SUGAR-BASED SURFACTANTS

Fundamentals and Applications

SURFACTANT SCIENCE SERIES

SUGAR-BASED SURFACTANTS

Fundamentals and Applications

Edited by
Cristóbal Carnero Ruiz
Universidad de Málaga
Málaga, Spain

CRC Press
Taylor & Francis Group
Boca Raton London New York

CRC Press is an imprint of the
Taylor & Francis Group, an **informa** business

CRC Press
Taylor & Francis Group
6000 Broken Sound Parkway NW, Suite 300
Boca Raton, FL 33487-2742

First issued in paperback 2019

© 2009 by Taylor & Francis Group, LLC
CRC Press is an imprint of Taylor & Francis Group, an Informa business

No claim to original U.S. Government works

ISBN-13: 978-1-4200-5166-7 (hbk)
ISBN-13: 978-0-367-38624-5 (pbk)

Library of Congress Cataloging-in-Publication Data

Sugar-based surfactants : fundamentals and applications / editor, Cristóbal Carnero Ruiz.
    p. cm. -- (Surfactant science series)
    Includes bibliographical references and index.
    ISBN 978-1-4200-5166-7 (alk. paper)
    1. Sugar-based surfactants. I. Ruiz, Cristóbal Carnero. II. Title. III. Series.

TP994.S776 2008
668'.1--dc22   2008028862

Visit the Taylor & Francis Web site at
http://www.taylorandfrancis.com

and the CRC Press Web site at
http://www.crcpress.com

Contents

Preface

Sugar-based surfactants are mainly characterized by having hydrophilic groups in their polar moiety. This structural feature, together with the many possibilities for linkage between the hydrophilic sugar head group and the hydrophobic alkyl chain, provides unique physicochemical properties to these surfactants, some of them substantially different from the common nonionic ethoxylated surfactants. Although interest in these amphiphiles was traditionally entirely academic, they have recently become the object of increasing attention for many researchers, thus opening new areas of research in surface and colloidal science from both fundamental and technological perspectives. Among the characteristic properties of these surfactants, a frequently remarked fact is that they can be produced from renewable resources and exhibit excellent ecological behavior. Certainly, there is currently a clear tendency to replace conventional surfactants with more environmentally benign compounds, though it should be recognized that sugar surfactants possess other properties that turn them into very advantageous products in view of their performance, for example, being healthier to consumers and for other technical applications. Therefore, many research groups today are getting involved in studies on the design and development of novel sugar-based surface-active molecules, the characterization of their solution and interfacial behavior, the establishment of structure–property relationships, as well as in the advance of new industrial applications for these materials.

When this book was being planned, we thought about the suitability of a new book on sugar-based surfactants taking into account the existence of related books, for instance, *Nonionic Surfactants: Alkyl Polyglucosides* (edited by Balzer and Lüders) and *Alkyl Polyglycosides: Technology, Properties, and Applications* (edited by Hill, von Rybinsky, and Stoll). However, after evaluating relevant information on different fundamental and applied aspects related to sugar-based surfactants as reported in the recent years, we concluded that the enterprise should be accomplished.

Although information about synthesis, production, chemical properties, and various applications in the fields of cosmetics, detergency, and manufacturing of alkyl polyglycosides, as well as the analytical chemistry of these surfactants and their behavior from ecological, toxicological, and dermatological points of view has been widely revised in the aforementioned volumes, many recent findings on the solution behavior and applications of systems involving these and other sugar-based surfactants remain to be revised. Among them, special attention must be devoted to those related to the adsorption of sugar-based surfactants, in both liquid–air and solid–liquid interfaces, the appearance of new kinds of these surfactants, and their applications in the field of biomembranes or in the preparation of pharmaceuticals, to mention only a few. Therefore, this book seeks to cover information about these research areas that are currently being developed.

Sugar-Based Surfactants: Fundamentals and Applications consists of 15 chapters. Chapter 1 provides a general perspective of the universe of amphiphiles, including several aspects on synthesis, fields of application, and production. Chapters 2 through 7 deal with the study of diverse physicochemical properties, including solution behavior, self-assembly, adsorption in the air–liquid and liquid–solid interfaces, and rheology, all of which have decisive implications for numerous industrial processes. Chapters 8 through 10 collect recent findings on synthesis, properties, and applications of new kinds of sugar-based surfactants, including gemini, ionic, and isoprenoid-type sugar-based surfactants, whose potential in new medical and biophysical applications is promising. Chapters 11 through 14 are dedicated to complex systems, where a sugar-based surfactant combines with another amphiphilic agent to form mixed micelles, microemulsions, vesicles, or protein–surfactant complexes. Chapter 15 presents remarkable applications of sugar-based surfactants in the solubilization of liposomes and cell membranes.

I would like to thank the collaboration and the effort of all the contributors who have participated in the making of this book, and also many thanks to Patricia Roberson, Taylor & Francis project coordinator, for her assistance.

Cristóbal Carnero Ruiz

Editor

Cristóbal Carnero Ruiz is an associate professor and head of the Department of Applied Physics II at the University of Málaga (Spain). He is a member of the Spanish Group of Colloids and Interfaces and of the editorial boards of the *Journal of Surface Science and Technology* and *Atomic, Molecular & Optical Physics Insights*.

Dr. Carnero Ruiz received his BS and MS in chemistry from the University of Granada (Spain) in 1976. Thirteen years later, he obtained a PhD (with honors) in chemistry from the University of Málaga and in 1998 a PhD (with honors) in physics from the University of Granada. In 1990, he joined the Department of Applied Physics II at the University of Málaga, where he is the leader of the structured fluids and amphiphilic systems research group, whose main interest lies in the field of association colloids. His contributions in this area (as author or coauthor of numerous professional papers) deal with mixed micellization and the effect of additives or cosolvents on micellization of surfactants.

Contributors

Juan Aguiar
Departamento de Física Aplicada II
Universidad de Málaga
Málaga, Spain

Emilio Aicart
Departamento de Química Física I
Facultad de Ciencias Químicas
Universidad Complutense de Madrid
Madrid, Spain

Patrizia Andreozzi
Department of Chemistry
SOFT-INFM-CNR Research Center
La Sapienza University
Rome, Italy

Lucyanna Barbosa-Barros
Departamento de Tecnología de
  Tensioactivos
IIQAB-CSIC
Barcelona, Spain

Cristóbal Carnero Ruiz
Departamento de Física Aplicada II
Universidad de Málaga
Málaga, Spain

Per Claesson
Department of Chemistry and Surface
  Chemistry
Royal Institute of Technology
Stockholm, Sweden

and

Institute for Surface Chemistry
Stockholm, Sweden

Mercedes Cócera
Departamento de Tecnología de
  Tensioactivos
IIQAB-CSIC
Barcelona, Spain

Luisa Coderch
Departamento de Tecnología de
  Tensioactivos
IIQAB-CSIC
Barcelona, Spain

Jan B.F.N. Engberts
Department of Organic and Molecular
  Inorganic Chemistry
Stratingh Institute
University of Groningen
Groningen, the Netherlands

Jan Christer Eriksson
Department of Chemistry and Surface
  Chemistry
Royal Institute of Technology
Stockholm, Sweden

Paula D. Galgano
Instituto de Química
Universidade de São Paulo
São Paulo, Brazil

Giacomo Gente
Department of Chemistry
La Sapienza University
Rome, Italy

Masakatsu Hato
Systems and Structural Biology Center
RIKEN Yokohama Institute
Yokohama, Japan

José M. Hierrezuelo
Departamento de Física Aplicada II
Universidad de Málaga
Málaga, Spain

Karlheinz Hill
Cognis GmbH
Monheim, Germany

Ingegärd Johansson
Akzo Nobel Surfactants Europe
Stenungsund, Sweden

Elena Junquera
Departamento de Química Física I
Universidad Complutense de Madrid
Madrid, Spain

Tadashi Kato
Department of Chemistry
Tokyo Metropolitan University
Tokyo, Japan

Jaap Klijn
Department of Organic and Molecular
  Inorganic Chemistry
Stratingh Institute
University of Groningen
Groningen, the Netherlands

Rumen Krastev
Max Planck Institute of Colloids
  and Interfaces
Potsdam, Germany

Atte J. Kumpulainen
Department of Chemistry and Surface
  Chemistry
Royal Institute of Technology
Stockholm, Sweden

Camillo La Mesa
Department of Chemistry
SOFT-INFM-CNR Research Center
La Sapienza University
Rome, Italy

Catherine LeHen-Ferrenbach
Cognis France
Saint Fargean-Ponthierry, France

Olga López
Departamento de Tecnología de
  Tensioactivos
IIQAB-CSIC
Barcelona, Spain

Shaohua Lu
Langmuir Center for Colloids
  and Interfaces
Columbia University
New York, New York

Alfonso de la Maza
Departamento de Tecnología de
  Tensioactivos
IIQAB-CSIC
Barcelona, Spain

Hiroyuki Minamikawa
Nanotube Research Center
National Institute of Advanced
  Industrial Science and
  Technology
Tsukuba, Japan

José A. Molina-Bolívar
Departamento de Física Aplicada II
Universidad de Málaga
Málaga, Spain

Gemma Montalvo
Departamento de Química Física
Universidad de Alcalá
Alcalá de Henares
Madrid, Spain

José L. Parra
Departamento de Tecnología de
  Tensioactivos
IIQAB-CSIC
Barcelona, Spain

José M. Peula-García
Departamento de Física Aplicada II
Universidad de Málaga
Málaga, Spain

Marco Scarzello
Department of Organic and Molecular
  Inorganic Chemistry
Stratingh Institute
University of Groningen
Groningen, the Netherlands

Omar A. El Seoud
Instituto de Química
Universidade de São Paulo
São Paulo, Brazil

Ponisseril Somasundaran
Langmuir Center for Colloids and
  Interfaces
Columbia University
New York, New York

Thomas Sottmann
Institut für Physikalische Chemie
Universität zu Köln
Köln, Germany

Marc C.A. Stuart
Department of Organic and Molecular
  Inorganic Chemistry
Electron Microscopy Group
Groningen Biomolecular Sciences
  and Biotechnology Institute
University of Groningen
Groningen, the Netherlands

Cosima Stubenrauch
School of Chemical and Bioprocess
  Engineering
University College Dublin
Dublin, Ireland

Eric C. Tyrode
Department of Chemistry, Surface
  Chemistry
Royal Institute of Technology
Stockholm, Sweden

and

Institute for Surface Chemistry
Stockholm, Sweden

Mercedes Valiente
Departamento de Química Física
Universidad de Alcalá
Alcalá de Henares
Madrid, Spain

Anno Wagenaar
Department of Organic and Molecular
  Inorganic Chemistry
Stratingh Institute
University of Groningen
Groningen, the Netherlands

Lei Zhang
Akzo Nobel Chemicals
Brewster, New York

1 Sugar-Based Surfactants for Consumer Products and Technical Applications*

Karlheinz Hill and Catherine LeHen-Ferrenbach

CONTENTS

1.1 INTRODUCTION

Sugar-based surfactants are the final result of a product concept, which is based on the greatest possible use of renewable resources. While the derivatization of fats

* Updated from Hill, K. and Rhode, O., *Fett/Lipid*, 101, 25, 1999. With permission.

TABLE 1.1
Availability of Carbohydrate Raw Materials

	Production Volume [t/a]	Average Price [€/kg][a]
Sucrose	150,000,000	0.25
Glucose	30,000,000	0.30
Sorbitol	650,000	1.80

Source: Lichtentaler, F.W., *Methods and Reagents for Green Chemistry*, Wiley-Interscience, 2007.

[a] Due to various reasons market prices of agricultural commodities are of high volatility–therefore the figures shown can only be indications.

and oils to produce a variety of different surfactants for a broad range of application has a long tradition and is well established [1], the production of surfactants based on fats, oils, and carbohydrates on a bigger industrial scale is relatively new. Today, the most important sugar-based surfactants are alkyl polyglycosides, sorbitan esters, and sucrose esters [1c].

Considering the amphiphilic structure of a typical surfactant with a hydrophilic head group and a hydrophobic tail, it has always been a challenge to attach a carbohydrate molecule as alternative to polyethylene glycol to a fat and oil derivative, such as a fatty acid or a fatty alcohol [2]. Although science has reported numerous ways of making such linkages and has also described a big number of different carbohydrates used in such reactions, it is clear from an industrial perspective that only a few carbohydrates fulfill the criteria of price, quality, and availability to be an interesting raw material source. Those are mainly sucrose from sugar beet or sugarcane, glucose, derived from starches, and sorbitol as the hydrogenated glucose derivative (Table 1.1) [2e]. More recently, lactose, xylose, and the carbohydrate-based residues from straw and hemicellulose processing have been used as well as the starting materials for the respective derivatives.

1.2 PRODUCTS AND APPLICATIONS

1.2.1 SORBITAN ESTERS

Sorbitan esters have been known for a couple of decades when the first industrial chemical processes were established for their manufacturing. One differentiates between a two-step and a one-step process (Scheme 1.1). In the first process, dehydration of sorbitol occurs in the presence of acid (e.g., NaH_2PO_3) to form 1,4-sorbitan as main isomer, which is subsequently esterified with fatty acids in a second reaction step with an alkaline catalyst (e.g., K_2CO_3) at 200°C–250°C. In the one-step process, both reactions are carried out simultaneously [3]. Both methods have been developed for industrial scale production. Depending on the type and amount of fatty acid used, different product compositions, consisting of mixtures of mono-, di-, or trisorbitan esters (e.g., laurates, oleates, or stearates)

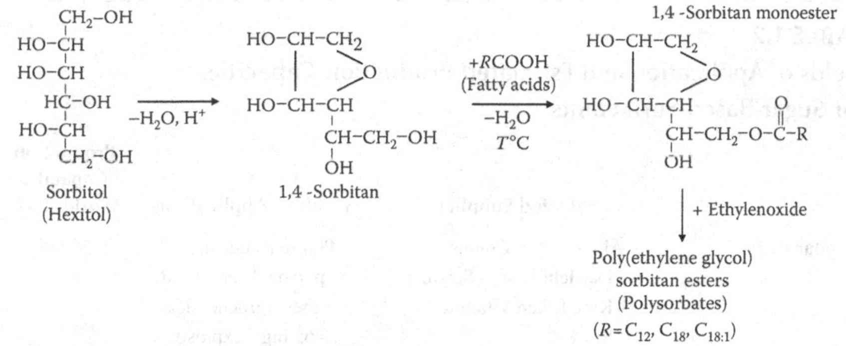

SCHEME 1.1 Synthesis of sorbitan esters by intramolecular dehydration of sorbitol and subsequent base-catalyzed esterification with fatty acids.

are produced with hydrophilic/lipophilic balance (HLB) values in a typical range of 1–8. To modify these relatively hydrophobic materials, a common technology is to further derivatize the sorbitan esters by reaction with ethylene oxide to produce sorbitan ester ethoxylates—or polysorbates—with HLB values of 10–17, depending on the number of ethylene oxide units attached (Figure 1.1) [3a].

The main manufacturers for sorbitan ester products today are listed in Table 1.2. The total market size for sorbitan esters (including the ethoxylated products) is estimated to be approximately 20,000 t/year. Mainly used as emulsifiers in pharmaceuticals, foods, cosmetic products, pesticides, for emulsion polymerization and explosives and other technical applications, the market size seems to be relatively stable and there is obviously no attempt and no need for further development of this mature technology.

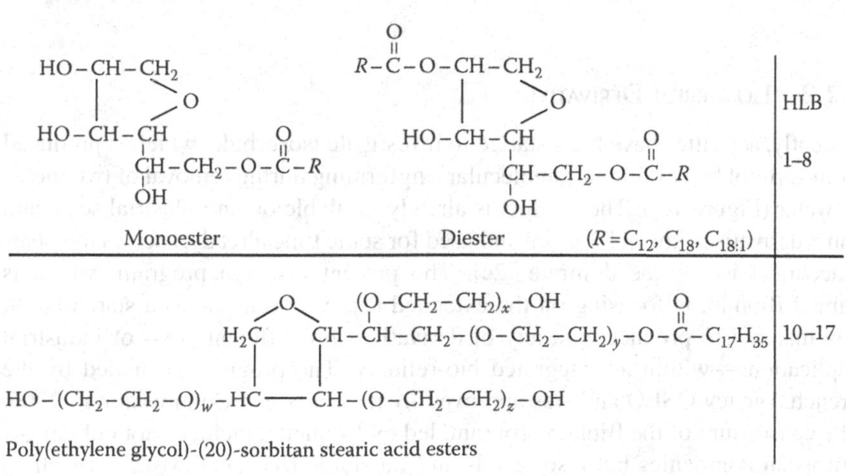

FIGURE 1.1 Hydrophilicity of sorbitan esters.

TABLE 1.2
**Fields of Application and Estimated Production Capacities
for Sugar-Based Surfactants**

	Selected Suppliers	Fields of Applications	Production Capacity, World [t/a][a]
Sorbitan esters	Akzo Nobel, Cognis, Dai-Ichi Kogyo Seiyaku, Kao, Riken Vitamin, SEPPIC	Pharmaceuticals, personal care, food, fiber, agrochemicals, coatings, explosives	20,000
Sucrose esters	Cognis, Croda, Dai-Ichi Kogyo Seiyaku, Evonik/ Goldschmidt, Mitsubishi-Kagaku, Sisterna, Jiangsu Weixi	Food, personal care, pharmaceuticals	<10,000
Alkyl polyglycosides	Akzo Nobel, BASF, China Research Institute of Daily Chemical Industry, Cognis, Dai-Ichi Kogyo Seiyaku, Kao, LG, SEPPIC	Personal care, detergents, agrochemicals, I + I	85,000
Others: Methylglucoside esters;	Lubrizol/Noveon	Personal care, pharmaceuticals	<10,000
Anionic alkyl polyglycoside derivatives	Cognis, Cesalpina	Personal care	

[a] Estimated figures based on private communications and literature data, references given in the text.

1.2.2 ISOSORBIDE DERIVATIVES

Recently, activities have been started to investigate isosorbide, which is produced from sorbitol by twofold intramolecular ringforming during removal of two moles of water (Figure 1.2). The product is already available on an industrial scale and some derivatives are well known and used for some time already, such as the pharmaceutical isosorbide dinitrate [2e]. The present research program, which is named Biohub, is focusing on the extended use of isosorbide as a starch-based raw material to produce a variety of derivatives for different areas of industrial applications—within an integrated bio-refinery. The program is funded by the French Agency OSEO and was approved by the European Commission in 2006. The consortium of the Biohub program, led by Roquette, includes not only major European companies but also small- and medium-sized enterprises, as well as public research laboratories [4].

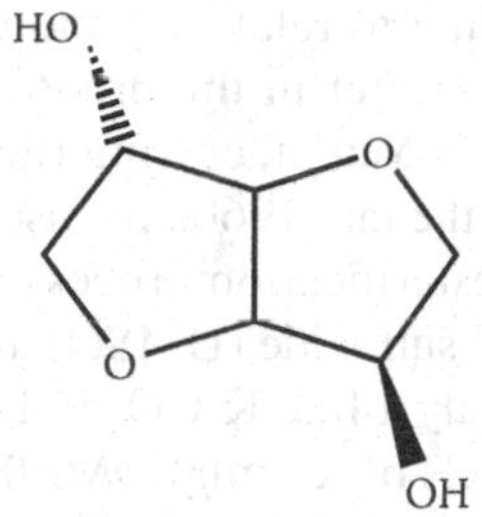

FIGURE 1.2 Structure of isosorbide.

1.2.3 SUCROSE ESTERS

Sucrose esters are described as very mild with regard to their dermatological properties and are approved as food additives in many countries. As a consequence, these products seem to be perfect raw materials for food and cosmetic formulations and their use in those applications as specialty emulsifier has long tradition [5a–c]. In Asia, one can find sucrose esters with a low degree of esterification (or degree of substitution, DS) as well in special detergent products. The octaesters have been developed by Procter & Gamble and are used as non-caloric fat substitute in food technology.

Sucrose is a nonreducing disaccharide containing multifunctionalities: three primary alcohols, five secondary alcohols, and two anomeric carbons. The manufacture of sucrose esters is quite challenging because the sucrose molecule is very temperature sensitive and, due to its high functionality with eight hydroxyl groups, selectivity in the esterification reaction is difficult to achieve (Scheme 1.2). General chemical pathways are either direct esterification with fatty acid or *trans*-esterification of sucrose with fatty acid methyl ester. Typically complex product mixtures consisting of mono-, di-, tri-, tetra-, and pentaesters are formed. These products are very hydrophobic and of limited application potential, except for specific emulsification needs. Historically, the studies carried out by Sugar Research Foundation gave a real boost to the applied

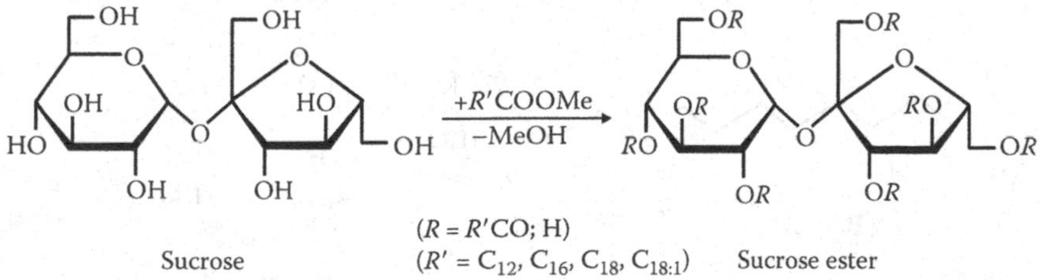

$(R = R'CO; H)$

$(R' = C_{12}, C_{16}, C_{18}, C_{18:1})$ Sucrose ester

SCHEME 1.2 Synthesis of sucrose esters by base catalyzed trans-esterification with fatty acid methyl esters ($R'COOMe$), usually carried out in solvents (e.g., dimethyl formamide), microemulsions or solvent free.

research of sugar esters, leading to relatively more hydrophilic products (i.e., higher content of mono- and diester in the product mixture). These were first commercialized by Dai-Nippon Manufacturing (now Mitsubishi-Kagaku Food Corporation, Ryoto esters) in the late 1960s for use in the food industry. Manufacturing is based on a trans-esterification process in solvents, such as dimethyl formamide (DMF) or dimethyl sulfoxide (DMSO), to solubilize the sucrose, and in presence of an alkaline catalyst like K_2CO_3 [5d,e]. The purification in order to achieve the permissible levels of residual DMF/DMSO is a critical point and requires a relatively complex process. Therefore, several methods have been developed to achieve a higher selectivity in the reaction or provide economical purification procedures using liquid–liquid extraction and crystallization technologies.

Several improvements of the process, such as the reaction with acyl chlorides, or the application of two phase reaction systems with propylene glycol and an emulsifier in order to build a microemulsion, have been described in literature and patents [5f]. Another approach for the synthesis of sugar esters is the use of enzymes. Enzymatic catalysis in the field of carbohydrate chemistry has been actively explored over years in laboratory scale, resulting in a higher degree of selectivity. Whether optimized enzymes, e.g., that can operate in low-water environment, can revolutionize this type of synthesis to make it industrially feasible, is still not known [5g–k]. Therefore, solvent-free processes based on trans-esterification of the molten sucrose and fatty acid ester at temperatures of about 130°C and in the presence of an alkaline catalyst and an emulsifier (fatty acid soap or the sucrose esters itself) are still preferred [5g]. However, the purification step to eliminate the residual soap content remains as a difficult process step here. An optimized solvent-free and water-free process was described recently by Cognis [5i]. The synthesis is carried out in an organized medium under heterogeneous catalysis (K_2CO_3) with sucrose ester as emulsifier [5g]. In the first step, so the theory, the continuous phase is composed of methyl esters. These methyl esters are adsorbed on the catalyst surface (Figure 1.3) and react with the sucrose esters already present to form a sucrose polyester as intermediate [5g,i]. In the second step, the sucrose polyesters

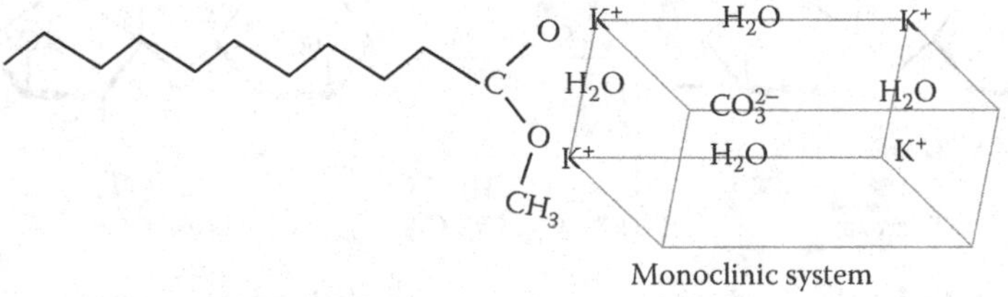

Monoclinic system

FIGURE 1.3 Surface reaction with preadsorption of fatty acid methyl esters on K_2CO_3. (From Mouloungui, Z., *Réactions en milieu hétérogène solide/liquide faiblement hydrate*, Thèse d'état INPT, Toulouse, 1987.)

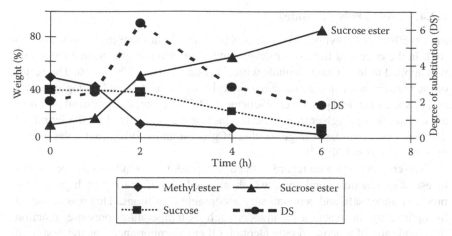

FIGURE 1.4 Mass balance of reagents and products in the esterification of sucrose with fatty acid methyl ester under solvent- and water-free conditions (125°C; K_2CO_3; sucrose ester as emulsifier).

react with the molten sucrose in an acyl transfer reaction to obtain the final sucrose ester composition with an optimized DS of about 2.5 (Figure 1.4). The final product is mixed with a cosmetic emollient, which enables a final refining by filtration. Applied as a cosmetic emulsifier, it improves the stability of the final product formulation and the sensorial performance by forming lamellar liquid crystal structures [5h].

Today, the major producers for sucrose esters are Dai-Ichi Kogyo Seiyaku and Mitsubishi in Japan, Croda and Procter & Gamble in the United States, Sisterna in the Netherlands, and Cognis in Germany (Table 1.2). It seems that the existing production capacities are much higher than the actual market potential, which is estimated to be less than 10,000 t/year. However, demand and market volume could increase substantially, if reaction processes—especially for the synthesis of high mono- and diester products—can be further optimized.

1.2.4 GLUCOSE-DERIVED SURFACTANTS

The first step in overcoming the problem of nonselective derivatization of carbohydrates was achieved when Emil Fischer discovered the reaction of glucose with alcohol to form alkyl glucosides [6a]. The glucosidation reaction is highly selective due to the hemiacetal function in the glucose molecule and the resulting high reactivity of the hydroxyl group at C1. The same is true for the synthesis of fatty acid glucamides. Here the glucose molecule reacts initially with methylamine, which, after hydrogenation, gives selectively the glucamine as an intermediate [6b,c]. Further derivatization with fatty acid methyl ester leads to the desired product.

1.2.4.1 Alkyl Polyglycosides

First syntheses of alkyl polyglycosides were carried out more than 100 years ago [6a]. In the course of the further development, reaction of glucose with alcohols was applied to long chain alcohols with alkyl chains from C8 to C16. The result of the reaction is a complex mixture of alkyl mono-, di-, tri-, and oligoglycosides as a mixture of α- and β-anomers (Scheme 1.3). Therefore, the industrial products are called alkyl polyglycosides. The products are characterized by the length of the alkyl chain and the average number of glucose units linked to it—the degree of polymerization (DP) [7].

The crucial point with regard to the development of an industrial process was to establish reaction conditions, which allowed manufacturing of high-quality products under safe and economically acceptable conditions. This was achieved by optimizing the reaction parameters such as temperature, pressure, reaction time, and ratio of glucose to fatty alcohol. Of equal importance was the design of a special distillation technology to remove the excess fatty alcohol as smoothly as possible as well as an appropriate bleaching and stabilization in the final treatment step (Scheme 1.4). This so-called direct synthesis of alkyl polyglycosides nowadays is the preferred manufacturing mode.

The breakthrough in the production of mid-chain ($C_{12/14}$) alkyl polyglycoside occurred in 1992 with the inauguration of a large scale production plant for Alkyl Polyglycoside surfactants by Henkel Corporation in the USA (now Cognis Corporation) and in 1995 with the opening of a second plant of equal capacity by Henkel in Germany (now Cognis GmbH). Cognis was the first company to offer mid-chain ($C_{12/14}$) alkyl polyglycosides in industrial scale at

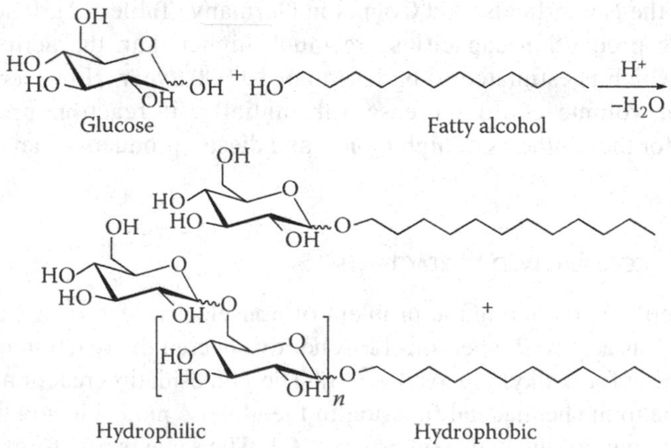

SCHEME 1.3 Synthesis of alkyl polyglycosides by acid-catalyzed acetalization of glucose in molar excess of fatty alcohol and removal of water under vacuum at 100°C–120°C.

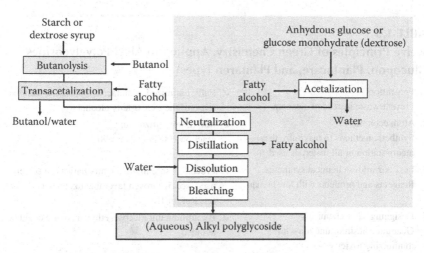

SCHEME 1.4 Manufacturing process for alkyl polyglycosides.

required quality. Currently, Cognis is the supplier with the largest capacity worldwide. Other manufacturers include Akzo Nobel, BASF, China Research Institute of Daily Chemical Industry, Dai-Ichi Kogyo Seiyaku, Kao, LG, and SEPPIC. The overall capacity available wordwide is estimated to be approximately 85,000 t/year (Table 1.2).

By combining vegetable oil and sugar as raw materials, it has for the first time become possible to offer commercially significant amounts of nonionic surfactants, which are completely based on renewable resources. The main applications for the $C_{12/14}$ alkyl polyglycosides are liquid dishwashing agents and detergents and personal care products; for the $C_{8/10}$ (or branched C_8) alkyl polyglycosides, it is hard surface cleaners, agrochemicals, and products for industrial and institutional cleaning (Table 1.2). The main characteristics of process and product regarding environmental impact are summarized in Table 1.3 by applying the "12 principles of green chemistry" according to Anastas and Warner for Glucopon, Plantaren, and Plantacare types [7h].

1.2.4.2 Fatty Acid Glucamides

The synthesis to produce the fatty acid glucamides involves the reaction of glucose with methyl amine under reductive conditions to form the corresponding N-methyl glucamine. In a subsequent reaction step, this intermediate is converted with fatty acid methyl ester to the corresponding fatty acid amide. Compared to the alkyl polyglycosides, fatty acid glucamides are composed of only one single carbohydrate molecule attached to the fatty acid chain. This is one reason why fatty acid glucamides are less soluble and tend to crystallize more easily from

TABLE 1.3
Twelve Principles of Green Chemistry: Applied to Alkyl Polyglycosides (Glucopon, Plantacare, and Plantaren Types)

1. Prevention
   Prevent waste instead of clean up
   ✓ Fully optimized manufacturing process with the reuse of the excess fatty alcohol

2. Atom economy
   Synthetic methods to maximize the incorporation of all materials used
   ✓ Maximum utilization:
   reaction 100% – water >90%

3. Less hazardous chemical syntheses
   Reagents and products with low human
   ✓ Process is safe including auxiliaries (no solvent used, see 5): proven favorable tox and ecotox data (cf. Ref. in 10)

4. Designing safer chemicals
   Guarantee desired function while minimizing toxicity
   ✓ By introducing glucose, ethylene oxide could be avoided

5. Safer solvents and auxiliaries
   Avoid wherever possible
   ✓ No solvent used in the process; only water for the dilution of the final product

6. Design for energy efficiency
   Preferably ambient pressure and temperature if possible
   ✓ Reaction under ambient pressure and continuous process for the distillation of excess fatty alcohol are minimizing the energy consumption

7. Use of renewable feedstocks
   ✓ Raw materials used are 100% renewable (glucose and vegetable fatty alcohol)

8. Reduce derivatives
   Avoid blocking/protection groups, if possible
   ✓ No protection groups are used

9. Catalysis
   ✓ Acid is used in catalytic quantities

10. Design for degradation
    Favorable biodegradation properties after use
    ✓ Aerobic and anaerobic degradation proven (cf. *Handbook of Detergents*, Marcel Dekker NY, part B, Chap 18 "Ecology and Toxicology of Alkyl Polyglycosides", pp. 487–521)

11. Real-time analysis for pollution prevention
    Process monitoring
    ✓ Process control via process information system PI

12. Inherently safer chemistry for accident prevention
    ✓ Process is inherently safe, no runaway possible due to raw materials

Source: Cognis, unpublished results.

aqueous solutions. Scheme 1.5 shows the manufacturing scheme for the production of fatty acid glucamides. To avoid significant amounts of unreacted *N*-methyl glucamine, which could be considered as potential precursors for nitrosamines, Procter & Gamble has developed an optional reaction with acetic anhydride in the product finishing. Free secondary amines can be acetylated in this step, the resulting acetates can remain in the final product [8].

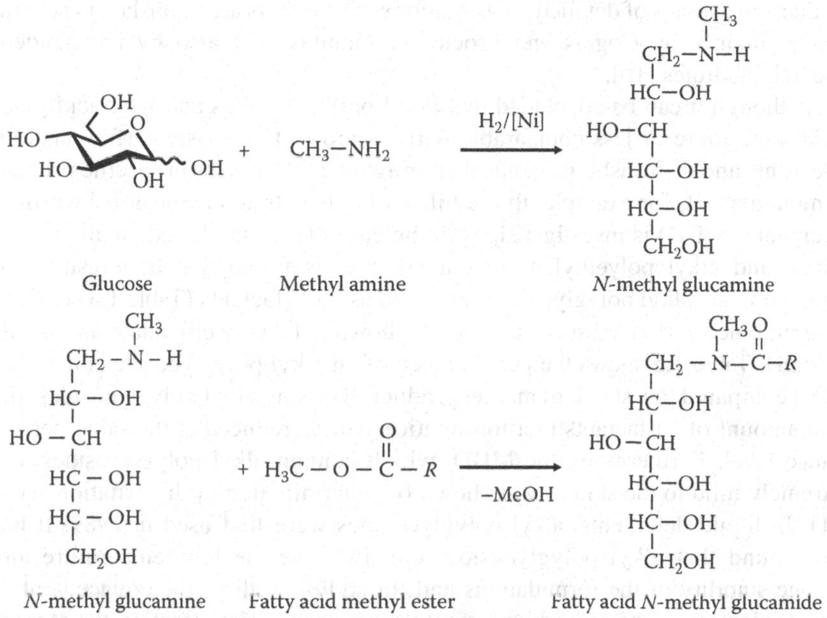

Glucose   **Methyl amine**   *N*-methyl glucamine

N-methyl glucamine   Fatty acid methyl ester   Fatty acid *N*-methyl glucamide

SCHEME 1.5 Two-step synthesis of fatty acid glucamides by reductive alkylation of methyl amine with glucose to obtain *N*-methyl glucamine, which is acylated by base catalyzed reaction with fatty acid methyl ester in a second step.

For several years, fatty acid glucamides with $C_{12/14}$ and $C_{16/18}$ alkyl chains have been used exclusively in powdered and liquid detergents and liquid dishwashing agents. A couple of years ago their use was limited to a few applications only and at present they are rarely found in any application. Basically they have disappeared from the market. Derivatives of glucamides, such as glucamine oxides and betaines or anionic glucamides, as well as bifunctional glucamides, have been described as well in literature [9], but are hardly found in any market product so far.

1.2.4.3 Properties of Alkyl Polyglycosides and Fatty Acid Glucamides

With regard to their basic physical chemical properties, such as surface and interfacial tension and critical micelle concentration, alkyl polyglycosides and fatty acid glucamides ($C_{12/14}$) are rather comparable. There are slight differences in the basic foam behavior for the pure sugar-based surfactants as well as in binary combinations. With regard to their ecological, toxicological, and dermatological properties, alkyl polyglycosides as well as fatty acid glucamides can be considered as surfactants with extraordinary product safety. This has been proven for both products in a series of detailed investigations. The results are published in several papers, mainly by Cognis and Procter & Gamble, but also by independent research institutes [10].

Although it can be concluded that alkyl polyglycosides and fatty acid gluc-amides are more or less comparable with regard to their basic performance in detergents and dishwashing agents, there might be differences in specific product formulations. If, for example, the stability of concentrated manual dishwashing detergents (MDD) is investigated, as in the case of a paste, based on alkyl ether sulfate and alkyl polyethylene glycol ether, it is found that best results are obtained when alkyl polyglycosides are used as cosurfactants (Table 1.4) [11]. In general, glucose-derived surfactants were shown to be very efficient components in MDD. Table 1.5 shows the performance of an alkyl polyglycoside containing MDD compared to a standard market product. By using alkyl polyglycosides, the total amount of surfactants in a formulation can be reduced at the same performance level. Furthermore, the MDD, which contain alkyl polyglycosides, are extremely mild to the skin as was shown by determination of the irritation score [11]. In liquid detergents, alkyl polyglycosides were first used in 1989. It has been found that alkyl polyglycosides can improve the low-temperature and storage stability of the formulations and, in addition, allow the replacement of triethanolamine soaps by sodium or potassium soaps. The effect on the storage stability of enzymes and, as a consequence, the washing performance, by using alkyl polyglycosides as a component in detergents has been described as well [11]. Table 1.6 summarizes the various applications for short- ($C_{8/10}$) and medium-chain ($C_{12/14}$) alkyl polyglycosides in the detergents field.

In personal care products, alkyl polyglycosides represent a new concept in compatibility and care. They may be combined with conventional components and can even replace them in new types of formulations, leading to a rich spectrum of

TABLE 1.4
Stability of Concentrated MDD (Pastes)

Ingredients	Product 1 (wt %)	Product 2 (wt %)	Product 3 (wt %)	Product 4 (wt %)
Alkyl ether sulfate	10	10	10	10
Alkyl polyethylene glycol ether	15	15	15	15
Fatty acid alkanolamide	18	—	—	—
Alkylamidobetaine	—	18	—	—
Fatty acid glucamide	—	—	18	—
Alkyl polyglycoside	—	—	—	18
Appearance at 68°F (20°C)	Cloudy	Gel	Clear	Clear
Pourpoint	—	—	54°F (12°C)	32°F (0°C)
Storage test 3 weeks, 41°F (5°C)	Solid	Solid	Solid	Clear liquid

TABLE 1.5

Examples for Conventional MDD

Ingredients	MDD without $C_{12/14}$ APG [wt%][a]	MDD with $C_{12/14}$ APG [wt%][a]
sec-Alkane sulfonate	15–20	—
Alkyl sulfate	—	2–10
Alkyl ether sulfate	2–10	5–15
Alkyl polyglycoside	—	1–5
Alkylamidobetaine	—	1–5
Ethanol	<10	<10
Fragrances	<1	<1
Colorants	<0.1	<0.1
Water	Balance	Balance
Total amount of surfactants[b]	25	20

[a] Typical amount of surfactants used.

[b] Total amount of surfactants needed to achieve comparable performance in a standardized dishwashing test (plate test) [11b].

supplementary effects. With regard to foam, they are comparable with betaines and sulfosuccinates and can stabilize the foam of anionics in hard water and in the presence of sebum. The alkyl polyglycoside foam consists of finer bubbles and can be creamier compared to other systems (Figure 1.5) [12a]. The extreme mildness of alkyl polyglycosides is demonstrated by the so-called arm flex test, where skin

TABLE 1.6

Use of Alkyl Polyglycosides in Detergents Products Depending on the Alkyl Chain Length

Type of Products	Alkyl Chain Length
MDDs	($C_{12/14}$)
Cleaners	($C_{12/14}$; $C_{8/10}$)
All purpose cleaners	($C_{12/14}$; $C_{8/10}$)
Bathroom cleaners	($C_{8/10}$)
Liquid toilet cleaners	($C_{8/10}$)
Window cleaners	($C_{8/10}$)
Floor care formulations	($C_{12/14}$)
Liquid laundry detergents	($C_{12/14}$)
Powder/extrudate detergents	($C_{12/14}$)

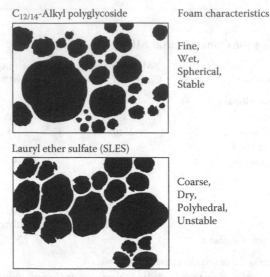

FIGURE 1.5   Foam structure of surfactant solutions.

is repeatedly treated with solutions of a standard surfactant for personal care products (SLES) and, for comparison, with a solution of a mixture of SLES and alkyl polyglycosides. Afterward, the skin roughness is measured by profilometry and the positive effect of the sugar surfactant is shown. The whole range of applications for alkyl polyglycosides is described in recent publications and examples for formulations are given in [12]. The broad spectrum of applications in the personal care field is briefly summarized in Table 1.7.

TABLE 1.7
Alkyl Polyglycosides in Personal Care Products

Properties[a]	Types of Applications
Nonionic surfactant based on renewable raw materials	Basic or cosurfactant for all kinds of personal care and cleansing products
Mild, readily biodegradable, ecotoxicologically safe, good foamer, viscosity enhancer, cleansing power, rheology modifier, substantivity to hair, influence on strength and interaction of hair fibers, stable at alkaline pH value, emulsifying properties, microemulsion formation, perfume solubilization	Shampoo, shower bath, baby products, oil bath, syndet bars, dental care products, blends and core concentrates, shampoo concentrate, setting lotions, styling gels, hair conditioners, shampoo for fine or damaged hair, perms, hair colors, creams and lotions for sensitive skin, refatting preparations, facial cleansers

[a] Relevant for the application in personal care products.

To demonstrate the large performance spectrum of alkyl polyglycosides, one more application should be mentioned as well. Alkyl polyglycosides ($C_{8/10}$ and $C_{12/14}$) have been shown to be substitutes for alkyl phenol ethoxylates in agrochemical formulations. They lead to higher salt tolerances and show good results as adjuvants in several postapplied herbicides, such as control of giant foxtail in soybeans with Assure II (DuPont) and control of common lambs quarters in soybean with Pursuit (American Cyanamid). By now, the short chain products ($C_{8/10}$ and C_{9-11}) are approved as inert ingredients by the United States Environmental Protection Agency, USEPA [13].

1.2.5 DERIVATIVES OF ALKYL POLYGLYCOSIDES

Since alkyl polyglycosides are available in sufficient quantities, their use as a raw material for the development of specialty surfactants is arousing considerable interest. The derivatization of alkyl polyglycosides is currently being pursued with great commitment with the goal to modify the surfactant properties [14]. A broad range of alkyl polyglycoside derivatives can be obtained by using relatively simple methods, for example, nucleophilic substitution. Besides the reaction to esters or ethoxylates, ionic products, such as sulfates and phosphates, can also be synthesized (Scheme 1.6). However, until now, there are only a few products that are established in the market: methyl glucoside esters and selected anionic derivatives of alkyl polyglycosides.

1.2.5.1 Methyl Glucoside Esters

Esterification of methyl glucoside (obtained from starch or glucose and methanol) with methyl esters of stearic or oleic acid leads to the desired products (Scheme 1.7).

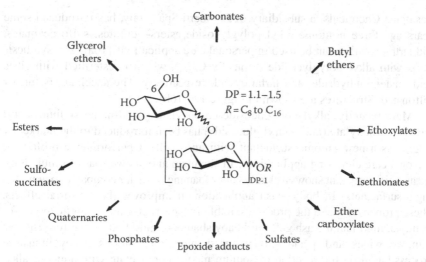

SCHEME 1.6 Overview on alkyl polyglycoside derivatives.

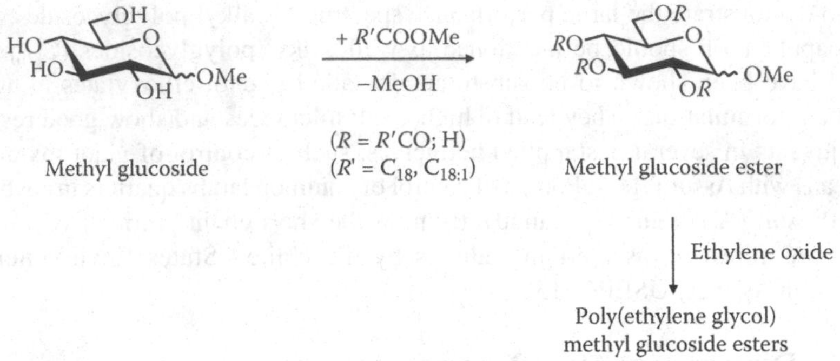

SCHEME 1.7 Synthesis of methyl glucoside ester by base-catalyzed (K_2CO_3) transesterification of methyl glucoside with fatty acid methyl ester ($R'COOMe$) at 120°C–160°C.

By subsequent ethoxylation, the respective polyethylene glycol methyl glucoside esters are obtained. Because of their relatively lipophilic character, methyl glucoside esters are, in contrast to alkyl polyglycosides with the same hydrophobic chain length, hardly soluble in water, but exhibit excellent emulsification properties [5a,b,15]. They have found application as emollients, moisturizing and emulsifying agents, and thickeners for cosmetics. The hydrocarbon length and DS can be varied to adjust the rheological properties and obtain a specific w/o emulsification behavior. A major manufacturer for methyl glucoside esters and ethoxylated derivatives is Lubrizol with its Noveon line. The total market size, including the ethoxylated products, is estimated to be less than 10,000 t/year (Table 1.2).

1.2.5.2 Anionic Derivatives of Alkyl Polyglycosides

Cesalpina Chemicals, a subsidiary of Lamberti Spa., Italy, has introduced some years ago three nonionic alkyl polyglycoside esters—citrates, sulfosuccinates, and tartrates—that can be used in personal care applications [16a]. The synthesis starts with alkyl polyglycoside (typically $C_{12/14}$), which is esterified with citric acid, maleic anhydride, and tartaric acid, respectively. The succinate is further sulfonated. Structures are shown in Figure 1.6 [16b].

More recently, alkyl polyglucoside carboxylate (INCI-name sodium lauryl glucose carboxylate (and) lauryl glucoside) has been introduced to the market by Cognis as a new anionic surfactant with an excellent performance profile for personal care cleansing applications. In shampoo and shower bath formulations, the anionic surfactant shows a clearly better foaming behavior compared to nonionic sugar-surfactants. In body wash applications it improves the sensorial effects. These properties make the product suitable for several cosmetic applications, for example, mild facial wash gel, mild baby shampoo, mild body wash for sensitive skin, wet wipes, and special sulfate-free shampoo applications. A new industrial process based on the reaction of sodium monochloroacetate with aqueous alkyl polyglycoside (without additional solvents) enables the manufacturing of this product in an economically and ecologically favorable way (Scheme 1.8) [17].

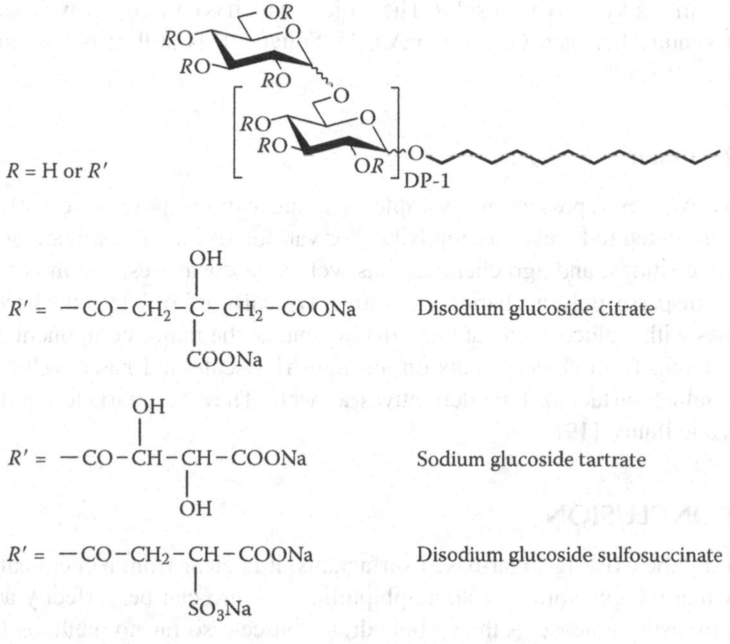

$R = H$ or R'

$R' = -CO-CH_2-\underset{\underset{COONa}{|}}{\overset{\overset{OH}{|}}{C}}-CH_2-COONa$ Disodium glucoside citrate

$R' = -CO-\underset{\underset{OH}{|}}{\overset{\overset{OH}{|}}{CH}}-CH-COONa$ Sodium glucoside tartrate

$R' = -CO-CH_2-\underset{\underset{SO_3Na}{|}}{CH}-COONa$ Disodium glucoside sulfosuccinate

FIGURE 1.6 Examples of anionic alkyl polyglycoside derivatives.

1.2.6 OTHER CARBOHYDRATE RAW MATERIALS IN THE PRODUCTION OF SUGAR-BASED SURFACTANTS

1.2.6.1 Hemicellulose

Recently, development activities for alkyl polypentosides have been described. The products are based on hemicellulose, which, after hydrolysis, leads to a mixture of glycosides (mainly pentoses such as xylose and arabinose). These pentoses are transformed into the desired alkyl polypentosides via the reaction with fatty alcohols under standard reaction conditions. The products are reported to have similar

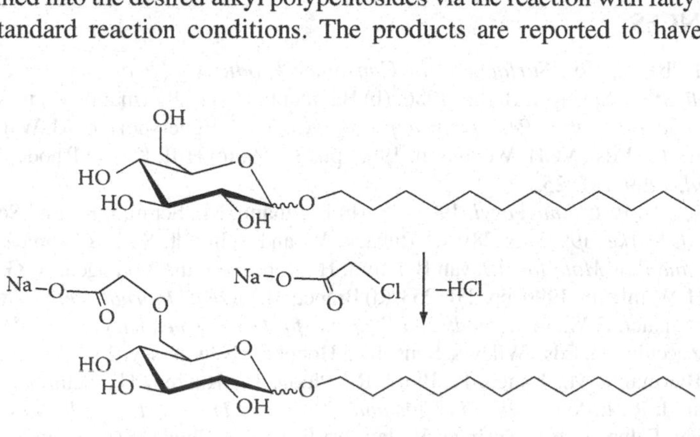

SCHEME 1.8 Synthesis of alkyl polyglycoside carboxylate.

properties than alkyl polyglycosides. The project, which is carried out by Wheatoleo (a joint venture between Oleon and A.R.D./Soliance), is still at precommercial scale and state [18].

1.2.6.2 Inulin

Inulin derivatives represent an example of a sugar-based polymeric surfactant, mainly promoted to be used as emulsifier for various technical applications, such as paints, coatings, and agrochemicals, as well as in cosmetics. Inulin is defined as a polydisperse polysaccharide consisting typically of $\beta(2\text{-}1)$-linked fructose molecules with a glucose unit at the reducing end as the major component. Orafti extracts inulin from chicory roots on an industrial scale and has developed the corresponding surfactant type derivatives as well. They are marketed under the brand name Inutec [19].

1.3 CONCLUSION

Comparing the existing sugar-based surfactants, it is clear from a chemical point of view that the generation of an amphiphilic structure can be perfectly accomplished by using glucose as the carbohydrate source—so far no methods for the selective derivatization of sorbitol and sucrose have been developed for industrial scale. This, combined with outstanding performance, multifunctionality, competitive price, high product safety, and an environmentally favorable manufacturing process, could be one reason why alkyl polyglycosides are the most successful sugar-based surfactants nowadays.

(Ryoto is a registered trademark of Mitsubishi-Kagaku Food; APG surfactants, Glucopon, Plantacare, and Plantaren are registered trademarks of the Cognis group; Noveon is a registered trademark of Lubrizol; Inutec is a registered trademark of Orafti.)

REFERENCES

1. (a) Falbe, J., Ed., *Surfactants in Consumer Products—Theory, Technology and Application*, Springer, Berlin, 1986; (b) Baumann, H. and Biermann, M., in *Nachwachsende Rohstoffe, Perspektiven für die Chemie*, Eggersdorfer, M.,Warwel, S., Wulff, G., Eds., VCH, Weinheim, 1993, pp. 33–55; (c) Hill, K. and Rhode, O., *Fett/Lipid*, 1999, *101*, 25.
2. (a) Schulz, P., *Chimica oggi*, 1992, 33; (b) Biermann, M., Schmid, K., and Schulz, P., *Starch/Stärke*, 1993, *45*, 281; (c) Ruback, W. and Schmidt, S., in *Carbohydrates as Organic Raw Materials III*, van Bekkum, H., Röper, H., and Vorhagen, A. G. J., Eds., VCH, Weinheim, 1996, pp. 231–253; (d) Brancq, B., *SÖFW-Journal*, 1992, *118*, 905; (e) Lichtentaler, F. W., in *Methods and Reagents for Green Chemistry*, Tundo, P., Perosa, A., Zecchini, F., Eds., Wiley & Sons, Inc., Hoboken, New Jersey, 2007, Chap. 2.
3. (a) Biermann, M., Lange, F., Piorr, R., Ploog, U., Rutzen, H., Schindler, J., and Schmid, R., in *Surfactants in Consumer Products-Theory, Technology and Application*, Falbe, J., Ed., Springer-Verlag, Berlin, 1987, Chap. 3; (b) Ropuszynski, S.,

and Sczesna, E., *Tenside Surf. Det.*, 1990, *27*, 350; (c) Ropuszynski, S. and Sczesna, E., *Tenside Detergents,* 1985, *22*, 190.

4. Rupp-Dahlem, C., Bio refinery concept from cereals: The BioHub® Program, in *Renewable Raw Materials for Industry: Contribution to Sustainable Chemistry*, Brussels, 2007.

5. (a) Desai, N. B., *Cosmetics & Toiletries*, 1990, *105*, 99; (b) Nakamura, S., *INFORM*, *8*, 866, 1997; (c) Cecchin, M., *Domaines d'application des sucroesters et sucroglyérides*, DESS Ingénierie Documentaire, ENSSIB de Lyon, 2001, p.87; (d) Osipow, L. I., Snell, F. D., and Finchler, W.C., *Ind. Eng. Chem.*, 1956, *48*, 1459; (e) Hass, H. B., in *Sugar Esters*, Snell, F. D., Ed., Sugar Research Foundation, New York, 1968, Chaps. 1–7; (f) Osipow, L. I. and Rosenblatt, W., *J. Am. Oil. Chem. Soc.*, 1967, *4*, 307; (g) Claverie, V., *Mise au point d'un nouveau procédé de synthèse d'esters de saccharose de degré de substitution contrôlé*, Thèse ENCT de Toulouse, 1998; (h) Pian, V., Le Hen-Ferrenbach, C., Beuché, M., and Roussel, M., *New Sucrose Polystearate Effective Emulsifier Combining Skin Touch with Moisturizing Properties*, ISFCC, Florence, 2005; (i) Le Hen-Ferrenbach, C., Beuché, M., and Roussel, M., DE 054432, 2005, Cognis; (j) Mouloungui, Z., *Réactions en milieu hétérogène solide/liquide faiblement hydraté*, Thèse d'état INPT, Toulouse, 1987; (k) Drummond, C. J., Fong, C., Krodkiewska, I., Boyd, B. J., and Baker, I. J., in *Novel Surfactants/114*, Marcel Dekker, New York, 2003, Chap.3.

6. (a) Fischer, E., *Ber. Dtsch. Chem. Ges.*, 1893, *26*, 2400; (b) Jürges, P. and Turowski, A., Vergleichende Untersuchung von Zuckerestern, *N*-Methylglucamiden und Glycosiden am Beispiel von Reinigungsprodukten, in *Perspektiven Nachwachsender Rohstoffe in der Chemie*, Eierdanz, H., Ed., VCH, Weinheim, 1996, pp. 61–70; (c) Kelkenberg, H., *Tenside Surf. Deter.*, 1988, *25*, 8.

7. (a) Hill, K., von Rybinski, W., and Stoll, G., Eds., *Alkyl Polyglycosides—Technology, Properties and Applications*, VCH, Weinheim, 1997; (b) von Rybinsky, W. and Hill, K., *Angew. Chem. Int. Ed.*, 1998, *37*, 1238; (c) von Rybinski, W. and Hill, K., in *Novel Surfactants, Preparations, Applications, and Biodegradability*, Holmberg, K., Ed., Marcel Dekker, New York 1998, Chap. 2; (d) Balzer, D. and Lüders, H., Ed., *Nonionic Surfactants, Alkyl Polyglucosides*, Marcel Dekker, New York, 2000; (e) Koeltzow, D. E. and Urfer, A. D., *J. Am. Oil Chem. Soc.*, 1984, *61*, 1651; (f) Straathof, A. J. J., van Bekkum, H., and Kieboom, A. P. G., *Starch/Stärke*, 1988, *40*, 438; (g) Böcker, Th. and Thiem, J., *Tenside Surf. Det.*, 1989, *26*, 318; (h) Anastas, P. T. and Warner, J. C., *Green Chemistry: Theory and Practice*, Oxford University Press, New York, 1998.

8. (a) Laughlin, R. G., Fu, Y.-C., Wireko, F. C., Scheibel, J. J., and Munyon, R. L., in *Novel Surfactants, Preparations, Applications, and Biodegradability*, Holmberg, K., Ed., Marcel Dekker, New York, 1998, Chap. 1; (b) Scheibel, J. J., Connor, D. S., Shumate, R. E., and St. Laurent, J. C. T. R. B., EP-B 0558515, US-A 598462, 1990, Procter & Gamble, *Chem. Abstr.*, 1992, *117*, 114045.

9. Beck, R., *Chimica Oggi*, 2002, *May*, 45.

10. (a) Aulmann, W. and Sterzel, W., in *Alkyl Polyglycosides—Technology, Properties and Applications*, VCH, Weinheim, 1997, Chap. 9; (b) Matthies, W., Jackwerth, B., and Kraechter, H.-U., in *Alkyl Polyglycosides—Technology, Properties and Applications*, VCH, Weinheim, 1997, Chap. 10; (c) Steber, J., Guhl, W., Stelter, N., and Schröder, F. R., in *Alkyl Polyglycosides—Technology, Properties and Applications*, VCH, Weinheim, 1997, Chap. 11; (d) Stalmans, M., Matthijs, E., Weeg, E., and Morris, S., *SÖFW-J.*, 1993, *119*, 794.

11. (a) Schmid, K., in *Perspektiven Nachwachsender Rohstoffe in der Chemie*, Eierdanz, H., Ed., VCH, Weinheim, 1996, pp. 41–60; (b) Andree, H., Hessel, J. F., Krings, P.,

Meine, G., Middelhauve, B., and Schmid, K., in *Alkyl Polyglycosides—Technology, Properties and Applications*, VCH, Weinheim, 1997, Chap. 6.

12. (a) Tesmann, H., Kahre, J., Hensen, H., and Salka, B. A., in *Alkyl Polyglycosides— Technology, Properties and Applications*, VCH, Weinheim, 1997, Chap. 5; (b) Ansmann, A., Busch, B., Hensen, H., Hill, K., Kraechter, H.-U., and Müller, M., in *Handbook of Surfactants, Part D: Formulation, Surfactant Science Series*, Showell, M. S., Ed., CRC Press, Taylor & Francis Group, Boca Raton, 2005, p. 207; (b) Weuthen, M., Kawa, R., Hill, K., and Ansmann, A., *Fat. Sci. Technol.*, 1995, *97*, 209.

13. Garst, R., in *Alkyl Polyglycosides—Technology, Properties and Applications*, VCH, Weinheim, 1997, Chap. 7.

14. Rhode, O., Weuthen, M., and Nickel, D., in *Alkyl Polyglycosides—Technology, Properties and Applications*, VCH, Weinheim, 1997, Chap. 8.

15. Anonymous, *Parfums Cosmetiques Actualites*, 1998, *139(Feb–Mar 1998)*, 44.

16. (a) Bernardi, P., Fornara, D., Paglino L., and Verzotti T., *The 1st Concise Surfactants Directory*, 15, 1996; (b) *Chem. Marketing Rep.* 1996 *(June 17)*, 23.

17. Behler, A., Hensen, H., and Seipel, W., in Proceedings 6th World Surfactant Congress, CESIO June, 2004, Berlin, 2004.

18. *ICIS Chemical Business*, March 8, 2007.

19. (a) Booten, K. and Levecke, B., *Specialty Chemicals Magazine*, 2004 *(March)*, 14; (b) Tharwat, T., Levecke, B., and Booten, K., *Chimica Oggi*, 2006, 55.

2 Solution Behavior of Alkyl Glycosides and Related Compounds

Patrizia Andreozzi, Giacomo Gente, and Camillo La Mesa

CONTENTS

2.1 INTRODUCTION

New generation surfactants exhibit a variety of physical–chemical properties, making them more performance oriented compared to the classical ones. Their characters arise from the need to get products having properties useful for ad hoc applications. According to the rules required by national or international standards, new generation surfactants are obtained from renewable sources, using green and low-cost chemicals. Biodegradable, high added value, and versatile materials are, thus, produced. Among them alkyl glycosides (indicated as AGs), alkyl thio-glycosides (ATGs), alkyl maltosides (AMs), their homologues and derivatives, betaines (zwitter-ionic surfactants), Gemini ("twin" species), and Bolas (bearing two polar groups linked to both ends of long alkyl chains) are the most promising new generation surfactants. Bolas and Gemini may have glycosidic polar head groups as well [1].

AGs, ATGs, AMs, and related sugar-based species can be chemically functionalized, giving esters, amines, sulfonic and carboxylic acid derivatives, etc, and may allow getting different ionic species (Chapter 9) [2]. The physical–chemical properties of such ionic derivatives are described in Chapter 9 and shall not be dealt with here.

Glycosides and structurally related species fulfill most of the following requirements:

1. High biocompatibility
2. Low, if any, protein denaturant action
3. Nonirritant character toward the derma
4. Good solvent capacity with respect to lipids and sterols
5. High surface-active efficiency
6. Relevant adsorption onto solid surfaces
7. Lubricant activity

Nonionic glycosides fulfill the desired performances in presence of co-solutes and from low to moderate temperatures in solution or in complex matrices (such as gels, creams, and pastes). The nonionic character of commercially available AGs allows modulating their solution properties in a wide range of experimental conditions and avoids drawbacks associated with the control of pH and ionic strength, which are, conversely, pertinent to many mild surfactants.

Compared to commercial nonionic surfactants, such as alkyl polyoxyethylene glycols [3,4], AGs do not show the presence of critical phenomena or consolute boundaries (with relevant exceptions) [5,6]. Aggregates formed by glycosides are characterized by hydrophilic surfaces, consisting in more or less compact layers of hydrated sugar units. The polar head groups are bulky and may form extended hydrogen-bond networks.

An extended thermodynamic characterization on glycosides in solution was performed in the past, since the pioneering work of Shinoda [7–11], Jones [12,13], Rosen [14,15], and others [16–25]. Studies reported so far span from adsorption at the air–solution interface [26,27] to micelle formation and growth [28–31]. AGs are also used in studies on the adsorption onto solids [32,33], and may have a different behavior, depending on whether hydrophilic or hydrophobic surfaces are considered [34,35]. They play a key role in the rheology of complex fluids [36], form thermotropic and lyotropic liquid crystals [37–39], are used in emulsion stabilization [40], to mind but a few selected properties. Hence, possible applications are many fold.

Formulations based on glycosides show performances-related properties in detergency [41], antifoaming agents [42], template chemistry [43], household, body care and cosmetics [44,45], biomimetic sciences [46], and biochemical separation [47]. There are several good reasons, thus, to have an exhaustive knowledge of their physical, chemical, technological, and biological properties. This is documented by many review articles issued on specialist journals and books in the last years [48–61].

Studies reported so far deal with AGs, ATGs, AMs, linear and branched polysaccharides (schematically defined as alkyl polyglycosides, APGs) [62,63], more exotic species, such as lactosides and galactosides [64], and their fluoro-alkane derivatives, as well [65,66]. Many contributions focus on their solvent capacity with respect to lipids and proteins. An exhaustive knowledge of their thermodynamic properties offers the way to estimate lipid-surfactant, polymer-surfactant, and protein-surfactant interactions [13,67], and to quantify their recovering capacity from complex biological matrices with respect to proteins or lipids.

In this chapter, attempts are made to report on current thermodynamic knowledge, to draw the behavior of some AGs, and, in very few cases, of APGs, and hydrophobically modified polysaccharides (HMPSs). The former products are linear block copolymers, consisting of alkyl chains linked to polysaccharide chains. The latter, conversely, are graft polysaccharides, consisting in a series of alternating glycosidic domains separated by alky glycoside groups (Figure 2.1).

Comparing the latter species with simple AGs may give relevant information on the contribution played by saccharide units in the association features. This line

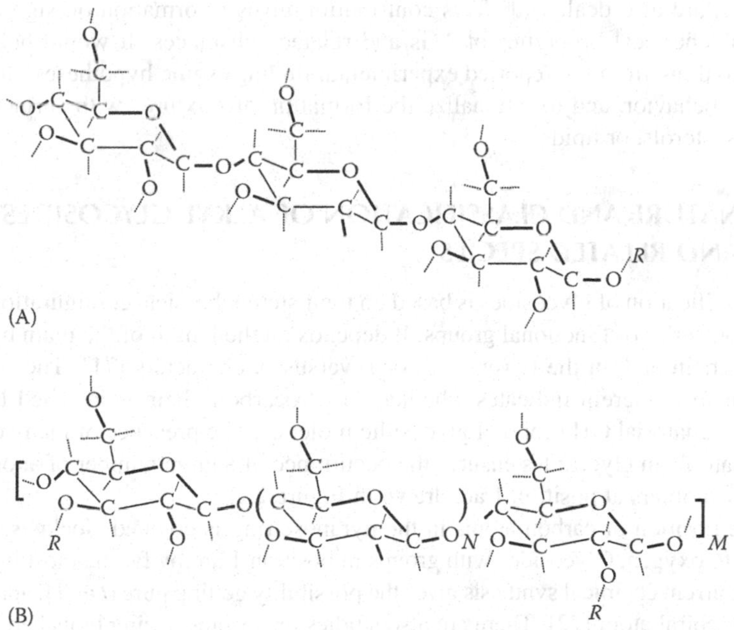

FIGURE 2.1   Schematic view of linear alkyl poly-glycosides, PAGs, characterized by the presence of a hydrocarbon chain, R, at one extremity of a poly-glycoside unit, A, and of (randomly grafted) hydrophobically modified polysaccharides, HMPSs, B. The symbols $\langle N \rangle$ and $\langle M \rangle$ in B indicate the number of sugar units in the polymeric sequences.

finds analogy with current work on alkyl polyethylene oxide glycols, indicated as AEO_N [68]. Knowledge of glycoside properties may predict the adsorption performances of AGs compared with APGs and HMPSs onto solid surfaces and particles [69]. The latter strategy seems promising in surface functionalization and stabilization of large colloid particles. It also includes new functional materials based on commercially available low-cost polysaccharides, such as inulines and the related HM derivatives.

2.2 SCOPE OF THE CONTRIBUTION

In the following, we report on the thermodynamics of micelle formation and on the nonideal mixing behavior of glycosides. Studies reported so far give a comprehensive view of a complex phenomenology, which is generally rationalized according to the hydrophilic–lipophilic balance (HLB). Alternative approaches take into account the compensation between temperature-dependent enthalpy and entropy contributions to the aggregate stability [70]. Modeling the above findings in terms of mass action or (pseudo) phase separation models is reported. Brief indications are also given on the possibility to treat their supramolecular association features in terms of the small systems thermodynamics (SST).

Modifications in the physical–chemical properties associated with changes in composition, in some thermodynamic properties and in the subsequent phase behavior, are also dealt with. This contribution gives information on significant physical–chemical properties of AGs and related substances. It would help the reader to draw from the reported experimental findings some hypotheses on their solution behavior, and to rationalize the formation of mixtures with surfactants, proteins, sterols, or lipids.

2.3 NATURE AND CLASSIFICATION OF ALKYL GLYCOSIDES AND RELATED SPECIES

The classification of glycosides is based on their stereochemical configuration and on the presence of functional groups. It depends on the length of the main hydrocarbon chain and on the α versus β, or L versus D, characters [71]. The α or β denomination therein indicates whether a hydrocarbon chain is attached to the axial or equatorial OH group of glycosidic moieties. The presence of many chiral carbon atoms in glycosides ensures the occurrence of a large number of anomers. α and β anomers at position 1 are drawn in Figure 2.2.

The position of carbon atoms in the pyranose ring is counted clockwise with respect to oxygen. Glycosides with groups in position 1 are by far the most investigated. Current chemical synthesis gives the possibility getting pure α and β anomers for each chiral atom [72]. There are also studies on species having branched sugar units [73], or alkyl groups located onto other positions in the sugar ring [74].

Glycosides can be conformationally pure or not, depending on the synthetic procedures; they differ from each other not only in absolute conformation but also in solubility, phase behavior, critical micelle concentration (CMC) values, and performances. There is a remarkable influence of carbohydrate stereochemistry

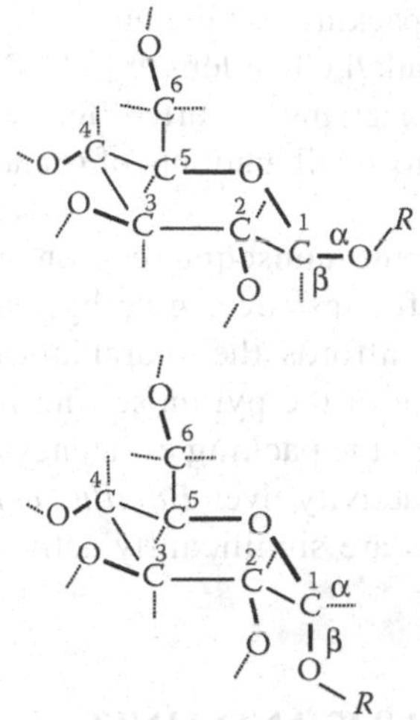

FIGURE 2.2 Absolute conformation of α- (up) and β-glycopyranoside anomers, when the alkyl groups are bound at position 1. C–O bonds are in bold, C–H and O–H bonds are indicated by dotted lines. Hydrogen atoms are not shown, for the sake of clarity; R indicates an alkyl chain. The position of different carbon atoms in the glycopyranoside unit is counted clockwise with respect to oxygen.

on glycoside properties at a molecular level. α and β anomers, or L and D isomers, have different bulk and surface properties. For instance, differences in the melting and clearing temperatures for lyotropic liquid crystalline phases of galactosides arise from a more extensive hydrogen-bond network between proximal head-groups with respect to glycosides [75]. This behavior is presumably due to the axial position of the four OH groups.

Compared to AGs, galactoside-based surfactants have a much higher hydrogen bonding degree, and water surrounding such micelles has long residence times, due to the configuration of OH groups facing toward the bulk [76]. This behavior is reflected in the strength of solute–solvent interactions, and some head groups are considerably more solvated compared to others. Head-group interactions with water are strong and modulate the entropy of micelle formation. Intermolecular hydrogen-bond networks may settle out between the sugar units, or with the solvent, depending on their stereochemistry [36,77].

The links between performances and stereochemistry become evident by comparing α-D and β-D-mono-alkyl glycopyranosides with 8 to 16 carbon atoms in the main chain. The above products are prepared according to Fischer's, Koenigs–Knorr's, or Schmidt's reaction routes [78], dynamic light scattering (DLS), differential scanning calorimetry (DSC), Fourier–transform IR spectroscopy (FTIR), cross polarization magic angle spinning (CP-MAS) and x-ray diffraction investigated the physicochemical properties of the resulting anomers in solution and in solid form. Above the CMC, micelles or bilayer-like aggregates (disks) can be observed, depending on the configuration of the anomeric center.

In solid state, the molecular packing and the number of hydrogen bonds between head groups depend on the alkyl chain length [37]. Presumably, the differences observed in some phase diagrams reported for alkyl β-D-glycopyranosides [38,79,80] may be ascribed to small amounts of α anomers in the mixture, or to the presence of L derivatives.

Stereochemistry has some consequences on biochemical activity too. Arabino-hexopyranosides, for instance, give by reaction α or β glycosides. Deacetylating the anomers affords the L-arabino-hexopyranosides. In such systems, the stereochemistry of the pyranose ring is related to the surfactant hydrophobic character and to the packing efficiency at the air–water interface. In addition, the fungicidal activity over *Bacillus cereus* and *Bacillus subtilis* indicates that L-pyranosides are significantly active, while D-ones much less active [81].

2.4 PHASE EQUILIBRIUMS AND LINKS WITH OTHER PROPERTIES

Mixtures composed of water and glycosides (with, eventually, other species) [82] undergo a series of phase transitions, depending on composition and temperature. Binary, ternary, and higher-order systems are reported in the literature. The presence of different phases is reflected by significant changes in some macroscopic properties, in solute (or solvent) activity, and in many effects related to thermal transitions as well [83]. The selected macroscopic appearance and the optical textures of different liquid crystalline phases obtained by selected AGs and AMs in mixtures with water are reported in Figure 2.3.

Thermodynamic information on liquid crystalline phases is based on DSC [73,80] and solvent activity measurements [84,85]. Water sorption isotherms indicate that the activity of each component (in isothermal mode) changes at the molecule–micelle equilibrium, at the solution–liquid crystalline, and at the lamellar–cubic phase transitions, respectively [86].

Mixtures made of *n*-octyl-β-D-glycopyranoside and water are strongly nonideal and Raoult's law, accordingly, is not fulfilled in such systems. Changes in the chemical potentials of the components, μ_i, occur in proximity of each phase transition. This behavior, shown in Figure 2.4, allows one to determine whether first- or second-order phase transitions are dealt with.

In particular, the derivative of the chemical potential for the solute (or the solvent), $\left(\frac{\partial \mu_i}{\partial X_i}\right)$, is significantly different in each phase. This behavior is reflected in the respective slopes of the function below or above the transition concentration. It also accounts for changes in the relative weight of different contributions to the chemical potential and depends on phase structure. μ_i values of the surfactant, in particular, contain information on the weight of different contributions to the overall system stability, including hydration, curvature, or size and shape effects. In supramolecular association, in fact, the chemical potential for the surfactant is the sum of different energy contributions, ascribed to [87,88]

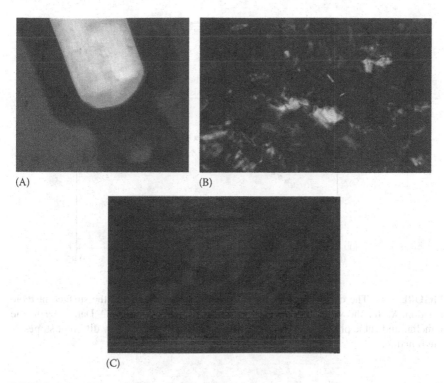

FIGURE 2.3   Selected examples of the phase behavior observed in concentrated aqueous glycoside solutions. In (A), the optical birefringence is reported in a 71.0 wt% octyl-β-D-thio-glycoside sample, (located in a 1.0 mm glass tube), at 20°C. In (B), lamellar textures is shown in an 82.5 wt% dodecyl-maltoside system, at 40°C. In (C), the fanlike textures are reported relative to a 67.0 wt% octyl-β-D-glycopyranoside, at 15°C.

1. Transfer energy of a surfactant molecule from water to a nonpolar, micelle-like, environment, ΔG_{tr}
2. Work required forming an interface, ΔG_{int}
3. Energy required to built up an electrical double layer, ΔG_{el}
4. Some constraints on the alkyl chain conformation, ΔG_{conf}
5. Effect of polar head groups hydration, ΔG_{hydr}
6. Steric repulsions, ΔG_{st}
7. Hydrogen bond contributions, ΔG_{HB}
8. Other terms

The relative weight of each contribution determines the overall system stability. As to hydrogen bonds, it is cumbersome estimating their stabilizing role from simple models. It is well known that their dependence on distance between adjacent groups obeys a complex, and strongly directional, power law behavior, which is approximately close to two [89]. Such considerations do not hold only for

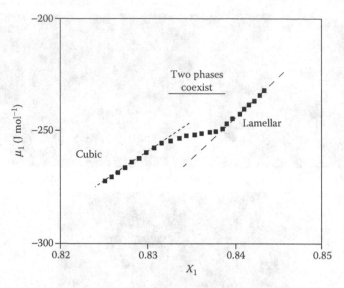

FIGURE 2.4   The chemical potential of water, μ_1 (J mol^{-1}) versus the surfactant mole fraction, X_i, for the water/octyl-β-D-glycopyranoside system, at 60°C. Data refer to the lamellar and cubic phase, respectively. Each phase is characterized by different slopes of the function.

micelles and apply also to the formation of microemulsions and, particularly, of liquid crystalline phases.

Calculations based on chemical potentials are useful in modeling size and shape transitions. Since the pioneering work of Mazer et al. [90,91], efforts were devoted to rationalize such transitions on thermodynamic grounds. In the sphere to rod transition, for instance, aggregates grow in one direction and both ends of the rod still keep a semi-spherical shape, as indicated in Figure 2.5. μ_i values can be easily calculated according to such an hypothesis.

Among many others, Nagarajan and Blankschtein recently proposed thermodynamic developments of theories on micelle size and shape [92–96]. Their contributions allow one to determine the driving forces responsible for association, to detect structural modifications of the aggregates, or to predict phase transitions. Such theories, in addition, are relevant in evaluating the preferred self-assembly mode for a given species.

In thermodynamic terms, the chemical potential is defined according to

$$\mu_i = \left(\frac{\partial \Delta G}{\partial n_i} \right)_{P,T} = \bar{L}_i - T\bar{S}_i \qquad (2.1)$$

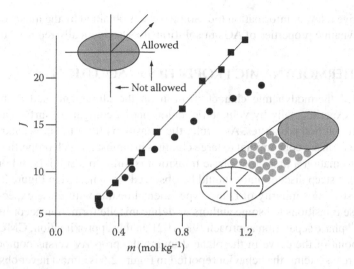

FIGURE 2.5 Growth mechanism of micellar aggregates normal to their section, to give cylindrical micelles. In the right hand side is reported the dependence of micelle axial ratios, J, on octyl-β-D-glycopyranoside molality, m_i (mol kg⁻¹), at 25.0°C. Plots refer to soft and deformable cylindrical micelles, circles, or stiff and rigid ones, squares.

where $\bar{L}_i \left(= \left(\frac{\partial \Delta H}{\partial n_i} \right) \right)$ and $\bar{s} \left(= \left(\frac{\partial \Delta S}{\partial n_i} \right) \right)$ indicate the partial molal enthalpy and entropy, respectively. Surface and electrical work contributions also may be introduced into Equation 2.1.

It can also be demonstrated that the partial molal volume, $\bar{V}_i$, fulfills the relation

$$\left(\frac{\partial \mu_i}{\partial P} \right)_T = \bar{V}_i \tag{2.2}$$

All partial molal quantities can be obtained by elaborating the data inferred by selected thermodynamic experiments. A large body of information available on the thermodynamic properties of AGs and similar substances is based on volumetric, calorimetric, activity coefficients, and surface activity. The aforementioned quantities (and the derived ones, as well) may define the role of different contributions to the overall system stability in more detail. Structural information on size and shape transitions can be obtained from heat capacity [97]. The latter quantity, in fact, is extremely sensitive to structural changes, being the second derivative with respect to the system Gibbs energy. Partial molal heat capacities, expansibility, and compressibility (which are somehow linked to the quantities mentioned above) give relevant information too. Some aspects obtained by the most common thermodynamic properties of AGs in solution are systematically reported below.

2.5 THERMODYNAMIC PROPERTIES IN SOLUTION

A detailed thermodynamic characterization on the aforementioned systems is obtained experimentally by volumetric, calorimetric, colligative, surface activity, and other solution properties. As a rule, the transitions from the molecular to the associate form are concomitant to large changes in almost all such properties. Were micelle formation a first-order phase transition (similar to a solid–liquid one, for instance), a steep discontinuity should be observed, as indicated in Figure 2.6.

The striking similarity of many experimental results with those expected for true phase transitions led some authors to define micelle formation according to a "pseudo" phase separation approach [98–102]. In this approximation, CMC is the salient point of the curve in the plane drawing the property versus composition plots. In real systems, the behavior reported in Figure 2.6 is almost never observed; it is approached only when very long alkyl chains are being used (i.e., when the number of molecules into micelles is extremely high). The behavior depicted by the dotted line in Figure 2.6, conversely, holds when micelle size is moderate and the alkyl chains contain less, or about 10 carbon atoms.

An important consequence associated to the above statements implies that a thermodynamically consistent definition of CMC fulfills the following requirement, expressed in differential form as [103]

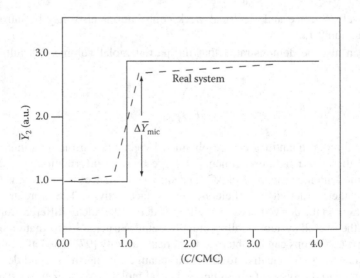

FIGURE 2.6 Plot of a generic partial molal quantity, $\overline{A}_i$, in arbitrary units, versus the a-dimensional ratio (C/CMC). The vertical arrow indicates changes in $\overline{A}_i$ at the CMC. The dotted line indicates the behavior occurring in real systems, when micelle formation is moderately cooperative.

$$\left(\frac{\partial^3 \bar{A}_i}{\partial X_i^3}\right)_{T,P} = 0 \qquad (2.3)$$

where

$\bar{A}_i$ is any experimentally determined partial molal quantity

X_i the surfactant mole fraction in the medium

Thermodynamic properties are almost always elaborated from apparent molal quantities, defined as $\Phi_{Y,i}$. The latter are linked to the corresponding partial molal ones, $\bar{A}_i$, by the relation [104]

$$\bar{A}_i = \left(\frac{\partial\left(\Phi_{Y,i} X_i\right)}{\partial X_i}\right) = \Phi_{Y,i} + X_i\left(\frac{\partial\Phi_{Y,i}}{\partial X_i}\right) \qquad (2.4)$$

In most instances, it is not possible getting partial molal quantities directly from experiments, but it is easy getting them from calculations based on Equation 2.4. Details on the derivation procedures and transformations are reported in Table 2.1.

The correct definition of CMC in Equation 2.3 is coincident with the salient point of the plot in Figure 2.6, where changes in partial molal quantities, $\Delta\bar{A}_i$, must be calculated. Other possible definitions of critical concentration currently are used. In some cases, for instance, the CMC is considered the inflection point at the first change in slope of the curve in Figure 2.6 [105–108]. In all cases reported so far, micelle formation is always concomitant with changes in slope or

TABLE 2.1
Transformation of Some Apparent Molal Quantities, $\Phi_{Y,i}$, into Partial Molal Ones, $\bar{A}_i$

$\Phi_{Y,i}$	$\bar{A}_i$	Dimensions	Transformation
Practical osmotic coefficient, $\Phi_{O,i}$	Activity coefficient, γ_i	A-dimensional	$\ln\gamma_i = \left(\Phi_{O,i}-1\right)\left[1+\left(\frac{1}{M^\circ}\right)\int_0^{m_i}\left(\frac{\partial m_i}{m_i}\right)\right]$
Volume, $\Phi_{V,i}$	Volume, $\bar{V}_i$	cm³ mol⁻¹	$\bar{V}_i = \Phi_{V,i} + X_i\left(\frac{\partial\Phi_{V,i}}{\partial X_i}\right)$
Heat of dilution, $\Phi_{L,i}$	Heat of dilution, $\bar{L}_i$	J mol⁻¹	$\bar{L}_i = \Phi_{L,i} + X_i\left(\frac{\partial\Phi_{L,i}}{\partial X_i}\right)$

Notes: [a] The term $(1/M^\circ)$ before the integral indicates the solvent molar mass. In mixed systems, it is the weight average value of mole fractions and molar mass of the components, and is calculated according to $\langle M^\circ\rangle = \sum_j\left(M_j^\circ X_j\right)$.

[b] For thermodynamic consistency, the lower limit in the integral must be replaced with a reference m_i^0 value. For more details, see Ref. [100].

in activity, volume, heat of dilution, compressibility, and heat capacity versus concentration plots, as indicated in the forthcoming sections.

Significant differences in thermodynamic properties between the molecular and the micellar regimes are almost always observed. In the former concentration range, partial molal quantities are generally expressed in terms of a power law equation for composition, according to the virial-like expression

$$\overline{Y}_i = \overline{Y}_i^0 + A_1 X_i + B_2 X_i^2 + C_3 X_i^3 + \cdots = \sum_{j=0} A_j X_i^j \qquad (2.5)$$

where

$$A_0 = \overline{Y}_i^0 \qquad (2.6)$$

The constants in Equation 2.5 represent the second, third, and higher virial coefficients of the power law equation. There, X_i is the mole fraction of the surfactant solute in the medium and $\overline{Y}_i^0$ the limiting value of the partial molal quantity at infinite dilution. Such term is a reference state for most thermodynamic quantities, except free energies and entropies.

All concentration scales can be used in Equation 2.5 and molality ($m_i = $ mol kg^{-1}), or molarity ($M_i = $ mol dm^{-3}), may replace X_i, which is, perhaps, currently used in concentrated regimes. It is worth noticing that m_i does not depend on temperature and, therefore, is extensively used in thermodynamic calculations on dilute solutions.

Equation 2.5 applies to almost all thermodynamic quantities (activity, enthalpy, volume, etc.), each having its own A_1 constants. In water, such constants are well known in a wide temperature and pressure range [109]. Other terms in the summation reported in Equation 2.5 are obtained by the polynomial fits. Usually, second- or third-order polynomials fit the data with an appreciable accuracy.

All the above statements are valid also in saline media or in mixed solvents, provided the relative ratios between the components are kept fixed. In salt solutions or mixed solvents, perhaps, the A_1 constant in Equation 2.5 must be recalculated. The reasons for this find their justification in classical thermodynamics, which imposes a strict relation between A_1 and the medium permittivity [110]. The value of A_1, in fact, is related to the transfer energy for the solute from water to another solvent medium. Consider, as an example, the dilution enthalpy for a nonionic solute in water at 298.15 K. According to this theory, A_1 is 29,805 J mol$^{-3/2}$ kg$^{1/2}$, when the concentration is in m_i scale. When the solvent is not water, the following relation holds [110]

$$A = \left(\frac{1.67 * 10^7}{\varepsilon}\right)\left(\frac{1}{\sqrt{T\varepsilon}}\right)\left[1 + \left(\frac{T}{\varepsilon}\right)\left(\frac{\partial \varepsilon}{\partial T}\right)\right] \approx \left(\frac{1.67 * 10^7}{\varepsilon}\right)\left(\frac{1}{\sqrt{T\varepsilon}}\right)\left[1 + \left(\frac{\partial \ln \varepsilon}{\partial \ln T}\right)\right] \quad (2.7)$$

where
　ε is the medium permittivity
　T is the temperature, K

(Note: For ionic species, the summation in Equation 2.5 is a power law function of $\sqrt{X_i}$ and A_1; the limiting slope of the quantity under test is predicted by Debye–Hückel's theory for the thermodynamic properties of electrolytes in solution.)

When micelles are formed, they coexist with the surfactant in molecular form. Hence, it is not possible getting relevant thermodynamic information from Equation 2.5. In this case, iterative procedures are required to extract thermodynamic quantities from the data. Suppose, as an example, that a dilute micellar solution, at concentration slightly above the CMC, C_i, is diluted with an equal amount of water, so that the final solute concentration, $C_i/2$, is below the critical threshold. The partial molal enthalpy at $C_i/2$ fulfills Equation 2.5, whereas the one at C_i does not. In such case, the integral dilution enthalpy of the process, $\Delta H_{\text{int,dil},i}$ is defined as

$$\Delta H_{\text{int,dil},i} = \Phi_{L,i,\text{mol}} - \Phi_{L,i,\text{mic}} \tag{2.8}$$

where

$\Phi_{L,i,\text{mol}}$ refers to molecular
$\Phi_{L,i,\text{mic}}$ to the micellar form, respectively

Since

$$\Phi_{L,i,\text{mol}} = A_1 X_i + B_2 X_i^2 + C_3 X_i^3 + \cdots \tag{2.9}$$

combination of Equation 2.8 with Equation 2.9 gives the $\Phi_{L,i,\text{mic}}$ value, to be used in further computational stages. A sketch of that procedure is drawn in Figure 2.7.

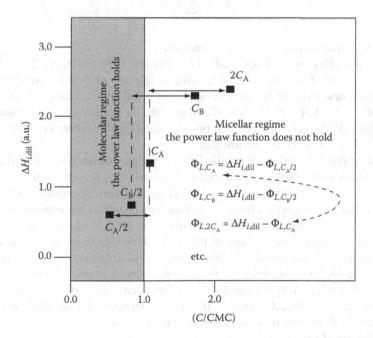

FIGURE 2.7 Procedure required to calculate the apparent molal enthalpies of dilution in micellar systems. In the regime, at a concentration, $C_i > \text{CMC}$, $\Phi_{L,i,\text{mic}}$ is equal to the difference $\Delta H_{\text{int,dil},i} - \Phi_{L,\text{mol}}$. The latter quantity, taken at a concentration $C_i/2 < \text{CMC}$, is obtained by fitting the data according to Equation 2.8.

2.5.1 VOLUMETRIC PROPERTIES

The possibility to get precise density values, ρ, from vibration densimeters greatly improved the quality of volumetric data to concentrations as low as $10^{-5}\ X_i\ (2.0 \times 10^{-4}\ m\ kg^{-1})$. Thermostatting to $0.01°C$ gives density values accurate to $1.0 \times 10^{-5}\ g\ cm^{-3}$ or less.

Apparent molal volumes, $\Phi_{V,i}$, are directly obtained by density ones through well-known equations. In molality scale, they are obtained by [111]

$$\Phi_{V,i} = \left(\frac{1}{\rho}\right)\left[MW_i - \left(\frac{10^3\left(\rho - \rho^0\right)}{\rho^0 m_i}\right)\right] \qquad (2.10)$$

where
ρ is the solution density
ρ^0 is that of the solvent
MW_i is the solute molecular mass

In concentrated regimes, the transformation to mole fraction units is required. In any case

$$V_{tot} = \sum_{i=1} \bar{V}_i X_i \qquad (2.11)$$

where V_{tot} is the overall volume of the mixture.

Thanks to Desnoyers [18], Fukada [112], Harada [113], and others [23,31,114], a large amount of volumetric data relative to selected glycosides are currently available. These indicate the occurrence of significant volume changes in proximity of the CMC. For a given class of compounds, the volume change for micelle formation, $\Delta \bar{V}_{i,mic}$, regularly depends on temperature and number of carbon atoms in the main alkyl chain, N_C, Figure 2.8.

The volumetric behavior of alkyl β-D-glycopyranosides reported therein is consistent with other nonionic species, such as alkyl dimethylamine oxides [105], alkyl dimethylphosphine, and alkyl dimethylarsine oxides [115]. The above behavior is expected from considerations based on the transfer of alkyl chains from water to a micelle-like environment [105,116]. According to the literature, additivity rules occur as far as the dependence of $\bar{V}_i^0$ on the alkyl chain length is concerned, but departures from linearity are observed in $\Delta \bar{V}_{i,mic}$ values. This behavior is in agreement with studies by Sesta, who showed that $\Delta \bar{V}_{i,mic}$ is not related only to the transfer of surfactant molecules from water to a micellar environment. She suggested that other effects, particularly those due to a partial dehydration of the polar head groups, are also relevant [117].

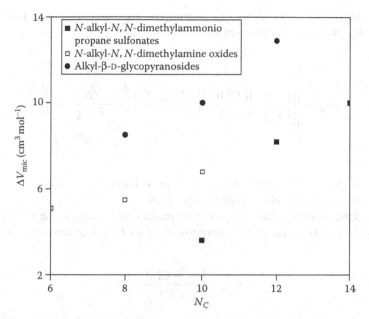

FIGURE 2.8 Volume change upon micelle formation, $\Delta V^0_{mic,i}$, (cm³ mol⁻¹) versus the number of carbon atoms in the surfactant alkyl chain, N_C, for N-alkyl-N,N-dimethylammonio propane sulphonates, full squares, N-alkyl-N,N-dimethylamine oxides, empty squares, and alkyl-β-D-glycopyranosides, circles at 25.0°C. (Values taken from Refs. [111,112,114,115]).

Pressure plays a significant role on both CMC and $\Delta \bar{V}_{i,mic}$ values. The latter quantity decreases on increasing the applied pressure, P, and changes in sign from positive to negative [118,119]. This behavior is not unexpected. It can be demonstrated, in fact, that

$$RT \left(\frac{\partial \ln CMC}{\partial P} \right)_T = \Delta \bar{V}_{mic,i} \qquad (2.12)$$

The point at which $\Delta \bar{V}_{mic,i}$ is zero is known as Tanaka's pressure. It is a technologically relevant quantity and indicates how applied pressure shifts the molecular–solid–micellar triple point (also termed Krafft point) to high temperatures [9]. Despite its relevance in lubricants and hydraulic fluids, unfortunately, no studies have been performed along this line on alkyl glucosides.

It is possible estimating the volume change of micelle formation from ultrasonic relaxation. This method gives the exchange (or relaxation) time associated with the entry–exit of surfactant molecules from micelles [120]; at the same time, the relaxation amplitude is grossly proportional to $\Delta \bar{V}_{mic,i}$ [121]. The absorption and propagation of sound waves in a given medium, in fact, depend on P and T. Therefore, the relaxation amplitude depends on $\Delta \bar{V}_{mic,i}$ and/or $\Delta \bar{H}_{mic,i}$ [122], as a consequence of Le Chatelier's principle.

At the relaxation frequency, f_{rel}, the maximum sound absorption per wavelength, $\mu_{max} = (\alpha\lambda)_{max}$, is expressed as

$$\mu_{max} = \left\{ \left[\left(\frac{\pi}{\kappa_{S,i,\infty}} \right) \left(\frac{CMC}{RT} \right) \left(\frac{(\Delta V_{mic,i})^2}{V} \right) \right] \left[\frac{\dfrac{\sigma^2 X_{i,red}}{\langle N_i \rangle}}{1 + \left(\dfrac{\sigma^2 X_{i,red}}{\langle N_i \rangle} \right)} \right] \right\} \qquad (2.13)$$

where
$\kappa_{S,i,\infty}$ is the high frequency adiabatic compressibility of the solute
σ^2 accounts for micelle polydispersity [120]
$\langle N_i \rangle$ is the average number of surfactant molecules, or ions, in the aggregate
$X_{red,i}$, finally, is a reduced concentration scale for the surfactant, defined as

$$X_{red,i} = \left(\frac{m_i - CMC_i}{CMC_i} \right) \qquad (2.14)$$

$\Delta \bar{V}_{mic,i}$ values obtained by ultrasonic relaxation are slightly higher (10%–15%) than those obtained by classical volumetric methods [121,123]. The reasons for such discrepancies are at yet unknown. Very presumably, they arise by neglecting enthalpic or heat capacity terms in the equation for the maximum sound absorption per wavelength. It is well acquainted, in fact, that the propagation of acoustic waves in a given medium results in an energy dispersion and implies subsequent changes in $\bar{V}$, $\bar{H}$, and $\bar{C}_P$. The small variations with respect to volumetric data are presumably due to the fact that $\Delta \bar{H}_{mic,i}$ for ionic and nonionic species is close to zero at room temperature [119,124].

Recent efforts were made to detect volumetric properties in mixed surfactant systems. Aicart et al. determined the volumetric properties in octyl-β-D-glycopyranoside/dodecyl- (or tetradecyl)- trimethylammonium bromide mixtures [125,126]. The above results indicate that $\Delta \bar{V}_{mic,i}$ is not a linear function of composition. This behavior is not unexpected, since $\Delta \bar{V}_{mic,i}$ contains terms due to the transfer of not only surfactant molecule into micelles but also head group hydration ones.

When the alkyl chains of the components have different length, in addition, $\Delta \bar{V}_{mic,i}$ depends on the surfactants packing into micelles in at yet unspecified way. More details on the thermodynamic elaboration and modeling based on volumetric properties, calculated according to the pseudo-phase or mass action approaches, are discussed in Section 2.6.

2.5.2 Calorimetric Properties

Compared to batch and flow methods, isothermal titration calorimeters (ITC) and similar equipments have greatly improved the use of solution calorimetry. Information inferred from calorimetry spans from the detection of CMC values to the determination of micellization enthalpy and, eventually, to heat capacity changes. The sensitivity given by calorimetric methods ensures the possibility to get reliable $\Delta H_{int,dil,i}$ values at concentrations as low as $10^{-4} m_i$. Therefore, surfactants having CMC in the millimolal range can be safely investigated. The CMC regularly decreases with alkyl chain length and the heat change associated to micelle formation becomes, generally, sharper.

Studies reported so far deal with dilute solutions, where thermal effects due to inter-micellar interactions are immaterial. In such regime, micelles are immersed in a medium containing molecular surfactant, at concentration close to the CMC and continuously exchanging molecules with the bulk. Micelle–micelle interactions are moderate, particularly in nonionic species. Hence, changes in $\Delta H_{int,dil,i}$ are moderate, if any, above the CMC.

$\Delta \overline{H}_{mic,i}$ values significantly depend on T and change in sign above a temperature that is peculiar to the surfactant [119]. This effect is related to the compensation between enthalpic and entropic contributions to micelle formation. At concentrations well above the CMC, the observed heat effects level off, unless micelles modify their size and shape. Such behavior is also reflected by minor (second order) changes in heat capacity values [97].

Two distinct calorimetric regimes hold in solution. Operationally, the CMC is the concentration value above which calorimetric data cannot be fitted according to the classical theory for molecular solutions. On going from the molecular to the micellar range, data interpolation is cumbersome, unless proper calculus procedures are used. Details on such procedures have been reported in an exhaustive review by Desnojers and Perron [109].

The calorimetric behavior of AGs, ATGs, and AMs is based on experimental results on pure aqueous systems (and in mixed systems) [23,114,124,127,128], in concentrated solution regimes, through DSC measurements [73,74,80], in mixed glycoside-lipid systems [129,130], and on glycosides adsorption onto surfaces [131,132]. As to binary systems, data indicate the occurrence of a marked increase (or decrease) of the dilution enthalpy on concentration, the salient point being located at the CMC. Above that limit the integral heat of dilution, $\Delta H_{int,dil,i}$, levels off.

Particularly relevant are the studies on N-decanoyl-N-methyl glucamine–water mixtures [124]. Experiments on the above system were performed in a wide temperature range and indicate a significant change in sign of $\Delta H_{int,dil,i}$ on increasing T (Figure 2.9). A peculiar behavior, which deserves further investigation, arises from some discrepancies between experimentally determined $\Delta \overline{H}_{mic,i}$ values and the values calculated by van't Hoff equation. This effect is presumably ascribed to changes in the number of molecules participating to micelle formation. In the above system, the enthalpy change, $\Delta \overline{H}_{mic,i}$, decreases with increasing T, reaches zero at 35.0°C, and becomes progressively more negative above that value. Hence, the thermodynamics of micelle formation can be expressed in terms of compensation between enthalpic and entropic terms. The above behavior also shows nice correspondence with that observed in ionic surfactants [70,119,133].

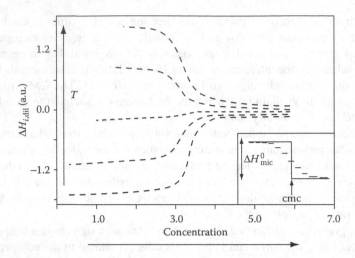

FIGURE 2.9 Integral enthalpy of dilution, $\Delta H_{int,dil,i}$ (kJ mol^{-1}) versus the millimolarity (m mol l^{-1}) of decanoyl-N-methylglucamine in water at different temperatures. Note the inversion in sign of $\Delta H_{int,dil,i}$ and of $\Delta \bar{H}_{mic,i}$ by raising T.

According to standard thermodynamic modeling, enthalpy–entropy compensation effects can be detected from $\Delta \bar{H}_{mic,i}$ versus T plots. The enthalpic contribution plays a major role in the Gibbs energy of micelle formation at low temperatures, when the entropic term, conversely, becomes dominant at high T values [134,135].

The interactions between molecular octyl-β-D-thioglycopyranoside and lecithin such as POCP, 1-palmitoyl-2-oleoyl-sn-glycero-3-phosphocholine, were also reported [136–138]. Mixing of the two species is not ideal and the (mixed) vesicle–micelle transition implies significant partitioning of the surfactant between the bulk and the lipid bilayers. The corresponding enthalpy is a linear function of octyl-β-D-thioglycopyranoside mole fraction in the membrane. Compensation between enthalpic and entropic contributions occurs in this system too. Comparison with the solution behavior of octyl-β-D-glycopyranoside indicates that thio derivative is much more hydrophobic compared to the former, as can be also inferred from $\Delta \bar{H}_{mic,i}$ for the pure species (Table 2.2).

The relevance of $\Delta \bar{H}_{mic,i}$ values in thermodynamic analysis is noticeable, since many other quantities are obtained from it. In the pseudo-phase separation approach, for instance,

$$\Delta G^0_{mic,i} = RT \ln X_{i(CMC)} \tag{2.15}$$

where $X_{i(CMC)}$ is the solute mole fraction at the critical threshold. Accordingly,

TABLE 2.2
Standard Enthalpy of Micelle Formation, $\Delta H^0_{mic,i}$, (kJ mol⁻¹), for Alkyl-β-D-Glycopyranosides, Octyl-β-D-Thioglycopyranoside, and Decyl-β-D-Maltoside, at 25.0°C and 30.0°C, in Water

Substance	$\Delta H^0_{mic,i}$ 25.0°C	$\Delta H^0_{mic,i}$ 30.0°C
Octyl-β-D-glycopyranoside	6.4	6.0
Decyl-β-D-glycopyranoside	7.5	6.9
Dodecyl-β-D-glycopyranoside	10.1	—
Octyl-β-D-thioglycopyranoside	6.8	6.7
Decyl-β-D-maltoside	8.9	—

Source: Fukada, K., Kawasaki, M., Seimiya, T., Abe, Y., Fujiwara, M., and Ohbu, K., *Colloid Polym. Sci.*, 278, 576, 2000.

Note: The uncertainty on $\Delta H^0_{mic,i}$ values is to ±0.3 kJ mol⁻¹. It slightly decreases on increasing the alkyl chain length.

$$\Delta H^0_{mic,i} = \frac{\partial\left(\dfrac{\Delta G^0_{mic,i}}{T}\right)}{\partial\left(\dfrac{1}{T}\right)} \qquad (2.16)$$

The entropy and heat capacity changes, $\Delta S^0_{mic,i}$ and $\Delta C^0_{P,mic,i}$, respectively, are obtained by the relations

$$\Delta S^0_{mic,i} = \frac{\Delta H^0_{mic,i} - \Delta G^0_{mic,i}}{T} \qquad (2.17)$$

and

$$\Delta C^0_{P,mic,i} = \left(\frac{\partial\left(\Delta H^0_{mic,i}\right)}{\partial T}\right) \qquad (2.18)$$

Equation 2.18 is particularly suitable when differential calorimetric techniques are not available or the determination of $\Delta C^0_{P,mic,i}$ values is cumbersome. In such cases, care must be taken to get calorimetric measurements at very close temperatures, to avoid the overlapping of classical thermal effects with relaxational contributions (due to the shift of CMC values on increasing T or P) [109]. These imply additional problems in the analysis of experimental data and the subsequent elaboration requires significant data manipulation, which may mask the observed solution behavior.

2.5.3 COLLIGATIVE PROPERTIES

Colligative properties can be obtained by many experimental methods. Those involved with cryoscopy, ΔT_{cr} [19,106], and vapor pressure, $\Delta P_{vap\,pres}$ [139], are by far the most common. Modifications of the latter method measure the differential heat of vaporization for the solution with respect to the solvent, in presence of vapor-saturated air. This differential heat change gives rise to a difference in temperature on the electrodes onto which the liquids are stratified, is transformed into an electric signal and properly elaborated. The data quality can be high and makes possible getting accurate values below the millimolal range. Isopiestic methods are only of historical interest.

The osmotic coefficients, $\Phi_{cr,i}$, are inferred by cryoscopic data according to [106]

$$\Phi_{cr,i} = \left[\frac{2.303 \times 55.5 \times \Delta T \left[4.207 \times 10^{-3} + \left(2.1 \times 10^{-6} \right) \times \Delta T \right]}{m_i} \right]$$

(2.19)

where ΔT is the freezing point lowering and other symbols have their usual meaning. The numerical constants are inferred by proper use of the thermodynamic quantities relating the equality of chemical potentials for the solvent to that in solid form.

Activity coefficients, γ_i, are inferred by practical osmotic coefficients, as indicated in Table 2.1. The integration reported in the above table must be transformed into finite differentials, to ensure numerical and physical consistency to calculus procedures. A few studies have been reported on the colligative properties of aqueous alkylglycosides [20,21,23,31,114]. They define CMCs, the solid–solution equilibrium, and other effects associated with micelle formation. Small, but significant, humps of γ_i values close to the CMC are sometimes observed; the increase with respect to unity can be more or less significant [20]. Generally, micelle formation is concomitant to significant departures from the (nearly) ideal low concentration regimes and a steep decrease in activity coefficients is observed above the CMC. In Figure 2.10, the practical osmotic coefficients, $\Phi_{O,i}$, of a selected glycoside are reported. The curve relative to the activity coefficients (not shown) is quite similar. The reasons for that behavior arise from the small effect played by integration, calculated according to Table 2.1.

As to γ_i values, significant changes in slope are observed when micelles are formed. Above the CMC, the practical osmotic coefficients (and the activity coefficients too) rapidly level off. Drastic departures from the ideal behavior observed in dilute, molecular, regimes are concomitant to supramolecular formation. Thus, micellar solutions are strongly not ideal.

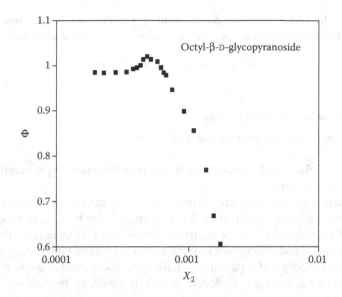

FIGURE 2.10 Practical osmotic coefficients, $\Phi_{O,i}$, versus the surfactant mole fraction, X_i, for octyl-β-D-glycopyranoside in water. The data, obtained by freezing point depression, are in semi-logarithmic scale. The CMC is located at the maximum of the curve.

Activity coefficients are sometimes used to determine micelle aggregation numbers in terms of the pseudo-phase separation model [109], or according to a mass action approach [105].

2.5.4 SURFACE ACTIVITY

Surface tension is of everyday use in the determination of CMCs and areas per molecule, $A_{mol,i}$ (nm^2). Balance methods are by far the most used, but dynamic, spinning drop, and stalagmometric ones are useful too. All such methods are well known and shall not be discussed. Methods based on the du Noüy ring facility are user friendly, simple, and fast. Drawbacks inherent to this technique are the amount of surfactant and the time required getting the equilibrium between bulk and surface phases. It can be several minutes long, because of the diffusion-controlled motion of surfactant molecules from the bulk to the surface phase. The effect is particularly relevant in close proximity of the CMC, because of the relatively long times (some minutes) required getting the equilibrium value. A kinetic analysis of surface adsorption based on the du Noüy method may be questionable, unless very long adsorption times are used. For more details, look at specialized books [140].

The work required to increase the air–solution interface area is expressed according to the well-known relation

$$\left(\frac{\partial G}{\partial A}\right)_{T,P} = \sigma \qquad (2.20)$$

where σ is the surface tension. Such quantity, indicating the work required to increase the air–water interface, can be related to the solute chemical by

$$\partial\sigma = -\Gamma_i^1 \partial\mu_i = -\Gamma_i^1 RT \partial \ln a_i \tag{2.21}$$

where
   a_i is the surfactant solute activity
   Γ_i^1 the excess surface concentration of the solute

In very dilute solutions, the solute activity is currently replaced by the corresponding concentration scale.

The surface excess concentration indicates the occurrence of a proportionality between bulk and surface properties. It is written as Γ_i^1, to indicate that the excess of the surfactant species is taken relative to the solvent (component 1). Formally, it is an operator transforming bulk to surface activity. Another interesting property of Γ_i^1 is its similarity with a partial molal quantity. According to Equation 2.21, in fact, the area occupied per molecule, which is inversely proportional to Γ_i^1, depends on concentration.

The minimum area per molecule, $A_{i,\min}$ (nm^2), is obtained from Γ_i^1 values by the relation

$$A_{i,\min} = \left(\frac{10^{18}}{N_A \Gamma_{i,\max}^1} \right) \tag{2.22}$$

where
   N_A is Avogadro's number
   $\Gamma_{i,\max}^1$ refers to the maximum slope of surface tension curve versus the solute activity in proximity of the CMC

In dilute concentration regimes, Γ_i^1 increases with surfactant amount and levels off when no more room is available for insertion of molecules in the interface. Hence, micelle formation is concomitant to surface saturation. This is exactly the meaning of CMC, given in a classical compilation of data [141].

Another quantity of interest is the Gibbs energy of adsorption, $\Delta G_{ads,i}^0$, defined as

$$\Delta G_{ads,i}^0 = \Delta G_{mic,i}^0 - \left(\frac{\Pi_{CMC,i}}{\Gamma_{i,\max}^1} \right) \tag{2.23}$$

where
   $\Pi_{CMC,i}$ is the surface pressure
   $\Pi_{CMC,i} = \sigma^\circ - \sigma_{CMC}$, at the CMC
   $\Delta G_{ads,i}^0$ is proportional to the Gibbs energy of transfer from water to micelles, indicated as $\Delta G_{mic,i}^0$, and is modulated by the surface pressure term in the right-hand side of Equation 2.23

The latter is the work required forming an interface. Thus, $\Delta G^0_{ads,i}$ is an important parameter in thermodynamic modeling, because it contains terms relative to different forces responsible for micelle formation, i.e., the transfer Gibbs energy and the surface work ones, respectively. On these grounds, it is also expected that the performances of different surfactants, characterized by very close CMC values, can be cogently evaluated by $\Delta G^0_{ads,i}$ terms.

Surface tension data relative to AGs are reported in Figure 2.11. The CMC therein is the intersection point of two straight lines. More correctly, it is the endpoint of surface activity, i.e., the concentration at which surface activity ends and micelles begin to form [141]. From that point onward, the surface tension remains roughly constant, and proper use of Equation 2.21 is a good approximation of micelle formation as the onset of a separate phase. It can be demonstrated, in fact, that

$$\left(\frac{1}{\Gamma^1_i RT}\right)\partial\sigma = \partial\ln a_i \approx 0 \qquad (2.24)$$

which is an implicit formulation of the pseudo-phase separation approach.

Above that point, the solute activity remains nearly constant and a new phase is formed. In real systems the situation is more complex. The nonconstancy of surface tension data above the CMC, in particular, was used to demonstrate that micelle formation fulfills a mass action model [21].

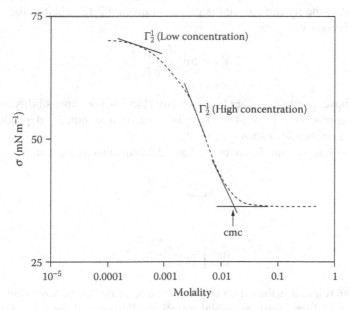

FIGURE 2.11 Surface tension, σ (mN m^{-1}) versus the surfactant molality, m_i (mol kg^{-1}), for octyl-β-D-glycopyranoside at 30.0°C. The CMC is the intersection point of the two branches of the curve. Γ^1_i, the slope of the function, depends on m_i.

Many contributions have been reported on the surface activity of AGs and related species [7–9,14,15,21–23,26,27,114]. Some of them account for the behavior in mixtures with other surfactants [14,142]. Such studies indicate the occurrence of surfactant association to give micelles and are used to get the area per polar head group in pure and mixed systems. AGs and AMs show no significant interaction with each other, or with nonionic surfactants, but interact more strongly with AEO_N. AMs, conversely, weakly interact with cationic, anionic, or zwitter-ionic surfactants. Interactions are more pronounced when AGs and AMs are mixed with non-glycosidic surfactants. Similar considerations apply to polyglycosides and HM ones.

Quite interesting, on this regard, is the behavior of hydrophobically modified (HM) pullulan, (α 1-4 polysaccharide), grafted with lauroyl groups [143]. This system shows an unusually slow kinetics of adsorption at the air–solution interface (some days long). The effect, observed at concentrations well below the CMC, depends on HM degree. The adsorption kinetics was explained by taking into account surface spreading, self-diffusion, and elastic properties of the polymer chains. It was rationalized by assuming that very long polymer chains need time to get an optimal arrangement at the air–aqueous solution interface and to pack into compact monolayers. This hypothesis has been recently confirmed by measuring the adsorption kinetics of similar grafted pullulan-based polymers onto solid surfaces [69,144].

2.5.5 OTHER SOLUTION PROPERTIES

Thermodynamic information on the association modes of glycosides in water can also be obtained by heat capacity, expansibility, and adiabatic compressibility. Definition of the former quantity is given in Equation 2.18, whereas the expansibility is defined as

$$\overline{Y}_{i,\zeta} = \lim_{\partial T \to 0} \left(\frac{\partial \overline{Y}_{V,i}}{\partial T} \right)_P \tag{2.25}$$

All the above quantities, either heat capacity [145,146] or expansibility and adiabatic compressibility [147,148], share in common a significant dependence on concentration, below or above the CMC.

Taking into account Equations 2.4 and 2.5 and considering that

$$\lim_{Xi \to 0} \overline{A}_i = \overline{Y}_i \tag{2.26}$$

results in

$$\lim_{Xi \to 0} \left(\frac{\overline{A}_i - \overline{Y}_i^0}{\overline{Y}_i - \overline{Y}_i^0} \right) = 2 \tag{2.27}$$

The above relation is the direct consequence of using power law series in Xi to express partial (or apparent) molal quantities. For more details see Appendix.

Heat capacity is obtained by differential methods, or performing calorimetric measurements at very close temperatures ($\Delta T \approx 1.0°C$, or less). The fit quality depends, obviously, on the width of ΔT windows. In most cases reported so far, the relaxational contributions, due to a shift of CMC values on increasing the temperature, are minimized by proper procedures. Similar considerations apply to expansibilities. C_P and expansibility values can also be obtained by extrapolation from the temperature dependence of dilution heats, or volumes, respectively. Their plots versus m_i are characterized by significant humps, or changes in slope, close to the critical concentration.

Compressibility data are obtained by combining density and sound velocity, v (cm s^{-1}), into

$$\kappa_{S,i} = \left(\frac{1}{\rho v^2} \right)$$ (2.28)

As a consequence of the surfactant packing into micelles, changes in slope occur in the $\kappa_{S,i}$ versus concentration plots close to the CMC.

2.6 THERMODYNAMIC MODELING

The aforementioned features obtained from different thermodynamic methods are a direct consequence of the surfactant packing into micelles or other supramolecular aggregates, such as disks or vesicles. The above behavior is modeled on thermodynamic grounds in the forthcoming section, where phase separation approaches or simplified mass action models for micelle formation are reported. Preliminary information is also given in the SST approach. In the latter case, the average number of surfactant molecules in an aggregate, $\langle N_i \rangle$, is considered an additional variable to be introduced into the equations for the solute chemical potential [149].

The more refined approaches developed so far are based on equilibrium conditions [145,150,151]. Thermodynamic properties are expressed in terms of the ratio between molecular and associate forms, depending on surfactant content. For partial molal quantities (enthalpy, entropy, volume, etc.), the following equality holds:

$$\overline{Y}_i = \theta_i \left[\overline{Y}^0_{\text{mol},i} + \theta_i B_Y m_i \right] + \left(1 - \theta_i \right) \overline{Y}^0_{\text{mic},i}$$ (2.29)

where

$\overline{Y}^0_{\text{mol},i}$ is the lower limit of the investigated property in molecular form the constant B_Y accounts for pair-wise interactions between nonassociated species

$\overline{Y}^0_{\text{mic},i}$ is the high concentration limit of $\overline{Y}_i$ well above the CMC

θ_i is defined as

$$\theta_i = \left(\frac{m_{\mathrm{mol},i}}{m_i}\right) \approx \left(\frac{\mathrm{CMC}_i}{m_i}\right)$$

(2.30)

At the CMC, Figure 2.12, the following equality holds

$$\left(\frac{\partial^2 \theta_i}{\partial m_{i,\mathrm{tot}}^2}\right) = 0$$

(2.31)

Obviously, such definition is somehow reminiscent of the requirements dictated by Phillips [103] in Equation 2.3.

The equations reported above can be applied to the mass action model for micelle formation or to the pseudo-phase one. In the two-site approximation, molecular and associate species formed by, in the average, $\langle N_i \rangle$ units coexist above the CMC. No size and shape transitions of the aggregates are supposed to occur and no further association stages or micelle growth are allowed. In such hypotheses, micelles are considered as a "non deformable reservoirs" for surfactant molecules. Such behavior implies that the forces responsible for micelle formation are the same in the whole range of concentration and do not depend on composition. This is a very special case of a more complex behavior. As a matter of fact, micelles often change in size and shape on increasing the surfactant content. This statement holds to be true in simple surfactant solutions as cetyltrimethylammonium salicylate ones [152] or in more exotic species such as the bile acids [153]. In all such cases, micelle formation leads to the formation of large colloidal aggregates, whose average sizes depend on concentration in an yet unspecified way. As a rule, micelle

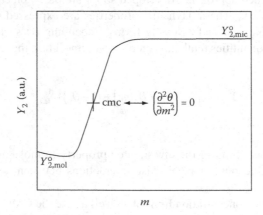

FIGURE 2.12 The limiting values of partial molal quantities for the molecular, $Y_{\mathrm{mol},i}^0$, and micellar, $Y_{\mathrm{mic},i}^0$, forms. The CMC is the salient point of the curve, defined by Equation 2.31.

growth in the above systems is regularly dependent on composition, micelles grow in one direction, and rod-like, ellipsoidal, aggregates are usually found.

Proper elaboration of θ_i based on the above relations leads to

$$\theta_{i,\text{CMC}} = \left(\frac{\langle N_i \rangle}{\langle N_i \rangle - 1}\right)\left[1 - \left(\frac{1}{2\langle N_i \rangle}\right)^{\frac{1}{2}}\right] \tag{2.32}$$

where $\theta_{i,\text{CMC}}$ is the relative amount of molecular to the overall surfactant content at the CMC. Use of Equations 2.31 and 2.32 relates, thus, $\theta_{i,\text{CMC}}$ to the skewness of a generic $\overline{Y}_i$ function in proximity of the critical threshold. In general, θ_s is controlled by the cooperativity of the association process, i.e., by $\langle N_i \rangle$ (Figure 2.13).

Proper use of the above relations allows one getting the number average micelle aggregation numbers. A near constancy of $\langle N_i \rangle$ values from different thermodynamic methods is expected to occur. Perhaps, this hypothesis is not strictly true and some scattering occurs, depending on the experimental method used. This is because quantities obtained by deriving chemical potentials (and related quantities) are extremely sensitive to small variations in composition, temperature, or pressure. Unfortunately, the computational procedures amplify the errors on $\langle N_i \rangle$ values. Hence, caution is recommended when the quality of data is questionable or the number of points low. In any case, extensive smoothing of the resulting plots can be necessary.

According to theoretical considerations by Desnoyers [109], T plays only a minor role in aggregation features and that is the reason for a slight decrease in

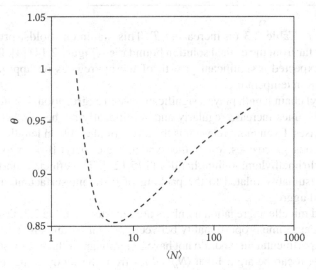

FIGURE 2.13 Dependence of θ_i at the CMC, defined by Equation 2.29, on the number of molecules in the micelle, $\langle N_i \rangle$. Data are in semilogarithmic scale.

TABLE 2.3
Number Average Micelle Aggregation Numbers, $\langle N_i \rangle$, Obtained by Different Partial Molal Quantities, for OBG, Octyl-β-D-thioglycopyranoside (OTG), Decyl-β-D-glycopyranoside (DeBG), Decyl Maltoside (DeMAL), and Dodecylmaltoside (DMAL) at Different Temperatures

Method	OBG	OTG	DeBG	DeMAL	DMAL
Cryoscopy	28±6	26±5	42±10	47±13	58±16
Vapor pressure	26±5[a]	23±3[a]	49±16[a]	55±18[a]	
	23±4[b]	21±2[b]			
Surface tension	26±5[c]	24±4[c]	40±11[c]	47±17[c]	61±20[c]
Calorimetry	25±5[c]	23±3[c]		43±12[c]	64±21[c]
	24±4[d]	21±2[d]			
Volumetry	25±5[c]	24±4[c]	41±12[c]		64±22[c]
Heat capacity	24±4[e]	22±2[e]			
Compressibility	28±6[c]	26±4[c]	43±14[c]		68±25[c]
Expansibility	25±4[e]	23±3[e]			

Note: The uncertainty on $\langle N_i \rangle$ values is indicated too.
[a] = 35.00°C.
[b] = 45.00°C.
[c] = 25.00°C.
[d] = 30.00°C.
[e] = 27.50°C.

$\langle N_i \rangle$ values in Table 2.3 on increasing T. This statement holds, provided the systems are far from the critical solution boundaries (Figure 2.14) [4]. In the latter case, it is expected a significant growth of the aggregates on approaching the critical solution temperature.

The alkyl chain length plays a significant role in aggregation. For homologous species $\langle N_i \rangle$ values increase regularly and significantly as the number of carbon atoms increases. Even more relevant is the effect of alkyl chain length in mixtures of homologous glycosides, or in the system water/octyl-β-D-glycopyranoside (OBG)/alkyltrimethylammonium halides [125,126]. The effects reported therein are very presumably related to the packing of different surfactant species into nonspherical aggregates.

Selected micelle aggregation numbers are reported in Table 2.3. Data reported therein indicate some proportionality between $\langle N_i \rangle$ values and alkyl chain length. Different experimental methods do not have a significant influence on such values. Accordingly, it can be argued that $\langle N_i \rangle$ values from thermodynamic methods are self-consistent of each other. In general, perhaps, results from thermodynamic methods give slightly lower $\langle N_i \rangle$ values compared to those obtained from light scattering [154], and close to those inferred by relaxation methods [155].

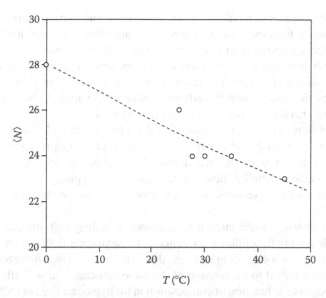

FIGURE 2.14 The number average micelle aggregation numbers of OBG, $\langle N_i \rangle$, as a function of temperature, T, in °C. Data are mean values taken from different experimental methods at each temperature.

Surfactant solutions are strongly nonideal, because micelle formation is concomitant to significant changes in the solute activity close to the CMC. In almost all cases, the pair-wise surfactant–surfactant interaction parameter, β, is different from zero. In terms of the regular solution theory, the solute activity is expressed as [156,157]

$$\gamma_i = \exp\left[\beta\left(X_i\right)^2\right] \tag{2.33}$$

Accordingly, the excess Gibbs energy of mixing is [158]

$$\left[\frac{\Delta G_{exc,mix}}{RT}\right] = \left[\frac{\Delta G_{mix,mic} - \Delta G^{\circ}_{mix.mic}}{RT}\right] = \ln\left[\frac{\prod\limits_{i=2}\gamma_i}{\sum\limits_{i=2}\gamma_i}\right] \tag{2.34}$$

where the excess function is taken relative to that pertinent to the ideal mixture behavior. In ternary systems, and when β does not depend on composition, developments of the above equation lead to

$$\ln\prod\limits_{i=2}\gamma_i - \ln\sum\limits_{i=2}\gamma_i = \left[\frac{\prod\limits_{i=2}\beta X_i^2}{\sum\limits_{i=2}\beta X_i^2}\right] = \beta\left[\frac{X_i^2 X_{i+1}^2}{X_i^2 + X_{i+1}^2}\right] \tag{2.35}$$

In multicomponent surfactant mixtures, thus, the nonideality of mixing is more relevant than in the binary ones. In addition, quaternary systems are by far less ideal than ternary ones and so forth. This is a direct consequence of Equation 2.35. The above hypothesis is in fair good agreement with the theory originally proposed by Rubingh [159,160], and recently revised by some of us [158]. It is worth noticing that the assumptions underlying Equation 2.35 apply to both surface and bulk phases, provided proper transformations are made.

Quite different is the theoretical model leading to SST [149,161–164]. The above approach is used in alternative to the pseudo-phase or the mass action one. In some aspects, it is considered a statistical mechanics approach applied to ensembles made of small colloid particles. It can be applied, with small formal changes, to micelles, vesicles, polymer molecules, and surface-functionalized nanoparticles.

In the following, some interesting approaches dealing with micelle formation and growth are briefly outlined. In general, the equations defining the chemical potential of a given species contain terms due to the bulk, surface free energy terms, contributions ascribed to the rotation of the same particles, and so forth. Since the chemical potential is function of composition in the aggregate (i.e., of $\langle N_i \rangle$), the relative weight of the surface free energy and rotation terms becomes more and more significant as the number of molecules decreases. For the above reasons it is suggested to start with the application of macroscopic thermodynamics to a large bunch of independent small systems and to introduce the number of molecules (or particles) and/or their volume as extensive variables. Finally, the resulting energy of the system is derived according to classical thermodynamic routes.

The set of variables to be used in SST contains terms related to the difference between microscopic and macroscopic thermodynamics. Depending on the variables of interest, the difference between two such terms is zero or not. Extra energy terms are peculiar to the small system and, obviously, are different for different environments. Some special cases are reported in Table 2.4, where the small system energy, $\wp$, and the corresponding characteristic function, ℓ, are reported. The SST approach applies to the calculation of entropy, as well.

Comparison with classical, macroscopic thermodynamics is, thus, at hand. In classical thermodynamics, micelles and other association colloids are considered a separate phase (i.e., $\langle N_i \rangle$ is not accounted for) or the result of a chemical equilibrium. The former approach finds its theoretical justification in Equation 2.23. Formally, the aggregation numbers calculated in the pseudo-phase approach are indefinite, even though determinations of $\langle N_i \rangle$ values based on that model have been sometimes reported [145].

In the SST approach, conversely, the environmental variables are chosen in such a way that the number of independent variables, fulfilling the phase rule, is $(F + 1)$. Accordingly, the integration of the thermodynamic function of interest can be expressed as

$$\wp_i(X_i) = \int_a^b \left[\langle N_i(X_i) \rangle \overline{Y}_i(X_i) \right] \partial X_i \qquad (2.36)$$

TABLE 2.4
Links between Residual Energy Functions, $\wp$, and Characteristic Thermodynamic Functions, ℓ, for Different Sets of Variables Links Are Elaborated According to Small Systems Thermodynamics

Set of Variables	Residual Energy Functions $\wp$ (kJ mol^{-1})	Characteristic Functions ℓ (kJ mol^{-1})
μ_i, V, T	$\wp = (P - \hat{P})V$	$\ell = -\hat{P}V = \wp - PV$
N_i, P, T	$\wp = (\hat{\mu}_i - \mu_i)N_i$	$\ell = \hat{\mu}_i N_i = \mu_i N_i + \wp$
$N_1, N_2, N_3, \ldots, P, T$	$\wp = G - \sum_i \mu_i N_i$	$\ell = G = \sum_i \mu_i N_i + \wp$
N_i, V, T	$\wp = A + PV - \mu_i N_i$	$\ell = A = \mu_i N_i - PV + \wp$
N_i, V, E		$\ell = \mu_i N_i - PV + \wp - E$
μ_i, P, T		$\ell = \wp$

Note: Links are elaborated according to small systems thermodynamics.

where the term in square brackets obeys Euler's theorem, i.e., both $\langle N_i \rangle$ and $\overline{Y}_i$ are implicit functions of composition. The results are also quite different compared to those inferred by the mass action approach, where $\langle N_i \rangle$ is a stoichiometric constant.

The integral in Equation 2.36 is such that all thermodynamic constraints on the original function, and its derivatives, are fulfilled. In particular, continuity of the function is required to deal with the above equation. Unfortunately, no detailed results based on SST have been reported on glycosides. Any serious attempt to get the results of such elaboration procedures requires the independent determination of thermodynamic properties, such as volumes, and light scattering ones. In this way, a cogent application of the theory could be possible.

2.7 SUMMARY

In the present chapter, attention is focused on the solution behavior of mild surfactants of the AGs family. They are nonionic and can be obtained by classical synthetic routes. The reasons for so much interest toward the above compounds arise from their mild and not denaturant character, making them excellent candidates to be used in fundamental and applied studies of biochemical relevance. Investigation on protein-surfactant systems are an example [67]. Other relevant aspects imply the possibility to get extended hydrogen bond networks (Figure 2.15). That peculiar contribution to the system energy ensures a relatively high stability

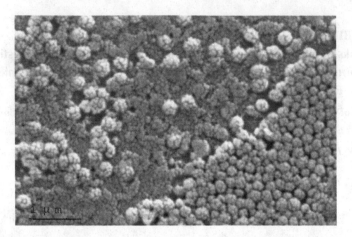

FIGURE 2.15 Coverage of silica nanoparticles (200 nm in diameter, according to the top bar) by a layer of HMPSs, a graft copolymer made of polymethyl-metacrylic acid, PMMA, chains linked to pullulan, as inferred by SEM. (From Andreozzi, P., La Mesa, C., Masci, G., and Suber, L., *J. Phys. Chem. C*, 111, 18004, 2007. With permission.)

and bond directionality to micelles, liquid crystals, and other association colloids formed by AGs. This behavior is quite unfamiliar in surfactant systems and deserves, thus, further attention. The mildness of hydrogen bond interactions toward biological macromolecules is a good reason for using them.

On this purpose, we reported and discussed a series of thermodynamic properties for the above compounds, spanning from binary to multicomponent systems. The conclusions presented here are essentially based on information dealing with micellar systems, but can be easily extended to micro-emulsions and lyotropic phases. Examples based on the thermodynamics of micelle formation and on the nonideality of mixing of glycosides with other surface-active species have been reported. The same holds for some information on the links between thermodynamics and phase equilibria in concentrated regimes, when different liquid crystalline phases are found. Some links between the solution behavior and physical–chemical performances have been briefly discussed.

ACKNOWLEDGMENTS

Thanks to Professors Antonio Capalbi and Marta Letizia Antonelli (Department of Chemistry, Sapienza University in Rome) for the revision of the chapter and many useful suggestions. Thanks are due also to Professors Giuseppe Antonio Ranieri, Luigi Coppola, and Dr. Rita Muzzalupo, at Calabria University, for cogent criticism.

The present research line was developed under the auspices of the European Community, through a COST D-35 Action Project on Interfacial Chemistry and Catalysis, for the years 2006–2010.

Thanks to the Ministry of University and Scientific Research, MIUR, for supporting us through a financing project (PRIN n° 2006 8-121-24-20) for the years 2006–2008.

APPENDIX

Apparent molal quantities depend on solute concentration according to a power-law equation, such as

$$\Phi_{Y,i} = \sum_{i=0} A_i X^i \tag{2.A.1}$$

where A_i are proper constants and the exponent is an integer, or semi-integer in electrolyte solutions. The partial molal quantities derived by Equation 2.A.1 are defined as

$$\overline{Y}_i = \left(\frac{\partial \left(X \Phi_{Y,i} \right)}{\partial X_i} \right) = \left(\frac{\partial \sum_{i=0} A_i X^{i+1}}{\partial X_i} \right) = \sum_{i=0} (i+1) A_i X^i \tag{2.A.2}$$

The above series converge to the same value as X_i approaches zero. In addition, the ratio

$$\lim_{X_i \to 0} \left[\frac{\overline{Y}_i - A_0}{\Phi_{Y,i} - A_0} \right] = 2 \tag{2.A.3}$$

Although apparent and partial molal quantities have the same limiting value, the ratio inferred by Equation 2.A.3 can be used as a further constraint in the data analysis. This procedure finds use when significant changes in slope occurs at moderate concentrations, as usually happens in association colloids.

NOMENCLATURE

AGs	alkyl glycosides
AMs	alkyl maltosides
ATGs	alkyl thio-glycosides
APGs	alkyl polyglycosides
HMPSs	hydrophobically modified polysaccharides
AEO_N	alkyl polyoxyethylene glycols
HLB	hydrophilic–lipophilic balance
CMC	critical micelle concentration
α or β	chiral anomers of carbon atoms
L or D	absolute configuration of carbon isomers
μ_i	chemical potential of the surfactant species. That of the solvent is referred to as μ_1

X_i, m_i, M_i	mole fraction, molality, and molarity of the surfactant component in the mixture. More generally, concentration of the ith component in the mixture
$\Delta G_{tr}, \Delta G_{int}, \Delta G_{el}, \Delta G_{conf}, \Delta G_{hydr}, \Delta G_{HB}$	Gibbs energy of transfer, of forming an interface, related to electrostatic work, conformational, of hydration, steric contributions, due to hydrogen bonds, etc.
$\bar{L}_i, \bar{S}_i, \bar{V}_i, \bar{C}_{P,i}$	partial molal enthalpy, entropy, volume, heat capacity of the surfactant species
n_i	number of moles for the ith component
$\bar{A}_i$	generic partial molal quantity for the ith component
T, P	temperature and pressure
$\Phi_{Y,I}$	apparent molal quantity for the ith component
ΔA_i	changes in partial molal quantities
$\bar{Y}_i, \bar{Y}_i^0$	generic partial molal quantity and its limit at high dilution
A_1, B_2, C_3	first, second, third constants into polynomial equations
ε	static dielectric permittivity of the medium
C_i	generic concentration of the ith species
$\Phi_{L,i,mol}, \Phi_{L,i,mic}$	apparent molal quantity relative to the molecular or micellar form
ρ, ρ^o	density, solvent density
$\mu_{max} = (\alpha\lambda)_{max}$	maximum sound absorption per wavelength
$\kappa_{S,i,\infty}$	adiabatic compressibility
σ^2	micelle polydispersity
$X_{i,red}$	reduced micellar concentration
$\langle N \rangle$	number average micelle aggregation number
$\Delta H_{int,dil,I}$	integral enthalpy of dilution for the ith species
POCP	1-palmitoyl-2-oleoyl-sn-glycero-3-phosphocholine
$\Delta G_{mic,i}^0, \Delta H_{mic,i}^0, \Delta S_{mic,i}^0, \Delta C_{P,mic,i}^0$	standard Gibbs energy, enthalpy, entropy, and heat capacity for micelle formation
ΔT_{cr}	freezing point depression
$\Delta P_{vap\ pres}$	vapor pressure lowering
γ_i	solute activity coefficient
$A_{mol,i}$	area per molecule at the air–aqueous solution interface
σ	surface tension of the solution
a_i	solute activity
Γ_i^1	surface excess concentration (taken relative to the solvent)
$A_{i,min}$	minimum area per molecule at the air-aqueous solution interface

N_A	Avogadro's number
$\Gamma^1_{i,\text{max}}$	maximum surface excess concentration
$\Delta G^0_{\text{ads},i}$, $\Delta G^0_{\text{mic},i}$	standard Gibbs energy of adsorption, and of micelle formation
$\Pi_{\text{CMC},i}$	surface pressure at the critical micelle concentration
HM	hydrophobically modified
$\bar{Y}_{i,\zeta}$	partial molal expansibility
v	sound velocity
$\langle N_i \rangle$	average number of molecules per aggregate

REFERENCES

1. Castro, M.J.L., Kovensky, J., and Fernandez Cirelli, A., *Langmuir*, 2002, *18*, 2477.
2. Focher, B., Savelli, G., and Torri, G., *Chem. Phys. Lipids*, 1990, *53*, 141.
3. Kenkare, P.U., Hall, C.K., and Kilpatrick, P.K., *J. Colloid Interface Sci.*, 1996, *184*, 456.
4. Weckstrom, K. and Zulauf, M.J., *Chem. Soc. Faraday Trans. I*, 1985, *81*, 2947.
5. Whiddon, C., Soderman, O., and Hansson, P., *Langmuir*, 2002, *18*, 4610.
6. Whiddon, C. and Soderman, O., *Langmuir*, 2001, *17*, 1803.
7. Shinoda, K., Yamanaka, T., and Kinoshita, K., *J. Phys. Chem.*, 1959, *63*, 648.
8. Shinoda, K., Yamaguchi, T., and Hori, R., *Bull. Chem. Soc. Jpn.*, 1961, *34*, 237.
9. Shinoda, K., in *Solvent Properties of Surfactant Solutions*, Shinoda, K., Ed., Marcel Dekker, New York, 1967, Chap. 1.
10. Fukuda, K., Söderman, O., Lindman, B., and Shinoda, K., *Langmuir*, 1993, *9*, 2921.
11. Shinoda, K., Carlsson, A., and Lindman, B., *Adv. Colloid Interface Sci.*, 1996, *64*, 253.
12. Jones, M.N., *Chem. Soc. Rev.*, 1992, *21*, 127.
13. Jones, M.N., in *Food Polymers, Gels and Colloids*, Dickinson, E., Ed., Royal Society of Chemistry, London, 1991, p. 65.
14. Rosen, M.J. and Sulthana, S.B., *J. Colloid Interface Sci.*, 2001, *239*, 528.
15. Li, F., Rosen, M.J., and Sulthana, S.B., *Langmuir*, 2001, *17*, 1037.
16. Warr, G.G., Drummond, C.J., Grieser, F., Evans, D.F., and Ninham, B.W., *J. Phys. Chem.*, 1986, *90*, 4581.
17. Platz, G., Thunig, C., Poelike, J., Kirchhoff, W., and Nickel, D., *Colloids Surf. A: Physicochem. Eng. Asp.*, 1994, *88*, 113.
18. Brown, G.M., Dubreuil, P., Ichhaporia, F.M., and Desnoyers, J.E., *Can. J. Chem.*, 1979, *48*, 2525.
19. Nilsson, F., Söderman, O., Hanson, P., and Johansson, I., *Langmuir*, 1998, *14*, 4050.
20. La Mesa, C., *Colloid Polym. Sci.*, 1990, *268*, 959.
21. La Mesa, C. and Ranieri, G.A., *Ber. Bunsen-Ges. Phys. Chem.*, 1993, *97*, 620.
22. D'Aprano, A., La Mesa, C., Proietti, N., Sesta, B., and Tatone, S., *J. Solution Chem.*, 1994, *23*, 1331.
23. Antonelli, M.L., Bonicelli, M.G., Ceccaroni, G.F., La Mesa, C., and Sesta, B., *Colloid Polym. Sci.*, 1994, *272*, 704.
24. Valente, A.J.M., Nilsson, M., and Söderman, O., *J. Colloid Interface Sci.*, 2005, *281*, 218.
25. Lam, H., Kavoosi, M., Haynes, C.A., Wang, D.I.C., and Blankschtein, D., *Biotechnol. Bioeng.*, 2005, *89*, 381.

26. Luders, H., in *Nonionic Surfactants, Surfactant Science Series*, Balzer, D. and Luders, H. Ed., Marcel Dekker, New York, 2000, Vol. 91, p. 77.
27. Shinoyama, H., Gama, Y., Nakahara, H., Ishigami, Y., and Yasui, T., *Bull. Chem. Soc. Jpn.*, 1991, *64*, 291.
28. Stradner, A., Glatter, O., and Schurtenberger, P., *Langmuir*, 2000, *16*, 5354.
29. Bonincontro, A., Briganti, G., D'Aprano, A., La Mesa, C., and Sesta, B., *Langmuir*, 1996, *12*, 3206.
30. Thiyagarajan, P. and Tiede, D.M., *J. Phys. Chem.*, 1994, *98*, 10343.
31. La Mesa, C., Bonincontro, A., and Sesta, B., *Colloid Polym. Sci.*, 1993, *271*, 1165.
32. Matsson, M.K., Kronberg, B., and Claesson, P.M., *Langmuir*, 2004, *20*, 4051.
33. Zhang, L., Somasundaran, P., and Maltesh, C.J., *Colloid Interface Sci.*, 1997, *191*, 202.
34. Johnson, R.A. and Nagarajan, R., *Colloids Surf., A: Physicochem. Eng. Asp.*, 2000, *167*, 21.
35. Johnson, R.A. and Nagarajan, R., *Colloids Surf., A: Physicochem. Eng. Asp.*, 2000, *167*, 37.
36. Schulte, J., Enders, S., and Quitzsch, K., *Colloid Polym. Sci.*, 1999, *277*, 827.
37. Aveyard, R., Binks, B.P., Chen, J., Esquena, J., Fletcher, P.D.I., Buscall, R., and Davies, S., *Langmuir*, 1998, *14*, 4699.
38. Sakya, P., Seddon, J.M., and Templer, R.H., *J. Phys. II, (Paris)*, 1994, *4*, 1311.
39. Jeffrey, G.A., *Acc. Chem. Res.*, 1986, *19*, 168.
40. Ryan, L.D. and Kaler, E.W., *Colloids Surf. A: Physicochem. Engin. Asp.*, 2001, *176*, 69.
41. Nieendick, C. and Schmid, K.H., *SOFW J.*, 1995, *121*, 412.
42. Marinova, K.G. and Denkov, N.D., *Langmuir*, 2001, *17*, 2426.
43. Stangar, U.L. and Huesing, N., *Silicon Chem.*, 2003, *2*, 157.
44. Tai, L.H.T. and Nardello-Rataj, V., *Oleagineux, Corps Gras, Lipides*, 2001, *8*, 141.
45. Tesmann, H., Kahre, J., Hensen, H., and Salka, B.A., in *Alkyl Polyglycosides*, Hill, K., von Rybinski, W., and Stoll, G., Eds., VCH, Weinheim, Germany, 1997, p. 71.
46. Suzuki, K., Ando, T., Susaki, H., Mimori, K., Nakabayashi, S., and Sugiyama, Y., *Pharm. Res.*, 1999, *16*, 1026.
47. Okada, T., Takeda, K., and Kouyama, T., *Photochem. Photobiol.*, 1998, *67*, 495.
48. Luders, H., in *Nonionic Surfactants, Surfactant Science Series*, Balzer, D. and Luders, H. Ed., Marcel Dekker, New York, 2000, Vol. 91, p. 7.
49. Luders, H., in *Nonionic Surfactants, Surfactant Science Series,* Balzer, D. and Luders, H. Ed., Marcel Dekker, New York, 2000, Vol. 91, p. 19.
50. Schmid, K. and Tesmann, H., in *Detergency of Specialty Surfactants, Surfactant Science Series*, Friedly, H. Ed., Marcel Dekker, New York, 2001, Vol. 98, p. 1.
51. Hoffmann, B. and Platz, G., *Curr. Opin. Colloid Interface Sci.*, 2001, *6*, 171.
52. Schoberl, P. and Scholz, N., in *Nonionic Surfactants, Surfactant Science Series*, Balzer, D. and Luders, H. Ed., Marcel Dekker, New York, 2000, Vol. 91, p. 331.
53. Balzer, D., in *Specialist Surfactants*, Robb, I.D., Ed., Blackie, London, 1997, p. 169.
54. Thiem, J. and Boecker, T., *Spec. Publ. Roy. Soc. Chem., Ind. Appl. Surfactants III*, 1992, *107*, 123.
55. Akoh, C.C., in *Fatty Acids in Foods and Their Health Implications*, 2nd ed., *Food Science and Technology*, Marcel Dekker, New York, 2000, Vol. 96, p. 375.
56. Dembitsky, V.M., *Lipids*, 2004, *39*, 933.
57. Hill, K. and Rhode, O., *Fett/Lipid*, 1999, *101*, 25.
58. von Rybinski, W. and Hill, K., in *Novel Surfactants, Surfactant Science Series*, Holmberg. K. Ed., Marcel Dekker, New York, 2003, Vol. 114, p. 35.

59. von Rybinski, W. and Hill, K., in *Novel Surfactants, Surfactant Science Series*, Holmberg. K. Ed., Marcel Dekker, New York, 1998, Vol. 74, p. 31.
60. von Rybinski, W. and Hill, K., *Angew. Chem., Intern. Ed.*, 1998, *37*, 1328.
61. von Rybinski, W., *Curr. Opin. Colloid Interface Sci.*, 1996, *1*, 587.
62. Jin, X., Zhang, S., Yang, J., and Lu, R., *Tenside, Surfactants, Detergents*, 2004, *41*, 126.
63. Balzer, D., *Tenside, Surfactants, Detergents*, 1996, *33*, 102.
64. Jung, J.H., Do, Y., Lee, Y.-A., and Shimizu, T., *Chem. Eur. J.*, 2005, *11*, 5538.
65. Hein, M., Miethchen, R., and Schwaebisch, D., *J. Fluorine Chem.*, 1999, *98*, 55.
66. Greiner, J., Riess, J. G., and Vierling, P., in *Organofluorine Compounds in Medicinal Chemistry and Biomedical Applications, Studies Org. Chem*, Filler, R., Kobayashi, Y., and Yagupolskii, L.M., Ed., Amsterdam, 1993, Vol. 48, p. 339.
67. La Mesa, C., *J. Colloid Interface Sci.*, 2005, *286*, 148.
68. Conroy, J.P., Hall, C., Leng, C.A., Rendall, K., Tiddy, G.J.T., Walsh, J., and Lindblom, G., *Progr. Colloid Polym. Sci.*, 1990, *82*, 253.
69. Andreozzi, P., Suber, L., Masci, G., and La Mesa, C., COST Action Chemistry D36, Coimbra, Pt, 2007.
70. Del Castillo, J.L., Czapkiewicz, J., Rodriguez, J.R., and Tutai, B., *Colloid Polym. Sci.*, 1999, *277*, 422.
71. Abe, Y., Fujiwara, M., Ohbu, K., and Harata, K., *Perkin II*, 2000, 2, 341.
72. Sharma, G.V.M. and Krishna, P.R., *Curr. Org. Chem.*, 2004, *8*, 1187.
73. Nilsson, F., Söderman, O., and Johansson, I., *Langmuir*, 1997, *13*, 3349.
74. Coppola, L., Gordano, A., Procopio, A., and Sindona, G., *Colloid Surf. A: Physicochem. Eng. Aspect*, 2002, *196*, 175.
75. Dawson, R.M., Alderton, M.R., Wells, D., and Hartley, P.G., *J. Appl. Toxicol.*, 2006, *26*, 247.
76. Chong, T.T., Hashim, R., and Bryce, R.A., *J. Phys. Chem. B*, 2006, *110*, 4978.
77. Matsumura, S., Imai, K., Yoshikawa, S., Kawada, K., and Uchibori, T., *J. Am. Oil Chem. Soc.*, 1990, *67*, 996.
78. Boyd, B.J., Drummond, C.J., Krodkiewska, I., and Grieser, F., *Langmuir*, 2000, *16*, 7359.
79. Nilsson, F., Söderman, O., and Johansson, I., *Langmuir*, 1996, *12*, 902.
80. Bonicelli, M.G., Ceccaroni, G.F., and La Mesa, C., *Colloid Polym. Sci.*, 1998, *276*, 109.
81. Rauter, A.P., Lucas, S., Almeida, T., Sacoto, D., Ribeiro, V., Justino, J., Neves, A., Silva, F.V.M., Oliveira, M.C., Ferreira, M.J., Santos, M.-S., and Barbosa, E., *Carbohydrate Res.*, 2005, *340*, 191.
82. Sottmann, T., Kluge, K., Strey, R., Reimer, J., and Söderman, O., *Langmuir*, 2002, *18*, 3058.
83. Kocherbitov, V. and Söderman, O., *Langmuir*, 2004, *20*, 3056.
84. Danielsson, I., *Adv. Chem. Ser.*, 1976, *152*, 13.
85. Aspler, J.S. and Gray, D.G., *Macromolecules*, 1981, *14*, 1546.
86. Kocherbitov, V., Söderman, O., and Wadso, L., *J. Phys. Chem. B*, 2002, *106*, 2910.
87. Nagarajan, R., *Langmuir*, 2002, *18*, 31.
88. Puvvada, S. and Blankschtein, D., *J. Chem. Phys.*, 1990, *92*, 3710.
89. Israelachvili, J.N., *Intermolecular and Surface Forces*, 2nd ed., Academic Press, London, 1991.
90. Missel, P.J., Mazer, N.A., Benedek, G.B., and Carey, M.C., *J. Phys. Chem.*, 1983, *87*, 1264.
91. Missel, P.J., Mazer, N.A., Benedek, G.B., Young, C.W., and Carey, M.C., *J. Phys. Chem.*, 1980, *84*, 1044.

92. Nagarajan, R. and Ruckenstein, E., *Langmuir*, 1991, *7*, 2934.
93. Nagarajan, R., *Langmuir*, 1985, *1*, 331.
94. Stephenson, B.C., Beers, K., and Blankschtein, D., *Langmuir*, 2006, *22*, 1500.
95. Blankschtein, D., Thurston, G.M., and Benedek, G.B., *J. Chem. Phys.*, 1986, *85*, 7268.
96. Blankschtein, D., Thurston, G.M., and Benedek, G.B., *Phys. Rev. Lett.*, 1985, *54*, 955.
97. Quirion, F. and Desnojers, J.E., *J. Colloid Interface Sci.*, 1986, *112*, 565.
98. Shinoda, K. and Hutchinson, E., *J. Phys. Chem.*, 1962, *66*, 577.
99. Kale, K.M. and Zana, R., *J. Colloid Interface Sci.*, 1977, *61*, 312.
100. Benjamin, L.J., *Phys. Chem.*, 1964, *68*, 3575.
101. Mukerjee, P., Mysels, K.J., and Kapuan, P., *J. Phys. Chem.*, 1967, *71*, 4166.
102. Hoiland, H., and Vikingstad, E., *J. Colloid Interface Sci.*, 1978, *64*, 126.
103. Phillips, J.N., *Trans. Faraday Soc.*, 1955, *51*, 561.
104. Lewis, G.N. and Randall, M., in *Thermodynamics*, Ed., McGraw-Hill, New York, 1965, Chap. XX, p. 241.
105. Desnojers, J.E., Caron, G., De Lisi, R., Roberts, D., Roux, A., and Perron, G., *J. Phys. Chem.*, 1983, *87*, 1397.
106. La Mesa, C. and Sesta, B., *J. Phys. Chem.*, 1987, *91*, 1450.
107. Wennerström, H. and Lindman, B., *Phys. Rep.*, 1979, *52*, 1.
108. Lindman, B. and Wennerström, H., *Topics Curr. Chem.*, 1980, *87*, 1.
109. Desnojers, J.E. and Perron, G., in *Surfactant Solutions: New Methods of Investigation*, *Surfactant Science Series*, Zana, R., Ed., Marcel Dekker, New York, 1987, Vol. 22, p. 1.
110. Debye, P. and Pauling, L., *J. Am. Chem. Soc.*, 1925, *47*, 2129.
111. *Handbook of Chemistry and Physics*, 61th ed., Weast, R.C. and Astle, M.J., Ed., CRC Press, Boca Raton, FL, 1981, Table III.
112. Fukada, K., Kawasaki, M., Seimiya, T., Abe, Y., Fujiwara, M., and Ohbu, K., *Colloid Polym. Sci.*, 2000, *278*, 576.
113. Harada, S. and Sahara, H., *Langmuir*, 1994, *10*, 4073.
114. Capalbi, A., Gente, G., and La Mesa, C., *Colloid Surf. A: Physicochem. Eng. Asp.*, 2004, *246*, 99.
115. Perron, G., Yamashita, F., Martin, P., and Desnoyers, J.E., *J. Colloid Interface Sci.*, 1991, *144*, 222.
116. Tanford, C., *J. Mol. Biol.*, 1972, *67*, 59.
117. Sesta, B., *J. Phys. Chem.*, 1989, *93*, 7677.
118. Sugihara, G. and Mukerjee, P., *J. Phys. Chem.*, 1981, *85*, 1612.
119. La Mesa, C., *J. Phys. Chem.*, 1990, *94*, 323.
120. Aniansson, E.A.G., Wall, S.N., Almgren, M., Hoffmann, H., Kielmann, I., Ulbricht, W., Zana, R., Lang, J., and Tondre, C., *J. Phys. Chem.*, 1976, *80*, 905.
121. Teubner, M., *J. Phys. Chem.*, 1979, *83*, 2917.
122. Eigen, M. and De Maeyer, L., in *Techniques of Organic Chemistry*, Friess, S.L., Lewis, E.S., and Weissberger, A., Eds., InterScience, New York, 1963, Vol. VII, p. 788.
123. D'Aprano, A., La Mesa, C., and Persi, L., *Langmuir*, 1997, *13*, 5876.
124. Prasad, M., Chakraborty, I., Rakshit, A.K., and Moulik, S.P., *J. Phys. Chem. B*, 2006, *110*, 9815.
125. Del Burgo, P., Junquera, E., and Aicart, E., *Langmuir*, 2004, *20*, 1587.
126. Lainez, A., Del Burgo, P., Junquera, E., and Aicart, E., *Langmuir*, 2004, *20*, 5745.
127. Ikawa, Y., Tsuru, S., Murata, Y., Okawauki, M., Shigematsu, M., and Sugihara, G., *J. Solution Chem.*, 1988, *17*, 125.
128. Sugihara, G., Yamamoto, K., Wada, Y., Murata, Y., and Ikawa, Y., *J. Solution Chem.*, 1988, *17*, 225.
129. Opatowski, E., Kozlov, M.M., Pinchuk, I., and Lichtenberg, D., *J. Colloid Interface Sci.*, 2002, *246*, 380.

130. Opatowsky, E., Kozlov, M.M., and Lichtenberg, D., *Biophys. J.*, 1977, *73*, 1448.
131. Paula, S., Sues, W., Tuchtenhagen, J., and Blume, A., *J. Phys. Chem.*, 1995, *99*, 11742.
132. Majhi, P. and Blume, A., *Langmuir*, 2001, *17*, 3844.
133. Tanford, C., *The Hydrophobic Effect and the Formation of Micelles, Vesicles and Biological Membranes*, 2nd ed., Wiley, New York, 1980.
134. Pilcher, G., Jones, M.N., Espada, L., and Skinner, H.A., *J. Chem. Thermodyn.*, 1969, *1*, 381.
135. Pilcher, G., Jones, M.N., Espada, L., and Skinner, H.A., *J. Chem. Thermodyn.*, 1970, *2*, 1.
136. Wenk, M.R. and Seelig, J., *Biophys. J.*, 1997, *73*, 2565.
137. Wenk, M.R., Alt, T., Seelig, A., and Seelig, J., *Biophys. J.*, 1997, *72*, 1719.
138. Keller, M., Kerth, A., and Blume, A., *Biochim. Biophys. Acta, Biomembr.*, 1997, *1326*, 178.
139. Gente, G., Iovino, A., and La Mesa, C., *J. Colloid Interface Sci.*, 2004, *274*, 458.
140. Adamson, A.V., in *Physical Chemistry of Surfaces*, 5th ed., John Wiley, New York, 1990, Chap. 3, p. 53.
141. Mukerjee, P. and Mysels, K.J., *Critical Micelle Concentrations of Aqueous Surfactant Systems*, NSR-NBS-36, Washington, DC, 1971.
142. Kiraly, Z. and Findenegg, G.H., *Langmuir*, 2000, *16*, 8842.
143. Sallustio, S., Galantini, L., Gente, G., Masci, G., and La Mesa, C., *J. Phys. Chem. B*, 2004, *108*, 18876.
144. Andreozzi, P., La Mesa, C., Masci, G., and Suber, L., *J. Phys. Chem. C*, 2007, *111*, 18030.
145. Desnojers, J.E., Caron, G., De Lisi, R., Roberts, D., Roux, A., and Perron, G., *J. Phys. Chem.*, 1983, *87*, 1397.
146. De Lisi, R., Perron, G., Paquette, J., and Desnoyers, J.E., *Can. J. Chem.*, 1981, *59*, 1865.
147. Desnojers, J.E., De Lisi, R., and Perron, G., *Pure Appl. Chem.*, 1980, *52*, 443.
148. Vikingstad, E., Skauge, A., and Hoiland, H., *J. Colloid Interface Sci.*, 1978, *66*, 240.
149. Hill, T.L., *Thermodynamic of Small Systems*, Parts I and II, Ed., Dover, New York, 1994.
150. Burchfield, T.E. and Woolley, E.A., *J. Phys. Chem.*, 1984, *88*, 2149.
151. Deanden, L.V. and Woolley, E.A., *J. Phys. Chem.*, 1987, *91*, 2404.
152. Olsson, U., Söderman, O., and Guering, P., *J. Phys. Chem.*, 1986, *90*, 5223.
153. Calabresi, M., Andreozzi, P., and La Mesa, C., *Molecules*, 2007, *12*, 1731.
154. Sonnino, S., Cantù, L., Corti, M., Acquati, D., and Venerando, B., *Chem. Phys. Lipids*, 1994, *71*, 21.
155. Zana, R., in *Surfactant Solutions. New Methods of Investigation, Surfactant Science Series*, Zana, R., Ed., Marcel Dekker, New York, 1987, Vol. 22, p. 1.
156. Shinoda, K. and Nomura, T., *J. Phys. Chem.*, 1980, *84*, 365.
157. Takasugi, K. and Esumi, K., *J. Phys. Chem.*, 1996, *100*, 18802.
158. Muzzalupo, R., Gente, G., La Mesa, C., Caponetti, E., Chillura-Martino, D., Pedone, L., and Saladino, M.L., *Langmuir*, 2006, *22*, 6001.
159. Rubingh, D.N., in *Solution Chemistry of Surfactants*, Mittal, K.L., Ed., Plenum Press, New York, 1979, Vol. I, p. 337.
160. Holland, P.M. and Rubingh, D.N., *J. Phys. Chem.*, 1983, *87*, 1984.
161. Hill, T.L., *Statistical Thermodynamics*, Ed., Addison Wesley, Reading, MA, 1960.
162. Hill, T.L., *J. Chem. Phys.*, 1962, *36*, 3182.
163. Hall, D.G. and Pethica, B.A., in *Nonionic Surfactants*, Schick, M.J., Ed., Marcel Dekker, New York, 1967, p. 516.
164. Hall, D.G. and Pethica, B.A., *J. Phys. Chem.*, 1981, *85*, 2429.

3 Self-Assembly and Micellar Structures of Sugar-Based Surfactants: Effect of Temperature and Salt Addition

José A. Molina-Bolívar
and Cristóbal Carnero Ruiz

CONTENTS

3.1 INTRODUCTION

The term nonionic surfactant is frequently associated with ethylene oxide based amphiphiles. However, there are other nonionic surfactants, and one such group consists of those derived from carbohydrates (sugar-based surfactants), which constitute a new and interesting class of surfactants. Although they have been known for more than 100 years, it is not until during the two last decades that the interest in them has increased. The reasons for this state of affairs reside in the fact that sugar-based surfactants show environmental properties that in many instances are superior compared with other surfactants. The existing commercial surfactants are mostly based on slowly degradable compounds (ethylene oxide based surfactants) and only in a few cases they are, or at some point during their degradation become, harmful to the environment or to human beings. Sugar-based surfactants are being produced on an industrial scale and both the hydrophobic and hydrophilic moieties of the surfactant stem from cheap renewable raw materials, and as a consequence they are biodegradable and nontoxic. These properties make them interesting as substitutes of other surfactants, which are potentially damaging to the environment. Industrially produced sugar-based surfactants often have complex structures as a result of the many possible isomers formed when two or several sugar molecules are linked together. Much of the research output has involved these commercial mixtures, sometime with added cosurfactant. In fact, Lorber et al. [1] discovered that sugar-based surfactants are often contaminated by significant amounts of UV-absorbing or ionic compounds that considerably influence the micellar properties. The sugar-based surfactants have a broader application spectrum. Due to the favorable environmental profile and the "mild" character of these surfactants, they have found application in personal care products since they exhibit favorable dermatological properties, as liquid dishwashing agents, hard surface cleaners, agrochemicals, and for industrial and office cleaning [2,3].

Thermodynamic studies on sugar-based surfactant solutions suggest that the hydrogen bonding that exists between sugars and water is stronger or more extensive than that between the water molecules themselves [4]. It has also been well established that the nature and extent of the hydration of a sugar depends upon the stereochemistry of the particular sugar molecule [5]. This unique interaction that certain sugars have with water is believed to be responsible for the operation of a number of biological processes. For example, it has been suggested that the lattice of strongly hydrogen-bonded water molecules that surrounds the sugar residues of glycoproteins located at the outer surface of biological membranes serves as a medium that allows selective passage of other types of molecules into cells [6]. The micelle is an ideal model system for comparing with sugar residues at biological membrane surfaces, since micelles provide the same hydrophobic/hydrophilic interface as the membrane surface [7].

An understanding of membrane protein crystallization is of central importance for structural biology. Integral membrane proteins require surfactant solutions in order to be solubilized in aqueous media. The presence of the surfactant greatly complicates the phase map for protein crystallization. The crystallization of integral membrane proteins has proven to be critically dependent upon the chemical nature of the surfactant, its polar head group, and the alkyl chain. In this context, sugar-based surfactants are preferred (instead of other kinds of nonionic surfactants) because of their significantly high critical micelle concentration (cmc), which permits rapid removal by dialysis, and their small micelle size that is compatible with gel-filtration procedures. On the other hand, this class of surfactants has high solubilizing power, has high solubility in water, and do not denaturalize membrane protein [8,9]. Sugar-based surfactants are incorporated into liposomes for increasing their colloidal stability or their resistance against macrophages (fuzzy liposomes as drug carriers). Also they are extensively used in the alimentary industry as food emulsifiers, and in the biochemical and pharmacological industries [10,11].

Not a great deal is known about the behavior in a system with pure sugar-based surfactants, relative to say ethylene oxide based surfactants (C_iE_j). It is important to obtain an understanding on self-aggregation, surface properties, and general phase behavior of these molecules on the same level that we presently have for other surfactants. This is necessary if they are to be used as efficiently as possible in modern applications. In particular, the influence of electrolyte and temperature on micelle formation has a direct bearing on the dispersing and solubilizing capacity of a given surfactant in various applications. For ethylene oxide-based surfactants, the effect of the temperature and other factors on micelle size and morphology are well-understood and may, in the general case, be directly linked to effects on ethylene oxide solubility [12]. However, for sugar-based surfactants, the information and understanding of the micelle formation are much less extensive. Furthermore, many of the features of the micelle formation of ethylene oxide based surfactants are difficult or impossible to apply to sugar-based surfactants in a direct and meaningful manner [13].

Sugar-based surfactants show a very versatile behavior in a solution. Contrary to other amphiphiles, there are many different possibilities for the linkage between the hydrophilic sugar head group and the hydrophobic alkyl chain [14]. Furthermore, the hydrophilic head group could consist of innumerable different sugar units, i.e., saccharide derivates, whereas the hydrophobic part could consist of one, two, or even more chains. Sugar surfactants with two or more hydrophilic chains are usually referred to as glycolipids [15]. Derivates with alkyl chains longer than 12 C-atoms are generally not included in the studies because these surfactants do not dissolve well in water due to the high Krafft temperature [16].

Sugar-based surfactants can be classified into sucrose esters, sorbitane esters, N-methylglucamides, and alkyl polyglucosides. The selective functionalization of saccharose and sorbitol with fatty acids for the construction of a perfect amphiphilic structure cannot be realized in simple technical processes because of the poly-functionality of the molecule. The ideal raw material for selective derivatization is glucose. Reaction with fatty alcohol produces alkyl glucosides, whereas N-methylglucamides are prepared by reductive amination with methylamine and subsequent acylation [17,18]. Both products have a marked amphiphilic character, and it is for this reason that they are preferred. The hydrophilic part of alkyl polyglucosides consists of one or several glucose molecules, and the hydrophobic parts are connected via an ether bond (glucoside linkage). The usual abbreviation is C_iG_j with i referring to the number of C-atoms in the hydrophobic chain and j to the polymerization degree of the head group. In nature, the breakdown of glucoside linkage is enzymatically controlled by different glucosidases [19]. This is the main reason for the biodegradability of the alkyl glucosides.

There are two possibilities to attach the hydrocarbon chain to the glucose molecule, either via an α-linkage or via a β-linkage. The hydrocarbon chain is linked to the glucose molecule at the hydroxyl group at carbon number one of the glucose ring. For the case of a β-linkage, the hydrocarbon chain is attached to the glucose molecule in an equatorial way, while in the α-linkage molecules it is attached axially [20]. It seems clear that the two cases have rather marked differences with regard to the conformation in the head group region. As a consequence, the two surfactants are expected to differ markedly with regard to their packing possibilities in aggregates.

N-methylglucamides have been known since 1982, when they were synthesized by Hildreth [21]. Because of their wide use in membrane proteins solubilization, N-alkanoyl-N-methylglucamides (MEGA-n surfactants) have been extensively studied. These surfactants are easily synthesized at low cost. The nondenaturing properties, ease of removal by dialysis, and well-defined chemical composition of this class of sugar-based surfactant make them attractive reagents for membrane research. These sugar-based surfactants have a characteristic feature, they do not show a cloud point even if their solutions are heated to boiling and a large amount of sodium chloride is added. This indicates that there is a considerable difference between the hydration of hydroxyl groups of MEGA-n surfactants and that of oxyethylene groups of ethylene oxide based surfactants in aqueous medium. This chapter summarizes the latest findings about the surface activity, micellization process, and structures in dilute aqueous solutions of alkyl polyglucosides (mono and di) and N-methylglucamides. After analyzing their behavior in pure water, we focus on the influence of the temperature and salt addition.

3.2 SURFACE ACTIVITY AND MICELLAR FORMATION

Surfactants experiment two important physical processes in solution, they self-assembly to form a wide variety of structures and adsorb in the interface lowering their surface tension. These processes must be well understood for further technological applications. Due to their amphipathic structure, surfactants in aqueous solutions tend to form thermodynamically stable molecular aggregates called micelles. Micelles begin to form at a specific concentration called cmc, which is dependent on the surfactant structure and experimental conditions. Below the cmc, the surfactants are solubilized as monomers in the solution. Once the cmc is reached, all additional surfactant that have been added are employed either in the formation of new micelles or for promoting the growth of the aggregates. Detergency, the removal of a soil from a substrate by using surfactants, is an extremely complex process, which can involve mechanisms of surfactant interfacial adsorption and solutilization. This last process may involve the direct absorption of soil into micelles or the formation of intermediate phases [22]. The ability of alkyl glucosides to remove a solid organic soil from a hard surface is critically dependent on the cmc, with very low performance at concentrations less than the cmc, and optimal performance above the cmc. In comparison with ionic and some nonionic surfactants, these reagents lower surface and interfacial tension more effectively.

The surface activity of a surfactant, as expressed in moles of adsorbed surfactant per unit area of interface, can be characterized by three parameters: the cmc, the required area per head group of the molecule, and the interfacial tension at concentrations greater than or equal to the cmc [23]. The surface activity mainly depends on the chain length and the polarity of the head group of the surfactant molecule. An effective measure of the adsorption of a surfactant at the air–liquid interface is usually obtained by means of the surface excess concentration, Γ_{max}, which can be determined, for dilute dilutions, by the Gibbs equation

$$\Gamma_{max} = -\frac{1}{RT}\left(\frac{\partial \gamma}{\partial \ln c}\right)_{T,P} \tag{3.1}$$

where
 γ is the corrected surface tension
 c is the concentration of surfactant

From the surface excess concentration, it is possible to calculate the minimum area per surfactant molecule, A_{min}, at the air–liquid interface, by the equation

$$A_{min} = \frac{1}{N_A \Gamma_{max}} \tag{3.2}$$

where N_A is the Avogadro's number. The effectiveness of a surface-active molecule is measured by the surface pressure at the cmc, Π_{cmc}, which can be obtained from the relationship

$$\Pi_{cmc} = \gamma_0 - \gamma_{cmc} \tag{3.3}$$

here γ_0 and γ_{cmc} are the surface tension of pure solvent and of the micellar solution at the cmc, respectively. The standard free energy of adsorption, ΔG_{ads}^0, can be determined using the equation

$$\Delta G_{ads}^0 = \Delta G_m^0 - \frac{\Pi_{cmc}}{\Gamma_{max}} \tag{3.4}$$

The glucose molecule has a highly hydrophilic head group, and when the degree of glucosidation increases, the surface-active properties will decrease. The focus of this section is on the question of how molecular architecture of sugar-based surfactants influences surface activity and micellization process. Detailed understanding of the micellization of long-chain glucosides such as n-tetradecyl-β-D-maltoside (β-$C_{14}G_2$) is important from an applied point of view. When solubilizing bulky aromatic molecules in micellar systems, it is advantageous to use surfactants comprising long hydrocarbon chains for the obvious reason that a longer hydrocarbon chain results in a larger hydrocarbon region in the micelle. Furthermore, the cmc decreases as the alkyl chain length increases, so that at a given total surfactant concentration, a longer chain surfactant generally has a larger concentration of surfactant in the micellar state. The cmc of alkyl glucosides determined by different methods, particularly by surface tension measurements, are collected in Table 3.1 for

TABLE 3.1
Surface and Micellar Properties of *n*-Alkyl Glucosides

Surfactant	cmc (mM)	A_{min} (Å^2/ molecule)	ΔG_m^0 (kJ/mol)	ΔG_{ads}^0 (kJ/mol)	References
Octyl-β-D-glucoside (β-C_8G_1)	25[a], 19[a]	—			[48,38]
	18.2[a]	42	−19.9		[53]
	25[a]	41	−19.1		[95]
	20[a]	48			[30]
	25[b], 18[b], 25.3[c]				[61,116,62]
	23.2[b], 25.8[b]				[118,82]
Octyl-α-D-glucoside (α-C_8G_1)	6.3[a], 10[a]				[38,48]
	12.0[a]	42	−21.0		[53]
	10[a]		−21.4		[35]
Octyl-β-D-thioglucoside (β-SC_8G_1)	7.9[a]	52.5	−21.9	−35.3	[84]
	8.7[b], 9.2[b]				[84,36]
Nonyl-β-D-glucoside (β-C_9G_1)	6.9[a], 6.5[a]				[13,118]
Decyl-β-D-glucoside (β-$C_{10}G_1$)	2.0[a]	41		−32.0	[83]
	2.2[a]	47			[114]
	1.96[a]	42	−25.4		[53]
	1.95[a]	39.7			[117]
	2.2[a]				[119]
Decyl-α-D-glucoside (α-$C_{10}G_1$)	0.85[a]	44		−31.9	[83]
	0.35[a]	49			[115]
Dodecyl-β-D-glucoside (β-$C_{12}G_1$)	0.19[d]	36	−31.2	−38.3	[114]
	0.17[a]				[116]
	0.13[a]	34			[34]
Dodecyl-α-D-glucoside (α-$C_{12}G_1$)	0.072[a]				[52]
	0.042[a]	38			[34]
Tetradecyl-β-D-glucoside (β-$C_{14}G_1$)	0.015[a]				[52]

[a] Surface tension.
[b] Fluorescence.
[c] Light scattering.
[d] Drop weight method.

TABLE 3.2
Surface and Micellar Properties of *n*-Alkyl Maltosides

Surfactant	cmc (mM)	A_{min} (Å²/ molecule)	$\Gamma_{max} \times 10^6$ (mol/m²)	ΔG_m^0 (kJ/mol)	ΔG_{ads}^0 (kJ/mol)	References
Octyl-β-D-maltoside (β-C$_8$G$_2$)	19.1[a]	42		−19.8		[53]
Decyl-β-D-maltoside (β-C$_{10}$G$_2$)	2.0[a]	56			−32.2	[83]
	1.99[a]	44		−25.3		[53]
	2.0[a]					[24]
	1.95[a]	49.3	3.4			[117]
Dodecyl-β-D-maltoside (β-C$_{12}$G$_2$)	0.15[a]	50	3.32	−31.8	−42.6	[7]
	0.24[a]					[38]
	0.13[a]	45		−32.1		[53]
	0.16[b], 0.16[b]					[118,120]
	0.18[b], 0.18[a], 0.20[a]					[61,78,39]
	0.13[a]	42				[34]
	0.16[a]	43.6	3.8			[117]
Dodecyl-α-D-maltoside (α-C$_{12}$G$_2$)	0.15[a]					[39]
	0.12[a]	49				[34]
Tetradecyl-β-D-maltoside (β-C$_{14}$G$_2$)	0.015[a]					[52]
	0.014[a]	36	4.6			[71]

[a] Surface tension.
[b] Fluorescence.

n-alkyl-monoglucosides and Table 3.2 for *n*-alkyl-diglucosides (maltoside). Figure 3.1 corresponds to the structure and nomenclature of surfactants that appear in this chapter. As can be seen, these sugar-based surfactants have higher cmc values compared to ethylene oxide based surfactants. For example, the cmc of β-C$_{10}$G$_2$ and β-C$_{10}$E$_8$ are 2 and 1.03 mM, respectively; the cmc of β-C$_{12}$G$_1$ and C$_{12}$E$_8$ are 0.18 and 0.071 mM, respectively [24]. The ΔG_{ads}^0 values are more negative than their corresponding ΔG_m^0 values, indicating that the adsorption of the surfactant at the air–liquid interface is a more favorable process than the micellar formation. In other words, when a micelle is formed, work has to be done to transfer the surfactant molecules in the monomeric form at the surface to the micellar stage through the aqueous medium [25,26].

3.2.1 α- AND β-LINKAGE

Table 3.1 shows the general observation that alkyl glucosides with α-linkage between the hydrophilic and hydrophobic parts have lower cmc values than the

APGs	n-alkyl-α-D-glucosides	n-alkyl-β-D-glucosides	n-alkyl-β-D-maltosides	
Structure				
R = n	α-C_nG_1	β-C_nG_1	β-C_nG_2	
Example				
	n-dodecyl-β-D-maltoside (β-$C_{12}G_2$)			

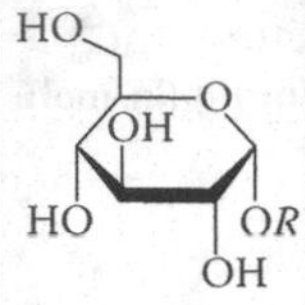

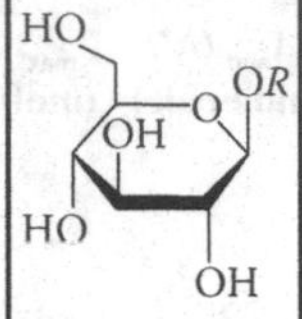

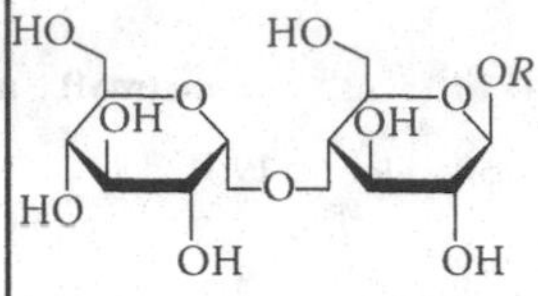

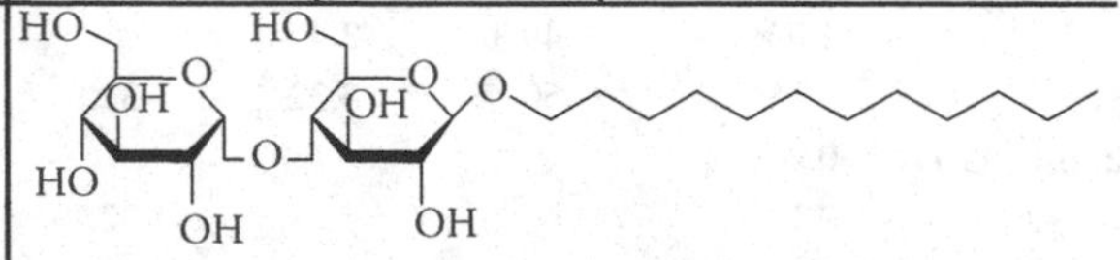

FIGURE 3.1 Structures and abbreviations for the n-alkyl polyglucosides (APGs) presented in this chapter.

alkyl glucosides with β-linkage. This suggests that α-anomers are slightly more hydrophobic in nature [27–30]. This result has been attributed to the fact that in the structure of the β-anomer, the different orientation of the head group results in the primary alcohol in the C-6 position being bent over to slightly shield the first methylene group in the alkyl chain. One of the basic characteristics of mono- and disaccharides in an aqueous environment is their ability to form a number of hydrogen bonds between the hydroxyl groups of the monosaccharide unit and the neighboring water molecules. In general, sugars locally increase the structure of water surrounding the sugar molecule. This entropically unfavorable fact is compensated by the enthalpy gain to the formation of hydrogen bonds between the sugar and water molecules [31]. The exact number of hydrogen bonds formed depends on the number and position of the hydroxyl groups on the sugar molecule, i.e., the exact stereochemistry is important. For instance, intramolecular hydrogen bonds may be formed that reduce the number of possible hydrogen bonds with solvent molecules and increase the rigidity of the sugar [32,33]. From Table 3.1, it is easy to notice that the differences between the cmc for α and β-anomers of glucosides are higher than for the maltosides. Assuming that the hydration of the hydrophilic moiety is an important factor in determining the ease of packing of the head groups, a higher degree of hydration of maltose compared to that of glucose could account for this difference [30]. Indeed, if the partial molar heat capacity at infinite dilution is taken as an indication of the hydration of the sugar, maltose monohydrate and glucose have partial molar heat capacity of 756 and 375 J/K mol at 35°C, respectively [31]; i.e., maltose tends to be much more highly hydrated than glucose. The contribution of the hydration around the anomeric center to the overall hydration is thus less in the case of maltose [34].

On the other hand, the minimum area per molecule (A_{min}) of different anomers in the interfacial layer is nearly the same. Compared with medium-chain ethylene oxide based surfactants such as C_8E_6 (A_{min} = 39.6 Å^2), the values for the corresponding glucosides are larger [3]. Brown et al. [35] have proposed that the different values of ΔG_m^0 for β-C_8G_1 anomers (α, β) are due to the fact that the hydrocarbon chain is in closer proximity to a hydroxyl group, and that probably increases the hydration of the octyl group resulting in an increase in ΔG_m^0 (i.e., less negative). In this respect, the lower value of ΔG_m^0 for octyl-β-D-thioglucoside (β-SC_8G_1) in comparison with β-C_8G_1 surfactant could reflect the fact that the former surfactant is less hydrated. In fact, Frindi et al. [36] have justified that the cmc of β-SC_8G_1 is less than half that of β-C_8G_1 because the -S- group is more hydrophobic than the -O- group.

An impressive example about the influence of the surfactant structure in the micellar formation is given by Fukuda et al. [37]. In this work, the properties of the α- and β-anomers of methyl 6-O-octanoyl-D-glucoside (α-MC_8G_1 and β-MC_8G_1) are investigated and compared with the properties of octyl-β-D-glucoside (β-C_8G_1). One of the results is that the aggregation number of α-MC_8G_1 is 1.5 times larger than that of β-MC_8G_1, whereas those of β-MC_8G_1 and β-C_8G_1 are nearly equal. On the other hand, thermodynamic properties such as the partial molar compressibility and the partial molar volume are similar for α-MC_8G_1 and β-MC_8G_1. These results are worth mentioning as they illustrate that small changes in the surfactant head group can greatly affect the micellar size and shape, which is not evident from the thermodynamic behavior [14]. Note that the anomerism at the acetal carbon (C-1) seems to have a stronger effect on the aggregation and, hence, on the packing behavior than the position of the hydrophobic chain link to the hydrophilic head (C-1 or C-6).

Focher et al. [38] have shown that the structures formed by the self-assembly of α-C_8G_1 and β-C_8G_1 depend upon the configuration at the anomeric center. The two anomers have identical hydrophilic–lipophilic balances but significant differences in molecular architecture. This different conformation assumes the hydration and the water structuration. The β-anomer forms spherical micelles with size comparable to C_iE_j surfactants while the α-glucoside forms very large and nonspherical aggregates. The axial conformation of α-C_8G_1 would permit the monomers to pack in extended structures. On the other hand, Dupuy et al. [39] have also observed, using small-angle x-ray and neutron scattering, that the structures formed by the self-assembly of α- and β-$C_{12}G_2$ in water depend on the configuration at the anomeric center. The α-anomer forms quasi-spherical aggregates, while the β-anomer forms larger oblate ellipsoidal micelles. This difference in behavior suggests that the configuration of the head group influences the orientation of the polar residue, and consequently, the packing of monomers in self-assemblies. On the other hand, the α- and β-anomers display a quite different crystallization behavior that arises from their crystal structures. The α-anomers form more stable crystal lattices than the β-anomers [40]. The melting enthalpy and Krafft points of the α-anomers are essentially higher than those of the β-anomers. The surfactants with β-linkage are much more soluble in water than those with α-linkage [41].

3.2.2 ALKYL CHAIN LENGTH

As is typical of all nonionic surfactants, the length of the alkyl chain in glucose-based surfactants has a large effect on the adsorption from solution at interfaces and on the micellization process of these surfactants. The effects of changes of length and structure of alkyl chain on the adsorption properties of series of surfactants with fixed head group have been well studied for some sugar-based surfactants (see references cited in the tables). Generally speaking, the length of the alkyl chain determines the order of magnitude of the cmc. This is shown for n-alkyl-glucosides and n-alkyl-maltosides in Figure 3.2. It has been established that the hydrophobic driving force for interfacial adsorption and micellization of surfactants increases as the chain length is increased, as there are more hydrocarbon units requiring minimal contact with the aqueous solution. This manifests itself with lower cmc and greater free energy of micellization (i.e., more negative). It is also expected to promote the phase separation phenomenon displayed by some of these glucose-based surfactants. On the other hand, the minimum area per molecule is relatively invariant with the change in the alkyl chain length. This demonstrates that in the case of n-alkyl glucosides, the presence of the hydrocarbon tail does not seem to affect the packing of surfactant molecules at the air–water interface. The molecular area is essentially determined by interrelations between the glucoside polar head groups. However, in the case of C_iE_j surfactants, the minimum area per molecule decreases with increasing the carbon number in the alkyl chain. For example, the A_{min} value for C_9E_8 is 75.5 Å^2/molecule, whereas the value for $C_{15}E_8$ is 45.2 Å^2/molecule [42]. Probably the addition of the carbon number in the alkyl chain increases their mutual attraction in the monolayer, so that the reduction of A_{min} as the chain length increases

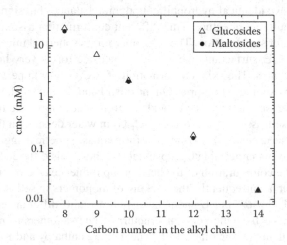

FIGURE 3.2 Dependence of the cmc on the alkyl chain length for glucosides and maltosides.

is expected. The Γ_{max} values that can be calculated from the A_{min} data presented in Tables 3.1 and 3.2 are lower than those reported for C_iE_j surfactants [43,44]. The difference may arise from different hydration by water molecules of the oligooxyethylene chain, on the one hand, and the hydroxyl groups of the sugar moiety on the other hand.

Coppola et al. [45] have investigated the properties of three nonionic surfactants from a series of 6-O-acyl-α,β-D-glucoside, differing from one other by the number of carbon atoms in their alkyl chain: 6-O-decanoyl-, 6-O-stearoyl-, and 6-O-oleyl-. A significant decrease in cmc is observed as the alkyl chain length increases. For all the surfactants studied, the minimum area per molecule at the air–water interface seems to fall in the range 35–37 Å^2/molecule. The relevant influence of alkyl chain length on the cmc of sugar-based surfactant was related by Boullanger et al. [46]. These authors investigated the micellar aggregation of a series of alkyl 2-amino-2-deoxy-β-D-glucosides having n-alkyl chain lengths of 8, 9, and 12 carbon atoms. An expected decrease of the cmc was observed with the increase of the alkyl chain, and the A_{min} values were of the order of 50 Å^2/molecule, irrespective of the alkyl chain length. The experimental results suggest that the head groups are closely packed at the air–water interface. This is possible because sugar head groups form hydrogen bonds with their neighboring molecules better than with water. This fact has been shown by molecular dynamic simulations [47]. Branching the alkyl chains in adopting secondary alkyl glucosides increases the cmc [3,41]. The surfactants with branched hydrocarbon chains do not form hexagonal or cubic phases. This is in contrast to the straight chain analogs. This is most likely caused by difficulties of packing the branched chains into the aggregates of the hexagonal cubic phases [41]. A systematic study on the physical-chemistry properties of branched alkyl glucosides was published by Nilsson et al. [48]. By using dynamic light scattering, these authors have determined the cmc of 2-ethylhexyl glucoside (2EHG) and isooctyl glucoside (IOG). Both surfactants, having eight carbons in the alkyl chain and similar degree of glucosidation, have similar values of cmc, 44 and 43 mM, respectively. These values are higher in comparison with the cmc of β-G_8C_1, probably due to the effect of steric hindrance in the aggregation process when branchings are introduced into the alkyl chains.

In respect of MEGA-n surfactants, Walter et al. [49] have determined the cmc of the MEGA-n series of surfactants by fluorescence. At 25°C the cmc values are 51.5 ± 5.6, 16.0 ± 1.0, and 4.8 ± 0.2 mM for MEGA-8, MEGA-9 and MEGA-10, respectively. As expected, the cmc values are inversely proportional to the carbon atoms number of hydrophobic chain. The cmc values of these surfactants obtained by surface tension measurements are 54.8, 17.1, and 4.12 mM for MEGA-8, MEGA-9, and MEGA-10, respectively [50]. In regard to surface tension at cmc, it is lowered in the order of MEGA-8 (35.3 mN m^{-1}), MEGA-9 (34.8 mN m^{-1}), and MEGA-10 (32.4 mN m^{-1}). On the other hand, the minimum area occupied by the surfactant molecule at the air–liquid interface, A_{min}, decreases with increasing chain length (50.4, 47.9, and 38.9 Å^2 molecule^{-1} for MEGA-8, MEGA-9, and MEGA-10, respectively).

3.2.3 HEAD GROUP

The effect of head group polymerization on the air–aqueous solution interfacial adsorption of n-alkyl glucoside surfactants has been investigated by different authors [51,52]. In contrast to the influence of the alkyl chain, the degree of polymerization of the head group has a very small effect on the adsorption behavior and the micellization process. The cmc values of the alkyl glucosides are similar regardless of the different size of the head group (see Figure 3.2). This is also quite different from the behavior of the ethylene oxide based surfactants, for which the cmc is clearly increased by increasing the degree of ethoxylation [42]. This behavior is an important structural trend of theses sugar-based surfactants [53]. The values for the occupied area at the interface are practically independent of the head group size, despite the much larger size of the maltoside head group. As can be seen in Tables 3.1 and 3.2, the ΔG_m^0 values are similar for glucosides and maltosides with the same chain length. Therefore, both β-maltoside and β-glucoside head groups must have a similar influence on the hydration of the alkyl chain with the same length. The ΔG_m^0 results are also consistent with the idea that glucose and maltoside moieties retain their hydration in the interfacial microenvironment of the micelle. It is evident from the ΔG_m^0 results that the structure of the hydration of the sugar head group is a very important factor in determining the thermodynamic properties of an n-alkyl glucoside surfactant. The presence of the thermotropic liquid crystalline phase in these systems has been attributed to this hydrogen bonding ability of the glucose head group [54].

The mentioned trends about the micellization of n-alkyl glucoside surfactants are extensive to sugar alcohol esters as deduced from a study by Piao et al. [55]. These authors have measured the cmc of a series of sugar alcohol esters in order to examine how the number and orientation of the hydroxyl groups of the sugar alcohol moiety affect the surfactant properties as well as the acyl chain length. It has been observed that the cmc of the sugar alcohol esters decreased as the acyl chain length increased. The hydrophilic moiety with different numbers and orientation of the hydroxyl groups of the sugar alcohols did not affect the cmc, while the surface tension at the cmc increased with the number of hydroxyl groups. The minimum area per molecule was independent of the type of hydrophilic and hydrophobic moieties, and was approximately $35 \pm 6 \ Å^2$. The cmc values did not depend on the temperature (from 25°C to 39°C) for any of the tested monoacyl sugar alcohols, indicating that the enthalpy of the micellization was nearly zero. The free energy of micellization was negative for every ester. Capalbi et al. [56] have determined by calorimetry the enthalpies of micelle formation (ΔH_m^0) for some n-alkyl-monoglucosides and n-alkyl-maltosides. They deduced that ΔH_m^0 regularly depends on the carbon numbers in the alkyl chain. The enthalpic contribution ascribed to the de-solvation of the polar head unit is positive and depends on the head group.

It is noteworthy that in the case of sugar esters, some differences in the self-assembly behavior with respect to alkyl glucosides has been observed. Söderberg et al. [57] have investigated the micellization of mono-dodecyl esters of glucose, sucrose, raffinose, and stachyose. They observed that the minimum area per surfactant molecule at the air–water interface increases with the sequential addition of galactose structural units, sucrose < raffinose < stachyose. On the other hand, the free energy of micellization increases linearly with the number of galactose units. In the case of sugar monoesters of xylose, galactose, sucrose, and lactose with different hydrophobic chain lengths, it has been corroborated that surfactants with more hydrophilic head groups exhibited higher cmc, though this trend was less pronounced as the alkyl chain length increased [34]. A reduction in the cmc with increasing carbon chain length was observed. Niraula et al. [58] have studied the interfacial properties of dodecyl-β-D-maltoside and dodecyl-α-D-fructofuranosyl-α-D-glucopyranoside (DFG), showing that the head group polarity plays a decisive role in interfacial properties and micellar formation. For both surfactants, the structure of their alkyl tail is identical with the same number of $-CH_2-$ groups in it. Therefore they differ slightly in head group polarity. For the same bulk surface concentration, β-$C_{12}G_2$ with less polar head group showed lower cmc and higher degree of surface/interfacial tension reduction as opposed to DFG. Finally, Milkereit et al. [59] have recently studied the effect of alkyl chain length and carbohydrate head group on micelle formation in aqueous solution of synthetic alkyloxyethyl glucosides containing an ethyl spacer with different conformations of the disaccharide head groups (glucose, lactose, and galactose).

3.2.4 Effect of the Temperature

The information and understanding about the effect of the temperature on the adsorption and micellization process of sugar-based surfactants are much less extensive than in the case of C_iE_j surfactants. For these surfactants, the increase of the temperature leads to larger micelles sizes and lower cmc, being very sensitive to temperature change [60]. For example, the cmc of $C_{10}E_8$ changes from 1.4 to 0.76 mM when the temperature is raised from 15°C to 40°C [42]. In the case of $C_{12}E_8$, the cmc changes from 0.097 to 0.058 mM in the same temperature range [42].

Among isomerically pure sugar surfactants, the β-C_8G_1 has received considerable attention in view of its extensive use in membrane research. The influence of the temperature on the aggregation behavior of β-C_8G_1 and β-$C_{12}G_2$ has been investigated by Aoudia et al. [61] using fluorescence probing and time-resolved fluorescence quenching (TRFQ). Figure 3.3 illustrates the variations of the cmc for these surfactants with the temperature. The cmc of β-$C_{12}G_2$ increases slowly with the temperature, a variation opposite to that generally reported for C_iE_j surfactants [44]. On the other hand, the cmc of β-C_8G_1 decreases fairly rapidly upon increasing the temperature, up to about 30°C. The results at higher temperature do not permit to evaluate with certainty whether the cmc levels off or goes through a flat minimum around 35°C and then increases very slowly with the temperature. Similar results have been obtained by Kameyama et al. [62] and Antonelli et al. [63]. Standard enthalpies of micellization (ΔH_m^0), Giggs energies of micellization

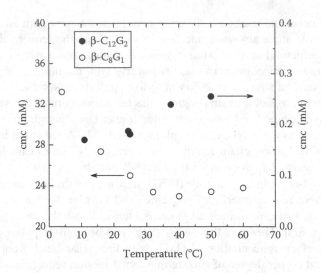

FIGURE 3.3 Effect of temperature on the cmc of β-$C_{12}G_2$ and β-C_8G_1. (Adapted from Aoudia, M. and Zana, R., *J. Colloid Interface Sci.*, 206, 158, 1998. With permission.)

(ΔG_m^0), and entropies of micellization (ΔS_m^0) are important in understanding micelle formation in aqueous solution. In the theory treating the micelle as a phase, the temperature dependence of cmc is correlated with the enthalpy of micellization by the Gibbs–Helmholtz equation:

$$\Delta H_m^0 = -R T^2 \left(\frac{\partial \ln \text{cmc}}{\partial T} \right)_P \tag{3.5}$$

In this way, the enthalpy of micellization can be evaluated from the slope of the plot of ln cmc versus temperature. Applying this treatment to the data obtained by the previously cited authors, the existence of two linear regions at lower and higher temperatures is observed. The micellization enthalpies were calculated to be around 13 and 1.3 kJ/mol in the region of lower and higher temperatures, respectively. This indicates a remarkable change in the enthalpy of micellization at 35°C. The former value is in agreement with the enthalpies of micellization of C_iE_j surfactants ranged from 6.7 to 18 kJ/mol in a temperature range between 15°C and 45°C [64]. The above behavior of aqueous β-C_8G_1 solution in the higher temperature region thus seems more unique than that in lower temperature region. The precise mechanism of this discontinuous change of the enthalpy of micellization is not clear, but it should be based on an entropic effect [3]. The influence of temperature on micelle formation of octyl-β-D-thioglucoside (β-SC_8G_1) was presented by Molina-Bolívar et al. [65]. The cmc values, thermodynamic parameters, and adsorption parameters of micellization obtained by these authors are collected in Table 3.3. The cmc of β-SC_8G_1 decreases as the system temperature increases in the studied temperature range (20°C–40°C). This trend could be a consequence

TABLE 3.3
cmc Values, Thermodynamic Parameters of Micellization, and Adsorption of Octyl-β-D-Thioglucoside at Different Temperatures

T(°C)	cmc (mM)	$-\Delta G_m^0$ (kJ/mol)	ΔH_m^0 (kJ/mol)	$T\Delta S_m^0$ (kJ/mol)	$\Gamma_{max} \cdot 10^3$ (mmol/m²)	A_{min} (Å²/ molecule)	Π_{cmc} (mN/m)	$-\Delta G_{ads}^0$ (kJ/mol)
20	9.3	21.18	14.35	35.53	2.84	58.56	44.99	37.05
25	8.5	21.76	14.84	36.60	3.06	54.33	43.69	36.05
30	7.5	22.44	15.34	37.78	3.24	51.30	42.86	35.68
35	6.9	23.02	15.85	38.87	3.41	48.70	42.21	35.40
40	6.2	23.66	16.37	40.03	3.44	48.27	41.51	35.73

Source: Molina-Bolívar, J.A., Aguiar, J., Peula-Garcia, J.M., and Carnero Ruiz, C., *J. Phys. Chem. B*, 108, 12813, 2004.

of the reduction in hydrophilicity of the surfactant molecules, which is due to a smaller probability of hydrogen bond formation at higher temperatures. In other words, the increase in temperature produces a decrease in hydration of the hydrophilic head group, which favors the micellar formation. Therefore, the micellization process occurs at lower concentrations as the temperature increases. According to the phase separation model, the standard Gibbs energy of micelle formation per mole of monomer, in the case of a nonionic surfactant, is given by

$$\Delta G_m^0 = RT \ln x_{cmc} \tag{3.6}$$

where x_{cmc} is the mole fraction of surfactant at the cmc. Whereas, the standard entropy contribution to the micellization process can be calculated from

$$T\Delta S_m^0 = \Delta H_m^0 - \Delta G_m^0 \tag{3.7}$$

As can be seen from Table 3.3, the free energy of micellization is negative and becomes more negative as the temperature increases, indicating that the formation of micelle becomes more spontaneous at higher temperatures. The values of ΔH_m^0 indicate that the micellization process is endothermic. This magnitude shows a weak dependence on temperature, indicating some variation in the environment surrounding the hydrocarbon chain of the surfactant molecule with temperature variation. Positive values of ΔH_m^0 are usually attributed to the release of structural water from the hydration layers around the hydrophobic parts of the surfactant molecule [66]. These hydrophobic interactions become significantly smaller with the partial disruption of the structure of the water as the temperature is increased. The entropic contribution of the micelle formation, $T\Delta S_m^0$, presents positive values that are larger than ΔH_m^0. In general, the entropy term is the main driving force for micelle formation by nonionic surfactants [67]. As micelles are formed, the hydrophobic hydration layers around the alkyl chains are broken down. This process is accompanied by a gain in entropy (more disorder in the structure of water) and favors micellization at a lower concentration [68]. More positive ΔS_m^0 values indicate that the micellization process is more spontaneous. The so-called enthalpy–entropy compensation [69] has been observed in many processes, including micellization of surfactants, which is reflected by a linear relationship between the enthalpy change and the entropy change, its slope being the so-called compensation temperature. The value found for the compensation temperature of β-SC_8G_1 was 302 K. The compensation temperature for C_iE_j surfactants is 322 K, for alkyl sulfates 304 K, and for alkyl trimethyl ammonium bromides 308 K [69]. This fact suggests that the hydrophilic group of β-SC_8G_1 behavior is more similar to that of the mentioned ionic surfactants than to the nonionic ones. In fact, the existence of a small but significant charge in alkyl polyglucosides micelles has been established [7,70]. The adsorption parameters presented in Table 3.3 suggest that surface excess concentration, Γ_{max}, increases slightly with the temperature. This behavior is due to the fact that hydration of the hydrophilic group of β-SC_8G_1 decreases as the temperature increases, and, therefore, increases the tendency to be located at the air–water interface. The ΔG_{ads}^0 values are negative and become less negative as the temperature increases, indicating that the adsorption of the surfactant at the air–water interface occurs spontaneously and becomes more spontaneous at lower temperatures. The difference between ΔG_{ads}^0 and ΔG_m^0 values is called effective Gibbs free energy change [68]. It is observed from Table 3.3 that this parameter decreases with increasing the temperature, indicating minimum energy requirement for micellization with increase of the temperature. As a result, micellization is favored over adsorption with the increase of the temperature. A priori, an increase in the adsorption of the surfactant should produce a decrease in the surface tension, and, therefore, an increase in the surface pressure (Π_{cmc}). However, the reduction rate in γ_0 is greater than in γ_{cmc}, resulting in a decrease of Π_{cmc}. Probably, this fact can be interpreted in the sense that the temperature has a greater effect on the water structure than on the dehydration of the hydrophilic groups of the surfactants. On the other hand, the decreases of A_{min} with the temperature could be attributed to more favored dehydration of the hydrophilic groups with the increase of temperature.

The effect of the temperature on the self-aggregation of nonyl-β-D-glucoside (β-C_9G_1) has been reported by Ericsson et al. [13]. They observed a decrease of Π_{cmc} from 39 mN/m at 10°C to 37 mM/m at 40°C, with the cmc values going from 9.1 to 6.0 mM in that temperature range. The slope of the isotherm (i.e., $d\gamma/d\ln c$) does not change appreciably with the temperature. This suggests that the packing area per molecule is fairly insensitive to temperature and close to 45 Å in the range of 10°C–40°C. Ericsson et al. [71] have also studied the surfactant tetradecyl-β-D-maltoside (β-$C_{14}G_2$). In this case, the cmc shows a modest increase with increasing temperature in the studied range (10°C–40°C). The calculations suggest that the area per molecule at the air–water interface increases from 32 ± 4 Å²/molecule to 45 ± 4 Å²/molecule, when the temperature is raised from 10°C to 40°C. Based on surface tension measurements, Boullanger et al. [46] have concluded that the cmc of a series of alkyl 2-amino-2-deoxy-β-D-glucosides showed little dependence on the temperature in the studied range 25°C–60°C.

Majhi et al. [72] have recently determined, by differential scanning calorimetry, the cmc of β-C_8G_1, β-C_9G_1, and β-$C_{10}G_2$ in an extended temperature range (up to 60°C). For the three studied surfactants the cmc-temperature profile shows a shallow minimum at 40°C. As deduced from published studies, an important topic of sugar-based surfactants is the relative temperature-insensitivity of their physicochemical properties. The temperature dependence of the cmc was found to be rather small. Typically the strongest effect was observed for β-C_8G_1. This insensitivity results from the strength of the hydrogen bonds between the hydroxyl groups of the sugar unit and water, which prevents any significant dehydration of the head group in the experimental relevant temperature range. The importance of this point can be understood if one compares the sugar-based surfactants with the ethylene oxide based surfactants. There is extensive water content in the head groups of both sets of micelles, but the ethylene oxide segments are hydrogen bond acceptor, and the sugar head groups, with their hydroxyl groups, are both acceptor and donors. This difference in the donor/acceptor balance allows the sugars head groups, or their attached water molecules, to be more effective than the C_iE_j head groups in hydrogen bonding [73]. One area where the relative temperature insensitive properties of the sugar-based surfactants may be an advantage is in energy-efficient low-temperature applications. To achieve this, attention has to be focused on ways to lower the Krafft boundary of sugar-based surfactants.

The effects of the temperature on the micellization process of MEGA-8, MEGA-9, and MEGA-10 have been studied by Walter et al. [49]. These authors have determined cmc in the range of biological interest, 5°C–35°C. In all cases, the cmc decreased very slightly with increasing temperature over this range. Similar results have been obtained for MEGA-10 in the temperature range 30°C–45°C by Sulthana et al. [74]. However, the experimental results of other researchers using the static light scattering method [75] and isothermal titration calorimetry [76] suggest a very shallow minimum in the cmc–temperature profile for the MEGA-n series. This minimum appears at 50°C for MEGA-8, at 45°C for MEGA-9, and at 40°C for MEGA-10. An explanation for this discrepancy is not at hand, but probably the sensitivities of the methods employed might be the reason for the different behavior.

A suggestion regarding the solution properties of MEGA-n surfactants has resulted from a comparison with those of ethylene oxide based surfactants. The partial specific volume of a molecule consists of two contributions, which are the intrinsic volume of the solute molecule and that of its hydration shell. It should be noted that the information obtained from the partial specific volume about the micelle hydration is entirely thermodynamic, not hydrodynamic, in nature. From density and viscosity measurements it is possible to access the partial specific volume. Molina-Bolívar et al. [77] have observed that the partial specific volume of MEGA-10 was roughly constant within the temperature range from 30°C to 50°C. This result indicates that the amount of water thermodynamically linked to micelle via hydrogen bonding is unaltered by the temperature. This can be ratio-nalized by the possible not/or little breaking of H-bonds between the sugar head and the surrounding water molecules at higher temperature, which can be sup-ported by the fact that this surfactant has not a cloud point. It is widely assumed that the phase separation in the cloud point of ethylene oxide based surfactants is due to the reduction of the micellar repulsions as a result of the dehydration of the oxyethylene groups, as the temperature is increased. At the cloud point, the water molecules get totally detached from the micelles [61,74]. On the other hand, it has been proposed that, due to the hydrophilicity, there arise two hydrogen bonds per oxyethylene alkyl ether with water whereas a hydroxyl group of the hydrophilic part of MEGA-n possibly forms three hydrogen bonds with water [75]. It may be concluded that there is a considerable difference between the interaction of hydroxyl groups of MEGA-n with water and that of the oxyethylene groups of ethylene oxide based surfactants with water.

Thermodynamic studies [74,76] have demonstrated that with an increase of the temperature, the micellization process of MEGA-10 becomes relatively more spontaneous, and that ΔH_m^0 changes sign from positive to negative at the temperature where the minimum in the cmc–temperature profile appears. The micellization is entropy dominated. As the temperature increases, ΔG_{ads}^0 values become more and more negative, suggesting that the adsorption process is relatively more favorable at higher temperatures. This can be ascribed to the fact that dehydration of the hydrophilic group is required for the adsorption to take place at the air–water interface and that, since at higher temperature the surfactant is comparatively less hydrated, adsorption at the air–water interface becomes easier. The A_{min} value also decreases with increasing temperature.

The slight sensitivity to temperature for MEGA-n surfactants is in contrast with the β-C_8G_1 cmc which decreases from 33 to 22 mM from 5°C to 40°C [61], whereas cmc of MEGA-8 changes just a few millimolars. The difference presumably is in the nature of the head group interaction. On the other hand, differences between the methylglucamide and glucoside moiety must count for the somewhat greater solubility of MEGA-n series of surfactants com-pared to the alkyl glucosides. For example, the cmc of β-C_8G_1 is closer to the cmc of MEGA-9 than to the cmc of MEGA-8. The linear head group on the methylglucamides and amino substitution may both contribute to differences in steric packing constraints or hydrogen bonding possibilities. Frindi et al. [36] have obtained by spectrofluorimetry that the cmc of MEGA-8 at 25°C is 69 mM and 62 mM at 40°C, in concordance with the values published by Okawauchi et al. [75].

3.2.5 Effect of the Salt Addition

Since electrolytes are often present together with surfactants in systems of practi-cal interest, it is important to have an understanding of the way in which salt effects arise in micelle formation and adsorption at interfaces. Sugar-based surfac-tants show a greater tolerance for electrolyte than ionic and C_iE_j surfactants.

Zhang et al. [78] have studied the influence of different electrolyte on the micellization of dodecyl-β-D-maltoside. This study revealed that the packing of the molecules at the air–water interface was not affected by the nature of the salt. In Table 3.4, the cmc values obtained in different salt solutions are shown. The surface tension and the cmc of β-$C_{12}G_2$ are reduced by the salts. The most accepted explanation of this trend is given in the context of the salting-out effect [79]. It has been suggested that the decrease in cmc for nonionic surfactants with electrolyte addition is related to the work required to disrupt the structure of the aqueous solvent by the insertion of a surfactant molecule [79]. Because the added salt acts to enhance the water structure, the introduction of the monomeric surfactant molecule will require an additional amount of work to overcome the added structural energy. As a consequence, it decreases the solubility of the surfactant and the cmc is reduced. Furthermore, other possible explanation for the decrease of the surface tension and cmc with electrolyte concentration is that the hydration of salt ions will decrease the "free" water molecules and increase the effective concentration of the surfactant [78].

Zhang et al. [78] rationalized the high salt tolerance of β-$C_{12}G_2$ surfactant in terms of the strong hydration of the sugar head group. According to this idea, the strong hydration would make difficult for electrolyte to interact with the head group. The explanation builds on earlier studies, which suggests that the maltoside moiety is strongly hydrated and that it is difficult for neighboring molecules

TABLE 3.4
cmc of Dodecyl-β-D-Maltoside in Different Salt Solutions

Salt	Concentration (M)	cmc (mM)
H_2O	—	0.17
LiCl	0.75	0.109
NaCl	0.75	0.0833
KCl	0.75	0.0787
RbCl	0.75	0.0955
NaF	0.75	0.0683
NaBr	0.75	0.114
NaI	0.75	0.141
NaSCN	0.75	0.136
$AlCl_3$	0.125	0.130
Na_3PO_4	0.125	0.106
$Na_3C_6H_5O_7$	0.125	0.108
Na_2SO_4	0.25	0.0938
$CaCl_2$	0.25	0.118

Source: Zhang, L., Somasundaran, P., and Maltesh, C., *Langmuir*, 12, 2371, 1996. With permission.

to approach the head group [7]. In the light of their differential scanning calorimetry results, Boyd et al. have suggested that there is strong binding of the two-sugar head group of maltoside with water [53]. Another reason for the salt tolerance has been claimed to be the rigidity of the sugar moieties, which is proposed to make it difficult for the electrolyte to penetrate the head group region [80]. Both explanations essentially suggest the same thing: the electrolytes cannot penetrate the head group region thereby making the surfactant more tolerant towards the electrolyte. The micellization of maltosides will be affected by the effect that the salts may have on the water structure. Clearly, this would serve to rationalize the difference in the salt effect between ethylene oxide based surfactants and sugar-based surfactants. From Table 3.4 it can be seen that, at the same ionic strength, different salts have different effects on cmc of the β-$C_{12}G_2$ surfactant. The effects of cations on the surface activity and cmc of the surfactant were found to be different from those of anions. The effects of the salts follow the well-know Hofmeister series [81]. This shows the salt order to basically reflect the intrinsic differences in the interaction of various ions with the aqueous solvent molecule. At the same ionic strength for the same cation Na^+, the effect of anions on the surface tension and cmc follows the order $F^- > Cl^- > SO_4^{2-} > Br^- > PO_4^{3-} >$ citrate $> I^- > SCN^-$. Similarly, for the same anion Cl^-, the effect of cations follows the order $K^+ > Na^+ > Rb^+ > Li^+ > Ca^{2+} > Al^{3+}$. Ions with a high charge/radius ratio, e.g., Li^+, Ca^{2+}, and Al^{3+}, are called structure-making ions, since they induce a more coherent structure of the water. Conversely, ions with a low charge/radius radio, e.g., Rb^+, are called structure-breaking ions. Because of the more intense electrostatic field in their vicinity, the structure-making ions are more highly hydrated than the structure-breaking ions. Consequently, the amount of distortion in the structure of the "free" water surrounding the hydrated structure-making ions is much less than that in the case of the weakly hydrated structure-breaking ions. In the case of a surfactant solution, water molecules are oriented around the structure-making ions, and this leads to a salting-out effect due to the reduction of hydration of the surfactant. The structure-breaking ions increase the ratio of monomeric water molecules in the bulk and promote the hydration of the surfactant, i.e., a salting-in effect. In Figure 3.4, the cmc values of β-$C_{12}G_2$ are plotted as a function of the charge/radius ratio of the ions. The effect of anions is explained on the basis of their structure-making and structure-breaking properties, while the effect of cations is exactly the reverse.

A thermodynamic and fluorimetric study of the micellar formation process of β-C_8G_1 in the presence of salts containing divalent cations has been carried out by Pastor et al. [82]. The addition of Cl_2Ca or Cl_2Zn leads to a loose decrease of cmc of β-C_8G_1. It can be noticed that, for a given salt concentration, $CaCl_2$ and $ZnCl_2$ decrease the cmc in a similar fashion. Aveyard et al. [83] have observed that the presence of 1 M NaCl has a significant effect on the surface tension of decyl-β-D-glucoside (β-$C_{10}G_1$). The salt causes a reduction in the cmc from 2 to 0.9 mM and, for a given surfactant concentration below the cmc, lowers the surface tension over much of the concentration range. On the other hand, the minimum area per molecule at the air–water interface does not change in the presence of salt. These

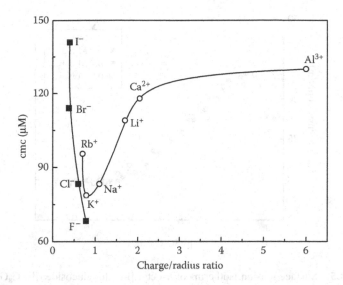

FIGURE 3.4 Dependence of the cmc of β-$C_{12}G_2$ on the charge to radius ratio of the ions. (From Zhang, L., Somasundaran, P., and Maltesh, C., *Langmuir*, 12, 2371, 1996. With permission.)

authors have pointed out that the groups of –OH moieties on carbohydrates can apparently compete with water for coordination to cations. A suitably arranged group of two or three –OH moieties appears to be necessary for extensive compl- exation. The complexation of sodium ions with the glucose moiety on a surfactant molecule could provoke a salting-in of head groups. Nilsson et al. [80] have stud- ied the effect of NaCl and NaSCN on β-C_9G_1 by tensiometry. Both salts were found to decrease the cmc, although the effect was more pronounced for the salting- out anion Cl$^-$. The chloride anion decreases the head group area at the air–water interface, whereas the SCN$^-$ anion increases the head group area. Both salts were found to affect the surface tension above cmc. Both effects follow the Hofmeister series. A recent paper deals with the influence of salt on the self-aggregation of octyl-β-D-thioglucoside [84]. The surface tension curves as a function of the β-SC_8G_1 concentration at different NaCl concentrations are shown in Figure 3.5. From these plots, it is seen that the surface tension value decreases with increasing NaCl concentration for a constant surfactant concentration. The obtained cmc values are listed in Table 3.5. Again, the cmc decreases as the salt concentration increases. The Gibbs energy of micellization is negative and becomes more negative as the electrolyte concentration increases, indicating that the micellar formation becomes more favorable in the presence of increasing NaCl concentrations. The observed increase of Γ_{max}, as well as the reduction of A_{min} when salt is presented in solution, can be related to the dehydration of the hydrophilic group. The standard free energy of adsorption is negative in all cases but undergoes small variations in the presence of NaCl. The ΔG^0_{ads} values are more negative than their

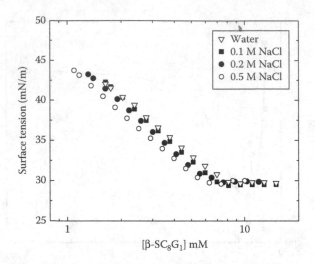

FIGURE 3.5 Surface tension isotherms of n-octyl-β-D-thioglucoside (β-SC$_8$G$_1$) at different electrolyte concentrations. (From Molina-Bolívar, J.A., Hierrezuelo, J.M., and Carnero Ruiz, C., *J. Phys. Chem. B*, 110, 12089, 2006. With permission.)

corresponding ΔG_m^0 values, indicating that the adsorption of the surfactant in the air–water interface is a more favorable process than the micellar formation, this situation being almost unaffected by the presence of NaCl.

Miyagishi et al. [85] have examined the effects of 16 kinds of salts on cmc of MEGA-n series determined by surface tension and fluorescent probe method. They have observed that the cmc of MEGA-8, MEGA-9, and MEGA-10 were lowered by the addition of salt, but this lowering was always much less compared

TABLE 3.5

Effect of NaCl Addition on Micellization and Adsorption Parameters of Octyl-β-D-Thioglucoside at 25°C

[NaCl] (M)	cmc (mM)	$-\Delta G_m^0$ (kJ/mol)	$\Gamma_{max} \cdot 10^3$ (mmol/m^2)	A_{min} (Å^2/ molecule)	Π_{cmc} (mN/m)	$-\Delta G_{ads}^0$ (kJ/mol)
0	7.86	21.96	3.16	52.54	42.27	35.34
0.1	7.09	22.21	3.49	47.57	42.47	34.38
0.2	6.37	22.48	3.41	48.69	42.63	34.98
0.5	5.59	22.76	3.38	49.12	43.44	35.61

Source: Molina-Bolívar, J.A., Hierrezuelo, J.M., and Carnero Ruiz, C., *J. Phys. Chem. B*, 110, 12089, 2006. With permission.

with that of the ionic surfactants. This decrease is explained in terms of dehydration of the hydrophilic groups. The salt added to micellar solution squeezes out some number of water molecules from the hydrated hydrophilic shell of the micelle and consequently destabilizes the micelle. However, it has been claimed that the salting-out of the hydrocarbon chains in the aqueous solvent by the electrolyte contributes significantly to the decrease in cmc [85]. It has been proposed that the depression of the cmc of nonionic surfactants with the electrolyte addition can be expressed by [43]

$$\log(\text{cmc}) = \log(\text{cmc})_{\text{no salt}} + k\,C_{\text{salt}} \tag{3.8}$$

where

    $k(k < 0)$ is a constant, which quantify the power of salt to decrease the cmc

    C_{salt} is the electrolyte concentration

The value of k can be determined from the slope of $\log(\text{cmc})$ versus the electrolyte concentration plot. Miyagishi et al. [85] have observed that the addition of divalent ions largely promotes decrease in the cmc, and that the values of k at 35°C were the same as those at 25°C within experimental error. The longer the acyl chain of MEGA-n became, the more effective at decreasing the cmc it became; i.e., the longer acyl chain gave the larger k. For example, k values of NaCl were 0.53, 0.56, and 0.59 for MEGA-8, -9, and -10, respectively. The ability to decrease cmc was observed in the order $Ca^{2+} > Na^+ > K^+ > Cs^+ > Li^+$ with respect to the cation and $SO_4^{2-} > CO_3^{2-} > SO_3^{2-} > HPO_4^{2-} > F^- > Cl^- > Br^- > NO_3^- > I^- > SCN^-$ with respect to the anion. This order was also similar to that of the Hofmeister series, except for Li. A possible explanation for the exceptional results of Li in MEGA-n surfactants may be the specific complexation of Li^+ with the hydroxyl group of the surfactant. Molina-Bolívar et al. [77] have observed that the micellization process of MEGA-10 becomes more favored as the NaCl concentration increases. In addition, the standard free energy of adsorption, ΔG^0_{ads} values, is negative and becomes more negative in the presence of increasing NaCl concentration. It is to be noted that ΔG^0_{ads} values are more negative than their corresponding ΔG^0_{m} values, indicating that the adsorption of the surfactant in the air–liquid interface is a more favorable process than the micellar formation, this situation being a little more pronounced with the NaCl addition.

3.3   SIZE AND STRUCTURE OF THE AGGREGATES

The determination of the basic structural parameters of micelles is necessary for understanding the physical mechanisms that drive the formation of their molecular assemblies in solution and that they are critical for applications. For example, accurate measurements of the properties of micelles formed by a single surfactant species can serve as a starting point to correlate surfactant packing preferences with trends in protein–surfactant interactions. The presence of the protein component in a protein–surfactant complex will in general alter the surfactant packing with respect to the surfactant-only micelles [86]. More specifically, recent results suggest that micelle geometry can influence the conformation of proteins buried in their hydrophobic core [86].

Since micelles are nonrigid dynamic structures with a liquid-like core, it is probably unrealistic to regard them as rigid structures with a precise shape [42]. Nevertheless, it is useful to consider an average micellar shape. Micelles consist of a rather limited number of surfactants, typically 50–150, forming a closed structure in order to minimize the contact between the surfactants hydrophobic part and the water. The mechanism behind this is called the hydrophobic effect. According to the amphiphilic nature of surfactant molecules, the tail groups will constitute the liquid-like hydrophobic interior of the aggregates while the head groups form an outer hydrophilic layer toward the water phase. The shape, size, and degree of polydispersity of the micelles are dependent on the surfactant structure (typically the relative size of the head group and tail group), concentration, temperature, and solution composition. The size/structure of a surfactant micelle is a compromise between the repulsion between the head groups, which tends to promote spherical aggregates, and the urge to minimize the unfavorable hydrocarbon/water contact, which tends to drive the structure towards a bilayer. Micellar shape is often described with the critical packing parameter (CPP), defined as

$$CPP = \frac{v}{al} \tag{3.9}$$

where

a is the optimal head group area

v and l are the volume and length of the surfactant hydrophobe, respectively

Thus, CPP is a geometrical parameter that does not take electrostatic, or other long-ranged forces, into account. The CPP heavily influences the phase behavior of the surfactant. Spherical micelles will be formed if $CPP < 1/3$. As CPP increases, meaning that the relative size of the hydrophobic part increases, the curvature of the aggregates will decrease and disc-, tablet-, and rodlike micelles are formed. As the concentration of surfactant increases, the micelles often grow more or less in size and at even higher concentrations various types of phases can be formed, e.g., hexagonal ($1/3 < CPP < 1/2$), lamellar ($CPP \approx 1$), and cubic ($CPP \geq 1$) phases. The different phases can have very different physicochemical properties, e.g., viscosity and the ability to scatter light.

Octyl-β-D-glucoside is the alkyl glucoside more extensively studied by a range of methods. A sedimentation equilibrium study realized by Roxby et al. [87] indicates that the micelles of β-C_8G_1 are homogeneous in size at the cmc and have an aggregation number of 70. As total concentration increases, micelles become larger and somewhat heterodisperse. By combined analysis of dielectric properties and viscosities, La Mesa et al. [88] have found a slight and continuous dependence of micellar shape on surfactant content for β-C_8G_1. An initial small-angle neutron scattering (SANS) study, based on relatively noisy data and measured over a small scattering vector (q) range, modeled β-C_8G_1 micelles as spheres [89], but subsequent works found them to be nonspherical. From measurements of molecular weight and partial specific volume for β-C_8G_1 micelles, Kameyama et al. [62] have deduced that a spherical shape is impossible and anisotropic shape is needed. They hypothesized the shape to be a one-component ellipsoid. This model was later used to fit SANS data for β-C_8G_1 micelles [90], and they found that the micelles grow significantly in the long dimension with increasing surfactant concentration. Furthermore, they observed interparticle repulsion effects for surfactant concentrations $\geq 0.6\,M$.

The self-diffusion coefficients of β-C_8G_1 have been obtained by ^{1}H-NMR in order to draw conclusions about the surfactant aggregation behavior [91]. The surfactant self-diffusion data at 15°C as a function of surfactant concentration (up to 12 wt %) are presented in Figure 3.6. The maximum size of a spherical aggregate is determined by the length of the extended hydrocarbon chain. In the case of a C-8 chain, this length can be calculated using the Tanford's formula: $l_c = 1.5 + 1.265n_c$, obtaining a value of 11.6 Å. [92]. A glucose moiety can be estimated to be 5 Å in length. Finally, adding on another 5 Å for the water of hydration, it is calculated that the maximum value for the hydrodynamic radius of a sphere is 21.6 Å. Using this value and taking into account the contributions of aggregate obstruction effects and upon the assumption that β-C_8G_1 micelles size does not change with concentration, the dashed curve in Figure 3.6 is obtained. This clearly shows that the assumption of spherical micelles is not in accordance with the experimental data. However, as is evident in this figure, if the micelles are modeled as hemisphere capped rods (cylindrical micelles) with a cylindrical radius of 21.6 Å and an axial ratio of 4.5 Å, there is a good agreement with the experimental data (solid line). In this prediction, the translational diffusion of a hemisphere capped rod at infinite dilution is given by

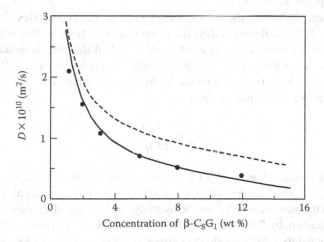

FIGURE 3.6 Experimentally determined diffusion coefficients of β-C_8G_1 (●). Dashed and solid lines are the predictions for spheres and for hemisphere-capped rods, respectively. (From Nilsson, F. and Söderman, O., *Langmuir*, 12, 902, 1996. With permission.)

$$D_0 = \frac{k_B T}{6\pi\eta b} \frac{\bar{s}}{p} \qquad (3.10)$$

where

k_B is the Boltzmann constant

T is the temperature

η is the viscosity of the medium

b is the radius of the hemisphere (or, equivalently, the radius of the cylindrical section)

p is the axial ratio of the micelle

$\bar{s}$ is the so-called reduced sedimentation coefficient, whose expression can be found in reference [93]

On the other hand, the obstruction effects for hemisphere capped rods are given by

$$D = D_0 \left(1 - \frac{k2\theta_{agg} p^3}{\bar{s}^3 (3p-1)} \right) \qquad (3.11)$$

where

k is an interaction constant

θ_{agg} is the volume fraction of aggregates

The analysis of x-ray diffraction (small-angle x-ray scattering [SAXS]) data gives proof of a rather short-ranged interaction between the micelles of β-C_8G_1. This short-ranged hydration force sets in at a distance of a few water layers [91]. It was finally suggested that the micelles of β-C_8G_1 are strongly hydrated with six water molecules per head group [91].

It has been shown that the diffusion coefficient of β-C_8G_1 micelles (D) obtained by dynamic light scattering (DLS) measurements decreases monotonically with micellar concentration, showing a tendency to flatten toward a constant value at the highest values [94]. At a very dilute limit, D approaches to a self-diffusion coefficient D_0 which is related to the hydrodynamic equivalent radius R_H for a sphere by the Stokes–Einstein equation:

$$R_H = \frac{k_B T}{6\pi\eta D_0} \qquad (3.12)$$

The hydrodynamic radius R_H of β-C_8G_1 estimated by D'Aprano et al. was 2.7 nm at 25°C [94]. This value agrees with the value reported by Shinoda et al. [95], but is higher than the maximum value for the hydrodynamic radius of a sphere. D'Aprano et al. have studied, by SANS, a micellar solution of β-C_8G_1 and the results obtained indicate that the micelles are cylinders also at concentrations very close to the cmc. They suggested that the radius and length of the cylinders are 12 Å and 60 Å, respectively. The size of the cylinder gives a gyration radius, R_g, of about 1.9 nm.

In an excellent study, the structure of micelles of different n-alkyl-β-D-glucosides in aqueous solution has been determined as a function of alkyl chain length (β-C_7G_1, β-C_8G_1, β-C_9G_1, and β-$C_{10}G_1$), using a combination of x-ray and SANS techniques over an extended q-range [96]. In this case, the authors interpreted their scattering data in terms of a cylindrical core–shell form factor, with the core representing the hydrocarbon chain and the shell representing the hydrated head group region. The length of the cylinders depends strongly on the number of carbons in the alkyl chain as well as on the concentration and isotopic composition of the solvent, while the cylinder radius grows in proportion to changes in the alkyl chain length only. The drastic increase in micelle length with increasing surfactant chain length is interpreted to reflect imperfect packing of the surfactant in the cap regions as solvent exposure of the alkyl chain. Figure 3.7 shows the effect of alkyl chain length on x-ray scattering patterns of n-alkyl-β-D-glucosides. The position of the first minimum is shifted to lower scattering vector (q) values when the alkyl chain length increases, and this is accompanied by significant increases in the slopes of the scattered intensities in the low q region ($q < 0.1\,\text{Å}^{-1}$). The changes of slope qualitatively indicate that the micelle size increases as a function of the chain length, with the largest size increase occurring between the β-C_8G_1 and β-C_9G_1 micelles. This trend for increasing micelle size with chain length was also seen with β-$C_{10}G_1$ solutions. However, in this case, solutions in the 0.1%–1% concentration range were visibly turbid, indicating extremely large structures. The large secondary broad peak seen in each x-ray scattering patterns (see Figure 3.7) is characteristic of scattering arising from a core–shell structure, in which the core and shell regions have scattering length densities that are

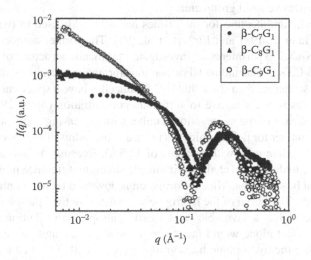

FIGURE 3.7   Effect of alkyl chain length on the x-ray scattering patterns for different n-alkyl-glucosides. (From Zhang, R., Marone, P.A., Thiyagarajan, P., and Tiede, D.M., *Langmuir*, 15, 7510, 1999. With permission.)

large and of opposite sign [97]. The x-ray and neutron scattering data confirm a concentration-dependent change in β-C_8G_1 micelle size, which is restricted to changes in micelle length, not diameter [95]. An essentially equivalent core radius of 10.7 ± 0.1Å was found by fitting β-C_8G_1 micelle scattering at all concentrations. This value is smaller than the maximum chain length for the fully extended alkyl chain. It should be emphasized that this core–shell model is a simplified model of a complex micelle structure. Probably, the real micelles have an irregular and dynamic structure, in which there is extensive mixing of core, shell, and solvent. The shell is especially not well defined with such mixing. In fact, the observed large fluctuations results in diffuse scattering at high q region, which is much higher than the expected from static cylindrical micelle structure, were modeled in terms of fluctuations in the position of the surfactant in the micelle [96]. The proposition put forth by Nilsson et al. [80] that β-C_9G_1 and β-$C_{10}G_1$ form a network structure between the micelles is worth mentioning in this context. These authors obtained information about the micellar size of β-C_9G_1 and β-C_9G_2 by using ^{1}H-NMR self-diffusion and TRFQ. Both techniques have shown that discrete nonspherical aggregates are formed at concentration immediately above the cmc (probably a prolate geometry with an axial ratio of 11 to 1) and there was micellar growth in one or two dimensions as the concentration increases. When the surfactant concentration was increased above 17 wt%, the micelles formed a network structure. The aggregation of β-C_9G_1 has been also studied by Ericsson et al. [13]. The main conclusions drawn from this work are that β-C_9G_1 micelles can be described as relatively stiff, elongated structures with a circular cross section and that substituting D_2O for H_2O was found to induce substantial micellar growth, without affecting cmc. This deuterium effect can be rationalized on the basis of the shorter length of the O–D bond, as compared with the O–H one, and the resulting decrease in effective head group area.

The micelle architecture for maltosides has been addressed in two contributions from He et al. [98] and Lipfert et al. [99]. The former authors combined SANS and SAXS experiments to investigate the micelle structure of octyl-β-D-maltoside (β-C_8G_2). The indirect Fourier transformation method and the model fitting of the scattering data show that β-C_8G_2 micelles have a spherical shape. The size of micelles is not sensitive to surfactant concentration (up to 188 mM) and there is no increase of the aggregation number with increasing concentration. The aggregation number for β-C_8G_2 micelles is 26, and the radius of the sphere is 23.7 Å (with a hydrocarbon core with a radius of 11.5 Å). Because β-C_8G_2 and β-C_8G_1 have the same hydrocarbon chain, their micelle structure difference must be due to their different head groups. Micelle formation is governed by two contributions to the free energy. The burial of the hydrophobic moieties in the lipid-like hydrophobic micelle core has a favorable (and mostly entropic) contribution to the free energy. This effect alone would favor very large aggregate and phase separation. However, since the hydrophilic head groups are covalently attached to the hydrophobic tails, micelle formation brings head groups into close proximity, which is energetically unfavorable and balances the hydrophobic effect. The balance of "opposing forces" determines the size and shape of the micelle [92]. It is known that, due to the strong binding of water to the sugar head group, there are strong short-range repulsive hydration forces between them [91,100] that can prevent the formation of large micelles [92]. To decrease the repulsive energy, a large optimal head group area is required. However, the attractive hydrophobic force between the hydrocarbon chains requires a small head group area in order to decrease the attractive energy. Together, these two forces determine the optimal head group area, which gives a minimum value to the total interaction energy per glucoside molecule [100]. Although β-C_8G_1 and β-C_8G_2 have the same head group area at the air–water interface (see Tables 3.1 and 3.2), β-C_8G_2 has a larger optimal head group area than does β-C_8G_1 because the repulsive hydration force of the two-sugar heads group is higher than that of one-sugar head group (the estimated hydration of the head group of β-C_8G_2 and β-C_8G_1 is eight and four molecules of water per molecule, respectively). Thus, β-C_8G_2 forms a spherical micelle as a result of the high hydration force of its head group whereas β-C_8G_1 forms a cylindrical micelle [98,99]. The formation of nonspherical shape is a reflection of weak steric hindrance of the polar head groups, whereas when the head groups are more hindered sterically the micelles are spheres. Then, hydration forces and steric repulsion play an important role in the determination of micelle structure. In a recent paper, Lipfert et al. [99] have investigated by means of SAXS measurements the shape of micelles of β-C_8G_1, β-C_9G_1, β-$C_{10}G_2$, and β-$C_{12}G_2$. The SAXS intensities for the glucosides are reasonably fitted by a prolate two-component ellipsoid model. This model features an ellipsoidal core with one semiaxis a, two semiaxis b, representing the hydrophobic interior of the micelle, and an outer shell corresponding to the surfactant head groups. For β-C_8G_1 and β-C_9G_1 the short dimension of the micelle is approximately constant with increasing surfactant concentration; however, the apparent length increases from 40Å at 50mM to 50–55Å at 150mM for β-C_8G_1, in reasonable agreement with the dimension determined in other studies [90,96], and from 55Å at 10mM to 105Å at 50mM for β-C_9G_1. Figure 3.8 shows the apparent aggregation number, as determined by Lipfter et al. [99] from SAXS measurements, as a function of the surfactant concentration for β-C_8G_1, β-$C_{10}G_2$, and β-$C_{12}G_2$ micelles. It is to be noted that the apparent aggregation number is not necessarily the same as the actual aggregation number, because the former, at high surfactant concentration, can be affected by the interactions between micelles, in such a way that the so-called interparticle interference effects can change the apparent aggregation number as a function of surfactant concentration without the micelle necessarily changing size. Data in Figure 3.8 indicate that whereas the apparent aggregation number of β-C_8G_1 increases with surfactant concentration, in the case of β-$C_{10}G_2$ and β-$C_{12}G_2$ it decreases. Data of β-C_8G_1 must be interpreted in the sense that the micelles of β-C_8G_1 actually do grow with increasing surfactant concentration. However, in the cases of β-$C_{10}G_2$ and β-$C_{12}G_2$ micelles, the observed decrease in apparent aggregation number is very likely due to interparticle repulsion effects; therefore, the size of these micelles seems to be approximately constant as a function of surfactant concentration [99]. For a surfactant concentration of 50mM, the aggregation number is 85, in good agreement with the value of 87 obtained by Kameyama et al. [62]. In the case of maltosides, the SAXS intensities are well fitted by oblate ellipsoid models. The results for β-$C_{12}G_2$ are in excellent agreement with the findings of Dupuy et al. [39], who modeled these micelles as a

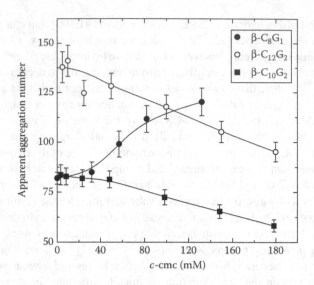

FIGURE 3.8   Apparent aggregation number as a function of micellar concentration for
β-C_8G_1, β-$C_{10}G_2$ and β-$C_{12}G_2$. (From data in Lipfert, J., Columbus, L., Chu, V.B., Lesley,
S.A., and Doniach, S., *J. Phys. Chem. B*, 111, 12427, 2007.)

two-component oblate ellipsoid with a 14.1Å semiminor axis and two 28.2Å
semimajor axes for the core and a 6.2Å outer shell of uniform thickness. For
β-$C_{10}G_2$, the core is smaller than for β-$C_{12}G_2$, the short dimension by about 1.8Å,
and the long one by about 5Å, while the thickness of the head group layer is
unchanged. This is to be expected as the two surfactants have the same head group,
but β-$C_{10}G_2$ has a shorter alkyl chain than β-$C_{12}G_2$. For β-$C_{10}G_2$ and β-$C_{12}G_2$, the
aggregation number is around 85 and 140, respectively. The micelle structure difference between β-C_8G_2 and β-$C_{10}G_2$ or β-$C_{12}G_2$ originates from the hydrocarbon
chain difference. The 12-carbon chain of β-$C_{12}G_2$ exerts a higher hydrophobic
force than does the 8-carbon chain of β-C_8G_2, although they have the same head
group; hence, β-$C_{12}G_2$ forms a relative large micelle (compared to that of β-C_8G_2)
and has ellipsoid shape. The literature provides data about the micellar shape
of β-$C_{14}G_2$ [101], which indicate the formation of rodlike micelles for this
surfactant.

The micellar shape of nonionic micelles is determined to a large extent by the
aggregation number (N_{agg}). A variety of experimental methods may be applied to
the determination of the aggregation number. The results obtained for different
n-alkyl glucosides and *n*-alkyl maltosides are summarized in Table 3.6. Some of
the reported values have been determined with the assumption that the micelles do
not grow with surfactant concentration. As observed, glucosides exhibit larger
aggregation numbers compared to maltosides for the same chain length. A likely
explanation is that the larger head group of the maltosides as compared to the
glucosides leads to stronger steric repulsion, and, therefore, to smaller micelles.

TABLE 3.6

Values of the Micelle Aggregation Numbers (N_{agg}) for *n*-Alkyl Glucosides and *n*-Alkyl Maltosides

Surfactant	N_{agg}	Method of Determination[a]	References
Heptyl-β-D-glucoside (β-C$_7$G$_1$)	58 at 1%	Sedimentation	[121]
Octyl-β-D-glucoside (β-C$_8$G$_1$)	92 at 60 mM	TRFQ	[36]
	113 at 0.5%	SE	[121]
	85 at 50 mM	SAXS	[99]
	87 at cmc	SLS	[62]
	84 at cmc	SE	[62]
	105 at 99.2 mM	TRFQ	[61]
	54 at 29 mM and 105 at 55 mM	SSFQ	[82]
	82	SLS	[123]
	70 at cmc	Sedimentation	[87]
Octyl-β-D-thioglucoside (β-SC$_8$G$_1$)	114 at 30 mM	TRFQ	[36]
	217	Sedimentation	[121]
Nonyl-β-D-glucoside (β-C$_9$G$_1$)	202 at 0.5%	Sedimentation	[121]
	240	SAXS	[99]
Octyl-β-D-maltoside (β-C$_8$G$_2$)	26	SAXS	[98]
Decyl-β-D-maltoside (β-C$_{10}$G$_2$)	85	SAXS	[99]
	69	—	[99]
Dodecyl-β-D-maltoside (β-C$_{12}$G$_2$)	132 at 40 mM	SAXS	[39]
	140	SAXS	[99]
	129±4 at 150 mM	SANS	[122]
	125±10 in the range 5–20 mM	TRFQ	[61]
	110±10 in the range 4–200 mM	TRFQ	[107]
	140	—	[124]
Tetradecyl-β-D-maltoside (β-C$_{14}$G$_2$)	>1000	—	[101]

[a] TRFQ, time-resolved fluorescence quenching; SSFQ, steady-state fluorescence quenching; SE, sedimentation equilibrium; SLS, static light scattering; SAXS, small-angle x-ray scattering; SANS, small-angle neutron scattering.

Also, the repulsive hydration is higher for maltosides than for glucosides. For a given head group, increasing the alkyl chain length favors large aggregation numbers, as the hydrophobic effect from packing the tail groups in the micelle interior becomes stronger (by ≈0.8 kcal/mol per CH_2 group) [92]. In agreement with this prediction, the observed aggregation numbers increase with increasing chain length for the glucoside and maltoside sequences. An important constraint on micelle shape is the maximum possible extension of the hydrocarbon chain (l_c given by the Tanford's formula). Since there cannot be a "hole" in the middle of the micelle, one dimension of the micelle is always limited by this extension. As emphasized by Israelachvili [100], the maximum aggregation number consistent with spherical geometry for micelles with a certain alkyl chain length is given by

$$N_{max,sph} = \left(\frac{4\pi}{3}\right)\frac{l_c^3}{V_{tail}}$$ (3.13)

where V_{tail} is the volume occupied by an alkyl chain of n_c carbon atoms ($V_{tail} = 27.4 + 26.9 n_c$). For all the surfactants included in Table 3.6 (except for β-C_8G_2), $N_{agg} > N_{max,sph}$. Glucosides and maltosides micelles should therefore be anisotropic, in agreement with the observation that spherical models are unable to account for the SANS and SAXS data.

Size and shape of MEGA-10 micelles were investigated by means of dynamic light scattering in combination with rheological measurements [102]. The micellar radius proved to be dependent on the concentration, increasing from 22 Å ($N_{agg} = 73$) up to 35 Å ($N_{agg} = 160$) in the surfactant concentration range studied. From the experimental results, the authors suggested that the micelles were at least nearly spherical at surfactant concentration close to the cmc. The formation of rodlike structures or long flexible micelles induces significant changes in the viscosity of the surfactant solution. The dynamic viscosity (η) of the surfactant solution is highly sensitive to changes in temperature. The temperature dependence of the viscosity follows the Arrhenius law:

$$\eta = B \exp\left(\frac{E_a}{RT}\right)$$ (3.14)

where
  B is the pre-exponential factor
  R is the gas constant
  E_a is the shear activation energy

The activation energy describes the energy that is necessary to move individual micelles in an environment of surrounding micelles. This parameter can be determined from the slope of the plot of ln η versus $1/T$. The reported values of E_a in water for MEGA-10 are 17.5 kJ/mol [77] and 15.7 kJ/mol [102]. Activation energy values between 15.9 and 17.5 kJ/mol are characteristic for aqueous solutions of molecularly dissolved substances and spherocolloids [103]. For rod-shaped micelles, a wide range of E_a values (30–300 kJ/mol) are reported for various micellar systems [104]. Probably for MEGA-10 micelles, the bulky head groups (sugar moiety which is highly hydrated) in turn favor the formation of spherical micelles due to the space requirements.

3.3.1 EFFECT OF THE TEMPERATURE

In many applications, the selection of a particular surfactant as well as suitable medium (temperature, electrolyte concentration, pH) is often achieved on a trial-and-error basis. It is therefore of interest to investigate the influence of experimental conditions on the micelle structure and it is hoped that the results will contribute to improve the utility of sugar-based surfactants. In general, the micelles of nonionic surfactants grow in size with the temperature, and such growth is usually accompanied by a change of micellar shape. For some ethylene oxide based surfactants, the size of the micelles dramatically increases as a function of the temperature (e.g., $C_{15}E_5$) [12]. In the case of sugar-based surfactants, there is not much known about the effect of the temperature on the size and shape of micelles. For β-C_8G_1 it has been obtained that the aggregation number and also the hydrodynamic radius of micelles show a flat maximum at 36°C, corresponding to a hydrodynamic radius of approximately 28 Å [38,62] (see Figure 3.9). Similar trend was obtained by Aoudia et al. [61], who observed that the aggregation number goes through a rather flat maximum at around 35°C. This temperature is close to that where the cmc appears to go through a shallow minimum (see Figure 3.3). This moderate growth of β-C_8G_1 micelles contrasts with the strong increases of the size of ethylene oxide based surfactants with the temperature [3]. He et al. [105] have performed SANS experiments at 15°C and 25°C in order to elucidate the structure of the β-C_8G_1 micelle. The SANS data suggest that micelle shape can be described by a cylindrical model. The radius of the cylinder is practically constant (13.2 ± 0.2 and 12.7 ± 0.2 at 15°C and 25°C, respectively), whereas its length is more affected by the temperature (68 ± 2 and 96 ± 2 at 15°C and 25°C, respectively). The gyration radius was 21.7 Å at 15°C and 29.1 Å at 25°C.

By means of DLS measurements, Ericsson et al. [13] have studied the effects of the temperature on the micellar size of β-C_9G_1. These authors have found that

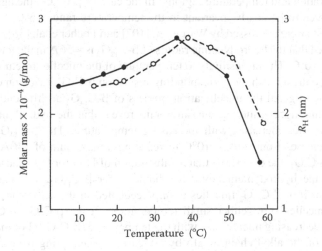

FIGURE 3.9 Effect of temperature on the molar mass (●) and hydrodynamic radius (○) of β-C_8G_1. (From Kameyama, K. and Takagi, T., *J. Colloid Interface Sci.*, 137, 1, 1990. With permission.)

the effective hydrodynamic diameter of the micelles decreases, almost linearly, from 195 to 110 Å when the temperature increases from 10°C to 60°C, suggesting that the effective head group size increases with the temperature. However, they observed that the unimer solubility of β-C_9G_1 increases with decreasing temperature. This behavior contrasts with that of ethylene oxide based surfactants, in which the temperature effects on both CPP and unimer solubility favor increased aggregate size upon increasing temperature. In order to explain the above behavior, Ericsson et al. [13] invoked the rigid and strongly hydrated head group of β-C_9G_1 as opposite to the flexible polyoxyethylene (POE) chain at higher temperatures as a consequence of the fact that water becomes a less good solvent for PEO with increasing temperature. The aforementioned authors proposed that this behavior can be explained if increased temperature is assumed to increase the effective head group size by influencing hydration and hydrogen bonding, then this effect would tend to favor decreased micelle size at higher temperature, thus competing with the temperature effects on unimer solubility. The fact that the micellar size of β-C_9G_1 is extremely sensitive to small changes in head group size is supported by the observation that an exchange of D_2O for H_2O has a profound influence on micelle size (for a given surfactant concentration, the size of micelles is higher in the presence of D_2O) [13]. An exchange of O–D bonds for the longer O–H bonds leads to a decrease in head group size and thus favors larger aggregates [106]. A comparison of the results previously discussed for β-C_8G_1 and β-C_9G_1 indicates that the temperature has much more dramatic effects on the micelle morphology of the latter surfactant.

The influence of the temperature on the aggregation of β-C_8G_2 has been investigated by the SANS technique [98]. As shown in Figure 3.10, there are small differences in the SANS patterns for 188 mM β-C_8G_2 from 20°C to 50°C. This indicates that the size of β-C_8G_2 micelles is temperature insensitive within the studied concentration and temperature regions. In the case of β-$C_{12}G_2$, the aggregation number is seen to be nearly invariant in the temperature range 16°C–60°C [63]. Similar results were addressed by Warr et al. [107] and Focher et al. [38]. The latter group showed that the hydrodynamic radius of β-$C_{12}G_2$ is ≈ 36 Å in the temperature range 15°C–40°C. The insensitivity to temperature of the micelle structure of these maltosides contrasts with the corresponding results on β-C_9G_1. Ericsson et al. [71] have also investigated the micellization process of β-$C_{14}G_2$ at different temperatures. Dynamic light scattering measurements reveal that the hydrodynamic diameter of micelles increases with increasing temperature. The β-$C_{14}G_2$ micelle diameter increases from 60 Å at 10°C to reaching a maximum of 340 Å at 70°C. Similar to β-C_9G_1, the data show that a substitution of D_2O for H_2O causes a large increase in the hydrodynamic micellar diameter for β-$C_{14}G_2$. The temperature induced growth of β-$C_{14}G_2$ micelles is unprecedented in the sense that the alkyl glucoside micelles show either small temperature effects (β-C_8G_1, β-C_8G_2, and β-$C_{12}G_2$) or decreasing micelle size with the temperature (β-C_9G_1). For maltosides, an increase of the alkyl chain length by six carbons (i.e., going from β-C_8G_2 to β-$C_{14}G_2$) leads to a transition from spherical micelles with little temperature dependence to rodlike micelles that grow dramatically with the temperature. In the case

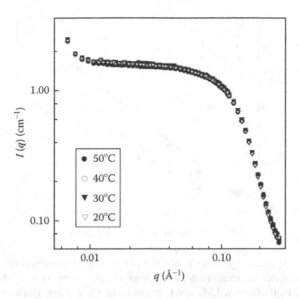

FIGURE 3.10 SANS data at different temperatures for 188 mM β-C_8G_2. (From He, L.Z., Garamus, V.M., Funari, S.S., Malfois, M., Willumeit, R., and Niemeyer, B., *J. Phys. Chem. B*, 106, 7596, 2002. With permission.)

of glucosides, an increase of one carbon (i.e., going from β-C_8G_1 to β-C_9G_1) provokes a decrease in the micellar size with the temperature. As can be seen, a relatively minor increase of hydrophobicity within a given homologous surfactant series may lead to a dramatic amplification of the effect of the temperature. The conformation of the alkyl chains changes with the temperature and promotes micellar growth at higher temperatures (the CPP increases and the curvature of the micelles decreases). Also, it is known that the hydrophobic force increases with the temperature, and consequently the micelles grow with increasing temperature. On the other hand, it has been suggested that the short-ranged repulsive force between sugar head groups increases with the temperature, which would result in an increase in the effective head group area with the temperature (the value of CPP decreases and smaller micelles are formed) [98]. Hence, temperature effects on the alkyl chain and hydrophobic force tend to counteract those on the head group size for sugar-based surfactants. The relative importance of head group effects on micelle size is diminished as the alkyl chain length increases. For long-chain surfactants (e.g., β-$C_{14}G_2$), the effects on the alkyl chain dominate and the micellar size decreases with the temperature. However, for short-chain surfactants, the effect of the temperature on the head group size balances the effect on the hydrocarbon chain. In this case, the micellar size does not change with the temperature (e.g., β-C_8G_1 or β-C_8G_2) or the micellar size decreases with increasing temperature (e.g., β-C_9G_1). In conclusion, the micelle morphology is very sensitive to changes in the surfactant molecular structure.

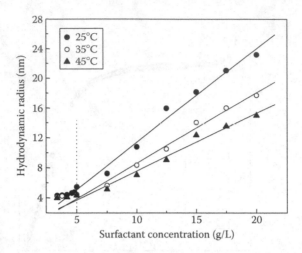

FIGURE 3.11   Apparent hydrodynamic radius of octyl-β-D-thioglucoside micelles as a function of surfactant concentration at different temperatures. (From Molina-Bolívar, J.A., Aguiar, J., Peula-Garcia, J.M., and Carnero Ruiz, C., *J. Phys. Chem. B*, 108, 12813, 2004. With permission.)

The structures of the aggregates of octyl-β-D-thioglucoside (β-SC$_8$G$_1$) through a temperature range have been reported by Molina-Bolívar et al. [65]. In Figure 3.11, the apparent hydrodynamic radius of β-SC$_8$G$_1$ micelles versus the surfactant concentration at various temperatures is plotted. As can be seen, the results showed a well-pronounced transition in the micelle size at a certain surfactant concentration. Below this concentration (around 5 g/L), the micelle size presents a weak dependence on the surfactant concentration. In this region of low concentration the micelles are small. After the break point in the curve, a strong concentration dependence of the micelle size, which rapidly increases, is observed. Although this light scattering study does not provide any explicit information about the micelle shape, it is accepted in the literature that spherical micelles undergo a transition to larger rodlike aggregates with the surfactant concentration [100]. Figure 3.11 shows that in the low surfactant concentration region, the micellar size is independent of the temperature (hydrodynamic radius of 35 Å with an aggregation number of 116). However, after the break point, a micellar growth is evidenced, which is more significant at low temperature. This fact suggests that the formation of rodlike micelles is enhanced by the decrease of the temperature. This decrease of micellar size with the temperature is in concordance with the behavior showed by β-C$_9$G$_1$. Additional experiments, particularly intramolecular excimer forming of 1,3-dipyrenylpropane and static light scattering measurements, evidenced the existence of this growth of the micelles from globular to rodlike structures [65].

There are reports on the micellar aggregation number of MEGA-8, MEGA-9, and MEGA-10 as a function of the temperature [75,76]. The N_{agg} values may be

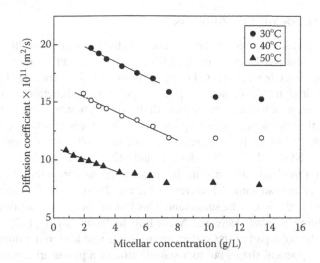

FIGURE 3.12 Apparent diffusion coefficient (D_c) of MEGA-10 versus micellar concentration (c-cmc) at different temperatures. (From Molina-Bolívar, J.A., Hierrezuelo, J.M., and Carnero Ruiz, C., *J. Colloid Interface Sci.*, 313, 656, 2007. With permission.)

regarded as being almost constant, irrespective of the temperature change, which differs remarkably from the strikingly increasing trend of N_{agg} of ethylene oxide based surfactants. This insensitivity of N_{agg} to the temperature can be presumably ascribed to the nature of the head group of MEGA-n surfactants, which is mainly characterized by the presence of the methylamine moiety, an open conformation, and, particularly, by a great capability to form hydrogen bonds [49]. Figure 3.12 illustrates the variation of the apparent diffusion coefficient D_c of MEGA-10 with the micellar concentration (c-cmc) at 30°C, 40°C, and 50°C [77]. It is evident from this figure that there is an initial decrease in the apparent diffusion coefficient with increasing concentration. This decrease is fairly independent on the temperature. The linear fits to the data in Figure 3.12 at low concentrations give as intersection the apparent diffusion coefficient at infinite dilution, D_0, from which the hydrodynamic radius can be calculated by using the Stokes–Einstein relation. The so obtained values were 26Å, 22Å, and 21Å at 30°C, 40°C, and 50°C, respectively. This slight decrease of the hydrodynamic radius with increasing temperature can be due to a reduction in the micellar aggregation number or a decrease in hydration. In respect to the dependence of the aggregation number with the temperature, it has been observed that this parameter, within experimental accuracy, remains almost invariant with the temperature [36,76,77] which differs remarkably from the strikingly increasing trend of the aggregation number of the C_iE_j surfactants. This implies that the decrease in the hydrodynamic radius with the temperature can be attributed to the reduction in the quantity of water nonspecifically associated to the micellar periphery (it is noteworthy to remember that, as deduced from density measurements, the partial specific volume of MEGA-10 is not affected by the temperature) [77].

3.3.2 Effect of the Salt Addition

Short papers that are focused on the influence of electrolytes on micelle size can be found in the literature. Ericsson et al. [13] have investigated the effects of cations and anions on micelle size of β-C_9G_1 using dynamic light scattering. Figure 3.13 shows the hydrodynamic diameter of β-C_9G_1 micelles as a function of the electrolyte concentration for five different salts with the same cation but different anion. As can be seen, the micelle size shows a pronounced dependence on both salt type and concentration, and the salting-out and salting-in effects follow the Hofmeister series $SO_4^{2-} > Cl^- > NO_3^- > I^- > SCN^-$. Here, I^- and SCN^- are salting-in anions, which gives rise to a moderate decrease in the micelle size as compared to pure water, whereas the salting-out anion Cl^- increases the micelle size. The Cl^- anion induces a reduction of hydration of the surfactant. This loss of solvation is accompanied by a corresponding volume decrease. According to Israelachvili [100], a volume decrease of the polar part of the surfactant molecule will lead to a reduction of the stability of a spherical shape, but to a stabilization of a planar arrangement of the surfactant molecule. A reduction of curvature, i.e., a tendency toward a more planar arrangement, will of course favor the association and the subsequent growth of micelles. In such a nonspherical arrangement, the packing of the surfactant molecules can be denser as a consequence of the smaller head group area. The opposite behavior is induced by the SCN^- anion, which promotes the hydration of the surfactant (salting-in effect). A similar investigation was performed for cations by studying the effects of chloride, nitrate, and sulfate salts of the alkali metals Li, Na, K, and

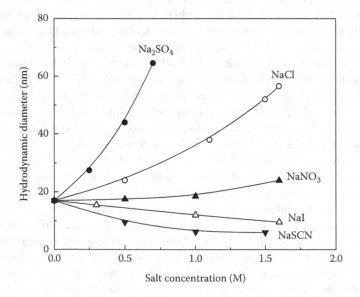

FIGURE 3.13 The effect of different sodium salts on the micellar hydrodynamic diameter for 10 g/L solution of β-C_9G_1 at 25°C. (From Ericsson, C.A., Söderman, O., Garamus, V.M., Bergström, M., and Ulvenlund, S., *Langmuir*, 20, 1401, 2004. With permission.)

Cs [13]. In terms of micelle size, the effects of the salts decrease with increasing the polarizability (increasing ionic radius) of the cation. The micelle hydrodynamic radius in the presence of 1.5 M LiCl is 550 nm whereas it is 280 nm in 1.5 M CsCl. These authors have pointed out that the micelle size correlated with head group area, rather than with monomer solubility [13]. Similar results have been observed for the maltoside β-$C_{14}G_2$ [71]. The aggregation of β-C_8G_1 in the presence of divalent cations has been investigated by Pastor et al. [82]. In this study, the speed of sound data together with the fluorescence results provided the aggregation number of the micelles. This parameter increased with increasing the content of $CaCl_2$ in the medium. These authors suggested that the $CaCl_2$ provokes a contraction of the surfactant hydration shell, which is accompanied by a higher packing, and that the β-C_8G_1 micelles turn from spherical (at low salt concentration) to ellipsoidal (at $[Ca^{2+}] \geq 0.1 M$).

The apparent hydrodynamic radius of octyl-β-D-thioglucoside (β-SC_8G_1) micelles in the presence of NaCl has been measured by dynamic light scattering (see Figure 3.14) [84]. The results showed a well-pronounced transition in the micelle size at certain surfactant concentration. At surfactant concentration lower than this transition surfactant concentration (around 4.5 g/L), the micelle size presents a weak dependence on surfactant concentration. After the break point in the curve, a strong increase of the micelle size with the concentration is observed. The data indicate a significant micellar growth with the corresponding change in the micelle shape from sphere to rodlike. The hydrodynamic radius of the micelles at infinite dilution increases from 35 Å ($N_{agg} = 114$) in water up to 62 Å ($N_{agg} = 537$)

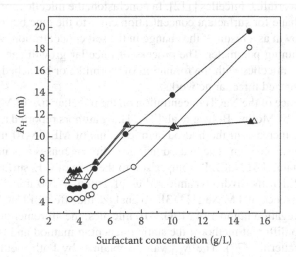

Surfactant concentration (g/L)

FIGURE 3.14 Apparent hydrodynamic radius of n-octyl-β-D-thioglucoside (β-SC_8G_1) micelles as a function of surfactant concentration at different NaCl concentrations. ($\bullet$) water, ($\circ$) 0.1 M NaCl, ($\blacktriangle$) 0.2 M NaCl, and ($\triangle$) 0.5 M NaCl. (From Molina-Bolívar, J.A., Hierrezuelo, J.M., and Carnero Ruiz, C., *J. Phys. Chem. B*, 110, 12089, 2006. With permission.)

in 0.5 M NaCl. On the other hand, it is noteworthy that two different patterns may be deduced in the nonspherical shape region from Figure 3.14. First, for 0 and 0.1 M NaCl, the micelle size monotonically increases with the β-SC$_8$G$_1$ concentration, with the micellar size bigger in the presence of 0.1 M NaCl. However, dynamic light scattering measurements performed in 0.2 and 0.5 M NaCl showed that the radius initially increases and then becomes practically constant for surfactant concentration higher than 7 g/L (for both electrolyte concentrations). By analogy with solutions of semiflexible polymers, it has been suggested that the surfactant concentration where the apparent hydrodynamic radius is constant may be identified as an overlap threshold concentration, $c*$, between a dilute and semidilute region [108,109]. According to this model, the micelles can be seen as individual aggregates that are diffusing almost independently of each other below $c*$ in the dilute region. Above $c*$, in the semidilute region, the micelle growth has progressed to such an extent that the aggregates start to overlap together and become entangled. The network so formed has an average mesh size ξ, called the correlation length, which is the average length between two points of entanglements [110]. This correlation length dominates the dynamic light scattering rather than the size of individual micelles. Molina-Bolívar et al. [84] have observed the existence of a strong increase in the Debye plots, obtained from static light scattering, above a minimum value. This behavior probably stems from the overlapping of micelles.

The overlapping or entanglement of micelles was first evidenced by Hoffmann et al. [111]. Rheological and viscometric studies of β-C$_9$G$_1$, β-C$_{10}$G$_1$, β-C$_{12}$G$_2$, and β-C$_{14}$G$_2$ have demonstrated that these surfactants form entangled networks of polydisperse wormlike micelles [112]. In conclusion, the micelles are often quasi-spherical in shape for surfactant concentration close to the cmc but may undergo a dramatic growth as a result of the change in the salt concentration, which can be defined as a tuning parameter. The process of micellar growth can result in the overlapping of micelles with the formation of wormlike or branched micelles, or even interconnected micellar networks.

The influence of the NaCl concentration on the micellar size of MEGA-10 has been analyzed by Molina-Bolívar et al. [77]. These authors reported a progressive, though slight, increase in the hydrodynamic radius of MEGA-10 micelles with increasing the salt content. The hydrodynamic radius increases around 20% when the water contains 1 M NaCl. In comparison to the case of C$_i$E$_j$ surfactants, it has been reported that the hydrodynamic radius of Triton X-100 micelles increases 40% in the presence of 1 M NaCl [113]. To analyze the effect of NaCl addition on the mean aggregation number of MEGA-10 micelles, the aforementioned authors have used two different methods: the static quenching method and that based on static light scattering [77]. The N_{agg} values obtained by both methods were in excellent agreement, indicating that the micellar aggregation number showed a similar trend to the hydrodynamic radius. Whereas the N_{agg} of MEGA-10 micelles increases around 50% in 1 M NaCl, in the case, for example, of Triton X-100 the aggregation number in 1 M NaCl is around 120% [113]. As can be seen, the effect of the electrolyte on the solution properties of MEGA-10 is much less pronounced than on the ethylene oxide based surfactants. In order to obtain information about the hydration of MEGA-10 micelles, the dependence of micellar solution density with the NaCl concentration was also investigated [77]. The partial specific volume of MEGA-10 calculated from these experimental data were practically constant for all NaCl concentrations studied; this suggests that the hydration shell of the micelle remains practically unaltered.

ACKNOWLEDGMENTS

We are grateful to the Spanish Ministry of Education and Science (project CTQ2005–04513) for supporting our research on sugar-based surfactants. We thank Dr. Lipfert for providing us the apparent aggregation number data in Figure 3.8 and for valuable comments about the interpretation of these data.

REFERENCES

1. Lorber, B., Bishop, J.B., and DeLucas, L.J. *Biochim. Biophys. Acta.*, 1990, *1023*, 254.
2. Hill, K. and Rhode, O., *Fett/Lipid.*, 1999, *101*, 25.
3. Balzer, D. and Lüders, H., *Nonionic Surfactants: Alkyl Polyglucosides*. Marcel Dekker, New York, 2000.
4. Suggett, A., Polysaccharides, in *Water-A comprehensive Treatise*, Franks, F., Ed. Plenum Press, New York, 1975; Vol. 4, pp. 519–567.
5. Barone, G., Cacace, P., Catronuovo, G., and Elia, V., *Carbohydr. Res.*, 1981, *91*, 101.
6. Marshall, R.D., *Biochem. Soc. Trans.*, 1984, *12*, 513.
7. Drummond, C.J., Warr, G.G., Grieser, F., Ninham, B.W., and Evans, D.F., *J. Phys. Chem.*, 1985, *89*, 2103.
8. Bonincontro, A., Briganti, G., D'Aprano, A., La Mesa, C., and Sesta, B., *Langmuir*, 1996, *12*, 3206.
9. Saito, S. and Tsuchiya, T., *Biochem. J.*, 1984, *222*, 829.
10. De pino, V., Benz, R., and Palmire, F., *Eur. J. Biochem.*, 1998, *179*, 188.
11. Hamada, T., Wakagi, T., Shiba, H., and Koyama, N., *Arch. Microbiol.*, 1999, *171*, 237.
12. Holmberg, K., Jönsson, B., Kronberg, B., and Lindman, B., *Surfactants and Polymers in Aqueous Solution*, 2nd ed. Wiley, New York, 2003.
13. Ericsson, C.A., Söderman, O., Garamus, V.M., Bergström, M., and Ulvenlund, S., *Langmuir*, 2004, *20*, 1401.
14. Stubenrauch, C., *Curr. Opin. Colloid Interface Sci.*, 2001, *6*, 160.
15. Hato, M., Minamikawa, H., Tamada, K., Baba, T., and Tanabe, Y., *Adv. Colloid Interface Sci.*, 1999, *80*, 233.
16. Laughlin, R.G., *The Aqueous Phase Behavior of Surfactants*. Academic Press, London, 1994.
17. von Rybinski, W. and Hill, K., *Angew. Chem. Int. Ed.*, 1998, *37*, 1328.
18. Hill, K., *Pure Appl. Chem.*, 2000, *72*, 1255.
19. Stryer, L., *Biochemistry*, 3rd ed. W.H. Freeman and Company, New York, 1988.
20. Sanders, C.R. and Prestegard, J.H., *J. Am. Chem. Soc.*, 1992, *114*, 7096.
21. Hildreth, J.E.K., *Biochem. J.*, 1982, *363*, 207.
22. Weerawardena, A., Boyd, B.J., Drummond, C.J., and Furlong, D.N., *Colloids Surf. A*, 2002, *169*, 317.
23. Hoffmann, H., *Prog. Colloid Polym. Sci.*, 1990, *83*, 16.

24. Liljekvist, P. and Kronberg, B., *J. Colloid Interface Sci.*, 2000, *222*, 159.
25. Sulthana, S.B., Bhat, S.G.T., and Rakshit, A.K., *Langmuir*, 1997, *13*, 4562.
26. Sulthana, S.B., Rao, P.V.C., Bhat, S.G.T., and Rakshit, A.K., *J. Phys. Chem. B.*, 1998, *102*, 9653.
27. Kutschmann, E., Findenegg, G., Nikel, D., and von Rybinski, W., *Colloid Polym. Sci.*, 1995, *273*, 565.
28. Strasthof, A., van Bekkum, H., and Kieboom, A.P.G., *Starch*, 1988, *40*, 438.
29. Focher, B., Savelli, G., and Torri, G., *Chem. Phys. Lipids*, 1990, *53*, 141.
30. Matsumura, S., Imai, K., Yoshikawa, S., Kawada, K., and Uchibori, T., *J. Am. Oil Chem. Soc.*, 1990, *67*, 996.
31. Banipal, P.K., Banipal, T.S., Lark, B.S., and Ahluwalia, J.C., *J. Chem. Soc. Faraday, Trans.*, 1997, *93*, 81.
32. Galema, S.A., Howard, E., Engberts, J.B.F.N., and Grigera, J.R., *Carbohydr. Res.*, 1994, *265*, 215.
33. Engelsen, S.B. and Pérez, S., *Carbohydr. Res.*, 1996, *292*, 21.
34. Garofalakis, G., Murray, B.S., and Sarney, D.B., *J. Colloid Interface Sci.*, 2000, *229*, 391.
35. Brown, G.M., Dubreuil, P., Ichhaporia, F.M., and Desnoyers, J.E., *Can. J. Chem.*, 1970, *48*, 2525.
36. Frindi, M., Michels, B., and Zana, R., *J. Phys. Chem.*, 1992, *96*, 8137.
37. Fukuda, K., Kawasaki, M., Seimiya, T., Abe, Y., Fujiwara, M., and Ohbu, K., *Colloid Polym. Sci.*, 2000, *278*, 576.
38. Focher, B., Savelli, G., Torri, G., Vecchio, G., McKenzie, D.C., Nicoli, D.F., and Button, C.A., *Chem. Phys. Lett.*, 1989, *158*, 491.
39. Dupuy, C., Auvray, X., Petipas, C., Rico-Lattes, I., and Lattes, A., *Langmuir*, 1997, *13*, 3965.
40. Nilsson, F., Söderman, O., and Johansson, I., *J. Colloid Interface. Sci.*, 1998, *203*, 131.
41. Hoffmann, B. and Platz, G., *Curr. Opin. Colloid Interface Sci.*, 2001, *6*, 171.
42. Schick, M.J., *Nonionic Surfactants: Physical Chemistry*. Marcel Dekker, New York, 1987.
43. Rosen, M.J., *Surfactants and Interfacial Phenomena*, 2nd ed. Wiley, New York, 1989.
44. Meguro, K., Takasawa, Y., Kawahashi, N., Tabata, Y., and Ueno, M., *J. Colloid Interface Sci.*, 1981, *83*, 50.
45. Coppola, L., Gordano, A., Procopio, A., and Sindona, G., *Colloids Surf. A*, 2002, *196*, 175.
46. Boullanger, P. and Chevalier, Y., *Langmuir*, 1996, *12*, 1771.
47. vanBuuren, A.R. and Berendsen, H.J.C., *Langmuir*, 1994, *10*, 1703.
48. Nilsson, F. and Söderman, O., *Langmuir*, 1997, *13*, 3349.
49. Walter, A., Suchy, S.E., and Vinson, P.K., *Biochim. Biophys. Acta*, 1990, *1029*, 67.
50. Ko, J.S., Oh, S.W., Kim, K., Nakahima, N., Nagadome, S., and Sugihara, G., *Colloids Surf. B*, 2005, *45*, 90.
51. Thiëm J. and Böcker, T., *Ind. Appl. Surfactants III*, 1992, 123.
52. Bëcker, T. and Thiem, J., *Tenside Surf. Det.*, 1989, *26*, 318.
53. Boyd, B.J., Drummond, C.J., Krodkiewska, I., and Grieser, F., *Langmuir*, 2000, *16*, 7359.
54. Jeffrey, G. and Bhattacharjee, S., *Carbohydr. Res.*, 1983, *115*, 53.
55. Piao, J., Kishi, S., and Adachi, S., *Colloids Surf. A*, 2006, *277*, 15.
56. Capalbi, A., Gente, G., and La Mesa, C., *Colloids Surf. A*, 2004, *246*, 99.
57. Söderberg, I., Drummond, C.J., Furlong, D.N., Godkin, S., and Matthews, B., *Colloids Surf. A*, 1995, *102*, 91.

58. Niraula, B.B., Chun, T.K., Othman, H., and Misran, M., *Colloids Surf. A*, 2004, *248*, 157.
59. Milkereit, G., Garamus, V.M., Gerber, S., and Willumeit, R., *Langmuir*, 2007, *23*, 11488.
60. Strey, R., *Curr. Opin. Colloid Interface Sci.*, 1996, *1*, 402.
61. Aoudia, M. and Zana, R., *J. Colloid Interface Sci.*, 1998, *206*, 158.
62. Kameyama, K. and Takagi, T., *J. Colloid Interface Sci.*, 1990, *137*, 1.
63. Antonelli, M.L., Bonicelli, M.G., Ceccaroni, G., La Mesa, C., and Sesta, B., *Colloid Polym. Sci.*, 1994, *272*, 704.
64. Corkill, J.M., Goodman, J.F., and Harrold, S.P., *Trans. Faraday Soc.*, 1964, *60*, 202.
65. Molina-Bolívar, J.A., Aguiar, J., Peula-Garcia, J.M., and Carnero Ruiz, C., *J. Phys. Chem. B*, 2004, *108*, 12813.
66. Kresheck, G.C., Surfactants, in *Water—A Comprehensive Treatise*, Franks, F., Ed. Plenum Press, New York, 1975. Vol. 4, pp. 95–167.
67. Pestman, J.M., Kevelam, J., Blandamer, M.J., van Doren, H.A., Kellogg, R.M., and Engberts, J.B.F.N., *Langmuir*, 1999, *15*, 2009.
68. Borse, M.S. and Devi, S., *Adv. Colloid Interface Sci.*, 2006, *123*, 387.
69. Chen, L.J., Lin, S.Y., and Huang, C.C., *J. Phys. Chem. B*, 1998, *102*, 4350.
70. Balzer, D., *Langmuir*, 1993, *9*, 3375.
71. Ericsson, C.A., Söderman, O., Garamus, V.M., Bergström, M., and Ulvenlund, S., *Langmuir*, 2005, *21*, 1507.
72. Majhi, P.R. and Blume, A., *Langmuir*, 2001, *17*, 3844.
73. Whiddon, C.R., Bunton, C.A., and Söderman, O., *J. Phys. Chem. B*, 2003, *107*, 1001.
74. Sulthana, S.B., Rao, P.V.C., Bhat, S.G.T., Nakano, T.Y., Sugihara, G., and Rakshit, A.K., *Langmuir*, 2000, *16*, 980.
75. Okawauchi, M., Hagio, M., Ikawa, Y., Sugihara, G., Murata, Y., and Tanaka, M., *Bull. Chem. Soc. Jpn.*, 1987, 60, 2718.
76. Prasad, M., Chakraborty, I., Rakshit, A.K., and Moulik, S.P., *J. Phys. Chem. B*, 2006, *110*, 9815.
77. Molina-Bolívar, J.A., Hierrezuelo, J.M., and Carnero Ruiz, C., *J. Colloid Interface Sci.*, 2007, *313*, 656.
78. Zhang, L., Somasundaran, P., and Maltesh, C., *Langmuir*, 1996, *12*, 2371.
79. Myers, D., *Surfactant Science and Technology*. VCH, New York, 1992.
80. Nilsson, F., Söderman, O., and Johansson, I., *Langmuir*, 1998, *14*, 4050.
81. Collins, K.D. and Washabaugh, M.W., *Q. Rev. Biophys.*, 1985, *4*, 323.
82. Pastor, O., Junquera, E., and Aicart, E., *Langmuir*, 1998, *14*, 2950.
83. Aveyard, R., Binks, B.P., Chen, J., Esquena, J., and Fletcher, P.D.I., *Langmuir*, 1998, *14*, 4699.
84. Molina-Bolívar, J.A., Hierrezuelo, J.M., and Carnero Ruiz, C., *J. Phys. Chem. B*, 2006, *110*, 12089.
85. Miyagishi, S., Okada, K., and Asakawa, T., *J. Colloid Interface Sci.*, 2001, *238*, 91.
86. Lipfert, J., Columbus, L., Chu, V.B., and Doniach, S., *J. Appl. Crystallogr.*, 2007, *40*, S229.
87. Roxby, R.W. and Mills, B.P. *J. Phys. Chem.*, 1990, *94*, 456.
88. La Mesa, C., Bonincontro, A., and Sesta, B., *Colloid Polym. Sci.*, 1993, *271*, 1165.
89. Thiyagarajan, P. and Tiede, D.M., *J. Phys. Chem.*, 1994, *98*, 10343.
90. Giordano, R., Maisano, G., and Teixeira, J., *J. Appl. Crystallogr.*, 1997, *30*, 761.
91. Nilsson, F. and Söderman, O., *Langmuir*, 1996, *12*, 902.
92. Tanford, C., *The Hydrophobic Effect: Formation of Micelles and Biological Membranes*, 2nd ed. Wiley, New York, 1980.

93. Yoshizaki, T. and Yamakawa, H., *J. Chem. Phys.*, 1980, *72*, 57.
94. D'Aprano, A., Giordano, R., Jannelli, M.P., Magazu, S., Maisano, G., and Sesta B., *J. Mol. Struc.*, 1996, *383*, 177.
95. Shinoda, K., Yamanaka, T., and Kinishita, K., *J. Phys. Chem.*, 1959, *63*, 648.
96. Zhang, R., Marone, P.A., Thiyagarajan, P., and Tiede, D.M., *Langmuir*, 1999, *15*, 7510.
97. Schurtenberger, P., Jerke, G., Cavaco, C., and Pedersen, J.S., *Langmuir*, 1996, *12*, 2433.
98. He, L.Z., Garamus, V.M., Funari, S.S., Malfois, M., Willumeit, R., and Niemeyer, B., *J. Phys. Chem. B*, 2002, *106*, 7596.
99. Lipfert, J., Columbus, L., Chu, V.B., Lesley, S.A., and Doniach, S., *J. Phys. Chem. B*, 2007, *111*, 12427.
100. Israelachvili, J.N., *Intermolecular and Surface Forces*, 2nd ed. Academic Press, London, 1992.
101. von Minden, H.M., Brandenburg, K., Seydel, U., Koch, M.H.J., Garamus, V.M., Willumeit, R., and Vill, V., *Chem. Phys. Lipids*, 2000, *106*, 157.
102. Lippold, H., Findeisen, M., Quitzsch, K., and Helmstedt, M., *Colloids Surf. A*, 1998, *135*, 235.
103. Ferry, J.D., *Viscoelastic Properties of Polymers*, 3rd ed. Wiley, New York, 1980.
104. Raghavan, S.R. and Kaler, E.W., *Langmuir*, 2001, *17*, 300.
105. He, L.Z., Garamus, V., Niemeyer, B., Helmholz, H., and Willumeit, R., *J. Mol. Liquids*, 2000, *89*, 239.
106. Whiddon, C.R. and Söderman, O., *Langmuir*, 2001, *12*, 2371.
107. Warr, G.G., Drummond, C.J., Grieser, F., Ninham, B.W., and Evans, D.F., *J. Phys. Chem.*, 1986, *90*, 4581.
108. Imae, T. and Ikeda, S., *J. Phys. Chem.*, 1986, *90*, 5216.
109. Kato, T., Terao, T., and Seimiya, T., *Langmuir*, 1994, *10*, 4468.
110. de Gennes, P.G., *Scaling Concepts in Polymer Physics*. Cornell University Press, Ithaca, New York, 1979.
111. Hoffmann, H., Platz, G., Rehage, H., and Schorr, W., *Adv. Colloid Interface Sci.*, 1982, *17*, 275.
112. Ericsson, C.A., Söderman, O., and Ulvenlund, S., *Colloid Polym. Sci.*, 2005, *282*, 1313.
113. Molina-Bolívar, J.A., Aguiar, J., and Carnero Ruiz, C., *J. Phys. Chem. B.*, 2002, *106*, 870.
114. Shinoda, K., Yamaguchi, T., and Hori, R., *Bull. Chem. Soc. Jpn.*, 1961, *34*, 237.
115. Matsumura, S., Imai, K., Yoshikawa, S., Kawada, K., and Uchibori, T., *Yukagaku*, 1991, *40*, 11.
116. Nikel, D., Nitsch, C., Kurzendörfer, P., and von Rybinski, W., *Prog. Colloid Polym. Sci.*, 1992, *89*, 249.
117. Rossen, M.J. and Sulthana, S.B., *J. Colloid Interface Sci.*, 2001, *239*, 528.
118. De Grip, W.J. and Bovee-Geurts, P.H.M., *Chem. Phys. Lipids*, 1979, *23*, 321.
119. Helenius, A., McCaslin, D.R., Fries, E., and Tanford, C., *Methods Enzymol.*, 1979, *56*, 734.
120. Van Aken, T., Foxall-VanAken, S., Castleman, S., and Ferguson-Miller, S., *Methods Enzymol.*, 1986, *125*, 27.
121. Tiefenbach, K.J., Durchschlag, H., and Jaenicke, R., *Progs. Colloid Polym. Sci.*, 1999, *113*, 135.
122. Bucci, S., Fagotti, C., Degiorgio, V., and Piazza, R., *Langmuir*, 1991, *7*, 824.
123. Esumi, K., Arai, T., and Takagusi, K. *Colloid. Surf. A*, 1996, *111*, 231.
124. Strop, P. and Brunger, A.T., *Protein Sci.*, 2005, *14*, 2207.

4 Thin Film and Foam Properties of Sugar-Based Surfactants

Per Claesson, Cosima Stubenrauch, Rumen Krastev, and Ingegärd Johansson

CONTENTS

4.1 INTRODUCTION

Foam generation first leads to the formation of wet foam that consists of spherical air bubbles surrounded by a liquid. While draining, the structure gradually changes as the liquid content of the foam decreases and the air bubbles are transformed into polyhedral air cells separated by thin liquid films that are stabilized e.g. by surfactants adsorbed at the water/air interface. Although the application of liquid foams is widespread (cleaning agents, beverages, fire-fighting, flotation, and oil recovery to mention just a few), too little is yet understood about the parameters with which their stability can be controlled. Thus, the development of new products is often based on "trial and error." In order to learn more about the properties of well-drained, dry foams, the investigation of the foam building blocks, i.e. the foam films, is generally regarded as promising [1–4]. With respect to low molecular weight (LMW) surfactants, we usually distinguish between sterically and electrostatically stabilized films. The former are called Newton black films (NBFs), whereas the latter are referred to as common black films (CBF) or common thin films (CTFs). For the sake of clarity, we will solely use the abbreviation CBF (the difference between CBFs and CTFs is explained in Ref. [5]). The CBFs are relatively thick and water rich whereas the NBFs are thin films with a very low water content.

One way of studying foam films is to measure the disjoining pressure Π as a function of the film thickness h (see Section 4.2). In doing so, one obtains two important pieces of information. First, the mechanism stabilizing the film against further drainage can be determined and quantified. For instance, a film of 20 nm thickness can be stabilized (1) by steric forces in the case of an amphiphilic polymer (reviewed in Ref. [6]), (2) by structural forces in the case of surfactant/polyelectrolyte mixtures (reviewed in Refs. [7,8]), or (3) simply by long-range electrostatic forces. A distinction is easily possible when the $\Pi(h)$ curve is known. With this knowledge the properties can be tuned in a controlled way. For example, the thickness and stability of an electrostatically stabilized foam film is conveniently tuned by adding electrolyte, whereas addition of electrolyte will not influence the properties of a sterically stabilized film to any large extent. The type of surfactant, the surfactant concentration, surface-active additives, electrolyte concentration, and pH determine if a CBF or an NBF is formed (see Section 4.2). The second important piece of information that is obtained from $\Pi(h)$ curves is the maximum pressure sustainable by the film. This pressure is a measure of the film stability, and it may be expected that the stability will increase with increasing repulsive interaction forces. This is, however, only part of the truth. It has been found experimentally for both ionic [9] and nonionic surfactants [10–12] that equal electrostatic repulsion does not automatically result in equal foam film stability. Furthermore, long-range steric repulsion between amphiphilic polymers does not guarantee the formation of stable foam films [13]. Thus, the stability of thin foam films cannot be explained solely by the magnitude of the repulsive interactions operating normal to the film surfaces. What is needed is a surface that is able to prevent the film from rupturing by dampening external vibrations. This ability is believed to be mirrored in the surface viscoelasticity of the monolayer [9,11,14–17] (see Section 4.3), which affects the probability of hole formation and thus rupture.

The stability of thin foam film lamellas is one of the important parameters, which govern the short- and the long-term stability of the foams. Another factor which has to be considered when the long-term stability of the foams is predicted is the exchange of gas between the foam bubbles. Formation of long-lived foams is not possible when the process of Ostwald ripening is fast. The exchange of gas between the bubbles is controlled by the gas permeability of the films, which is strongly related to the concentration and the chemical structure of the surfactant used to stabilize the foam. The gas permeability of foam films is strongly affected by the presence of molecular scale defects (holes) in the surface layer since these holes form an easy pathway for gas transport across the interfacial layer. The gas permeability of a foam film also depends on the film thickness and the interfacial interactions in the films. Thus, studies on the gas permeability of foam films do not only provide us with information on the foam stability, but also supply valuable information about the structure of the surfactant layers that stabilize a foam film.

The last question that is addressed in this chapter is whether or not the properties of single surfaces and/or isolated foam films can be compared with those of foams. The mechanical strength of interfacial layers and their response to dilational and shear deformations in the lateral direction are definitely important for the foam stability. However, the exact correlation between surface rheological parameters and foam stability is far from being clear [16,18]. Moreover, specifying the correlation between properties of isolated foam films and those of foams (see, for example, Refs. [1,2,19,20] and references therein) is also quite a challenge. Isolated films used in model studies are either horizontal or vertical and of fixed sizes whereas foams consist of a three-dimensional network of interconnected films, which are of various orientations and sizes. Further, foam films are usually studied under defined pressures while foams are studied under gravity, i.e. the pressure is not constant but a function of the height of the foam column. In this chapter we summarize the current state of the art regarding the properties of foams stabilized by sugar-based surfactants (Section 4.5) and compare these data with those obtained for single surfaces (Section 4.3) and isolated foam films (Sections 4.2. and 4.4).

TABLE 4.1

Critical Micelle Concentration (cmc), Surface Tension at cmc (γ_{cmc}), Surface Excess at cmc (Γ_{cmc}), and Area Per Molecule at the Air–Water Interface at cmc (A_{cmc}) for Some Alkyl Glucosides

Surfactant	cmc/mM	γ_{cmc}/mN m^{-1}	Γ_{cmc}/µmol m^{-2}	A_{cmc}/nm^2
β-C_8G_1	22	30.8	4.40	0.380
β-$C_{10}G_1$	2.0	28.3	4.43	0.375
β-$C_{10}G_2$	2.2	36.7	3.45	0.481
β-$C_{12}G_2$	0.17	35.3	3.56	0.467

Note: The data for *n*-octyl-β-glucoside, β-C_8G_1 are from Ref. [21], for *n*-decyl-β-glucoside, β-$C_{10}G_1$ from Ref. [22], for *n*-decyl-β-maltoside, β-$C_{10}G_2$ from Ref. [22], and for *n*-dodecyl-β-maltoside β-$C_{12}G_2$ from Ref. [23].

The surprisingly complex structural changes in layers of sugar-based surfactants adsorbed at the air–water interface, which appear to occur with increasing surfactant concentration, are addressed in Chapter 5 of the present book. Some important characteristics of their adsorption to air–water interfaces are provided in Table 4.1. We note that for a given head group the adsorption density increases slightly and the surface tension at cmc decreases slightly with increasing hydrocarbon chain length. The cmc decreases by roughly a factor of 10 when the hydrocarbon chain is increased by two methylene units. For a given chain length an increase in the head group size results in a slight increase in cmc, a significant increase in the surface tension at the cmc, and in a significant decrease of the packing density at the interface.

4.2 DISJOINING PRESSURE IN FOAM FILMS

By a convention introduced by Derjaguin [24] the interaction between two air–liquid interfaces across a foam film is called the disjoining pressure, emphasizing that the forces that prevent the film from collapsing are repulsive, i.e. they disjoin the interfaces. The disjoining pressure (Π) as a function of film thickness (h) is known as the disjoining pressure isotherm, which emphasizes that the measurements are done at constant temperature. In the following we will call these isotherms $\Pi(h)$-curves. Equation 4.1 relates the disjoining pressure (Π) to the change in film tension (γ), with film thickness (h), at constant temperature (T), and chemical potential (μ).

$$\Pi(h) = -\left(\frac{\partial\gamma}{\partial h}\right)_{T,\mu} \tag{4.1}$$

Experimental details on how $\Pi(h)$-curves are measured and evaluated can be found in numerous books, articles, and reviews [2,7,20,25–27].

4.2.1 EFFECT OF SURFACTANT CONCENTRATION

Interactions across foam films stabilized by sugar-based surfactants have been intensely investigated during the last 10–15 years. In an early study it was shown that the equilibrium thickness of foam films stabilized by n-octyl-β-glucoside, β-C_8G_1, decreases with salt concentration and surfactant concentration [28], in agreement with the trends observed for other classes of nonionic surfactants [29]. The data strongly suggested that the long-range interaction in foam films stabilized by this nonionic surfactant was due to repulsive electrostatic double-layer forces. That this indeed is the case was shown in 1996 when the $\Pi(h)$-curves for β-C_8G_1 were reported, see Figure 4.1 [30]. The long-range repulsion can in all cases be well described by a repulsive double-layer force using the Debye-length expected for the given ionic strength. The double-layer forces were calculated within the nonlinear Poisson–Boltzmann (PB) model, using the surface potential as a variable until a good fit between theory and experiments was obtained. We note that the nonlinear PB model is an approximation due to the neglect of ion–ion correlation and ion size effects [31]. Thus, surface potentials and surface charge densities deduced from these measurements should be regarded as approximate values. Nevertheless, it is clear that the surface charge density at the interface is low. For foam films stabilized by a 1 mM β-C_8G_1 solution the area per charge at the interface is about 100 nm², while it increases to ~450 nm² with increasing surfactant

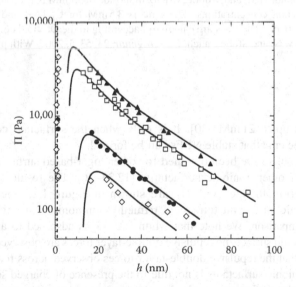

FIGURE 4.1 Disjoining pressure (Π) versus thickness (h) curves for n-octyl-β-glucoside in 0.1 mM KBr solution at different surfactant concentrations: 3 mM (filled triangles), 10 mM (unfilled squares), 21 mM (filled circles), and 25 mM (unfilled diamonds). The cmc is 22 mM. The lines are calculated according to the DLVO theory. (Reprinted from Bergeron, V., Waltermo, Å., and Claesson, P.M., *Langmuir*, 12, 1336, 1996. With permission.)

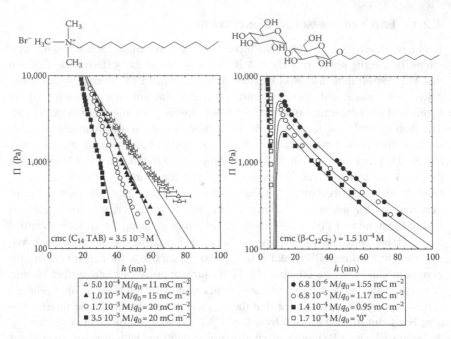

FIGURE 4.2 Disjoining pressure (Π) versus thickness (h) curves for tetradecyl trimethyl ammonium bromide (left) and n-dodecyl-β-D-maltoside in 0.1 mM NaCl solution (right) at different surfactant concentrations. The cmc is 3.5 mM for $C_{12}TAB$ and 0.17 mM for β-$C_{12}G_2$, respectively. The lines are calculated according to the DLVO theory. (Reprinted from Buchavzov, N. and Stubenrauch, C., *Langmuir*, 23, 5315, 2007. With permission.)

concentration up to 21 mM [30]. It is only when the surfactant concentration approaches the cmc that stable NBFs can be formed.

Similar results have been obtained for other sugar-based surfactants [11,32], as well as for other nonionic surfactants [12,33–36]. The results obtained for n-dodecyl-β-D-maltoside, β-$C_{12}G_2$, are shown in Figure 4.2, where also data for the cationic surfactant tetradecyl trimethyl ammonium bromide are illustrated for comparison. We note that when β-$C_{12}G_2$ is adsorbed to an uncharged hydrophobic solid surface, no repulsive double-layer forces are observed [37]. Thus, we conclude that the repulsive double-layer forces observed across foam films stabilized by nonionic surfactants is not due to the presence of charged surface active impurities in the surfactant sample, but the available data rather suggest that the charge originates from the intrinsic properties of the air–water interface. However, it is of great importance to use surfactants of high purity for these studies [38].

In general an increase in nonionic surfactant concentration decreases the stabilizing long-range repulsion since the surface charge density decreases. On the other hand, the stability against rupture of CBFs and NBFs increases with increasing surfactant concentrations. Thus, the stability of these films is imposed by other, short-range interactions acting both normal and tangentially to the film surface. This is further discussed in Section 4.3.

4.2.2 Effect of Electrolyte Concentration

Once it has been established that the long-range interaction between nonionic surfactant layers adsorbed at the air–water interface is due to repulsive double-layer forces, it becomes natural to control these forces by the electrolyte concentration. The PB model of the double-layer force predicts that the decay rate of the force at large separation should equal the Debye-length (κ^{-1}) given by [39]:

$$\kappa^{-1} = \sqrt{\frac{\varepsilon_0 \varepsilon_r kT}{e^2 \sum_i C_i z_i^2}} \tag{4.2}$$

where
  ε_0 is the permittivity of vacuum
  ε_r is the dielectric constant of the solution
  k is the Boltzmann constant
  T is the absolute temperature
  e is the elementary charge
  C is the ion concentration (expressed as number of ions per m^3)
  z is the valency of the ion

The summation goes over all ion types present in solution. This prediction is confirmed by experiments. Both for foam films stabilized by β-C_8G_1 and by β-$C_{12}G_2$ the range of the repulsion decreases with increasing salt concentration as expected and the exponential decay of the force at large separation is consistent with that of a double-layer force at the given ionic strength [30]. A second finding, with a less obvious explanation, is that the charge density of the interface increases with increasing salt concentration. For instance, for 21 mM β-C_8G_1 it was found that the area per charge at the interface decreased from 320 nm^2 to 130 nm^2 as the KBr concentration was increased from 0.1 to 1 mM [30]. At even higher salt concentrations the range of the double-layer force was too small to allow formation of any stable CBF and thus the charge density at the interface could not be determined at higher ionic strengths. Similarly, for 0.068 mM β-$C_{12}G_2$ (Figure 4.3) the area per charge at the interface decreased from 130 nm^2 to 45 nm^2 as the NaCl concentration was increased from 0.1 to 5 mM. If it is assumed that the charge originates from binding of ionic species (e.g. OH$^-$ ions) to certain sites at the air–water interface [35], an increased charge density at higher ionic strength follows naturally. However, if this is indeed the case, what is the nature of these adsorption sites? This question will be further addressed in Section 4.2.3.

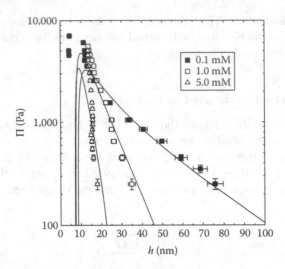

FIGURE 4.3 Disjoining pressure (Π) versus thickness (h) curves for 0.068 mM n-dodecyl-β-D-maltoside solutions at three salt concentrations: 0.1 mM (filled squares), 1 mM (unfilled squares), and 5 mM (unfilled triangles). The lines are calculated according to the DLVO theory. (Data from Stubenrauch, C., Schlarmann, J., and Strey, R., *Phys. Chem. Chem. Phys.*, *4*, 4504, 2002. With permission; *Phys. Chem. Chem. Phys.*, 5, 2736, 2003 [erratum].)

4.2.3 EFFECT OF PH

Purified water typically has a "natural" pH of 5.5–5.7 due to dissolution of carbon dioxide, i.e. the formation of H_2CO_3 that dissociates to H^+ and HCO_3^- ions. Measurements of $\Pi(h)$-curves at different pH values can shed light on the sign of the surface charge at the air–water interface in the presence of nonionic surfactants [40,41]. One of the most extensive studies in this area is that of Karraker and Radke [35], who systematically investigated pH effects on foam films stabilized by $C_{12}E_4$ and $C_{10}E_8$, respectively. They concluded that the surface charge vanishes at about pH 3–4 and reaches saturation at about pH 8. A similar trend with an increasing surface charge with increasing pH has been reported for 10 mM β-C_8G_1 solutions in 1 mM KBr [30]. At pH 4, the double-layer force was too weak to allow the $\Pi(h)$-curve to be measured. The double-layer force was found to be significant at pH 5, increasing further at pH 6 whereas virtually identical $\Pi(h)$-curves were found at pH 6 and pH 9–10. (In this latter study, the pH was not controlled during measurements but CO_2 absorption may have shifted the pH-value to lower values during the measurements). Also for 6.8×10^{-5} M β-$C_{12}G_2$ an increase in surface charge density with increasing pH has been found [41]. Hence, an increasing surface charge density with increasing pH has been observed for all nonionic surfactants studied so far, but the details of this pH dependence are influenced by the surfactant type as can be seen in Figure 4.4.

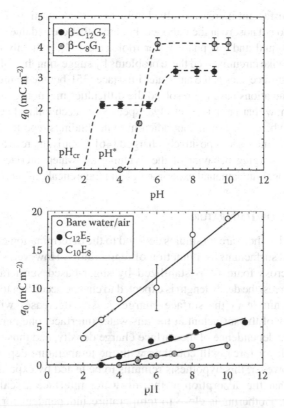

FIGURE 4.4 Variation of the surface charge density q_0 with pH for two sugar-based surfactants at a concentration of about 0.5 cmc in 1 mM electrolyte solution (top) and for two ethoxylated surfactants, namely: $C_{12}E_5$ at 1/7 cmc in 1 mM electrolyte and $C_{10}E_8$ at 2/3 cmc in 0.3 mM electrolyte solution (bottom). For the sake of comparison data for the bare air–water interface are also provided. (Reprinted from Stubenrauch, C., Cohen, R., and Exerowa, D., *Langmuir*, 23, 1684, 2007. With permission.)

From the results above it seems clear that the sign of the surface charge of the air–water interface in the presence of nonionic surfactants is negative, as has also been determined to be the case for the pure air–water interface [7,42]. The widely accepted hypothesis for the origin of the surface charge is that it is due to adsorption of OH^- ions [7,35,40,43], but it has also been suggested that HCO_3^- ion adsorption may play a role in the surface charging process as this would more easily rationalize the pH dependence of the surface charge [30]. However, the HCO_3^- adsorption hypothesis was not supported by the measurements of Karraker et al. [35] who studied foam films at equal pH but different carbonate ion content and found that the resulting double-layer repulsion was the same. All explanations based on ion adsorption fail to explain why the surface charge drops rapidly when the surfactant cmc is approached, since in this concentration range the surface excess of the surfactant increases only slowly with concentration. The hypothesis that OH⁻ ions do not adsorb at the interface, but are rather created due to surface reactions between liquid and gas phase water molecules [44], is hardly tenable either. Karraker and Radke circumvented these problems by suggesting that OH⁻ ions adsorb only to certain surface sites at the air–water interface [35], but the nature of these sites is not clear. Simulations have not resolved the difficulties mentioned above. Some of these studies show that anionic electrolyte species are accumulated at the air–water interface [45], whereas some, in contradiction to the findings reported above, predict that the air–water interface is positively charged [46]. Clearly, there is a need to better understand the charging behavior of the dynamic air–water interface, and there is room for additional fundamental experimental and theoretical efforts in this area.

4.2.4 Effect of Temperature

To our knowledge there are no studies devoted to the determination of $\Pi(h)$-curves for sugar-based surfactants as a function of temperature. However, since the long-range force across foam films stabilized by sugar-based surfactants is due to double-layer forces, the decay length is expected to change according to Equation 4.2. Since the magnitude of the surface charge density decreases with increasing packing density of the surfactant at the air–water interface, one may suggest that the temperature dependence of the surface charge density, and thus the magnitude of the double-layer force, will correlate with the temperature dependence of the adsorption. However, this hypothesis remains to be tested by experiments.

We note that the adsorption at the air–water interface as calculated from surface tension isotherms is close to temperature independent for n-decyl-β-D-glucoside (β-$C_{10}G_1$). For n-decyl-β-D-maltoside (β-$C_{10}G_2$) hardly any change in adsorption was observed between 8°C and 22°C, while a further increase in temperature to 29°C resulted in increased adsorption [47]. This latter finding was interpreted as being due to a change in orientation of the maltoside head group. No higher temperatures were investigated so it remains to be tested if the expected decrease in adsorption with temperature due to the increased thermal motion will occur at even higher temperatures. Similarly, for β-$C_{12}G_2$ an increase in adsorption with temperature has been observed between 15°C and 25°C, and in this case the adsorption decreases again as the temperature is increased further [48]. Clearly, there is a need for systematic investigations in this area covering a range of sugar-based surfactants and a large temperature interval before the general picture is clarified. On the other hand, for alkyl ethoxylate (C_nE_m) surfactants it has been established that the adsorption increases with temperature [49] due to the decreased solvent quality of water for oligo(ethylene oxide) chains at higher temperatures.

For NBFs, it has been reported that the film thickness increases only weakly with temperature [2,50]. For instance, for β-$C_{12}G_2$ above the cmc in 0.2 M NaCl the NBF thickness increased from 6.5 to 7.2 nm as the temperature was increased from 15°C to 40°C [50]. On the other hand, the monolayer permeability showed a minimum at 25°C [48], which correlates with a maximum in adsorption density. We will come back to this point in Section 4.4.

4.2.5 Effect of Stereochemistry and Chain Branching

Even the simplest alkyl glucoside exists in two anomeric forms, namely α- and β-glucose. Chemically these forms differ in the orientation of the chemical bonds next to the hydrocarbon–sugar linkage. Interestingly, this difference results in significantly different physical properties [51,52]. For instance, β-C_8G_1 is highly soluble in water showing a cmc-value of 22 mM [21], whereas α-C_8G_1 at 25°C is soluble only to a concentration of about 10 mM. This has been attributed to the lower crystal energy of the α-form [53,54]. A comparison of the foamability and foam stability of the two anomers at concentrations below the solubility limit of the α-form, i.e. below 10 mM, shows that β-C_8G_1 generates larger foam volume and more stable foam compared to α-C_8G_1 [32]. This may be related to the slightly higher surface excess of β-C_8G_1 as compared to α-C_8G_1 [32], even though it is also possible that in the dynamic environment of the foam the local concentration of surfactant may exceed the solubility limit for α-C_8G_1, and crystallites may form and act as foam destabilizers. In contrast, for the β-C_8G_1 anomer such a local concentration increase would promote foam stability by increasing the adsorption at the interface to the level obtained close to the cmc.

The branched surfactant 2-ethylhexyl-α-glucoside (α-$C_6C_2G_1$) is an even less potent foaming agent than α-C_8G_1. This has nothing to do with the long-range double-layer interaction across the foam films since the $\Pi(h)$-curves of these surfactants are very similar (Figure 4.5), and in neither case do

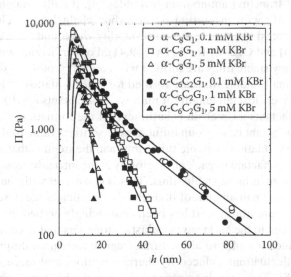

FIGURE 4.5 Disjoining pressure (Π) versus thickness (h) curves for α-C_8G_1 (unfilled symbols) and α-$C_6C_2G_1$ (filled symbols). The surfactant concentration was 10 mM and the KBr concentration was 0.1 mM (circles), 1 mM (squares), and 5 mM (triangles). The lines are calculated according to the DLVO theory. (Reprinted from Waltermo, Å., Claesson, P. M., Manev, E., Simonsson, S., Johansson, I., and Bergeron, V., *Langmuir*, 12, 5271, 1996. With permission.)

stable NBFs form [32]. The surface excess evaluated from surface tension isotherms is also very similar for these two compounds. Rather, it may be suggested that it is the surface organization of these three octylglucosides (β-C_8G_1, α-C_8G_1, and α-$C_6C_2G_1$) and possibly the formation of crystallites in the foam, which distinguish the foam properties of these surfactants. It is expected that the surface organization of the surfactants will be reflected in different surface viscoelastic properties. Indeed, surface laser light scattering shows such differences at the high frequencies probed by this technique, but no molecular interpretation for these differences has been offered [55]. Further studies of surface viscoelastic properties of these compounds, over a broad frequency range, would provide important information that would broaden our understanding of their different foam properties. We will come back to the correlation between foam film stability and viscoelasticity in Section 4.3.

4.2.6 Effect of Surfactant Chain Length

For n-alkyl ethoxylates the electrostatic component of the $\Pi(h)$-curve is not significantly affected by the alkyl chain length, but the effect of surfactant concentration, pH, and salt follows the trends discussed in the previous sections [35]. In contrast, the stability of the foam films, both CBFs and NBFs, is strongly affected by the surfactant chain length. For both n-alkyl ethoxylates [35] and alkyl trimethyl ammonium bromides [9], the film stability has been found to increase with increasing hydrocarbon chain length. The minimum chain length needed to promote thin film stability was found to be C10 for C_nE_4 surfactants [35] and C12–14 for C_nTAB [9,41]. For surfactants with maltoside head groups very stable films are formed when the tail group is n-dodecyl [12], whereas marginally stable films are formed for n-decyl maltoside [38]. Thus, it appears that the critical tail length for maltoside surfactants is C10, whereas for glucoside surfactants a C8 chain is sufficient for the formation of stable films. Clearly, the surfactant head group influences the critical value of the tail length needed for the formation of stable thin films, and the trend is that a head group that allows the surfactant to pack more closely at the interface requires a shorter tail group for formation of stable films. This is consistent with the observation that addition of a small amount of dodecanol dramatically increases the stability of C_{12}TAB foam films [9]. It has been convincingly argued [9] that the film stability is related to the surface viscoelastic properties that need to reach critically high values in order to allow film integrity to remain despite thickness and pressure fluctuations induced by thermal motions and mechanical disturbances [56–58]. Thus, the long-range repulsive double-layer forces that are accurately determined by measurements of $\Pi(h)$-curves should have a bearing on film drainage and thus influence foam properties. However, it is obviously not these long-range interactions that are decisive for foam stability but rather short-range interactions and surface viscoelasticity, which are discussed further in Section 4.3.

4.2.7 EFFECT OF HEAD GROUP POLYMERIZATION

The effect of head group polymerization on $\Pi(h)$-curves has most consistently been investigated for alkyl ethoxylates [34,35]. For instance, Karraker and Radke showed that for the $C_{10}E_n$ series, with n between 2 and 8, the long-range electrostatic repulsion was independent of the head group polymerization. On the other hand, stable NBFs were formed only when $n \geq 4$, indicating that smaller head groups do not generate sufficiently strong short-range repulsive forces to support these very thin films.

For alkylglucosides no similar systematic study is available, but we can compare the results reported for β-C_8G_1 [30] and β-$C_{12}G_2$ [11]. In both cases stable NBFs are formed when the surfactant concentration approaches cmc. Further, the repulsive double-layer forces encountered at concentrations below cmc are of similar magnitude and show similar dependence on surfactant concentration and salt addition. Thus, from these data it cannot be deduced that the head group polymerization of alkyl glucosides has any major effect on the $\Pi(h)$-curves. However, this conclusion may need to be modified once pure sugar-based surfactants with larger head groups have been investigated. Indeed, an influence of the head group size is indicated by the results obtained for technical grade sugar surfactants as described in the next section.

4.2.8 TECHNICAL COMPARED TO PURE SURFACTANTS

As in all surface studies small amounts of strongly surface active species can dramatically affect the outcome of an experiment, be it measurements of $\Pi(h)$-curves [9,38,59,60] or any other surface property. This should be kept in mind when comparing data obtained for pure surfactants with those of technical surfactants used in applications. Technical alkyl polyglucosides are normally produced by Fischer glucosidation that gives a mixture of the α- and β-anomers, with a thermodynamically determined α/β ratio of 2–3 [61]. As discussed above, there are distinct differences in solubility between these anomers, with the β-form being more soluble, and this alone will make the properties of the mixture different to those of the anomer pure substance. Furthermore, in the Fischer process a head group polydispersity will arise and there will be ring isomerism [62,63]. Thus, when the aim is to understand properties of technical surfactants for optimizing a particular application it is not sufficient to study only pure surfactants. Structure–property relationships can be determined for both technical and pure surfactant systems, but the molecular interpretation of the findings is more easily achieved for pure systems. Thus, studies of pure and technical surfactants complement each other and one can draw important conclusions by comparing results obtained with pure surfactants and with technical surfactants. It is a bit surprising that only very few studies are devoted to such comparisons [64,65].

In one report the surface forces between hydrophobic solid surfaces coated by a technical octyl glucoside was compared with the corresponding data for β-C_8G_1, and large differences were found [66]. The adsorbed layer of β-C_8G_1 was uncharged,

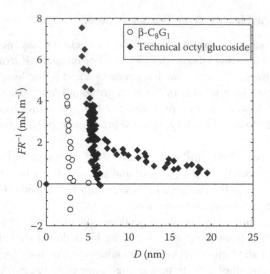

FIGURE 4.6   Force (F) normalized by radius (R) as a function of distance (D) for β-C_8G_1 (unfilled circles) and for a technical octyl glucoside (filled diamonds) adsorbed on hydrophobic solid surfaces. The surfactant concentration was in both cases 5.84 g/L, which is above the cmc. (Reprinted from Persson, C. M., Claesson, P. M., and Johansson, I., *Langmuir*, 16, 10227, 2000. With permission.)

very thin (≈ 1.5 nm), and nearly incompressible (Figure 4.6). In contrast, the adsorbed layer of the technical surfactant contained some charges (area/charge = 47 nm^2), which generated strong electrostatic double-layer forces. Thus, one can conclude that the technical process results in some charged surface-active impurities that affect the surface properties of the mixture. The technical surfactant also had a high degree of glucosidation and head group polydispersity, which resulted in the formation of thicker (≈ 2.5 nm) and more compressible layers. Thus, the technical mixture should be a better dispersing agent than β-C_8G_1, and the technical mixture was also found to be the better wetting agent [66].

In accordance with the force measurements using nonpolar solid substrates, the $\Pi(h)$-curve of the technical mixture was found to display a stronger double-layer force than that of β-C_8G_1 [66]. Thus, a more stable CBF was formed by the technical mixture. On the other hand, no stable NBF could be formed by the technical mixture, which we attribute to the presence of large head groups in combination with the small hydrophobic alkyl chain.

A branched technical alkyl glucoside has also been investigated. It was found that this surfactant generated long-range attractive forces between nonpolar solid surfaces [61], a feature not observed for any of the pure alkyl glucosides. The solution of the technical surfactant was in this case slightly turbid, showing that some of its components had a rather poor solubility in water. The long-range attraction was explained by a phase separation of the solution that was induced when the gap between the two surfaces became sufficiently small, i.e. by capillary condensation of poorly soluble components.

4.2.9 SHORT-RANGE INTERACTIONS: HYDRATION AND STERIC EFFECTS

The short-range interaction between surfactant layers results from a complex interplay of van der Waals, steric, and hydration forces, as discussed in detail in a recent review [67]. The distance determinations at small separations is, unfortunately, less accurate with the thin film pressure balance technique as compared to, e.g. the surface force apparatus [27]. Thus, detailed studies of short-range interactions are more accurately performed for solid–liquid interfaces than for air–liquid interfaces. We note that even though the surface excess at the nonpolar solid surface in many cases is close to that at the air–water interface [21,68,69], the interfacial mobility of the surfactants is likely to differ significantly with a higher mobility being found at the fluid interface. This and the presence of undulations at fluid interfaces are expected to affect the short-range interaction and these differences should be kept in mind when using data obtained at solid nonpolar surfaces to gain a better understanding of short-range interactions in foam films. When data on short-range interactions for the same system obtained with different techniques are compared, one finds good qualitative agreement, but as expected quantitative agreement is not always found. This issue is discussed and some examples are provided in the review by Claesson et al. [27], and the interested reader is referred to this paper and references therein.

We end this section by pointing out that it is not only the extent of hydration that is of importance for the short-range interactions, but also the nature of the hydration layer. The spectroscopic technique vibrational sum frequency spectroscopy, VSFS, allows a detailed picture of the hydration of surfactant layers in contact with bulk solution to be obtained without interference from the isotropic bulk phase [70,71]. Such studies have shown that the hydration of ethoxylated surfactants is very different to that of sugar-based surfactants. For alkyl ethoxylates the hydration appears to give rise to a clathrate-like water structure around the oligo(ethylene oxide) chain, whereas no similar feature is observed around a sugar head group. In addition, some strong hydrogen bonds are formed between water and both types of nonionic head groups [70,71].

4.3 SURFACE RHEOLOGY AND FOAM FILM STABILITY

Numerous studies of the surface rheology of LMW surfactants have been described in the literature. In this area it is important to distinguish between responses to changes in interfacial area (dilational rheology), and to changes in interfacial shape (shear rheology). In the case of LMW water soluble surfactants shear rheology can (in most cases) be neglected, and we focus on the correlation between surface dilational rheology and foam film stability. As most of these studies were performed at only one or two frequencies they are not very informative and might lead to wrong conclusions, as discussed below. More information can be extracted from studies where both the concentration and the frequency dependence of the surface rheological parameters have been investigated [14,15,17,19,55,72–85]. However, for sugar surfactants only four surface rheology studies exist [17,55,84,85], three of which deal with the surface rheology and the stability of foam films of the same solutions [17,84,85]. In these three studies the same surfactant was investigated, namely the nonionic sugar surfactant n-dodecyl-β-D-maltoside (β-$C_{12}G_2$), and for this reason this section summarizes only the results obtained for β-$C_{12}G_2$.

4.3.1 THEORETICAL BACKGROUND

"Equal disjoining pressures do not automatically result in equal foam film stabilities." This cannot be called into question because it has been corroborated experimentally for both ionic [9] and nonionic surfactants [6,17,85]. Thus, the stability of thin foam films cannot be explained solely by the magnitude of the repulsive interactions operating normal to the film surfaces. What is needed is a surface layer that is able to dampen external disturbances and thus preventing the film from rupturing. This ability is usually believed to be reflected in the surface dilational elasticity of the monolayer [6,9,14,16,38], although it was argued only recently that it is the surface viscosity rather than the surface elasticity that determines whether or not a foam film is stable [15,79,81,82]. As a definite answer to this controversy can only be given by additional measurements at high frequencies (i.e. $v > 100\,\text{Hz}$), which do not yet exist, we compare surface elasticities and film stabilities, keeping in mind, however, that we may need to revise the correlation between surface and film properties. In contrast to the disjoining pressure which operates normal to the surface, the surface dilational elasticity is related to processes which operate tangential to the surface. However, the exact relationship between surface dilational elasticity and film stability is still unclear and certainly there is no direct proportionality between the two. Moreover, experimental data are few and the surface rheological properties in most frequency ranges are virtually unknown. The dilemma is that of the systems investigated so far either the surface rheology or the film properties have been investigated in detail, but in most cases not both. We are aware of only three studies in which both the surface rheology and the stability of foam films were investigated for the same solutions [17,84,85].

The surface elasticity ε and viscosity η are the real and the imaginary parts, respectively, of the complex surface viscoelasticity $\bar{\varepsilon}$, which describes the linear response of a surfactant monolayer to a sinusoidal deformation of frequency v, i.e.

$$\bar{\varepsilon} = \varepsilon + i\omega\eta \qquad (4.3)$$

with $\omega = 2\pi v$. In simplified terms, the surface elasticity ε reflects the ability of the monolayer to adjust its surface tension in an instant of stress and is thus a measure of the energy stored in the surface layer as a result of external stress. The surface viscosity η is a measure of the energy dissipation in the surface layer and depends on the relaxation processes that restore the equilibrium after the disturbance. The surface viscosity represents the energy loss due to all relaxation processes that occur in the surface layer. They might derive from the exchange of molecules between the bulk and the surface phase or from rearrangements of molecular segments within a surface region or from other processes.

The surface rheological parameters depend on the frequency of the disturbance in relation to the relaxation rates, and these rates are affected by the surfactant concentration c. We will come back to this in connection with Figure 4.7. Equations for $\varepsilon(\nu,c)$ and $\eta(\nu,c)$ were first derived by van den Tempel and Lucassen [86,87]. According to their model, there are no adsorption/desorption barriers at the interface, and exchanges between bulk and surface are controlled by simple diffusion. With these assumptions one obtains:

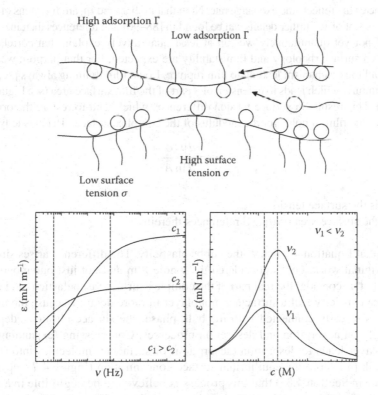

FIGURE 4.7 Schematic drawing (top) of a thin foam film. Spatial fluctuations lead to changes in the surfactant's surface concentration, which, in turn, results in local surface tension differences. These differences can be evened out by surfactant molecules diffusing along the surface and/or from the bulk to the surface. Schematic graphs (bottom) of the dilational surface elasticities ε as a function of the frequency ν (left) and the surfactant concentration c (right). (Reprinted from Stubenrauch, C. and Rippner-Blomqvist, B. In *Colloid Stability: The Role of Surface Forces, Part 1*; Tadros, T., (Ed.), Wiley-VCR: Weinheim, 2006, p 263. With permission.)

$$\varepsilon(v,c) = \varepsilon_0 \frac{1+\xi}{1+2\xi+2\xi^2} \tag{4.4}$$

and

$$\eta(v,c) = \frac{\varepsilon_0}{2\pi v} \frac{\xi}{1+2\xi+2\xi^2} \tag{4.5}$$

with

$$\xi = \sqrt{\frac{\omega_0}{4\pi v}} \tag{4.6}$$

Thus, the parameters ε_0 and ω_0 determine the shape of the $\varepsilon(v, c)$- and the $\eta(v,c)$-curves. The former is the high-frequency limit of the surface elasticity (according to Equation 4.4 one obtains $\varepsilon = \varepsilon_0$ for $v \to \infty$) and the latter, known as the molecular exchange parameter ω_0, reflects the exchange rate. Note that both ε_0 and ω_0 are functions of c but independent of v. Further details can be found in [88–90] and references therein.

If not yet quantitatively we can at least qualitatively explain that correlations between surface rheology and film stability are expected. For that purpose we have to recall the mechanism leading to film rupture. Local film thinning always precedes film rupture, which leads to extension of a part of the film surface area (see Figure 4.7 (top)). This results in surface tension differences which tend to restore the original state of the film. It holds for the modulus of the complex surface viscoelasticity

$$\bar{\varepsilon} = \frac{d|\varepsilon|\sigma}{d\ln A} \tag{4.7}$$

where
  σ is the surface tension
  A the surface area ([90] and references therein)

Note that Equation 4.7 is half the Gibbs elasticity. The difference arises since in his original work, Gibbs considered the whole film and not just one monolayer [91]. Let us consider the real part of the complex surface viscoelasticity $\bar{\varepsilon}$, i.e. the surface elasticity ε of a surfactant monolayer in more detail. Because the monolayer is directly connected with the bulk phase, the surface elasticity depends strongly on adsorption and desorption processes. Compressing (expanding) the monolayer leads to desorption (adsorption) of surfactant molecules into (from) the bulk to restore the equilibrium surface concentration (Figure 4.7 (top)). We will see in Section 4.3.3 that this process is believed to be negligible in a foam film. Surfactant molecules diffusing along the surface (Figure 4.7 (top)), thereby dragging liquid layers lying underneath back to the thinned area, is another mechanism to restore equilibrium. This process (the Gibbs-Marangoni effect) restores not only the surface concentration but also the shape of the film [91]. Whether or not the original state will be restored in the first place, and which of the two restoring processes that dominates, depends on the frequency of the disturbance. Two extreme cases are easy to understand: when the frequency is low, the monolayer has time to reach equilibrium and there is no resistance to the disturbance ($\varepsilon = 0$ for $v \to 0$). When the frequency is high, the monolayer has no time to respond and behaves as if it were insoluble ($\varepsilon = \varepsilon_0$ for $v \to \infty$). Thus ε increases with increasing v until a plateau value is reached which is schematically shown in Figure 4.7 (bottom, left).

The situation is, however, even more complex because the time needed to reach equilibrium strongly depends on the surfactant concentration. The higher the surfactant concentration, the faster the molecular exchange between bulk and surface and the Gibbs–Marangoni effect becomes less significant. It is because of this fast exchange between the bulk and the surface layer that surface tension differences $d\sigma$ are evened out very quickly at high surfactant concentrations, which may result in $\varepsilon = 0$ for a given frequency v. The experimental observation related to this exchange is a maximum in the $\varepsilon(c)$-curve which shifts towards higher concentrations with increasing frequency as is seen in Figure 4.7 (bottom, right). At infinite high frequencies ($v \to \infty$) the $\varepsilon(c)$-curve should reach a plateau as the frequency of the disturbance is faster than any restoring process. The interrelation between frequency and concentration is illustrated by the vertical lines drawn in Figure 4.7 (bottom). In the case of the $\varepsilon(v)$-curves it is seen that at some low frequency v_1 it holds $\varepsilon(c_1) > \varepsilon(c_2)$, while at v_2 it holds $\varepsilon(c_1) < \varepsilon(c_2)$. In the case of the $\varepsilon(c)$-curves, a concentration range exists in which the elasticity decreases with increasing c at v_1, while ε increases with c at the higher frequency v_2. These two examples illustrate the need for both frequency and concentration dependent measurements in order to obtain data with large information content. This, however, is not trivial as the available experimental techniques are time-consuming and restricted to certain frequency ranges depending on the method via which the surfaces are deformed. Moreover, the evaluation of the data is complicated and often results obtained by different techniques do not lead to the same quantitative results, as will be shown below.

4.3.2 FREQUENCY AND CONCENTRATION DEPENDENCE OF THE SURFACE ELASTICITY

In three consecutive studies the surface rheology of aqueous β-$C_{12}G_2$ solutions was studied over broad frequency (0.005–400 Hz) and concentration (1.02×10^{-5}–1.37×10^{-4} M) ranges. Three different techniques were used, namely the profile analysis tensiometer (PAT, frequency range 0.005–0.2 Hz) [17], the capillary pressure tensiometer (CPT, frequency range 0.2–80 Hz) [85], and excited capillary waves (CW, frequency range 100–400 Hz) [84]. To our knowledge it is only β-$C_{12}G_2$ that has been studied this extensively, so quantitative comparisons with other systems will not be made. Qualitatively, however, the general trends described in the following have been found for all LMW surfactants [14,15,17,19,55,72–85].

The surface elasticities of aqueous β-$C_{12}G_2$ solutions are shown as a function of the frequency in Figure 4.8 and as a function of the concentration in Figure 4.9, respectively. The theoretical curves were calculated according to Equation 4.4 and the corresponding fitting parameters are listed in Table 4.2.

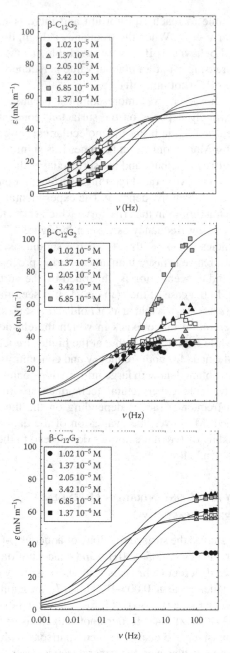

FIGURE 4.8 Dilational surface elasticities ε as a function of the frequency v for six different β-$C_{12}G_2$ concentrations c. Results obtained with three different techniques: profile analysis tensiometer, frequency range from 0.005 to 0.2 Hz (top); capillary pressure tensiometer, frequency range from 0.2 to 80 Hz (middle); capillary waves, frequency range from 100 to 400 Hz (bottom). The solid lines are calculated according to Equation 4.4. See text for further details.

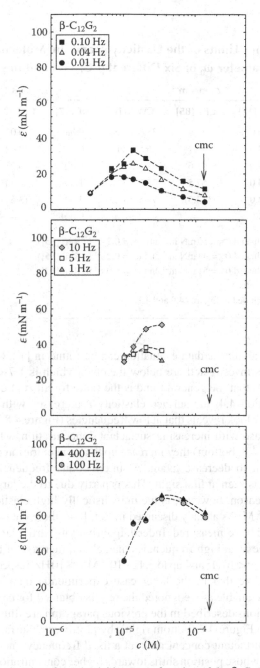

FIGURE 4.9 Dilational surface elasticities ε as a function of the β-$C_{12}G_2$ concentration c for eight different frequencies ν. Results obtained with three different techniques: profile analysis tensiometer, frequency range from 0.005 to 0.2 Hz (top); capillary pressure tensiometer, frequency range from 0.2 to 80 Hz (middle); capillary waves, frequency range from 100 to 400 Hz. The dashed lines are guide to the eye (bottom). See text for further details.

TABLE 4.2

High-Frequency Limits of the Elasticity ε_0 and the Molecular Exchange Parameter ω_0 of Six Different β-$C_{12}G_2$ Solutions

	ε_0/mN m^{-1}			ω_0/s^{-1}		
c/M	PAT [17]	CPT [85]	CW [84]	PAT [17]	CPT [85]	CW [84]
1.02 10^{-5}	36.0	35.6	35.0 ± 3.5	0.08	0.02	0.11
1.37 10^{-5}	48.0[a]	38.7	56.5 ± 5.6	0.17[a]	0.02	0.35
2.05 10^{-5}	51.0[b]	46.6	58.5 ± 5.9	0.40[b]	0.07	0.68
3.40 10^{-5}	53.0[c]	57.5[d]	72.7 ± 7.3	1.00[c]	4.50[d]	2.35
6.85 10^{-5}	58.0	(113.5)[e]	71.5 ± 7.2	3.60	(3.50)[e]	5.40
1.37 10^{-4}	64.0[d]	—	64.2 ± 6.4	9.00[d]	—	8.87

[a] Data was reevaluated ($\varepsilon_0 = 50$ mN m^{-1} and $\omega_0 = 0.2$ s^{-1} in Ref. [85]).
[b] Data was reevaluated ($\varepsilon_0 = 43$ mN m^{-1} and $\omega_0 = 0.2$ s^{-1} in Ref. [85]).
[c] Data was reevaluated ($\varepsilon_0 = 51$ mN m^{-1} and $\omega_0 = 1.0$ s^{-1} in Ref. [17]).
[d] Unpublished results.
[e] Data are not included in Figures 4.9 and 4.10.

More details about the data evaluation can be found in [17,84,85]. Note that all concentrations investigated are below the cmc, which is 1.7×10^{-4} M. Figure 4.8 clearly reveals that the general trend is the same for all $\varepsilon(v)$-curves. In agreement with Equation 4.4, the surface elasticity ε increases with increasing frequency v. A closer look reveals that at low frequencies (Figure 4.8, top) the surface elasticities decrease with increasing surfactant concentration, while at high frequencies (Figure 4.8, bottom) they increase up to a concentration of 3.42×10^{-5} M before they begin to decrease again. At intermediate frequencies (Figure 4.8, middle) no trend is seen at first sight. This is partly due to the large scatter of the data. The main reason, however, is the unrealistically high elasticities measured for $c = 6.85 \times 10^{-5}$ M. As already discussed in Ref. [84], these data seem to be inaccurate and should be re-measured. Indeed, ignoring this particular $\varepsilon(v)$-curve, one observes the same as at high frequencies, namely an increase of the elasticity up to 2.05×10^{-5} M at $v < 10$ Hz and up to 3.42×10^{-5} M $v > 10$ Hz, respectively, followed by a decrease. Note that in the latter case experimental data showing such a decrease are not available as we ignored the results obtained for $c = 6.85 \times 10^{-5}$ M.

The observations described in the previous paragraph are illustrated in Figure 4.9. According to Figure 4.7 (bottom right), by plotting the surface elasticity as a function of the surfactant concentration at a fixed frequency one obtains a curve with a maximum whose position shifts towards higher concentrations with increasing frequency. This is exactly what has been observed experimentally for low (Figure 4.9, top) and high frequencies (Figure 4.9, bottom). The same trend is also seen at intermediate frequencies (Figure 4.9, middle) if the values obtained for $c = 6.85 \times 10^{-5}$ M are ignored as was discussed above. It can be seen in Figure 4.9

that the concentration at which the elasticity ε reaches its maximum shifts from ~7 ×10⁻⁶ M at 0.01 Hz to ~3 × 10⁻⁵ M at 400 Hz. Simultaneously, the value of ε at this maximum increases from ~20 to ~70 mN m⁻¹.

The concentration-dependent behavior of the surface elasticities is easily understood if one recalls that two different frequencies play a role, namely the frequency of disturbance (ν) and the frequency of the molecular exchange (ω_0). When $\nu < \omega_0$, the elasticity is mainly determined by the molecular exchange: the higher the concentration, the faster the exchange and thus the lower the elasticity. On the other hand when $\nu > \omega_0$, very limited molecular exchange takes place and the elasticity is mainly determined by the surface excess concentration Γ: the higher the bulk concentration, the higher the Γ and thus the higher the elasticity. Theoretically the maximum of the $\varepsilon(c)$-curve should be located at the concentration for which $\omega_0 \sim \nu$. This is indeed the case at low frequencies but not observed at high frequencies. Additional experiments to quantify the ω_0-values at high concentrations would be needed to clarify this point. Based on what has been said so far it is clear that the high frequency limit of the elasticity ω_0 is expected to also increase with increasing bulk concentration and finally to level off once the maximum surface coverage is reached. That this is indeed the case will be shown in Section 4.3.3.

We conclude by taking a quantitative look at the results. It is obvious from both the experimental data and the fitting parameters ε_0 and ω_0 (see Table 4.2) that the $\varepsilon(\nu)$-curves measured with three different techniques agree perfectly at the lowest concentration (1.02 × 10⁻⁵ M), while comparatively large differences were observed at all other concentrations. The largest difference occurs at a surfactant concentration of $c = 6.85 \times 10^{-5}$ M for which completely different $\varepsilon(\nu)$-curves and ε_0 values have been obtained from the PAT and the CW measurements on the one hand and from the CPT measurements, on the other. It is argued and explained in Ref. [84] that the ε-data measured with the CPT for this particular concentration are not reliable. Indeed, ignoring the CPT results for $c = 6.85 \times 10^{-5}$ M the data obtained with the three different techniques complement each other convincingly. The remaining discrepancies are due to the different experimental and evaluation procedures but do not affect the general conclusion that a combination of PAT, CPT, and CW allows us to cover a frequency range from around 0.001 Hz up to 1000 Hz! As has been discussed, the question of what technique should be taken cannot be answered in general terms as the required frequency range depends mainly on the surfactant concentration. However, as a "rule of thumb" one can say that the frequency range should be chosen such that the data points reach the expected plateau value, or are at least close to it. To make measurements between 0.001 Hz and 1 Hz if the plateau value is expected to occur around 100 Hz will certainly not lead to a reliable fit—not because the theory is not working but because of the interplay of the fitting parameters ε_0 and ω_0, which makes it possible to fit the same $\varepsilon(\nu)$-curve with various couples of ε_0 and ω_0 values. The closer the data are at the high-frequency limit, the more reliable are the fitting parameters ε_0 and ω_0. It is explained in the following section that ε_0 is believed to be the parameter that determines the film stability and should thus be determined as accurately as possible.

4.3.3 Surface Elasticity and Foam Film Stability

Section 4.2 discusses the disjoining pressure of foam films stabilized by aqueous β-$C_{12}G_2$ solutions in detail. What we focus on in the following is the pressure at which films rupture, and the relation between surface elasticity and foam film stability. What is still under debate is whether film rupture occurs at one particular frequency and whether it is at this frequency that the surface elasticity has to be measured. The most convincing argument that it is the high frequency limit of the surface elasticity ε_0 that has to be considered is given in [92], where it was found that during the thinning of foam films surfactant diffusion is too slow to replenish the film surfaces. In other words, surfactant exchange with the bulk is ineffective in foam film thinning. Transferring this result to film rupture which is preceded by a local film thinning (see Figure 4.7) should lead to a correlation between ε_0 and the maximum pressure applicable to the film before it ruptures. Such a correlation could explain the experimental observation that films for which equal $\Pi(h)$–curves have been calculated can have completely different stabilities, as mentioned in Section 4.3.1. It is widely discussed that a stable film requires elastic surfaces to dampen external disturbances and thus to prevent the film from rupturing. Two examples of a direct correlation between the pressure above which the film ruptures and the elasticity ε_0 of the film surfaces are reported in [85].

1. The $\Pi(h)$–curves of foam films stabilized by the nonionic surfactant tetra-ethylene glycol monodecylether ($C_{10}E_4$) reveal that the stability of the common black film increases with increasing $C_{10}E_4$ concentration. However, the increase of the film stability is accompanied by a decrease of the surface charge density, which, in turn, is expected to destabilize the film. Thus, the fact that the stability of $C_{10}E_4$ films increases as the surface charge decreases cannot be explained in terms of interaction forces, while it can be explained in terms of surface elasticities. It was shown that the ε_0 value increases continuously within the investigated concentration range, which, in our view, is the reason for the increasing stability of the respective foam films.
2. The $\Pi(h)$–curves of a 6.85×10^{-5} M β-$C_{12}G_2$-solution and of a 1.1×10^{-4} M $C_{10}E_4$-solution do not differ with respect to the disjoining pressure; in both cases the surface charge density q_0 is calculated to be 1.17 mC m^{-2}. However, the corresponding film stabilities are far from being comparable. Whereas the β-$C_{12}G_2$ film is stable up to 5000 Pa, the $C_{10}E_4$ film ruptures already at pressures around 800 Pa. In other words, the β-$C_{12}G_2$ film is much more stable than the corresponding $C_{10}E_4$ film although both systems have the same surface charge density. Again, the low stability of the $C_{10}E_4$ film at $q_0 = 1.17$ mC m^{-2} cannot be explained in terms of interaction forces, while the different stabilities are clearly reflected in the respective ε_0-values, which is 68 ± 10 mN m^{-1} for 6.85×10^{-5} M β-$C_{12}G_2$ (see Table 4.2) and 36 mN m^{-1} for 1.1×10^{-4} M $C_{10}E_4$ (see Table 3 in Ref. [85]).

The second example includes only one β-$C_{12}G_2$ solution. What happens if we vary the β-$C_{12}G_2$ concentration? Will we observe the same as for $C_{10}E_4$, namely an

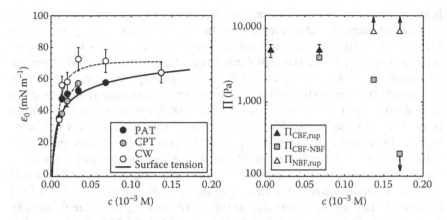

FIGURE 4.10 (left) High frequency limits of dilational surface elasticities ε_0 as a function of the β-$C_{12}G_2$ concentration c. Results obtained via fitting the $\varepsilon(\nu)$-curves seen in Figure 4.8. All data are listed in Table 4.2. The dashed lines are guide to the eye. For comparison surface elasticities calculated from surface tension data are also shown (solid line). See Ref. [84] for details. Note that the concentration axis is linear and not logarithm as was the case for Figure 4.9. (right) Disjoining pressure at which either the rupture of a CBF ($\Pi_{CBF,rup}$), the rupture of an NBF ($\Pi_{NBF,rup}$)* or a CBF–NBF transition ($\Pi_{CBF-NBF}$) was observed for a given β-$C_{12}G_2$ concentration c. Film stabilities were studied in a pressure range from 200 to 10,000 Pa (see Ref. [11] for details). The two upper arrows indicate that the NBF rupture was not observed up to 10,000 Pa. The lower arrow indicates that the CBF–NBF transition occurs at pressures below 200 Pa.

increasing stability with increasing surfactant concentration accompanied by an increase in ε_0 and a decrease in q_0? The answer to this question is given in Figure 4.10. As discussed in Section 4.2 an increasing β-$C_{12}G_2$ concentration indeed leads to a decrease of the surface charge density and thus to a decrease of the pressure $\Pi_{CBF-NBF}$ at which the transition from CBF to NBF occurs (Figure 4.10, right). However, the stability of the CBF does not change if we increase the concentration from 6.8×10^{-6} to 6.8×10^{-5} M although the surface charge density q_0 drops from 1.47 to 1.17 mC m^{-2}; at both concentrations film rupture occurs at $\Pi_{CBF,rup} = 5000 \pm 1000$ Pa (Figure 4.10, right). Thus, there is no correlation between the disjoining pressure (i.e. surface charge density) and the film stability. Unfortunately, a clear correlation between the surface elasticity and film stability cannot be seen either. Looking at the ε_0-values (Figure 4.10, left and Table 4.2), one clearly sees that ε_0 increases continuously in this concentration range. On the

---

* A comment has to be made regarding the $\Pi_{NBF,rup}$ values given in Figure 4.10. Measuring surface rheology, one studies single surfaces and one always has to keep in mind that the surface rheology of a single monolayer may not be the appropriate parameter to describe film stability. In case it was, it would only apply to CBFs as these films indeed consist of two separate monolayers. An NBF, however, is a bilayer rather than a monolayer and is often described as a two-dimensional liquid crystalline phase. In an NBF the monolayers are in direct contact to each other, which makes studies at single surfaces inappropriate to understand the stability of NBFs. Thus the stabilities for the NBF are only given for the sake of completeness.

basis of these observations one may conclude that there is a minimum elasticity $\varepsilon_{0,min}$ that is required to stabilize foam films. In other words, for $\varepsilon_0 < \varepsilon_{0,min}$ foam films are unstable, while they are expected to be stable for $\varepsilon_0 \geq \varepsilon_{0,min}$. As foam films stabilized by β-$C_{12}G_2$ are stable at $c \geq 6.8 \times 10^{-6}$ M, one may conclude from the ε_0-data in Table 4.2 that $\varepsilon_0 = 36$ mN m^{-1} is already above $\varepsilon_{0,min}$, i.e. that $\varepsilon_0 = 36$ mN m^{-1} is enough to stabilize a foam film. Unfortunately, looking at the data obtained for 1.1×10^{-4} M $C_{10}E_4$ reported in [85], one sees that the situation is much more complex. At this concentration ε_0-values of 36.2 mN m^{-1} were reported by Santini et al. [85] and 44 mN m^{-1} was measured by Stubenrauch and Miller [17], but a stable film did not form. Thus an $\varepsilon_{0,min}$-value that is the same for all surfactants cannot be deduced from the data obtained so far and it still needs to be clarified whether such a value exists at all. It was suggested in [85] not only to look at the ε_0 but also at the corresponding ω_0-value to explain film stabilities. For example, one finds for a 1.02×10^{-5} M β-$C_{12}G_2$ solution $\varepsilon_0 = 35$–36 mN m^{-1} and $\omega_0 = 0.02$–0.11 s^{-1} (Table 4.2), while for a 1.1×10^{-4} M $C_{10}E_4$ solution it is $\varepsilon_0 = 36.2$–44 mN m^{-1} and $\omega_0 = 1.93$–4.5 s^{-1} (Table 3 in Ref. [85]). Thus a possible explanation for the different film stabilities at similar ε_0-values could be the much lower ε_0-value, which means that for β-$C_{12}G_2$ the value of 36 mN m^{-1} is reached at a frequency much lower compared to $C_{10}E_4$. However, additional data are required before a final conclusion can be drawn.

4.4 GAS PERMEABILITY OF FOAM FILMS

The permeation of gas through foam films is of interest in many physical, chemical, and biological studies as well as in technological applications. Despite a lot of investigations of foam films, measurements of the gas permeability are relatively rare. The first measurements were performed in the middle of the twentieth century [93,94], while detailed studies on the basic relations between the gas permeability of foam films and the thermodynamic parameters of the system were carried out much later [95–97]. In these studies the influences of the film thickness, the surfactant adsorption density, the surfactant chain-length, and the temperature on the gas permeability of the foam films were investigated. While the short-term foam stability is governed by the dynamics of foam formation and drainage of the foam films, the long-term foam stability is administered by the properties of the well-drained foam films. The rupture of these films leads to a coalescence of the bubbles and thus to the destruction of the foam. In addition, the transfer of gas between two bubbles via the foam films (Ostwald ripening) also regulates the foam stability. It has been shown that measuring the gas permeability of foam films delivers valuable information on the structure and the interactions in the thin layers [98]. A detailed summary of the theoretical and experimental studies of the gas permeability of foam films has recently been provided by Farajzadeh et al. [99].

4.4.1 THEORETICAL BACKGROUND

An equilibrium foam film with thickness h, which consists of an aqueous core (thickness h_2) covered on each side with a monolayer of adsorbed surfactant molecules (thickness h_1) [2,25,100]. This type of foam film was called CBF in the previous sections. A measure of the gas permeability of such a foam film is the permeability coefficient K (cm s^{-1}) [93,94,101] defined by:

$$\frac{dN}{dt} = -K S \Delta C_g \tag{4.8}$$

with

$$K = \frac{D}{h}$$

Here, N is the number of moles of gas that permeates through the film, t is the time, S is the area of the film, ΔC_g is the concentration difference of the gas between the two sides of the film, and D is the diffusion coefficient of the gas through the film. The definition of the permeability coefficient K according to Equation 4.8 is valid only when the permeability obeys Fick's law and D is a constant. However, many studies [97,101–105] have shown that this usually is not the case because D varies with the number of CH_2 groups in the hydrophobic part of the surfactant molecules. Thus, commonly the results are described in terms of permeability rather than in terms of diffusion.

A detailed expression for the foam film permeability was proposed by Princen [94] applying Fick's first law. It takes into account the detailed "sandwich-like" structure of the film and the solubility of gas in the liquid:

$$K = \frac{D_w H}{h_2 + 2D_w / k_{ml}} \tag{4.9}$$

Here, D_w is the diffusion coefficient of the gas in the aqueous core of the film and H is the Henry's solubility coefficient, k_{ml} is the gas permeability of a single surfactant monolayer obtained from existing theories (for review see Refs. [99,101]). Equation 4.9 shows that the total foam film resistance against gas permeability is a sum of the resistance offered by the liquid core and the resistance due to the two monolayers. This is similar to the equations proposed in Refs. [106,107] for the permeability of rubber membranes for gases and of polymer membranes for water vapor. The equation shows that the permeability of the foam films depends on the thickness of the aqueous layer as well as on the solubility and diffusion of the gas in the aqueous phase. Higher values of D_w and H of the bulk liquid result in increased permeation rates and thus higher values of K. It appears from Equation 4.9 that for thick foam films ($h_2 \gg 2D_w/k_{ml}$) the rate of permeation is controlled by the liquid layer via D_w and H, giving $K = D_w H/h_2$.

Very thin NBFs consist of two strongly interacting surfactant monolayers i.e. a bilayer, where the hydrated head groups face each other. For such films $h_2 \ll 2D_w/k_{ml}$ and the film permeability is $K = Hk_{ml}/2$. This shows that the permeability of the monolayers is the rate determining process of gas transfer across NBF. As the monolayers do not behave as single layers, and the foam film permeability needs special models to be explained. The nucleation theory of hole formation due to fluctuations in the NBF [95,108] successfully describes the permeability and the stability of the films as a function of surfactant concentration. According to the theory (1) the NBF is a bilayer where the two surfactant monolayers are mutually adsorbed on each other; (2) the film is in contact and in equilibrium with the ambient gas phase, and its periphery with the bulk phase (meniscus); (3) the molecules within each of the monolayers are packed in a two-dimensional (2D) lattice; (4) some places in the lattice are not occupied by surfactant molecules due to thermal fluctuations. These free places are called vacancies. These vacancies move in the plane of the monolayers and aggregate in clusters (called holes) which consist of $i = 1, 2, 3$, etc. single vacancies. The number of surfactant vacancies increases with decreasing surfactant concentration. The gas transfer across the bilayer film occurs simultaneously through its hole-free area as well as through holes with different sizes i. K is a sum of the area weighted permeability coefficients K_o of the hole-free bilayer and K_i of holes of i vacancies:

$$K = K_o \frac{S_o}{S} + \sum_{i=1}^{\infty} K_i \frac{S_i}{S} \qquad (4.10)$$

Here S_i is the overall area occupied by holes of size i, while $S_o = S - \sum_i S_i$ is the hole-free film area. This area is accepted to be very close to the whole area of the film, S. The area S_i may be determined using the nucleation theory of hole formation and K can be presented as a function of the surfactant concentration c_s by

$$K = K_o + \sum_{i=1}^{\infty} a_i c_s^{-i} \qquad (4.11)$$

where a_i is defined as

$$a_i = K_i i c_e^i \exp(-P_i/kT) \quad \text{with} \quad P_i = (4\pi A)^{1/2} \kappa i^{1/2} \qquad (4.12)$$

In Equation 4.12, c_e is a critical equilibrium concentration at which the NBF and a large 2D phase of vacancies are in a thermodynamic equilibrium; P_i is the work necessary for formation of the perimeter of a hole in the film which consists of i single vacancies and which has a line tension κ at the three-phase contact between the hole and the surrounding film; A is the area per surfactant molecule. Equation 4.11 provides an expression for the dependence of K on the surfactant concentration in a general form. It shows that K increases with decreasing c_s due to the permeability through the holes in the foam bilayer, i.e. in the NBF. This is due to the decrease of the work associated with hole formation at lower c_s values, as well as to the increased number of holes present at lower surfactant concentrations. At a certain threshold surfactant concentration c_t, K becomes equal to K_o and a further increase of c_s does not change K, i.e. $K = K_o$ at $c_s > c_t$.

The theoretical approaches presented above are a suitable base for understanding the influence of different thermodynamic parameters on the gas permeability of foam films. In the following we summarize the experimentally found gas permeability of foam films stabilized by β-$C_{12}G_2$. We show how the surfactant concentration, the film thickness (regulated by the salt concentration), and the temperature influence the foam film permeability.

4.4.2 EFFECT OF THE SURFACTANT CONCENTRATION

Variations in surfactant concentration lead to changes in adsorption density and thus to changes in gas permeability. This is of no importance for the permeability of the thicker CBF where, according to the analysis of Equation 4.9, the gas diffusion through the central aqueous core of the film controls the permeability process. This conclusion was experimentally confirmed by Krustev et al. [109]. In contrast, NBFs are very thin and their thickness is usually hardly influenced by the surfactant concentration. In this case the gas permeability is controlled by the permeability of the single surfactant monolayers. Thus, it is strongly affected by the properties of the monolayer, which, in turn, depend on the surfactant packing density that is controlled by the bulk surfactant concentration.

The gas permeability measurements through films stabilized with β-$C_{12}G_2$ were performed at constant NaCl concentration of 0.2 M that ensured formation of NBFs [50]. These films consist of two surfactant monolayers with a thickness of around 2 nm each, and an aqueous core of around 1 nm thickness. The aqueous core is assumed to contain the hydration water of the surfactant head groups. A preposition for a successful gas permeability experiment is the stability of the foam film for at least 30 min, which is the time window for the experiment. Such stability is assured at β-$C_{12}G_2$ concentrations above 3.9×10^{-6} M [110].

The foam film permeability as a function of β-$C_{12}G_2$ concentration is shown in Figure 4.11 at two temperatures, 25°C and 35°C. The permeability was found to be constant at surfactant concentrations above cmc (1.7×10^{-4} M and 0.2 M NaCl). Foam films of β-$C_{12}G_2$ are much less permeable in this concentration range ($K \approx 0.012$ cm s^{-1}) compared to foam films stabilized with other surfactants under similar conditions [95–97,109] (e.g. $K \approx 0.033$ cm s^{-1} in the case of a NBF formed by sodium dodecylsulphate). The permeability was found to increase steeply at surfactant concentrations around and below cmc. It also increases with increasing temperature, i.e. the whole $K(c_s)$-curve is shifted towards higher K-values compared to the curve obtained at lower temperature.

The above-mentioned theory of hole formation was applied for understanding the experimental permeability results. The $K(c_s)$-curves were statistically treated using equations similar to Equation 4.11. The analysis routine included fitting of the equation to the experimental data for both temperatures including all possible combinations of the summands with $i = 1, 2, 3$, up to 6 and K_o. The theoretical equations were fitted only to the experimental points for $c_s < $ cmc, which is the range where the monomer surfactant concentration changes and K depends on c_s. The solid lines in Figure 4.11 are calculated according to the statistical treatment. The curves coincide well with the experimental points in the whole concentration range, even at higher surfactant concentrations where K is constant. The constant K-value obtained from the horizontal part of the fitted theoretical curve

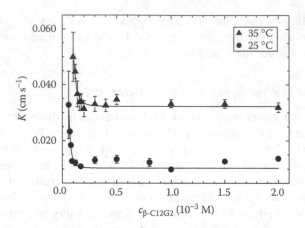

FIGURE 4.11   Gas permeability of foam films as a function of the β-$C_{12}G_2$ concentration at two temperatures (25°C and 35°C). The NaCl concentration is constant ($c_{NaCl} = 0.2\,M$) and assures formation of NBF. The solid lines denote the theoretical curves fitted to the experimental points for $c_{\beta\text{-}C12G2} < cmc$ according to Equation 4.11. (Reprinted from Muruganathan, R. M., Krastev, R., Müller, H.-J., and Möhwald, H., *Langmuir*, 22, 7981, 2006. With permission.)

was identified as the background permeability coefficient K_o, meaning that the permeability occurs through the hole-free (defect-free) area of the film. The statistical treatment led to significant values only for $a_i = a_4$ at both temperatures. All other coefficients a_i were not significant. Thus, it was concluded that holes consisting of four molecular vacancies contribute to the increase in the permeability at low surfactant concentrations. The population of such holes decreases with increasing surfactant concentration until it reaches the plateau value. The contribution of holes smaller than $i = 4$ to the increasing film permeability is minor because of the low permeability coefficient through such small holes. On the other hand, the gas flux through the larger holes ($i > 4$) is larger, but the number of such holes is low because the statistical probability to form larger defects is small. Note that formation of larger holes most likely leads to film rupture.

The specific film interaction free energy Δg^f is a convenient quantity for studying the interaction forces in foam films. It is related to the disjoining pressure and to the easily measurable contact angle θ between the film and its meniscus by the relation [111,112]:

$$2\sigma\left(\cos\theta - 1\right) = \Delta g^f = -\int_{\infty}^{h} \Pi(h)dh + \Pi h \approx -\int_{\infty}^{h} \Pi(h)dh \qquad (4.13)$$

The quantity Πh is usually a few orders of magnitude smaller than the values of the specific film interaction free energy and thus can be neglected. The information that can be extracted in this way is especially important in the case of higher salt concentrations where only NBFs are formed even at small capillary pressures. Since the repulsive part of the experimental $\Pi(h)$-curve is interrupted at the CBF to NBF transition and since the attractive branch of the $\Pi(h)$-curve cannot be determined with a thin film pressure balance, it is impossible to obtain Δg^{f} from direct integration of the $\Pi(h)$-curve in this case.

Our results show that the specific film interaction free energy is very sensitive to changes in the interactions in the foam films, which is demonstrated by the dependencies of the film thickness and the contact angle (respectively the specific film interaction free energy $|\Delta g^{\mathrm{f}}|$) on the surfactant concentration (Figure 4.12). The contact angle increases upon increasing the surfactant concentration up to the cmc, although in this range of concentrations the NBF has a constant thickness. Above cmc θ remains essentially constant. A comparison between Figures 4.11 and 4.12 (see the highlighted area in Figure 4.12) shows that the gas permeability of the NBF changes exactly in the range of surfactant concentrations where the contact angle changes significantly even though the film thickness is constant. Obviously, changes in the film structure occur which strongly influence the interactions between the film surfaces but do not change the film's thickness. This observation demonstrates that the contact angle is a more sensitive tool to register changes in the structure of NBFs than the film thickness.

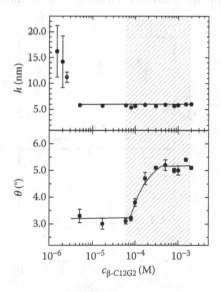

FIGURE 4.12 Dependencies of (top) foam film thickness h and (bottom) contact angle θ on the β-$C_{12}G_2$ concentration. The NaCl concentration is constant (0.2 M) and assures formation of NBF. The temperature is 25°C. All lines are only guides for the eye. The highlighted area shows the range of surfactant concentrations used for the permeability experiments (Figure 4.11). (Reprinted from Muruganathan, R. M., Krustev, R., Müller, H.-J., and Möhwald, H., *Langmuir*, 20, 6352, 2004. With permission.)

4.4.3 Effect of the Film Thickness

Equation 4.8 predicts a strong dependence of the gas permeability on the thickness of the foam film. At a given surfactant concentration, the thickness of a CBF can be varied either by applying pressure similar to thin film pressure balance measurements (see Section 4.2), or by adding salt to the film forming solutions. The addition of salt reduces the double-layer repulsion between the film surfaces and the films become thinner. One has to keep in mind that the salt may also change the adsorption density at the film surfaces. However, when nonionic surfactants are used and the surfactant at concentrations above the cmc one can expect that the changes in the adsorption density are minor. Even though gas permeability experiments at different applied pressures are possible, only few experiments with ionic surfactants have been performed [113]. Experiments with foam films stabilized by β-$C_{12}G_2$ were performed by changing the salt (NaCl) concentration while the surfactant concentration was fixed at 1×10^{-3} M [110], which is above the cmc. The results are presented on Figure 4.13 together with the dependence of the film thickness on the NaCl concentration. The observed dependence $K(c_{el})$ is weak. Although the increase in the electrolyte concentration leads to a decrease in the film thickness of the CBF, the film permeability is constant within the limits of the experimental error. The following explanation may describe the difference between the theoretical expectations (see Equation 4.8) and the observed results. Additional pressure from one side of the film has to be created in

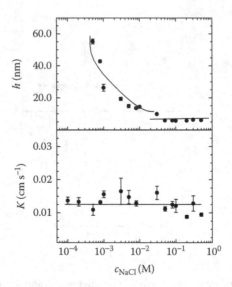

FIGURE 4.13 Dependencies of (top) foam film thickness h and (bottom) foam film permeability, K, on the NaCl concentration. The β-$C_{12}G_2$ concentration is constant (10^{-3} M) and above the cmc for the salt free solutions. The temperature was kept constant at 25°C. (Data from Muruganathan, R. M., Krustev, R., Müller, H.-J., and Möhwald, H., *Langmuir*, 20, 6352, 2004; Muruganathan, R. M., Krastev, R., Müller, H.-J., and Möhwald, H., *Langmuir*, 22, 7981, 2006. With permission.)

order to study the gas permeability. This pressure is used as a driving force for the permeability process.* It was shown by Muruganatahan et al. [50] that CBF films stabilized with β-$C_{12}G_2$ undergo a transition to a NBF at a capillary pressure of around 400 Pa, irrespective of the NaCl concentration. This low transition pressure is due to the low charge density of the film interfaces formed by the nonionic surfactant. Thus, due to the higher pressure in the bubbles we suggest that NBFs with constant thickness were formed under our experimental conditions. Since the thickness of the NBFs does not depend on the salt concentration, no change in the permeability was registered.

4.4.4 EFFECT OF THE TEMPERATURE

The gas permeability of NBFs as a function of temperature has also been investigated [48] for films prepared from a solution of 1×10^{-3} M β-$C_{12}G_2$ in 0.2 M NaCl. The temperature was varied in the range 15°C–35°C. The results are summarized in Figure 4.14. The film permeability decreases in the temperature range from 15°C to 25°C. It reaches a minimum at 25°C and then increases up to 35°C. The NBF thickness is constant in this temperature range or it increases slightly above 25°C [50]. This cannot explain the experimental observation.

It has been shown in other studies [96,99] that the gas permeability of foam films increases with temperature. The reason is that the average energy of the gas molecules increases with increasing temperature. Thus, the number of molecules

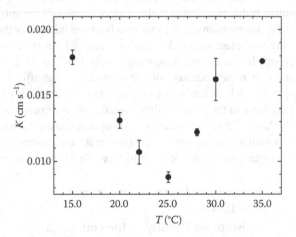

FIGURE 4.14 Dependence of the foam film permeability, K, on the temperature T. Films were prepared from a 1 mM β-$C_{12}G_2$ solution containing 0.2 M NaCl.

---

* For example in Muruganathan et al. (*Langmuir* 2006, 22, 798) microscopic bubbles were used to study the gas transport through the films. The additional pressure in this case was equal to the capillary pressure in the bubbles, around 700 Pa (bubble radius approximately 100 µm and surface tension approximately 35 mN m^{-1}).

that possess the necessary energy to overcome the energy barrier (in our case the foam film) increases, which leads to the observed increase in permeability. However, when we deal with NBFs the permeability is controlled solely by the permeability of the single monolayers that form the film (Equation 4.9), which, in turn, depends on the adsorption density of the surfactant molecules [101–104]. In many cases the adsorption density decrease monotonously with increasing temperature because of the enhancement of the thermal motion of surfactant molecules at the surface. Hence, at higher temperatures one would expect a larger area per surfactant molecule and, consequently, a higher permeability. The theory of fluctuation formation of holes in the foam bilayers also predicts an increase in the number of vacancies in the film with increasing temperature. All these factors predict that an increase in temperature should result in a continuous increase in the permeability of the foam films. The unexpected decrease in the gas permeability of foam films stabilized by β-$C_{12}G_2$ in the temperature range from 15°C to 25°C could be explained, if the monolayer adsorption density exhibits a non-monotonous temperature dependence. The surfactant densities at cmc were calculated from the surface tension isotherms measured at three temperatures (15°C, 25°C, and 35°C) (Table 4.3). Indeed, the surface density at 25°C was found to be larger than that at higher and lower temperatures, suggesting a non-trivial temperature dependence of the surface density of β-$C_{12}G_2$. An increase in the surface density with temperature has also been observed for nonionic surfactants of the type C_nE_m and is considered to be a result of dehydration of the surfactant molecules [49].

Thus, we can conclude that the temperature dependence of the gas permeability of β-$C_{12}G_2$ films is due to changes in the surface density of the surfactant. It can be understood as resulting from the competition between increasing thermal energy of the gas and the surfactant molecules, and a decreased effective size of the surfactant head group. In the temperature range from 15°C to 25°C the surfactant adsorption density increases because of the decrease in the effective size of the surfactant head groups, which can be due to dehydration or conformational changes. This leads to a decrease in the permeability even though the energy of the gas molecules increases. Above 25°C the surfactant adsorption density follows the general trend. It decreases with increasing temperature due to the enhancement of the thermal motion of the surfactant molecules at the interface. This leads to an increase in the

TABLE 4.3
Adsorption Density at the cmc (Γ_{cmc}) of β-$C_{12}G_2$ at the Air–Water Interface at Different Temperatures T

$T/°C$	$\Gamma_{cmc}/\mu mol\ m^{-2}$
15	4.08
25	4.25
35	3.75

film permeability supported by the increase in the energy of the gas molecules. The results presented above underline the importance of small variations of the adsorption density in the film surfaces for the gas permeability of foam films.

4.5 STABILITY OF FOAMS

Surface active glucosides have been known since the beginning of the last century, initially being found under the name "saponins" [114,115]. The most prominent surface chemical property quoted at that time was their ability to foam. Consequently, over time a considerable amount of information has been accumulated on the foamability and foam stability of these surfactants. This has led to a qualitative understanding on how the surfactant structure, temperature, surfactant, and salt concentration affect foam properties. Some examples are provided below. The data on foamability and foam stability that are discussed in the following have been determined using different methods, namely Ross–Miles [116], ASTM standard D 1173, a winding equipment (Vindan) [117], and a bubbling technique (FoamScan [118–121]). However, it goes beyond the aim of this section to describe these techniques in detail.

4.5.1 PURE ALKYL GLUCOSIDES

Some data on the foam characteristics of pure β-$C_{12}G_2$ will be discussed first as a continuation of the information on film stability, surface rheology, and gas permeability of this surfactant provided in previous sections. Results obtained for aqueous solutions of β-$C_{12}G_2$ with the Vindan and the FoamScan method, respectively, are presented in Figure 4.15. As for all surfactants the foamability (Figure 4.15 (top)) and foam stability (data not provided) increase with surfactant concentration up to some multiples of cmc. At sufficiently high concentrations, for example at 10 cmc, a huge amount of foam with a high stability can be generated with β-$C_{12}G_2$. Comparing these results with those obtained for the equally nonionic surfactant $C_{12}E_6$, one notices that the latter surfactant has a much lower foamability and foam stability at the same relative surfactant concentration (10 cmc). When the two surfactants β-$C_{12}G_2$ and $C_{12}E_6$ are mixed in a 1:1 ratio the resulting foam properties were found to be more similar to that of $C_{12}E_6$ than to that of β-$C_{12}G_2$, i.e. the properties of the mixture are dominated by the less foaming surfactant (Figure 4.15). This behavior has been observed for foam generation with the winding technique (Vindan, large energy input) as well as for foam generation with the bubbling technique (FoamScan, low energy input). Thus, the properties of the 1:1 mixture mimic that of the pure $C_{12}E_6$ independently on the amount of energy used in the foam generation process. On the other hand, when only traces of one surfactant are added to the second surfactant, for example β-$C_{12}G_2$:$C_{12}E_6$ = 50:1 (Figure 4.15 (top)) or β-$C_{12}G_2$:$C_{12}E_6$ = 1:50 (data not provided), the resulting foam properties mimic that of the main surfactant.

For C_8G_1 and $C_{10}G_1$, an increased foaming with increasing surfactant concentration as well as a better foaming profile compared to the corresponding alkyl polyglycol ethers has been demonstrated by Shinoda et al. [122] and by Matsumura et al.

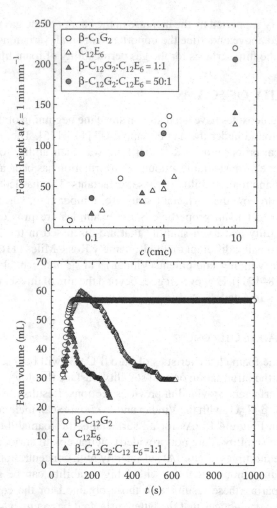

FIGURE 4.15 Foam height in mm obtained with the Vindan method for β-$C_{12}G_2$, $C_{12}E_6$ and mixtures of them as a function of the relative surfactant concentration c/cmc. The foam was created from a 200 mL solution in a 500 mL cylinder turned around at 40 rpm (top). Foam volume in milliliters produced with the FoamScan instrument as a function of time for β-$C_{12}G_2$, $C_{12}E_6$ and a 1:1 mixture of β-$C_{12}G_2$ and $C_{12}E_6$ at fixed total surfactant concentration of 10 cmc (bottom). The initial foam volume and the gas flow rate were fixed to 50 mL and 60 mL min^{-1}, respectively.

using the Ross–Miles method [123]. Shinoda et al. showed that the foam volume and the foam stability of β-C_8G_1 reach plateau values above the cmc. Similar features have also been observed for nonionic octyl derivatives with other sugar head groups. Matsumura et al. found that among the C_nG_1 surfactants $C_{10}G_1$ is the best foamer both in terms of foamability and foam stability. It was also found, in general, that better foam properties were obtained for the β anomer compared to the

α anomer. This stereochemical effect has also been pointed out and discussed in Section 4.2.5. The dramatic drop in foamability observed when increasing the hydrocarbon chain length from C10 to C12 was explained by the much lower water solubility of $C_{12}G_1$ as compared to that of $C_{10}G_1$. It was also noted that the foam properties depend on the type of sugar unit used as head group, and, for a given hydrocarbon chain length, the foamability and foam stability were found to be higher for galactose than for glucose.

The problem with low water solubility can be avoided by including more than one glucose unit, thus increasing the head group size [124]. It has been shown that if the molecules are adequately balanced with 2 or 3 glucose units the foam produced by longer hydrocarbon chain alkyl glucosides is both of high volume and high stability. For instance, Koeltzow and Urfer found by employing the Ross–Miles method [125], in agreement with the results presented above, that for $\beta\text{-}C_nG_1$ optimum foamability and foam stability were achieved with a straight C10 hydrocarbon chain. For $\beta\text{-}C_nG_2$ a broad maximum in performance for C10–C13 chains was found. An increase to a C15 chain resulted in significant reduced foamability, whereas the stability of the foam, measured after 5 min, remained high. An increase in head group polymerization to $\beta\text{-}C_nG_3$ resulted in similar foam properties as for $\beta\text{-}C_nG_2$, except that a slightly higher foamability for $\beta\text{-}C_{18}G_3$ compared with $\beta\text{-}C_{18}G_2$ was noted. It is clear that a high foamability requires high surfactant solubility, but this is not a sufficient criteria. It appears that the foam results depend on a delicate balance between the different factors influencing the organization or packing in the foam lamellae, which affects surface rheology and stabilization of possible holes formed in the lamellae. From a synthetic point of view this balance is reached by variations in hydrocarbon chain length and head group polymerization.

4.5.2 COMMERCIAL ALKYL POLY GLUCOSIDES

A general picture emanates from all the product descriptions of alkyl poly glucosides (APGs) published over the years [117,126–128]. APGs are always characterized as good to medium foamers giving stable foam, and they perform better than the corresponding ethoxylated nonionic surfactants in both foamability and foam stability tests. As explained in Section 4.2.5, the APGs consist of a mixture of both stereoisomers and oligomers. The commercial glycol ethers also contain a distribution of oligomers. It is not self-evident that the foams produced from these mixtures would be in any respect similar to the foams produced from the pure surfactants that we have discussed so far. However, as will be demonstrated below, the general trends on how foam properties are affected by surfactant structure and concentration appears to be the same for commercial APGs and pure alkyl glucosides.

One interesting, and for applications very important, feature is the synergistic effects that are seen in mixtures of APGs and anionic surfactants. Anionic surfactants produce a large amount of foam. However, this foam is rather loose and brittle and is easily destroyed by oily dirt that is incorporated during a cleaning procedure. When a cleaning liquid contains both anionic surfactants and APGs the foam becomes more stable against contaminants and can be used for a longer period of time for instance in hand dish washing [62]. This can, of course, be understood in terms of denser packing in the foam lamellae when the repulsion between the anionic head groups is shielded by the neutral glucose units of the APGs.

In the following an investigation of a set of 25 APGs with varying carbon chain lengths, carbon chain structures, and varying degrees of glucosidation are discussed. Further details can be found in the report by Johansson et al. [129]. The aim of this study was to identify APG structures that do not foam and still can be used for cleaning and wetting purposes. The reason for this is that foam generation is detrimental to new cleaning techniques using high-pressure jets or circulation of cleaning solutions. The set of surfactants studied included C8, C10, C12, and C14 hydrocarbon chains with three different hydrocarbon chain structures: straight, methyl branched, and so-called Guerbet branched. The latter structure has one defined branch, e.g. 2-ethyl hexanol, 2-propyl heptanol, 2-butyl octanol, or 2-pentyl nonanol. Two different solutions were investigated: (1) 0.05 wt% APG in water and (2) a formulated liquid containing 0.5 wt% APG, 0.6 wt% tetra potassium pyro phosphate, 0.4 wt% meta silicate solubilized with 0.4 wt% hydrotrope. In the following, we will call the formulated liquid a 0.5 wt% APG solution. All measurements were made with the winding equipment at room temperature.

4.5.2.1 Influence of Chain Structure

The data shown in Figure 4.16, obtained above the cmc of the APGs, demonstrate that both foamability (initial foam height) and foam stability (as judged from the time dependence of the foam height) depend on the branching. This holds true for both the C10 and the C12 homologues. The Guerbet structures with their defined branching give rise to the lowest foam, which is suggested to be due to formation of less well-organized surface structures. As expected, the formulations with higher content of surfactant generate larger foam volumes.

4.5.2.2 Influence of Chain Length

As can be seen in Figure 4.16, the straight and methyl branched C12 surfactants are better foamers than the C10 homologues. The same holds true for the Guerbet branched APGs at 0.05 wt%. However, the formulated liquid containing 0.5 wt% Guerbet branched surfactant behaves differently. In this case the foamability and the foam stability of the formulation containing the C12 homologue are smaller than those of the formulation containing the C10 homologue. The reason for this is not clear, but it demonstrates the point that results obtained for pure surfactants and technical surfactants cannot directly be extrapolated to the properties of formulated solutions. The results for the straight and the Guerbet branched surfactants are shown again in Figure 4.17 together with data obtained for C8 and C14. These results indicate that, depending on the isomer structure, there might be an optimal chain length around C10 or C12 where both foamability and foam stability reach a maximum. To obtain a molecular understanding for the effect of chain branching on foam properties would require an increased knowledge of the surface rheological properties of such structures over a broad frequency and concentration

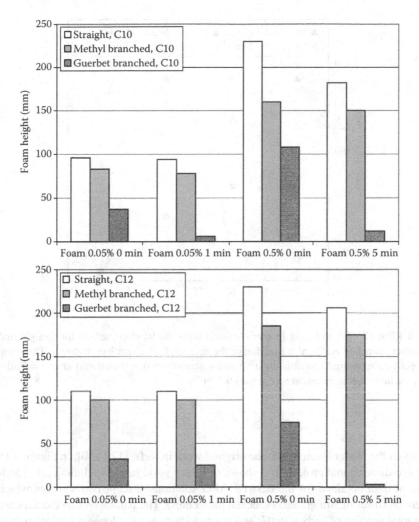

FIGURE 4.16 Foam height in mm obtained with the Vindan method for different isomers of $C_{10}APG$ (top) and $C_{12}APG$ (bottom). Two different surfactant concentrations were studied, namely 0.05 wt% and 0.5 wt%. In the former case the foam height was measured immediately after foam generation ($t = 0$ min) and after $t = 1$ min, while in the latter case it was studied at $t = 0$ min and $t = 5$ min.

range. In addition, an analysis based on chemometric multivariate techniques will be shown in the following.

4.5.2.3 Multivariate Structure–Property Relationship

Only two examples of the many multivariate analyses that have been made are discussed. For more details concerning results and computer programs used in the

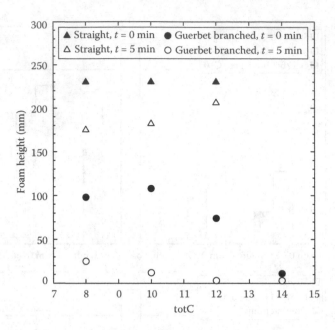

FIGURE 4.17 Foam height in mm obtained with the Vindan method for straight and Guerbet branched APGs as a function of the total carbon chain length (totC). The foam height was measured immediately after foam generation ($t = 0$ min) and after $t = 5$ min. The surfactant concentration was always 0.5 wt%.

analysis the reader is referred to the original work in Refs. [129,130]. In Figure 4.18 a three-dimensional projection is shown of a very good model including all experimental data obtained for the 0.5 wt% APG solutions (except two samples which were mixtures of straight and branched molecules). The parameters on the axes are the total amount of carbon (totC) and the ratio between the longest straight part in the carbon chain (redC) divided by totC, i.e. a measure of the branching. This means that the smaller the value of redC/totC is, the larger the branching is. Thus, the Guerbet structures have the lowest value of redC/totC, whereas redC/totC = 1 corresponds to the straight carbon chain. The foam height was measured 5 min after generation and is plotted versus totC and redC/totC, respectively. The amount of foam left after 5 min is a parameter of importance for the practical use. The optimum total chain length for high foamability and foam stability is just below 12, and the chain should be straight. Both C8 and C14 give lower foam, qualitatively in agreement with the findings for pure alkyl glucosides, and the Guerbet branched structures produce the lowest overall foam volume. The same trend (not shown) is obtained for the 0.05 wt% APG solutions though the amount of foam is generally lower.

The influence of the degree of glucosidation (Dp) is demonstrated in Figure 4.19. The interesting parameter here is the ratio Dp/totC, i.e. the amount of glucose

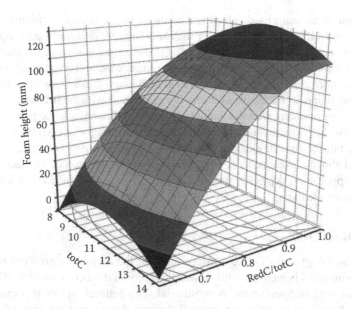

FIGURE 4.18 Multivariate plot of foam height after 5 min generated from 0.5 wt% APG in formulation against total carbon chain length (totC) and amount of branching (redC/totC).

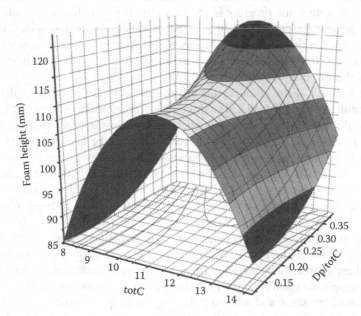

FIGURE 4.19 Multivariate plot of foam height after 5 min generated from 0.5 wt% APG in formulation against total carbon chain length (totC) and degree of polymerization/total carbon chain length (Dp/totC).

per carbon in the tail group, which is a measure of the actual hydrophilicity. The dependency of the foam properties of the ratio Dp/totC is somewhat complicated. Depending on where you are in the pseudo three-dimensional space the foamability and foam stability may increase (most often) or decrease with increasing hydrophilicity. The more glucosidated species seem to be the best foamers, and the low foamers can be found in the areas with low Dp combined with high branching, again qualitatively consistent with the limited amount of data available for pure alkyl glucosides. Further investigations revealed that rapid wetting was correlated with low foaming, but the overall cleaning got worse. The structure–property relationships described above make it possible to predict design rules for suitable surfactant structures for applications depending on specific demands, e.g. combinations of low foaming and good cleaning.

4.6 OUTLOOK

The methodology for accurate measurements of the disjoining pressure acting in single foam films is well-established and the body of data collected for MW ionic as well as nonionic surfactants is substantial. The general picture that emerges is that the disjoining pressure can be well-described by a combination of double-layer repulsion, van der Waals attraction, and a short-range repulsion due to dehydration and steric confinement. The situation is similar to that encountered between surfactant layers adsorbed at solid surfaces. The molecular mechanism behind the adsorption of OH⁻ ions (and other ions) at the air–water interface remains a bit of a puzzle, despite some progress in recent years. It seems likely that additional cleverly designed experiments and computer simulations are able to develop our understanding in this area. It also seems that a systematic use of contact angle measurements for NBFs would allow us to gain understanding of interactions between these very thin films.

The stability against film rupture is not related to the strength of the long-range repulsion, but rather linked to short-range forces acting normal and tangentially to the film surface. The hole nucleation theory appears to describe the process from the mechanistic point of view, and it can be used for interpretation of monolayer and thin film gas permeability data. However, a molecular model that links surfactant structure, adsorbed layer structure and hydration to the probability of hole nucleation is still lacking. Development of such a model would be beneficial for predicting structure–property relationships in foam films and foams.

It has become increasingly evident that surface rheological properties are very important for foam film and foam stability, and several techniques exist for measuring these properties over certain frequency ranges. Nevertheless, we feel that there is significant room for further development in this area. One needs to be able to measure both shear and dilational rheological properties over a (very) broad frequency range in order to obtain a detailed picture of all relevant relaxation processes in the adsorbed monolayers. It seems likely that we will see a large progress in this area during the coming years.

For alkylglucosides, a correlation between thin film stability on the one hand, and foamability and foam stability on the other hand can be found. Further, the structure–foam property relationships found for pure surfactants are well mimicked by those for technical surfactants, and it seems that the trends obtained for the evolution of foam properties with surfactant structure is the same independent on which foaming method is used. Most of the studies have been performed with glucose units as polar part of the surfactant. However, there are some indications that the properties may change when different sugar units are used in the surfactant structure. This area seems to be rather poorly explored, and considering the difference in hydration and solubility properties of different sugar units it can be expected that a systematic variation of the nature of the sugar unit would provide novel surfactants with interesting properties, not only for foam applications but in general terms.

ACKNOWLEDGMENTS

Part of the work was funded by the European Community's Marie Curie Research Training Network "Self-Organisation under Confinement (SOCON)," contract number MRTN-CT-2004-512331. We thank Lok Kumar Shresta for carrying out the FoamScan measurements.

REFERENCES

1. Pugh, R., *Adv. Colloid Interface Sci.* 1996, *64*, 67.
2. Exerowa, D. and Kruglyakov, P. M., *Foam and Foam Films—Theory, Experiment, Application*; Elsevier: Amsterdam, 1998, Vol. 5.
3. Bergeron, V., *J. Phys.: Condens. Matter* 1999, *11*, R215.
4. Stubenrauch, C., *Tenside Surf. Det.* 2001, *38*, 350.
5. Stubenrauch, C., Makievski, A. V., Khristov, K., Exerowa, D., and Miller, R., *Tenside Surf. Det.* 2003, *40*, 196.
6. Stubenrauch, C. and Rippner-Blomqvist, B. In *Colloid Stability: The Role of Surface Forces, Part 1*; Tadros, T., (Ed.); Wiley-VCR: Weinheim, 2006, p 263.
7. Stubenrauch, C. and Klitzing, R. V., *J. Phys. Condens. Matter* 2003, *15*, R1197.
8. Bergeron, V. and Claesson, P. M., *Adv. Colloid Interface Sci.* 2002, *96*, 1.
9. Bergeron, V., *Langmuir* 1997, *13*, 3474.
10. Karraker, K. PhD Thesis, University of California, Berkeley, 1999.
11. Stubenrauch, C., Schlarmann, J., and Strey, R., *Phys. Chem. Chem. Phys.* 2002, *4*, 4504 and *Phys. Chem. Chem. Phys.*, 2003, 5, 2736 (erratum).
12. Schlarmann, J. and Stubenrauch, C., *Tenside Surf. Det.* 2003, *40*, 190.
13. Rippner, B., Boschkova, K., Claesson, P. M., and Arnebrant, T., *Langmuir* 2002, *18*, 5213.
14. Monroy, F., Giermanska-Khan, J., and Langevin, D., *Coll. Surf. A* 1998, *143*, 251.
15. Fruhner, H., Wantke, K. -D., and Lunkenheimer, K., *Coll. Surf. A* 2000, *162*, 193.
16. Langevin, D., *Adv. Colloid Interface Sci.* 2000, *88*, 209.
17. Stubenrauch, C. and Miller, R., *J. Phys. Chem. B* 2004, *108*, 6412.
18. Bos, M. and Vliet, T., *Adv. Colloid Interface Sci.* 2001, *91*, 437.
19. Lucassen, J. In *Anionic Surfactants: Physical Chemistry of Surfactant Action*; Lucassen-Reynders, E., (Ed.); Marcel Dekker: New York, 1981, p 217.

20. Kruglyakov, P., In *Thin Liquid Films*; Ivanov, I. B., (Ed.); Marcel Dekker: New York, 1988.
21. Kjellin, U. R. M., Claesson, P. M., and Vulfson, E. N., *Langmuir* 2001, *17*, 1941.
22. Persson, C. M., Kumpulainen, A. J., and Eriksson, J. C., *Langmuir* 2003, *19*, 6110.
23. Persson, C. M., Kjellin, U. R. M., and Eriksson, J. C., *Langmuir* 2003, *19*, 8152.
24. Derjaguin, B. V., *Acta Physicochim URSS* 1939, *10*, 25.
25. Scheludko, A., *Adv. Colloid Interface Sci.* 1967, *1*, 391.
26. Sedev, R. and Exerowa, D., *Adv. Colloid Interface Sci.* 1999, *83*, 111.
27. Claesson, P. M., Ederth, T., Bergeron, V., and Rutland, M. W., *Adv. Colloid Interface Sci.* 1996, *67*, 119.
28. Waltermo, Å., Manev, E., Pugh, R., and Claesson, P. M., *J. Disp. Sci. Techn.* 1994, *15*, 273.
29. Manev, E. D. and Pugh, R. J., *Langmuir* 1991, *7*, 2253.
30. Bergeron, V., Waltermo, Å., and Claesson, P. M., *Langmuir* 1996, *12*, 1336.
31. Attard, P., Mitchell, J., and Ninham, B. W., *J. Chem. Phys.* 1988, *89*, 4358.
32. Waltermo, Å., Claesson, P. M., Manev, E., Simonsson, S., Johansson, I., and Bergeron, V., *Langmuir* 1996, *12*, 5271.
33. Kolarov, T., Cohen, R. C., and Exerowa, D., *Coll. Surf.* 1989, *42*, 49.
34. Stubenrauch, C. and Strey, R., *J. Phys. Chem. B* 2005, *109*, 19798.
35. Karraker, K. A. and Radke, C. J., *Adv. Colloid Interface Sci.* 2002, *96*, 231.
36. Buchavzov, N. and Stubenrauch, N., *Langmuir* 2007, *23*, 5315.
37. Rojas, O. J., Stubenrauch, C., Schultze-Schlarmann, J., and Claesson, P. M., *Langmuir* 2005, *21*, 11836.
38. Persson, C. M., Claesson, P. M., and Lunkenheimer, K., *J. Colloid Interface Sci.* 2002, *251*, 182.
39. Israelachvili, J. N., *Intermolecular and Surface Forces*; Academic Press: London, 1991.
40. Exerowa, D., *Kolloid-Z.* 1969, *232*, 703.
41. Stubenrauch, C., Cohen, R., and Exerowa, D., *Langmuir* 2007, *23*, 1684.
42. Li, C. and Somasundaran, P., *J. Colloid Interface Sci.* 1991, *146*, 215.
43. Marinova, K. G., Alargova, R. G., Denkov, N. D., Velev, O. D., Petsev, D. N., Ivanov, I. B., and Borwankar, R. P., *Langmuir* 1996, *12*, 2045.
44. Stubenrauch, C., Rojas, O. J., Schlarmann, J., and Claesson, P. M., *Langmuir* 2004, *20*, 4977.
45. Garrett, B. C., *Science* 2004, *303*(5661), 1146.
46. Buch, V., Milet, A., Vacha, R., Jungwirth, P., and Devlin, J. P., *PNAS* 2007, *104*, 7342.
47. Kumpulainen, A. J., Persson, C. M., and Eriksson, J. C., *Langmuir* 2004, *20*, 10534.
48. Muruganathan, R. M., Krustev, R., Ikeda, N., and Müller, H. J., *Langmuir* 2003, *19*, 3062.
49. Wongwailikhti, K., Ohta, A., Seno, K., Nomura, A., Shinozuka, T., Takiue, T., and Aratono, M., *J. Phys. Chem. B* 2001, *105*, 11462.
50. Muruganathan, R. M., Krustev, R., Müller, H. -J., Möhwald, H., Kolaric, B., and V. Klitzing, R., *Langmuir* 2004, *20*, 6352.
51. Nilsson, F., Söderman, O., and Johansson, I., *J. Colloid Interface Sci.* 1998, *203*, 131.
52. Claesson, P. M. and Kjellin, U. R. M., In *Encyclopedia of Surface and Colloid Science*; Hubbard, H., (Ed.); Marcel Dekker: New York, 2002, p 4909.
53. Dorset, D. L. and Rosenbusch, J. P., *Chem Phys Lipids* 1981, *29*, 299.

54. Straathof, A. J. J., van Bekkum, H., and Kieboom, A. P. G., *Starch/Stärke* 1988, *40*, 438.
55. Rojas, O. J., Neuman, R. D., and Claesson, P. M., *J. Phys. Chem. B* 2005, *109*, 22440.
56. Vrij, A., *Discuss. Faraday Soc.* 1966, *42*, 23.
57. Vrij, A. and Overbeek, J. T. G., *J. Am. Chem. Soc.* 1968, *90*, 3074.
58. Exerowa, D., Kashchiev, D., and Platikanov, D., *Adv. Colloid Interface Sci.* 1992, *40*, 201.
59. Schlarmann, J., Stubenrauch, C., and Strey, R., *Phys. Chem. Chem. Phys.* 2003, *5*, 184.
60. Monin, D., Espert, A., and Colin, A., *Langmuir* 2000, *16*, 3873.
61. Waltermo, Å., Claesson, P. M., and Johansson, I., *J. Colloid Interface Sci.* 1996, *183*, 506.
62. Balzer, D., *Tenside Surf. Det.* 1991, *28*, 419.
63. Hughes, F. A. and Lew, B. W., *J. Am. Oil Chem. Soc.* 1970, *47*, 162.
64. Black, I. J. PhD Thesis, University of Reading, United Kingdom, 1992.
65. Black, I. J. and Herrington, T. M., *J. Chem. Soc. Faraday Trans.* 1995, *91*, 4251.
66. Persson, C. M., Claesson, P. M., and Johansson, I., *Langmuir* 2000, *16*, 10227.
67. Claesson, P. M., Kjellin, M., Rojas, O. J., and Stubenrauch, C., *Phys. Chem. Chem. Phys.* 2006, *8*, 5501.
68. Bargeman, D. and Van Voorst Vader, F., *J. Colloid Interface Sci.* 1972, *42*, 467.
69. Gau, C. -S. and Zografi, G., *J. Colloid Interface Sci.* 1990, *140*, 1.
70. Tyrode, E., Johnson, M. C., Kumpulainen, A., Rutland, M. W., and Claesson, P. M., *J. Am. Chem. Soc.* 2005, *127*, 16848.
71. Tyrode, E., Johnson, C. M., Rutland, M. W., and Claesson, P. M., *J. Phys. Chem. C* 2007, *111*, 11642.
72. Lucassen, J. and Giles, D., *J. Chem. Soc. Faraday Trans. I* 1975, *71*, 217.
73. Lucassen, J., *Faraday Disc. Chem. Soc.* 1975, *59*, 76.
74. Stenvot, C. and Langevin, D., *Langmuir* 1988, *4*, 1179.
75. Jiang, Q., Chiew, Y., and Valentini, J., *J. Colloid Interface Sci.* 1993, *155*, 8.
76. Jiang, Q., Chiew, Y., and Valentini, J., *J. Colloid Interface Sci.* 1993, *159*, 477.
77. Jayalakshmi, Y., Ozanne, L., and Langevin, D., *J. Colloid Interface Sci.* 1995, *170*, 358.
78. Wantke, K. -D., Fruhner, H., Fang, J., and Lunkenheimer, K., *J. Colloid Interface Sci.* 1998, *208*, 34.
79. Wantke, K. -D. and Fruhner, H., *J. Colloid Interface Sci.* 2001, *237*, 185.
80. Liggieri, L., Attolini, V., Ferrari, M., and Ravera, F., *J. Colloid Interface Sci.* 2002, *255*, 225.
81. Koelsch, P. and Motschmann, H., *Langmuir* 2005, *21*, 6265.
82. Andersen, A., Oertegren, J., Koelsch, P., Wantke, D., and Motschmann, H., *J. Phys. Chem. B* 2006, *119*, 18466.
83. Aksenenko, E. V., Kovalchuk, V. I., Fainerman, V. B., and Miller, R., *Adv. Colloid Interface Sci.* 2006, *122*, 57.
84. Grigoriev, D. and Stubenrauch, C., *Coll. Surf. A* 2007, *296*, 67.
85. Santini, E., Ravera, F., Ferrari, M., Stubenrauch, C., Makievski, A., and Krägel, J., *Coll. Surf. A* 2007, *298*, 12.
86. Lucassen, J. and van den Tempel, M., *Chem. Eng. Sci.* 1972, *27*, 1283.
87. Lucassen, J. and van den Tempel, M., *J. Colloid Interface Sci.* 1972, *41*, 491.
88. Gaydos, J., In *Drops and Bubbles in Interfacial Research*; Möbius, D. and Miller, R., (Eds.); Elsevier: Amsterdam, 1998, p 1.

89. Loglio, G. E. A., In *Novel Methods to Study Interfacial Layers*; Möbius, D. and Miller, R., (Eds.); Elsevier: Amsterdam, 2001.

90. Lucassen-Reynders, E. H., Cagna, A., and Lucassen, J., *Coll. Surf. A* 2001, *186*, 63.

91. Gibbs, J. W., *The Scientific Papers Vol. 1*; Dover Publications: New York, 1961.

92. Sonin, A. -A., Bonfillon, A., and Langevin, D., *J. Colloid Interface Sci.* 1994, *162*, 323.

93. Brown, A. G., Thuman, W. C., and Mc Bain, J. W., *J. Coll. Sci.* 1953, *8*, 508.

94. Princen, H. M. and Mason, S. G., *J. Colloid Interface Sci.* 1965, *20*, 353.

95. Nedyalkov, M., Krustev, R., Kashchiev, D., Platikanov, D., and Exerowa, D., *Coll. Polym. Sci.* 1988, *266*, 291.

96. Nedyalkov, M., Krustev, R., Stankova, A., and Platikanov, D., *Langmuir* 1992, *8*, 3142.

97. Krustev, R., Platikanov, D., Stankova, A., and Nedyalkov, M., *J. Disp. Sci. Techn.* 1997, *18*, 789.

98. Krustev, R. and Müller, H.-J., *Langmuir* 1999, *15*, 2134.

99. Farajzadeh, R., Krastev, R., Pacelli, L., and Zitha, J., *Adv. Colloid Interface Sci.* 2008, *137*, 27.

100. Ivanov, I. B. E., Ed., *Thin Liquid Films*, Marcel Dekker: New York, 1988.

101. Barnes, G. T., *Adv. Colloid Interface Sci.* 1986, *25*, 89.

102. Langmuir, I. and Langmuir, D. B., *J. Phys. Chem.* 1927, *31*, 1719.

103. Langmuir, I. and Schaefer, V. J., *J. Frankin Inst.* 1943, *235*, 119.

104. Archer, R. J. and La Mer, V. K., *J. Phys. Chem.* 1955, *59*, 200.

105. Li, J., Schuring, H., and Stannarius, R., *Langmuir* 2002, *18*, 112.

106. van Amerongen, G. J., *J. Polymer Science* 1947, *2*, 381.

107. Korvezee, A. E. and Mol, E. A., *J. Polymer Sci.* 1947, *2*, 371.

108. Kashchiev, D. and Exerowa, D., *Biophys. Biochem. Acta* 1983, *732*, 133.

109. Krustev, R., Platikanov, D., and Nedyalkov, M., *Colloids and Surfaces A: Physicochemical and Engineering Aspects* 1993, *79*, 129.

110. Muruganathan, R. M., Krastev, R., Müller, H. -J., and Möhwald, H., *Langmuir* 2006, *22*, 7981.

111. de Feijter, J. A., Rijnbout, J. B., and Vrij, A., *J. Colloid Interface Sci.* 1978, *64*, 258.

112. de Feijter, J. A. and Vrij, A., *J. Colloid Interface Sci.* 1978, *64*, 269.

113. Krustev, R. and Müller, H. -J., *Rev. Sci. Instrum.* 2002, *73*, 398.

114. Kobert, *Sueddeutsche Apotheker-Zeitung* 1911, *51*, 158.

115. Fisher, E., *Berichte Deutshen Chemischen Gesellschaft* 1914, *47*, 1377.

116. Ross, J. and Miles, G. D., *Oil and Soap* 1941, *18*, 99.

117. Baltzer, D. In *Nonionic Surfactants, Alkyl Polyglucosides*; Baltzer, D. and Lueders, H., (Eds.); Marcel Dekker: New York, 2000.

118. Rippner Blomqvist, B., Ridout, M. J., Mackie, A. R., Wärnheim, T., Claesson, P. M., and Wilde, P., *Langmuir* 2004, *20*, 10150.

119. Schmitt, C., Bovay, C., and Frossard, P., *J. Agric. Food Chem.* 2005, *53*, 9089.

120. Acharya, D. P., Gutiérrez, J. M., Aramaki, K., Aratani, K., and Kuneida, H., *J. Colloid Interface Sci.* 2005, *291*, 236.

121. Ruiz-Henestrosa, V. P., Sanches, C. C., and Patino, J. M. R., *J. Agric. Food Chem.* 2007, *55*, 6339.

122. Shinoda, K., Yamanaka, T., and Kinoshita, K., *J. Phys. Chem.* 1959, *63*, 648.

123. Matsumura, S., Imai, K., Yoshikawa, S., Kawada, K., and Uchibori, T., *J. Am. Oil Chem. Soc.* 1990, *67*, 996.

124. Jönsson, B., Lindman, B., Holmberg, K., and Kronberg, B., *Surfactants and Polymers in Solution*; John Wiley & Sons, Chichester, 1998.
125. Koeltzow, D. E. and Urfer, A. D., *J. Am. Oil. Chem. Soc.* 1984, *10*, 1651.
126. Busch, P., Hensen, H., and Tesmann, H., *Tenside Surf. Det.* 1993, 2, 116.
127. Nieendick, C. and Schmid, K. -H., *Chemica Oggi* 1995, *13*, 42.
128. Hill, K. -H., von Rybinski, W., and Stoll, G., *Alkyl polyglucosides*; VCH: Weinheim, 1997.
129. Johansson, I., Strandberg, C., Karlsson, B., and Gustavsson, B., In *Proc. 4th World Surfactant Congr.* Barcelona, 1996, Vol. 2, p 283.
130. Lindgren, Å., PhD Thesis, Umeå University, Sweden, 1995.

5 Soluble Monolayers of Sugar-Based Surfactants at the Air–Solution Interface

Atte J. Kumpulainen, Eric C. Tyrode,
and Jan Christer Eriksson

CONTENTS

5.1 STUDIES OF THE SURFACE ACTIVITY OF SUGAR-BASED SURFACTANTS

At the core of the topic of sustainable development stands the use of petroleum-based chemicals. By replacing xenobiotic oil-derived chemicals with substitutes

153

originating from the biosphere a significant positive environmental impact might be achieved [1]. To this end a resumed interest in the study of sugar-based surfactants and their properties has emerged [2–8]. Furthermore, carbohydrate head groups are found to be essential for many biological processes [9,10], e.g., in glycolipids which have important, yet poorly understood, roles in cell signaling or host–guest interaction [11,12]. Toward this background it is realized that a better understanding of the basic physicochemical properties of carbohydrate-based surfactants and lipids is urgently needed.

Modern surface science and engineering relies on the exploration of mechanisms at the molecular level for designing novel materials with special properties [13,14]. Self-organizing molecular systems rely on both kinetic and thermodynamic factors for controlling growth and morphology [15,16]. Hence, to gain further knowledge of the basic causes of adsorption and interactions among surfactant molecules is a prerequisite for scientific progress as well as for the development of many engineering applications.

The issues raised in this chapter are related to the nature of surface-active molecules composed of a hydrocarbon tail and a carbohydrate head group and their interaction with water. We account for the main molecular features involved in the transfer of such surfactant molecules from a dilute water solution to the air–water interface.

Combining extensive surface tension measurements with the surface-sensitive spectroscopic technique known as vibrational sum frequency spectroscopy (VSFS) enables comprehensive testing of some of the thermodynamically inferred models. VSFS yields valuable information on the orientation of the molecules, but suffers in turn from the drawback that the recorded signal depends on a combination of both the molecular orientation and the packing density in the surface. This circumstance implies that an independent determination of the packing density is required for the interpretation of the spectra recorded. Hence, by means of combining surface tension and VSFS measurements one can obtain information about the overall structure of a surfactant monolayer for different surface packing densities, henceforth referred to as surface densities, of a surfactant. In addition, VSFS allows for simultaneous assessment of changes in the participating water molecules due to changes in surfactant surface density.

The surface tension and adsorption studies of sugar-based surfactants reported in the literature [17–37], frequently involve a limited number of collected data points, in particular in the low concentration range, and often focus on just deducing the (saturation) surface density at the cmc and, of course, the value of the cmc itself. Furthermore, the data presented for sugar-based surfactants may exhibit large discrepancies, especially with respect to the deduced surface densities. For e.g., n-decyl-β-D-maltopyranoside (C_{10}-Mal) we have 56 Å^2/molecule at the cmc at 298 K according to Aveyard et al. [36] and 44 Å^2/molecule at the cmc at the same temperature according to Boyd et al. [37], as was pointed out by Stubenrauch [38]. For the purpose of molecular modeling this is unsatisfactory, and can, of course, easily result in erroneous conclusions regarding the mechanisms behind the adsorption at different surface densities. Hence, our account will almost exclusively be based on the comparatively accurate and extensive surface tension and

VSFS sets of data obtained in our own laboratory. One chief purpose is to focus on the cooperative interactions at play for common sugar-based surfactants in the air–water interface, resulting in surface micelle formation upon passing from the dilute Henry range toward the liquid-expanded state [39].

For insoluble surfactants it is generally recognized that spontaneous cooperative surface organization is indeed of great significance from both a fundamental scientific point of view as well as a technical point of view [40]. There is hence a growing need of knowledge about self-organizing surfactant film systems and their potential use in fabrication of advanced microdevices [41,42]. Insoluble monolayers suffer, however, from the drawback that metastable states commonly occur. Many transitions are found to take place above the equilibrium spreading pressure, and are thus inherently unstable with respect to the formation of three-dimensional crystals [43,44]. On the other hand, for surfactants that are soluble in water, a constant surface tension value is usually reached quite rapidly for a certain concentration and represents for the most part a true equilibrium state. This favorable thermodynamic feature makes soluble surfactant systems ideal for studying phenomena at interfaces involving the interactions between hydrocarbon tails, water, and head groups at high surface densities. Unlike the case of an insoluble surfactant, however, the surface density cannot be derived right away from a mass balance but has to be deduced from the surface tension data using the Gibbs surface tension equation while taking into account that (for a nonionic surfactant) there is rarely any significant change of the activity factor below the cmc.

Generally, carbohydrate head groups are "surface-phobic" and tend to stay away from the air–water interface (most carbohydrates raise the surface tension to a modest degree). In addition they exhibit short-range mutual interactions; the head group repulsion is found to be rather accurately accounted for by the hard-disk equation of state. Hence, carbohydrate-based surfactants constitute in fact useful vehicles for testing the models of hydrocarbon tail interaction with surface water in the dilute regime as well as the onset of repulsive interactions upon increasing the surface density of surfactant molecules.

In Section 5.2, the basic thermodynamic relations for soluble monolayers are briefly presented, followed in Section 5.3 by a presentation of surface tension isotherms for sugar-based surfactants. Then we discuss in Section 5.4 the difficulties encountered when applying the old Langmuir–Szyszkowski model for describing the entire adsorption range from the dilute Henry to the fully developed liquid-expanded range. In Section 5.5 a brief introduction of the VSFS technique and some relevant results obtained from such studies of the air–water interface is presented. This is followed by an in-depth treatment of each adsorption range observed in Section 5.6, treating the Henry range and in Section 5.7, presenting the emergence and interactions among surface micelles. In Section 5.8, VSFS results comparing sugar-based and ethylene-oxide (EO)-based nonionic surfactants are presented. In Section 5.9 the fully developed liquid-expanded monolayer for sugar-based surfactants is treated, which is followed by a study of mixed sugar-based surfactant monolayers in Section 5.10. Thereafter the adsorption isotherms of typical sugar-based soluble surfactants are presented in Section 5.11, and a comparison is made based on headgroup differences in Section 5.12.

In addition, we discuss the temperature behavior of the surface tension, corresponding to the entropy and energy change due to the adsorption in Section 5.13. Finally we exemplify how the number of monomers of a spherical micelle formed in the bulk solution is estimated by means of surface tension measurements in Section 5.14. A summary of the findings is presented in Section 5.15. The appendix treats the spontaneous emergence of surface micelles in comparison to ordinary spherical bulk micelles. In total this yields a full account of the rich and complex phenomena occurring at the air–sugar-based surfactant solution interface with increasing chemical potential of the surfactant.

5.2 THERMODYNAMIC RELATIONS FOR SOLUBLE MONOLAYERS

Let us begin with some surface-thermodynamic relations needed for the analysis of equilibrium surface tension data of surfactant solutions. We choose to consider a dilute binary solution of a nonionic (sugar) surfactant (component 2, denoted by subscript 2) in water (component 1, denoted by subscript 1). By adhering to the Gibbs scheme [45], as the dividing surface between the solution phase and the gas phase we take the geometrical surface where the surface excess of water (Γ_1) equals zero. This obviously implies that the surface phase considered has zero volume ($V^s = 0$). Thus, we have the following Helmholtz free energy expression for the interface (denoted by superscript S) as our formal starting point,

$$dF^s = -S^s\, dT + \mu_2\, dn_2^s + \gamma\, dA \qquad (5.1)$$

Here all symbols have their usual meaning: F^s is the Helmholtz (free) energy, S^s its entropy, T the absolute temperature, μ_2 the chemical potential of the surfactant, n_2^s the number of surfactant molecules, γ the surface tension, and A the total surface area. In strictness, n_2^s represents (like F^s, S^s, etc.) a surface excess relative to the $\Gamma_1 = 0$ dividing surface chosen. However, since a surfactant as a rule is strongly adsorbed already at very low surfactant concentrations, $\Gamma_2 = n_2^s/A = 1/a$ closely represents the actual superficial density of the surfactant in the interface. At equilibrium, the chemical potential of the surfactant, μ_2 is the same, of course, in the surface and in the bulk solution.

In terms of the thermodynamic interfacial properties, according to Equation 5.1 the surfactant chemical potential is defined by the partial derivative

$$\mu_2 = \left(\frac{dF^s}{dn_2^s} \right)_{T,A} \qquad (5.2)$$

implying a differential addition of surfactant at constraints of fixed temperature, T, and constant surface area, A. Correspondingly, for the surface tension we have

$$\gamma = \left(\frac{dF^s}{dA} \right)_{T,n_2^s} \qquad (5.3)$$

This relation obviously involves a stretching of a (closed) surfactant monolayer.

Furthermore, integration of the starting expression, Equation 5.1, at constant T and chemical potential (surfactant concentration) yields

$$F^s = \mu_2 n_2^s + \gamma A \tag{5.4}$$

or

$$\Omega^s \equiv F^s - \mu_2 n_2^s = \gamma A \tag{5.5}$$

where Ω^s denotes the grand Ω-potential of the interface. Accordingly, the surface tension γ is the same as the Ω-potential per unit area. By writing Equation 5.5 per mole of surfactant we also have

$$\varepsilon \equiv \frac{\Omega^s}{n_2^s} \equiv f^s - \mu_2 = \gamma a \tag{5.6}$$

where ε stands for the Ω^s-potential per mole of adsorbed surfactant and $a \equiv A/n_2^s$. Upon combining Equations 5.1 and 5.5 we get

$$d\Omega^s = -S^s \, dT - n_2^s d\mu_2 + \gamma \, dA \tag{5.7}$$

or, for one mole of adsorbed surfactant

$$d\varepsilon = -S_2^s \, dT - d\mu_2 + \gamma \, da \tag{5.8}$$

Equation 5.8 gives the surface tension γ as a partial derivative with respect to the molar area, a

$$\left(\frac{d\varepsilon}{da}\right)_{T,\mu_2} = \gamma \tag{5.9}$$

Moreover, by making use of Equations 5.5 and 5.7 we obtain the Gibbs surface tension equation in its traditional form, viz,

$$d\gamma = -\frac{S^s}{A} dT - \Gamma_2 \, d\mu_2 \tag{5.10}$$

where γ is considered as a function of the two degrees of freedom, T and μ_2, we have (in agreement with the phase rule) at constant atmospheric pressure. Insofar as the surfactant solution is approximately ideal we can rewrite Equation 5.10 in the following way:

$$d\gamma = -\left(\frac{S^s}{A} - \Gamma_2 s_2^b\right) dT - \Gamma_2 kT d\ln c_2 \tag{5.11}$$

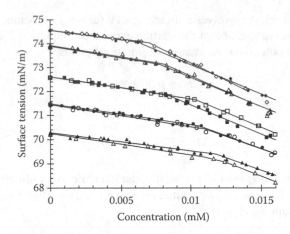

FIGURE 5.1 Surface tension data for C_{10}-Mal, n-decyl-β-D-maltopyranoside (filled symbols), and C_{10}-Glu, n-decyl-β-D-glucopyranoside (empty symbols) at the following temperatures: 8°C (highest surface tension values), 13°C, 22°C, 29°C, and 37°C (lowest surface tension values). The lines only serve as guides for the eye, dotted lines for C_{10}-Glu and solid lines for C_{10}-Mal. (From Kumpulainen, A.J., Persson, C.M., Eriksson, J.C., Tyrode, E.C., and Johnsson, C.M. *Langmuir*, 21, 305, 2005. With permission.)

where

c_2 denotes the surfactant concentration

s_2^b stands for the partial molar entropy of the surfactant in the bulk solution (denoted by superscript b)

k is the Boltzmann constant

According to Equation 5.11, the entropy change per mole due to passing from the bulk phase to the interface is given by the relation

$$a\left(\frac{d\gamma}{dT}\right)_{c_2} = -\left(s_2^s - s_2^b\right) \tag{5.12}$$

For low enough concentrations, that is in the dilute Henry range, the entropy associated with the uncovered water surface in between the surfactant molecules makes a predominant contribution to the right-hand side of Equation 5.12 and, independently of the concentration of the surfactant, the temperature coefficient of γ becomes approximately the same as for pure water [39] (cf. Figure 5.1).

5.3 SURFACE TENSION ISOTHERMS OF SUGAR-BASED SURFACTANTS

Here we survey the surface tension data and the molecular mechanisms that cause the surface tension lowering effect for sugar-based surfactants. Typically, a surface tension isotherm of a sugar-based surfactant can be subdivided into a series of concentration ranges corresponding to surface phases of different nature.

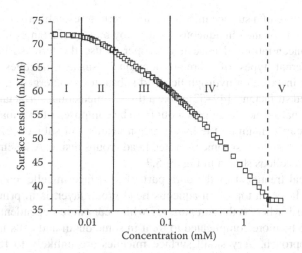

FIGURE 5.2 Surface tension isotherm for n-decyl-β-D-maltopyranoside (C_{10}-Mal) at 22°C recorded by Kumpulainen [46]. The isotherm can be subdivided into five main ranges, the Henry law range (I), the dilute surface micelle range (II), the granular range (III), and finally, the liquid-expanded range (IV). A fifth range is also included commencing at the cmc, the spherical micelle range (V). This range is not influenced, however, by any molecular events occurring in the surface.

In Figure 5.2 the surface tension data for n-decyl-β-D-maltopyranoside (C_{10}-Mal) at 22°C due to Kumpulainen [46] are displayed.

Below the cmc (above which the surface tension drops at a very slow rate), the surface tension isotherm for C_{10}-Mal can be split into four main ranges; in Figure 5.2 each transition between successive ranges is demarked by a vertical line. The first part of the isotherm, where the bulk phase concentration of surfactant still is very low, corresponds to ideal two-dimensional solution behavior. This range we refer to as the Henry law or just the Henry range. Here, the average surface area available to each surfactant molecule, a, is much larger than the molecular area proper, a_2, and the interactions between the adsorbed surfactant molecules are negligible [47,48]. This right away implies that the surface density of surfactant is proportional to the surfactant concentration in the bulk phase. Consistent with the Gibbs surface tension equation, this also means that the surface tension drops linearly with the surfactant concentration c_2 in the bulk solution.

At the high concentration end of the Henry range, clusters of molecules start to appear in the surface. To begin with the number of clusters is fairly small. However, by gradually raising the bulk concentration of surfactant, the surface density of clusters will rapidly grow. The quasi-planar clusters are mainly oriented parallel with the surface plane and arranged in such a way that the hydrocarbon tails give rise to a two-dimensional hydrocarbon core. The head groups are likely to be located at the edge and below the water face of a cluster, supposedly being fully immersed in the subjacent water phase.

This structure of a surface micelle bears a certain resemblance to an ordinary spherical micelle formed in aqueous solution by most surfactants at high enough surfactant concentrations. Hence it is proper to use the term surface micelle. Slightly different types of (three-dimensional) surface micelles have been observed, for instance on hydrophilic solid substrates submerged in solutions of e.g., EO-based surfactants [49–52]. Like a three-dimensional surfactant micelle, a two-dimensional surface micelle is thought to be composed of two main parts: the central core part containing the hydrocarbon chains and subjacent head groups and the rim part where just some hydrated head groups reside. A schematic picture of a surface micelle is shown in Figure 5.3.

Theoretical treatment of the core part of a surface micelle, composed of a hydrocarbon layer on top of an aqueous head group layer, is in principle rather straightforward, whereas the rim part, for which curvature-dependent effects sets in, appears to be more complicated to treat in some detail and calls for making a simplified approach. Very small surface micelles are unlikely to form as they would have an excess of hydrocarbon–water contacts. However, too large surface micelles tend to be disfavored due to head group repulsion and hydrocarbon chain stretching. By applying the Helfrich curvature free energy expression [39,53] for a short cylinder to account for the line tension effects to be anticipated for a surface micelle, we can readily generate a free energy minimum for a certain number of surfactant molecules participating in a surface micelle. On the basis of such a model we also find a maximum, indicating a gap in the size distribution just below the micelle size range. Typically, a small circular surface micelle is composed of 8–12 surfactant molecules. Upon increasing the bulk concentration, the surface density of surfactant micelles becomes high enough to make repulsive micelle–micelle interactions significant due to overlap of the rim parts. Eventually, the rim parts will be annihilated altogether followed by the formation of a surface-covering fluid hydrocarbon chain monolayer on top of a subjacent head group layer, the so-called liquid-expanded (LE) phase.

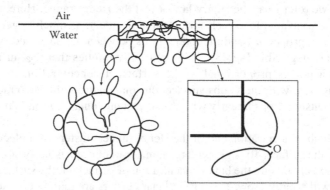

FIGURE 5.3 A schematic representation of a surface micelle of a sugar surfactant (Mal) at the air–water interface, the wiggly lines correspond to hydrocarbon chains and the ellipses to the glucose units of the head groups. (From Kumpulainen, A.J., Persson, C.M., Eriksson, J.C., Tyrode, E.C., and Johnsson, C.M. *Langmuir*, 21, 305, 2005. With permission.)

The (repulsive) interactions among the surface micelles have been success-fully treated by modeling them as hard disks and applying the hard-disk equation of state [54,55]. In the initial stage of surface micelle formation (range II in Figure 5.2) this generally accounts quite well for the interactions, but with increasing packing density shape deformations are likely to become involved. This brings us to range III, the "granular" range. The most striking feature of this range is the rapid decrease in surface tension at an almost constant area/molecule, i.e., an approximately linear slope of the surface tension curve when plotted against the logarithm of the bulk concentration. Below in this chapter, we apply a modified Langmuir–Szyszkowski model taking into account surfactant aggregation in the interface. In contrast to the hard-disk model, this model can account for the sur-face saturation behavior observed in the granular range very well, whereas it captures the mixed state with both monomers and surface micelles in the dilute surface micellar range less satisfactorily.

In the granular range (III) we suppose that the micelle rims still are intact, although no longer circular but distorted. Further, it is reasonable to assume that the micelles gradually adopt a quasi-hexagonal shape to eventually cover the surface in hexagonal packing. Raising the surfactant chemical potential in this range only yields a very slight increase in surfactant surface density. Presum-ably, in these distorted surface micelles that we refer to as grains, the central core parts (that are subject to the line tension pressure) are in much the same state as in the dilute surface micelle phase II. The increase in chemical potential results instead in a slight compression of the grain boundaries. As the chemical potential is increased to beyond where it is the same as in a surface-covering hydrocarbon/head group twin layer film, the extent of adsorption will increase in a stepwise fashion due to annihilation of the grain boundaries. At this point a "knee," in other words a fairly sharp change in slope, for the surface tension isotherm will appear.

Upon raising the chemical potential further within the liquid-expanded range, the thickness of the hydrocarbon tail layer tends to increase and likewise the den-sity of head groups in the subjacent head group–water layer. We may tentatively treat the hydrocarbon tail layer, and the mixed head group–water monolayer immediately below it, separately when modeling the free energies that affect the adsorption equilibrium. Hence, we realize that a major contribution to the surface pressure in the liquid-expanded range stems from the head group–water mixing that can be quantitatively assessed by means of a hard-disk equation of state.

A pictorial representation of the state of the surfactant-laden water–air inter-face through steps I–IV is given in Figure 5.4.

5.4　LANGMUIR–SZYSZKOWSKI MODEL OF SURFACTANT ADSORPTION

Since many years it has been common practice to employ the two-dimensional ideal Langmuir–Szyszkowski model to account for the surface tension drop rela-tive to pure water caused by a surfactant in the range below its cmc. In essence, no other feature is then invoked but a gradual variation of the configurational surface

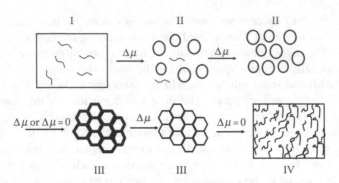

FIGURE 5.4 Schematic representation of the adsorption states for C_{10}-Mal through the phases I–IV. Wavy lines correspond to dispersed molecules (I and II), circles to surface micelles (II), hexagons to grains (III), and wiggly lines to hydrocarbon tails in a fluid state (IV). (From Kumpulainen, A.J., Persson, C.M., and Eriksson, J.C., *Langmuir*, 20, 10534, 2004. With permission.)

entropy of the top monolayer of adsorbed surfactant molecules. Thus, this model primarily implies just ideal mixing of surfactant and water in the interface on the length scale of a surfactant molecule that means, in turn, that the water surface is only completely covered by surfactant at saturation close to the cmc.

For a number of surface tension isotherms of different surfactants, however, Motomura and Aratono et al. [56,57] noted a fairly sharp break point at about 0.001–0.1 of the cmc, in particular when the data points were plotted on a linear concentration scale, indicating a (first-order) surface phase transition from the mixed "gaseous" to the surface-covering liquid-expanded state. This transition was found to be accompanied by a relatively large stepwise reduction of the molecular area of the surfactant.

For some sugar-based surfactants it was later on clarified that the passage from the gaseous to the liquid-expanded monolayer state is a more complex process. In the early stage it involves formation of circular, two-dimensional surface micelles/clusters [39]. As the surfactant concentration is raised, the concentration of surface micelles rapidly grows giving rise to repulsive interactions that cause a significant part of the surface tension lowering. In a final, true first-order step, the (distorted) border lines between crowded micelles are annihilated, resulting in the appearance of a laterally homogeneous liquid-expanded phase. This sequence of events that takes place in the interface implies that there is just a very minor exposure of water molecules to the adjacent gas phase in most of the concentration range where a great deal of the lowering in surface tension occurs, something that evidently disagrees with the very basis of the Langmuir–Szyszkowski model though it is in fair agreement with recent VSFS measurements [39,58]. In the following we further consider these fascinating observations in more detail, leaving behind the out-dated original Langmuir–Szyszkowski conceptions. To get the proper background for doing so, however, we include a recapitulation of the Langmuir–Szyszkowski model following in principle the Hill treatment of Langmuir adsorption of a gas onto a solid surface [59].

Let us consider the interface between a surfactant solution and a gas phase. We adopt a monolayer model and assume that there is mixing on the length scale of a surfactant molecule. By invoking a (hexagonal) surface lattice with mesh size, a_2 (corresponding to the surface area occupied by one surfactant molecule) and the Butler relations [60,61] we can write down the following equilibrium conditions.

$$kT \ln \theta_1 = a_2 (\gamma - \gamma_1^\circ) \tag{5.13}$$

$$kT \ln(\theta_2/\theta_2^{mic} y_2) = a_2(\gamma - \gamma_2^{mic}) \tag{5.14}$$

The above conditions (where k is the Boltzmann constant) are based upon an assumption of ideal mixing in the monolayer of surfactant molecules that constantly change place with packages of water molecules that occupy the same surface area, a_2, as a single surfactant molecule. As before, subscript "1" refers to water and subscript "2" to the surfactant component. Hence, γ_1° stands for the surface tension of pure water. Equation 5.13 also presupposes that a significant adsorption occurs already for very low surfactant concentrations in the bulk solution. As to Equation 5.14 we may note that $\theta_2 = 1 - \theta_1$ is the fraction of the monolayer surface that is covered by surfactant. The dimensionless variable y_2 used here is the surfactant concentration in the bulk solution, c_2, divided by the cmc, whereas γ_2^{mic} denotes the surface tension value recorded at the cmc, and θ_2^{mic} the fraction of the surface area that is covered by surfactant monomers at the cmc. We may also note that we tacitly assume in this context that the surface area per surfactant molecule a_2 is constant, in other words that it stays the same irrespective of the surface density of the surfactant.

Upon eliminating $a_2\gamma$ from Equations 5.13 and 5.14 we readily find the relation

$$\frac{\theta_2}{\theta_1 y_2} = \theta_2^{mic} \exp[a_2 (\gamma_1^\circ - \gamma_2^{mic})/kT] = k_a \tag{5.15}$$

from which the Langmuir isotherm immediately follows

$$\theta_2 = \frac{k_a y_2}{1 + k_a y_2} \tag{5.16}$$

where k_a is the adsorption constant. Furthermore, according to Equation 5.13 we have

$$\gamma_1^\circ - \gamma = -\frac{kT}{a_2} \ln \theta_1 \tag{5.17}$$

From this last relationship we get the surface pressure that supposedly arises due to surfactant adsorption as a consequence of ideal molecular mixing. By combining Equations 5.16 and 5.17 we finally obtain the Langmuir–Szyszkowski equation in the form

$$\gamma_1^o - \gamma = \frac{kT}{a_2} \ln(1 + k_a y_2) \qquad (5.18)$$

It includes two parameters: the surface mesh size, a_2, that is often assumed equal to the surface area occupied by a surfactant molecule close to the cmc (we shall come back to this matter), and the adsorption constant k_a. The Langmuir–Szyszkowski equation can also be written in exponential form as follows

$$\exp[a_2(\gamma - \gamma_1^o)/kT] + y_2\theta_2^{mic} \exp[a_2(\gamma - \gamma_2^{mic})/kT] = 1 \qquad (5.19)$$

This form is akin to the well-known Butler–Guggenheim relation [62] and, as Equation 5.18 also does, it accounts for the dependence of the surface tension γ on the concentration of the surfactant in the bulk solution.

It has turned out that in general one can generate quite satisfactory fits to surface tension data using Equation 5.18, particularly so for relatively high surfactant concentrations and considering the parameters involved as free parameters. However, this is no guarantee for the correctness of the underlying physical model which in the end has been proven to have a major flaw: by means of VSFS spectroscopic studies it has been experimentally established that only relatively few water molecules are actually exposed to the gas phase above the first break-point of the $\gamma(c_2)$ isotherm. Moreover, to assume ideal Langmuir behavior for a concentrated surfactant–water mixture in the interface appears unrealistic and the same can be said about the proposition that a surfactant molecule, irrespective of the packing density, always occupies the same surface area in the monolayer.

5.5 VIBRATIONAL SUN FREQUENCY SPECTROSCOPY (VSFS) PRINCIPLES: THE AIR–WATER INTERFACE

The development of techniques capable of recording the molecular events occurring at the liquid/vapor interface due to surfactant adsorption has been hampered by the general difficulties that probing this interface entails. Ideally, the preferred technique should be sensitive to chemical and structural changes and more importantly, sufficiently surface-sensitive to probe only surface molecules proper, even in the presence of a vast majority of the same molecules in the underlying solution phase. In this sense, vibrational sum frequency spectroscopy (VSFS or SFG spectroscopy) has proven to be a useful tool. VSFS is a second order, nonlinear spectroscopic technique with an exquisite surface sensitivity, which probes vibrational transitions and consequently provides a chemical fingerprint of the surface species. The intrinsic surface sensitivity of this technique arises from the second order nature of the sum frequency (SF) process, which can only occur in non-centrosymmetric media.

Molecules in the bulk of diluted surfactant solutions, as well as in a vast majority of other liquids and solids, are isotropic (centrosymmetric), and therefore SF inactive. In contrast, molecules at interfaces are distinguished by their net polar orientation (non-centrosymmetric), which allows the generation of a sum frequency signal.

In practice, the SF process involves overlapping two ultra-short laser beams at the surface, one fixed in the visible range (ω_{vis}) and the other tunable in the IR range (ω_{IR}), and detecting a third laser beam at the sum of the frequencies of the two incident fields ($\omega_{SF} = \omega_{vis} + \omega_{IR}$), which is generated in the interfacial region and carries the desired information. As schematized in Figure 5.5, when the frequency of the infrared beam approaches one of the resonant vibrations of the species present in the interface, the SF signal is resonantly enhanced giving rise to the SF spectra. Analyzing the positions of the bands in a VSFS spectrum gives direct chemical information about the species, as well as some information about the local molecular environment (hydrogen bonded, free vibrations, etc.). Moreover, varying the polarization of the different laser beams involved in the SF process, also allows the possibility of determining the orientation of specific bonds in a molecule. A detailed description of VSFS, as well as practical examples of its applicability in the study of interfaces can be found in a set of comprehensive reviews, which have appeared on the subject over the past decade [63–66]

An illustrative example that not only constitutes an important reference for the study of surfactant solutions, but also demonstrates the potential of VSFS as a surface spectroscopy tool, is the VSFS spectrum of pristine water at the liquid/vapor interface presented in Figure 5.6. The spectral range shown corresponds to the OH stretching region of water molecules (~3000 to 3800 cm^{-1}). The vibrational spectrum in this region is particularly useful for determining the local environment of the surface water molecules, because the resonant frequencies of the OH stretching modes are very sensitive to the number and strength of hydrogen bonds, red shifting (toward lower wave numbers) more than 600 cm^{-1} when going from isolated water molecules in the gas phase to tetrahedrally coordinated hydrogen bonded water molecules in ice. In the spectrum, two main features can be easily distinguished: A broad band extending for a few hundred wave numbers in the low frequency range, followed by a sharp peak centered at approximately

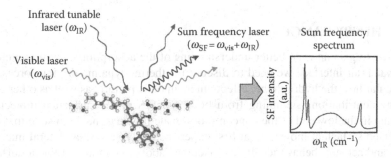

FIGURE 5.5 Schematics of the sum frequency process.

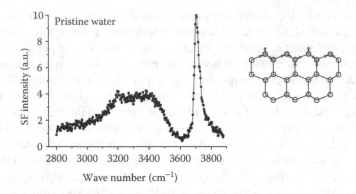

FIGURE 5.6 VSF spectra of pure water (polarization combination ssp). For reference right figure: Lateral view of an ideal basal surface of ice I_h. Note the (3D) tetrahedral arrangement of the water molecules. In the surface of liquid water the tetrahedral network is not complete and the surface structure is more disordered.

3700 cm^{-1}. This last narrow and strong peak is characteristic of the VSFS spectrum of water, and it is assigned to the stretching vibrations of water molecules with an OH bond protruding out in the gas phase and vibrating free from hydrogen bonds (free OH). In fact, only one-quarter of the water molecules present at the very top monolayer have a configuration that give rise to this intensity which demonstrates the surface sensitivity of VSFS [67]. The broader band at lower frequencies, which is usually subdivided into two contributing intensities, is associated with water molecules interacting with neighbors through hydrogen bonds of varying strength and number, which somehow support the free OH configuration. The first intensity is centered at ~3200 cm^{-1} and loosely referred to as "ice-like" because a prominent peak appears at this position in the ice spectrum (tetrahedrally coordinated water), while the second intensity is located at ~3450 cm^{-1} and called "liquid like" (associated with molecules in a more disordered local network). Overall the spectra of a liquid water surface display many features that resemble those of the surface of ice [68].

5.6 HENRY RANGE

In order to gain an even better understanding of the adsorption of a surfactant at the water–air interface we need to discuss the thermodynamic driving forces, in particular how the hydrophobic effect enters into the picture, as well as other free energy contributions stemming from both moieties of the surfactant molecule. Starting in the low end of the concentration scale it is reasonable to state that the dispersed surfactant molecules at first experience only very weak lateral interactions and actually behave ideally as discussed above, yielding the linear surface pressure expression:

$$\Pi \equiv \gamma_1^o - \gamma = bc_2 \tag{5.20}$$

where

γ_1^o is the surface tension of pure water

Π is the surface pressure

b is an adsorption constant, sometimes denoted as the Traube constant, that depends on the intrinsic properties of the surfactant [69,70]

Applying the Gibbs surface tension equation at constant temperature and using the notion of the equimolecular dividing surface, and assuming ideality (or at least a constant activity factor) we obtain at constant T,

$$\frac{1}{kT}\frac{d\Pi}{d\ln c_2} = \Gamma_2 = \frac{\Pi}{kT} = \frac{bc_2}{kT} \tag{5.21}$$

Thus, in the Henry region, both the surface density of the surfactant and the surface pressure are linear functions of the bulk concentration, c_2. Moreover, the gas-law-like expression $\Pi = kT/a$ is valid. In addition, from Equation 5.18 we also have the expression

$$\gamma_1^o - \gamma = \frac{kT\,k_a}{a_2} \times \frac{c_2}{cmc} \tag{5.22}$$

that should hold for concentrations well below the cmc. This means that the constant b which determines the rate of increasing the surface pressure upon raising the surfactant concentration, is given by

$$b = \frac{kT\,k_a}{a_2 \times cmc} \tag{5.23}$$

or, making use of Equation 5.15:

$$b = \frac{kT}{a_2 \times cmc}\,\theta_2^{mic}\,\exp[a_2(\gamma_1^o - \gamma_2^{mic})/kT] \tag{5.24}$$

Since in general, to a fairly crude approximation, we have that $\gamma_1^o - \gamma_2^{mic} \approx 35$ to $45\,mN/m$, according to Equation 5.24 we may expect the Traube constant b to vary approximately inversely with the cmc when comparing C_{10} and C_{12} Mal surfactants. In the present case the ratio of the cmc values is about 11 and the ratio of b-values about 7 (Figure 5.8). It is well known that the cmc's scale with the number of methylene groups in the hydrocarbon tail in such a way that the cmc's of nonionic surfactants diminish with about a factor 3 for each methylene group added. Hence, according to Equation 5.24 we may expect the Traube constant b to increase with a factor of about 9 upon switching from a C_{10} to a C_{12} hydrocarbon chain in conformity with the old Traube's rule [69,70].

In order to properly confront the Langmuir–Szyzskowski model with extensive sets of experimental data we may proceed as follows. We first make use of

Equation 5.24 to obtain the surface mesh area a_2 while assigning the experimental values to $\gamma_1^o - \gamma_2^{mic}$ and the Traube constant b. It is seen from Equation 5.14 that the surface area fraction θ_2^{cmc} at the cmc can be obtained from the expression

$$\theta_2^{mic} = 1 - \exp[-a_2(\gamma_1^o - \gamma_2^{mic})/kT] \tag{5.25}$$

Hence, Equation 5.24 can written in the more convenient form

$$b = \frac{kT}{a_2 \times cmc}\{\exp[a_2(\gamma_1^o - \gamma_2^{mic})/kT] - 1\} \tag{5.26}$$

From experimental values of b we can evaluate the corresponding a_2-values using the Equation 5.26. Then it is an easy matter to predict the surface tension as a function of the surfactant concentration making use of Equations 5.14 and 5.25. From Figure 5.7 it is seen that the agreement with experimental data (by necessity) is satisfactory in the Henry range (though for rather small values of the mesh size) but much less so in the isotherm ranges that we have denoted as II, III, and IV where most of surface pressure arises when the surfactant concentration is increased.

Conversely, if a_2 would be assumed equal to the molecular area as derived by applying the Gibbs surface tension equation to the data recorded just below the cmc, we would end up with predicting a much to large value for the b constant in comparison with what is found experimentally (Figure 5.8). This analysis hence substantiates our claim that the Langmuir–Szyzskowski model in its original form

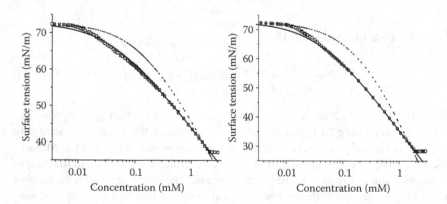

FIGURE 5.7 Langmuir-Szyszkowski fits optimized to account for the dilute behavior of the surface tension data for C_{10}-Mal (on the left-hand side) and C_{10}-Glu (on the right-hand side) using mesh sizes a_2 31.8 and 21 Å^2, respectively (dashed lines). Langmuir–Szyszkowski fitting functions chosen to account for the data at concentrations closer to the cmc using mesh sizes 47.9 Å^2 for C_{10}-Mal and 36.6 Å^2 for C_{10}-Glu (solid lines), that yield poor agreement at lower concentrations.

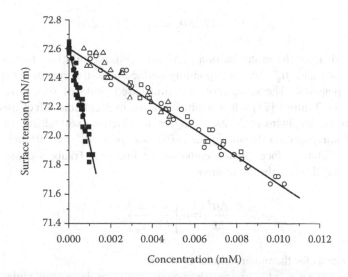

FIGURE 5.8 Surface tension data at 22°C for low concentrations of C_{10}-Mal (empty squares), C_{10}-Glu (empty circles), and C_{10}-S-Mal (empty triangles) all exhibiting the same negative slope with absolute value $b = 93 \pm 9$ mJ·m·mol⁻¹, and C_{12}-Mal (filled squares) and C_{12}-S-Mal (filled circles) both having negative slope with absolute value $b = 650 \pm 50$ mJ·m·mol⁻¹.

for the most part is an insufficient basis for unraveling the molecular mechanisms behind the surface tension lowering brought about by surfactants. It will be shown later on, however, that a Langmuir–Szyszkowski model where surface micelle formation is taken into account is capable of producing a more satisfactory agreement with the experimental surface tension data.

Our experimental observations in the Henry range made on n-decyl-β-D-maltopyranoside (C_{10}-Mal), n-decyl-β-D-glucopyranoside (C_{10}-Glu), n-decyl-β-D-thiomaltopyranoside (C_{10}-S-Mal), n-dodecyl-β-D-maltopyranoside (C_{12}-Mal), and n-dodecyl-β-D-thiomaltopyranoside (C_{12}-S-Mal) surfactant solutions indicate that for the C_{10}-surfactants we have a common slope, b, of 93 mJ·m·mol⁻¹ and for C_{12}-surfactants the common slope is found to be 650 mJ·m·mol⁻¹ at room temperature. Surface tension measurements at low concentrations at 22°C are displayed in Figure 5.8.

Interestingly, at room temperature the slopes, b, of $\gamma(c_2)$ for the measured systems are the same regardless of the circumstance that the head group consists of one or two glucose units, i.e., glucopyranoside and maltopyranoside, respectively. Neither does the exchange from oxygen- to sulfur-linkage between the head group and the hydrocarbon tail, i.e., exchange from maltopyranoside to thiomaltopyranoside, contribute to the free energy of adsorption in the Henry region at room temperature.

When it comes to estimating the various thermodynamic factors that influence the surfactant adsorption in the Henry range, we first recognize that within this dilute range the following expressions for the chemical potential are valid, in the interface and in the bulk, respectively.

$$\mu_2 = \mu_2^{so} + kT \ln \theta_2 = \mu_2^{bo} + kT \ln x_2 \qquad (5.27)$$

Here μ_2^{bo} denotes the mole fraction (x_2)-based (infinite dilution) bulk solution standard state and μ_2^{so} the corresponding surface area–fraction based standard chemical potential. The same mole–fraction based standard state was in fact employed by Tanford [71] when quantifying the hydrophobic effect associated with hydrocarbon chains of different lengths, to which we will adhere for convenience of interpretation. Denoting the difference $\mu_2^{bo} - \mu_2^{so}$ by $\Delta\mu^o$ and making use of the Gibbs surface tension equation we find that Traube's constant b is related to $\Delta\mu^o$ through the expression

$$\frac{\Delta\mu^o}{kT} = \ln\left[\frac{a_2}{kT} \cdot \frac{b}{v_1}\right] \qquad (5.28)$$

where v_1 stands for the molar volume of water.

Transferring e.g., a C_{12} hydrocarbon chain (with a projected molecular area of about 75 Å², due to the favored orientation of the hydrocarbon tail in the surface plane [39]) from the Tanford standard state in water solution to the corresponding hydrocarbon bulk phase, there is a free energy gain at room temperature amounting to 19.96 kT. Putting a surfactant molecule in the interfacial standard state at constant A implies eliminating a portion of the water surface, thus providing an driving force of about 13.5 kT which, however, is counterbalanced almost exactly by the work needed to generate a thin hydrocarbon film on top of the water surface [72,73]. Counteracting factors summing up to about 4.3 kT are related with residual, mostly lateral, hydrocarbon–water contacts and restricting the number of hydrocarbon chain conformations [74–78]. In this, admittedly somewhat loose, manner we can begin to understand why we are left with a net thermodynamic driving force of 15.7 kT as derived using Equation 5.28 and inserting the experimental value for b in the C_{12} case, 650 mJ·m·mol⁻¹. Comparing the results obtained for the C_{12}-surfactants with the C_{10}-surfactants, assuming a projected area of 65 Å² and by invoking a hydrophobic Tanford transfer free energy of −16.98 kT and an experimental b-value of 93 mJ·m·mol⁻¹, we find that the difference between the Tanford free energy contribution and the net thermodynamic driving force, $\Delta\mu^o$, is somewhat diminished, 3.4 kT. Hence, the difference between the two hydrocarbon tail lengths, assuming that the relatively inferior hydrocarbon tail configurational and head group transfer dependent free energy contributions are of similar proportions, is actually related to the decrease in the lateral hydrocarbon tail–surface water contacts for the shorter chain with a factor of about 0.4 kT per methylene group. In the following section, this particular hydrocarbon tail–surface water contribution emerges as the driving force for surface aggregation. Notably, the agreement between the Traube rule (and Equation 5.26) and the experimentally derived b-values is not fully satisfactory. However, the empirical Traube rule was based on measurement on substances carrying relatively short hydrocarbon tails and the Langmuir–Szyszkowski model clearly displays some severe flaws as discussed above. Moreover, the temperature coefficient of the Traube constant is related to the enthalpy of adsorption. The very slight change in b-value with temperature around room temperature (cf. Figure 5.1) indicates that the standard state free energy of adsorption, $\Delta\mu^{\circ}$, is chiefly entropically determined. This is in line with the circumstance that the hydrophobic Tanford free energy is controlling the adsorption.

Furthermore, one may also note that the model proposed by Rosen and Aronson [79,80], which admittedly relates to a different standard free energy state; assuming that saturation of the monolayer is at hand close to the cmc, does not converge with the infinite dilution standard state obtained by measurements in the Henry range (cf. Figure 5.7).

The change in free energy of adsorption upon lengthening the hydrocarbon tail by two methylenes is also notably different from some previous studies where figures on the order of -2.5 to -3 kT are presented. However, these studies suffer from either unreliable data, or from a poor theoretical framework [81]. On the other hand, data presented by Spaull and Nearn agree with respect to the magnitude of the shift in free energy of adsorption upon hydrocarbon tail lengthening by two methylenes (-2 kT) [82].

5.7 MICELLE FORMATION IN THE SURFACTANT SOLUTION–AIR INTERFACE

The extent of the adsorption at low surface densities in the Henry range is strongly correlated with the size of the surfactant molecule, or rather the size of the hydrocarbon tail. As we have discussed above, both the surface pressure and the adsorbed amount depend linearly on the bulk concentration. By analyzing the total free energy of adsorption we find that the most important contribution is the annihilation of air–water interface. Hence, a flat configuration of the hydrocarbon tail should be favored to maximize the reduction of the air–water interface.

At a certain concentration a pronounced change in slope of the $\gamma(c_2)$ curve is observed. At this point the adsorption starts increasing more rapidly. One may consider three different modes for the transition ending the Henry region. (1) Formation of a surface covering hydrocarbon film, i.e., a condensation; (2) formation of a loose network of surfactant molecules driven by replacing water–hydrocarbon contacts with hydrocarbon–hydrocarbon contacts; and (3) the formation of regions of higher surfactant density in equilibrium with a dilute surface solution phase of surfactant molecules. These three possible phases obtained after the transition schematically are shown in Figure 5.9. Both the condensation and the network model would yield regular (macroscopic) first-order transitions, whereas cluster formation in strictness would not imply a true first-order phase transition. Micelle/cluster formation in the surface has been widely discussed. Already Langmuir proposed that cluster formation may be behind the nonhorizontal behavior seen for surface pressure-molecular area isotherms at phase transitions for insoluble monolayers [83]. In later years the notion of cluster formation has been invoked for both insoluble, Langmuir [84–92], and soluble, Gibbs monolayers [93–96].

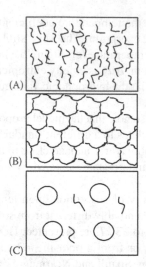

FIGURE 5.9 Three possible models of the transition that ends the Henry range. (A) The surface state after condensation of the hydrocarbon moieties. Wiggly lines correspond to hydrocarbon tails in a fluid state. The entire surface is covered and there is essentially no exposure of water to the air phase. This kind of surface phase requires a high surface densities of surfactant. (B) A network of hydrophobically bonded surfactant molecules is generated that forms a quasi-hexagonal pattern extending over the surface. For this surface phase to arise the superficial density of surfactant needed is obviously less than for case *A*. (C) Formation of flat circular surfactant micelles in the interface. Empty circles represent surface micelles (cf. Figure 5.9) and wavy lines surfactant molecules in the so-called gaseous (Henry) state.

On the basis of surface tension and VSFS measurements for sugar-based surfactants we have found that it is most likely that the transition from the Henry range actually involves formation of clusters of a limited size of surfactant molecules. Immediately after the transition from the Henry region, we note considerable differences in adsorption between C_{10}-Mal and C_{10}-Glu. As noted above, the nature of the head group does not to any measurable degree affect the adsorption in the Henry region, whereas the phase appearing after the Henry region exhibits sensitivity to the number of glucose units in the head group. The effect on the molecular analogue C_{10}-S-Mal is similar and even stronger.

The most compelling evidence for surface micelle formation relies, however, on estimating the molecular area, a, of each surfactant before and after the respective transitions. The molecular areas at the transition found for C_{10}-Mal, C_{10}-Glu, and C_{10}-S-Mal differ considerably. At room temperature the change in molecular area is from 440 to 150 Å^2 for C_{10}-Glu, from 550 to 200 Å^2 for C_{10}-Mal and from 900 to 250 Å^2 for C_{10}-S-Mal. These figures are estimated from linear fits of $\gamma(c_2)$ in the region around the transition ending the Henry range. In comparison a horizontally oriented decyl hydrocarbon tail covers around 65 Å^2 and a dodecyl chain about 75 Å^2. Apparently these data yield no firm support for the notion that a hydrocarbon chain condensation may occur. Instead, we are prompted to assume a heterogeneous film implying either formation of a loose network or formation of regions of locally higher surfactant density in equilibrium with a dilute surface phase containing isolated monomers. Although the network model cannot be discarded right away it is likely that such a structure might be unstable with respect to clustering/micelle formation. Hence, the transition actually observed will correspond to where the concentration of surface micelles gets high enough to give rise to a measurable surface pressure. Earlier we have referred to this range as the dilute surface micelle range (II). It corresponds to a rather narrow surfactant concentration interval, and is soon enough followed by the granular range (III), where a strong repulsion makes a predominant contribution to the reduction of the surface tension. The relation between the critical concentrations for formation of flat surface micelle in the interface and ordinary spherical micelles in the bulk solution is treated semiquantitatively in the appendix. Broadly speaking the csmc (critical surface micelle concentration) is two orders of magnitude below the ordinary cmc on the concentration scale.

In a preliminary manner we can account for the surface tension data recorded within the surface micelle and granular ranges as follows. We imagine flat circular surface micelles to form with an average radius of the hydrocarbon core of about the extended hydrocarbon chain length, i.e., about 16 Å for the C_{12} case and approximately 13.5 Å for the C_{10}-tail. The sugar head groups are supposedly located below the hydrocarbon core and in the rim part below and outside the hydrocarbon core. A typical aggregation number compatible with such a simplistic model is 10. We may note, however, that a rather large range of aggregation numbers of this order actually yield acceptable predictions of the surface tension data around the transition.

As a result of surface micelle formation, lateral hydrocarbon–water contacts are replaced by hydrocarbon–hydrocarbon contacts, in analogy with what we have as the main driving force for formation of ordinary surfactant micelles in the bulk solution. To have equality between the chemical potentials of the surfactant in monomer form in the bulk solution, μ_2, and in the form of surface micelles we apply the following condition that is based upon the Hill theory of Langmuir adsorption (subscript "smic" stands for "surface micelle"):

$$N_{smic}\mu_2 = G_{smic}^{\circ} + kT \ln\left(\frac{\theta_{smic}}{1-\theta_{smic}}\right) \qquad (5.29)$$

In other words, we imagine a surface lattice with sites that can harbor a single surface micelle, and estimate the configurational surface entropy contribution on this very basis. N_{smic} is the aggregation number and G_{smic}° the relevant standard state free energy of the aggregate. This expression is by the way in complete harmony with the corresponding expression for ordinary micelle formation, which can be written as follows:

$$G_{mic}^{\circ} - N_{mic}\mu_2 + kT \ln\phi_{mic} = N_{mic}\varepsilon_{mic} + kT \ln\phi_{mic} = 0 \qquad (5.30)$$

where ϕ_{mic} stands for the volume fraction of micelles. At the cmc, the omega potential, ε_{mic}, per surfactant molecule as a rule amounts to about 0.2 kT. In the surface, however, we need to take due account of the eigen-area of the micelle aggregates. Hence Equation 5.29 is in a more appropriate form than Equation 5.30 that only applies to a solution, which is dilute in micelles.

Furthermore, from Equation 5.29 we obtain

$$N_{smic}(\mu_2^{bo} + kT \ln c_2) = G_{smic}^{o} + kT \ln\left(\frac{\theta_{smic}}{1-\theta_{smic}}\right) \tag{5.31}$$

It follows that

$$kT \ln c_2^{N_{smic}} = G_{smic}^{o} - N\mu_2^{bo} + kT \ln\left(\frac{\theta_{smic}}{1-\theta_{smic}}\right)$$

$$= -kT \ln K_{smic} + kT \ln\left(\frac{\theta_{smic}}{1-\theta_{smic}}\right) \tag{5.32}$$

or

$$K_{smic} c_2^{N_{smic}} = \left(\frac{\theta_{smic}}{1-\theta_{smic}}\right) \tag{5.33}$$

which in turn yields

$$\theta_{smic} = \frac{K_{smic} c_2^{N_{smic}}}{1 + K_{smic} c_2^{N_{smic}}} \tag{5.34}$$

Simultaneously, however, for the water component (in packages corresponding to a surface micelle) we have

$$f_1^{s} - \mu_1 = a_{smic}\gamma$$
$$f_1^{so} - \mu_1^{o} = a_{smic}\gamma_1^{o} \tag{5.35}$$

From which by assuming $\mu_1 \approx \mu_1^{o}$ we get

$$kT \ln \theta_1 = a_{smic}(\gamma - \gamma_1^{o}) \tag{5.36}$$

which, using Equation 5.34, yields the Langmuir–Szyszkowski equation in a modified form

$$\gamma_1^{o} - \gamma = -\frac{kT}{a_{smic}} \ln(1 - \theta_{smic}) = \frac{kT}{a_{smic}} \ln(1 + K_{smic} c_2^{N_{smic}}) \tag{5.37}$$

that takes into account the formation of surface micelles. Apparently, the parameters involved here are the surface area covered by a surface micelle, a_{smic}, and the

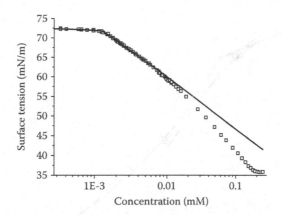

FIGURE 5.10 Application of Equation 5.37 to fit surface tension data for C_{12}-Mal in the surface micelle (II) and granular (III) ranges using a micelle aggregation number of 10 and a surface micelle size, a_{smic}, of 710 Å^2. Notably the fit obtained is satisfactory up to the end of the granular range, at about 63 mN/m.

equilibrium constant K_{smic}. Note that according to Equation 5.34, the surfactant coverage increases very rapidly around the concentration where $K_{smic}c_2^{N_{smic}} = 1$ which means that the constant K_{smic} can be estimated from the location of the transition that marks the end of the Henry region (cf. Figure 5.1).

In Figure 5.10 we have applied Equation 5.37 to C_{12}-Mal for the sake of simplicity assuming a Henry range type of ideal surface tension lowering up to the transition where the surface micelle mechanism takes over. It is seen that the agreement with the recorded surface tension data is quite satisfactory all the way up to the next transition, the one between the granular surface state and the liquid-expanded phase. This latter transition is thus bound to involve the disappearance of the water rims of the compressed surface micelles. Evidently, it marks the point where a laterally homogenous, surfactant-covered surface arises.

It is also noteworthy that in contrast to the original Langmuir–Szyszkowski equation, the modified Equation 5.37 in principle accounts for the rapid change of slope of the surface tension function toward the end of the Henry region for C_{12}-Mal as well as for the very high surface elasticity observed prior to the final transition to the liquid-expanded phase [97]. However, in its simplistic form so far used the model fails to correctly capture the adsorption as it predicts an unrealistically rapid increase of the surface density at the transition from the Henry range, which is particularly evident for C_{10}-Mal (cf. Figure 5.11). Previously we have attempted to devise a model description of the granular range on the basis of the hard-disk expression, with limited success, however [39]. This makes sense, of course, as we can envisage surface micelles to be rather easily deformed and that a hard-disk model might exaggerate the rise of the chemical potential resulting from concentrating the micelles. The hard-disk expression does not give a correct prediction of the very sharp surface tension decrease at almost constant surface density in the

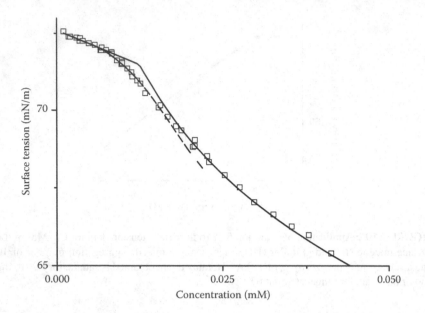

FIGURE 5.11 Surface tension data for C10-Mal at 22°C (empty squares) fitted with a hard-disk equation of state for a hard-disk micelle of size 610 Å^2 and an aggregation number equal to 10 (dashed line). A fit combining the Henry range behavior, up until approx. 0.015 mM, and Equation 5.37 at higher concentrations using a surface micelle size of 780 Å^2 and an aggregation number equal to 10 (solid line).

granular range, whereas the Langmuir–Szyszkowski model essentially captures the observed saturation behavior. However, the predictions obtained by applying the hard-disk equation for mixed states in the dilute surface micellar range (range II in Figure 5.2) are in better agreement with the data than the current Langmuir–Szyszkowski model. Hence, the true interfacial behavior in the surface micellar and granular range is, hardly surprisingly, not correctly captured neither by the hard disk nor the modified Langmuir–Szyszkowski model. Nevertheless, the models do yield complementary descriptions of the interactions between the surface micelles and grains (cf. Figure 5.11). It is worth noting, however, that so far the complicated contributions due to surface micelle rim stretching upon increasing the surface micelle area fraction in the range close to 1 are left out of consideration.

Granular ranges are observed for a considerable concentration interval in the isotherms for C_{10}-Mal, C_{10}-Glu, and the 1:1 mixed solution, corresponding to chemical potential increase on the order of 1.5, 0.7, and 0.8 kT, respectively [98]. Concurrently, the change in molecular surface area for these surfactants in the entire granular range is as low as 2–3 Å^2. Surface tension data highlighting the granular range fitted with a linear function $\gamma \sim \ln c$ representing surface tension lowering at a constant molecular area of 79.5 Å^2 are displayed in Figure 5.12. Interestingly, the change in surface tension with bulk fraction of C_{10}-Mal in a

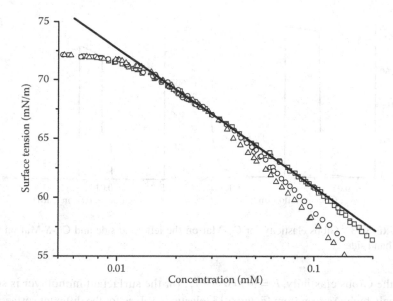

FIGURE 5.12 Surface tension data for C_{10}-Mal (squares), C_{10}-Glu (triangles) and the 1:1 mixed solution (circles) at 22°C fitted with a function representing surface tension lowering at a constant molecular area equal to 79.5 Å².

mixture of C_{10}-Mal and C_{10}-Glu in the granular range is nil, and consequently the contribution arising from an exchange of the head group from Mal to Glu, or vice versa, in the grains is also nil. The chief difference seen between C_{10}-Mal and C_{10}-Glu is related to the stability of the granular phase we find that the absence of the second glucose unit seems to diminish the concentration, and surface pressure, at which the grain, or surface micelle, rims are annihilated. This difference is thought to stem from the difference in head group size. The accommodation of a smaller head group in a fully developed liquid-expanded film is obviously less costly energetically. A discussion on the entropic contributions to the stability of the granular phase is deferred until later in the chapter.

For C_{10}-S-Mal a critical point can be observed at a concentration of 7 μM, where the molecular area changes from 200 Å² to approximately 80 Å² [46] (cf. Figure 5.24). Attractive forces between surface micelles might compensate for the free energy of stretching the micelle rim against the line tension to accommodate the micelles into the granular phase. Attractive forces are observed at head group–head group contact between macroscopic hydrophobic surfaces immersed in sugar-based surfactant solutions [99,100]. Phase transitions for sugar-based lipid monolayers have also been attributed to attractive forces between the head groups [101]. However, attractive forces between the surface micelles are difficult to account for due to the complex interplay with stretching of micelle rims and concurrent shape fluctuations.

The surface tension decrease being nearly linear with respect to the chemical potential change, i.e., the extent of adsorption being largely constant also indicates

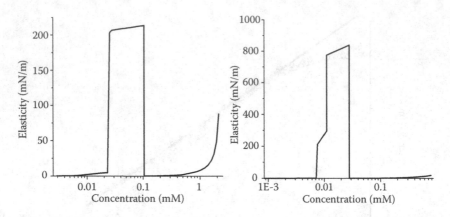

FIGURE 5.13   Gibbs elasticity for C_{10}-Mal on the left-hand side and C_{10}-S-Mal on the right-hand side.

that the Gibbs elasticity, $E = \left(\partial\gamma/\partial\ln a\right)_T$, for the surfactant monolayer is surprisingly high. As seen from Figure 5.13, elasticity values for the different surfactant solutions in the granular region at room temperature range between 200 and 800 mN/m, i.e. about the same as for a liquid-condensed monolayers [102].

Early on the liquid condensed type of film was considered by Adam [103] and Langmuir [83] to be a semisolid film. Notably, the elasticity increases somewhat with increasing chemical potential of the surfactant in the granular ranges as presented below, which can be understood by contemplating the elasticity derived on the basis of the modified Langmuir–Szyszkowski model:

$$E = \frac{kT}{a}\frac{1}{(a-a_{\text{smic}})} \tag{5.38}$$

assuming that the eigen-area, a_{smic}, is constant. The area/molecule in the granular range for C_{10}-Glu and C_{10}-Mal at room temperature is found to be around 79 Å^2, corresponding to a surface density of around 2.1 μmol/m^2 and for C_{10}-S-Mal the corresponding molecular area is approximately 70 Å^2, corresponding to a surface density of around 2.35 μmol/m^2. C_{10}-S-Mal does exhibit a short concentration interval between 7 and 10 μM in which the molecular area is almost constant around 80 Å^2. In addition, granular surface saturation is also observed for n-dodecyl-β-D-maltopyranoside (C_{12}-Mal), where the area/molecule is found to be around 70 Å^2. Interestingly, the densities are of similar magnitude despite the increase in hydrocarbon tail length. Morphologically similar phases have been reported for partially fluorinated alkanoic acids [104,105] and similarly for partially fluorinated alkanes [106].

5.8   VSFS ON MALTOPYRANOSIDE AND EO-BASED SURFACTANT SOLUTIONS

Upon addition of a surface-active compound, the surface water structure is seen to be distinctly perturbed. However, with the current instrument limitations, surface densities of at least 500 Å^2 per molecule are generally required to observe a change in the spectra of pure water, which makes VSFS somehow less sensitive than carefully performed surface tension or ellipsometric measurements. The presence of the surfactant is manifested by the appearance of vibrational modes characteristic of the surfactant molecules (i.e., CH stretches between 2700 and 3000 cm^{-1}), as well as by a decrease in the intensity of the characteristic spectral features of pure water, in particular the sharp free OH peak centered at ~3700 cm^{-1}. The spectrum of n-dodecyl-β-D-maltopyranoside, C$_{12}$-Mal, at 0.9 μM is shown as an example in Figure 5.14, which corresponds to the concentration regime in the initial stages of the dilute micellar range, where the average molecular area is around 300 Å^2. Though reduced, the fact that the sharp free OH peak can be observed in the spectra indicates that patches of unperturbed water are still present in the surface. This band is not seen to disappear until molecular areas of approximately 80 Å^2 for both C$_{10}$-Mal and C$_{12}$-Mal are reached [39]. These findings provide direct evidence that the surfactant molecules do not form a properly surface-covering phase until reaching fairly high surface densities, a statement which is consistent with the formation of surface micelles as the disappearance of the free OH band can be correlated the increase in adsorption of surface micelles, see Figure 5.15.

Notably, a decrease in intensity is observed in Figure 5.14 in the bonded OH region (3000–3500 cm^{-1}), particularly in comparison to the pure water spectrum. The reduced intensity in this interval (and this concentration range) does not imply a lack of water molecules hydrating the sugar head group, rather that the water

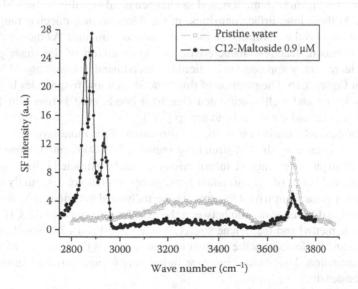

FIGURE 5.14 VSF spectra of 0.9 μM solution of n-dodecyl-β-D-maltopyranoside (C$_{12}$-Mal). The spectrum of water is included for reference. The concentration shown corresponds to the range where patches of unperturbed surface water are still observed (dilute surface micellar range, $a \approx 300$ Å^2). Polarization combination ssp.

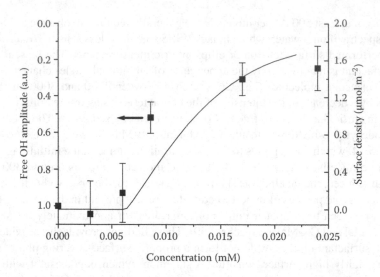

FIGURE 5.15 VSFS free OH-amplitude in relation to the surface micelle adsorption. Filled squares represent the free OH amplitude as a function of C_{10}-Mal bulk concentration (left-hand side y-scale, presented in reverse order). The solid line corresponds to the calculated adsorption of surface micelles, obtained by deducting the free monomer density from the total surface density (right-hand side y-scale). (From Kumpulainen, A.J., Persson, C.M., Eriksson, J.C., Tyrode, E.C., and Johnsson, C.M. *Langmuir*, 21, 305, 2005. With permission.)

molecules have no net orientation, and as such are not detectable in the VSF spectra [58]. At these low surface densities, in the dilute surface micellar range, the dominant effect is the disruption of the water surface structure. As the concentration is increased, features characteristic of the hydration of the sugar groups become increasingly apparent, in particular a broad band centered at $\sim 3150\,cm^{-1}$ shown in Figure 5.16. The position of this band at such low frequencies is indicative that strong and well-coordinated (ice like) bonds exist between hydrating water molecules and the sugar head group [107].

Additional information about the conformation of the monolayer can also be obtained by examining the CH stretching region ($2700-3000\,cm^{-1}$). However, in the case of sugar surfactants the information that can be extracted is limited, since the vibrational bands of the surfactant head group and tail are essentially convoluted, and it is far from trivial to separate the individual contributions. Nonetheless, from the difference in ratio between the two sharpest peaks in the CH region (symmetric methyl and methylene stretches at at ~ 2850 and $\sim 2880\,cm^{-1}$, respectively) it can be inferred that the number of gauche defects decreases with increasing concentration. The obvious increase in intensity is mainly due to an increase in surface density.

EO-based surfactants are the most common class of nonionic surfactants and it is instructive to compare their behavior with sugar surfactants at the liquid/air interface. Similarly to the behavior described previously for surfactants with a

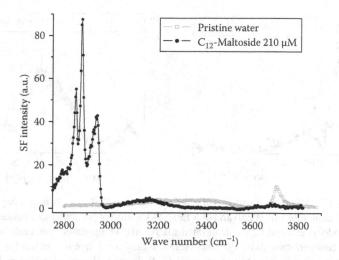

FIGURE 5.16 VSF spectra of 210 μM solution of n-dodecyl-β-D-maltopyranoside (C$_{12}$-Mal). The spectrum of water is also included for reference. The concentration shown is above the cmc. Note the OH stretching band at ~3150 cm^{-1} which corresponds to the water molecules strongly bound to the sugar head groups. Polarization combination ssp.

maltoside head group, a sharp "free OH" peaks is observed until relatively high surface densities indicating that patches of unperturbed surface water are present on the surface. In the particular case of penta(ethylene oxide) n-dodecyl ether (C$_{12}$-E$_5$), these patches disappear when surface densities around 65 Å^2 are reached. Actually the disappearance of the free OH peak at ~65 Å^2 coincides with a rapid change in the orientation of surfactant tails, which adopt a more upright configuration as the surface covering liquid-expanded layer is formed. Increasing the concentration also promotes a reduction in the number of gauche defects in the monolayer. However, a significant number of gauche defects remain even above the cmc. The observed results are consistent with the notion of formation of surface micelles of limited size with a flat orientation of the surfactants at low surface concentrations [39].

Moreover, a band common to both types of surfactants and associated with water molecules vibrating in close proximity to the hydrocarbon tails has also been reported at around ~3600 cm^{-1} in spectra recorded using different sets of polarization combinations (sps and ppp) [107].

The VSF spectra of EO-based surfactants also exhibit some clear differences compared to sugar-based counterparts. In contrast to the sugar surfactants, where the water structure is seen to be first disrupted by the adsorbing molecules (cf. Figure 5.4) the EO chains actually tend to enhance the structuring of water at low surface densities. This behavior is explained by the formation of a complex cage structure of water molecules with a preferred orientation around the

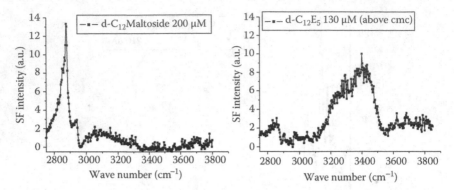

FIGURE 5.17 VSF spectra of a $200\,\mu M$ solution of $CD_3(CD_2)_{11}$-β-D-maltopyranoside (d-C_{12}-Mal, deuterated alkyl tail) and of a $130\,\mu M$ solution of perdeuterated penta(ethylene oxide) n-dodecyl ether (d-$C_{12}E_5$, deuterated alkyl tail). The solution concentration correspond to a concentration above the cmc. Note the only weak features stemming from the head group are observed in the spectra for d-$C_{12}E_5$ in the CH stretching region. Polarization combination ssp.

EO head group. The structured shell around the EO groups is, of course, not static and should be considered as a dynamic cage formation, which is constantly forming and rearranging [58]. Moreover, striking differences are observed with respect to preferred orientation of the surfactant head groups. Experiments using EO and maltoside surfactants carrying perdeuterated alkyl tails specifically allow targeting conformational changes in the head group, as the heavier mass of the deuterium isotope displaces the stretching vibrations of the methylene and methyl modes of the surfactant tail to considerably lower frequencies ($2050–2250\,cm^{-1}$). In the spectra presented in Figure 5.17 for surfactant solutions of d-C_{12}-E_5 and d-C_{12}-Mal the differences are patent. The interpretation of the lack of signal in the CH region for the EO surfactant is that the poly-EO head group is randomly oriented at the liquid–air interface, though the water molecules surrounding this polar group are in fact significantly oriented [48]. In contrast, the stiffer glucose units of the maltopyranoside head group display a preferred orientation along the whole concentration range to the cmc, while ordered water molecules hydrating the head group are only observed at concentrations close to the cmc.

5.9 FULLY DEVELOPED LIQUID-EXPANDED RANGE

LE monolayer state is often supposed to arise already at the transition that ends the Henry region and to prevail all the way up to the cmc and above, unless there are intermolecular forces favoring the formation of a crystalline monolayer at the temperature in question. Thus, this range might entail strongly stretched monolayer states as well as states with compressed hydrocarbon films in the vicinity of the cmc. In our opinion, however, such a definition of the LE surface state is somewhat inappropriate, at least for carbohydrate-based surfactants, as the special properties of the micelle and granular ranges tend be left out of consideration as a result of such a categorization. Thus, we wish to reserve the term "liquid-expanded" surface phase for a monolayer state that is distinguished by having a thin but essentially coherent surface-covering hydrocarbon film part with liquid-like properties.

Next, we discuss the LE surface phase in more specific terms. We recognize that the LE phase usually forms as a result of a transition from the granular range by which the water rims suddenly disappear that at first surround the compacted surface micelles. The molecular area varies from about 65 Å^2 for both C_{10} as well as C_{12} sugar surfactants down to 37 Å^2 for Glu (C_{10}) and 46–48 Å^2 for Mal and S-Mal (C_{10} and C_{12}) at the cmc (and beyond). Surface tension data for C_{10}-Mal, C_{10}-Glu, C_{10}-S-Mal, C_{12}-Mal, and C_{12}-S-Mal is presented in Figure 5.18.

Obviously, the thiomaltosides have lower cmc values than the maltoside analogues. This is usually interpreted in terms of a greater hydrophobicity of the

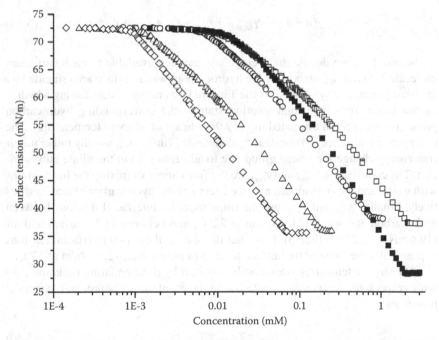

FIGURE 5.18 Surface tension data at 22°C for C_{10}-Mal (empty squares), C_{10}-Glu (filled squares), C_{10}-S-Mal (empty circles), C_{12}-Mal (empty triangles) and C_{12}-S-Mal (empty diamonds). Note that the surface tension level at the cmc, corresponding to the work of forming the interface from the solution, is much the same for the C_{10}- and C_{12}-Mal and S-Mal surfactant pairs. The main reason is that the chemical potential at the cmc is practically the same for the C_{10} and C_{12} pairs as the difference in monomer concentration is counterbalanced by the difference in standard state free energy.

thiomaltoside molecule. However, this is erroneous. As the slopes in the Henry range are identical (cf. Figure 5.8) the thiomaltoside molecules are in fact equally hydrophobic as the maltoside molecules. Hence, the cmc difference observed stems from better packing conditions in the micelles of the molecules carrying thiomaltoside head groups. This is probably related to the higher polarizability of the sulfur linkage as compared to the oxygen linkage, which makes coping with the amphiphilic environment in a micelle and in the LE interface less energetically costly.

The tensile properties of the monolayer may be derived by separately considering the hydrocarbon film part and the polar head group layer where head groups and water mix. Accordingly, the free energy of the monolayer can be obtained approximately as a sum of the contributions from the headgroups and the hydrocarbon tails, respectively.

Further, the monolayer is open to a huge reservoir kept at the same temperature which means that at equilibrium, the chemical potentials of its components are the same as in the bulk solution. For the excess free energy, or more precisely, the grand Ω-potential per surfactant molecule, we can write [108]

$$\frac{\Omega}{n_2^s} = \varepsilon = \gamma_{hc,w}(a - a_{pg}) + \gamma_{hc,air}a + \varepsilon_{Tan} + \varepsilon_{pg} + \varepsilon_{mix} + \varepsilon_{conf} = \gamma a \qquad (5.39)$$

As before, by a we denote the average surface area available to each surfactant molecule, whereas a_{pg} stands for the hydrocarbon–water surface area shielded by the head group. The symbol ε_{Tan} is the Tanford free energy of transferring a hydrocarbon chain from the actual solution state to the corresponding hydrocarbon phase (for a C_{12} chain estimated to $-(19.96 + \ln x_2)\, kT$ at room temperature). The polar group (subscript pg) constant ε_{pg} denotes the (intrinsic), usually rather minor free energy change for a head group due to adsorption from the dilute bulk solution. The contributions ε_{mix} and ε_{conf} are the free energy of mixing the head groups with water, and the conformational free energy of the hydrocarbon chains, respectively. Finally, $\gamma_{hc,w}$ and $\gamma_{hc,air}$ are the (macroscopic) interfacial tensions between hydrocarbon and water (51.0 mN/m at 22°C) and between hydrocarbon and air (19.6 mN/m at 22°C) [72,73]. Note that the sum of these two interfacial tensions is practically the same as the surface tension of pure water, 72.5 mN/m at 22°C.

The surface tension is most readily obtained by differentiating Equation 5.39 with respect to the area/molecule, a, at constant temperature and chemical potentials

$$\gamma = \gamma_{hc,w} + \gamma_{hc,air} + \gamma_{mix} + \gamma_{conf} \qquad (5.40)$$

This means that to a good approximation, just γ_{mix} and γ_{conf} actually contribute to the surface pressure $\gamma_1^\circ - \gamma$. Note that in this approximation there are no contributions pertaining to ε_{Tan} and ε_{pg} as they are supposed to be independent of the surface density of surfactant. It appears from Equation 5.39 that the surface tension can also be quantified by estimating all the various components of ε/a. However, insofar as the Gibbs surface tension equation is fulfilled the end result must necessarily be the same as when we instead take derivatives of ε with respect to the molecular area, a [108].

Fortunately, for decyl and dodecyl surfactants practically all the information needed is at hand enabling us to account for the surface tension lowering that arises partly in the hydrocarbon film and partly in the head group layer. For the head group–water mixing it has proven rather successful to employ the attractively simple hard-disk model [109–113] (validated by computer simulations [54,55]),

$$\frac{\varepsilon_{mix}}{kT} = \ln\left(\frac{\theta^{hd}}{\theta_1}\right) + \frac{\theta^{hd}}{\theta_1} + \text{const.} \tag{5.41}$$

where the water component is considered as a structureless medium and where θ^{hd} is the area fraction covered by hard disks and $\theta_1 = 1 - \theta^{hd}$ is the area fraction of exposed water surface. Here we insert $\theta^{hd} = a_{pg}/a = \theta_2$. The resulting surface pressure generated by the hard disks is simply,

$$\Pi_{mix} = \frac{kT}{a}\frac{1}{(1-\theta_2)^2} \tag{5.42}$$

It develops in the interfacial water to keep the water chemical potential the same as in the dilute solution phase in spite of the presence of the head groups. Evidently, when θ_2 is no longer small, the quadratic factor in the denominator causes the surface pressure to rise much quicker than for the dilute ideal case. The reason is, of course, that due to the surface area occupied by the disks themselves, the lateral mixing will be restricted, in fact even more so than for the Langmuir adsorption case.

As to the hydrocarbon chain contribution to the surface pressure we have gathered some experience from utilizing the results of Gruen's mean-field single-chain theory, first presented in 1984 [74] that was later on verified by Ben-Shaul et al. [76,77]. Accordingly, for a C_{12} chain in a planar bilayer, the configurational free energy passes through a minimum for a packing density close to 48 Å^2 implying that (disregarding surface effects) a thin hydrocarbon monolayer film is subject to a lateral tension related with the hydrocarbon chain packing for lesser packing densities and, conversely, to a lateral pressure for higher packing densities. In Figure 5.19 we show the calculated Gruen γ_{conf}-function for C_{12} chains that have been employed, and in the same figures we have also plotted the corresponding functions for C_{10} chains that could be inferred from analyzing surface tension data for mixed Mal and Glu surfactant solutions [114].

Finally, in Figure 5.20 we present plots of the experimentally and theoretically derived surface pressure versus molecular area functions. The agreement is seen to be rather satisfactory and it is noteworthy that there is just one adjustable parameter involved: the head group area fraction θ_{pg}, or the head group size a_{pg}. At higher packing densities, however, the approximations invoked become untenable, in the first place presumably the assumption that ε_{pg} is independent of the packing density. The repulsive (dehydrating) contacts among the more crowded head groups are then anticipated to gradually become more and more significant as the

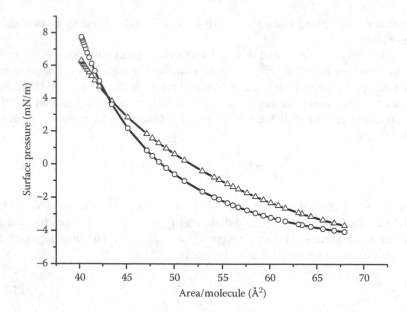

FIGURE 5.19 Calculated configurational surface pressure for a dodecyl tail, $-\gamma_{conf}(C_{12})$, represented by the solid line with empty circles and the experimentally deduced surface pressure of the decyl tail, $-\gamma_{conf}(C_{10})$, solid line with triangles. (From Kumpulainen, A.J., Persson, C.M., and Eriksson, J.C., *Langmuir*, 20, 10935, 2004. With permission.)

surface density is further raised. The surface pressure of C_{12}-Mal, C_{10}-Mal, C_{10}-Glu fitted with the hard-disk pressures and the corresponding hydrocarbon tail configurational pressures are displayed below.

The head group-related spread seen for the surface tension curves of the various surfactants studied (cf. Figures 5.18 and 5.21) is, according to the above treatment, largely connected with the terms ε_{mix} and $-\gamma_{hw}a_{pg}$ of Equation 5.39, out of which the first one becomes the more predominant at high surface densities. According to Equation 5.42, a given slope of the surface tension curve (i.e., a given a) will be realized at a higher surface tension value, when a_{pg} is small, relatively speaking.

5.10 MIXED SOLUTIONS OF SUGAR-BASED SURFACTANTS

For mixed solutions of nonionic surfactants the total adsorbed amount, Γ_2, is obtained as readily as for a pure surfactant:

$$\frac{1}{kT}\left(\frac{\partial \gamma}{\partial \ln c_2}\right)_{T,x_M,x_G} = -\frac{n_2^s}{A} = -\Gamma_G - \Gamma_M = -\Gamma_2 \qquad (5.43)$$

where M and G denote the two components and x_M and x_G are the bulk mol fractions of the nonionic surfactants with $x_M + x_G = 1$ fulfilled. Obviously, by applying

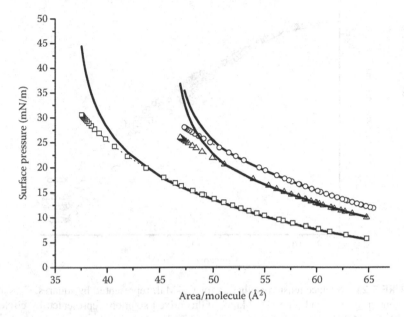

FIGURE 5.20 Surface pressure for C_{10}-Glu, represented by the solid line with the lowest surface pressure, fitted with a sum of a hard-disk size of 11.8 Å^2 and $-\gamma_{conf}(C_{10})$, C_{12}-Mal, represented by the solid line with the second highest surface pressure fitted with a combination of two hard disks states with sizes of 32 and 14 Å^2 and $-\gamma_{conf}(C_{12})$ and a deduction of 1.7 mN/m to account for the difference in macroscopic surface tension between C_{10} and C_{12} hydrocarbon chain phases, and C_{10}-Mal, represented by the solid line with the highest surface pressure fitted with a sum of two hard disks with sizes of 32 and 14 Å^2 and $-\gamma_{conf}(C_{10})$. The fitting functions are represented by the dotted lines with symbols.

Equation 5.43 we cannot separate the adsorption of M or G, what we obtain is an inseparable sum of the two components. This turns out to be one of the key problems when measuring on impure surfactants, since one of the components might be strongly surface-active and consequently might achieve high surface densities, even though the bulk fraction of the component may escape detection. Thus, it is of paramount importance that the surfactant sample is really very pure. Measurements on contaminated samples purified by various techniques [35,115] have repeatedly proven the deleterious effect of small amounts of impurities. Techniques for surfactant purification include foam fractionation [116,117] and the high performance surfactant purification apparatus [118].

In order to obtain the adsorbed amount of each of the components we need to measure the change in surface tension upon changing the bulk fraction of the surfactants [119] as we have

$$\left(\frac{\partial \gamma}{\partial x_M}\right)_{T,c_2} = -kT\Gamma_2 \left(\frac{x_M^s}{x_M} - \frac{x_G^s}{x_G}\right) \qquad (5.44)$$

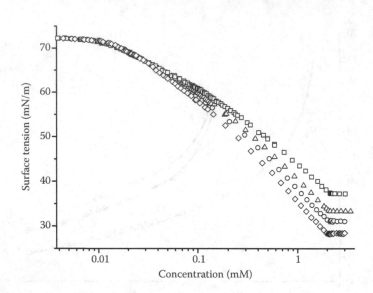

FIGURE 5.21 Surface tension isotherms for C_{10}-Mal, represented by squares, C_{10}-Glu, represented by diamonds, 1:1 (C_{10}-Mal:C_{10}-Glu) mixed solution, represented by circles, and 4:1 (C_{10}-Mal:C_{10}-Glu) mixture, represented by triangles.

where x_M^s and x_G^s are the surface fractions of one of the components M and G. Thus, by measuring the change in surface tension due to varying the concentration at constant bulk mole fractions we obtain the adsorbed amount. Then by measuring the change in surface tension due to changing the bulk mole fraction we obtain the surface fractions of the two components.

Surface tension data for mixed solutions of C_{10}-Mal and C_{10}-Glu were recorded by Persson et al. [120]. Isotherms included were: C_{10}-Mal, 90%, 80%, 65%, 50%, and 35% C_{10}-Mal and C_{10}-Glu at 22°C. Surface tension isotherms of C_{10}-Mal, C_{10}-Glu, 80% and 50% C_{10}-Mal are presented in Figure 5.21, the other isotherms are omitted for clarity. By applying the surface thermodynamic relations Equations 5.43 and 5.44 we obtain the surface density of each component. The adsorption of each component in the 1:1 mixed solution is presented in Figure 5.22, which indicates an interesting behavior. At low concentrations the larger head group of Mal is somewhat favored (not clearly visible in Figure 5.22), but with increasing adsorption Mal becomes less favored and finally it is slowly expelled from the interface. Notably, the two surfactants carry the same decyl-hydrocarbon tails, and consequently the state of the hydrocarbon tail film is the same at equal surface density.

By applying the derivative with respect to the surface fraction of one of the components of the molecular free energy for a mixture of hard disks and dropping the subscripts, $x_M^s = x^s$, we obtain the following x^s (a) expression [114],

$$0 = C + \ln\left[x^s/(1-x^s) \right] + \frac{\Delta a^{hd}(2a - a_1^{hd} - x^s \Delta a^{hd})}{(a - a_1^{hd} - x^s \Delta a^{hd})^2} \qquad (5.45)$$

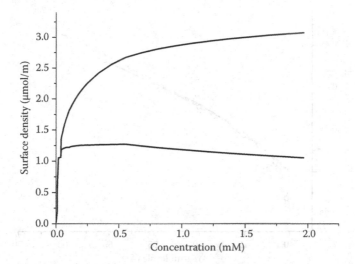

FIGURE 5.22 Component adsorption of C_{10}-Mal and C_{10}-Glu for the 1:1 mixed solution case. C_{10}-Glu displays clearly the higher extent of adsorption.

where Δa^{hd} is the size difference between hard disks M and G and C is a constant. C is well estimated by $-\gamma_{hw}\Delta a^{hd}/kT$ indicating the preference of adsorbing the larger disk at large molecular areas, due to the stronger shielding of direct hydrocarbon–water contact. At higher surface densities the larger disk gives rise to greater repulsion than the smaller one, whereby adsorption of the smaller disk becomes more favorable, which is in agreement with the experimental findings. Hence, we can use Equation 5.45 to predict $x^s(a)$ and compare with the thermodynamically derived surface fractions obtained by applying Equations 5.43 and 5.44 to the surface tension data of the mixed solutions. Using hard-disk sizes of 22.9 Å² to represent the Mal head group and 11.3 Å² for the Glu head group we achieve a fit accurate to within 1% from 62 to 42 Å² (the molecular area at the cmc is 40 Å²), which is presented in Figure 5.23.

The hard-disk model fails to predict the very rapid decline in Mal surface fraction in the final 2 Å² towards the cmc, indicating that the short-range forces are operating between the head groups at these molecular areas which are not accounted for by the hard-disk model. However, since the surface fraction can be obtained with some accuracy, then the surface pressure increase due to the head groups in the region between 63 to around 42 Å² is well accounted for by the hard-disk equation of state for a mixture of hard disks

$$-\frac{\gamma^{hd}}{kT} = \frac{a}{(a-\hat{a})^2} = \frac{1}{a}\frac{1}{(1-\theta_{hd})^2}$$
(5.46)

where $\hat{a} = x_M^s a_M^{hd} + x_G^s a_G^{hd}$ is the average hard-disk area. This proves to be a fruitful mean-field ansatz [114,121,122].

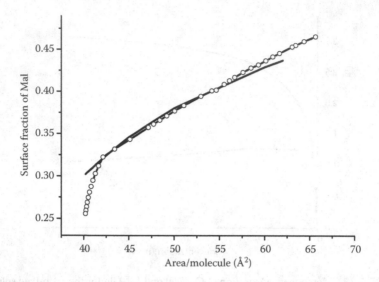

FIGURE 5.23 The surface fraction of Mal head groups in a 1:1 mixed solution represented by the solid line with empty circles and theoretical values using the hard-disk
model with hard-disk sizes 22.9 and 11.3 $Å^2$ to account for the Mal and Glu head groups,
respectively, solid line. (From Kumpulainen, A.J., Persson, C.M., and Eriksson, J.C.,
Langmuir, 20, 10935, 2004. With permission.)

5.11   ADSORPTION ISOTHERMS

The adsorption isotherms of sugar-based surfactants which can be deduced from
the corresponding surface tension isotherms by differentiation display rich and
complex phase behavior. Isotherms for C_{10}-Mal, C_{10}-Glu, and C_{10}-S-Mal in the
range from 0 to 0.2 mM are displayed in Figure 5.24. Note, in particular, the presence of vertical steps corresponding to phase transitions and nearly horizontal
portions corresponding to an almost fixed interfacial structure,

Unlike the surface pressure versus molecular area isotherms, the adsorption
isotherms, i.e., surface density versus bulk concentration, do depend on the free
energy contributions that are independent of molecular area, a [123]. This means
that the adsorption isotherms are shifted on the concentration scale depending on
for instance the magnitude of the hydrophobic Tanford contribution (cf. Equation
5.39). Full adsorption isotherms stretching all the way to the cmc for the decyl-
and dodecyl-surfactants examined are displayed in Figure 5.25.

5.12   COMPARISON BETWEEN THIOMALTOSIDE
AND MALTOSIDE HEAD GROUPS

In Figure 5.18 surface tension isotherms for C_{10}-Mal, C_{10}-S-Mal, C_{12}-Mal, and
C_{12}-S-Mal are shown. Quite evidently the C_{12}-S-Mal isotherm deviates from the
patterns seen for the other surfactants. Moreover, the adsorption isotherms
presented in Figure 5.25 indicate that the behavior of C_{12}-S-Mal is somewhat

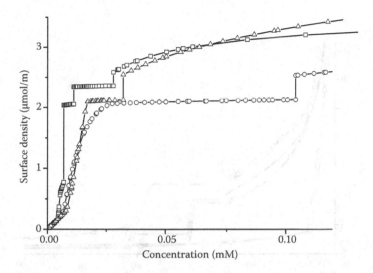

FIGURE 5.24 Surface densities at 22°C in the concentration range from zero through the granular surface phases and to the initial stages of the LE surface phase for C_{10}-Mal (solid line with circles), C_{10}-Glu (solid line with triangles), and C_{10}-S-Mal (solid line with squares) exhibiting two granular phases and the most quickly rising adsorption.

different from the behavior of the other surfactants investigated. In Figure 5.26 surface pressure versus molecular area isotherms for the thiomaltoside and maltoside pairs are presented. Obviously, for the decyl pair of Mal and S-Mal the surface tension difference in the liquid-expanded phase ($a < 65$ Å^2) is quite small. For the dodecyl pair, however, the surface tension difference is of very different magnitude. Interestingly, no granular phase or typical surface micellar phase is

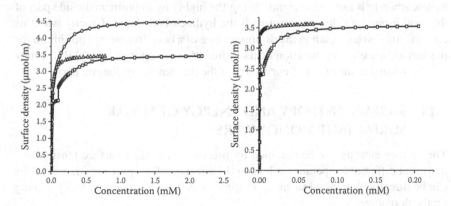

FIGURE 5.25 Adsorption isotherms for C_{10}-Mal (solid line with squares), C_{10}-Glu, (solid line with circles), and C_{10}-S-Mal, (solid line with triangles) are presented on the left-hand side, C_{12}-Mal (solid line with squares), C_{12}-S-Mal (solid line with triangles) displaying very rapid adsorption are presented on the right-hand side.

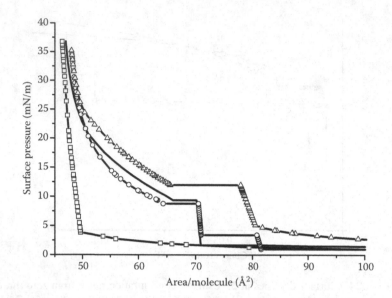

FIGURE 5.26   Surface pressure isotherms at 22°C for C_{10}-Mal (solid line with triangles), C_{10}-S-Mal (solid line with circles), C_{12}-Mal (solid line without symbols) and C_{12}-S-Mal (solid line with squares) displaying the clearly lowest surface pressure.

detected for the C_{12}-S-Mal surfactant, implying a rapid, perhaps first-order, phase change directly from the gaseous to the LE phase.

At a surface pressure of only 3.5 mN/m the area/molecule for C_{12}-S-Mal is already down at 50 Å². In comparison, for C_{12}-Mal 50 Å² is reached at a surface pressure of 23 mN/m. At the cmc, however, the molecular areas are much the same. Due to its greater polarizability the sulfur-linkage might tend to mix with the hydrocarbon tails and consequently bring the highly hydrophilic maltoside part of the head group into closer contact with the hydrocarbon chains. Exactly why this circumstance would result in the disappearance of a deep free energy minimum for the surface micelle as a function of its radius is so far difficult to say. But it may well be related to the effect of curvature on the line tension of the rim part.

5.13   SURFACE ENTROPY AND ENERGY OF SUGAR SURFACTANT MONOLAYERS

The surface entropy can be obtained by means of recording surface tension isotherms at different temperatures. It is well known that for a pure liquid, the results can be directly interpreted as an interfacial excess entropy quantity, S^s, by relying on the derivative:

$$\frac{d\gamma}{dT} = -\frac{S^s}{A} \tag{5.47}$$

Here S^s stands for the excess entropy estimated for the Gibbs equimolecular dividing surface for which the surface excess of the liquid vanishes. Written on a molar (or molecular) basis we thus have, e.g., for pure water (component 1):

$$a_1 \frac{d\gamma}{dT} = -s_1^s \tag{5.48}$$

Usually, for most simple liquids, s_1^s is independent of temperature, but not for water that shows a relatively low, though increasing surface entropy value as the temperature is raised. This is commonly attributed to a breakdown of the H-bond-dependent surface-induced structure present in the interfacial zone.

For a surfactant solution we can compute the excess surface entropy by applying Equation 5.11, preferably put in the form:

$$\left(\frac{d\gamma}{dT}\right)_{c_2} + \Gamma_2 k \ln x_2 = -\left(\frac{S^s}{A} - \Gamma_2 s_2^{bo}\right) \equiv -\frac{S_{x_2=1}^s}{A} \tag{5.49}$$

The entropy quantity introduced in this manner, $S_{x_2=1}^s$, corresponds to the surface excess in entropy relative to pure bulk water and surfactant in the $x_2 = 1$ standard state. As before, x_2 is the mole fraction of surfactant in the bulk solution, whereas s_2^{bo} denotes the corresponding standard state value (based on extrapolation to $x_2 = 1$) of the entropy per surfactant molecule. Evidently, both interfacial water and adsorbed surfactant molecules, inclusive of their mixing, will contribute to the value of $S_{x_2=1}^s$.

The number of data points it takes to precisely evaluate a differential quantity like $d\gamma/dT$ for a given temperature is, of course, relatively large. One would need to determine a whole set of isotherms for such a purpose. The isotherms for C_{10}-Mal and C_{10}-Glu at three different temperatures: 8°C, 22°C, and 29°C are presented in Figures 5.27 and 5.28, respectively. On this fairly limited basis we can only point out some trends.

For low surface densities of surfactant, i.e., in the dilute Henry range, the entropy associated with the free water in between the surfactant molecules is bound to make a predominant contribution to the surface entropy and energy. Thus, independently of the concentration, in this range the temperature coefficients of the surface tension of these sugar surfactant solutions are approximately the same as for pure water, about 0.15 mN/m·K (cf. Figure 5.1).

Both terms on the left-hand side of Equation 5.49 are negative and hence both of them contribute to a positive value of the surface entropy quantity $S_{x_2=1}^s$. The overall trend is seen to be, for the reason just mentioned, that $S_{x_2=1}^s$ at first stays fairly constant as the surfactant concentration is raised, then drops when surface micelles arise and attains a minimum in the granular range. After the transition to the LE state the value of $S_{x_2=1}^s$ grows nearly constant as the first of the contributing terms in Equation 5.49 increases in magnitude more than the second one decreases.

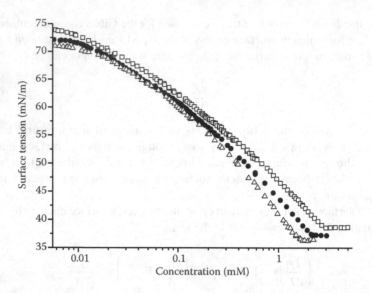

FIGURE 5.27 Surface tension data for C_{10}-Mal at 8°C (empty squares), 22°C (filled circles), and 29°C (empty triangles). (From Kumpulainen, A.J., Persson, C.M., and Eriksson, J.C., *Langmuir*, 20, 10534, 2004. With permission.)

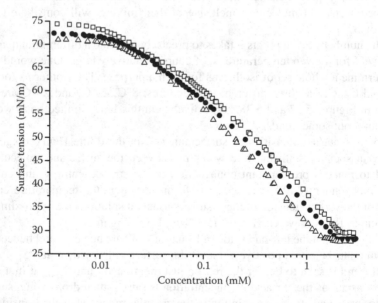

FIGURE 5.28 Surface tension data for C_{10}-Glu at 8°C (empty squares), 22°C (filled circles), and 29°C (empty triangles). (From Kumpulainen, A.J., Persson, C.M., and Eriksson, J.C., *Langmuir*, 20, 10534, 2004. With permission.)

Let us next denote the energy and free energy quantities that are analogous to $S^s_{x_2=1}$ by $U^s_{x_2=1}$ and $F^s_{x_2=1}$, respectively. It is readily proven that these thermodynamic quantities can be evaluated by using the following relations:

$$U^s_{x_2=1}/A = \gamma - T\left(\frac{\partial \gamma}{\partial T}\right)_{c_2} \tag{5.50}$$

$$F^s_{x_2=1}/A = \gamma + \Gamma_2 kT \ln x_2 \tag{5.51}$$

In particular, we may note that the first of the above relations is similar in form to the well-known relation that is commonly applied for pure liquids to compute the surface energy

$$U^s/A = \gamma - T\frac{d\gamma}{dT} \tag{5.52}$$

By means of Equations 5.50 and 5.51 and the data given in Figures 5.27 and 5.28 we have derived the $TS^s_{x_2=1}$ and $U^s_{x_2=1}$ values for C_{10}-Mal shown in Figure 5.29.

It appears that the surface excess energy drops quickly in the surface micelle/granular range down to a minimum and reaches a nearly constant level in the LE

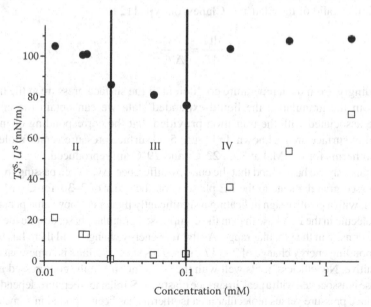

FIGURE 5.29 $TS^s_{x_2=1}$ and $U^s_{x_2=1}$ values for C_{10}-Mal. $TS^s_{x_2=1}$ (empty squares) and $U^s_{x_2=1}$ (filled circles). The ranges II, III, and IV demarked by the vertical lines denote the surface micelle, granular and fully developed liquid-expanded ranges, respectively.

range. Evidently, the courses of the $U^s_{x_2=1}$ functions are roughly parallel with the courses of the $S^s_{x_2=1}$ functions. Unfortunately, the resolution is not sufficiently high to be able to verify the presence of the steps in the curves to be anticipated at the break points in the surface tension curves.

As a result of surface micelle formation, the molecular organization in the interface changes a whole lot. Some interfacial water is likely to be present in the surface micelle/granular range but less, of course, than in the Henry range, and for the most part it should be in contact with unshielded hydrocarbon patches beneath the micelles. Yet another energy/entropy source is related with the hydrocarbon chains themselves. Due to the constraint of planarity their conformational entropy contribution is supposedly less than in a pure hydrocarbon liquid. Eventually, in the liquid-expanded state this constraint becomes less stringent, but at the same time the amount of interfacial (i.e., excess entropy) water successively decreases with an increasing surface density of head groups.

Obviously, the whole surface energy/entropy issue is a rather complex one, and much further work will be needed, experimentally as well as theoretically, to get the full picture of what is actually going on, and how the thermodynamic state of the surfactant solution/air interface varies with the extent of adsorption. Yet, a major driving force behind the surface tension lowering is likely to be a reduction of the number of free, non-hydrogen-bonded OH groups originally present in the interface, about one per 30 Å², corresponding to a surface energy of about 90 mJ/m².

For the case of a true first-order surface phase transition it can be shown that a relation is valid of the Clausius–Clapeyron type [124]:

$$\frac{d\Pi}{dT} = \frac{\Delta s^s_2}{\Delta a_2} \qquad (5.53)$$

Accordingly, from the temperature coefficient of the surface pressure at the transition from the granular to the liquid-expanded state we can obtain the entropy change associated with the transition provided that the corresponding change in molecular surface area is known. In Figure 5.30 surface pressure versus molecular area isotherms for C_{10}-Mal at 8°C, 22°C, and 29°C are reproduced.

In this way we have found that the entropy difference Δs^s_2 when passing from the compressed granular state to the LE phase is on the order of 2–3 k for C_{10}-Mal and C_{10}-Glu, with a positive sign indicating a significantly higher entropy value per surfactant molecule in the LE phase than in the compressed granular phase, despite the lower surface density in the granular range. As the free energy change is nil there has to be a corresponding energy change of 2–3 kT units. This state of affairs is somewhat counter-intuitive. Nonetheless, it fits well with the observation that the compressed granular phase is associated with a peak in surface elasticity. Similar temperature dependences for surface pressure versus molecular area isotherms has been reported in some cases, e.g., hydrogen bonding between head groups for the case of hexadecyl- and octadecyl-urea studied by Glazer and Alexander [125]. Furthermore, mixtures of the hexadecyl- and octadecyl-urea were studied in great detail by Hayami et al. [126]. For these systems a hydrogen-bonding pattern between the head groups was claimed to stabilize

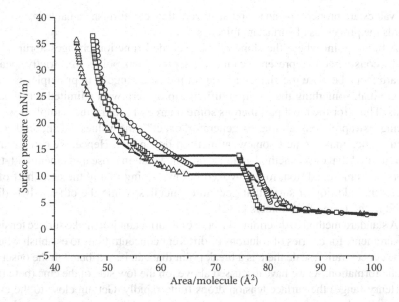

FIGURE 5.30 Surface pressure for C_{10}-Mal at 8°C (solid line with squares) with the highest transition pressure, 22°C (solid line with circles), and 29°C (solid line with triangles) displaying the lowest transition pressure.

a surface phase with a molecular area of approximately 25 Å², that upon compression collapses to a condensed phase with an average molecular area of approximately 20 Å². The energy change upon the phase transition is on the order of 7.5 kT, which is a figure consistent with disruption of hydrogen bonds. Moreover, measurements on diblock copolymer monolayers (poly(styrene-block-4-vinylpyridinium decyl iodide)) [127,128] offer further tantalizing indications that the reorganization of decyl-tails from a dilute surface micellar state at the air–water interface indeed can be responsible for transitions from ordered (at large molecular areas) to less ordered monolayers (at lower molecular areas), i.e., the transition is associated with an increase in entropy upon compressing the monolayer. These studies yield overall qualitative features that are surprisingly similar to the results obtained for sugar surfactants, despite the fact that the alkyl tails grafted to the vinylpyridine monomers is the only common denominator. Similar temperature behavior for transitions have also been detected for spin-labeled probe molecules [129], 17-methyloctadecanoic acid [130] and theoretically described in a one-dimensional lattice model [131].

5.14 DERIVING SOME PROPERTIES OF SUGAR SURFACTANT MICELLES FROM SURFACE TENSION MEASUREMENTS

Knowing the cmc of a surfactant is a prerequisite, of course, for many practical applications. For instance, it is only above the cmc that solubilization of water-insoluble compounds will take place. Moreover, insofar as the experiments to determine the cmc values are properly planned and analyzed, they can furnish valuable insights as regards the properties of surfactant micelles.

A break point where the slope of the recorded function changes abruptly is usually considered to represent the cmc. Such cmc break points are, strictly speaking, artefacts, because the change arising in the cmc range is in principle smooth and gradual, something that is often difficult to discern with a limited set of data points. Thus, for the most part there is some degree of arbitrariness implied when making extrapolations aiming to generate "exact" cmc values. In the end, this circumstance makes cmc's somewhat method-dependent. Hence, we cannot take it for granted that every method generates a cmc value in close agreement with the theoretical cmc-condition, to be invoked below, stating that at the cmc, half of a differential addition of surfactant goes into micelles, while the other half is dissolved in the form of monomers [132].

A standard method to determine the cmc of a surfactant is to make surface tension measurements for a series of solutions of different concentrations to establish where on the concentration scale there is a break point that can be attributed to the onset of micelle formation. As we have discussed above, on the low side of the cmc (but after the Henry range) the surface tension drops quite rapidly. Getting close to the cmc break point, the molecular area, a, certainly continues to diminish, but to an almost insignificant degree. Consequently, we may expect a to stay practically the same even after the cmc where the chemical potential of the surfactant increases very slowly upon adding more surfactant. In this range, the experimental concentration variable is, of course, no longer the surfactant monomer concentration, c_2, but the total surfactant concentration, $c_{tot} = c_{mic} + c_2$. From the Gibbs surface tension equation it thus appears that in the range from slightly below to well above the cmc, $-a\,d\gamma$ may in fact constitute a handy expression for the increase of the chemical potential, $d\mu_2$, of the surfactant, as we can put a equal to its value immediately before the cmc, i.e.,

$$\Delta\mu_2 = \mu_2 - \mu_2(\text{cmc}) = -a\,[\gamma - \gamma(\text{cmc})] = -a\,\Delta\gamma \qquad (5.54)$$

Here $\Delta\gamma$ is simply the change recorded of γ after the cmc. Thus, for a nonionic surfactant we can write:

$$c_2 \approx \text{cmc} \times e^{-a\Delta\gamma/kT} \qquad (5.55)$$

This relationship implies that c_2 is actually a known quantity which is easily evaluated to a good approximation. On the other hand, from the aggregation equilibrium condition we have

$$c_{mic} = \text{const} \cdot c_2^{\langle N \rangle} \qquad (5.56)$$

where $\langle N \rangle$ denotes the average number of monomers in the micelle. On this basis one can demonstrate that the following relation holds:

$$\frac{d\ln c_{tot}}{d\ln c_2} = \frac{\langle N \rangle c_{mic} + c_2}{c_{mic} + c_2} \qquad (5.57)$$

which upon combining with the Gibbs surface tension equation yields [133]

$$\frac{d\gamma}{d \ln c_{tot}} = -\frac{kT}{a}\left[\langle N\rangle\left(1-\frac{c_2}{c_{tot}}\right)+\frac{c_2}{c_{tot}}\right]^{-1} \tag{5.58}$$

showing that in the high-concentration limit, $c_{tot} \gg c_2$, the slope of the γ versus ln c_{tot} curve is equal to the slope immediately before the cmc, kT/a, divided by the average aggregate number $\langle N\rangle$ [117].

Furthermore, upon inserting the cmc criterion $c_{mic} = c_2/\langle N\rangle$ (that results from Equation 5.57 by invoking $dc_2/dc_{tot} = 0.5$) in Equation 5.58, we find the approximate relationship:

$$\frac{d\gamma}{d \ln c_{tot}} \approx -\frac{kT}{2a} \tag{5.59}$$

Consequently, the special cmc definition that we have favored actually corresponds to that particular point on the γ versus ln c_{tot} plot where the slope is half of the slope just prior to the cmc.

To evaluate $\langle N\rangle$ from surface tension data for concentrations above the cmc we can most conveniently go back to a couple of relations contained in Hill's small systems thermodynamics [134]:

$$\varepsilon = -kT \ln \phi_{mic} \tag{5.60}$$

$$\frac{d\varepsilon}{d\mu_2} = -\langle N\rangle \; (\text{constant } T \text{ and } p) \tag{5.61}$$

that are valid for systems which are dilute in noninteracting nonionic micelles. The symbol ε corresponds to the excess free energy of a micelle and ϕ_{mic} is the volume fraction of micelles (in addition ϕ_{mic} has also the nature of a partition function). By inserting the Gibbs surface tension equation put in the form $d\mu_2 + ad\gamma = 0$ we get:

$$\frac{d\gamma}{d \ln c_{mic}} = -\frac{kT}{a\langle N\rangle} \tag{5.62}$$

where now

$$c_{mic} = cmc\left(\frac{c_{tot}}{cmc} - e^{-a\Delta\gamma/kT}\right) \tag{5.63}$$

Accordingly, by plotting γ versus $\ln(c_{mic}/cmc)$ we obtain $\langle N\rangle$ from the slope upon inserting the value of a found right before the cmc.

It is evident however, that only rather crude estimates of $\langle N\rangle$ are feasible on the basis of surface tension measurements alone, as the estimated decrease of the surface tension upon for example a 10-fold increase of c_{tot} after the cmc merely amounts to 0.25–0.5 mN/m. The surface tension data of C_{12}-Mal in the proximity of the cmc are displayed in Figure 5.31.

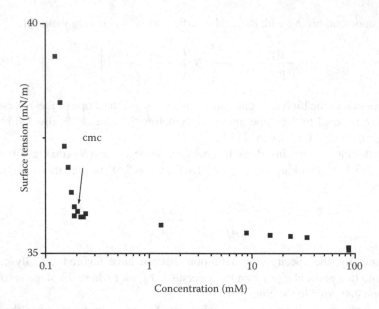

FIGURE 5.31 Surface tension data for C_{12}-Mal below and well above the cmc. The arrow marks the estimated cmc value, 0.19 mM. The average aggregation number of a micelle is ≈ 76.

For C_{12}-Mal surfactant micelles $\langle N \rangle = 76$, with a variation at 95% confidence between 45 and 200 monomers per micelle was obtained for solutions ranging from the cmc to 100 mM in concentration, in due agreement with other type of investigations [135]. Yet, one has to bear in mind that the assumption made here of a constant average aggregation number is only approximately valid. In fact $\langle N \rangle$ is expected to increase by several percent for a change of c_{tot} by one order of magnitude. This we can conclude from the relationship

$$\frac{d \ln \langle N \rangle}{d \ln c_{tot}} = \left(\frac{\sigma_N}{\langle N \rangle} \right)^2 \tag{5.64}$$

where σ_N is the standard deviation of the micelle size distribution. This relation follows from differentiating the subdivision potential, ε, a second time with respect to the chemical potential. Typically σ_N is on the order of 10.

Finally, it is worth noting that the temperature dependence of the cmc carries valuable thermodynamic information about the micelles formed. In particular, at constant pressure we have

$$\frac{\langle h_N^{mic} \rangle - h_2^b}{kT^2} = -\frac{d \ln cmc}{dT} \tag{5.65}$$

Here $\langle h_N^{mic} \rangle$ is the average value of the molecular enthalpy in a micelle and h_2^b the (partial) molecular enthalpy in the bulk solution. Using this relation and the cmc data in Figures 5.27 and 5.28, we can estimate $\langle h_N^{mic} \rangle - h_2^b$ at about 20°C to approximately 2.7 kT units for both C_{10}-Glu and C_{10}-Mal. This is in line with some published reports showing that this enthalpy difference becomes equal to zero at about 40°C where the cmc's for sugar-based surfactants have minima [136,137].

5.15 SUMMARY

In this chapter, we have brought up some salient features for discussion that emerge from recent studies of dilute solutions of simple sugar surfactants. In particular, we have focused on the analysis of surface tension data recorded for such solutions and, to a lesser extent, the results of parallel VSFS studies. In broad terms the pattern of behavior observed as the surfactant concentration is raised is as follows:

1. First there is a linear Henry, ideal solution range where the surface pressure grows rather slowly while obeying the gas-law-like relation $\Pi a = kT$.
2. Rather abruptly the surface tension then starts to drop at a much faster rate which is often, though not necessarily always, associated with the formation of flat surface micelles in the interface with an anticipated aggregation number on the order of 10.
3. A first-order transition ends the surface micelle/granular range. It takes place from a compressed surface micelle state to the laterally homogeneous liquid-expanded state resulting in an increase in entropy and energy despite the increase in surface density.
4. The surface tension lowering eventually almost stops as a consequence of the formation of ordinary spherical micelles.

Hydrophobic factors by and large cause the above sequence of adsorption events. For example, transferring a surfactant molecule from the dilute solution state to the liquid-expanded surface film implies that the number of unfavorable hydrocarbon chain–water contacts will diminish. Similarly, surface micelle formation will reduce the number of lateral hydrocarbon–water contacts in the interface.

Counteracting molecular factors are largely of entropic origin. Filling up the interface with surface micelles gives rise to repulsion because the configurational entropy is reduced. A repulsive mechanism of essentially the same kind sets in for the liquid-expanded state when the area fraction of head groups increases. We argue that simple models based on Langmuir adsorption of surface micelles and the hard-disk equation of state are promising theoretical starting-points for reaching a better understanding of the fundamental issue regarding the causes of the surface tension lowering effect in general shown by soluble surfactants, including sugar surfactants.

APPENDIX

Comparison between the Critical Concentrations of Forming Spherical Surfactant Micelles in Solution and Flat Surface Micelles in the Solution–Air Interface

Let us consider the surfactant solution–air interface toward the end of the Henry range where just a few noninteracting surface micelles have formed. In other words, the interface is dilute in terms of surface micelles as well as monomers. We picture a surface micelle as a short cylinder with a thickness of some 4 Å and a hydrocarbon core of a radius corresponding to the extended chain length, approximately 16 Å for a C_{12}-chain and with the polar head groups distributed beneath it in the water phase, and, additionally, some of them spreading to below the water rim that surrounds the circular core (cf. Figure 5.3). Focusing on the particular surface micelle of equilibrium size the condition expressed by Equation 5.6 must be fulfilled (per molecule):

$$f_2^{smic} - a_2^{smic}\gamma = \mu_2^{smic} = \mu_2^{bo} + kT \ln x_2 \text{(csmc)} \qquad (5.A.1)$$

where csmc stands for the critical surface micelle concentration, or approximately,

$$f_2^{smic} - a_2^{smic}\gamma_1^o - \mu_2^{bo} = kT \ln x_2 \text{(csmc)} \qquad (5.A.2)$$

By invoking the various contributions to the molecular free energy difference $f_2^{smic} - \mu_2^{bo}$ we can write

$$\Delta\mu_{Tan}^o + a_2^{smic}\gamma_{hc,air} + (a_2^{smic} - a_{pg})\gamma_{hc,w} - a_2^{smic}\gamma_1^o + \varepsilon_{conf} + \varepsilon_{mix} + \varepsilon_{rim} + \varepsilon_{pg}$$
$$= kT \ln x_2 \text{(csmc)} \qquad (5.A.3)$$

where $\Delta\mu_{Tan}^o$ is the free energy of bringing the hydrocarbon chain from the infinite dilution, $x_2 = 1$, standard state to liquid hydrocarbon ($-19.96\ kT$ for a C_{12} chain); a_2^{smic} ($\gamma_{hc,w} + \gamma_{hc,air}$) is the free energy required to form a hydrocarbon film from that hydrocarbon fluid; $-a_{pg}\gamma_{hc,w}$ is the relative gain in free energy due to attaching the head groups, ε_{conf} is the change in conformational free energy of the hydrocarbon chains; ε_{mix} is the free energy of mixing the head groups with water; ε_{rim} is the increase in surface free energy caused by the lateral hydrocarbon–water contact at the rim of the hydrocarbon core of the surface micelle; and ε_{pg} is the (minor) free energy change associated with the head group per se upon transferring it from the dilute surfactant solution to the surface micelle.

Similarly, for a spherical micelle of the same surfactant of equilibrium size, we have as a good approximation at the cmc:

$$\Delta\mu_{Tan}^o + (a_2^{mic} - a_{pg})\gamma_{hc,w} + \varepsilon_{conf} + \varepsilon_{mix} + \varepsilon_{pg} = kT \ln x_2 \text{(cmc)} \qquad (5.A.4)$$

Next, we form the difference between Equations 5.A.3 and 5.A.4 while assuming that $\Delta\varepsilon_{conf}$, $\Delta\varepsilon_{mix}$, and $\Delta\varepsilon_{pg}$ are all negligible and $a_2^{smic} = a_2^{mic} = a_2$ resulting in the expression:

$$a_2(\gamma_{hc,air} - \gamma_1^0) + \varepsilon_{rim} = kT\ln\left(\frac{x_2(\text{csmc})}{x_2(\text{cmc})}\right) = kT\ln\left(\frac{\text{csmc}}{\text{cmc}}\right) \quad (5.A.5)$$

A reasonable estimate of ε_{rim} assuming a line tension of about 10^{-11} N and multiplying with the circular perimeter length is $\varepsilon_{rim} = 2.46\ kT$ for an aggregation number of 10. Finally, inserting $a_2 = 60\ \text{Å}^2$ and $\gamma_1^0 - \gamma_{hc,air} = 50\,\text{mN/m}$ we get

$$\frac{\text{csmc}}{\text{cmc}} = 7.4 \cdot 10^{-3} \quad (5.A.6)$$

which is in broad agreement with our experimental observations for sugar-based surfactants.

REFERENCES

1. Holmberg, K. *Curr. Opinion Colloid Interface Sci.* 2001, *6*, 148.
2. Balzer, D. *Tenside Surf. Det.* 1991, *28*(6), 419.
3. von Rybinski, W. *Curr. Opinion Colloid Interface Sci.* 1996, *1*, 587.
4. von Rybinski, W. and Hill, K. *Angew. Chem. Int. Ed.* 1998, *37*, 1328.
5. Söderman, O. and Johansson, I. *Curr. Opinion Colloid Interface Sci.* 2000, *4*(6), 391.
6. Allen, D. K. and Tao, B. Y. *J. Surfactants Deterg.* 1999, 2, 383.
7. Weuthen, M., Kawa, R., Hill, K., and Ansmann, A. *Fat Sci. Technol.* 1995, *97*, 209.
8. Burczyk, B. *Novel Surfactants; Preparation, Applications and Biodegradability*, 2nd ed., Holmberg, K. (Ed.), Dekker: New York, 2003 Chapter 4, p. 129.
9. Dwek, R. A. *Chem. Rev.* 1996, *96*, 683.
10. Jelinek, R. and Kolusheva, S. *Chem. Rev.* 2004, *104*, 5987.
11. Hakomori, S., I. *Glycoconjugate J.* 2000, *17*, 143.
12. Mammen, M., Choi, S.-K., and Whitesides, G. M. *Angew. Chem. Int. Ed.* 1998, *37*, 2754.
13. Swalen, J. D., Allara, D. L., Andrade, J. D., Chandross, E. A., Garoff, S., Israelachvili, J., McCarthy, T. J., Pease, R. F., Rabolt, J. F., Wynne, K. J., and Yu, H. *Langmuir* 1987, *3*, 932.
14. Allara, D. L. *Nature* 2005, *437*, 638.
15. Barth, J. V., Costantini, G., and Kern, K. *Nature* 2005, *437*, 671.
16. Love, J. C., Estroff, L. A., Kriebel, J. K., Nuzzo, R. G., and Whitesides, G. M. *Chem. Rev.* 2005, *105*, 1103.
17. Shinoda, K., Yamanaka, T., and Kinoshita, K. *J. Phys. Chem.* 1959, *63*, 648.
18. Drummond, C. J., Warr, G. G., Grieser, F., Ninham, B. W., and Evans, D. F. *J. Phys. Chem.* 1985, *89*, 2103.
19. Lu, J. R., Thomas, R. K., and Penfold, *J. Adv. Colloid Interface Sci.* 2000, *84*, 143.
20. Cooke, D. J., Lu, J. R., Lee, E. M., Thomas, R. K., Pitt, A. R., Simister, E. A., and Penfold, J. *J. Phys. Chem.* 1996, *100*, 10298.
21. Eastoe, J., Rogueda, P., Howe, A. M., Pitt, A. R., and Heenan, R. K. *Langmuir* 1996, *12*, 2701.
22. Zhang, L., Somasundran, P., and Maltesh, C. *Langmuir* 1996, *21*, 2371.
23. Söderberg, I., Drummond, C. J., Furlong, D. N., Godkin, S., and Matthews, B. *Colloids Surf. A* 1995, *102*, 91.

24. Waltermo, Å., Claesson, P. M., Simonsson, S., Manev, E., Johansson, I., and Bergeron, V. *Langmuir* 1996, *12*, 5271.
25. Drummond, C. J. and Wells, D. *Colloids Surf. A* 1998, *141*, 131.
26. Garofalakis, G., Murray, B. S., and Sarney, D. B. *J. Colloid Interface Sci.* 2000, *229*, 391.
27. Kjellin, U. R. M., Reimer, J., and Hansson, P. *J. Colloid Interface Sci.* 2003, *262*, 506.
28. Kjellin, U. R. M., Claesson, P. M., and Vulfson, E. N. *Langmuir* 2001, *17*, 1941.
29. Coppola, L., Gordano, A., Procopio, A., and Sindona, G. *Colloids Surf. A* 2002, *196*, 175.
30. Molina-Bolivar, J. A., Aguiar, J., Peula-Garcia, J. M., and Carnero Ruiz, C. *J. Phys. Chem. B* 2004, *108*, 12813.
31. Molina-Bolivar, J. A., Hierrezuelo, J. M., and Carnero Ruiz, C. *J. Phys. Chem. B* 2006, *110*, 12089.
32. Rojas, O. J., Neuman, R. D., and Claesson, P. M. *J. Phys. Chem. B* 2005, *109*, 22440.
33. Piao, J., Satsuki, K., and Shuji, A. *Colloids Surf. A* 2006, *277*, 15.
34. Pilakowska-Pietras, D., Lunkenheimer, K., and Piasecki, A. *Langmuir* 2004, *20*, 1572.
35. Persson, C. M., Claesson, P. M., and Lunkenheimer, K. *J. Colloid Interface Sci.* 2002, *251*, 182.
36. Aveyard, R., Binks, B. P., Chen, J., Esquena, J., Fletcher, P. D. I., Buscall, R., and Davies, S. *Langmuir* 1998, *14*, 4699.
37. Boyd, B., Drummond, C. J., Krodkiewska, I., and Grieser, F. *Langmuir* 2000, *16*, 7359.
38. Stubenrauch, C. *Curr. Opinion Colloid Interface Sci.* 2001, *6*, 160.
39. Kumpulainen, A. J., Persson, C. M., Eriksson, J. C., Tyrode, E. C., and Johnsson, C. M. *Langmuir* 2005, *21*, 305.
40. Zasadizinski, J. A., Viswahathan, R., Madsen, L., Garnaes, J., and Schwartz, D. K. *Science* 1994, *263*(5154), 1726.
41. Shipway, A. N. and Willner, I. *Acc. Chem. Res.* 2001, *34*, 421.
42. Bjornholm, T., Hassenkam, T., and Reitzel, N. *J. Mater. Chem.* 1999, *9*, 1975.
43. Bell, G. M., Combs, L. L., and Dunne, L. J. *Chem. Rev.* 1981, *81*, 15.
44. Goddard, B. D. *J. Colloid Interface Sci.* 1979, *68*, 196.
45. Gibbs, J. W. *The Collected Works of J. Willard Gibbs vol.*1, *Thermodynamics*, Yale University Press: New Haven, 1928.
46. Kumpulainen, A. J. Thesis, Royal Institute of Technology, Stockholm, Sweden, 2004.
47. Meguro, K., Ueno, M., and Esumi, K. Non-ionic surfactants: Physical chemistry, in *Surfactant Science Series vol.* 23, Schick, M. (Ed.), Marcel Dekker, New York, pp. 109–183, 1987.
48. Adamson A. W. *Physical Chemistry of Surfaces*, 5th ed. Wiley: New York, 1990.
49. Brinck, J., Jönsson, B., and Tiberg, F. *Langmuir* 1998, *14*, 5863.
50. Tiberg, F. *J. Chem. Soc. Faraday Trans.* 1996, *92*(4), 531.
51. Kiraly, Z., Börner, P. L., and Findenegg, G. H. *Langmuir* 1997, *13*, 3308.
52. Grant, L., Tiberg, F., and Ducker, W. A. *J. Phys. Chem. B.* 1998, *102*, 4288.
53. Helfrich, W. *Zeitschrift Naturforsch.* 1973, c. *28*, 693.
54. Erpenbeck, J. J. and Luban, M. *Phys. Rev. A* 1985, *32*, 2920.
55. Nilsson, U. Thesis, University of Lund, Sweden, 1992.
56. Motomura, K., Iwanaga, S. -I., Hayami, Y., Uryu, S., and Matuura, R. *J. Colloid Interface Sci.* 1981, *80*, 32.
57. Aratono, M., Uryu, S., Hayami, Y., Motomura, K., and Matuura, R. *J. Colloid Interface Sci.* 1984, *98*, 33.
58. Tyrode, E., Johnsson, C. M., Rutland, M. W., and Claesson, P. M. *J. Phys. Chem. C* 2007, *111*, 11642.
59. Hill, T. L. Theory of Physical Adsorption, in *Advances in Catalysis*, Frankenberg, W. G. (Ed.), Academic: London, 1952, p. 211.
60. Butler, J. A. V. *Proc. R. Soc. Ser. A* 1932, *135*, 348.

61. Eriksson, J. C. *Arkiv Kemi* 1966, *33*, 343.
62. Guggenheim, E. A. *Trans. Faraday Soc.* 1945, *41*, 150.
63. Bain, C. D. *J. Chem. Soc., Faraday Trans.* 1995, *91*, 1281.
64. Shen, Y. R. and Ostroverkhov, V. *Chem. Rev.* 2006, *106*, 1140.
65. Richmond, G. L. *Chem. Rev.* 2002, *102*, 2693.
66. Shultz, M.J., Baldelli, S., Schnitzer, C., and Simonelli, D. *J. Phys. Chem. B* 2002, *106*, 5313.
67. Du, Q., Superfine, R., Freysz, E., and Shen, Y. R. *Phys. Rev. Lett.* 1993, *70*, 2313.
68. Shen, Y. R. *Solid State Comm.* 1998, *108*, 399.
69. Traube, I. *Liebigs Ann. Chem.* 1891, *265*, 26.
70. Langmuir, I. *J. Am. Chem. Soc.* 1917, *39*, 1848.
71. Tanford, C. *The Hydrophobic Effect*, 2nd ed., Wiley and Sons: New York, 1980.
72. Goebel, A. and Lunkenheimer, K. *Langmuir* 1997, *13*, 369.
73. Birdi, K. S. (Ed.), *Handbook of Surface and Colloid Chemistry*, CRC: Boca Raton, 1997.
74. Gruen, D. W. R. and de Lacey, E. H. B. Packing of amphiphile chains in micelles and bilayers, in *Surfactants in Solution Vol. 1*, Mittal, K. L. and Lindman, B. (Eds.) pp. 279–306. Plenium Press: New York, 1984.
75. Gruen, D. W. R. *J. Phys. Chem.* 1985, *89*, 146.
76. Ben-Schaul, A., Szleifer, I., and Gelbart, W. M. *J. Chem. Phys.* 1985, *83*, 3597.
77. Szleifer, I., Ben-Schaul, A., and Gelbart, W. M. *J. Chem. Phys.* 1985, *83*, 3612.
78. Szleifer, I., Ben-Schaul, A., and Gelbart, W. M. *J. Phys. Chem.* 1990, *94*, 5081.
79. Rosen, M. J. and Aronson, S. *Colloids Surf.* 1981, *3*, 201.
80. Ross, S. and Morrison, I. D. *Colloids Surf.* 1983, *7*, 121.
81. Posner, A. M., Anderson, J. R., and Alexander, A. E. *J. Colloid Interface Sci.* 1952, *7*, 623.
82. Spaull, A. J. B. and Nearn, M. R. *J. Phys. Chem.* 1946, *68*, 2043.
83. Langmuir, I. *J. Chem. Phys.* 1933, *1*, 756.
84. Stoeckly, B. *Phys. Rev. A* 1977, *15*, 2558
85. Smith, T. *Adv. Colloid Interface Sci.* 1972, *3*, 161.
86. Israelachvili, J. *Langmuir* 1994, *10*, 3774.
87. Ruckenstein, E. and Li, B. *Langmuir* 1995, *11*, 3510.
88. Ruckenstein, E. and Li, B. *J. Phys. Chem* 1996, *100*, 3108.
89. Ruckenstein, E. and Bhakta, A. *Langmuir* 1994, *10*, 2694.
90. Popielawski, J. and Rice, S. A. *J. Chem. Phys.* 1988, *88*, 1272.
91. Wang, Z.-G. and Rice, S. A. *J. Chem. Phys.* 1988, *88*, 1290.
92. Harris, J. and Rice, S. A. *J. Chem. Phys.* 1988, *88*, 1298.
93. Meister, A., Kerth, A., and Blume, A. *J. Phys. Chem. B* 2005, *109*, 6239.
94. Fainerman, V. B., Miller, R., Wüstneck, R., and Makievski, A. V. *J. Phys. Chem.* 1996, *100*, 7669.
95. Drach, M., Rudzinski, W., Warszynski, P., and Narkiewicz-Michlek, J. *J. Phys. Chem. Chem. Phys* 2001, *3*, 5035.
96. Fainerman, V. B. and Miller, R. *Langmuir* 1996, *12*, 6011.
97. Kizling, J., Stenius, P., Eriksson, J. C., and Ljunggren, S. *J. Colloid Interface Sci.* 1995, *171*, 162.
98. Kumpulainen, A. J., Persson, C. M., and Eriksson, J. C. *Langmuir* 2004, *20*, 10534.
99. Persson, C. M. and Kumpulainen, A. J. *Colloids Surf. A* 2004, *233*, 43.
100. Waltermo, Å., Claesson, P. M., and Johansson, I. *J. Colloid Interface Sci.* 1996, *183*, 506.
101. Tamada, K., Minamikawa, H., Hato, M., and Miyano, K. *Langmuir* 1996, *12*, 1666.
102. Harkins, W. D. *The Physical Chemistry of Surface Films*, Reinhold: New York 1952.
103. Adam, N. K. *The Physics and Chemistry of Surfaces*, 3rd ed., Oxford University Press: London 1941.
104. Kato, T., Kameyama, M., Ehara, M., and Iimura, K. -I. *Langmuir* 1998, *14*, 1786.

105. Ren, Y., Iimura, K.-I., and Kato, T. *J. Phys. Chem B* 2002, *106*, 1327.
106. Fontaine, P., Goldmann, M., Muller, P., Fauré, M. -C., Konovalov, O., and Krafft, M. P. *J. Am. Chem. Soc.* 2005, *127*, 512.
107. Tyrode, E., Johnson, C. M., Kumpulainen, A., Rutland, M. W., and Claesson, P. M. *J. Am. Chem. Soc.* 2005, *127*, 16848.
108. Eriksson, J. C. and Ljunggren, S. *Colloids Surf.* 1989, *38*, 179.
109. Lebowitz, J. L. *Phys. Rev.* 1964, *133*, 895.
110. Lebowitz, J. L., Helfand, J. L., and Praestegaard, E. *J. Chem. Phys.* 1970, *53*, 471.
111. Henderson, D. *Mol. Phys.* 1975, *30*, 971.
112. Barrio, C. and Solana, J. R. *Phys. Rev. E* 2001, *63*, 011210.
113. Santos, A., Yuste, S. B., and Lopez de Haro, M. *J. Chem. Phys.* 2002, *101*, 4622.
114. Kumpulainen, A. J., Persson, C. M., and Eriksson, J. C. *Langmuir* 2004, *20*, 10935.
115. Chang, C.-H. and Radke, C. J. *Colloids Surf. A* 1995, *100*, 1.
116. Schubert, K. V., Strey, R., and Kahlweit, M. *J. Colloid Interface Sci.* 1991, *141*, 21.
117. Elworthy, P. H. and Mysels, K. *J. Colloid Interface Sci.* 1966, *21*, 331.
118. Lunkenheimer, K. and Wantke, K.-D. *Rev. Sci. Instrum.* 1987, *58*, 2313.
119. Todoroki, N., Tanaka, F., Ikeda, N., Aratono, M., and Motomura, K. *Bull. Chem. Soc. Jpn.* 1993, *66*, 351.
120. Persson, C. M., Kumpulainen, A. J., and Eriksson, J. C. *Langmuir* 2003, *19*, 6110.
121. Manciu, M. and Ruckenstein, E. *Colloids Surf. A* 2004, *232*, 1.
122. Mansoori, G. A., Carnahan, N. F., Starling, K. E., and Leland, T. W. *J. Chem. Phys.* 1971, *54*, 1523.
123. Persson, C. M., Kjellin, U. R. M., and Eriksson, J. C. *Langmuir* 2003, *19*, 8152.
124. Eriksson, J. C. *J. Colloid Interface Sci.* 1971, *37*, 659.
125. Glazer, J. and Alexander, A. E. *Faraday Trans. Soc.* 1951, *47*, 401.
126. Hayami, Y., Kawano, M., and Motomura, K. *Colloid Polymer Sci.* 1991, *269*, 167.
127. Zhu, J., Eisenberg, A., and Lennox, R. B. *J. Am. Chem. Soc.* 1991, *113*, 5583.
128. Zhu, J., Lennox, R. B., and Eisenberg, A. *Langmuir* 1991, *7*, 1579.
129. Cadenhead, D. A. and Müller-Landau, F. *J. Colloid Interface Sci.* 1974, *49*, 131.
130. Asgharian, B. and Cadenhead, D. A. *J. Colloid Interface Sci.* 1990, *134*, 522.
131. Bell, G. M. and Dunne, L. J. *J. Chem. Soc. Faraday Trans. 2* 1978, *74*, 149.
132. Evans, D. F. and Wennerström, H. *The Colloidal Domain: Where Physics, Chemistry and Biology Meet* 2nd ed., Wiley-VCH: New York, 1999.
133. Rusanov, A. I. Micellization in surfactant solution, in *Chemistry Reviews Vol. 22:* Volpin, M. E. (Ed.), Taylor & Francis, Boca Raton, 1997, p 132.
134. Hill, T. L. *Thermodynamics of Small Systems Vol. II*, Benjamin: New York, 1964.
135. Aoudia, M. and Zana, R. *J. Colloid Interface Sci.* 1998, *206*, 158.
136. Mahji, P. R. and Blume, A. *Langmuir* 2001, *17*, 3844.
137. Kameyama, K. and Takegi, T. J. *J. Colloid Interface Sci.* 1990, *137*, 1.

6 Adsorption of Sugar-Based Surfactants at Solid– Liquid Interfaces

Ponisseril Somasundaran, Lei Zhang,
and Shaohua Lu

CONTENTS

6.1   INTRODUCTION

Sugar-based surfactants can be obtained from renewable materials such as fatty alcohols and sugars and are easily biodegradable. Due to their unique solution and interfacial properties as well as their benign environmental profile, they are finding more and more industrial applications for detergency, emulsification, dispersion, wetting, solubilization, etc. [1–4].

As nonionic surfactants, sugar-based surfactants are quite different from the common nonionic alkyl polyethyleneglycol ethers and alkylphenol polyethyleneglycol ethers (C_iE_j and $C_i\Phi E_j$). The solubility of sugar-based surfactants arises from the hydroxyl groups of sugar headgroup instead of the ether oxygens and increases with temperature. Therefore, unlike nonionic alkyl polyethyleneglycol ethers, they do not have cloud points and are not temperature sensitive. Sugar-based surfactants in solution and at air–water interface have been subjected to extensive studies and reviews in Refs. [5–14]. However, the study of them at solid–liquid interfaces is limited. In this chapter, the adsorption behavior of sugar-based surfactants n-alkyl-β-D-glucoside and n-alkyl-β-D-maltoside, on various solids and the mechanisms of adsorption are reviewed. Mixtures of these sugar-based surfactants with anionic, cationic, and other nonionic surfactants at solid–liquid interfaces are also discussed. Particular emphasis is placed on the synergism or antagonism between these surfactants.

6.2   ADSORPTION OF n-DODECYL-β-D-MALTOSIDE ON HYDROPHILIC SOLIDS

Adsorption of n-dodecyl-β-D-maltoside on oxides alumina, and silica at pH 7°C and 25°C is shown in Figure 6.1. It can be seen that the surfactant adsorbs on alumina but in negligible amounts on the silica surface. It also adsorbs on hematite and titania in a similar manner. The isotherm in Figure 6.1 suggests that there is a three-stage adsorption process for the sugar-based surfactants in the case of alumina. In the first stage, the surfactant adsorbs individually and sparsely on the surface. In the second stage, a sharp increase in the adsorption density occurs as a result of the incorporation of the surfactant species into solloids or hemimicelles due to the hydrophobic chain–chain interactions [15,16]. The adsorption isotherm reaches a plateau region at the onset of the third stage. The inflexion point between stages II and III corresponds to the critical micelle concentration of the surfactant.

In the plateau region, the adsorption density is about 5.5×10^{-6} mol/m^2 and the surface area per molecule adsorbed is calculated to be ~30 A^2 in this region.

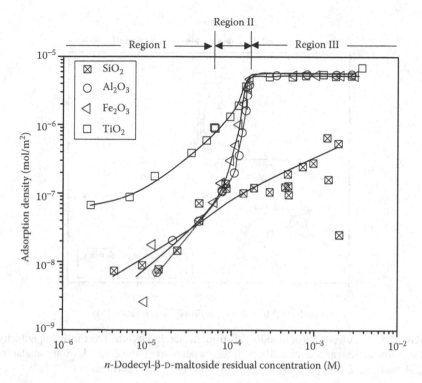

FIGURE 6.1   Adsorption isotherms of n-dodecyl-β-D-maltoside on hydrophilic solids. (From Zhang, L., Somasundaran, P., Maltesh, C., *J. Colloid Interface Sci.*, 191, 202, 1997. With permission.)

Since the area per molecule for n-dodecyl-β-D-maltoside is ~48.5 A^2 (derived from the surface tension data), the amount of surfactant adsorbed on the solids surpasses the amount required to form a theoretical monolayer and is close to that necessary to form a bilayer. However, this value can also be explained by considering the heterogeneous adsorption of spherical micelles.

To acquire information on the nature of the adsorbed layer conformation of surfactant on a solid surface, the changes in relative hydrophobicity of solid has been examined as a function of surfactant adsorption. The effect of adsorption of n-dodecyl-β-D-maltoside (DM) on the wettability of the alumina is illustrated in Figure 6.2 along with the adsorption isotherm. In the absence of the surfactant the alumina exhibits complete hydrophilicity. With an increase in adsorption, however, the surface becomes hydrophobic due to the surfactant adsorption with their hydrophobic tails oriented towards the bulk solution. The hydrophobicity reaches a maximum at the beginning of region II and then drops. The drop in the hydrophobicity after the maximum suggests that hydrophilic groups orient themselves toward the aqueous phase. The hydrophobicity of the alumina drops further as the adsorption reaches the plateau region. Further variation in hydrophobicity in the plateau region is possibly related to the effect of high residual DM that interferes with the wettability measurements [17]. Due to the bulky nature of the maltose

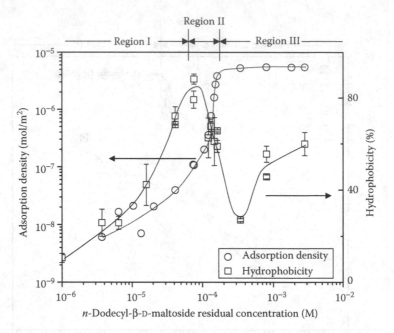

FIGURE 6.2   Adsorption of *n*-dodecyl-β-D-maltoside and its effect on the hydrophobicity of alumina particles as determined by two phase separation. (From Zhang, L., Somasundaran, P., Maltesh, C., *J. Colloid Interface Sci.*, 191, 202, 1997. With permission.)

headgroup, close packing of the adsorbed layer will require the hydrocarbon chains of the opposing surfactant to interpenetrate each other so as to create one layer of hydrocarbon chains with two layers of headgroups on each side. The proposed bilayer is a strong possibility given that the cross-sectional area of the headgroup is 48.5 A² while that of the paraffin chain is about 20 A².

The effects of the hydrocarbon chain length and the glucose units on headgroups on the adsorption of alkyl glucosides and maltosides on alumina are shown in Figure 6.3 for octyl- and decyl-β-D-glucosides and octyl-, decyl-, dodecyl, tetradecyl-β-D-maltosides. All the isotherms are similar in shape with the inflection point for each surfactant corresponding to its cmc value [18]. It is to be noted that all isotherms have a very small slope in the low concentration range, indicating a weak driving force for adsorption. Stronger interactions such as electrostatic interactions would result in a slope of 1 at low concentrations under constant ionic strength conditions [15]. The S shape of the adsorption isotherm indicates that there are strong interactions among the adsorbate species, in this case hydrophobic chain, and comparatively weaker interactions between the adsorbate and the adsorbent [19,20].

It can also been seen from these figures that there is a small increase in the adsorption density in the plateau region with an in increase in the chain length. This suggests that the packing of the longer chain surfactants is more compact than that of the shorter chain ones. This is likely to be due to the increase in hydrophobicity

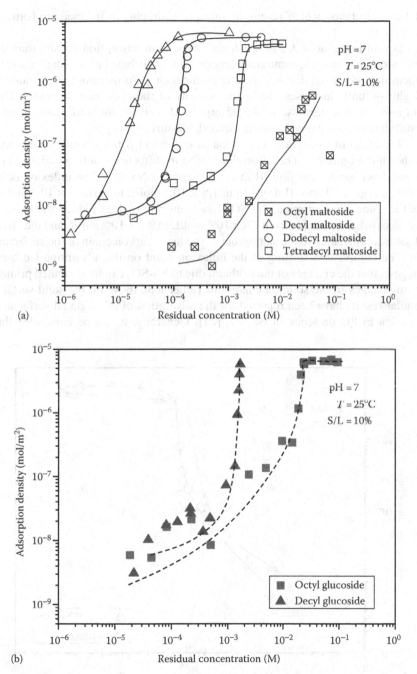

FIGURE 6.3 (a) Adsorption of n-alkyl-β-D-maltosides on alumina. (b) Adsorption of n-alkyl-β-D-glucosides on alumina.

of the surfactants, which results in stronger hydrophobic interactions between surfactants.

In addition, glucosides have a higher maximum adsorption density than the maltosides that have the same chain length. This is attributed to the smaller size of glucosides, which requires less area for their packing. Furthermore, as the number of glucose units increases, the hydrophobicity of the surfactant decreases. This can also cause a decrease in the adsorption density of the surfactant. Similar results have been obtained also for ethoxylated surfactants [21].

The effect of salt on the adsorption of n-dodecyl-β-D-maltoside on alumina is shown in Figure 6.4. The adsorption isotherm shifts to the left in regions I and II and downwards in region III in the presence of Na_2SO_4. The inflexion point between regions II and III drops from 1.8×10^{-5} mol/L to about 9×10^{-6} mol/L. At the same concentration, Na_2SO_4 has been reported to reduce the cmc of n-dodecyl-β-D-maltoside from 1.8×10^{-5} mol/L to 9.4×10^{-6} mol/L and the effect of salt was attributed to the salting-out of the hydrocarbon chain of the surfactant [9]. The comparable shifting of the inflexion point on the adsorption isotherm suggests that the changes of the isotherm due to Na_2SO_4 can be attributed primarily to changes in the solution conditions rather than those on the solid surface. Similar results have been reported for the adsorption of ethoxylated surfactants on silica in the presence of Na_2SO_4 [22]. Generally, it can be concluded that

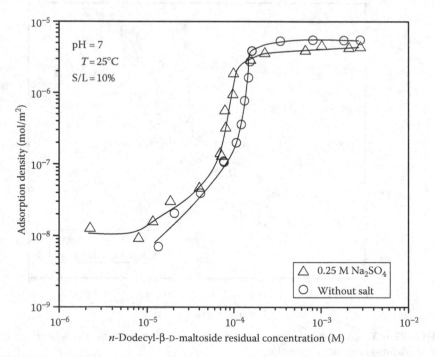

FIGURE 6.4 Effect of salt (Na_2SO_4) on the adsorption of n-dodecyl-β-D-maltoside on alumina. (From Zhang, L., Somasundaran, P., Maltesh, C., *J. Colloid Interface Sci.*, 191, 202, 1997. With permission.)

n-dodecyl-β-D-maltoside is fairly tolerant to salt with respect to adsorption on hydrophilic solids.

It is interesting to note that n-dodecyl-β-D-maltoside adsorbs on alumina, hematite, and titania. However, it absorbs much less on the silica surface. This behavior is very similar to that of polysaccharide polymers such as dextrin and starch, which adsorb only on alumina and hematite and not on silica [23,24]. This absorption preference can be explained by the fact that interactions between n-dodecyl-β-D-maltoside, which has an oligosaccharide headgroup, and oxides are similar to those between polysaccharide polymers and oxides. Nonionic ethoxylated surfactants, on the other hand, readily adsorb on silica but much less so on alumina and hematite [25–27].

To explore the mechanism by which sugar-based surfactants adsorb, the effect of pH on the adsorption of n-dodecyl-β-D-maltoside on alumina has been investigated and the results are shown in Figures 6.5 and 6.7 [17,28]. As the isoelectric point of alumina is 8.9, it is positively charged at pH 7 and negatively charged at pH 11. The identical adsorption isotherms obtained at these two pH conditions suggest that the surface charge of alumina does not govern the adsorption, indicating that the electrostatic interaction is not a factor in determining the adsorption of this surfactant on alumina. Electrokinetic measurements also show that the zeta-potential of alumina and titania after n-dodecyl-β-D-maltoside adsorption changes very little (Figure 6.6), suggesting again that the driving force for the adsorption is not electrostatic. The relatively small reduction in the

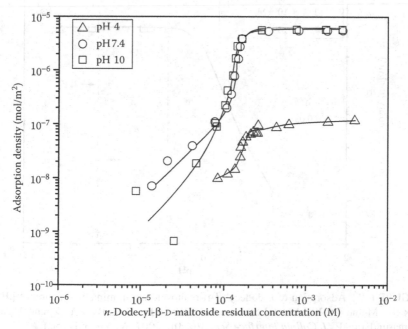

FIGURE 6.5 Effect of pH on the adsorption of n-dodecyl-β-D-maltoside on alumina.

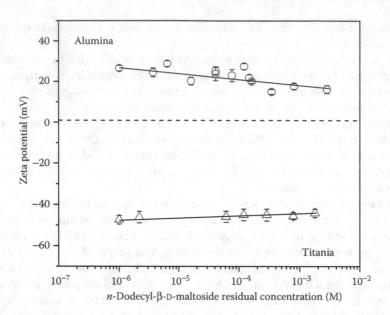

FIGURE 6.6 Zeta-potential of alumina and titania after n-dodecyl-β-D-maltoside adsorption at neutral pH. (From Zhang, L., Somasundaran, P., Maltesh, C., *J. Colloid Interface Sci.*, 191, 202, 1997. With permission.)

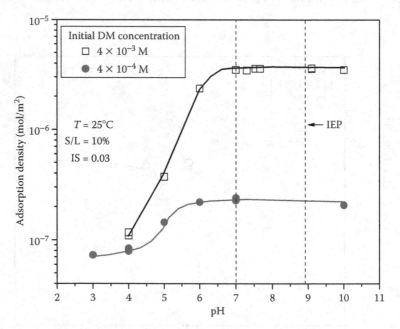

FIGURE 6.7 Adsorption of n-dodecyl-β-D-maltoside on alumina as function of pH at 4×10^{-3} M and 4×10^{-4} M total concentration. (From Lu, S.H., Bian, Y., Zhang, L., and Somasundaran, P., *J. Colloid Interface Sci.*, 316, 310, 2007. With permission.)

zeta-potential for both solids is proposed to be due to the masking of the solid surface by the adsorbed surfactant species. Surprisingly, there is drastic difference between adsorption at pH 7 and that at pH 4 shown in Figure 6.5. The saturation adsorption at pH 4 is ~1.1 × 10^{-7} mol/m^2, only 2% of that at pH 7. However, the adsorption isotherm at pH 4 shows a similar "S" shape as at pH 7, suggesting that there is a similar three-stage adsorption process. The pH effects on the adsorption of n-dodecyl-β-D-maltoside on alumina are further illustrated in Figure 6.7 for adsorption at constant initial surfactant concentrations of 4 × 10^{-3} M and 4 × 10^{-4} M in the region of pH 3 to 10 (in the case of 4 × 10^{-3} M, the concentration depletion was too small to be detected below pH 4). The adsorption decreases significantly with a decrease in pH below pH 7 but remains constant from pH 7 to 10, despite the fact that n-dodecyl-β-D-maltoside is a nonionic surfactant. This pH-dependant adsorption behavior can be accounted for by considering the driving force for adsorption as being mainly due to the hydrogen bonding between the surface hydroxyl groups and the hydroxyl groups of the surfactant sugar ring. Smith et al. [29] have proposed that the hydroxyl groups of the sugar-based surfactants are slightly acidic in nature and that they can hydrogen bond with the basic OH groups on the surface of titania particles. In the case of alcohol adsorption on alumina, it has been demonstrated that the alcohol behaves as an electronic donor, while the basic OH groups on the surface act as acceptors [30]. In addition, it has been revealed using analytical ultracentrifugation technique that the micellization of n-dodecyl-β-D-maltoside is pH dependent from pH 3 to 10 [28]. Therefore, the adsorption density changes can be attributed to the concentration of hydroxyl groups of the surfactant and/or the OH groups on solid surface.

The role of these surface groups on the surfactants and solids is examined by correlating the surface OH group concentration and adsorption density. Surface speciation of oxide surface is expected to depend on the solution parameters such as pH and ionic strength. For instance, the surface species of alumina in the aqueous solutions can be described by the following surface reactions [31].

$$-AlOH_2^+ \leftrightarrow -AlOH + H_s^+$$

$$-AlOH \leftrightarrow -AlO^- + H_s^+$$

The surface concentration of hydrogen ion can be replaced with the solution concentration using Boltzmann's equation, in which H_s^+ is a function of solution concentration H^+ and the surface potential of the solid. The surface ionization constants can be obtained from the zeta potential data [32]. The ionization constants pK_{a1} and pK_{a2} used are 5.8 and 12. By assuming that $H_s^+ \approx H^+$ and that the effect of ionic strength is negligible, the surface ion concentrations are estimated. The total hydroxide content is assumed to be twice that of the site density of gamma alumina (6.1 × 10^{-6} mol/m^2) [31]. The concentrations of alumina surface

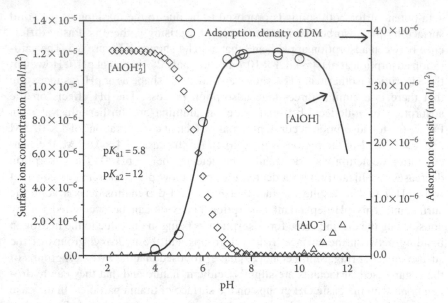

FIGURE 6.8 Surface ion concentration of alumina and adsorption of DM on alumina. (From Lu, S.H., Bian, Y., Zhang, L., and Somasundaran, P., *J. Colloid Interface Sci.*, 316, 310, 2007. With permission.)

species, $-AlOH$, $-Al(OH_2)^+$, and $-AlO^-$ are plotted as a function of pH in Figure 6.8 along with the adsorption density. When plotted on the appropriate scales, the adsorption density of n-dodecyl-β-D-maltoside follows the same trend as the concentration of $-AlOH$. Consequently, the findings clearly demonstrate that the adsorption density of n-dodecyl-β-D-maltoside is determined by the concentration of surface hydroxyl groups on alumina. To quantify the correlation, the adsorption density on alumina is plotted in Figure 6.9 as a function of $-AlOH$ group concentration. It can be seen that the adsorption density increases linearly with the $-AlOH$ concentration.

Similar phenomena were observed in the case of adsorption on hematite. Literature values were used for the surface ionization constants: pK_{a1}, 2.9 and pK_{a2}, 10.5 for hematite. The surface $-FeOH$ group concentration changes with pH and in turn affects the adsorption of the surfactant, as shown in Figure 6.10. The adsorption density of n-dodecyl-β-D-maltoside on hematite follows the surface $-FeOH$ group concentration.

The experimental results show an excellent linear relationship between the adsorption density of the sugar-based surfactant and the estimated surface hydroxyl group concentration, suggesting an adsorption mechanism for sugar based surfactant on the oxide surface as illustrated in Figure 6.11. It is proposed that the hydrogen-bonding interaction occurs only between the $-AlOH$ groups, the $-OH$ and oxygen groups of the surfactant and the other surface $-AlO^-$ and $-Al(OH_2)^+$ groups do not contribute due to the different electronic properties. This mechanism

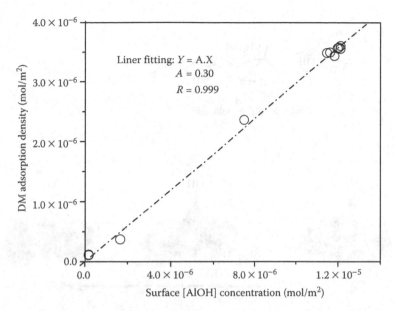

FIGURE 6.9 Correlation between surface ion concentration of alumina and adsorption of dodecyl maltoside on alumina. (From Lu, S.H., Bian, Y., Zhang, L., and Somasundaran, P., *J. Colloid Interface Sci.*, 316, 310, 2007. With permission.)

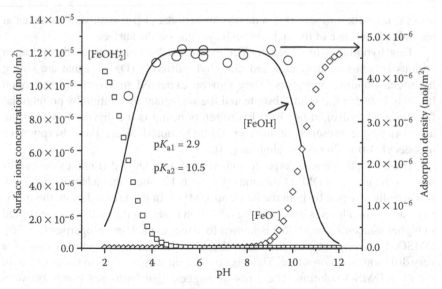

FIGURE 6.10 Surface ion concentration of alumina and adsorption of n-dodecyl-β-D-maltoside on hematite. (From Lu, S.H., Bian, Y., Zhang, L., and Somasundaran, P., *J. Colloid Interface Sci.*, 316, 310, 2007. With permission.)

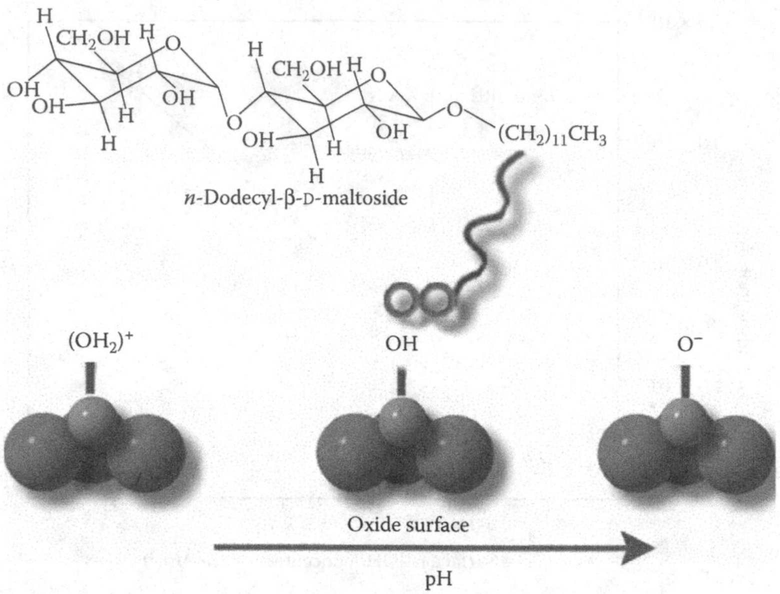

FIGURE 6.11 Illustration of hydrogen bonding between hydroxyl groups of sugar-based surfactant and oxide surface. (From Lu, S.H., Bian, Y., Zhang, L., and Somasundaran, P., *J. Colloid Interface Sci.*, 316, 310, 2007. With permission.)

also explains the low adsorption density of *n*-dodecyl-β-D-maltoside on silica at neutral pH, because of the lack of –SiOH groups on the surface.

This hypothesis has been further examined by conducting adsorption in various solvents such as urea and dimethyl sulfoxide (DMSO) that are strong hydrogen-bonding acceptors. They can be expected to affect the hydrogen bonding between the solid substrate and the surfactant in solution by preferential formation of hydrogen bonds. If hydrogen bonding is the driving force for the adsorption, the presence of urea or DMSO should affect the adsorption of *n*-dodecyl-β-D-maltoside on alumina [33].

The adsorption and desorption isotherms of *n*-dodecyl-β-D-maltoside on alumina in the urea and DMSO solutions (Figure 6.12) show that adsorption density is markedly affected by both the urea and DMSO. In the urea solution, the maximum adsorption level is lower and the inflection points of the isotherms are shifted to higher concentrations. Urea is known to affect the solvent properties [34–36]. DMSO, however, not only decreased the adsorption density, but also yielded a very different isotherm shape. There is no clear inflection point on the isotherm in the 85% DMSO solution. These results suggest that hydrogen bonds between solids and surfactants are weakened in the presence of urea and DMSO. In addition, the reversible adsorption of *n*-dodecyl-β-D-maltoside on alumina in water and in the presence of urea and DMSO also suggests that there is no chemical

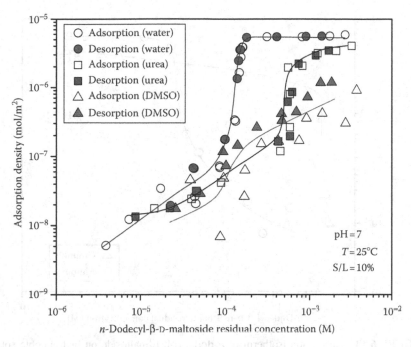

FIGURE 6.12   Adsorption and desorption of n-dodecyl-β-D-maltoside in the presence of 5 M urea and 85% wt DMSO.

interaction between the two. No chemical interaction has been found in the Fourier transform infrared spectroscopy (FTIR) study of the n-dodecyl-β-D-maltoside/alumina system [37,38].

6.3 ADSORPTION OF n-DODECYL-β-D-MALTOSIDE ON HYDROPHOBIC SOLIDS

Adsorption isotherm of n-dodecyl-β-D-maltoside on a hydrophobic solid graphite, is shown in Figure 6.13. The adsorption isotherm on graphite is quite different from that on hydrophilic solids. The adsorption density rapidly increases at a very low DM concentration and reaches a plateau far below the cmc of the surfactant. Evidently aggregation of the surfactant in the bulk solution itself does not affect the adsorption process. These results are in accordance with the literature on the adsorption of nonionic surfactants on hydrophobic surfaces [39]. Due to the hydrophobic nature of the solid, surfactants adsorb on the solid with the hydrophobic groups attached to the graphite surface by hydrophobic interactions [40]. The sharp increase of the isotherm at low DM concentration suggests very strong interaction of the surfactant with the solids in solutions. Above the inflexion point, the slow increase in adsorption density occurs due to the reconstruction of the adsorbed layer, where the chains are closely packed and stand vertically.

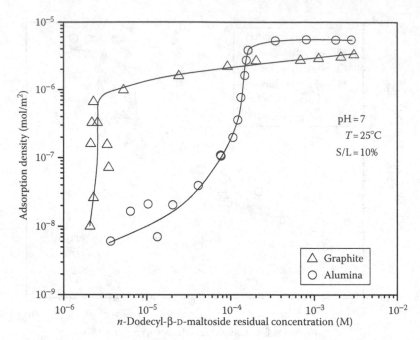

FIGURE 6.13 Adsorption isotherm of n-dodecyl-β-D-maltoside on hydrophobic solid graphite. (From Zhang, L., Somasundaran, P., and Maltesh, C., *J. Colloid Interface Sci.*, 191, 202, 1997. With permission.)

In the plateau region, the adsorption density is 3×10^{-6} mol/m² and the surface area per molecule in this region is estimated to be 55 A². This value is close to the value of 48.5 A² that was derived from the surface tension experiments and it is also half of that for alumina. The amount of surfactant adsorbed on the graphite is close to that of a monolayer.

The conformation of the adsorbed layer on graphite is illustrated in the form of the effect of surfactant adsorption on the wettability and stability. Both tests give the relative hydrophobicity of the solid. In the wettability tests (Figure 6.14), the graphite surface exhibits complete hydrophobicity in the absence of the surfactant. With the adsorption on graphite, the surface becomes increasingly hydrophilic due to the surfactant orientation with their hydrophobic heads attached to the surface and the hydrophilic parts dangling into the solution. The effect of the surfactant on the stability of graphite particles, as measured by the settling rate, is shown in the Figure 6.15. The graphite suspension is unstable without the surfactant. As DM adsorbs, the settling rate is found to decrease sharply. Complete dispersion of graphite occurred at a concentration corresponding to the onset of plateau on the adsorption curves. The stabilization of hydrophobic suspensions by nonionic surfactants has been observed for many other systems [41,42].

The above observations suggest that the adsorption of n-dodecyl-β-D-maltoside on graphite is monomolecular with the nonpolar groups attached to the solid surface and the headgroups oriented toward the bulk solution. The adsorption of surfactant

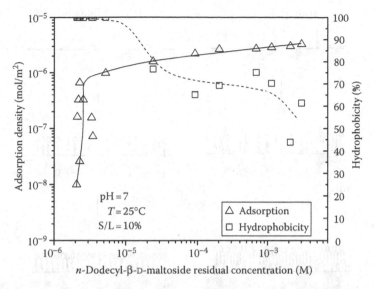

FIGURE 6.14 Adsorption isotherm of *n*-dodecyl-β-D-maltoside on graphite and the effect on its hydrophobicity. (From Zhang, L., Somasundaran, P., Maltesh, C., *J. Colloid Interface Sci.*, 191, 202, 1997. With permission.)

on hydrophobic solid demonstrates a drastically different behavior when compared with the hydrophilic solids due to the markedly different interactions between the surfactant and the solid surface. Hydrophobic interactions between

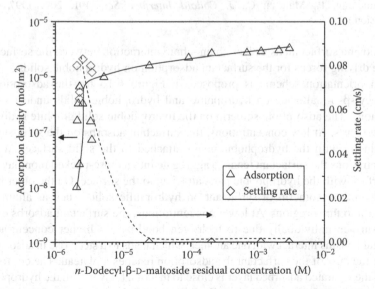

FIGURE 6.15 Adsorption isotherm of *n*-dodecyl-β-D-maltoside on graphite and its effect on the stability of graphite suspension. (From Zhang, L., Somasundaran, P., Maltesh, C., *J. Colloid Interface Sci.*, 191, 202, 1997. With permission.)

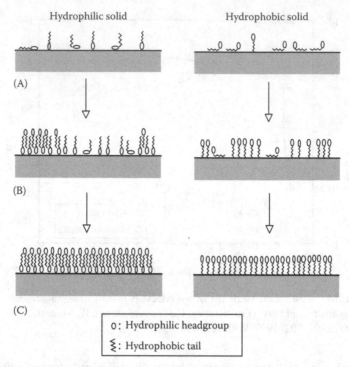

FIGURE 6.16   Proposed orientation model for the adsorption of *n*-dodecyl-β-D-maltoside on hydrophilic and hydrophobic surfaces in aqueous media, showing the conformation of surfactant at the surface. A–C indicate the successive stages of adsorption. (From Zhang, L., Somasundaran, P., Maltesh, C., *J. Colloid Interface Sci.*, 191, 202, 1997. With permission.)

the solid and surfactant as well as chain–chain interaction between the surfactants are the driving forces for the surfactant adsorption on hydrophobic solids.

An orientation scheme is proposed in Figure 6.16 for the adsorption of *n*-dodecyl-β-D-maltoside on hydrophilic and hydrophobic solids under various conditions. The adsorption isotherm on the hydrophobic solid graphite is divided into two parts. At low concentrations, the surfactant adsorbs on the solid surface individually with the hydrophobic chain attached to the solid surface. At high concentrations the surfactant tends to aggregate into a close-packed monolayer on the surface with the hydrophobic parts attached to the surface. On the other hand, the adsorption isotherm of surfactant on hydrophilic solids such as alumina is divided into three regions. At lower concentrations, the surfactant adsorbs on the solid surface individually due to hydrogen bonding. At higher concentrations surfactant chains interact with one another, which leads to a steep rise in adsorption. Above the cmc of the surfactant the adsorption reaches a plateau. The conformation of the saturated adsorbed layer is close to that of a bilayer with the hydrophobic chains interpenetrating each other. On both types of solids the surface becomes hydrophilic in the final adsorption state.

6.4 ADSORPTION OF MIXTURES OF n-DODECYL-β-D-MALTOSIDE WITH OTHER SURFACTANTS ON SOLIDS

The use of mixed surfactants for the modification of solid surfaces is important for many applications since the occurrence of beneficial synergism depends on the surfactant type and mixing conditions. The solid/liquid interfacial behavior of surfactant mixtures has been explored extensively as it is important for practical applications of surfactants [43–50].

Sugar-based surfactants have been studied in mixtures with other surfactants to determine whether or not there is possible synergism or antagonism under various conditions. In solutions, sugar-based surfactants interact synergistically with cationic and anionic surfactants. The magnitude of such interactions generally follows the order anionic/nonionic > cationic/nonionic > nonionic/nonionic [12]; however, at solid/liquid interface, the synergy or antagonism between the mixed surfactants is more complicated, since it is governed not only by intermolecular interactions but also by the surfactant/solid interactions. Here we use a general active/passive model [51] developed to discuss the adsorption of sugar-based surfactant in mixed system.

Generally, ionic surfactants adsorb only on the oppositely charged solid surfaces through electrostatic attraction, while the adsorption of nonionic surfactants is usually not affected by the surface charge. In a surfactant mixture, we define a surfactant that adsorbs on a solid as an active species and the one that coadsorbs as a passive species. For instance, ionic surfactants are passive in the case of solids with same charge. As mentioned above, n-dodecyl-β-D-maltoside is passive in the case of silica, but becomes active in the case of alumina from pH 7 to 10. The interaction between the mixed surfactant at solid/liquid interfaces follows a general rule as shown in Table 6.1.

1. Active nonionic + active ionic/nonionic surfactants
   When both components in the mixed surfactant system adsorb strongly on the solids, the mixture generally exhibits synergism in the preplateau region, whereas competition for adsorption sites occurs in the plateau region.

TABLE 6.1
Interactions between Mixed Surfactants at Solid/Liquid Interfaces

Nonionic	Anionic/Cationic/ (Nonionic)	Interaction		
		Pre-Plateau Region	Plateau Region	References
Active	Active	Synergism	Antagonism	[52–54]
Active	Passive	Synergism	Antagonism	[54]
Passive	Active	Synergism	Synergism	[55–60]

Note: Active, a surfactant that adsorbs; passive, a surfactant that coadsorbs in a mixture system.

For instance, in the case of mixed sodium dodecyl sulfate (SDS)/$C_{12}EO_8$ on kaolinite [52], strong synergistic interaction has been reported in the pre-plateau region, while competition for adsorption sites was observed in the plateau region. The composition of the adsorbed layer in the plateau region is governed by the balance between the adsorption driving forces for each surfactant. Generally, ionic surfactants adsorb more in this region than non-ionic surfactants, because the electrostatic attraction is stronger than the driving forces for nonionic surfactants, such as hydrogen bonding.

2. Active nonionic + passive ionic surfactants

Usually, anionic or cationic surfactants are passive when it comes to simi-larly charged solid surfaces; however, in the presence of active nonionic sur-factants, they can coadsorb through hydrophobic chain–chain interactions. Also, the nonionic surfactants screen the electrostatic repulsion and favor molecular packing in the adsorbed layer. In the preplateau region synergistic interaction often promotes the adsorption of the ionic components, while at the plateau region the adsorption of the mixture is lower than that of the nonionic surfactant alone due to the antagonistic interaction in this region.

3. Passive nonionic + active ionic/nonionic surfactants

When a nonionic surfactant is passive, a strong synergism is often expected in the mixture with an active ionic or nonionic surfactant in both preplateau and plateau regions. The synergism is attributed to the hydrophobic chain–chain interaction between the surfactants, which increases the adsorption of the nonionic surfactant significantly. For example, the adsorption of n-dodecyl-β-D-maltoside on silica was enhanced by two orders of magni-tude by the presence of cationic dodecyldimethylamine oxide (DDAO) [55] and Gemini surfactant [57]. Similar phenomena have been reported in the case of mixed nonionic $C_{12}EO_8$/sodium p-octylbenzene sulfonate [58], and mixed pentadecylethoxylated nonyl phenol (NP.15)/tetradecyl trimethylam-monium chloride (TTAC) systems on alumina [59].

As discussed in the previous section, the adsorption of n-dodecyl-β-D-maltoside is not only solid selective but is also pH dependent [61]. The investigation of the surfactant mixtures adsorption that contains n-dodecyl-β-D-maltoside is important for understanding the synergistic/antagonistic interaction in the adsorbed layer of mixed surfactants. The adsorption of n-dodecyl-β-D-maltoside in the mixtures with anionic sodium dodecyl sulfate (SDS) and sodium dodecyl sulfonate (SDS_1), cat-ionic and nonionic dodecyl trimethylammonium bromide (DTAB), and nonionic nonyl phenol ethoxylated decyl ether (NP-10) on alumina and silica are discussed below to understand these interactions between the surfactants.

6.4.1 Adsorption of N-Dodecyl-β-D-Maltoside with Anionic Surfactant on Alumina

Adsorption of nonionic sugar-based surfactant, n-dodecyl-β-D-maltoside, in mixtures with anionic sodium dodecylsulfate and sodium dodecyl sulfonate on alumina was studied from pH 4 to 11. In this range, the adsorption of *n*-dodecyl-β-D-maltoside and the anionic surfactant on alumina is pH-dependent with opposite trends.

6.4.1.1 Adsorption of *n*-Dodecyl-β-D-Maltoside/Sodium Dodecylsulfate Mixtures at pH 6 (Active Nonionic/Active Anionic)

The isoelectric point of alumina used in this study is 8.9. At pH 6, alumina is positively charged. Anionic sodium dodecylsulfate (SDS) by itself can adsorb on alumina due to electrostatic interaction. Nonionic *n*-dodecyl-β-D-maltoside (DM) can also adsorb on alumina through hydrogen bonding. The adsorption isotherms of DM/SDS 3:1, 1:1, and 1:3 mixtures on alumina at pH 6 are shown in Figure 6.17, together with those of dodecyl-β-D-maltoside and SDS alone. The adsorption of the mixtures is higher than that of either of the components in the sharply rising part of the isotherm, showing a strong synergy between DM and SDS. This is the region where hydrophobic chain–chain interactions dominate the adsorption process, because the surface is not yet saturated with the surfactants. At lower surfactant concentrations, SDS adsorbs more than DM, because the electrostatic interactions are stronger than hydrogen bonding. In this region, the adsorption takes place mainly due to the electrostatic attraction between the negatively charged dodecylsulfate and positively charged alumina. However, some adsorption of the sugar-based surfactant is due to hydrogen bonding. At higher concentrations, the adsorbed SDS forms mixed aggregates with DM through hydrophobic

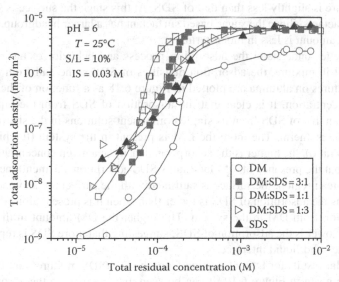

FIGURE 6.17 Adsorption of DM, SDS, and DM/SDS mixtures on alumina. (From Zhang, L., Zhang, R., and Somasundaran, P., *J. Colloid Interface Sci.*, 302, 25, 2006. With permission.)

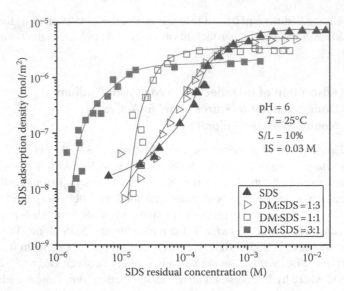

FIGURE 6.18 Adsorption of SDS on alumina: Adsorption for SDS alone and from DM/SDS 3:1, 1:1, and 1:3 mixtures. (From Zhang, L., Zhang, R., and Somasundaran, P., *J. Colloid Interface Sci.*, 302, 25, 2006. With permission.)

chain–chain interactions and promotes the DM adsorption. The low critical micellar concentration of DM causes the aggregates to form at lower concentrations promoting total adsorption as well. In the plateau region, the adsorption density of the mixture is slightly less than that of SDS. At this stage the surface is saturated with surfactants. Since the sugar-based surfactant has a larger headgroup, the total adsorbed amount is less in molar terms.

To better understand the adsorption process and behavior of an individual surfactant in mixture, the adsorption densities of SDS alone and from the DM/SDS mixtures on alumina are plotted in Figure 6.18 as a function of the residual SDS concentration. It is clear that the adsorption of SDS from the mixtures is higher than that of SDS from its single component solutions in the sharply rising part of the isotherm. The more the DM is present in the system (the higher the DM/SDS ratio), the higher is the adsorption density at a given concentration, suggesting that the presence of DM facilitates SDS adsorption. As mentioned above, in the plateau region, the surface is saturated with the surfactant, and under these conditions the adsorption of SDS is lower than when it is present alone due to the competition from DM in the system. The higher the DM amount in the mixing ratio, the lower is the adsorption of SDS, suggesting that more SDS is replaced by DM at the solid/liquid interface.

Similar results for DM adsorption from the DM/SDS mixtures and from DM alone are given in Figure 6.19. It can be seen that in this case the adsorption of DM is enhanced by SDS in the entire concentration range. Interestingly, the more the SDS is present in the system, the higher the adsorption is of DM in the rising part,

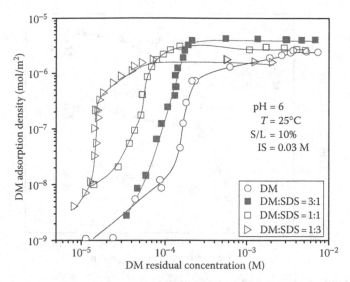

FIGURE 6.19 Adsorption of DM on alumina: Adsorption for DM alone and from DM/SDS 3:1, 1:1, and 1:3 mixtures. (From Zhang, L., Zhang, R., and Somasundaran, P., *J. Colloid Interface Sci.*, 302, 25, 2006. With permission.)

suggesting that there are synergistic effects between SDS and DM adsorption. It should be noted that the adsorption activity of DM on alumina at pH 6 is limited as shown in Figure 6.7, while that of SDS is fully active. This explains the fact that the plateau adsorption of DM is promoted while that of SDS is reduced.

To further understand the interaction between the two surfactants, the composition of the adsorbed layer is plotted in Figure 6.20 as a function of the total adsorption density. In the case of DM/SDS 1:1 and 1:3 mixtures, the DM ratio in the adsorbed layer is very close to the bulk mixing ratio. However, in the case of 3:1 mixture, the DM ratio is small at low concentrations and reaches a maximum at the on-set of the adsorption plateau and then decreases again. This phenomenon is attributed to the stronger interaction between SDS and the solid, due to the fact that the electrostatic interaction between SDS and alumina is stronger than the hydrogen bonding between DM and alumina. The results suggest that the low molar ratio component is favored in the preplateau region. In addition, it is apparent that the DM ratio in the adsorbed layer decreases in the plateau regions at the three mixing ratios tested. This suggests that the replacement of DM molecules by SDS molecules in the surfactant mixtures increase when the alumina surface is saturated. This compositional change in the adsorbed layer can be attributed to the decrease in DM monomer concentration in bulk. The DM monomers concentration is expected to decrease with concentration above cmc, since DM is more surface active than SDS in bulk. For this system, the monomer concentration was estimated using the regular solution theory and a correlation between the DM ratio in the adsorbed layer and that in solution was observed [62].

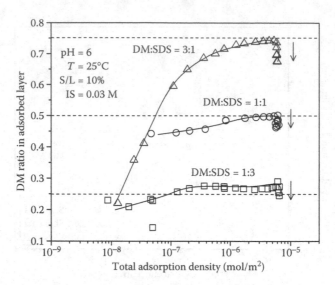

FIGURE 6.20 DM ratios in the adsorption layer.

6.4.1.2 Adsorption of *n*-Dodecyl-β-D-Maltoside/Sodium Dodecylsulfate Mixtures at pH 11 (Active Nonionic/Passive Anionic)

At pH 11, alumina are negatively charged and *n*-dodecyl-β-D-maltoside can adsorb under these conditions. However, sodium dodecylsulfate cannot adsorb due to its anionic nature. The adsorption of *n*-dodecyl-β-D-maltoside (DM), SDS, and their 3:1, 1:1, and 1:3 mixtures is illustrated in Figure 6.21. The adsorption behavior can be regarded as a mixed active nonionic/passive anionic system. At this pH, the adsorption of negatively charged SDS on the similarly charged alumina is very low, while the adsorption of DM is similar to that at pH 6. Surprisingly, the adsorption of the mixtures under these conditions is between that of DM and SDS. The more the SDS is present in the system, the lower is the total adsorption density of the surface mixtures. The presence of SDS in the system reduces the adsorption of the sugar-based surfactants under these conditions due to the antagonistic interaction in the plateau region.

To explore the adsorption behavior of individual surfactant in the mixtures at the solid/liquid interface, the adsorption of SDS from SDS solution and from the DM/SDS mixtures on alumina is plotted in Figure 6.22 as a function of the residual SDS concentration. Adsorption of SDS from the mixtures is enhanced by the presence of DM except in the very high concentration region. This is proposed to be due to the adsorbed DM functioning as anchor molecule for the SDS through hydrophobic chain–chain interactions. Therefore at least for SDS, there are some synergistic effects in the surfactant mixtures with DM.

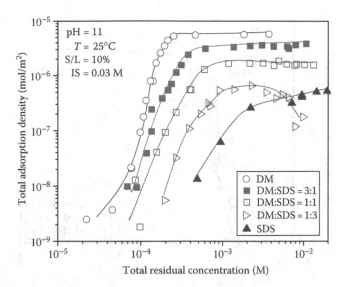

FIGURE 6.21 Adsorption of DM, SDS, and their 3:1, 1:1, and 1:3 mixtures on alumina at pH 11. (From Zhang, L., Zhang, R., and Somasundaran, P., *J. Colloid Interface Sci.*, 302, 25, 2006. With permission.)

In contrast, it can be seen from Figure 6.23 that the DM adsorption is enhanced by SDS only in the rising part of the plateau, but is depressed in the plateau region. Figure 6.21 demonstrates that the total adsorption of DM + SDS is decreased as SDS in the mixture increased. Thus it can be concluded that in this system at pH 11,

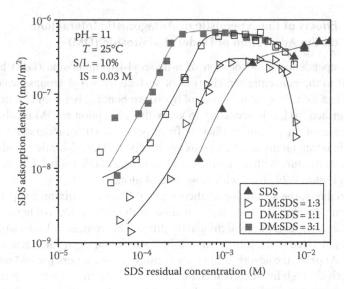

FIGURE 6.22 Adsorption of SDS on alumina at pH 11: Adsorption for SDS alone and from DM/SDS 3:1, 1:1, and 1:3 mixtures. (From Zhang, L., Zhang, R., and Somasundaran, P., *J. Colloid Interface Sci.*, 302, 25, 2006. With permission.)

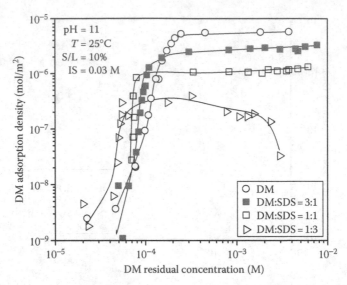

FIGURE 6.23 Adsorption of DM on alumina at pH 11: Adsorption for DM alone and from DM/SDS 3:1, 1:1, and 1:3 mixtures. (From Zhang, L., Zhang, R., and Somasundaran, P., *J. Colloid Interface Sci.*, 302, 25, 2006. With permission.)

there are mainly antagonist effects between DM and SDS, because the adsorption of the mixtures of DM/SDS is markedly lower than that of DM alone.

6.4.1.3 Effects of the Synergistic or Antagonistic Interactions on the Adsorption of Dodecyl Maltoside (DM)

The pH dependence of adsorption of n-dodecyl-β-D-maltoside (DM) has been correlated to the concentration changes of surface hydroxyl groups, which were proposed to account for the formation of hydrogen bonding between DM molecules and the surface [61]. It is necessary to know the adsorption of DM on alumina in the presence of an anionic surfactant from pH 4 to 10 to understand whether synergistic or antagonistic interactions are involved. The adsorption isotherms of DM from its mixture with sodium dodecyl sulfonate (SDS_1) at pH 4, 7, and 10 are plotted in Figure 6.24 along with those of DM alone.

Surprisingly, the adsorption isotherms of DM from its mixtures at pH 4 and 7 on alumina are almost identical, even though the adsorption isotherms of DM alone under same conditions significantly differ. The mixture exhibits strong synergism at pH 4 due to the formation of mixed surfactant aggregates at solid/liquid interface. At pH 4, the adsorbed SDS_1 molecules act as anchors for DM molecules to coadsorb through hydrocarbon chain–chain interaction, even though the direct association between DM and the surface is very weak. Furthermore, the presence of DM molecules screens the electrostatic repulsion between SDS_1 molecules and

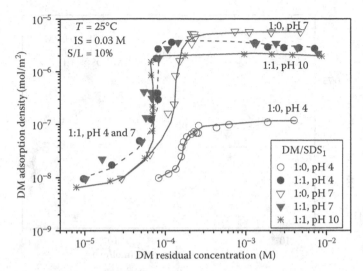

FIGURE 6.24 Adsorption of DM on alumina from its mixtures with sodium dodecyl sulfonate at various pH values. (IS: ionic strength, Mol/L; S/L: solid/liquid ratio, g/g; DM/SDS$_1$, mixing ratio Mol/Mol.) (From Lu, S., and Somasundaran, P., *Langmuir*, 24, 3874, 2008. With permission.)

promotes the formation of mixed aggregates and thus enhancing the total adsorption. At pH 10, the saturation adsorption of DM from its mixture at a mixing ratio 1:1 is 1/3 of that of DM alone, suggesting an antagonistic interaction.

6.4.1.4 Effects of pH on the Composition of the Adsorbed Layer of Mixed DM/SDS$_1$ on Alumina

The pH dependence of the adsorption of mixed DM/SDS$_1$ on alumina makes this system a unique mixture in terms of the interactions involved. In other words, the interaction between these two surfactants at solid/liquid interface is tunable with pH. At pH 4, this mixture acts as a passive nonionic/active anionic; at pH 7 it becomes an active nonionic/active anionic; and at pH 10 it behaves as an active nonionic/passive anionic. In other words, it exhibits all the three types of mixtures in the passive/active model.

The DM molar ratios in the adsorbed layer at pH 4, 7, and 10 are shown as a function of total adsorption density in Figure 6.25 for a mixing ratio 1:1. Although the mixing ratio remains constant, the DM ratios in the adsorbed layer are very different. At pH 4, the DM molar ratio in the adsorbed layer approaches 0.5 and then decreases in the plateau region. The DM ratio is lower than 0.5 in the whole isotherm region, since the adsorption of DM depends on that of SDS$_1$ adsorption. At pH 7, the DM ratio is close to 0.5 in the preplateau region, because both DM and SDS are active. At pH 10, the DM ratio in the adsorbed layer is greater than 0.5 in the preplateau region, due to the fact that the adsorption of SDS$_1$ depends

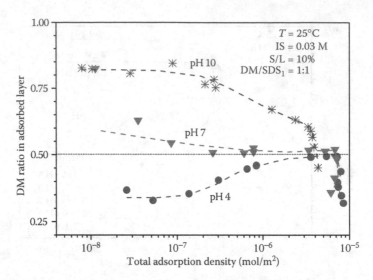

FIGURE 6.25 DM molar ratio in adsorbed layer on alumina. (IS: ionic strength, Mol/L; S/L: solid/liquid ratio, g/g; DM/SDS1, mixing ratio Mol/Mol.) (From Lu, S., and Somasundaran, P., *Langmuir*, 24, 3874, 2008. With permission.)

on DM. Interestingly, the DM molar ratios in the adsorbed layer decrease in the plateau region at all the three pHs, suggesting that this compositional change is independent of the interaction whether it is synergistic or antagonistic. As DM is more surface active than SDS_1, the DM monomer concentration is expected to be lower than that of SDS_1 in the concentration above cmc. It is hypothesized that the molar ratio in the adsorbed layer in the plateau region is determined mainly by the monomer concentrations of the components in the mixture.

Due the difference in the adsorption density and the interactions involved, the structure of the adsorbed layers is expected to vary significantly with the pH (Figure 6.26). At an acidic pH, DM is passive, while SDS is active. In this pH range the mixture exhibits strong synergism, as the pre-adsorbed SDS molecules act as anchors for DM to adsorb through hydrophobic chain–chain interaction. At neutral pH, both DM and SDS are active. In the preplateau region, the formation of mixed aggregates promotes the adsorption with synergistic interactions, while in the plateau region, competition for the adsorption sites occurs and the mixture exhibits antagonism. At pH above isoelectric point (IEP) of alumina, DM is active, whereas SDS is passive. The adsorption of SDS depends on the adsorbed DM, which was achieved through hydrophobic chain–chain interaction.

In addition, the DM molar ratio in the adsorbed layer was observed to decrease in the plateau region at the pH and mixing ratios tested. The compositional changes in the adsorbed layer can be attributed to the monomer concentration change in the bulk, which is related to the surface activity. In this mixture system, the DM molar ratios in monomers are expected to decrease due to their higher surface activity at concentrations above cmc and this decrease accounts

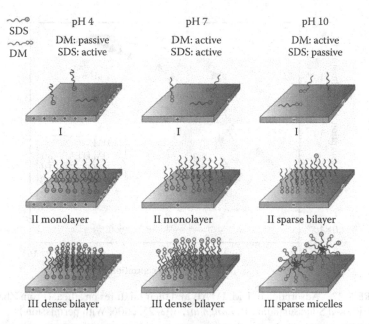

FIGURE 6.26 Illustration of adsorption of DM/SDS mixture on alumina at pH 4, 7, and 10 in the three adsorption region. I, the low adsorption region; II, the sharply rising region; III, the plateau region. (From Lu, S., and Somasundaran, P., *Langmuir*, 24, 3874, 2008. With permission.)

for the compositional change in the adsorbed layer. Our study yields unique information for understanding the synergistic and antagonistic interactions in adsorption of surfactant mixtures.

6.4.2 ADSORPTION OF *N*-DODECYL-β-D-MALTOSIDE WITH CATIONIC SURFACTANT ON SILICA (PASSIVE NONIONIC/ACTIVE CATIONIC)

The adsorption of nonionic–cationic mixtures of *n*-dodecyl-β-D-maltoside/dodecyltrimethyl ammonium bromide (DM/DTAB) on silica was studied as a function of their total concentrations. *n*-dodecyl-β-D-maltoside does not adsorb on silica at the tested pH due to a lack of hydrogen bonding, whereas under the same conditions the cationic DTAB adsorbs strongly on silica. The adsorption isotherms of 3:1, 1:1, and 1:3 DM/DTAB mixtures on silica at pH 9 are shown in Figure 6.27, along with those of for the single surfactants. The adsorption of DM alone on silica is very low. This is in accord with our previously reported finding, that DM adsorbs strongly on alumina but very weakly on silica. DTAB, on the other hand, exhibits strong adsorption on the negatively charged silica due to the electrostatic attractions between the surfactant and the solid.

The adsorption of the mixtures is lower than that of DTAB at low concentrations because at these concentrations DM does not adsorb on silica. The isotherms show

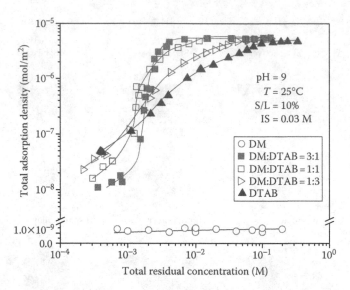

FIGURE 6.27 Adsorption of DM, DTAB and their mixtures on silica. (From Zhang, L., Zhang, R., and Somasundaran, P., *Langmuir*, 302, 25, 2006. With permission.)

a sharp increase at 0.1–0.2 mM, indicating the onset of the region where hydrophobic chain–chain interaction begins to dominate the adsorption process. In this region, the adsorption of the mixtures is higher than that of DM or DTAB alone. Two factors contribute to this increase. First, the highly surface active DM reduces the critical micelle and hemimicelle concentration of the surfactant, leading to a high adsorption at low concentrations. Next, DM can coadsorb with DTAB on silica due to hydrophobic chain–chain interactions. Therefore, there is a strong synergistic effect between dodecylmaltoside and dodecyltrimethyl ammonium bromide, especially in the sharp rising portion of the isotherms.

To elucidate the behavior of each surfactant component in the mixture, the adsorption densities of dodecyltrimethyl ammonium bromide alone and from the DM/DTAB mixtures on silica are plotted in Figure 6.28 as a function of residual DTAB concentration. It is clear that the presence of DM promotes DTAB adsorption and the adsorption of DTAB from the mixtures is higher than that of DTAB from its single component solutions. The higher the percentage of DM in the mixture, the higher is the adsorption of DTAB in regions below the plateau. In the plateau region, the surface is saturated with surfactants and under these conditions the adsorption of dodecyltrimethyl ammonium bromide (DTAB) is less than when it is present alone due to the competition from dodecyl maltoside (DM) for adsorption sites on the solid.

Similarly, to identify the behavior of dodecyl maltoside in the mixture, adsorption of DM from the DM/DTAB mixtures and DM is given in Figure 6.29. It can be seen that while DM does not adsorb on silica by itself, its adsorption is greatly facilitated by the presence of DTAB. The adsorption in mixtures is proposed to be due to the hydrophobic chain–chain interactions between DM and DTAB.

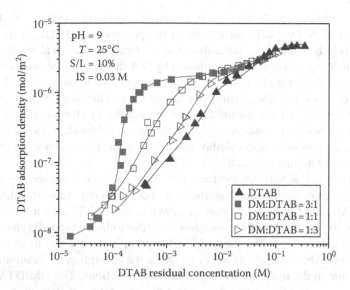

FIGURE 6.28 Adsorption of DTAB on silica: Adsorption alone and from DM/DTAB mixtures. (From Zhang, L., Zhang, R., and Somasundaran, P., *J. Colloid Interface Sci*, 302, 25, 2006. With permission.)

The DTAB in the hemimicelles, which form at the solid/liquid interface, acts as anchor species for the DM molecules. This situation is similar to that of mixed DM/SDS$_1$ on alumina at pH 4.

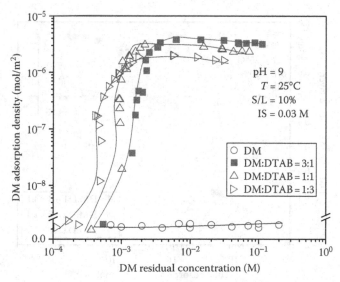

FIGURE 6.29 Adsorption of DM on silica: Adsorption alone and from DM/DTAB mixtures. (From Zhang, L., Zhang, R., and Somasundaran, P., *J. Colloid Interface Sci*, 302, 25, 2006. With permission.)

Thus, it can be concluded that sugar-based dodecyl maltoside, which does not adsorb on silica by itself, can adsorb in the presence of DTAB. In the mixed system, DTAB also adsorbs more from the mixture than it does alone, due to the reduction of critical micelle concentration by DM and an increase in the monomer activity. In a similar study, Huang et al. has reported that in the case of adsorption of the cationic tetradecyl trimethyl ammonium chloride/nonionic pentadecylethoxylated nonyl phenol mixtures at the negatively charged alumina/liquid interface, presence of tetradecyl trimethyl ammonium chloride in the system leads to the adsorption of pentadecylethoxylated nonyl phenol on alumina, where the latter does not normally adsorb by itself [59].

The interaction between the two surfactants can be clearly seen by plotting the adsorption of the mixtures together with DM ratios in the adsorbed layer. The DM/DTAB ratio for the 1:1 mixture is shown in Figure 6.30. At a low adsorption region ($<10^{-7}$ mol/m^2), DM shows a low ratio due to lack of specific chain–chain interactions between it and the solid. Adsorption is mostly due to electrostatic interactions. Above the low adsorption region, the adsorption of the mixture rises sharply due to the hydrophobic chain–chain interactions. The DM/DTAB ratio also increases rapidly at this concentration, suggesting that there are strong hydrophobic chain–chain interactions that cause more DM to coadsorb on the surface. In a certain adsorption region, the DM ratio is more than 0.5, suggesting that DM adsorbs more than DTAB itself, although it is coadsorbing through DTAB. The DM/DTAB ratio reaches a maximum and then decreases. The decrease is attributed to the competition between the micelles in solution and the hemimicelles at the solid/liquid interface for the more surface active DM. Interestingly,

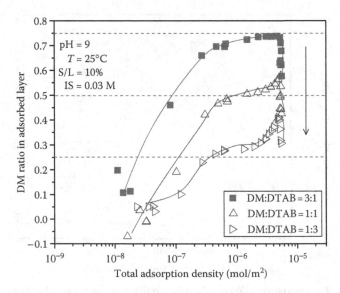

FIGURE 6.30   Adsorption of DM/DTAB mixtures and the DM/DTAB ratios on silica.

the maximum ratio corresponds to the onset of the plateau region. Therefore, one can conclude that interactions between the surfactants in solutions and at solid/liquid interfaces are dominated by different forces in different regions.

Similar trends are also found for the DM/DTAB ratios in 3:1 and 1:3 mixtures. For both mixtures, the DM/DTAB ratio is low at low concentrations, but increases rapidly at the concentration where hydrophobic chain–chain interaction dominates the adsorption process. It reaches a maximum at the onset of the plateau region and then proceeds to decrease. It should be noted that the decrease of DM ratio in the plateau adsorption region is similar to the case of mixed DM/SDS on alumina, which is attributed to the decrease of DM monomer concentration. Since composition of the adsorbed layer will determine to a large extent the interfacial behavior of the solids such as wettability and dispersion, the practical application of the above finding is to be noted.

6.4.3 Adsorption of N-Dodecyl-β-D-Maltoside with Nonionic Surfactant in Solids

6.4.3.1 Adsorption of n-Dodecyl-β-D-Maltoside/Nonylphenol Ethoxylated Decyl Ether Mixtures on Silica (Passive Nonionic/Active Nonionic)

The adsorption of n-dodecyl-β-D-maltoside (DM) and nonyl phenol ethoxylated decyl ether (NP-10) mixtures on silica is investigated next. Although, both surfactants are nonionic and their liquid/air interfacial behaviors are similar, they behave quite differently at solid/liquid interfaces. Sugar-based surfactants adsorb on alumina but not on silica, whereas ethoxylated nonionic surfactants exhibit an opposite behavior. The behavior of mixtures of the two at solid/liquid interfaces is further examined below.

The adsorption isotherms of n-dodecyl-β-D-maltoside (DM), nonyl phenol ethoxylated decyl ether (NP-10), and their mixtures on silica are shown in Figure 6.31. NP-10 adsorbs on silica strongly due to hydrogen bonding, while DM does not. Their mixtures adsorb less than NP-10 alone at low concentrations and approximately the same as NP-10 alone at higher concentrations. This is different for DM/DTAB mixtures, where the adsorption densities of the mixture are higher than those of either DM or DTAB in the region where hydrophobic interactions dominate the adsorption process. This can be explained by the fact that NP-10 is more surface active than DM, and that DM plays a minor role in the adsorption of mixtures.

Comparable to DM/DTAB mixtures, where DM promotes DTAB adsorption in mixtures, it was found that DM increases the adsorption of NP-10, especially at low concentrations. The dodecyl maltoside in the DM/NP-10 mixture, on the other hand, readily adsorbs on the surface than DM alone as a result of the hydrophobic chain–chain interactions. This phenomenon again is very similar to the DM adsorption from DM/DTAB mixtures. The variation of DM/NP-10 ratio in the adsorbed layer as a function of adsorption was found to be very similar to those of DM/DTAB mixtures as well.

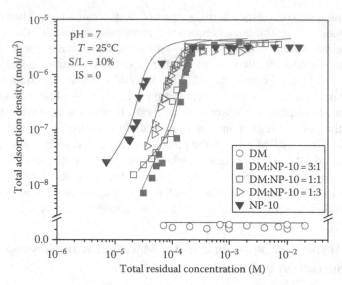

FIGURE 6.31 Adsorption of DM, NP-10 and their mixtures on silica. (From Zhang, L., Zhang, R., and Somasundaran, P., *J. Colloid Interface Sci*, 302, 25, 2006. With permission.)

6.4.3.2 Adsorption of *n*-Dodecyl-β-D-Maltoside/Nonylphenol Ethoxylated Decyl Ether Mixtures on Alumina (Active Nonionic/Passive Nonionic)

The adsorption isotherms of *n*-dodecyl-β-D-maltoside (DM), nonylphenol ethoxylated decyl ether (NP-10), and their mixtures on alumina are shown in Figure 6.32. In this case, DM adsorbs strongly on alumina, while NP-10 does not adsorb. At all mixing ratios, the adsorption of the mixtures is considerably lower than that of DM alone. The adsorption density data shows that the surface is not fully covered by the surfactant. The presence of NP-10 in the systems reduces the adsorption of the sugar-based DM, suggesting that there are antagonistic interactions between the two surfactants at the alumina–water interfaces.

The adsorption of NP-10 from solutions with and without DM on alumina is plotted in Figure 6.33 as a function of residual NP-10 concentration. Adsorption of NP from the mixtures is facilitated by the presence of DM. This is proposed to be due to the adsorbed DM functioning as anchor molecules for NP-10 through hydrophobic chain–chain interactions. Therefore, NP-10 appears to have clear synergistic effects when mixed with DM.

In contrast it can be seen from Figure 6.34 that the DM adsorption is depressed by NP-10 in the plateau region. From Figure 6.32, it can be seen that the total adsorption of DM + NP-10 is much less than that of DM alone. Therefore it can be noted that in this system as a whole, there are mainly antagonistic effects between DM and NP-10. This is very similar to the adsorption of *n*-dodecyl-β-D-maltoside/sodium dodecylsulfate on negatively charged alumina.

In the case of nonionic–cationic mixtures of *n*-dodecyl-β-D-maltoside (DM) and dodecyltrimethyl ammonium bromide (DTAB) on silica, DM does not adsorb

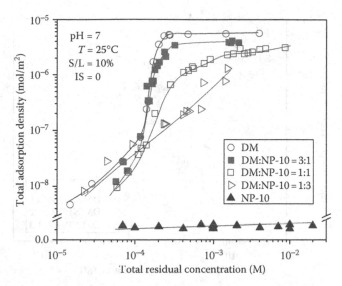

FIGURE 6.32   Adsorption of DM, NP-10 and their mixtures on alumina. (From Zhang, L., Zhang, R., and Somasundaran, P., *J. Colloid Interface Sci*, 302, 25, 2006. With permission.)

on the silica by itself. However, in the mixtures, DM can adsorb on silica through hydrophobic chain–chain interactions with the adsorbed DTAB species. The DM adsorption is characterized by a sharp increase in density at a given concentration

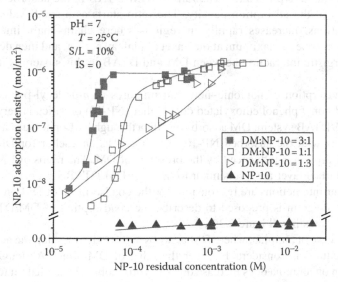

FIGURE 6.33   Adsorption of NP-10 on alumina: Adsorption alone and from mixtures. (From Zhang, L., Zhang, R., and Somasundaran, P., *J. Colloid Interface Sci*, 302, 25, 2006. With permission.)

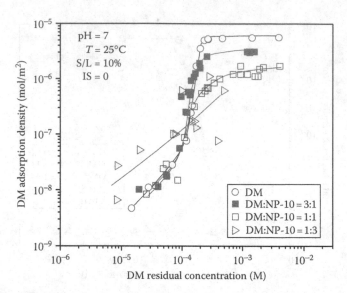

FIGURE 6.34 Adsorption of DM on alumia: Adsorption alone and from mixtures. (From Zhang, L., Zhang, R., and Somasundaran, P., *J. Colloid Interface Sci*, 302, 25, 2006. With permission.)

In these mixed systems, DTAB acts as anchor molecules for DM. Its adsorption is markedly affected by the presence of DM. As long as the surface is not saturated, DTAB adsorption is increased by the presence of DM. When the surface is saturated, DTAB adsorption density is reduced due to competition from DM for adsorption sites. The ratio of DM/DTAB in the adsorbed layer is a function of the adsorption density. The ratio starts as a small value at low concentrations, increases rapidly in regions where chain–chain interactions dominate, reaches a maximum at the onset of plateau region, and then decreases. The synergistic interaction between DM and DTAB in this system is therefore obvious.

The adsorption of nonionic–nonionic mixtures of *n*-dodecyl-β-D-maltoside (DM) and nonyl phenol ethoxylated decyl ether (NP-10) on silica is very similar to the DM/DTAB system: DM adsorbs on silica through hydrophobic chain–chain interactions with the adsorbed NP-10, which acts as an anchor for DM, while NP-10 adsorption is influenced by the presence of DM. The ratios of DM/NP-10 in the adsorbed layer are also similar to those of DM/DTAB systems, suggesting that similar interactions are responsible for the coadsorption of DM on silica. A schematic diagram is proposed to describe the coadsorption of DM/DTAB and DM/NP-10 on silica (Figure 6.35).

In the case of alumina, where DM adsorbs and NP-10 does not, the adsorption of the mixtures is considerably lower than that of DM alone. Although NP-10 adsorption on alumina does increase due to hydrophobic chain–chain interactions with adsorbed DM, the DM adsorption is reduced. For the DM/NP-10 mixtures at high concentrations the surface is not fully covered by the surfactants. The presence

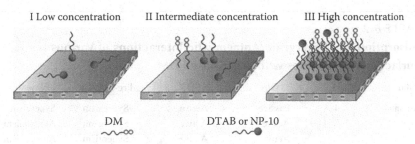

FIGURE 6.35   Adsorption models for DM/DTAB and DM/NP-10 mixtures on silica.

of NP-10 in the system reduces the adsorption of the DM and in turn decreases the total adsorption. Therefore there are antagonistic interactions between the two surfactants in this case.

6.5   SUMMARY

Adsorption of sugar-based surfactant n-alkyl-β-D-glucosides and n-alkyl-β-D-maltosides on solids is discussed in this chapter. The surfactants were found to absorb on alumina, hematite, and titania, but much less on silica. The adsorption behavior on hydrophilic solids is opposite to that of nonionic ethoxylated surfactants, and this unique behavior of the sugar-based nonionic surfactants have practical implications. An increase in the hydrocarbon chain length of the surfactants in adsorption causes a shift of the adsorption isotherms to lower concentrations. The magnitude of the shift corresponds to a change in the cmc of these surfactants. Furthermore, an increase in glucose units on the hydrophilic head causes a decrease in adsorption density due to steric effect.

pH is found to have no effect on the adsorption of n-dodecyl-β-D-maltoside on alumina, while salt appears to have only a minor effect. This is likely to be due to the changes in water structure rather than the modifications of the adsorption process itself.

The adsorption isotherms of n-dodecyl-β-D-maltoside on alumina and hematite show a typical three-stage adsorption of nonionic surfactants on hydrophilic solids. Hydrophobicity tests suggest that there is a bilayer formation at solid/liquid interfaces at higher concentrations.

The adsorption of n-dodecyl-β-D-maltoside on hydrophobic solid graphite showed fairly different behavior from that obtained from hydrophilic solids. The driving force for the adsorption is exclusively hydrophobic interaction. Surfactants adsorb on graphite with their hydrophobic chain attached to the surface and the adsorption is monolayer. This is confirmed by the results of the hydrophobicity and settling tests.

The adsorption mechanism of n-dodecyl-β-D-maltoside on alumina and hematite was investigated. The effect of pH on adsorption and zeta-potential measurements suggests that electrostatic interaction is not the driving force for the adsorption.

TABLE 6.2

Adsorption and Synergistic/Antagonistic Interactions of Various Surfactants in Mixtures with Sugar-Based DM

Solid	pH	DM	SDS	Pre-Plateau	Plateau
Alumina	4	Passive	Active	Synergism	Synergism
	6	Active	Active	Synergism	Antagonism
	7	Active	Active	Synergism	Antagonism
	10	Active	Passive	Synergism	Antagonism
	11	Active	Passive	Synergism	Antagonism
Solid	**pH**	**DM**	**DTAB**	**Pre-plateau**	**Plateau**
Silica	9	Passive	Active	Synergism	Synergism
Solid	**pH**	**DM**	**NP-10**	**Pre-plateau**	**Plateau**
Silica	7	Passive	Active	Synergism	Synergism
Alumina	7	Active	Passive	Synergism	Antagonism

However, it is proposed that hydrogen bonding between hydroxyl groups on surfactants and alumina surface hydroxyl species on solid surface is the driving force for the adsorption of n-dodecyl-β-D-maltoside on alumina. The presence of urea and DMSO, both strong hydrogen-bonding acceptors, decreases the adsorption of n-dodecyl-β-D-maltoside on alumina, suggesting that the hydrogen bonds between the surfactant and the solid are weakened by the presence of urea and DMSO.

In addition, it was found that the adsorption of n-dodecyl-β-D-maltoside on alumina and silica was significantly affected by the presence of another surfactant that is from anionic, cationic to nonionic. The adsorption of the mixtures is governed by the interactions between the surfactants at solid/liquid interface. Synergistic/antagonistic effects of sugar-based DM in mixtures with other surfactants at solid/liquid interfaces are summarized in Table 6.2. Some general trends are observed:

1. When the other surfactant adsorbs on the solid by itself, no matter whether DM adsorbs on the solid or not, the mixture of the two usually shows synergy. The two surfactants assist each other in adsorption. They compete only when the surface is saturated with surfactants and reduce the adsorption of each.
2. When the other surfactant does not adsorb on the solid by itself, and DM adsorbs, the mixture of two usually shows antagonism. The other surfactant prevents DM adsorption, resulting in low total adsorption.

The various synergistic/antagonistic interactions between sugar-based surfactant and other surfactant have implications for designing surfactant combinations for controlled adsorption to optimize their performance.

ACKNOWLEDGMENTS

The authors acknowledge the support of the Department of Energy, National Science Foundation, and industrial members of the IUCR Center.

REFERENCES

1. Hill, K., Rybinski, W.V., and Stoll, G. (Eds.), *Alkyl Polyglycosides: Technology, Properties and Applications*, VCH Verlagsgesellschaft mbH, Weinheim, 1997.
2. Rybinski, W.V. and Hill, K., *Angew. Chem. Int. Ed.*, 1998, *37*, 1328.
3. Balzer, D., *Tenside Surf. Det.*, 1991, *28*, 419.
4. Balzer, D. and Luders, H. (Eds.), *Nonionic Surfactants: Alkyl Polyglucosides (Surfactant Science)*, Marcel Dekker, New York, 2000.
5. Shinoda, K., Yamaguchi, T., and Hori, R., *Bull. Chem. Soc. Jpn.*, 1961, *34*, 237.
6. Drummond, C.J., Warr, G.G., Grieser, F., Ninham, B.W., and Evans, D.F., *J. Phys. Chem.*, 1985, *89*, 2103.
7. Kameyama, K. and Takagi, T., *J. Colloid Interface Sci.*, 1990, *137*(1), 1.
8. Focher, B., Savelli, G., Torri, G., Vecchio, G., McKenzie, D.C., Nicoli, D.F., and Button, C.A., *Chem. Phys. Lett.*, 1989, *158*(6), 491.
9. Zhang, L., Somasundaran, P., and Maltesh, C., *Langmuir*, 1996, *12*(10), 2371.
10. Boyd, B.J., Drummond, C.J., Krodkiewska, I., and Grieser, F., *Langmuir*, 2000, *16*(19), 7359.
11. Stubenrauch, C., *Curr. Opin. Colloid Interface Sci.*, 2001, *6*(2), 160.
12. Zhang, R., Zhang, L., and Somasundaran, P., *J. Colloid Interface Sci.*, 2004, *278*(2), 453.
13. Soderman, O. and Johansson, I., *Curr. Opin. Colloid Interface Sci.*, 1999, *4*(6), 391.
14. Hato, M., *Curr. Opin. Colloid Interface Sci.*, 2001, *6*(3), 268.
15. Somasundaran, P. and Fuerstenau, D.W., *J. Phys. Chem.*, 1966, *70*(1), 90.
16. Kunjappu, J.T. and Somasundaran, P., *J. Colloid Interface Sci.*, 1995, *175*(2), 520.
17. Zhang, L., Somasundaran, P., and Maltesh, C., *J. Colloid Interface Sci.*, 1997, *191*(1), 202.
18. Zhang, L., Interfacial properties of N-alkyl-B-D-glucosides and N-alkyl-B-D-maltosides at air-liquid and solid-liquid interfaces, PhD Thesis, Columbia University, New York, 1998.
19. Giles, C.H., Smith, D., and Huitson, A., *J. Colloid Interface Sci.*, 1974, *47*(3), 755.
20. Giles, C.H., D'Silva, A.P., and Easton, I.A., *J. Colloid Interface Sci.*, 1974, *47*(3), 766.
21. Portet, F., Desbene, P.L., and Treiner, C., *J. Colloid Interface Sci.*, 1997, *194*(2), 379.
22. Doren, A., Vargas, D., and Goldfrab, J., *Inst. Min. Metall. Trans., Sect. C*, 1975, *84*, 34.
23. Ravishankar, S.A., Pradip, and Khosla, N.K., *Intl. J. Mineral Process.*, 1995, *43*(3–4), 235.
24. Weissenborn, P.K., Warren, L.J., and Dunn, J.G., *Colloids Surf. A*, 1995, *99*(1), 29.
25. Somasundaran, P., Snell, E.D., and Xu, Q., *J. Colloid Interface Sci.*, 1991, *144*(1), 165.
26. Fu, E., Adsorption of anionic-nonionic surfactant mixtures on oxide minerals, PhD Thesis, Columbia University, New York, 1987.
27. Lawrence, S.A., Pilc, J.A., Readman, J.R., and Sermon, P.A., *J. Chem. Soc., Chem. Commun.*, 1987, *13*, 1035.
28. Lu, S.H., Bian, Y., Zhang, L., and Somasundaran, P., *J. Colloid Interface Sci.*, 2007, *316*(2), 310.

29. Smith, G.A., Zulli, A.L., Grieser, M.D., and Counts, M.C., *Colloids Surf. A*, 1994, *88*(1), 67.
30. Knozinger, H. and Stubner, B., *J. Phys. Chem.*, 1978, *82*(13), 1526.
31. Sprycha, R., *J. Colloid Interface Sci.*, 1989, *127*(1), 1.
32. Sprycha, R. and Szczypa, J., *J. Colloid Interface Sci.*, 1984, *102*(1), 288.
33. Weissenborn, P.K., Warren, L.J., and Dunn, J.G., *Colloids Surf. A*, 1995, *99*(1), 11.
34. Mukerjee, P. and Ray, A., *J. Phys. Chem.*, 1963, *67*(1), 190.
35. Souza, S.M.B., Chaimovich, H., and Politi, M.J., *Langmuir*, 1995, *11*(5), 1715.
36. Asakawa, T., Hashikawa, M., Amada, K., and Miyagishi, S., *Langmuir*, 1995, *11*(7), 2376.
37. Zhang, L. and Somasundaran, P., *J. Colloid Interface Sci.*, 2002, *256*(1), 16.
38. Mielczarski, E., Mielczarski, J.A., Zhang, L., and Somasundaran, P., *J. Colloid Interface Sci.*, 2004, *275*(2), 403.
39. Corkill, J.M., Goodman, J.F., and Tate, J.R., *Trans. Faraday Soc.*, 1966, *62*(520P), 979.
40. Ben-Naim, A.Y., *Hydrophobic Interactions*, Springer, New York, 1980.
41. Glazman, Y. and Blashchuk, Z., *J. Colloid Interface Sci.*, 1977, *62*(1), 158.
42. Mathai, K.G. and Ottewill, R.H., *Trans. Faraday Soc.*, 1966, *62*(519P), 750.
43. Somasundaran, P., Snell, E.D., Fu, E., and Xu, Q., *Colloids Surf.*, 1992, *63*(1–2), 49.
44. Somasundaran, P. and Huang, L., *Adv. Colloid Interface Sci.*, 2000, *88*(1–2), 179.
45. Thibaut, A., Misselyn-Bauduin, A.M., Grandjean, J., Broze, G., and Jérôme, R., *Langmuir*, 2000, *16*(24), 9192.
46. Portet-Koltalo, F., Desbene, P.L., and Treiner, C., *Langmuir*, 2001, *17*(13), 3858.
47. Huang, Z.Y. and Gu, T., *Colloids Surf.*, 1989, *36*(3), 353.
48. Lokar, W.J. and Ducker, W.A., *Langmuir*, 2004, *20*(11), 4553.
49. Brinck, J. and Tiberg, F., *Langmuir*, 1996, *12*(21), 5042.
50. Huang, L. and Somasundaran, P., *Langmuir*, 1997, *13*(25), 6683.
51. Lu, S. and Somasundaran, P., *Langmuir*, 2008, *24*(8), 3874.
52. Qun, X., Vasudevan, T.V., and Somasundaran, P., *J. Colloid Interface Sci.*, 1991, *142*(2), 528.
53. Rao, P.H. and He, M., *Chemosphere*, 2006, *63*(7), 1214.
54. Zhang, L. and Somasundaran, P., *J. Colloid Interface Sci.*, 2006, *302*(1), 20.
55. Matsson, M.K., Kronberg, B., and Claesson, P.M., *Langmuir*, 2005, *21*(7), 2766.
56. Zhang, L., Zhang, R., and Somasundaran, P., *J. Colloid Interface Sci.*, 2006, *302*(1), 25.
57. Zhou, Q. and Somasundaran, P., Synergistic adsorption of cationic gemini and sugar-based nonionic surfactant mixtures on silica, submitted to *J. Colloid Interface Sci.*, 2008.
58. Somasundaran, P., Fu, E., and Qun, X., *Langmuir*, 1992, *8*(4), 1065.
59. Huang, L., Maltesh, C., and Somasundaran, P., *J. Colloid Interface Sci.*, 1996, *177*(1), 222.
60. Zhang, R. and Somasundaran, P., *Langmuir*, 2005, *21*(11), 4868.
61. Lu, S. and Somasundaran, P., *J. Colloid Interface Sci.*, 2007, *316*, 310.
62. Lu, S. and Somasundaran, P., Micellar evolution in mixed nonionic/anionic surfactant mixture, in submission, 2008.

7 Rheological Properties of Systems Containing Sugar-Based Surfactants

Gemma Montalvo and Mercedes Valiente

CONTENTS

7.1   INTRODUCTION

Surfactant systems can self-assemble into various structures that show a rich variety of rheological properties [1]. Newtonian and non-Newtonian behavior can occur even in the micellar region, and the viscosity (η) of highly dilute aqueous solutions of spherical micelles increases with the volume fraction of the globular aggregates (φ_m) according to Einstein's equation

$$\eta = \eta_{CMC}(1 + 2.5\varphi_m) \tag{7.1}$$

where η_{CMC} is the Newtonian continuous phase viscosity at the critical micellar concentration. This equation does not include interactions, however, and is strictly limited to the low-concentration regime.

Steady-state flow measurements allow the flow curve of a fluid to be determined. Newtonian fluids display a linear relationship between the shear stress (σ) and the shear rate ($\dot\gamma$) that takes the form $\sigma = \eta\dot\gamma$; the curve resulting from plotting σ or η as a function of $\dot\gamma$ is known as the flow curve. When the viscosity is shear-dependent, the fluids are denoted as non-Newtonian (see Figure 7.1 for typical flow curves of non-Newtonian fluids). The viscous properties of surfactant solutions are described by curves of types (a)–(c):

- Curve (a) are the flow curves of a Newtonian fluid, i.e., an aqueous solution of spherical micelles
- Curve (b) characterize shear-thinning fluids, i.e., aqueous solutions of rod micelles
- Curve (c) characterize fluids with a limiting shear stress (yield stress) at which the material starts to flow

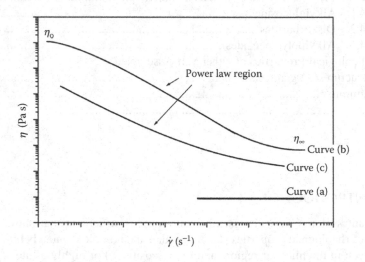

FIGURE 7.1 Typical flow curves of viscosity (η) as a function of shear rate ($\dot\gamma$): (a) Newtonian behavior; (b) shear-thinning behavior; and (c) shear-thinning behavior with yield stress.

The flow curve of a shear thinning fluid with no yield stress has three regions:

- Newtonian region at low shear rates, which is characterized by a zero-shear-rate viscosity, η_0
- Region of intermediate shear rates, which is characterized by a power law-type shear-dependent viscosity
- Second Newtonian region at high shear rates, which is characterized by a constant infinite-shear-rate viscosity, η_∞

This kind of flow curve is well described by a viscosity function of the following general form

$$\frac{\eta - \eta_\infty}{\eta_0 - \eta_\infty} = f(\gamma) \qquad (7.2)$$

Non-Newtonian behavior and elastic effects usually appear at high concentrations, although many surfactant solutions are very viscous at low concentrations. A great many papers have been published concerning viscoelastic surfactant solutions [2,3]. The micelles in these cases are elongated and the contour length is much greater than the persistence length, which means that they are often described as wormlike micelles. The elastic properties of these micelles are due to the existence of a transient network-like structure built up from contacts between the anisometric micelles (entanglements) or from intermicellar branching processes. This domain is known as the semidilute domain. The transient network can be characterized by a shear modulus (G_0) and a structural relaxation time (τ), and its zero-shear viscosity is given by

$$\eta_0 = G_0\tau \qquad (7.3)$$

The value of the shear modulus is related to the entanglement density of the mesh, and shear thinning is observed under steady-state flow rheological conditions due to deformation of the network and alignment of the micelles in the shear flow.

The rheological properties of a viscoelastic material can also be investigated by applying a sinusoidal stress of angular frequency ω in what is known as a "dynamic" or "oscillatory" experiment. Such nondestructive analyses provide information regarding the intermolecular and inter-particle forces within the system. For example, the elastic (storage) modulus G', the viscous (loss) modulus G'', and the magnitude of the complex viscosity modulus $|\eta^*|$ can be calculated from the phase angle between the sinusoidally varying stress and strain signals. The dynamic rheological behavior of a viscoelastic micellar solution can be described by a mechanical model known as the Maxwell model. This model consists of an elastic spring with a Hookean constant G_0 and a dashpot with viscosity η_0 connected in series. When a sudden strain is applied to the system for a short time, the stress (σ) relaxes exponentially with a time constant τ:

$$\sigma = \sigma_0 \exp(-t/\tau) \qquad (7.4)$$

As semidilute solutions of wormlike micelles frequently show monoexponential relaxation properties, these surfactants can be described by simple equations such as those shown below for a Maxwell fluid:

$$G'(\omega) = \frac{\eta_0 \tau \omega^2}{(\tau \omega)^2 + 1} \tag{7.5}$$

$$G''(\omega) = \frac{\eta_0 \omega}{(\tau \omega)^2 + 1} \tag{7.6}$$

$$|\eta^*| = \frac{\sqrt{(G'^2 + G''^2)}}{\omega} = \frac{\eta_0}{\sqrt{(\tau \omega)^2 + 1}} \tag{7.7}$$

It can be seen from these equations that G' approaches a constant limiting plateau equal to the shear modulus G_0 when $\tau \omega \gg 1$. The shear modulus is related to the entanglement density (υ) according to the following equation

$$G_0 = \upsilon k T \tag{7.8}$$

As υ usually decreases when the micelles become shorter, the value of G_0 also decreases.

Many rheological results for viscoelastic surfactant solutions can be represented over a large frequency range with a simple structural relaxation time and a single shear modulus G_0. These fluids are called Maxwell fluids. However, there are other situations where the rheograms cannot be fitted to the Maxwell model, especially where deviations occur at elevated frequencies.

Shear-induced structures in surfactant solutions have been described for wormlike micelles. Surfactant solutions can change their structure during flow, although the new phase formed is only stable under the action of a velocity gradient. This metastable phase is known as the shear-induced structure (SIS).

Surfactants form lyotropic liquid crystals in the high-concentration regime. Nematic, lamellar, and hexagonal liquid crystals are anisotropic materials whose rheological properties are much more complex than micelles; cubic liquid crystals are isotropic. The rheology of liquid crystal lamellar phases has been studied for a wide range of systems [4,5], and significant progress has been made in characterizing the microstructural changes that occur in the lamellar phase on going from sheet-like bilayers to dispersed multilamellar vesicles under shearing.

In this chapter, we will discuss the application of rheological methods to surfactants containing sugar-based derivatives. Interest in sugar-based surfactants, such as alkylglucosides, is due to a number of reasons [6], including their good surface-active and emulsifying properties, biocompatibility, nontoxic nature, and low skin irritation. Furthermore, their hydrophile–lipophile balance (HLB) values can be varied from 6 to 14 simply by varying the length of the alkyl tail and the nature of the sugar head-group. This allows a wide variety of colloidal structures to be prepared over a broad range of pHs and ionic strengths. The high biocompatibility of these systems means that they have no adverse effects on human health and are not a serious environmental threat, even at relatively high concentrations. Another important advantage of sugar-based surfactants is that they are much less sensitive to temperature variations than other typical nonionic surfactants. This chapter will provide a brief review of the rheological data for sugar surfactants, ranging from dilute (micelles) to highly concentrated (liquid crystals) systems.

7.2   RHEOLOGICAL PROPERTIES OF MICELLES

7.2.1   ALKYLGLUCOSIDES

Among the alkylglucoside systems, the phase behavior of n-octyl-β-D-glucopyranoside (C_8G_1) in water has been extensively studied by various authors, who found minor differences in the phase borders [7–10]. This system is very interesting due to its rich phase-behavior whereby it shows different liquid-crystalline phases: a fluid lamellar phase at low hydrations and bicontinuous cubic and hexagonal H_1 phases at higher hydrations. The flow behavior of C_8G_1 + water has been characterized by Quitzsch et al. [11], who found that the samples behave as Newtonian liquids at very low concentrations and that their viscosity increases linearly with concentration (Figure 7.2). In the middle range of surfactant weight fraction,

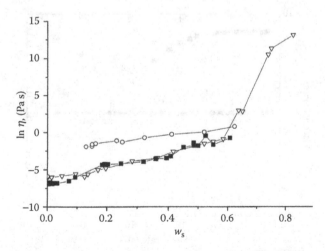

FIGURE 7.2   Viscosity (η) as a function of surfactant weight fraction (w_s) at 298 K. (∇: C_8G_1/water system; $\bigcirc$: $C_{10}G_1$/water system; ■: $C_{10}G_2$/water system). The lines are a guide for the eyes and have no other significance. (From Kahl, H., Enders, S., and Quitzsch, K., *Colloids Surf. A: Physicochem. Eng. Aspects*, 661, 183, 2001. With permission; Häntzschel, D., Schulte, J., Enders, S., and Quitzsch, K., *Phys. Chem. Chem. Phys.*, 1, 895, 1999.)

the Newtonian viscosity increases logarithmically, whereas the flow behavior does not follow the Newtonian law at high surfactant concentrations (higher than 20 wt%). This flow behavior can be described by the Ostwald law $\dot{\gamma} = K\sigma''$ where K and n are constants and σ is the shear stress. A decrease in viscosity with shear rate ($\dot{\gamma}$) is only visible at higher shear rates, and increasing the chain length of the hydrophobic tail ($C_8G_1, C_{10}G_1$) leads to a higher viscosity (Figure 7.2). The solutions show viscoelastic behavior in this region. Oscillatory experiments in the low-frequency range follow the simple Maxwell model and a plot of the structure relaxation time as a function of the surfactant concentration passes through a maximum.

Viscosimetric and density measurements on C_8G_1 + water have also been carried out well above the CMC by Sesta et al. [12], although the surfactant concentration was always less than 26 wt%. This concentration range corresponds to the middle range studied by Quitzsch et al., where the samples behave as Newtonian fluids. Sesta et al. explained the relative viscosities by considering the formation of anisometric aggregates, where intermicellar interactions are negligible. They estimated from these results that the micellar axial ratio increases with surfactant concentration by assuming that the micellar aggregates become more elongated as the concentration increases. This behavior is typical of surfactant solutions with fibre-like structures [13].

Aqueous solutions of different glycosides such as n-nonyl-β-D-glucoside (C_9G_1) and n-decyl-β-D-glucoside ($C_{10}G_1$), and their mixtures, have been characterized by capillary and rheology [14,15]]. The solutions investigated by Söderman et al. in the shear-rate range from 0.1 to 785 s^{-1} show Newtonian behavior at surfactant contents higher than 20 wt% (Figure 7.3). The dependence of the viscosity

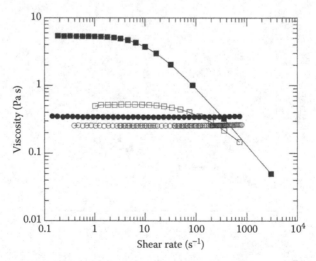

FIGURE 7.3 Viscosity (η) as a function of shear rate ($\dot{\gamma}$). ($\square$: 20.0 wt% $C_{14}G_2$ in water; $\blacksquare$: 18.5 wt% $C_{14}G_2$ in heavy water; $\bigcirc$: 42.5 wt% C_9G_1 in water; $\bullet$: 42.4 wt% C_9G_1 in heavy water). The lines are a guide for the eyes and have no other significance. (From Ericsson, C.A., Söderman, O., and Ulvenlund, S., *Colloid Polym. Sci.*, 283, 1313, 2005. With permission.)

on concentration follows a power law with an exponent equal to 2.0 for C_9G_1. The authors consider that the low numerical value of the exponent is indicative of the presence of branched micelles. They also point out that the micellar phase of C_9G_1 is followed by a bicontinuous cubic phase, which suggests that the microstructure of the two phases is similar. It should be noted that these experiments were carried out at a lower shear rate than for the micellar phase of C_8G_1 studied by Quitzsch et al., which could explain the different Newtonian and non-Newtonian flow behaviors observed. The viscoelasticity of C_9G_1 solutions is weak, and their relaxation is so fast that no crossover point between G' and G'' can be determined in the frequency spectrum. The rheological data for concentrated solutions of the micellar phase of $C_{10}G_1$ are similar to those for C_9G_1. It is not possible to compare dilute solutions of these two systems as phase separation occurs in the $C_{10}G_1$ + water system.

In $C_9G_1/C_{10}G_1$ mixtures, the scaling exponent of viscosity versus concentration decreases monotonously from 2.0 to 1.3 with increasing $C_{10}G_1$ content. Moreover, the relaxation time obtained from the frequency spectra suggests that this decrease follows the sequence $C_9G_1 > C_9G_1/C_{10}G_1 > C_{10}G_1$. Ericsson et al. have concluded from all these viscosimetric and rheological data that the increased packing parameter of the glycosides on going from C_9G_1 to $C_{10}G_1$ gives rise to a transition from a branched, interconnected micellar system to a proper, space-filling micellar network.

Taken together [16], these results suggest that micellar solutions of glucosides in water contain elongated aggregates that give rise to branched micelles at high surfactant concentrations, although discrete micelles can form in ternary systems that contain oil such as octane. This oil induces structural changes in the micellar shape that are directly related to the solubilized site of the oil in the aggregates [17,18].

Mixtures of sugar-based surfactants and other types of surfactants have also been investigated. For example, a viscometric study of mixed n-octyl-β-D-glucoside (C_8G_1) and tetraethylene glycol monododecyl ether micelles has been carried out with systems containing a fixed amount of sugar-based surfactant [19]. An abrupt viscosity change, due to a transition from spherical to rod-shaped micelles, was observed as the tetraethylene glycol monododecyl ether concentration was increased. Viscosity measurements have also been applied to determine the CMC of micelles, and the zero-shear viscosity of C_8G_1 and water-soluble polymer systems at different C_8G_1 contents has been used to study the effect of the polymer on the CMC of the sugar-based surfactant micelles [19].

7.2.2 DISACCHARIDES

Viscosity measurements have also been utilized to investigate the effect of head-groups on aggregate size in the low-concentration regime [15]—more hydrophilic head-groups at a constant tail length ($C_{10}G_2$ + water and $C_{10}G_1$ + water) lead to a

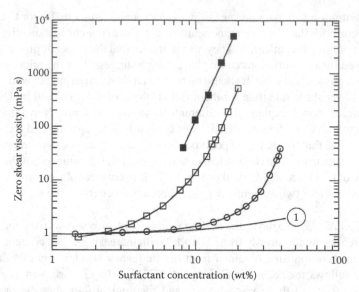

FIGURE 7.4 Zero-shear viscosity (η_0) as a function of surfactant weight fraction. ($\square$: $C_{14}G_2$ in water; $\blacksquare$: $C_{14}G_2$ in heavy water; both of them at 35°C; $\bigcirc$: $C_{12}G_2$ in water at 22°C. The solid line denoted by 1 is based on Einstein's equation (Equation 7.1) for hard spheres at 22°C). (From Ericsson, C.A., Söderman, O., and Ulvenlund, S., *Colloid Polym. Sci.*, 283, 1313, 2005. With permission.)

lower viscosity (Figure 7.2), and Newtonian fluid-type behavior was observed for the system $C_{10}G_2$ + water up to a surfactant concentration of around 20 wt%. For *n*-dodecyl-β-D-maltoside ($C_{12}G_2$) and *n*-tetradecyl-β-D-maltoside ($C_{14}G_2$) [14], the micellar solutions are shear thinning (Figure 7.3) with a zero-shear viscosity that scales with concentration according to a power law with an exponent of about 5.8, which is about three times higher than for glucosides (Figure 7.4). Concentrated $C_{14}G_2$ solutions display a viscoelastic behavior with a characteristic relaxation time of approximately 0.17 s. The frequency dependence of the $C_{14}G_2$ system, however, does not follow the Maxwell model, and a similar non-Maxwellian behavior has been observed in mixtures of cetylpyridinium chloride and sodium salicylate at low and moderate salicylate concentrations [20]. These systems are comprised of wormlike, nonbranched micelles, and the fact that the Maxwell model is unable to describe their rheological behavior has been explained in terms of a pronounced polydispersity, which gives rise to a continuous spectrum of relaxation times rather than a single, well-defined one. The rheometric and viscosimetric data are thus consistent with the proposal that $C_{14}G_2$ forms wormlike micelles at high concentrations and are also in agreement with previous small angle neutron scattering (SANS) measurements, which have shown that $C_{14}G_2$ micelles are elongated and flexible aggregates at low concentration. If the micelles also remain disconnected at higher concentrations, a transition from a dilute solution of wormlike micelles at low concentrations to a semidilute regime with entangled wormlike micelles at a crossover concentration (c^*) could be observed. Furthermore, the viscosity ought to scale with concentration with an exponent of more than 3.5 in the semidilute region, where the dependence of the viscosity on concentration follows a power law with an exponent for H_2O of 5.8 (Figure 7.4). The large numeric value of the exponent is not consistent with the existence of a micellar network, however, and strongly suggests that the micelles are wormlike. Moreover, the micellar phase in the $C_{12}G_2$ and $C_{14}G_2$ systems is followed by a hexagonal phase that is built up from cylindrical micelles in a hexagonal-packed array. Thus, in light of the phase transition rule, this hexagonal phase reinforces the proposal of wormlike micelles.

The study of the head-group effect can be completed with other sugar-based surfactants that contain sucrose as head-group. Thus, rheology studies have been carried out for sucrose stearate ($C_{18}SE$) and sucrose hexadecanoate ($C_{16}SE$). Systems with up to 2 wt% sucrose stearate [21] exhibit a power-law-type decrease in the steady-state apparent viscosity with shear rate. At higher sucrose stearate concentration, however, the flow curves exhibit two well-defined regions that depend on shear rate and the apparent viscosities can be fitted to the Carreau model. The range of shear rates over which the viscosity (η_0) remains constant becomes wider as the surfactant concentration increases. This increase was attributed to the existence of an increasingly strong micellar structure due to the fact that the size and number of micelles increases with surfactant concentration up to a concentration of about 10 wt%. This means that intermicellar interactions play an increasingly important role. A significant weaker micellar structure was found for systems containing 1 wt% sucrose stearate as the limiting viscosity could not be reached at the lower shear rates available and the lower viscosity and dynamic viscoelastic values obtained. The micellar shape must be similar within this composition range, as can be deduced from the fact that the limiting viscosity values do not significantly vary over the range 3–10 wt% sucrose stearate. It was found for the micellar region in binary mixtures of sucrose hexadecanoate ($C_{16}SE$) + water (up to 42 wt% surfactant) [18] that the zero-shear viscosity increases with surfactant concentration according to a power law, which indicates that there is micellar growth.

The addition of small amounts of a lipophilic cosurfactant (triethylene glycol monododecyl ether ($C_{12}E_3$) and monolaurin) to sucrose hexadecanoate induces micellar growth and leads to the formation of wormlike micelles, although these micelles are disrupted when the cosurfactant fraction reaches some critical value, whereupon the system returns to Newtonian-type behavior. The wormlike micelles in sucrose hexadecanoate can solubilize oil although, as we have mentioned previously, this solubilization depends very much on the type of oil. Micellar growth is inhibited to a greater extent as the solubility of the cosurfactant in oil increases. Additionally, the disruption of wormlike micelles is more severe as the solubilization site tends to shift towards the micellar core due to an increase in its curvature. On the other hand, oils that tend to penetrate into the surfactant palisade layer near the interface induce micellar growth at a certain cosurfactant fraction, probably due to a decrease in the curvature of the aggregates.

7.2.3 ALKYLPOLYGLUCOSIDES

Alkylpolyglucosides (APGs) $(C_Y G_X)$ are industrial surfactants that contain mixtures of a hydrocarbon chain with Y carbon atoms linked to X sugar residues. Several phase behavior studies of technical grade APGs in aqueous systems $(C_{8-10}G_x, C_{12-14}G_x,$ and $C_{8-16}G_x)$ [22] or in the presence of a hydrocarbon [23] have been reported in the literature, although the rheological behavior of aqueous APGs has not often been studied. It is therefore interesting to note that the phase diagrams of $C_{8-16}G_x$ have been estimated by combining rheological methods with other techniques such as optical microscopy with crossed polarizers, light scattering or small angle x-ray scattering (SAXS) studies, amongst others [24]. The rheological behavior of APGs is rather complicated as their rheological properties [24] are dependent on the alkyl chain length, Y, and on the number of glucose units, X. Samples in the semidilute concentration range behave as viscoelastic liquids that can be described by the Maxwell model in the low-frequency range. In general, the relaxation times pass through a maximum surfactant content (Figure 7.5a), which occurs at higher concentrations for pure surfactant systems. The variation of shear modulus, G_0, with concentration reported by Quitzsch et al. shows that the dynamic properties are much more dependent on chain length than on the number of glucose units (Figure 7.5b) [24], whereas the viscosity shows an Arrhenius-like dependence on temperature. In summary, a comparison between glucosides and polyglucosides shows that the rheological properties of the latter are more sensitive to a change in chain length than in the degree of polymerization.

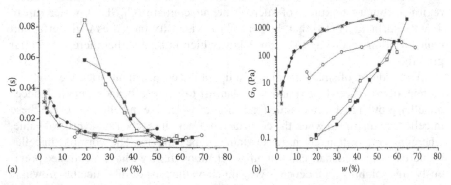

(a) w (%)   (b) w (%)

FIGURE 7.5 Rheological parameters as a function of surfactant weight fraction (w_s): (A) structure relaxation time (τ) and (B) plateau modulus (G_0) calculated from the Maxwell model (Equations 7.5 and 7.6). (■: $C_{8-10}G_x$ in water; ●: $C_{12-14}G_x$ in water; *: $C_{8-16}G_x$ in water; □: C_8G_1 in water [11]; ○: $C_{10}G_1$ in water [11]. (From Schulte, J., Enders, S., and Quitzsch, K., *Colloid Polym. Sci.*, 277, 827, 1999. With permission.)

7.3 RHEOLOGICAL PROPERTIES OF LIQUID CRYSTAL PHASES

7.3.1 ALKYLGLUCOSIDES

Liquid crystals prepared from alkylglucosides have been the focus of a great many papers due to the potential significance of these mixtures in commercial formulations with a good synergic effect. The rheology of these liquid-crystalline systems has not yet been studied in detail. Several nondestructive analyses of these materials by means of oscillatory experiments have been reported in the literature. In particular, they have been used for phase-transition determination due to the large difference between the rheological moduli values of an isotropic micelle and a liquid-crystalline phase and between the various types of liquid crystals. For example, phase transitions have been characterized during a cooling ramp at different concentrations in the n-octyl-β-D-glucopyranoside (C_8G_1) + water system, where a hexagonal liquid-crystalline phase causes an increase in the viscoelastic parameter G' of more than four orders of magnitude and higher weight fractions lead to a further increase of the viscosity up to about six orders of magnitude due to the formation of a cubic phase (Figure 7.6).

A comparative study of the aqueous systems n-octyl (C_8G_1) and n-decyl-β-D-glucopyranoside ($C_{10}G_1$) has been performed by rotational rheometry at constant stress and 25°C (Figure 7.2) [11,15]. In general, the viscosity increases abruptly due to a phase transition at high amounts of surfactant. This viscosity increase is more pronounced for the C_8G_1 surfactant solution than for the $C_{10}G_1$

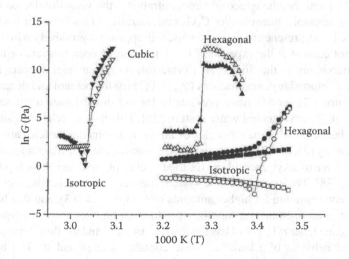

FIGURE 7.6 Storage (G', open symbols) and loss moduli (G'', solid symbols) as a function of reciprocal temperature at different surfactant weight fractions for the C_8G_1/water system. (Square, $w_s = 0.55$; circle, $w_s = 0.6$; up-triangle, $w_s = 0.637$; down-triangle, $w_s = 0.751$). (From Häntzschel, D., Schulte, J., Enders, S., and Quitzsch, K., *Phys. Chem. Chem. Phys.*, 1, 895, 1999. With permission.)

solution due to the formation of different liquid-crystalline phases. For example, the lamellar liquid-crystalline phase formed in the $C_{10}G_1$ system leads to a lower viscosity than the hexagonal structure in the C_8G_1 system at weight fractions of about 0.6 (Figure 7.2). Although these results are interesting for determining the phase boundaries as a function of temperature or concentration, they are not an original contribution since similar method has been used previously for different systems and different surfactants [21,25]. Although the rheological behavior is related to the type of component (i.e., type of alkylglucoside or surfactant), it is more strongly dependent on the three-dimensional structure of the sample (i.e., type of liquid-crystalline or micellar phase).

Phase-transition determinations have also been performed for mixtures of C_8G_1 with the nonionic surfactant tetraethylene glycol monododecyl ether ($C_{12}E_4$) in the very highly dilute regime [19] and aqueous mixtures of methyl octade-canoate D-glucopyranoside (glucate SS)/ethoxylated methyl glucose ether deriv-ative (Glucamate SSE-20) [26]. The lamellar region of both these systems has a higher viscosity than the micellar or two-phase region. The plastic behavior (i.e., shear-thinning), yield stress values and optical microscopy observations with crossed polarizers correlate well with the latter system being thixotropic. A big thixotropic loop means that a long time is required to recover the deformation of this lamellar structure.

The rheological behavior of the dilute lamellar phase of $C_8G_1/C_{12}E_4$ mixtures in the presence of an alcohol has also been investigated. The total surfactant con-centration was kept constant at 0.1 M while varying the $C_8G_1/C_{12}E_4$ molar ratio (χ_{OG}) in the presence of hexanol, octanol, decanol, and benzyl alcohol [27]. The flow behavior of these systems was found to depend on both the type of alcohol (Figure 7.7) and the alkylglucoside concentration—the viscosity decreases with increasing hexanol content and/or C_8G_1 concentration. This behavior was attrib-uted to the lower presence of vesicles, which disappear progressively with increas-ing alcohol content at the expense of the formation of open bilayers or micelles [28]. A maximum in the flow curves (viscosity vs. shear rate) appears for low n-octyl-β-D-glucoside concentrations ($\chi_{OG} < 0.2$) for the octanol and decanol sys-tems (Figure 7.7), as described previously for the dilute lamellar phase of the ternary $C_{12}E_4$/benzyl alcohol/water system [25]. This may be related to structural changes that occur when new vesicles or larger multilamellar vesicles are formed under shearing [25,29–37]. Consequently, the same orientation diagram is obtained in the presence of alkylglucoside as for the lamellar phase under shear reported by Diat et al. [38]. The viscosity of the system increases with increasing octanol (or decanol) concentration for higher amounts of C_8G_1 ($\chi_{OG} > 0.3$), and this has been correlated to an open bilayer topology predominating over a vesicle-type topol-ogy. In contrast, benzyl alcohol has the lowest viscosity and the flow curves exhibit the typical behavior of a lamellar phase containing open and stacked bilayers, with no zero-shear viscosity and with a sharper decrease in the viscosity at lower rather than at higher shear rates (Figure 7.7).

Dilute lamellar phases of $C_8G_1/C_{12}E_4$/water systems in the presence of alcohol show a "weak gel" behavior with an infinite relaxation time, as determined by

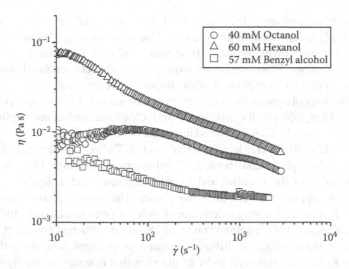

FIGURE 7.7 Viscosity (η) as a function of shear rate ($\dot{\gamma}$) for dilute lamellar phases of $C_{12}E_4$/alcohol/water systems in the presence of n-octyl-β-D-glucoside. Total surfactant concentration: 0.1 M; C_8G_1/$C_{12}E_4$ molar ratio: 0.1. The different alcohols are given in the insert.

dynamic experiments [27]. The viscoelastic moduli (G', G'') are dependent on both the angular frequency and the mole fraction of the glucoside [27]. A long structural recovery time has also been proposed from the results of flow experiments in other lamellar systems with glucopyranoside derivatives [26]. The elastic behavior is also affected by the different location of the distinct alcohols (hexanol, octanol, decanol, and benzyl alcohol) in the bilayer, which depends on the hydrophobicity of the alcohol. Thus, the most elastic samples are found for systems containing more hydrophobic alcohols (decanol and octanol), and the smallest χ_{OG} for octanol. These samples may correspond to the lamellar samples formed by larger and more flexible vesicles [39].

The concentrated lamellar regime has also been investigated for sugar-based surfactant systems. The incorporation of a small amount of C_8G_1 (1 wt%) produces small changes in the boundary of the lamellar phase of the lauryl sulfobetaine/1-butanol/water system, which is formed from both lamellar sheets and vesicles. However, this lamellar phase disappears completely in the presence of 2 wt% of C_8G_1. A complete study of the dependence of viscoelasticity on composition has been performed at high lauryl sulfobetaine surfactant concentrations with 1 wt% C_8G_1 [27]. These mixtures behave as plastic fluids with the yield stress values of between 18 and 270 Pa. The viscosity of these systems decreases with increasing shear rate according to a power law with an exponent close to −0.8, which has been attributed to a multilamellar vesicle topology [40,41], and strong elastic properties have been found for this concentrated lamellar phase by means of oscillatory experiments. The elastic (G') and viscous (G'') moduli are parallel and practically independent of the frequency due to a "strong gel" behavior, which is in good agreement with the yield stress values [42–45]. In summary, the different results obtained from oscillatory experiments with concentrated and dilute lamellar systems can be explained by the different bilayer topologies.

The rheology of concentrated systems containing n-octyl-β-D-glucopyranoside (C_8G_1) has been reported for mixtures with the cationic surfactant cetyltrimethylammonium bromide (CTAB). The addition of a minimum amount of C_8G_1 (1 wt%) to ternary CTAB/glyceraldehyde/water and CTAB/glycerol/water systems affects both their phase and rheological behaviors significantly [46]: the sugar surfactant enlarges the micellar and hexagonal phases and a liquid-crystalline nematic phase appears between these two phases. The nematic phase is built up of rod-like micelles with a long-orientational order. Cortes et al. have studied the rheological behavior of the nematic phase at 25% total surfactant content (CTAB + 1% C_8G_1) in the presence of different amounts of glycerol. The shear-thinning flow curves are consistent with rod-like micelles that orientate in the flow direction. At low shear rates there is a Newtonian region in the flow curve, while the plateau disappears upon increasing the glycerol content. There is no difference in the flow behavior between nematic and micellar phases at the highest amount of glycerol for this system. The elastic behavior of the nematic phase is only important for high angular frequencies (with a structural relaxation time of around 50 ms), and the hexagonal liquid-crystalline phase is more viscous than the nematic phase.

7.3.2 DISACCHARIDES

A drastic temperature-induced change in the lamellar phase morphology has been detected in disaccharide systems. This change occurs in a gel phase-to-lamellar liquid-crystalline phase-transition in a commercial sucrose stearate (SS-110)/water system [47]. Sucrose stearate consists of 72% monoester and 22% diester and its lamellar phase is built up from a combination of multilamellar vesicles and stacked bilayers. These samples undergo shear-thinning due to their decreasing viscosity as the shear rate increases. The viscosities also change significantly upon heating. These thermotropic results differ from those obtained by Calahorro et al. in their studies of the same system, where this phase transition is not observed. These authors proposed the existence of micelles with strong intermicellar interactions [21]. This difference may be due to a different monoester/diester ratio in the commercial blend used. Kahl et al. have also reported a thermally induced phase transition from a liquid-crystalline phase into an isotropic solution that is accompanied by a decrease in the viscosity values for a system containing n-decyl-β-D-maltopyranoside ($C_{10}G_2$) [15].

A viscosity hysteresis between the increasing and decreasing temperature ramps has been reported for a dilute lamellar phase containing 4 wt% SS-100 in water at a constant shear rate of $30 s^{-1}$ [47]. This thixotropic phenomenon may indicate that the dilute lamellar phase suffers a topology transformation from lamellar sheets to multilamellar vesicles (gel state). The induced vesicles did not relax after more than 3 weeks at 20°C. A similar long time to recover the lamellar structure has already been reported for lamellar phases of nonionic sugar-based surfactants like glucate SS and glucamate SSE-20 blends at different weight ratios [26]. These nonionic mixtures also undergo shear thinning, show thixotropic behavior, and manifest yield stress values. Larger thixotropic regions mean that longer times are required for the structure to recover, which is a good indication of the elasticity properties. Similarly, the more the system is structured in the gel stage (vesicles), the more it is sensitive to shear, which could induce an alteration and/or a high degree of structural deterioration. Topological transformations of a surfactant bilayer under shearing have also been clearly established by the SAXS technique for different surfactant types [34,39,48].

7.3.3 ALKYLPOLYGLUCOSIDES OR GLUCOPONE

APGs, which are amphiphilic molecules, can form liquid-crystalline phases upon addition of a solvent [49]. For example, a lamellar liquid-crystalline phase exits in the range 75–85 wt% in the binary $C_{8-10}G_x$/water system at temperatures below 20°C [50], while three lamellar regions with different bilayer topologies have been observed in binary concentrated $C_{12-14}G_x$ solutions from their corresponding polarization microscope textures. An increase in the $C_{12-14}G_x$ content means that the lamellar topology changes via vesicles and mixed vesicles/open bilayers to finally form open bilayers. The same sequence of lamellar topologies is found in dilute systems of zwitterionic and nonionic surfactants containing a fatty alcohol [22,25]. Optical microscopy studies have shown that samples containing vesicles and open bilayers together are transformed into vesicles upon shearing and that these vesicles return to the initial mixed topologies several hours after shear.

As described above (Section 7.3.1), the shear viscosity (η) and shear moduli change suddenly near the phase-transition boundary, which is very useful for phase determination. Thus, for low concentrations of $C_{8-16}G_x$, the isotropic micellar phase undergoes a phase transition to a lamellar liquid-crystalline phase upon cooling, according to the rheological behavior. The storage modulus of this liquid-crystalline phase is higher than the loss modulus, as would be expected from the elastic behavior of the lamellar structure. The storage modulus is weakly dependent on the applied frequency, while the loss modulus shows fluctuations in the lower frequency range. A similar behavior has also been described for the lamellar structures of different surfactants [51,52], which again confirms that the rheological behavior is due to the material's microstructure.

One of the few systematic rheological studies of APG systems was performed by Siddig et al. for systems containing a hydrocarbon [23,53]. The APG used (Glucopone 215 CSUP) was studied in the presence of four different hydrocarbons: heptane, octane, dodecane, and tetradecane. The isotropic micellar solution formed initially is transformed into hexagonal, lamellar, and bicontinuous cubic (*Ia3d* space group) liquid-crystalline phases upon increasing the

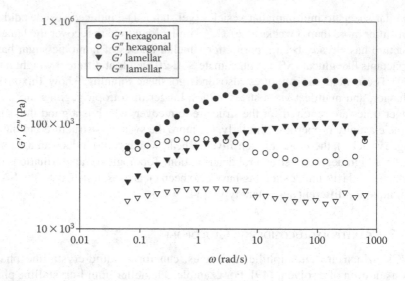

FIGURE 7.8 Storage (G', solid symbols) and loss moduli (G'', open symbols) as a function of frequency for hexagonal (circles) and lamellar (down-triangles) liquid-crystalline samples of the glucopone 215 CSUP (APG-type)/heptane/water system at 298 K. (From Siddig, M.A., Radiman, S., Jan, L.S., and Muniandy, S.V., *Colloids Surf. A: Physicochem. Eng. Aspects*, 276, 15, 2006. With permission.)

surfactant concentration. Varying the hydrocarbon produces a shift in the respective liquid crystal phase boundaries. The rheograms of the lamellar and hexagonal phases are similar, according to the authors (Figure 7.8). Both of these structures are more elastic than viscous, and the values of the elastic parameters (e.g., G', G'', etc.) of the hexagonal phase are higher than those of the lamellar phase. A shear-thinning behavior was observed for the complex viscosity moduli for both the lamellar and hexagonal phases, which are quasi-Maxwellian systems with a tendency for G' and G'' to cross. A terminal relaxation time ($\tau = 1/\omega$) may be estimated from the crossover frequency, ω, although the distorted circular shape seen in the Cole–Cole plots (represented by a plot of G'' against G') indicates a process with some distribution of relaxation times rather than a single relaxation time (Figure 7.9). The dynamic response of the hexagonal phases was found to be similar for all the hydrocarbon systems (terminal relaxation times of about 1 s) except for tetradecane, which was one order of magnitude higher. This was attributed to a more viscous behavior of the latter. The relaxation times for the lamellar phases were found to be of the same order of magnitude as those for the hexagonal phases [53]. The existence of a relaxation time for lamellar structures is not common. Indeed, many different surfactant systems with different lamellar topologies (vesicles and stacked bilayers) display a gel behavior with an infinite relaxation time [51,52,54–58], which means there is no crossing of the G' and G'' moduli.

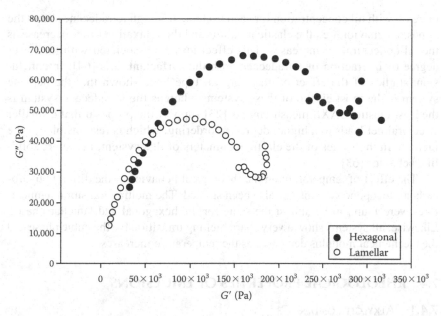

FIGURE 7.9 Cole–Cole plot (loss modulus, G'', as a function of storage modulus, G') for the glucopone 215 CSUP (APG-type)/heptane/water system at 298 K. (●: hexagonal phase; ○: lamellar liquid-crystalline phase). The semicircular shape indicates that the sample behaves as a Maxwellian fluid. (From Siddig, M.A., Radiman, S., Jan, L.S., and Muniandy, S.V., *Colloids Surf. A. Physicochem. Eng. Aspects*, 276, 15, 2006. With permission.)

The viscoelastic parameters of the hexagonal phases were found to be unaffected by this series of alkanes, while their values for the lamellar phases were found to decrease with an increase in alkyl chain length. In general, the effect of the different types of hydrocarbon on the rheological behavior was found to be greater for the lamellar than for the hexagonal phases.

The cubic liquid-crystalline regions of the different glucopone/hydrocarbon systems behave as Maxwell-type fluids with a finite structural relaxation time. A similar behavior has been described for the cubic crystalline phases of a large number of different types of surfactant systems [45,59–62]. In general, cubic phases with a bicontinuous structure have a finite structural relaxation time whereas cubic phases made up of individual spherical aggregates have an infinite relaxation time [43]. This is in good agreement with the structure determined by SAXS experiments for these glucopone systems [23]. Moreover, the instantaneous elastic modulus values (G_0) are similar (around 0.7×10^6 Pa) for all compositions in the different systems, with the exception of tetradecane. This G_0 value is similar to other values reported in the literature for the cubic phase of surfactant systems [59,62,63]. The elastic properties of the cubic phase increase with oil concentration for each system, although tetradecane shows the opposite behavior, i.e., the elastic modulus and the relaxation times decrease as the oil concentration increases. This effect may be associated with the lower degree of interaction of tetradecane with the surfactant tails [64]. In conclusion, studies of the effect of changing oil type have shown that the heptane system is the most elastic of these systems whereas the tetradecane system is the least elastic. SAXS measurements [23] support the proposal that a smaller structural net leads to a higher degree of ordering, which is responsible for the increase in the values of the elastic parameters of this system, i.e., more solid-like behavior [63].

The effect of temperature on the rheological behavior of the different hydrocarbon/glucopone systems has also been studied. The melting transition temperatures were found to be almost the same for the hexagonal and lamellar phases. Likewise, all systems show a very sharp melting transition for the cubic phase and the mechanical modulus decreases as the temperature increases.

7.4 RHEOLOGICAL PROPERTIES OF EMULSIONS

7.4.1 ALKYLGLUCOSIDES

Alkylglucosides are commonly used in food, cosmetic, and drug-delivery emulsions. Rheology is becoming increasingly important for studying the properties of these emulsions, especially their structure and storage stability, since rheological properties can often be correlated with physical stability. The linear viscoelastic properties of alkylglucosides depend in a complex manner on a number of variables whose effect may be most notable during the emulsification process or once the emulsion has formed. These variables should therefore be controlled in order to get a better applicability of the emulsions. For example, both the shear-thinning effect and the degree of viscoelasticity have been found to depend on the volume fractions of the dispersed phase, the average droplet-size distribution, and the polydispersity index, amongst others.

The rheological properties of oil-in-water (o/w) emulsions containing a mineral oil (complex commercial mixture of naphthonic and aromatic hydrocarbons) and a sugar-based surfactant as emulsifier have been studied [65]. The effect of concentration and alkyl chain length were considered using a homologous series of glucopyranosides that possess identical polar head-groups but alkyl chain lengths ranging from C_7G_1 to $C_{10}G_1$. All these emulsions were found to exhibit a shear-thinning response with a yield stress value when subjected to a flow test. However, they do not show a zero-shear viscosity (η_0). Their shear-thinning behavior was characterized by a power law index, n, in the flow equation $\eta \propto 1/\dot{\gamma}^n$. Smaller values of n therefore lead to a greater degree of droplet deformation, which in turn leads to a greater degree of shear thinning or plastic behavior. The shear viscosity, degree of shear thinning (n), and yield stress increase with an increase in both the length of the alkyl chain and the glycopyranoside concentration, i.e., with a decrease in the droplet size of the dispersed phase. The reason for this is that large droplets suffer a greater degree of deformation than smaller droplets, which tend to be more rigid, thereby conferring greater viscosity. The yield stress values were found not to change with storage time, which suggests a good storage stability and the absence of flocculation.

In the frequency sweep experiment, the elastic modulus was found to be higher than the loss modulus, which suggests that the elastic properties predominate over viscous properties in these emulsions and that very little energy is dissipated. G' is frequency independent while G'' increases slowly [65] and these two values do not cross, which implies that the characteristic relaxation times of these emulsions cannot be estimated by the Maxwell model. The values of the elasticity parameters increase with decreasing average droplet size of the dispersed phase.

7.4.2 DISACCHARIDES

Disaccharide-based surfactants have also been used in food emulsions in the search for new alternatives for improving the physical, organoleptic, and nutritional properties of emulsions. For example, different sugar-based surfactants have been tested to replace egg and milk derivatives as they may confer certain advantages, such as a decrease in cholesterol content or an increase in microbiological stability. Gallegos and coworkers have intensely investigated the effect of total or partial substitution of egg yolk by other emulsifiers, in particular two sucrose esters (sucrose palmitate and sucrose stearate) [21,66–71], and a sucrose distearate. Both the sucrose esters have a HLB of 15 and a monoester content of around 70 wt%, while the sucrose distearate (HLB = 8) consists of the sucrose esters of palmitic and stearic acids in an approximate 1:2 molar ratio [72]. These authors studied several variables that influence the emulsification process, such as composition, temperature, and aging effects [73].

Sucrose palmitate forms very stable o/w emulsions with sunflower oil. The flow curves were qualitatively similar for all samples (range of compositions: 1–10 wt% for sucrose palmitate and 55–85 wt% for the oil), and they exhibited a shear-thinning behavior over a wide range of shear rates and tended to a constant high-shear viscosity, η_∞ [68]. These flow curves were well fitted to the Sisko model given by the equation $\eta = \eta_\infty + k\dot{\gamma}^{n-1}$, except those samples with an oil content higher than 80%. These latter also tended to show a Newtonian region at low shear rates (η_0), and these flow curves were better fitted by the Carreau equation:

$$\frac{(\eta - \eta_\infty)}{(\eta_0 - \eta_\infty)} = \frac{1}{\left[1 + \left(\dfrac{\dot{\gamma}}{\dot{\gamma}_c}\right)^2\right]^s}$$

where $\dot{\gamma}_c$ is the critical shear rate for the onset of the shear-thinning region and the exponents n and s of the respective models are related to the slope of the power-law region that may represent the sensitivity of the emulsion to shear. However, several years later the same authors presented slightly different rheological results for the same emulsions obtained following the same sample preparation protocol, preshearing conditions, and storage [69]. In this case the flow curves were recorded with a decreasing shear rate ramp. These shear-thinning curves clearly exhibit a constant viscosity at low shear rates (η_0) independently of the oil concentration. The η_0 has also been described for emulsions stabilized by sucrose stearate [67], sucrose stearate/egg yolk mixtures [71], and sucrose distearate/egg yolk blends [72]. Those flow data were collected with an increasing shear-rate ramp and were fitted to the Carreau model. The "plateau" viscosity (η_0) was considered to be metastable equilibrium between restructuring kinetics and the shear rate decrease [69].

The effect of composition on the flow properties of sucrose palmitate/sunflower oil emulsions was studied on the basis of their Carreau model parameters (e.g., $\dot{\gamma}_c$, η_∞, η_0, and s). Thus, an increase in sucrose palmitate content was found to lead to an exponential increase in both η_∞ and η_0. A linear increase of s with sucrose palmitate concentration may suggest either that the emulsion becomes more sensitive to the variation of shear rate or that the emulsion restructuring rate is higher [69]. The influence of the oil was found to be qualitatively similar to the influence of the sucrose palmitate content. Thus, for the salad-dressing-type emulsions stabilized by sucrose distearate/egg yolk blends, the viscosity values clearly decrease with sucrose concentration in the emulsifier blend [72].

The two limiting viscosities (η_∞ and η_0) were found to show an Arrhenius-like dependence on temperature for the sucrose palmitate and sunflower oil system [69], whereas the temperature has less influence on s and does not influence the value of $\dot{\gamma}_c$ obtained from the Carreau model. The influence of processing temperature has also been studied for sucrose stearate/egg yolk blends [71]. For this system, the zero-shear viscosity (η_0) was found to increase with temperature and longer emulsification times. However, $\dot{\gamma}_c$, as fitted by Carreau model, showed two different behaviors: an initial increase in the processing temperature resulted in a higher values for the critical shear rate while the use of water at 50°C led to a decrease in this parameter.

A viscoelastic response has been reported for highly concentrated food emulsions with disaccharides such as commercial or model mayonnaises [66,74], salad dressings [71,75], or mixed emulsifiers like sucrose stearate/egg yolk [71] and sucrose distearate/egg yolk [72]. A predominantly elastic response was established by frequency sweep experiments (Figure 7.10), although a crossover point tended to appear for some compositions in the low-frequency regime. This crossover is related to a characteristic relaxation time for the onset of the terminal or flow region. There was also a tendency for the elastic modulus to exhibit a plateau regime at intermediate frequencies (G_0), which may be related to the formation of an elastic structural network due to an extensive flocculation process [71,76]. An unusual oscillatory behavior has been reported, however, for sucrose stearate/sunflower oil emulsions where no linear viscoelastic region was observed in an amplitude sweep test [67].

In general, the viscoelastic properties of highly concentrated o/w emulsions depend on the state of flocculation. Thus, (1) an extensive flocculation process provides enhanced stability to the emulsion given by the plateau regime of G' (i.e., G_0) [72,77] and (2) unflocculated or weakly flocculated emulsions show a crossover

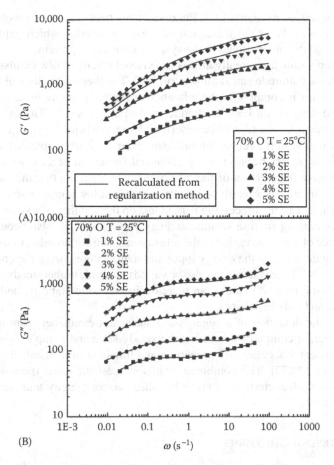

FIGURE 7.10 Rheological parameters as a function of frequency for sucrose palmitate-stabilized emulsions containing 70 wt% sunflower oil at 25°C: (A) Storage (G') and (B) loss moduli (G''). The sample composition is given in the insert. (From Guerrero, A., Partal, P., and Gallegos, C., *J. Rheol.*, 42, 1375, 1998. With permission.)

point between G' and G''. Consequently, it is important to note that the state of flocculation is a sugar surfactant composition effect. Thus, an increase in the sucrose palmitate content leads to an increase in the values of the viscoelastic functions (Figure 7.10), a higher value for the modulus G_0, and a broader plateau region in the relaxation time spectrum [66]. Taken together, these facts corroborate the enhancement of the elastic network. Similar, but less pronounced, results were obtained upon increasing the oil content, whereas the dynamic response of o/w emulsions stabilized by sucrose distearate and egg yolk blends depends on the composition ratio. Thus, a higher proportion of egg yolk enhances the elasticity of the emulsion in almost the whole frequency range studied due to the formation of a highly flocculated emulsion [72]. Entanglements between protein molecules at the interface may favor the formation of elastic networks, which explains the smaller droplet diameters found in the systems containing protein.

The temperature also greatly affects the viscoelasticity of o/w emulsions containing sucrose palmitate and sunflower oil [66]. The thermorheological behavior of these systems is complex as the influence of the temperature on G' and G'' is unequal and depends on both the frequency and composition. The influence of absolute temperature on G_0 can be described by an Arrhenius-type equation and the plateau region in the relaxation time spectrum is clearly expanded when the temperature decreases because of the enhanced formation of an elastic network due to extensive flocculation of the oil droplets. This means that the stability is substantially improved when the emulsion is stored at low temperature.

The influence of the processing parameters on the stability of a salad dressing emulsion containing sucrose stearate and egg yolk blends has also been studied. The existence of a plateau region in the relaxation spectrum reveals extensive flocculation that depends on the energy input and temperature during the emulsification process [71]. These factors can also be varied to provide higher steady-viscosity and viscoelastic (G', G'') values, an increase in emulsion stability, and a lower droplet size and polydispersity.

Finally, the deviation of a sugar-based surfactant emulsifier from the Cox–Merz rule (i.e., a comparison between the steady-state and complex viscosities) and its nonlinear viscoelasticity properties and transient flow behavior have also been reported [70,72]. The combined results suggest the same rheological and microstructural characterization given by linear viscoelasticity and steady-state flow studies.

7.4.3 ALKYLPOLYGLUCOSIDES

Oil-in-water cosmetic or pharmaceutical creams are complex, multicomponent preparations that combine a number of surfactants, polymers, and other additives. In particular, APG mixtures have been proposed as an alternative to polyoxyethylene glycol (PEG) derivatives for use as emulsifier blends. The microstructure of one type of o/w cream has been proposed to involve lamellar liquid-crystalline bilayers surrounding the oil droplets [78,79]. The swelling capacity of the lamellar bilayers in these systems should be an important control parameter, and the flow properties and stability of these emulsions provide significant information since the viscosity of creams is the numerical result of the forces acting within the structure.

The mixed APGs cetearyl glucoside and cetearyl alcohol as emulsifier blend and medium-chain triglycerides (Miglyol 812) as the oil phase have been tested as a drug-delivery medium for the treatment of skin diseases [78]. Systematic studies have been carried out for binary emulsifier/water systems and cream samples with increasing oil content (5–25 wt%). All samples were found to exhibit shear-thinning flow behavior with thixotropy and yield stress values. This is the most desirable type of flow behavior in pharmaceutical or cosmetic products since they should be "thin" during application and "thick" afterwards. The yield stress values were found to increase with the sample's oil content, which is related to the stronger network and uniform packaging of the oil droplets. Thus, the greater the structural forces holding the oil phase together, the greater the viscosity of the cream. The same trend is found for the apparent viscosity recorded by flow experiments and the viscoelastic parameters obtained by oscillatory experiments. All these creams have a much more pronounced elastic component than a viscous component in the whole frequency range, which correlates well with the existence of yield stress. The elastic modulus shows a weak dependence on the applied frequency while the viscous modulus fluctuates at lower frequency. This behavior has already been described for the lamellar structure of different surfactants [51,80].

The rheological behavior also appears to influence the *in vitro* and *in vivo* permeation profiles (bioavailability) of several complex systems containing cetearyl glucoside/cetearyl alcohol as emulsifier blend and the topical drug hydrocortisone [78,79]. Savic et al. determined the yield stress values and elasticity behavior, amongst other physicochemical properties (by polarization and transmission electron microscopy, wide-angle x-ray diffraction, and thermal analysis), in order to find the most convenient pharmaceutical formulation. Kónya et al. have also evaluated the flow behavior of different o/w emulsion creams with potential pharmaceutical applications. The complex emulsions studied contained cetostearyl alcohol as fatty amphiphile, isopropyl myristate [81], or white petrolatum [82] as oil, ionic APG derivatives (alkylpolyglucoside citrate and alkylpolyglucoside tartrate) as emulsifier, distilled water, and preservatives. Although other types of emulsifiers were also tested, systems containing these APGs showed better rheological and stability characteristics. All these creams were found to behave as plastic (shear thinning with yield stress)-thixotropic fluids [81] and their low yield stress values did not change significantly with surfactant concentration or during storage. Their small thixotropic region indicates the good structure-recovery properties of those creams. Further studies (in combination with other techniques such as transmission electron microscopy and x-ray diffraction) have shown that these APGs form mixed micelles rather than bilayers with cetostearyl alcohol. This is a different microstructure from those proposed in the literature for other o/w emulsion creams with pharmaceutical applications [78,79].

7.5   RHEOLOGICAL PROPERTIES OF OTHER SELF-ASSEMBLED STRUCTURES: ORGANOGELS

Organogels consist of a three-dimensional network of entangled reverse cylindrical wormlike micelles in an organic solvent. These systems are currently of interest in the pharmaceutical field because of their structural and functional benefits as they facilitate transport when applied topically and thereby lead to a dermal or transdermal effect [83]. Organogels containing alkylglucosides have been described for systems with lecithin [84–86], which forms spherical reverse micelles when dissolved in a nonpolar medium [87]. However, the addition of small amounts of strongly polar solvents (for instance water, glycerol, ethylene glycol, or formamide) can result in the transformation of these spherical micelles into cylinders due to the formation of hydrogen bonds between the phospholipid molecules and the polar solvent [88]. The presence of an alkylglucoside, which can also form hydrogen bonds between the molecules in the micellar aggregates, favors a jelly-like phase, which means that sugar-derived surfactants can modify the rheological properties of lecithin organogels in different ways and can provide an opportunity to control the properties of organogel matrices. In addition, the rheological behavior of any potential vehicle must be determined before it can be used for topical drug-delivery applications.

Oscillatory experiments have been used to study the viscoelastic properties of lecithin organogels containing α- and β-anomers of $C_{10}G_1$. In general, alkylglucosides with a β-linkage between the hydrophilic and hydrophobic parts are much more soluble in water, have higher CMC values, lower Kraft points, and form smaller aggregates at lower concentrations than alkylglucosides with an α-linkage [89–91]. Their rheograms are common to all organogel systems containing alkylglucosides that form wormlike micelles [92]. At low frequencies, G' and G'' have a slope of 2 and 1, respectively, whereas the complex viscosity modulus is constant. The elastic modulus increases slowly at frequencies above the crossover of G' and G'', whereas the loss modulus decays or passes through a minimum and the complex viscosity modulus decreases. However, a deviation from Maxwellian behavior is clearly observed at high frequencies. Both types of alkylglucoside linkages cause an equal decrease of zero-shear viscosity that tends to a constant value at around 4 wt% of $C_{10}G_1$. A pronounced difference is found between their effects on the plateau modulus (G_0), which decreases with increasing surfactant concentration to reach a minimum, which is lower for the α-linkage type (Figure 7.11) [85].

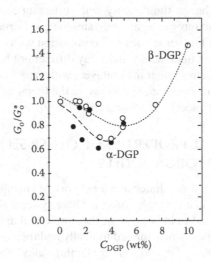

FIGURE 7.11   Plateau modulus (G_0) as a function of the sugar surfactant weight fraction (c_{DGP}). (●: n-decyl-α-D-glycopyranoside; ○: n-decyl-β-D-glycopyranoside). G_0^* is the plateau modulus of the organogel without alkylglycoside. (From Shchipunov, Y.A., Shumilina, E.V., and Hoffmann, H., *J. Colloid Interface Sci.*, 199, 218, 1998. With permission.)

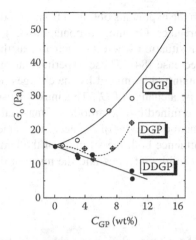

FIGURE 7.12   Plateau modulus (G_0) as a function of n-alkyl-β-D-glycopyranoside weight fraction (c_{DG}). (OGP: n-octyl-β-D-glycopyranoside; DGP: n-decyl-β-D-glycopyranoside; DDGP: n-dodecyl-β-D-glycopyranoside). (From Shchipunov, Y.A., Shumilina, E.V., and Hoffmann, H., *J. Colloid Interface Sci.*, 199, 218, 1998. With permission.)

The rheological study of lecithin/β-alkylglucoside mixtures has been extended to derivatives containing alkyl chains of various lengths (from 8 to 12 carbons): sugar derivatives with the shortest alkyl chain lead to an increase in the plateau modulus whereas derivatives with the longest chain decrease it, and G_0 reaches a minimum for n-decyl-β-D-glycopyranoside (medium chain length; Figure 7.12) [85]. The different behaviors were evaluated by taking into account the classical theory of rubber-like elasticity (Equation 7.8). In the general case, G_0 is determined by the number of contacts between micellar aggregates instead of the micelle length. The entanglement density, v, may decrease if the micelles become shorter, thereby making the G_0 value smaller. As lecithin organogels containing C_8G_1 and $C_{10}G_1$ behave differently, their dynamic properties resulting from micellar breaking (scission/recombination) were investigated. Complementary studies by Fourier-transform infrared (FTIR) spectroscopy showed that there is a difference in the alkylglucoside effect on hydrogen bonds formed by the lecithin phosphate group, which allowed Shchipunov et al. to conclude that the changes observed in the molecular interactions within micelles are responsible for the variation of the rheological parameters through the relaxation time (τ) for micellar breaking/recombination.

The effect of differences in the molecular shape of the sugar-based surfactant on the viscoelastic properties has also been studied. The viscosity, structural relaxation time, and shear moduli decrease with increasing n-dodecyl-β-D-glucopyranoside content whereas they increase for n-dodecyl-β-D-lactobionamide (n-dodecyl-β-D-galactopyranosyl-D-gluconamide) [84,93]. Despite their different effects, both the sugar-based surfactants change their scaling behavior in a similar way and agree well with the exponents obtained from a model involving branched connected micellar aggregates. Organogels containing $C_{12-14}G_{1.3}$ alkylpolygluco-side have an in-between situation where G_0 increases as the viscosity and structural relaxation time decrease [84]. These experimental results suggest that the viscosity changes are mainly determined by the changes in structural relaxation times and not so much by a change of G_0. This makes sense as the entanglement density, v, is mainly determined by the amount of material in the entangled network and not by the dynamic behavior of the network, as mentioned above. Sugar derivatives therefore influence both the strength of the hydrogen bonds between the molecules in the wormlike micelles and the micellar dynamics, even at low concentrations [93].

7.6 SUMMARY

The rheological study of aqueous solutions of glucosides has mainly been carried out for β-D-glucopyranoside surfactants with different chain lengths. The results have shown that they are elongated aggregates that become branched micelles at high surfactant concentration. The addition of oil induces a structural change in the micellar shape that is closely related to the solubilized oil site in the aggregates.

The effect of head-groups on aggregation size in the dilute regime has also been studied by viscosity measurements. The hydrophilicity of the head-group affects the viscosity: micelles with more hydrophilic head-groups (i.e., disaccharides) are less viscous. The viscoelastic behavior of these systems agrees with the presence of wormlike micelles.

The rheology of liquid-crystalline systems containing glucosides has not yet been studied in detail, although viscoelastic parameters have been used to determine phase boundaries as they differ from one type of liquid crystal to another. n-Octyl-β-D-glucopyranoside (C_8G_1)-water systems, and their mixtures, are the most studied amongst all the sugar surfactants. Liquid-crystalline phases are generally very elastic and thixotropic shear-thinning fluids with yield stress. Topological transformation under shearing has been established for many of these lamellar systems and different results have been reported for dilute and concentrated lamellar samples. The latter behave as strong gels with viscoelastic moduli that are practically independent of the frequency, while the viscoelastic moduli of dilute systems are frequency-dependent. This different behavior has been attributed to the different bilayer topologies (vesicles or bilayers) although, generally speaking, larger and more flexible vesicles are more elastic. In the presence of an alcohol, the rheological behavior depends on the type of alcohol and the alkylglucoside concentration. Lamellar, hexagonal, and bicontinuous-cubic liquid-crystalline regions form for APG systems in the presence of a hydrocarbon and all these liquid crystals have similar dynamic responses of the Maxwellian type.

As regards o/w emulsions, sugar-based surfactants have been used in food, cosmetic, and drug-delivery formulations whose viscoelastic properties depend on a huge number of variables, such as the emulsification process, composition, flocculation degree, temperature, storage conditions, and aging effects. These emulsions commonly show a shear-thinning response with a yield stress value and predominant elastic properties. In general, however, the viscoelastic properties depend on the state of flocculation: an extensive degree of flocculation results in a plateau regime for the elastic modulus (G_0) while weakly flocculated emulsions exhibit terminal relaxation time. The sugar content obviously also has a huge influence. The influence of the oil and sugar surfactant is qualitatively similar for disaccharide systems employed as food emulsifiers and the viscosity values increase with their concentration. For egg yolk/sucrose distearate blends, however, the viscosity decreases with sugar surfactant concentration.

Pharmaceutical creams have recently begun to include APGs in their formulations as an alternative to PEG derivatives, and rheological determinations have been used to evaluate the *in vitro* and *in vivo* permeation profiles of the topical drug hydrocortisone. These o/w emulsions are thixotropic, shear-thinning fluids with yield stress and a pronounced elastic contribution. Shear-thinning behavior is the most desirable flow type for topical applications.

Glucosides are also important components in lecithin organogels, which are networks of entangled reverse wormlike micelles in an organic solvent that are commonly used as topical drug-delivery vehicles. These organogels behave as Maxwellian fluids, although they deviate from Maxwellian-type behavior at high frequencies. The elastic plateau modulus (G_0) behaves differently for different alkyl chain lengths and for different hydrophilic character. This is due to the fact that sugar derivatives influence both the strength of the hydrogen bonds between the molecules in the wormlike micelles and the micellar breakdown dynamics.

Finally, it should be pointed out that the great possibilities opening up following the new and important applications of sugar-based surfactants suggest that their rheological studies should be updated. Rheological characterization of a product is crucial for a successful finished product as well as for resolving rheological problems encountered during manufacturing such as mixing, pumping, flow, heat exchangers, etc. A new rheological review of sugar-based surfactants that includes new, high-value-added products will therefore soon be necessary.

REFERENCES

1. Herb, C. A. and Prud'homme, R. K., (Eds.), *Structure and Flow in Surfactant Solutions*, ACS Symposium Series 578, Washington DC, 1994.
2. Hoffmann, H., Viscoelastic Surfactant Solutions, in *Structure and Flow in Surfactant Solutions*, Herb, C. A., Prud'homme, R. K., Ed., ACS Symposium Series, Washington, DC, 578, 1994, p 2.
3. Rehage, H., in *Rheological Properties of Viscoelastic Surfactant Solutions: Relationship with Micelle Dynamics*, CRC Press, Boca Raton, 2005, 125, p 419.
4. Berni, M. G., Lawrence, C. J., and Machin, D., *Adv. Colloid Interface Sci.*, 2002, *98*, 217.
5. Rodriguez-Abreu, C., Acharya, D., Aramaki, K., and Kunieda, H., *Colloids Surf. A: Physicochem. Eng. Aspects*, 2005, *269*, 59.

6. Balzer, D. and Lüders, H., (Eds.), *Nonionic Surfactants. Alkyl Polyglucosides*, Marcel Dekker, New York, 2000.
7. Chung, Y. and Jeffrey, G., *Biochim. Biophys. Acta (BBA)–Biomembranes*, 1989, *985*, 300.
8. Nilsson, F. and Söderman, O., *Langmuir*, 1996, *12*, 902.
9. Sakya, P., Seddon, J. M., and Templer, R. H., *J. Phys. II France*, 1994, *4*, 1311.
10. Loewenstein, A. and Igner, C., *Liq. Cryst.*, 1991, *10*, 457.
11. Häntzschel, D., Schulte, J., Enders, S., and Quitzsch, K., *Phys. Chem. Chem. Phys.*, 1999, *1*, 895.
12. Bonincontro, A., Briganti, G., D'Aprano, A., La Mesa, C., and Sesta, B., *Langmuir*, 1996, *12*, 3206.
13. Antonelli, M. L., Bonicelli, M. G., Ceccaroni, G., La Mesa, C., and Sesta, B., *Colloid Polym. Sci.*, 1994, *272*, 704.
14. Ericsson, C. A., Söderman, O., and Ulvenlund, S., *Colloid Polym. Sci.*, 2005, *283*, 1313.
15. Kahl, H., Enders, S., and Quitzsch, K., *Colloids Surf. A: Physicochem. Eng. Aspects*, 2001, *183–185*, 661.
16. Reimer, J., Nilsson, M., Álvarez Chamorro, M., and Söderman, O., *J. Colloid Interface Sci.*, 2005, *287*, 326.
17. Sato, T., Acharya, D. P., Kaneko, M., Aramaki, K., Singh, Y., Ishitobi, M., and Kunieda, H., *J. Disp. Sci. Technol.*, 2006, *27*, 611.
18. Rodriguez-Abreu, C., Aramaki, K., Tanaka, Y., Lopez-Quintela, M. A., Ishitobi, M., and Kunieda, H., *J. Colloid Interface Sci.*, 2005, *291*, 560.
19. Sanz, M., Granizo, N., Gradzielski, M., Rodrigo, M., and Valiente, M., *Colloid Polym. Sci.*, 2005, *283*, 646.
20. Rehage, H. and Hoffmann, H., *J. Phys. Chem.*, 1988, *92*, 4712.
21. Calahorro, C., Muñoz, J., Berjano, M., Guerrero, A., and Gallegos, C., *J. Am. Oil Chem. Soc.*, 1992, *69*, 660.
22. Platz, G., Pölike, J., and Thunig, C., *Langmuir*, 1995, *11*, 4250.
23. Siddig, M. A., Radiman, S., Muniandy, S. V., and Jan, L. S., *Colloids Surf. A: Physicochem. Eng. Aspects*, 2004, *236*, 57.
24. Schulte, J., Enders, S., and Quitzsch, K., *Colloid Polym. Sci.*, 1999, *277*, 827.
25. Montalvo, G., Rodenas, E., and Valiente, M., *J. Colloid Interface Sci.*, 1998, *202*, 232.
26. Ismail, Z., Kassim, A., Suhaimi, H., and Ahmad, S., *J. Disp. Sci. Technol.*, 2002, *23*, 769.
27. Granizo, N., Alvarez, M., and Valiente, M., *J. Colloid Interface Sci.*, 2006, *298*, 363.
28. Granizo, N., Thunig, C., and Valiente, M., *J. Colloid Interface Sci.*, 2004, *273*, 638.
29. Diat, O., Roux, D., and Nallet, F., *J. Phys. II Fr.*, 1993, *3*, 1427.
30. Gulik-Krzywicki, T., Dedieu, J. C., Roux, D., Degert, C., and Laversanne, R., *Langmuir*, 1996, *12*, 4668.
31. Versluis, P., van de Pas, J. C., and Mellema, J., *Langmuir*, 1997, *13*, 5732.
32. Schmidt, G., Müller, S., Schmidt, C., and Richtering, W., *Rheol. Acta*, 1999, *38*, 486.
33. Zipfel, J., Lindner, P., Tsianou, M., Alexandridis, P., and Richtering, W., *Langmuir*, 1999, *15*, 2599.
34. Escalante, J. I. and Hoffmann, H., *J. Physics: Condensed Matter*, 2000, *12*, A483.
35. Partal, P., Kowalski, A. J., Machin, D., Kiratzis, N., Berni, M. G., and Lawrence, C. J., *Langmuir*, 2001, *17*, 1331.
36. Zipfel, J., Nettesheim, F., Lindner, P., Le, T. D., Olsson, U., and Richtering, W., *Europhys Lett.*, 2001, *53*, 335.

37. Coppola, L., Nicotera, I., and Oliviero, C., *Appl. Rheol.*, 2005, *15*, 230.
38. Diat, O., Roux, D., and Nallet, F., *J. Phys. II Fr*, 1983, *3*, 1427.
39. Le, T. D., Olsson, U., Mortensen, K., Zipfel, J., and Richtering, W., *Langmuir,* 2001, *17*, 999.
40. Roux, D., Nallet, F., and Diat, O., *Europhys. Lett.*, 1993, *24*, 53.
41. Bergmeier, M., Gradzielski, M., Hoffmann, H., and Ortensen, K., *J. Phys. Chem. B*, 1999, *103*, 1605.
42. Hoffmann, H., Munkert, U., Thunig, C., and Valiente, M., *J. Colloid Interface Sci.*, 1994, *163*, 217.
43. Gradzielski, M., Hoffmann, H., Panitz, J., and Wokaun, K., *J. Colloid Interface Sci.*, 1995, *169*, 103.
44. Montalvo, G., Valiente, M., and Rodenas, E., *Langmuir*, 1996, *12*, 5202.
45. Montalvo, G., Valiente, M., and Khan, A., *Langmuir*, 2007, *23*, 10518.
46. Cortés, A. B. and Valiente, M., *Colloid & Polym. Sci.*, 2003, *281*, 319.
47. Sadtler, V. M., Guely, M., Marchal, P., and Choplin, L., *J. Colloid Interface Sci.*, 2004, *270*, 270.
48. Le, T. D., Olsson, U., and Mortensen, K., *Physica B: Condensed Matter*, 2000, *276–278*, 379.
49. Stubenrauch, C., *Curr. Opin. Colloid Interface Sci.* 2001, *6*, 160.
50. Platz, G., Thunig, C., Policke, J., Kirchhoff, W., and Nickel, D., *Colloids Surf. A: Physicochem. Eng. Aspects*, 1994, *88*, 113.
51. Robles-Vasquez, O., Corona-Galvan, S., Soltero, J. F. A., Puig, J. E., Tripodi, S. B., Valles, E., and Manero, O., *J. Colloid Interface Sci.*, 1993, *160*, 65.
52. Nemeth, Z., Halasz, L., Palinkas, J., Bota, A., and Horanyi, T., *Colloids Surf. A: Physicochem. Eng. Aspects*, 1998, *145*, 107.
53. Siddig, M. A., Radiman, S., Jan, L. S., and Muniandy, S. V., *Colloids Surf. A: Physicochem. Eng. Aspects*, 2006, *276*, 15.
54. Lauger, J., Weigel, R., Berger, K., Hiltrop, K., and Richtering, W., *J. Colloid Interface Sci.*, 1996, *181*, 521.
55. Montalvo, G., Rodenas, E., and Valiente, M., *J. Colloid Interface Sci.*, 1998, *202*, 233.
56. Warriner, H. E., Davidson, P., Slack, N. L., Schellhorn, M., Eiselt, P., Idziak, S. H. J., Schmidt, H., and Safinya, C. R., *J. Chem. Phys.*, 1997, *107*, 3707.
57. Bergmeier, M., Gradzielski, M., Hoffmann, H., and Mortensen, K., *J. Phys. Chem. B*, 1998, *102*, 2837.
58. Abdel-Rahem, R., Gradzielski, M., and Hoffmann, H., *J. Colloid Interface Sci.*, 2005, *288*, 570.
59. Radiman, S., Toprakcioglu, C., and McLeish, T., *Langmuir*, 1994, *10*, 61.
60. Jones, J. L. and McLeish, T. C. B., *Langmuir*, 1995, *11*, 785.
61. Jones, J. L. and McLeish, T. C. B., *Langmuir*, 1999, *15*, 7495.
62. Montalvo, G., Valiente, M., and Rodenas, E., *Langmuir*, 1996, *12*, 5202.
63. Gradzielski, M., Hoffmann, H., Panitz, J., and Wokaun, A., *J. Colloid Interface Sci.*, 1995, *103*, 189.
64. Maddaford, P. J. and Toprakcioglu, C., *Langmuir,* 1993, *9*, 2868.
65. Niraula, B., King, T. C., Chun, T. K., and Misran, M., *Colloids Surf. A: Physicochem. Eng. Aspects*, 2004, *251*, 117.
66. Guerrero, A., Partal, P., and Gallegos, C., *J. Rheol.*, 1998, *42*, 1375.
67. Bower, C., Gallegos, C., Mackley, M. R., and Madiedo, J. M., *Rheol. Acta,* 1999, *38*, 145.
68. Partal, P., Guerrero, A., Berjano, M., Muñoz, J., and Gallegos, C., *J. Texture Studies,* 1994, *25*, 331.

69. Partal, P., Guerrero, A., Berjano, M., and Gallegos, C., *J. Am. Oil Chem.*, 1997, *74*, 1203.

70. Partal, P., Guerrero, A., Berjano, M., and Gallegos, C., *J. Food Eng.*, 1999, *41*, 33.

71. Franco, J. M., Guerrero, A., and Gallegos, C., *Rheol. Acta,* 1995, *34*, 513.

72. Riscardo, M. A., Moros, J. E., Franco, J. M., and Gallegos, C., *Eur. Food Res. Technol.*, 2005, *220*, 380.

73. Gallegos, C. and Franco, J. M., *Rheology of Food Emulsions.* Elsevier, Amsterdam, 1999.

74. Gallegos, C., Berjano, M., and Choplin, L., *J. Rheol.,* 1992, *36*, 465.

75. Muñoz, J. and Sherman, P., *J. Texture Studies*, 1990, *21*, 411.

76. Dickinson, E., *Colloid Surf.* 1989, *42*, 191.

77. Franco, J. M., Berjano, M., and Gallegos, C., *J. Agric. Food Chem.*, 1997, *45*, 713.

78. Savic, S., Vuleta, G., Daniels, R., and Müller-Goymann, C. C., *Colloid Polym. Sci.,* 2005, *283*, 439.

79. Savic, S., Savic, M., Tamburic, S., Vuleta, G., Vesic, S., and Muller-Goymann, C. C., *Eur. J. Pharm. Sci.*, 2007, *30*, 411.

80. Németh, Z., Hasálsz, L., Pálinkás, J., Bóta, A., and Horányi, T., *Colloids Surf. A: Physicochem. Eng. Aspects*, 1998, *145*, 107.

81. Kónya, M., Dékány, I., and Erös, I., *Colloid Polym. Sci.*, 2007, *285*, 657.

82. Kónya, M., Bohus, P., Paglino, L., Csóka, I., Csányi, E., and Erös, I., *Progress Colloid Polym. Sci.,* 2004, 125, 161.

83. Kumar, R. and Katare, O. P., *AAPS Pharm. Sci. Tech.*, 2005, *6(2)*, 298.

84. Shchipunov, Y. A., Shumilina, E. V., and Hoffmann, H., *Colloid Polym. Sci.*, 1998, *276*, 368.

85. Shchipunov, Y. A., Shumilina, E. V., and Hoffmann, H., *J. Colloid Interface Sci.*, 1998, *199*, 218.

86. Shchipunov, Y. A., Shumilina, E. V., Ulbrich, W., and Hoffmann, H., *J. Colloid Interface Sci.*, 1999, *211*, 81.

87. Shchipunov, Y. A., *Russ. Chem. Rev.,* 1997, *66*, 301.

88. Shchipunov, Y. A. and Shumilina, E. V., *Mater. Sci. Eng.*, 1995, *3*, 43.

89. Shinoda, K., Yamanaka, T., and Kinoshita, K., *J. Phys. Chem.*, 1959, *63*, 648.

90. Shinoda, K., Yamaguchi, T., and Hori, R., *Bull. Chem. Soc. Jpn.*, 1961, *34*, 237.

91. Focher, B., Savelli, G., Torri, G., Vecchio, G., McKenzie, D. C., Nicoli, D. F., and Bunton, C. A., *Chem. Phys.Letters*, 1989, *158*, 491.

92. Hoffmann, H., in *Structure and Flow in Surfactants Solutions*, Herb, C. A., and Prud'homme, R. K., (Eds.), ACS Symposium Series 578, Washington DC, 1994, Chapter 1.

93. Shchipunov, Y. A., Shumilina, E. V., Ulbricht, W., and Hoffmann, H., *J. Colloid Interface Sci.,* 1999, *211*, 81.

8 Sugar-Based Gemini Surfactants with pH-Dependent Aggregation Properties

Marco Scarzello, Marc C.A. Stuart, Jaap Klijn,
Anno Wagenaar, and Jan B.F.N. Engberts

CONTENTS

8.1 SUGAR-BASED GEMINI SURFACTANTS: A UNIQUE MIX OF STRUCTURAL ELEMENTS IN THE AMPHIPHILE

Amphiphilic molecules possess parts, which are distinctively lipophilic and hydrophilic. This ambivalence lies at the basis of the intriguing properties of these molecules, selected by nature as the building-blocks of architectures essential for life (i.e., biological membranes), and chosen by scientists for several applications in fields as diverse as food industry, cleaning technologies, agrochemicals, and bio-medicine [1].

Amphiphile self-assembly can lead to a variety of structures, differing in shape and size, depending on the molecular structure of the amphiphile, and on the solution conditions (such as temperature, ionic strength, pH) [2]. Several studies showed that the ability of modulating the morphology of amphiphilic aggregates is of outmost importance for practical applications, particularly in the biomedical field [3–5].

A relatively new and particularly interesting class of amphiphilic molecules, the gemini surfactants (GS), is obtained by connecting two single-tailed surfactants via a spacer at the level of or near to the headgroups [6,7]. GSs generally possess a lower cmc and a higher ability to decrease the surface tension of water with respect to the parent surfactants [8,9], which constitute advantageous characteristics for practical applications. The sugar-based GSs, object of this chapter, exploit a unique combination of structural elements in the molecule in order to obtain self-assemblies with the desired properties [10–12] (Scheme 8.1):

1. First, the tertiary amino moieties in the headgroups, whose degree of protonation is a function of pH, confer a pH-dependent aggregation behavior. In fact, an increase in the degree of protonation of the headgroup nitrogen atoms [12–14] corresponds to an increase of the cross-sectional headgroup area a_0 (i.e., a decrease in the packing parameter P), as a consequence of the increased hydration and of stronger electrostatic repulsion between the positive charges. The effect of the pH on the molecular shape of GS1 (1,8-bis(N-octaden-9-yl-1-deoxy-D-glucitol-1-ylamino)3,6-dioxaoctane) is shown in Scheme 8.2. The pK_a values estimated for this compound are 6.0 and 8.1 [13].
2. Variations in the spacers allow modifications of the characteristics of the aggregates, by varying a_0 (also by modifying the hydration properties in the proximity of the headgroups) and the volume of the hydrophobic part V [15,16], as discussed in Section 8.4.1.

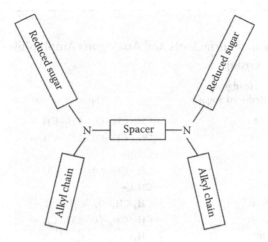

SCHEME 8.1 Schematic representation of a sugar-based GS.

3. Finally, the sugar moieties have a threefold function. First, they increase the solubility of the amphiphile in water. In fact, as we will see in Section 8.4.2, in the absence of the sugar or in the presence of short sugar moieties in the molecule, the electrically neutral compound behaves like an oil and not like an amphiphile [17]. Second, the presence of different reduced sugars determines different values of a_0 and, consequently, different aggregate morphologies [15,17]. Finally, the importance of carbohydrates in biomolecular recognition processes, which can potentially enhance the performances of amphiphiles in biomedical applications, has been recognized and has triggered a renewed interest in the study of their physical–organic chemistry.

The structural features of the different compounds which have been synthesized in the Engberts laboratory are summarized in Table 8.1.

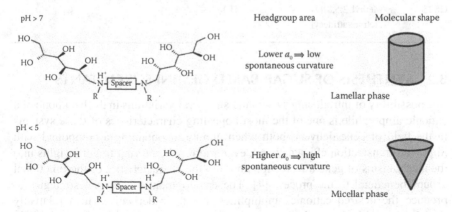

SCHEME 8.2 Schematic representation of the pH-dependence of the packing parameter of a sugar-based gemini surfactant.

TABLE 8.1
Sugar-Based Gemini Surfactants and Analogous Amphiphiles Synthesized in the Engberts' Group

Compound	Headgroup (Reduced Sugar)	Spacer	n-Alkyl Tails
GS1	Glucose	$-(CH_2-CH_2-O)_2-CH_2-CH_2-$	$C_{18:1}$
GS2	Mannose	$-(CH_2-CH_2-O)_2-CH_2-CH_2-$	$C_{18:1}$
GS3	Mannose	$-(CH_2)_6-$	$C_{18:1}$
GS4	Glucose	$-(CH_2-CH_2-O)_2-CH_2-CH_2-$	$C_{17}-C=O$
GS5	Glucose	$-(CH_2)_6-$	C_{18}
GS6	Mannose	$-(CH_2-CH_2-O)_2-CH_2-CH_2-$	C_{18}
GS7	Galactose	$-(CH_2-CH_2-O)_2-CH_2-CH_2-$	$C_{18:1}$
GS8	Galactose	$-(CH_2)_6-$	$C_{18:1}$
GS9	Talose	$-(CH_2)_6-$	$C_{18:1}$
GS10	Lactose	$-(CH_2-CH_2-O)_2-CH_2-CH_2-$	$C_{18:1}$
GS11	Lactose	$-(CH_2)_6-$	$C_{18:1}$
GS12	Melibiose	$-(CH_2)_6-$	$C_{18:1}$
GS13	Glucose	$-(CH_2)_2-$	$C_{18:1}$
GS14	Glucose	$-(CH_2)_4-$	$C_{18:1}$
GS15	Glucose	$-(CH_2)_6-$	$C_{18:1}$
GS16	Glucose	$-(CH_2)_8-$	$C_{18:1}$
GS17	Glucose	$-(CH_2)_{10}-$	$C_{18:1}$
GS18	Glucose	$-(CH_2)_{12}-$	$C_{18:1}$
GS19	Glucose	$-(CH_2)_3-$	C_{12}
GS20	Glucose	$-(CH_2)_3-O-(CH_2-CH_2-O)_2-(CH_2)_3-$	$C_{18:1}$
GS21	Arabinose	$-(CH_2)_6-$	$C_{18:1}$
GS22	Erythrose	$-(CH_2)_6-$	$C_{18:1}$
GS23	2,3–dihydroxy–1–glyceryl	$-(CH_2)_6-$	$C_{18:1}$
GS24	Methyl	$-(CH_2)_6-$	$C_{18:1}$
GS25	Methyl-2,5,8,11-tetraoxatridecyl	$-(CH_2)_6-$	$C_{18:1}$

8.2 SYNTHESIS OF SUGAR-BASED GEMINI SURFACTANTS

The possibility of introducing systematic structural variations in the backbone of a cationic amphiphile is one of the most appealing characteristics of these systems in the field of gene delivery, both when aiming at obtaining a compound with improved transfection efficiency and, even more, when trying to get insights into the mechanisms of gene transfection and into the characteristics of the functional groups beneficial to the process [4]. The development of efficient strategies to produce the desired cationic amphiphiles and their derivatives in a relatively straightforward and reproducible way is therefore essential. We will here briefly

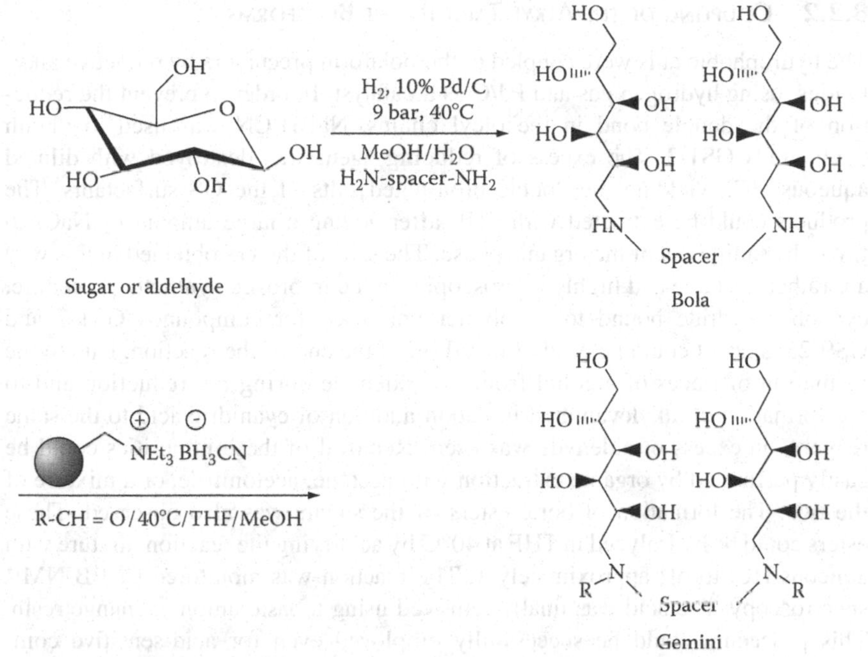

SCHEME 8.3 Synthesis of sugar-based GS and analogues. (From Wagenaar, A. and Engberts, J.B.F.N., *Tetrahedron*, 63, 10622, 2007.)

summarize a recently developed synthetic protocol for the synthesis of sugar-based GS and analogues compounds, which provide analytically pure compounds in good yields [18] and represents a substantial improvement to previously published procedures [11–13,19]. The synthetic route involves the preparation of the bola-forms and their subsequent alkylation to obtain the GS surfactants (Scheme 8.3).

8.2.1 SYNTHESIS OF THE BOLAFORM INTERMEDIATES

In order to obtain the intermediate bola amphiphile, a reductive amination is carried out in a solution of the desired sugar in a methanol/water mixture. The composition of the solvent mixture depends on the solubility of the sugar. The purification is the easiest when the product is a solid; crystallization is, in fact, sufficient to obtain the analytically pure precursors of GS1-9 and G13-19. In the case of GS8, the catalyst Pd/C had to be filtered off in acidic medium. For the precursors of GS10-12, the excess sugar had to be removed by using ion-exchange materials.

Liquid (precursor of GS25) or sticky (precursor of GS20) compounds were purified by both crystallization and ion-exchange. The precursor of GS21 was obtained in low yield due to the formation of several products arising from mono-, di-, tri-, and tetra-substitutions at the 1,6-diaminohexane nitrogen atoms. The different products could be separated and identified by liquid chromatography/mass spectrometry (LC/MS).

8.2.2 COUPLING OF THE ALKYL TAILS TO THE BOLAFORMS

The hydrophobic tails were coupled to the bolaform precursors by reductive alkylations, using hydrogen gas and Pd/C as a catalyst. In order to prevent the reduction of the double bond in the oleyl chains, $NaBH_3CN$ was used to obtain compounds GS1-3. The excess of reducing agent was destroyed with diluted aqueous HCl, yielding the double protonated salts of the GS surfactants. The products could be extracted with THF after adding a large amount of NaCl to favor the salting out of the organic phase. The salts of the GS obtained in this way are rather unstable and highly hygroscopic. In the improved synthetic procedure, cyanoborohydride bound to a polymer was used for compounds GS1-7 and GS9-25, since it could be easily filtered off at the end of the reaction. Due to the formation of traces of alcohol from the aldehyde during the reduction and to the formation of alkylcyanohydrine upon addition of cyanidric acid to the same reagent, an excess of aldehyde was used. Removal of these impurities could be easily performed by organic extraction with acetone, acetonitrile, or a mixture of the two. The formation of boric esters of the sugars was also observed. These esters could be hydrolyzed in THF at 40°C by acidifying the reaction mixture with aqueous HCl to pH approximately 1. The reaction was monitored by ^{11}B-NMR spectroscopy. The acid was finally removed using a basic anion-exchange resin. This procedure could be successfully employed even for acid-sensitive compounds (containing lactose and melibiose) and with 1,4 and 1,6 ether bonds (GS10-12). 0.5–2.0 moles of water present in the preparations of GS5-8, GS11-14, and GS22-23 could not be removed even by freeze-drying techniques, probably due to the strong binding of the water molecules to the reduced sugars of the headgroups. Only for GS21, a well-defined melting temperature could be measured.

8.3 pH-DEPENDENT AGGREGATION BEHAVIOR OF SUGAR-BASED GEMINI SURFACTANTS

In order to illustrate the pH-dependence of the phase behavior of sugar-based GS, we will describe the phase behavior of GS1 (Table 8.1) as a function of pH. GS1 is one of the first sugar-based GS to be synthesized and the one most extensively studied. Its behavior (together with that of GS15, see Section 8.4.1) will be used throughout this chapter as a reference in the rationalization of the effects induced by structural variations to its basic framework and is here described in detail.

8.3.1 VESICLE-TO-MICELLE TRANSITIONS

It was observed that upon decreasing the pH of a vesicular dispersion of the sugar-based GS1 prepared around neutral pH, the system undergoes transitions from a lamellar phase to worm-like micelles and ultimately to spherical micelles [12]. The pH values corresponding to the phase transitions can be inferred by static light scattering (SLS) and Nile Red fluorescence measurements [12–14,20]. The plots of the intensities of the scattered radiation from SLS measurements and of the wavelength

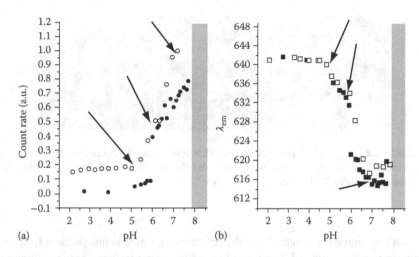

FIGURE 8.1 Normalized count rates (a) from SLS experiments and wavelength (b) of maximum Nile Red emission in water. Open and closed symbols refer to independent measurements. Arrows indicate the onset of the different transitions (see text for explanation). (Reprinted from Scarzello, M., Klijn, J.E., Wagenaar, A., Stuart, M.C., Hulst, R., and Engberts, J.B.F.N., *Langmuir*, 22, 2558, 2006. With permission.)

of maximum emission from Nile Red fluorescence measurements as a function of pH are consistent with the formation of aggregates of progressively smaller sizes and higher surface polarity with decreasing pH values (Figure 8.1).

The appearance of the samples between pH 7.7 and 7.0 (turbid/bluish) and the values of the Nile Red fluorescence emission maxima (about 615 nm) are consistent with the presence of vesicular aggregates (L_α phase). At pH < 5.9, the samples appear optically clear, indicating that the aggregates assume a micellar morphology. Moreover, the samples at pH values around 5.6 are more viscous relative to the vesicular dispersions found in less acidic conditions and are characteristic for the presence of cylindrical micelles. Consistently, dynamic light scattering (DLS) experiments show a broad size distribution in the pH interval between 5.9 and 5.2. Further evidence for the formation of worm-like micelles comes from the strong angular dependence of the intensity of the scattered light [12]. In the limits of the Guinier approximation, the dependence of the scattered intensity on the scattering angle θ is described by the following equation:

$$I(\theta) \propto \exp(-q^2 R_g^2/3) \quad (qR_g \ll 1) \tag{8.1}$$

where q is the scattering vector, $q = (4\pi n_s/\lambda_0)\sin(\theta/2)$, where n_s is the refractive index of the solution) and R_g is the radius of gyration. The value of R_g obtained by fitting the experimental intensity data obtained at pH 5.3 to Equation 8.1 is 42 nm. At the same pH value, the hydrodynamic radius measured by DLS is $R_h = 22$ nm. The ratio $R_g/R_h \approx 1.9$ is consistent with the presence of cylindrical aggregates.

FIGURE 8.2   Cryo-TEM pictures of GS1 dispersions at pH 2 (a), 5.4 (b), and 7 (c). (Reprinted from Johnsson, M., Wagenaar, A., Stuart, M.C., and Engberts, J.B.F.N., *Langmuir*, 19, 4609, 2003. With permission.)

On the contrary, the intensity of the scattered radiation is independent from the scattering angle at pH < 5.2 and the sizes of the aggregates are reduced to about 5–6 nm. These data indicate the presence of spherical micelles in this pH range. The conclusions drawn from visual inspections, spectroscopic, and diffraction data are confirmed by cryo-TEM pictures taken at different pH values (Figure 8.2).

8.3.2 FLOCCULATION AND REDISPERSION AT BASIC pH: THE INTRIGUING PHENOMENON OF OH⁻ BINDING

When the pH of a vesicular suspension prepared around neutral pH is increased, rapid flocculation is observed. ζ-potential measurements show that the colloidal stability decreases when the surface charge approaches neutrality (< +15 mV). This behavior is in agreement with the classical Derjaguin–Landau–Verwey–Overbeek (DLVO) theory: in the pH interval associated with a low surface charge, the vesicles aggregate as a result of attractive van der Waals interactions which overcame the modest electrostatic repulsion [12].

Interestingly, a further increase in pH leads to charge reversal and redispersion of the flocculated material as negatively charged vesicles are observed at ζ-potentials values < −15 mV.

DLS data demonstrate that when redispersion upon charge reversal is induced immediately after flocculation, the size distribution of the negatively charged vesicles is the same as that of the positively charged ones, indicating that fusion does not occur upon aggregation [13]. The cryo-TEM in Figure 8.3 confirms the light scattering data showing how the flocs consist of loosely aggregated vesicles.

In order to get insights into the requirements for GS1 vesicle fusion, we studied the effect of the equilibration time of the flocs on the sizes of the redispersed vesicles (Figure 8.4). GS1 vesicles were prepared at pH 7.5, flocculated upon addition of NaOH to pH 8.4, and redispersed after increasing equilibration times.

The redispersion was performed by adding either $NaOH_{(aq)}$ to a final pH of 8.9 or $HCl_{(aq)}$ to a final pH of 7.5. The hydrodynamic radii obtained from DLS measurements are plotted against the equilibration time of the flocculated material

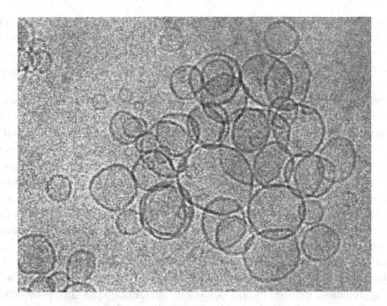

FIGURE 8.3 Cryo-TEM picture of flocculated GS1 vesicles. (Reprinted from Johnsson, M., Wagenaar, A., Stuart, M.C., and Engberts, J.B.F.N., *Langmuir*, 19, 4609, 2003. With permission.)

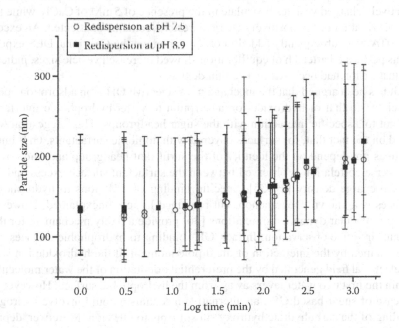

FIGURE 8.4 Hydrodynamic diameters measured for GS1 vesicles in water after flocculation–equilibration–redispersion at pH 7.5 and 8.9 as a function of the equilibration time after flocculation. The bars represent the width of the size distributions at half height.

(x-axis in Figure 8.4). It is clear that the size distribution of the redispersed aggregates depends on the equilibration time of the flocs and that redispersion by adding either acid or base leads to dispersed particles with comparable sizes. However, the size of the redispersed aggregates increases with respect to the original size of the vesicles when the flocs are equilibrated for more than about 15 min.

The charge reversal of the vesicular surfaces observed at basic pH is not due to adsorption at the vesicular interface of negatively charged buffer substances, as proven by the observation that an identical behavior is observed in 15 mM $NaCl_{(aq)}$ [12]. The only reasonable explanation for the presence of a negative surface potential at basic pH is selective OH^--ion adsorption at the interface. Consistently, the pH dependence of the ζ-potential can be well simulated by a simple Poisson–Boltzmann (PB) model in which the binding of hydroxide and protons to the vesicular surface is described by a number of surface equilibrium reactions [13].

The ability of Ca^{2+} cations to induce fusion of negatively charged phospholipid vesicles has been since long recognized [21,22]. Upon binding to the negatively charged phosphate headgroups, Ca^{2+} diminishes the surface charge of the vesicles and induces the formation of hydrophobic domains at the surface, which favor aggregation, induced by van der Waals interactions [23–28], provided that the concentration of the divalent cation is sufficiently high.

We studied the influence of Ca^{2+} ions on GS1 vesicular dispersions in order to verify if effects comparable to those just described could be observed even for particles which are negatively charged due to selective OH^- adsorption. Indeed, negatively charged vesicles flocculate in the presence of 5 mM of $CaCl_2$, while the size distribution of the positively charged vesicles remains unaffected. An excess of EDTA was subsequently added in order to remove the Ca^{2+} ions. DLS experiments performed after 1 h of equilibration showed increased vesicle sizes, indicating that fusion had occurred to a certain degree.

It has been argued that the mechanism of selective OH^--ion adsorption could be related to that of hydroxide-ion adsorption to water/hydrophobic interfaces and not to a specific interaction with the sugar headgroups. This suggestion was based on the fact that, for surfaces covered with nonionic surfactants, OH^- binding does not depend on the identity of the surfactant headgroup and, moreover, an inverse correlation is observed between the surfactant surface excess and the surface charge density [29–35]. Specific binding of OH^- ions to hydrophobic surfaces is a known phenomenon, but remained largely unexplained. However, recent molecular dynamics simulations [36] provide a likely mechanism for this counter-intuitive observation. In fact, OH^- binding to hydrophobic surfaces can be explained by the interaction of the dipole moment of the hydroxide ion with the electrical field generated by the preferential orientation of the water molecules within the first two water layers away from the hydrophobic surface. However, in the case of sugar-based GSs, an alternative mechanism could involve hydrogen bonding of the carbohydrate hydroxy substituents to OH^- ion. Moreover, deprotonation of one of the carbohydrate hydroxy groups would also lead to a negative surface charge. This last possibility is, however, unlikely due to the high pK_a values of carbohydrates (as an example, the pK_a of glycerol is about 14). As described in the following, in order to discriminate between the above-mentioned mechanisms and in order to obtain further insights into the requirements for selective hydroxide-binding to GS vesicular surfaces, a systematic comparative study has been carried out on: (1) GSs obtained by varying the length and nature of the spacer, (2) GSs obtained by substituting the sugar moieties with methyl groups, (3) GSs obtained by substituting the sugar moieties with short methoxy-terminated polyethylene oxide chains, (4) GSs obtained by varying the length and the stereochemistry of the sugar moieties, and (5) GS mixtures with single- and double-tailed amphiphilic additives.

8.4 EFFECTS OF STRUCTURAL VARIATIONS ON THE pH-DEPENDENT AGGREGATION BEHAVIOR OF SUGAR-BASED GEMINI SURFACTANTS

8.4.1 EFFECTS OF VARIATIONS OF THE LENGTH AND NATURE OF THE SPACER

The molecular structure of GS15 differs from that of GS1 only for the spacer (Table 8.1): while GS1 possesses an ethylene oxide spacer ($-(CH_2CH_2O)_2-CH_2CH_2-$), GS15 is characterized by a slightly shorter and more hydrophobic alkyl spacer ($-(CH_2)_6-$).

The morphologies of the aggregates formed by GS15, as determined by SLS (Figure 8.5a) and Nile Red fluorescence measurements as a function of pH, are analogous to those observed for GS1. Vesicles (Figure 8.5d) prepared around pH 7.5 undergo a transition toward worm-like micelles (Figure 8.5c) at pH 7.

A further decrease of the pH leads to the formation of spherical micelles (pH 5.8) and solely spherical micelles (Figure 8.5b) are present at pH < 4.7. The pH values of the transitions of GS1 and GS15 can be compared. The pH at which the transition toward spherical micelles is complete is lower for the compound with the shorter and most hydrophobic spacer (GS15). This finding is consistent with a higher cross-sectional headgroup area (i.e., lower packing parameter) of GS1 as compared to that of GS15, as expected due to the presence of a longer and more strongly hydrated spacer rather close to the headgroup. In fact, the value of the packing parameter associated with spherical micelle stability is reached at a higher degree of protonation of the amino moieties of the headgroup in the compound with a more hydrophobic spacer. The onsets of transitions toward worm-like micelles and spherical micelles are similar for GS1 and GS15.

The rationalization of the differences in the colloidal stability of GS1 and GS15 is complicated by the fact that the variation in the nature of the spacer is expected to affect both the ability of the vesicular surface to adsorb OH⁻ ions and the hydration of the headgroups. We will anyway attempt to discuss these differences, conscious of the fact that an unambiguous explanation would require a more detailed investigation. The pH value corresponding to vesicle flocculation is lower for GS1 (pH 7.7) than for GS15 (pH 8.2). Nevertheless, also the intensity of the scattered radiation of GS15 vesicular dispersions steadily increases going from pH 7.5 to 8.2 (Figure 8.5a). Cryo-TEM pictures confirm that the process of

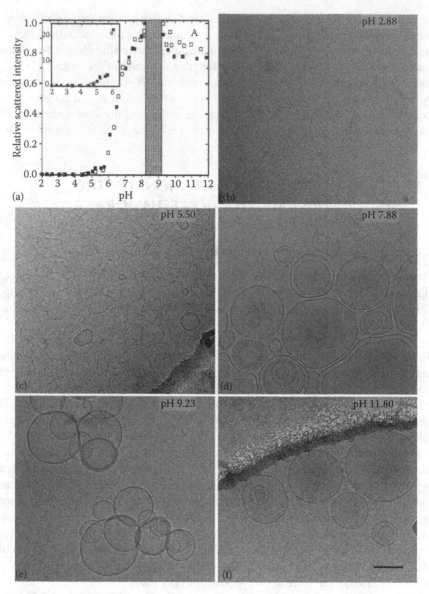

FIGURE 8.5 Normalized count rates (a) from SLS experiments of dispersions of GS15 in water (different symbols refer to independent measurements). Cryo-TEM pictures of GS15 dispersions taken at pH 2.88 (b), 5.50 (c), 7.88 (d), 9.23 (e), and 11.80 (f).

flocculation and redispersion is more gradual for GS15 than for GS1, since aggregated and multilamellar vesicles are observed at pH 7.9 (Figure 8.5d), before macroscopic aggregates are formed, consistent with the increase of the scattered intensities close to the pH of flocculation. The pH at which the aggregation of

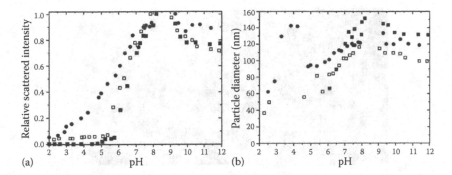

FIGURE 8.6 Relative intensities of the scattered radiation (a) and size distributions (b) of dispersions of GS15 (■), GS14 (□), and GS13 (●) as a function of pH. No points are present in the flocculation interval between pH 8 and 9. (Reprinted from Klijn, J.E., Stuart, M.C.A., Scarzello, M., Wagenaar, A., and Engberts, J.B.F.N., *J. Phys. Chem. B*, 110, 21694, 2006. With permission.)

GS15 vesicles starts is close to the pH at which GS1 flocculates. For this reason, we think that the surface charge of GS1 and GS15 vesicles is similar around pH 7.7, although ζ-potential measurements would be necessary to confirm this hypothesis. The pH of redispersion (Figure 8.5f) of the GS containing the hydrophilic spacer (GS1) is lower than the corresponding pH for GS15. We suggest that this effect might be due to the higher hydration energy of the hydrophilic spacer, which increases the colloidal stability of the dispersion.

GS14 (Table 8.1) is characterized by a spacer with four methylene units. As evident from the SLS and DLS data in Figure 8.6, this structural modification does not affect the pH of vesicle flocculation as compared to that of the parent compound GS15 [16].

This observation is consistent with the fact that in these variations in the headgroups electrostatic interactions arising from the decrease in the length of the spacer are negligible at these high pH values, since the distances between the few remaining charges are relatively high. On the contrary, the effect of the length of the spacer is significant for the transitions observed at acidic pH, where the aggregates possess higher surface charges [16]. While GS15 forms spherical micelles at pH 4.9, for GS14 dispersions the intensity of the scattered light levels off at pH lower than 5.5 to values consistent with the presence of worm-like micelles. At pH lower than 3.7, the presence of slower morphological transitions is indicated by the slow decrease in time of the scattered intensity. Cryo-TEM pictures taken on GS14 dispersions at pH 4.9 and 2.1 (Figure 8.7a) confirm that GS14 is predominantly aggregated in worm-like micelles. Moreover, the pictures show that small vesicles are also present even at pH 2.1.

The preference of GS14 for the formation of low curvature aggregates with respect to GS15 can be explained by a decrease in the packing parameter (lower cross-sectional surface area) and increased intramolecular electrostatic repulsion (lower pK_a of the double protonated headgroup) associated with the decrease

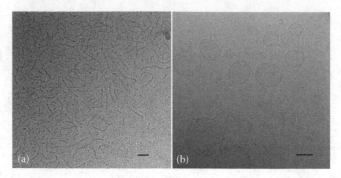

FIGURE 8.7   Cryo-TEM pictures of GS14 at pH 2.1 (a) and GS 13 at pH 1.1 (b) The bar represents 100 nm.

in the spacer length. Also the behavior of the amphiphiles obtained by a further decrease in the length of the hydrophobic spacer to two methylene units, GS 13 (Table 8.1) is consistent with what is expected on the basis of the observations for GS14: the pH of flocculation remains unaffected (Figure 8.6) while, at acidic pH, the cryo-TEM pictures indicate the presence of vesicles characterized by higher values of the packing parameter as compared to GS15 and GS14 at the same pH [16].

When increasing the number of methylene units in the spacer to eight (GS16, Table 8.1), an increase of the pH of flocculation (8.6) and redispersion (9.9) is observed. The transitions toward aggregates with higher curvature are only slightly affected. SLS and DLS experiments performed on the GS obtained by a further increase in the spacer length to 10 methylene units (GS17, Table 8.1, Figure 8.8) indicate a transition from vesicles to worm-like micelles (pH 7.2) and to spherical micelles (onset at pH 6.1, complete at pH 5.4).

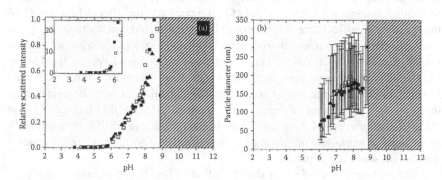

FIGURE 8.8   Intensities of the scattered radiation (a) and size distributions (b) of dispersions of GS17. (Reprinted from Klijn, J.E., Stuart, M.C.A., Scarzello, M., Wagenaar, A., and Engberts, J.B.F.N., *J. Phys. Chem. B*, 110, 21694, 2006. With permission.)

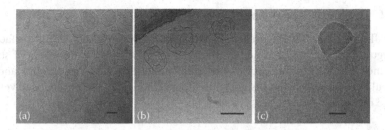

FIGURE 8.9 Cryo-TEM pictures of aqueous dispersions of GS17 under slightly basic conditions. pH 7.4 (a), 8.0 (b), and 8.5 (c). Bars represent 100nm.

Cryo-TEM pictures of GS17 dispersions taken at pH 7.4 (Figure 8.9) demonstrate the vesicular nature of the aggregate obtained at this pH [16].

In these pictures, intermediate morphologies in the process of vesicle fusion are visualized, together with elongated vesicular structures, indicating the instability of the morphologies with higher curvatures. The related high value of the packing parameter (around 1) can be rationalized assuming that the long hydrophobic spacer leads to a considerable increase of the hydrophobic volume of the surfactant. Consistently, it was reported that the plot of the surface areas of a family of GS with bisquaternary ammonium headgroups with increasing length of the alkyl spacers $-(CH_2)_n-$ versus the number of methylene units in the spacer (n) goes through a maximum for $n = 10-12$ [37]. This trend was explained on the basis of a change of the location of the spacer from the air–water interface (headgroup region) to the air side of the interface (hydrophobic region) for $n > 10$.

At more basic pH values, around 8, the intensity of the scattered radiation starts to increase sharply up to the pH of flocculation (Figure 8.8). Cryo-TEM pictures taken around this pH value (Figure 8.9) show vesicular structures and other lamellar aggregates with irregular shapes. In particular, some particles are recognizable, which exhibit the characteristics of the sponge phase (L_3), a morphology in which bilayers interconnect in a complex fashion without long-range order. The sponge phase has been previously observed in the phase diagram of ternary systems in regions close to the lamellar and the cubic phases [38]. Consistently, at pH 8.3, vesicles coexist with larger aggregates in the cubic phase (Figure 8.9). These data are once again consistent with folding of the C_{10} hydrophobic spacer towards the hydrophobic tails, leading to a higher increase of the hydrophobic volume than of the cross-sectional headgroup area (i.e., leading to an increase in the packing parameter).

Unexpectedly, obtaining reliable data on dispersions of GS18 (Table 8.1) turned out to be more difficult [16]. Since the reproducibility of the data obtained is at the moment unsatisfying, we only mention that the trends observed are anyway consistent with the expected further increase in the packing parameter with respect to GS17.

8.4.2 Effect of Variations of the Sugar Moieties

We will now discuss the aggregation properties of sugar-based gemini surfactant analogues [15,17], designed and synthesized in order to study the effect of structural modifications at the level of the sugar moieties on the headgroups and, in particular, to provide better insights into the mechanism of the pH-induced vesicle charge reversal.

8.4.2.1 Effect of the Number of Hydroxyl Groups in the Headgroup

A comparison between the aggregation properties of GS15 with the derivatives obtained by decreasing the length of the headgroup-reduced sugars is summarized in Figure 8.10.

The compounds carrying reduced sugars with 5 and 4 carbon atoms, GS21 and GS22 (Table 8.1) respectively, present similar aggregation behavior. The reductions of the cross-sectional headgroup areas with respect to GS15 do not lead to significant variations in the aggregation behavior at acidic pH and the

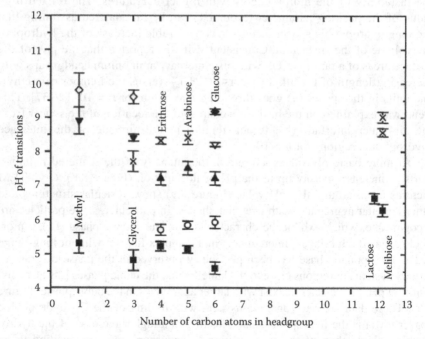

FIGURE 8.10 Plot of the pH of the transitions as a function of the number of carbon atoms in the headgroup. ▲ indicates the pH at which worm-like micelle formation from vesicles starts, ○ the pH at which the formation of spherical micelles from worm-like micelles starts, ■ the highest pH at which only spherical micelles are observed, ▽ the pH of flocculation, ◊ the pH of redispersion and × a different transition (see text for explanation). (Reprinted from Klijn, J.E., Stuart, M.C., Scarzello, M., Wagenaar, A., Engberts, J.B.F.N., *J. Phys. Chem. B*, 111, 5207, 2007. With permission.)

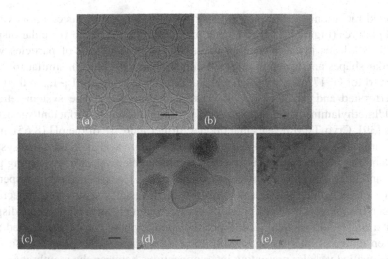

FIGURE 8.11 Cryo-TEM pictures of dispersions of (a) GS21 at pH 8.2 and (b) 8.6; (c) GS 22 at pH 7.80, (d) at pH 8.4, (e) at pH 8.64.

transitions towards worm-like micelles and toward spherical micelles occur at similar pH values as for GS15. These results indicate that the value of the cross-sectional headgroup area is mainly determined by the electrostatic repulsions between the positive charges at acidic pH and that the variation in the nature of the sugar does not significantly affect the value of the packing parameter. On the contrary, the aggregation behavior of GS21 and GS22 at basic pH is consistent with the expected reduction of the optimal cross-sectional headgroup area with respect to GS15. Cryo-TEM pictures of GS21 at pH 8.2 (Figure 8.11a) show vesicles with irregular structures, which undergo an unexpected transition toward flat bilayers (packing parameter is about 1, higher than that characteristic for vesicles) in the form of micrometers long fibers (Figure 8.11b).

The intensity of the scattered light and the particle size of GS22 dispersions increases slightly between pH 7.2 and 8.0. Vesicular aggregates are observed in cryo-TEM pictures of dispersions at pH 7.8 (Figure 8.11c). Interestingly, together with closed vesicles of irregular shapes similar to those formed by GS21, perforated vesicles are clearly visible in the pictures. Perforated vesicles have been described in the literature as possible intermediate structures in the transition from vesicular to micellar aggregates [38–44] and they can be formed by surfactants with a packing parameter between 1/2 and 2/3 [45]. During vesicle solubilization by detergents, direct micelle-to-vesicle transitions are observed for detergents with a lower packing parameter while intermediate formation of perforated vesicles is observed using detergents with a slightly higher packing parameter [41]. Consistently, we do not observe perforated vesicles in aqueous dispersions of GS15 and GS21 while they are present in dispersions of GS22, characterized by a slightly higher value of the packing parameter. Around pH 8, the intensity of the scattered radiation starts to increase sharply up to the pH of flocculation. Cryo-TEM pictures (Figure 8.11d) show that increasing the pH of the vesicular dispersion to 8.4 leads to the formation of tubular aggregates and of particles with irregular shapes and lacking inner periodic structure (L_3 phase), similar to those observed for GS17 (*vide supra*). An analogous phase sequence L_1–L_α in the form of perforated and intact vesicles–L_3 has been observed in the system tetra-*n*-decyldimethylamine oxide/heptanol/water with increasing cosurfactant/surfactant ratios [39]. Cryo-TEM reveals that upon a further increase of the pH (8.65), most of the amphiphile is organized in a cubic phase (Figure 8.11e). In the same system also lamellar phases can be observed, as double-walled vesicles, presenting passages and interconnections between the membranes. Coexistence of the dispersed sponge phase, cubic phase, and double-walled vesicles with interconnections between the bilayers has been previously observed in glyceryl monooleate dispersions at high ionic strength [38]. Since the sponge phase can be considered as a "molten lattice of interpassages" [38], the coexistence of the sponge phase with double-walled vesicles presenting interconnections between the membrane is not surprising. Above pH 8.6, the colloidally unstable inverted cubic phase dominates (Figure 8.11e).

Unfortunately, due to the different colloidal properties of lamellar and cubic phase, the data do not allow any conclusion about the ability of cubic phases to adsorb OH^- ions.

8.4.2.2 Aggregation Properties of Sugar-Based Gemini Surfactants Analogues: Implications for the Mechanism of OH^- Binding

As mentioned in the previous sections, GS1 vesicles flocculate upon charge neutralization and can be redispersed at sufficiently basic pH values. The colloidally stable vesicles at basic pH are negatively charged, as inferred from the negative values of the ζ-potentials, due to the selective adsorption of hydroxide ions. The comparison between the aggregation behaviors of GS23, GS24, and GS25, derivatives which present more radical modifications with respect to the parent compound GS15 than those we have already described (Figure 8.12), allows to discriminate between the possible mechanism of OH-binding introduced in Section 8.3, in particular between binding to hydrophobic surfaces and hydrogen bonding to hydroxyl moieties of the reduced sugar.

GS25 (liquid at room temperature) carries methoxy-terminated tetra(ethylene oxide) groups instead of reduced sugar units at both nitrogen atoms. Although the methoxy-terminated ethylene oxide moieties cannot be deprotonated nor act as hydrogen-bond donors, GS25 aggregates undergo morphological changes similar to those observed for GS1. In fact, at acidic pH spherical micelles are formed. Between pH 2 and 6.6 ± 0.2, the low intensities of the scattered radiation indicate the presence of spherical micelles, while larger aggregates are formed at higher pH values (Figure 8.13).

Cryo-TEM pictures taken at pH 7.4 show the presence of elongated lamellar structures (Figure 8.14a).

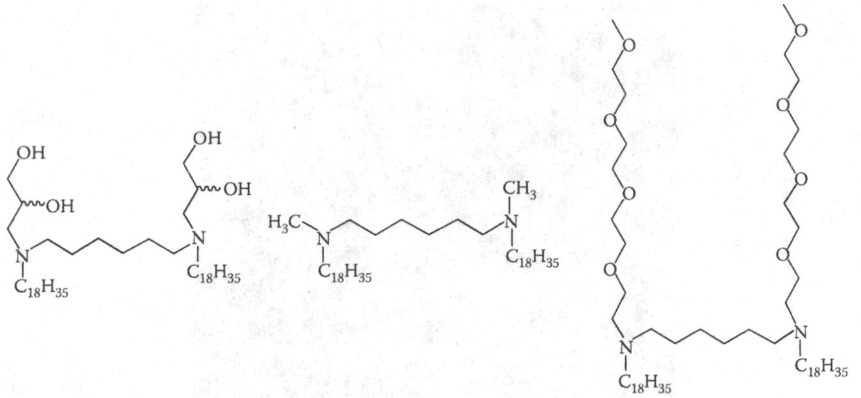

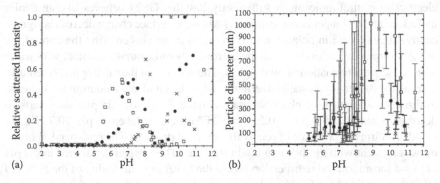

GS23   **GS24**   **GS25**

FIGURE 8.12 Molecular structures of sugar-based gemini surfactant analogues: GS23, 24, and 25.

Phase separation (arbitrarily defined here as formation of aggregates larger than 1 µm) is observed at pH 8.4 ± 0.2. Aggregates with sizes smaller than 1 µm are again observed upon a further increase in pH (9.2 ± 0.2). Cryo-TEM pictures taken at pH 11.2 demonstrate the lamellar nature of the redispersed aggregates (Figure 8.14b).

SLS data obtained on a dispersion of GS24, characterized by the presence of two methyl substituents on the nitrogen atoms, indicate the presence of spherical micelles at low pH values (Figure 8.13). The intensity of the scattered light dramatically increases around pH 5.3. Interestingly, cryo-TEM pictures taken at pH 6.8 (Figure 8.14c) show the presence of oil droplets rather than vesicles, as indicated by the homogenous appearance of the particles.

FIGURE 8.13 Normalized count rates from SLS experiments (a) and particle sizes (b) of dispersions of GS23 (●), GS24 (□), and GS25 (×) in water as a function of pH. (From Klijn, J.E., Scarzello, M., Stuart, M.C., and Engberts, J.B.F.N., *Org. Biomol. Chem.*, 4, 3569, 2006. With permission.)

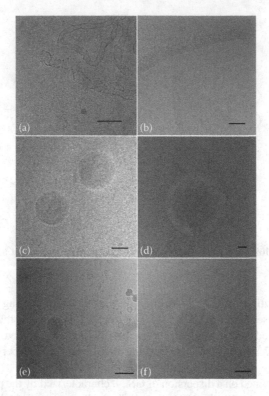

FIGURE 8.14   Cryo-TEM pictures of aqueous dispersions of GS25 at pH 7.4 (a) and pH 11.2 (b); GS24 at pH 6.8 (c) and pH 10.7 (d); GS23 at pH 6.9 (e) and pH 11.1 (f). Bar represents 100nm. (From Klijn, J.E., Scarzello, M., Stuart, M.C., and Engberts, J.B.F.N., *Org. Biomol. Chem.*, 4, 3569, 2006. With permission.)

This observation indicates that the polarity of the methylated headgroup of the electrically neutral molecule is sufficiently low that GS24 behaves like an apolar molecule (and no longer like a surfactant) when the surface charge decreases significantly. The decrease in polarity results in GS24 molecules entering the core of the initially formed micelles to reduce the entropically unfavourable contact with water with consequent formation of oil droplets. The stability of the oil droplets is remarkable, as the dispersion is stable overnight. DLS data indicate a continuous increase in droplet size until they phase separate (Figure 8.13). The droplet size starts to decrease again at pH > 9.8 ± 0.5. Cryo-TEM pictures taken at pH 10.7 (Figure 8.14d) confirm the presence of relatively small oil droplets. The redispersion of the oil is consistent with its ability to bind OH^- ions. The large uncertainties in the pH of phase separation and redispersion reflect the lower reproducibility of the SLS and DLS data obtained for GS24. A lower reproducibility of the data for this system with respect to those relative to vesicle-forming surfactants is not surprising, since it is known that the size and stability of oil droplets depend in a complex fashion on several parameters including, for example, the amount of dissolved gas [46–48].

The last derivative used in this study, GS23, is characterized by 2,3-dihydroxypropyl substituents and, like GS25, is a liquid at room temperature. The aggregation behaviour of this compound is intermediate between that of GS25 and GS24. In fact, while at acidic pH spherical micelles are present, at pH 6.9 both oil droplets and lamellar structures are observed by cryo-TEM (Figure 8.14e).

Aggregates displaying the characteristics of the sponge phase (*vide supra*) can also be observed. The formation of these structures is not surprising due to the similarity between the structures of GS23 and GS22 (Table 8.1). The reduction in the headgroup polarity of GS23 as compared with GS22 (carrying one more hydroxy substituent) results in the formation of oil droplets by the unprotonated surfactants instead of an inverted phase. Upon a further increase in pH, phase separation is observed at pH 8.4 ± 0.1 and the size of the aggregates decreases again at pH > 9 (Figure 8.13). Only oil droplets were observed by cryo-TEM for GS23 dispersions at pH 11.1 (Figure 8.14f).

From the ability displayed by all the derivatives described in this section to adsorb OH⁻ ions, we conclude that the characteristic orientation of the water molecules in the hydration layers that is associated with hydroxide binding can be generated also at rather hydrophilic interfaces, like those provided by GS23 and GS25, and not only at purely hydrophobic surfaces as those provided by GS24.

8.5 EFFECTS OF DOUBLE-TAILED PHOSPHOLIPIDS AND SINGLE-TAILED SURFACTANTS ON THE pH-DEPENDENT AGGREGATION BEHAVIOR OF SUGAR-BASED GEMINI SURFACTANTS

8.5.1 EFFECTS OF DOUBLE-TAILED PHOSPHOLIPID ADDITIVES

DOPC is a double-tailed phospholipid contained in natural membranes. The vesicular dispersions of this bilayer-forming amphiphile are commonly used as a model for biological membranes. The understanding of the effects of increasing DOPC contents in mixed dispersions with GS1 is obviously relevant in view of cell-biological or biomedical applications of this family of cationic gemini amphiphiles. Moreover, the aggregation properties of these mixtures with DOPC determined as a function of pH will contain information about the effects of mixing amphiphiles with several different packing parameters at each composition. The results inferred by the SLS and Nile Red fluorescence measurements on the mixed systems (Figure 8.15a) indicate analogous morphological transitions as for pure GS1 up to DOPC contents of 30 mol%. The data clearly show how the phase transitions of GS1 associated with the formation of aggregates of different curvatures are affected to a different extent by the presence of the bilayer-forming amphiphile.

Around neutral pH, the packing parameter of GS1 is close to 1 (slightly lower), as inferred by its ability to form vesicles, as is the packing parameter of DOPC, which assembles in the L_α phase in the whole pH range under investigation. Consistently, at these pH values the introduction of DOPC in mixed bilayers

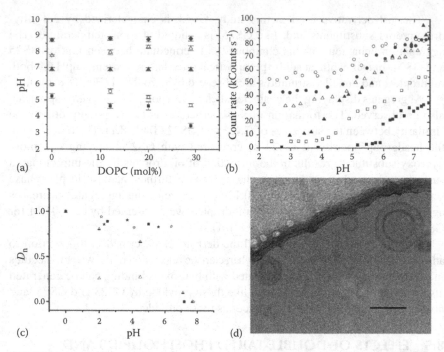

FIGURE 8.15 (a) Phase transitions of mixtures of GS1/DOPC in water. ▲ indicates the pH at which worm-like micelles formation from vesicles starts, □ the pH at which the formation of spherical micelles from worm-like micelles starts, ■ the highest pH at which only spherical micelles are observed, Δ the pH of flocculation, and ♦ the pH of redispersion (see text for explanation). (b) Intensity of the scattered light of GS1/DOPC dispersions as a function of pH. The symbols refer to DOPC contents of: ■ 0 mol%; □ 40 mol%; ▲ 50 mol%; Δ 60 mol%; ● 66 mol%; ○ 75 mol%. (c) Reduced water self-diffusion coefficients (D_n) for ■ GS1, ▲ GS1/DOPC = 60/40 mol%, Δ GS1/DOPC = 40/60 mol%, ☆ GS1/Lyso-PC = 75/25 mol%, ★ GS1/OTAC = 75/25. The value obtained for the pure solvent is shown at pH 0. See text for explanation. (d) Cryo-TEM picture of GS1/DOPC = 40/60 mol% at pH 2. Bar indicates 100 nm. (Reprinted from Scarzello, M., Klijn, J.E., Wagenaar, A., Stuart, M.C., Hulst, R., and Engberts, J.B.F.N., *Langmuir*, 22, 2558, 2006. With permission.)

with GS1 does not induce significant variations in the phase behavior with respect to the pure cationic gemini (Figure 8.15a).

As discussed previously, the packing parameter of GS1 decreases upon decreasing the pH of the dispersion, which results in a higher difference with the packing parameters of the two components of the mixture. The effect of this unbalance can be clearly seen in Figure 8.15a as a decrease of the pH values of the transitions toward highly curved aggregates: due to the presence of DOPC, lower pH-values (i.e., higher degree of protonation of GS1) are required in order to obtain the average packing parameters associated with micelle stability. The decrease of the inter-headgroup electrostatic repulsions, upon dilution of GS1 with the zwitterionic phospholipid DOPC, is also expected to favor the double protonation of the GS headgroup; however, the decrease in the pH of the transitions to micelles indicates that the dilution of the charge is not a dominant effect. Moreover, the linearity of the decrease of the pH associated with micelle formation with the DOPC content can be interpreted as an indication of the homogeneous mixing of the two components.

Similar conclusions concerning the effect of DOPC on the pH of the transitions can be drawn for DOPC contents up to 60 mol%. Nevertheless, while at DOPC contents lower than 40 mol% the intensity of the scattered radiation in the pH regions associated with spherical micelle stability is low and similar to that measured for pure GS1 dispersions, SLS experiments performed on dispersions with higher DOPC contents lead to higher residual count rates. This observation suggests the presence of residual vesicles at acidic pH in mixtures with a sufficiently high DOPC content (*vide infra*). Consistently, DLS data obtained under these conditions confirmed the presence of aggregates with sizes around 100 nm. At DOPC contents higher than 60 mol%, the data indicate that the compositions of the worm-like micelles and of the spherical micelles remain constant, since no further variation of the pH of the transitions with the DOPC content is observed.

Further insights into the nature of the aggregates formed by GS1/DOPC mixtures were obtained by diffusion-NMR. Due to the fact that from the NMR perspective, the water contained in the aqueous compartment of a vesicle diffuses with a similar diffusion coefficient as that of the vesicle, vesicle disruption is detected as an increase of the diffusion coefficient of water [49]. The pH dependence of the relative diffusion coefficients of the dispersions (D_n), normalized both with respect to the diffusion coefficient of the buffer in the absence of surfactant ($D_n = 0$) and with respect to the maximum amount of water entrapped in the vesicles ($D_n = 1$) is shown in Figure 8.15c. The values of D_n obtained for GS1 demonstrate that vesicular disruption is essentially complete at pH 4.5 since the diffusion coefficient measured in these conditions resembles that of bulk water. While the variations of the diffusion coefficient of water for the dispersions of 60/40 mol% GS1/DOPC are similar to those obtained for pure GS1 dispersions, the value of the diffusion coefficient obtained for DOPC contents of 60 mol% at acidic pH is only slightly bigger than that obtained around neutral pH (where vesicles are stable). These observations confirm that, at high DOPC contents, a significant amount of water is still trapped in vesicular pockets at pH 2. Consistent with the SLS and diffusion-NMR results, cryo-TEM pictures taken at acidic pH show the presence of a few small vesicles (about 30 nm diameter) for preparations with 40 mol% of DOPC and larger vesicles, with sizes around 100 nm, for a DOPC content of 60 mol% (Figure 8.15d). The equimolar mixtures of GS1 with DOPC have been studied also by broad line ^{31}P NMR. All the samples give rise to spectra in which the features characteristic of the L_α phase can be recognized. The absence of a quantitative transition toward the isotropic micellar phase at pH 2 is evident in the spectra, in accordance with the SLS, diffusion NMR and cryo-TEM results.

The analysis of the aggregation behavior at basic pH provides insights into the mechanism of OH$^-$ binding to GS vesicular surfaces. DOPC vesicles are colloidally stable in the whole pH range under investigation. The vesicles formed by this phospholipid are unable to bind hydroxide ions, as proven by an approximately neutral ζ-potential even at pH 10 [13]. The stability of the DOPC vesicles is due to the fact that fusion requires dehydration of the phospholipid headgroups at the site of closest approach, a process which has a unfavorable Gibbs energy due to the high hydration energies of the zwitterionic PC headgroups [50]. Consistently, the pH of flocculation slightly increases with increasing DOPC content. Surprisingly, due to the inability of DOPC vesicles to adsorb OH$^-$ ions, the pH of redispersion after flocculation remains about constant up to the highest DOPC content tested, 60 mol%. On the basis of the above mentioned mechanism for hydroxide binding, which attributes the cause of the phenomenon to an attractive electric field generated by the characteristic orientation of the first water layers at the interface, we suggest that the structure of the water of hydration of mixed GS1/DOPC vesicles is dominated by the cationic amphiphile even at DOPC contents up to 60 mol%.

As a second phospholipid for testing, DOPE was selected since it is one of the most commonly used helper lipids in amphiphile-mediated gene transfection. A helper lipid is an electrically neutral molecule that increases the transfection efficiency of a cationic amphiphile. DOPE resides in the inverted hexagonal phase (H_{II}) at temperatures higher than 10°C and its beneficial effect in transfection is related to its low packing parameter. In fact, genes in lipoplexes which can adopt the H_{II} morphology under physiological conditions are generally better transfected [3–5]. ^{31}P broad-line NMR experiments demonstrated that mixing GS1 with DOPE prevents the formation of the H_{II} phase even at neutral pH, physiological ionic strength conditions, and DOPE contents of 50 mol%. The consequences of this observation on the transfection efficiency will be discussed in Section 8.7.

The effect of DOPE on the pH-induced transitions of GS1 is similar to that observed for DOPC. The variations of the pH values associated with transitions towards highly curved aggregates are more affected by the presence of DOPE than by the presence of DOPC, consistent with the lower packing parameter of the latter. For example, the slope of the line obtained by linear fitting of the pH transitions from vesicles to worm-like micelles is about 30% higher for DOPE than for DOPC. As observed for DOPC-containing mixtures, the pH of redispersion of GS1/DOPE vesicles is independent of the amount of DOPE.

8.5.2 Effects of Single-Tailed Surfactant Additives

We finally tested the effect of two single tailed surfactants, lyso-PC (1-palmitoyl-2-hydroxy-sn-glycero-3-phosphocholine) and OTAC (n-octadecyltrimethylammonium chloride), on the morphology of GS1 dispersions. Lyso-PC is a natural phospholipid involved in several biological processes [51–53]. The decrease in the hydrophobic volume of lyso-PC with respect to its double-tailed counter part, DOPC, leads to a smaller packing parameter and to the formation of micelles instead of vesicles (cmc 4.0×10^{-5} M) [54] in pure dispersions. In mixed dispersions with GS1, the propensity of lyso-PC to form aggregates with high curvature

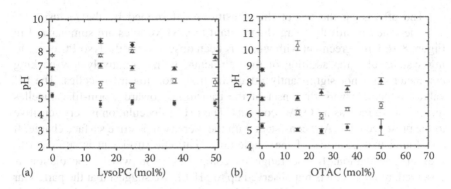

FIGURE 8.16 Phase transitions of mixtures of GS1/lyso-PC (a) and GS1/OTAC (b) in water. ▲ indicates the pH at which worm-like micelles formation from vesicles starts, □ the pH at which the formation of spherical micelles from worm-like micelles starts, ■ the highest pH at which only spherical micelles are observed, △ the pH of flocculation, and ◆ the pH of redispersion (see text for explanation). (Reprinted from Scarzello, M., Klijn, J.E., Wagenaar, A., Stuart, M.C., Hulst, R., and Engberts, J.B.F.N., *Langmuir*, 22, 2558, 2006. With permission.)

results in smaller particles, as indicated by the less turbid appearance of the samples (bluish/transparent) compared to the DOPC- or DOPE-containing systems (milky); the impression deduced by visual inspection is confirmed by DLS measurements. As expected, lyso-PC does not significantly modify the pH of the transitions of GS1 under acidic conditions (Figure 8.16a).

On the contrary, the single-tailed phospholipid affects the pH values of flocculation and redispersion more than DOPC. In fact, the pH of flocculation increases only slightly at lyso-PC contents up to 25 mol% but flocculation is not observed at still higher phospholipid contents (36 mol%). We suggest that the higher colloidal stability of the electrically neutral lyso-PC-containing vesicles is associated with higher hydration energies as compared with the corresponding DOPC-containing systems. This difference could arise from hydrogen bonding of the hydroxyl group at the *sn*-2 position of the glycerol backbone of lyso-PC compared to the ester function present in its double-tailed analogue. Addition of Ca^{2+} ions to the vesicles at basic pH leads to flocculation also in the absence of a pH-induced flocculation, indicating that the vesicles are negatively charged even in this case. In the composition range in which flocculation is observed, the pH of redispersion decreases with increasing lyso-PC content. This observation is consistent with an increase of the hydration energy for increasing phospholipid content, which would allow redispersion at lower levels of OH^- binding; an alternative possible explanation is a higher hydroxide binding constant to lyso-PC containing interfaces.

The diffusion NMR data for this system (Figure 8.15c) are in agreement with the SLS data, showing a behavior similar to that of GS1, i.e., vesicle disruption leading to formation of worm-like micelles.

The effects on the pH of the transitions determined by the inclusion of a single-tailed positively charged surfactant to GS1 vesicles are summarized in Figure 8.16b. In agreement with what has been observed for the lyso-PC containing system, also the addition of this surfactant with a relatively low packing parameter does not significantly affect the transition toward micelles. The pH values associated with the onset of the transitions toward worm-like micelles increase with increasing OTAC contents. The pH of flocculation is very sensitive to the presence of OTAC, consistent with the increased positive surface charge. It was found that 25 mol% of the single-tailed cationic surfactant is sufficient to prevent flocculation. In the range of compositions in which flocculation is observed, redispersion is not observed up to pH 12. We suggest that the particular water structure which gives rise to OH⁻ binding in the absence of the OTAC is disturbed by the presence of the positive charge of the single-tailed amphiphile. We cannot anyway exclude that redispersion would occur at extremely basic pH values (pH > 12). Since the cmc of lyso-PC is lower than the cmc of OTAC (3.5 × 10^4 M) [55], the amount of OTAC that leaves the vesicular bilayer to partition into the aqueous phase is expected to be higher than that of lyso-PC [56]. For this reason, additional experiments were performed using a double-tailed analogue of OTAC, N-hexadecyl-N-octadecene-9-yl-N-dimethylammonium chloride, which is expected to have a lower tendency for partitioning into the aqueous phase. The analogy between the results obtained with the single- and double-tailed cationic surfactants allows the conclusion that, also in the case of OTAC, the amount of surfactant in the aqueous phase is not significant.

As expected, the diffusion-NMR results for the OTAC-containing system (Figure 8.15c) resemble those for GS1 and GS1/lyso-PC.

8.6 EFFECTS OF PHYSIOLOGICAL IONIC STRENGTH CONDITIONS ON THE pH-DEPENDENT AGGREGATION BEHAVIOR OF SUGAR-BASED GEMINI SURFACTANTS

Samples containing GS1 mixed with 25 mol% of the additives introduced in the previous sections have been prepared using a saline buffer at physiological ionic strength (NaCl$_{aq}$ 150 mM). The main effect of an increase in ionic strength is the higher screening of the inter-headgroup electrostatic interactions, which can affect the colloidal stability and, in some cases, can even induce phase transitions toward aggregates with lower positive curvature [4,5].

The variations in the pH of the transitions induced by the changes of the ionic strength are in general rather modest in the systems under investigations. For pure GS1, the pH of flocculation is lower than that in water, due to the decrease in the electrostatic repulsion between vesicles. Interestingly, an increase in the size of the aggregates up to 500 nm is observed when decreasing the pH of a vesicular dispersion at physiological ionic strength from pH 6.9 to 6.2. The size of the aggregates decreases suddenly in correspondence with the pH values associated with spherical micelle formation. Cryo-TEM pictures recorded between pH 6.9 and 6.2 suggest that the large particles consist of aggregates of worm-like micelles.

For all the mixtures, vesicle flocculation is the transition, which presents the most significant variations upon an increase in ionic strength. The OTAC-containing vesicles display the most pronounced sensitivity to the increase in ionic strength, in accordance with the additional positive charge carried by the single-tailed surfactant. This system is the only one for which the pH of redispersion is sensitive to the variation in salt concentration. Under these conditions, flocculation is not observed in the lyso-PC containing system. We speculate that this observation could be due to a conformational change of the surfactants headgroups, which results in a higher density of polar residues at the interface, thereby increasing the Gibbs energy of hydration. The results suggest that the presence of lyso-PC in the bilayer facilitates this conformational change, in contrast with the phospholipids with a higher packing parameter. A further difference with respect to the aqueous dispersion is the absence of a quantitative transition toward spherical micelles for the mixture containing the double-tailed zwitterionic phospholipids. Cryo-TEM pictures taken at pH 3.1 reveal the presence of large aggregates, probably branched micelles. While in the presence of OTAC, aggregation of worm-like micelles is observed in analogy with GS1, the phenomenon is absent or strongly reduced in zwitterionic phospholipid-containing systems, probably due to their higher hydration energies. The effects of the ionic strength described in this section for DOPC- and DOPE-containing mixtures are also confirmed by ^{31}P-NMR results [20].

8.7 SUGAR-BASED GEMINI SURFACTANTS AS GENE DELIVERY VEHICLES

Some of the sugar-based GS herein described qualify among the most efficient chemical carriers available for *in vitro* gene transfection. Transfection experiments on COS-7 cell lines allowed to identify different GSs with transfection efficiencies higher than that of the commercial reagent Lipofectamine 2000 [57].

The first necessary requirement to achieve transfection is binding of the amphiphilic carrier to the hexogenous DNA (lipoplex formation). Gel electrophoresis experiments carried out on different sugar-based GS demonstrated that the presence of positive charges at the headgroup is the most important parameter for effective binding, irrespective of the structural details of the spacer and of the sugar moieties [57]. For example, GS4 which carries neutral amido groups instead of tertiary amines in the headgroup, is unable to bind DNA. Fluorescence microscopic experiments indicate that also cell internalization of the lipoplex is mainly driven by an excess of positive charge in the complex while interactions of specific sugar moieties with the cell surface do not seem to play a significant role, at least for transfection of COS-7 cell lines [57]. Consistently, while the use of GS4 does not lead to significant transfection, about 70% of the surviving cells after transfection using GS1 or GS2 (at the optimal charge ratio of +/− = 8) express the reporter gene green fluorescent protein (GFP). For comparison, only about 50% GFP-positive cells could be detected after transfection with the commercial reagent Lipofectamine 2000. An estimate of the cytotoxicities of these preparations was obtained by examining the cell monolayer 48 h after transfection.

Interestingly, among the compounds tested (GS1, GS2, GS3, GS4, and GS5), the ones carrying purely hydrophobic spacers (GS3 and GS5) showed the highest toxicity. The toxicity of GS1 and GS2 is comparable to that of Lipofectamine 2000 [57]. Preliminary experiments performed on Chinese Hamster Ovary (CHO) cell lines (A. Lima, M. Scarzello, L. van Vliet, F. Hollfelder, to be published), using a wider selection of compounds, confirmed that a hydrophilic spacer is a beneficial structural element as compared to a purely hydrophobic one. The screening carried out at the optimal charge ratio of +/− = 2 showed how GS1 is able to rival commercial products like Lipofectamine 2000 and JetPEI also on these cell lines and allowed to identify GS20 as the most efficient of the sugar-based GS. The nature of the spacer in GS20 is similar to that of the spacer in GS1, but slightly longer (Table 8.1). As we will see in the next section, a high transfection efficiency can be expected for this surfactant on the basis of considerations about the optimal morphology of the lipoplexes of sugar-based GS (*vide infra*).

The ability of GS1 and GS2 to transfect genes *in vivo* has been tested as well [57]. Three groups of three naked mice were transfected with a gene codifying for the firefly luciferase enzyme, which can be detected in situ in the presence of its substrate luciferin, oxygen, and ATP, by luminescence imaging (Figure 8.17) [58,59].

The images suggest that the expression of the reporter gene is mainly located in the lower abdominal region and, interestingly, neither for GS1 nor for GS2 formulations significant expressions in the lungs are observed. Sequestration of the lipoplexes in the small-diameter lung capillaries [60–62], which limits the circulation times and the possibility of targeting different organs, is often observed in amphiphile-mediated transfection. This phenomenon is due to the formation of large amphiphilic aggregates induced by the high ionic strength conditions and by the presence of negatively charged proteins under physiological conditions. The absence of this hurdle, when using GS1 or GS2 as DNA carriers, can be rationalized in terms of the morphological and colloidal stabilities data discussed in the next section.

Some expression is also observed in the mouth of the mice treated with GS1 and GS2 lipoplexes (but not in those treated with naked plasmid DNA), probably due to the animal licking the injection site (i.e., penile vein).

8.7.1 MORPHOLOGY OF THE LIPOPLEXES

As previously mentioned, the good transfection ability of several amphiphilic systems has been associated with their ability to undergo a transition toward an inverted hexagonal phase. The fusogenic characteristics of this phase, as compared to a lamellar morphology, are expected to increase the interactions of the lipoplexes with the endosomal membranes and facilitate endosomal release, commonly considered one of the major barriers in the transfection mechanism. On the other hand, due to the low colloidal stability of inverted phases, intravenous injection of a lipoplex, which resides in the inverted hexagonal phase is unlikely to be useful for therapeutic applications and, for *in vivo* transfection, a transition from the lamellar phase is desirable only once the lipoplex has reached the target tissue.

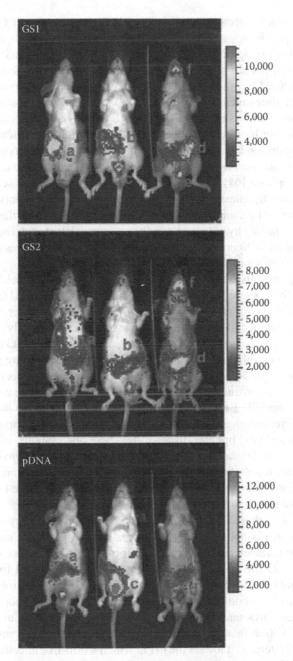

FIGURE 8.17 *In vivo* transfection mediated by GS1 and GS2. Male nude mice were injected with GS1 or GS2 lipoplexes or with naked plasmid DNA. The plasmid codes for the luciferase enzyme and its activity is visualized by luminescence after injection of luciferin using a cooled charged coupled device (CCD) camera. (From Wasungu, L., Scarzello, M., van Dam, G., Molema, G., Wagenaar, A., Engberts, J.B.F.N., and Hoekstra, D., *J. Mol. Med.*, 84, 774, 2006. With permission.)

SAXS measurements demonstrated the lamellar morphologies of GS1 and GS2 lipoplexes at physiological pH [10,57]. Turbidity measurements demonstrated that GS1 and GS2 lipoplexes do not aggregate upon an increase in ionic strength, in contrast with what has been observed for other effective transfection cocktails (SAINT2/DOPE) [57,63]. Although in serum aggregation is observed to a certain extent, the turbidities measured for GS1 and GS2 preparations are significantly lower than those observed for SAINT2/DOPE and large aggregates can be observed only after long equilibration times. Interestingly, the extent of aggregation is lower in dispersion containing 50% of serum than in dispersion containing 10% of serum. Similar observations have been previously reported for DOTAP/cholesterol lipoplexes [64]. The authors suggest that while at low serum contents bridging between lipoplexes might occur, at higher serum contents aggregation might be prevented by a uniform coating. The lower propensity of the GS preparations to cluster under physiological conditions is consistent with the observation that accumulation of lipoplexes in the capillaries of the lungs does not occur, as discussed in the previous section.

SAXS experiments performed on GS1 lipoplexes at mildly acidic pH values demonstrated the hexagonal symmetry of the complex [10]. Even if GS1 vesicular dispersions undergo a transition toward worm-like micelles under mildly acidic conditions, the inverted nature of this hexagonal phase was initially postulated on the basis of the analogy with several effective transfection cocktails. Further insights into the morphology of these lipoplexes were obtained by means of the solvatochromic probe Nile Red [63]. While for H_{II}-forming systems like SAINT2/DOPE mixtures, a transition toward the inverted hexagonal phase corresponds to a blue shift of the Nile Red fluorescence emission maximum, an increase in the wavelength corresponding to the maximum emission is observed upon decreasing the pH of GS1 and GS2 lipoplex dispersions. These observations, in combination with the high colloidal stability of the GS lipoplexes and the *in vivo* transfection results, suggest that these complexes reside in a normal hexagonal phase (H_I, Figure 8.18a) and not in the most commonly observed H_{II} phase (see SAINT2/DOPE example in Figure 8.18b).

In the light of these findings, the high transfection efficiency measured for GS20 (*vide supra*) is not surprising due to its similarity to GS1 and its slightly higher packing parameter, which favors the formation of worm-like micelles.

Lipoplexes, in the H_I phase, have been previously observed for the single-tailed surfactant CTAB [65–67]. It is not surprising that the behavior of GS1 lipoplexes under acidic conditions presents similarities with that of lipoplexes formed by single-tailed surfactants (*vide supra*). An H_I morphology is also consistent with the observation that the helper lipid DOPE inhibits the transfection efficiency of this system, as it lowers the pH at which worm-like micelles are formed (see Section 8.5.1).

The efficiencies of these transfection systems suggest that also lipoplexes in the H_I phase can induce the destabilization of the endosomal compartment necessary to allow the release of the gene into the cytosol and to avoid lysosomal degradation of the gene. Moreover, resonance energy transfer experiments,

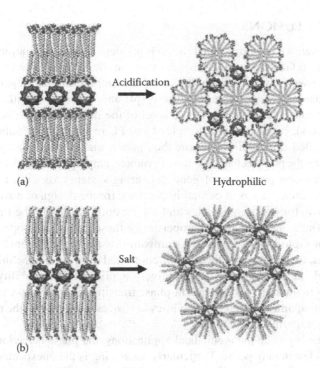

FIGURE 8.18 Model for the phase transition of lipoplexes formed from sugar-based GS (a); comparison with SAINT-2/DOPE (b). The model depicted here for the gemini lipoplexes (a) illustrates a transition from a lamellar phase L_α to a normal hexagonal H_I phase as presumably occurs in the endosomal compartment upon acidification. In such a H_I phase, the plasmid DNA is intercalated between micelles where the polar headgroup of the amphiphile is exposed on the outside, giving rise to externally hydrophilic particles. In (b) the lamellar L_α organization of SAINT-2/DOPE in the absence of salt and the inverted hexagonal H_{II} phase in its presence is displayed. In this H_{II} phase the polar headgroups of the amphiphiles interact with the plasmid DNA and the hydrophobic tails are exposed on the outside, giving rise to externally hydrophobic particles that will tend to aggregate. (Reprinted from Wasungu, L., Stuart, M.C., Scarzello, M., Engberts, J.B.F.N., and Hoebstra, D., *Biochim. Biophys. Acta*, 1758, 1677, 2006. With permission.)

performed to study the effect of GS1 lipoplexes on phospholipid model membranes, showed a strong ability of the GS1 lipoplex to destabilize phosphatidylserine (PS)-containing membranes, particularly at pH values (around 6) characteristic of the endosomal compartment [63]. Moreover, a fluorescence assay using the fluorescent probe PicoGreen showed that the interaction of the lipoplex with PS-containing model membranes at slightly acidic pH is also associated with DNA release from the complex. Since only naked DNA can be expressed in the cellular nucleus, DNA release is a further critical requirement in the transfection mechanism, which appears to be readily accomplished by these systems.

8.8 CONCLUSIONS

In this chapter, we have described the rich pH-dependent aggregation behavior of sugar-based GSs designed for applications in the field of gene transfection. We have shown how the morphologies of the amphiphilic aggregates can be tuned by changing the solution conditions (pH and/or ionic strength), by introducing structural modifications at the level of the headgroups (reduced sugars and/or spacers), by mixing with single-(Lyso-PC or OTAC) or double-(DOPC or DOPE) tailed amphiphiles. Despite the fundamental interest of analyzing and understanding the phase behavior of novel synthetic amphiphiles, the determination of the aggregation properties of gene delivering systems associated with high transfection efficiencies is of pivotal importance for the design of novel and still more effective transfection vehicles and for the optimization of the transfection cocktails. The elucidation of the properties of these compounds opens up possibilities for exploiting their good biocompatibility also for different biomedical applications, for example, in the field of controlled drug delivery. Both for gene therapy and for more general drug delivery applications, the ability of these compounds to undergo pH-dependent phase transitions might allow optimizing both circulation times and targeting abilities *in vivo*, as suggested by the promising, preliminary experiments on mice.

Even leaving aside the biomedical applications, the phase behavior of sugar-based GS is fascinating per se. Particularly interesting is the unexpected observation of vesicle charge reversal at basic pH, which was considered worth of further investigation despite the fact that it occurs under conditions which are not physiologically relevant. The comparison between the aggregation behavior of several GSs and of their derivatives supports the hypothesis of the analogy to the mechanism of OH⁻ binding to hydrophobic surfaces. Nevertheless, a definitive and detailed molecular picture of this phenomenon (and even of the well-known mechanism of OH⁻ binding to hydrophobic surfaces [36,68]) certainly requires new molecular dynamics simulations and further experimental work.

REFERENCES

1. Tadros, T. F., *Applied Surfactants*, Wiley-VCH, Weinheim, 2005.
2. Fennel Evans, D. and Wennerstrom, H., *The Colloidal Domain*, Wiley-VCH, New York, 1999.
3. Koltover, I., Salditt, T., Radler, J., and Safinya, C. R., *Science,* 1998, *281*, 78.
4. Scarzello, M., Smisterova, J., Wagenaar, A., Stuart, M. C. A., Hoekstra, D., Engberts, J. B. F., and Hulst, R., *J. Am. Chem. Soc.*, 2005, *127*, 10420.
5. Scarzello, M., Chupin, V., Wagenaar, A., Stuart, M. C. A., Engberts, J. B. F. N., and Hulst, R., *Biophys. J.*, 2005, *88*, 2104.
6. Menger, F. M. and Littau, C. A., *J. Am. Chem. Soc.*, 1991, *113*, 1451.
7. Menger, F. M. and Littau, C. A., *J. Am. Chem. Soc.*, 1993, *115*, 10083.
8. Esumi, K. and Ueno, M., *Structure-Performance Relationships in Surfactants*, Marcel Dekker, New York, 1997.
9. Menger, F. M. and Keiper, J. S., *Angew. Chem., Int. Ed. Engl.*, 2000, *39*, 1907.
10. Bell, P. C., Bergsma, M., Dolbnya, I. P., Bras, W., Stuart, M. C., Rowan, A. E., Feiters, M. C., and Engberts, J. B. F. N., *J. Am. Chem. Soc.*, 2003, *125*, 1551.

11. Fielden, M. L., Perrin, C., Kremer, A., Bergsma, M., Stuart, M. C., Camilleri, P., and Engberts, J. B. F. N., *Eur. J. Biochem.*, 2001, *268*, 1269.
12. Johnsson, M., Wagenaar, A., and Engberts, J. B. F. N., *J. Am. Chem. Soc.*, 2003, *125*, 757.
13. Johnsson, M., Wagenaar, A., Stuart, M. C., and Engberts, J. B. F. N., *Langmuir*, 2003, *19*, 4609.
14. Johnsson, M. and Engberts, J. B. F. N., *J. Phys. Org. Chem.*, 2004, *17*, 934.
15. Klijn, J. E., Stuart, M. C., Scarzello, M., Wagenaar, A., and Engberts, J. B. F. N., *J. Phys. Chem. B*, 2007, *111*, 5204.
16. Klijn, J. E., Stuart, M. C. A., Scarzello, M., Wagenaar, A., and Engberts, J. B. F. N., *J. Phys. Chem. B*, 2006, *110*, 21694.
17. Klijn, J. E., Scarzello, M., Stuart, M. C., and Engberts, J. B. F. N., *Org. Biomol. Chem.*, 2006, *4*, 3569.
18. Wagenaar, A. and Engberts, J. B. F. N., *Tetrahedron*, 2007, *63*, 10622.
19. Pestman, J. M., Terpstra, K. R., Stuart, M. C. A., van Doren, H. A., Brisson, A., Kellogg, R. M., and Engberts, J. B. F. N., *Langmuir*, 1997, *13*, 6857.
20. Scarzello, M., Klijn, J. E., Wagenaar, A., Stuart, M. C., Hulst, R., and Engberts, J. B. F. N., *Langmuir*, 2006, *22*, 2558.
21. Papahadjopoulos, D., Vail, W. J., Jacobson, K., and Poste, G., *Biochim. Biophys. Acta*, 1975, *394*, 483.
22. Papahadjopoulos, D. and Poste, G., *Biophys. J.*, 1975, *15*, 945.
23. Ravoo, B. J., Stuart, M. C. A., Brisson, A. D. R., Weringa, W. D., and Engberts, J. B. F. N., *Chem. Phys. Lipids*, 2001, *109*, 63.
24. Fonteijn, T. A. A., Engberts, J. B. F. N., and Hoekstra, D., *Biochemistry*, 1991, *30*, 5319.
25. Streefland, L., Wagenaar, A., Hoekstra, D., and Engberts, J. B. F. N., *Langmuir*, 1993, *9*, 219.
26. Ravoo, B. J. and Engberts, J. B. F. N., *Langmuir*, 1994, *10*, 1735.
27. Leckband, D. E., Helm, C. A., and Israelachvili, J., *Biochemisty*, 1993, *32*, 1127.
28. Ohki, S., *Biochim. Biophys. Acta*, 1982, *689*, 1.
29. Baba, T., Zheng, L. Q., Minamikawa, H., and Hato, M., *J. Colloid. Interface Sci.*, 2000, *223*, 235.
30. Zheng, L. Q., Shui, L. L., Shen, Q., Li, G. Z., Baba, T., Minamikawa, H., and Hato, M., *Colloids Surf., A*, 2002, *207*, 215.
31. Marinova, K. G., Alargova, R. G., Denkov, N. D., Velev, O. D., Petsev, D. N., Ivanov, I. B., and Borwankar, R. P., *Langmuir*, 1996, *12*, 2045.
32. Bergeron, V., Waltermo, Å., and Claesson, P. M., *Langmuir*, 1996, *12*, 1336.
33. Karraker, K. A. and Radke, C. J., *Adv. Colloid Interface Sci.*, 2002, *96*, 231.
34. Stubenrauch, C., Schlarmann, J., and Strey, R., *Phys. Chem. Chem. Phys.*, 2002, *4*, 4504.
35. Pashley, R. M., *J. Phys. Chem. B.*, 2003, *107*, 1714.
36. Zangi, R. and Engberts, J. B. F. N., *J. Am. Chem. Soc.*, 2005, *127*, 2272.
37. Alami, E., Beinert, G., Marie, P., and Zana, R., *Langmuir*, 1993, *9*, 1465.
38. Almgren, M., *Biochim. Biophys. Acta*, 2000, *1508*, 146.
39. Hoffmann, H., Thunig, C., Munkert, U., Meyer, H. W., and Richter, W., *Langmuir*, 1992, *8*, 2629.
40. Edwards, K., Gustafsson, J., Almgen, M., and Karlsson, G., *J. Colloid. Interface Sci.*, 1993, *161*, 299.
41. Silvander, M., Karlsson, G., and Edwards, K., *J. Colloid. Interface Sci.*, 1996, *179*, 104.
42. Gustafsson, J., Oradd, G., and Almgren, M., *Langmuir*, 1997, *13*, 6956.
43. Gustafsson, J., Oradd, G., Lindblom, G., Olsson, U., and Almgren, M., *Langmuir*, 1997, *13*, 852.
44. Holmes, M. C., *Curr. Opin. Colloid Interface Sci.*, 1998, *3*, 485.

45. Hyde, S. T., Andersson, S., Larsson, K., Blum, Z., Landh, T., Lidin, S., and Ninham, B. W., *The Language of Shape, the Role of Curvature in Condensed Matter: Physics, Chemistry and Biology*, Elsevier Science, Amsterdam, 1997.
46. Maeda, N., Rosenberg, K. J., Israelachvili, J. N., and Pashley, R. M., *Langmuir*, 2004, *20*, 3129.
47. Pashley, R. M., Rzechowicz, M., Pashley, L. R., and Francis, M. J., *J. Phys. Chem. B*, 2005, *109*, 1231.
48. Francis, M. J. and Pashley, R. M., *Colloids Surf., A*, 2005, *260*, 7.
49. Söderman, O., Herrington, K. L., Kaler, E. W., and Miller, D. D., *Langmuir*, 1997, *13*, 5531.
50. Cevc, G., *J. Chem. Soc., Faraday Trans.*, 1991, *87*, 2733.
51. Yuan, Y. P., Schoenwalder, S. M., Salem, H. H., and Jackson, S. P., *J. Biol. Chem.*, 1996, *271*, 27090.
52. Mori, S., Nakata, Y., and Endo, H., *Cell. Mol. Biol.*, 1991, *37*, 421.
53. Gopfert, M. S., Siedler, F., Siess, W., and Sellmayer, A., *J. Vas. Res.*, 2005, *42*, 120.
54. Hoyrup, P., Davidsen, J., and Jorgensen, K., *J. Phys. Chem. B*, 2001, *105*, 2649.
55. Kang, K. H., Kim, H. U., and Lim, K. H., *Colloids Surf., A*, 2001, *189*, 113.
56. Heerklotz, H. and Seelig, J., *Biochim. Biophys. Acta*, 2000, *1508*, 69.
57. Wasungu, L., Scarzello, M., van Dam, G., Molema, G., Wagenaar, A., Engberts, J. B. F. N., and Hoekstra, D., *J. Mol. Med.*, 2006, *84*, 774.
58. McElroy, W. D. and De Luca, M. A., *J. Appl. Biochem.*, 1983, *5*, 197.
59. Wilson, T. and Hastings, J. W., *Annu. Rev. Cell. Dev. Biol.*, 1998, *14*, 197.
60. Mahato, R. I., Tagliaferri, F., Meaney, C., Leonard, P., Wadhwa, M. S., Logan, M., French, M., and Rolland, A., *Hum. Gene Ther.*, 1998, *9*, 2083.
61. Niven, R., Pearlman, R., Wedeking, T., Meckeigan, J., Noker, P., Simpson-Herren, L., and Smith, J. G., *J. Pharm. Sci.*, 1998, *87*, 1292.
62. Li, S., Tseng, W. C., Stolz, D. B., Wu, S. P., Watkins, S. C., and Huang, L., *Gene Ther.*, 1999, *5*, 930.
63. Wasungu, L., Stuart, M. C., Scarzello, M., Engberts, J. B. F. N., and Hoekstra, D., *Biochim. Biophys. Acta*, 2006, *1758*, 1677.
64. Simberg, D., Weisman, S., Talmon, Y., and Barenholz, Y., *Crit. Rev. Ther. Drug. Carr. Syst.*, 2004, *21*, 257.
65. Krishnaswamy, P., Mitra, P., Raghunathan, A. K., and Sood, A. K., *Europhys. Lett.*, 2003, *62*, 357.
66. Ghirlando, R., Wachtel, E. J., Arad, T., and Minsky, A., *Biochemistry*, 1992, *31*, 7110.
67. Krishnaswamy, P., Raghunathan, A. K., and Sood, A. K., *Phys. Rev., E Stat. Nonlinear Soft Matter Phys.*, 2004, *69*, 031905.
68. Valcha, R., Zangi, R., Engberts, J.B.F.N., and Jungwirth, P., *J. Phys. Chem. C*, 2008, 112, 7689.

9 Sugar-Based Ionic Surfactants: Syntheses, Solution Properties, and Applications

Omar A. El Seoud and Paula D. Galgano

CONTENTS

9.1 INTRODUCTION

Sugar-based surfactants rank high on the list of compounds that are produced according to the principles of green chemistry [1,2]. The raw materials, mostly monosaccharides, disaccharides, fatty acids, and fatty alcohols are available from renewable sources; their production is economic because they are obtained either directly, e.g., sucrose by crystallization from concentrated sugarcane juice, or by simple processes, e.g., glucose by the enzymatic hydrolysis of starch (α-amylase and glucoamylase) [3], and fatty acids by steam-splitting of oils and fats [4]. Other raw materials that are obtained by employing straightforward, readily established processes include fatty esters, by transesterification of oils and fats with low molecular weight alcohols (mostly methanol and ethanol) [4]; fatty alcohols, by the catalytic reduction of the corresponding acids under high temperature/high pressure [5]; and N-alkylglucoamines by the reaction of glucose with alkylamines, followed by catalytic reduction of the intermediate imine. The economy of the production of sugar-based surfactants is assured because the reactions involved are relatively simple. Examples are the esterification of the sugar by a fatty acid, by using chemical or enzymatic pathways [6,7], the Fischer and Koenigs–Knorr syntheses of alkyl glycosides from glucose and fatty alcohols [8,9], and the synthesis of N-alkanoyl-N-alkyl-1-glucamines from the corresponding fatty acids and N-alkylglucoamines [10]. The relative simplicity of these reactions assures "atom economy" and reduction of waste, because all reactants are incorporated in the final products. The ready biodegradability of the products, their compatibility with the skin, and efficiency as surfactants, either alone or as mixtures with other "classic" surfactants are the most relevant examples of the favorable properties of sugar-based surfactants [11]. The demand for the latter is most likely to increase because the public is becoming increasingly concerned about the environmental impact of the products consumed. Additionally, the relative price advantage of petroleum-based raw materials that are employed in the detergent industry (alkenes, benzene, ethylene oxide, etc.) will probably not hold for long, because of the inexorably increased demand on crude oil, fueled by global economic expansion, not matched by an increase in oil and natural gas production. In summary, sugar-based surfactants are here to stay, and expand!

The transformation of nonionic sugar-based surfactants into their ionic counterparts is, in principle, relatively simple. Synthesis of anionic surfactants may include direct sulfonation of the alkylglycoside, sorbitan ester, or sucrose ester with HSO_3Cl or SO_3-pyridine complex [12–15]. The cationic counterparts can be produced by substituting a leaving group (e.g., halide or tosylate) for one of the OH groups of the nonionic surfactant, followed by (nucleophilic substitution) reaction with a tertiary amine, e.g., trimethylamine [16]. Nevertheless, the literature published on sugar-based ionic surfactant (SBIS) is far less than that published on the corresponding nonionic ones. The reasons are practical: Large-scale production of the latter is more recent than that of classic surfactants; SBNS perform well, e.g., in personal care products, hard-surface cleaners, washing formulations, and fabric softeners, either alone, or as mixtures with classic anionic and cationic surfactants [9]. Additionally, SBNSs are polyfunctional compounds; derivatization leads to several products, e.g., mono-, di-, and trisulfates; this requires (laborious) chromatographic purification, in order to obtain the individual compounds pure [17,18].

It is hoped that this chapter contributes to remedy this imbalance between the amount of work done on sugar-based ionic and nonionic surfactants. The literature, scattered in journals of many fields, on the syntheses, properties, and applications of SBIS, shows that they posses some interesting properties, distinct from those of classic ones, because of their multifunctionality and presence of a large, relatively rigid (sugar) head-ion. Where available, we include data on head-groups with three or more carbon atoms, including derivatives of glycerol, mono- and disaccharides, and sugar oligomers. In addition to the ionic surfactants proper, the discussion also covers *in situ* generated charged micelles, i.e., where the charge results from the complexation of an ion, e.g., the borate or boronate ion with the hydroxyl groups of the sugar [19]; or a change in solution pH, resulting in protonation of a nitrogen atom [20].

9.2 SYNTHESES OF SUGAR-BASED IONIC SURFACTANTS

9.2.1 PRELIMINARY REMARKS

Before discussing the syntheses of SBIS proper, it is worthwhile to consider the following:

1. *The nomenclature of the starting materials, the sugars, and their derivatives*:
   Unless stated otherwise, all alcohols, alkyl halides, and acyl groups have normal chain. For example, butyl bromide, octyl alcohol, and dodecanoyl stand for 1-bromobutane, octan-1-ol, and the straight-chain dodecanoyl group, respectively.

   The nomenclature of sugars, in particular their derivatives, are lengthy. For brevity and ease of reading, therefore, we have opted for shorter, self-explaining names, e.g., 2-aminoglucose instead of 2-amino-2-deoxy-D-glucopyranose and $C_n\alpha$ GlucoPy instead of *n*-alkyl-α-D-glucopyranoside. *A list of all abbreviations, acronyms, and symbols is given at the end of this chapter.*

   Figure 9.1 shows the structures and names of the D-family of aldoses, starting with D-triose and progressing systematically to the D-hexoses, as well as the disaccharides that have been employed in the synthesis of SBIS. For further discussion of the structural details of mono- and disaccharides and their derivatives, the reader is referred to any of the textbooks on carbohydrate chemistry [21,22].

   The conversion of an OH group into sulfate is sometimes referred to as sulfation. This leads to uncertainty, because the reaction involves incorporation of SO_3 into the molecule. Therefore, we use the term sulfonation all through, i.e., where SO_3 is introduced at carbon or oxygen. Finally, the anomeric composition of the SBIS will be given, where available.

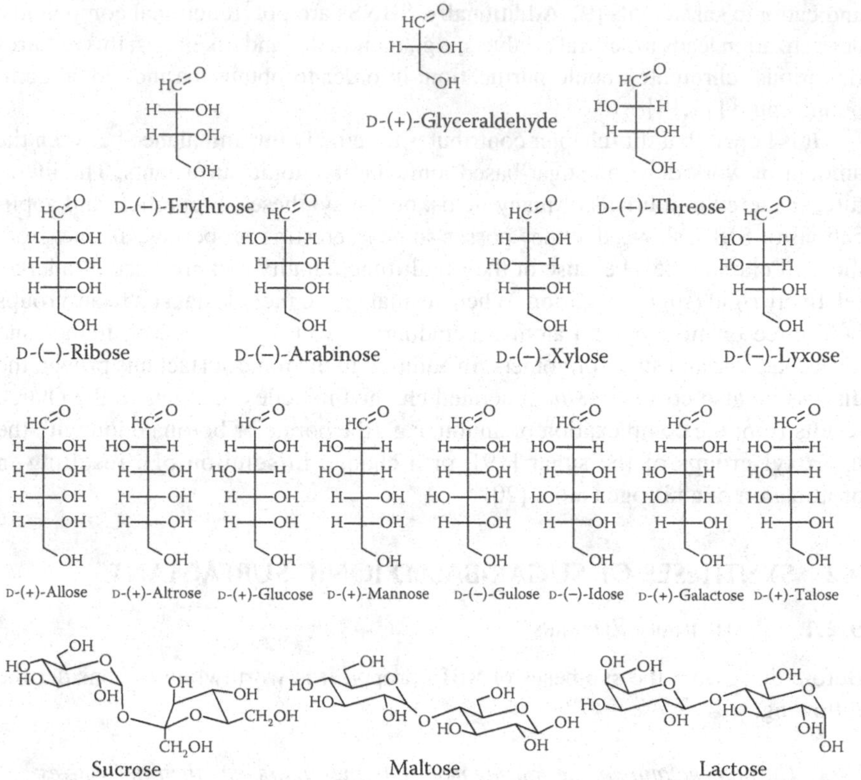

FIGURE 9.1 Structures and names of the D-family of aldoses, as well as those of some disaccharides.

2. *Inclusion of some nonionic, zwitterionic, and catanionic surfactants*:
   Under certain experimental conditions, some nonionic or formally neutral surfactants may acquire charge; it is useful to give a rational for their inclusion in our discussion. In an ionic surfactant, the charged head-group is covalently bonded to the hydrophobic tail. A charge, however, can be generated *in situ* on a nonionic molecule, e.g., by complexation or by change in solution pH. Both approaches are considered here, with more emphasis on the former, because the latter is discussed in detail in Chapter 8. Likewise, examples of formally neutral sugar-based surfactants are presented, in particular zwitterionic ones and the so-called catanionic surfactants. The surface charge of zwitterionic micelles is controlled by the pH of the solution, and the pK_a of the head ions. For example, alkylamine-*N*-oxides are transformed into cationic surfactants, e.g., by the reaction: $RN^+(CH_3)_2-O^- + HX \rightarrow RN^+(CH_3)_2-OH\ X^-$, at relatively low pH values [23]. Note that zwitterionic micelles condense ions from aqueous solutions, this indicates that there is charge fluctuation at the (formally neutral) micellar interface [24,25].

The term "catanionic" is employed to describe a surfactant that is formed by the pairing of two oppositely charged surfactant chains (the parent surfactants) with removal of the inorganic counterions (e.g., sodium, halide, sulfate, etc.). The resulting surfactant is thus uncharged, with a long-chain organic ion acting as a counterion of the other. If the two surfactants are single chained then the resulting catanionic can be considered as a pseudo-double-chain surfactant, e.g., dodecylammonium dodecanoate and dodecylammonium dodecylsulfate. The catanionic surfactants bear an obvious similarity with double-chain amphiphiles, particularly of the zwitterionic type, since both are formally neutral and possess two long alkyl chains. One important difference, however, is that the distance between charges is fixed in zwitterionic surfactants whereas it may vary in catanionic surfactants. Moreover, for the catanionics, factors like departure from equimolar ratio (charge neutrality), and asymmetry in chain length can break the symmetry of the system and induce more or less dramatic changes in phase behavior. That is, compared with typical double-chain zwitterionics, e.g., lecithin, the catanionics have new degrees of freedom [26]. Finally, although vesicles form spontaneously by catanionic; precipitation occurs when the component ions are present in equimolar concentrations. This implies that the stability of these catanionic vesicles is ensured by employing one of the surfactants in excess [27]; the aggregate formed has a net charge. These considerations justify inclusion of the above-mentioned surfactants in the present account.

9.2.2 GENERATION OF *IN SITU* CHARGED- AND SYNTHESES OF ZWITTERIONIC- AND CATANIONIC-SUGAR-BASED SURFACTANTS

One class of *in situ* charged micelles is formed via the complexation of borate or boronate anions with 1,2 or 1,3 diols. This complexation occurs at alkaline or neutral pH, as shown by equilibria (1–3), were $C_n = 0$ or 1, Figure 9.2 [19]:

In Figure 9.2, equilibrium (1) represents the complexation of boric acid with the hydroxide ion to form the tetrahydroxyborate anion, the only relevant species at low acid concentration (≤ 0.01 mol/L) [28], and equilibrium (2) is the complexation of the anion formed with the diol, with elimination of water. The same anionic species forms, less readily, under neutral conditions, as shown by equilibrium (3). The complex stability depends on the diol employed, being much higher for 1,2-diols (five membered ring) than for 1,3-diols (six membered ring) [29]. Results of ^{11}B NMR spectroscopy have shown that SBNS differ in their affinities toward borate. $C_n\beta$GlucoPy surfactants complex through O-4 and O-6 of the head-group; $C_8\beta$MaltoPy complex with borate mainly via the nonreducing glucose residue and have binding affinities very similar to that of $C_n\beta$GlucoPy [30]. The MEGA surfactants have much higher affinity toward borate than alkyl glucosides because the former posses an acylic, i.e., more flexible head group [31]. The interesting point is that the charge density at the micellar interface is a function of the ratio borate/surfactant and the pH of the solution, as depicted in Figure 9.3.

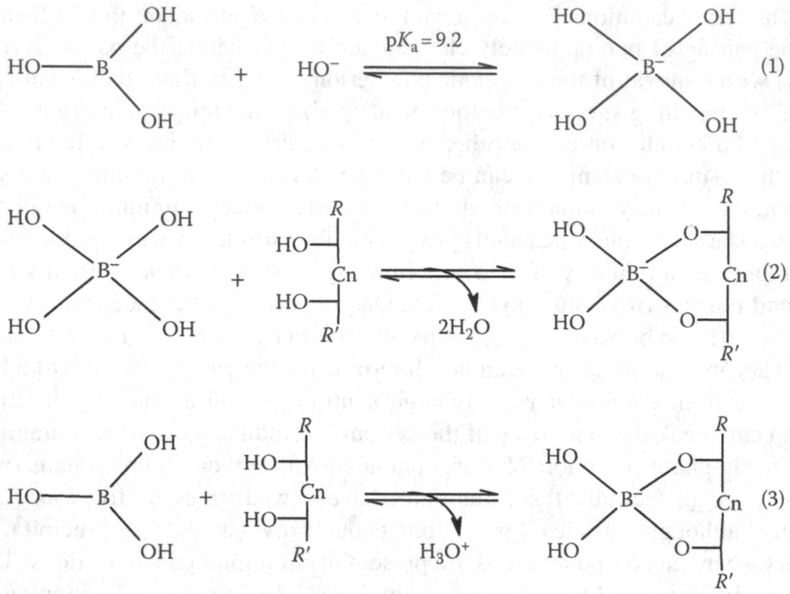

FIGURE 9.2 Representation of the equilibria established between boric acid and the hydroxide ion (1); between the tetrahydroxyborate anion and a 1,2-diol (2); and between boric acid and a 1,2-diol (3). (Redrawn from Smith, J.T. and El Rassi, Z., *J. Chromatogr. A*, 685, 131, 1994. With permission.)

This borate-based complexation scheme has been applied to $C_8\beta GlucoPy$, C_8Sucro, and $C_{12}\beta MaltoPy$ under alkaline conditions [19,32]; MEGA-C_n ($n = 7$, 8, 9, and 10) under alkaline conditions [31,33]; MEGA-9 under acidic, neutral, and alkaline pHs [34]; and N,N-bis-(3-gluconamidopropyl)-cholamide and -deoxycholamide under neutral and alkaline conditions [35].

In situ charged micelles can be generated from SBNS by changing the pH of the medium; this applies to single-chain and to "gemini" surfactants. The latter term was coined by Menger and Littau in order to describe a novel type of surfactant consisting of two identical conventional surfactants connected via a "spacer" at the level of the head groups [36]. The pH-induced change of the aggregates present is reversible, and can be readily detected by scattering techniques. Thus measurements of solution turbidity (at 600 nm), quasi-elastic light scattering, and ξ potential of solutions of the synthetic glycolipid, 1,3-di-O-phytanyl-2-O-(maltotriosyl) glycerol, Figure 9.4, have shown that the average size of the vesicles formed in the acidic region (pH 4–6) was well above 1000 nm, an indication of vesicle aggregation; in the alkaline region (pH 8–10), the aggregates became much smaller (110–130 nm), and much less polydisperse, indicating pH-induced breakdown of the vesicles. As a function of increasing the pH, the ξ potential measured becomes more negative, reaching a constant value of approximately −40 mV. The surface charges of these

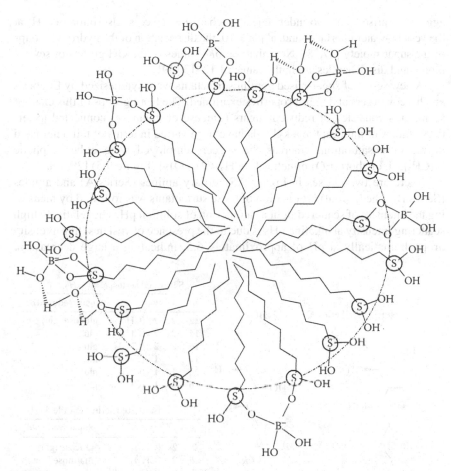

FIGURE 9.3 Schematic representation of a micellar aggregate formed between a sugar-based surfactant and the borate anion. (Redrawn from Smith, J.T. and El Rassi, Z., *J. Chromatogr. A*, 685, 131, 1994. With permission.)

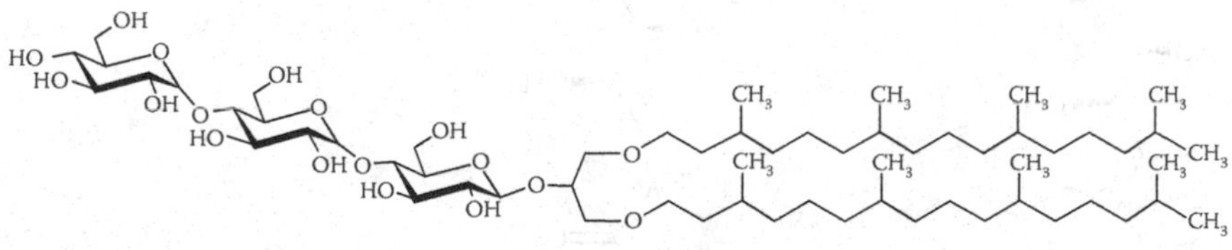

FIGURE 9.4 Structure of the synthetic glycolipid, 1,3-di-*O*-phytanyl-2-*O*-(β-D-maltotriosyl) glycerol. (Redrawn from Baba, T., Zheng, L.Q., Minamikawa, H., and Hato, M., *J. Colloid Interface Sci.*, 223, 235, 2000. With permission.)

aggregates arise from two independent mechanisms: Excess adsorption of OH$^-$ at the vesicle–water interfaces and, at pH > 10, the dissociation of the hydroxyl groups of the sugar moiety [37,38]. Note that the pK_a values of the OH groups of several mono- and disaccharides are in the range of 12–13 [39].

A new series of sugar-based gemini surfactants were synthesized by Engberts and his coworkers and their properties examined. The head-groups of this class of surfactants consisted of reduced sugars (glucose or mannose) connected to tertiary amines or amides; the alkyl tails have been varied in terms of tail length and degree of unsaturation, whereas the spacers employed were either aliphatic [–(CH$_2$)$_n$–], or short (EO) moieties [–(CH$_2$–CH$_2$–O)–] (Figure 9.5) [20,40].

There are two classes in Figure 9.5, tertiary amines (Series A) and amides (Series B). The (partial) protonation of these surfactants was followed by measuring the intensity of scattered light as a function of solution pH. The relatively high scattering intensity at alkaline pHs is due to the presence of vesicles; this intensity drops dramatically at pH of approximately 6.5, indicating a large decrease in

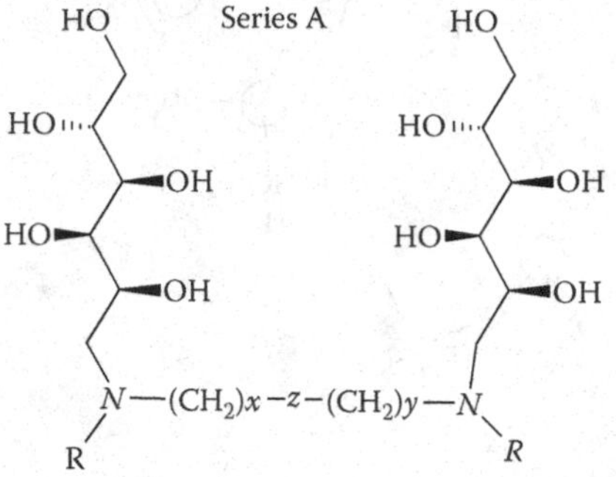

Saturated series, $R = C_mH_{2m+1}$

m	x	y	z	Sugar (reduced)
12	2	2	(CH$_2$)$_2$	glucose (glu)
14	2	2	(CH$_2$)$_2$	glu
16	2	2	(CH$_2$)$_2$	glu
12	1	1	(CH$_2$)$_2$	glu
14	1	1	(CH$_2$)$_2$	glu
16	1	1	(CH$_2$)$_2$	glu

Unsaturated series, R = oleyl

x	y	z	Sugar (reduced)
2	2	(CH$_2$)$_2$	glucose (glu)
2	2	(CH$_2$)$_2$	mannose (man)
2	0	(OCH$_2$CH$_2$)$_2$	glu
2	0	(OCH$_2$CH$_2$)$_2$	man

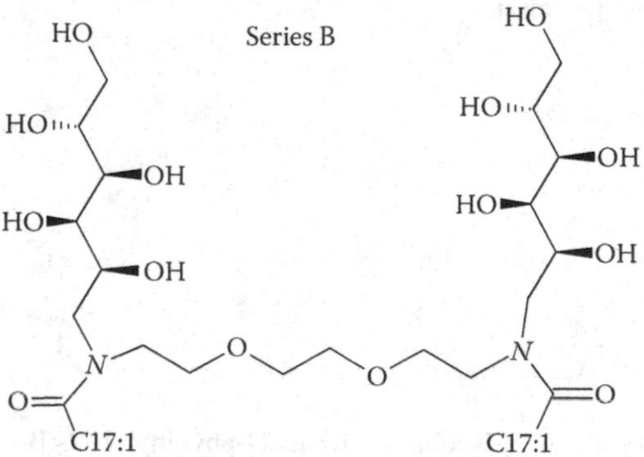

FIGURE 9.5 Structures of sugar-based gemini surfactants, including amine-based (Series A), and amide-based structures (Series B). (Redrawn from Johnsson, M. and Engberts, J.B.F.N., *J. Phys. Org. Chem.*, 17, 934, 2004. With permission.)

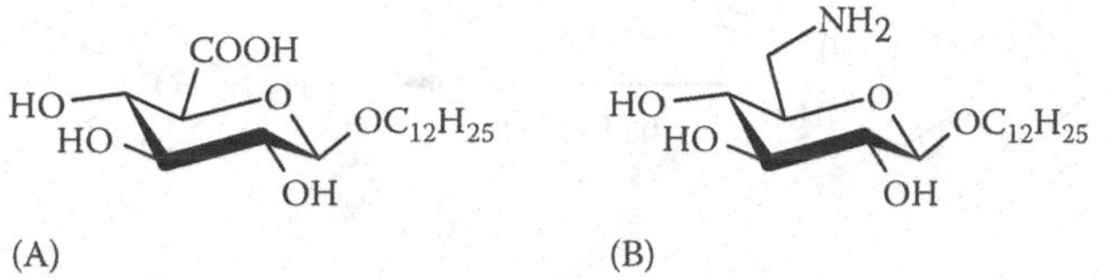

FIGURE 9.6 The structures of some SBNS, dodecyl-β-D-glucopyranoside uronic acid (A) and dodecyl-6-deoxy-6-amino-β-D-glucopyranoside (B), whose micellar behavior depends on solution pH. (Redrawn from Milkereit, G., Morr, M., Thiem, J., and Vill, V., *Chem. Phys. Lipids*, 127, 47, 2004. With permission.)

particle size, due to the formation of *in situ* generated cationic micelles. This behavior is general, with slight variation of the threshold pH at which the vesicles are transformed into micelles. Finally, it is worthwhile to note that the lyotropic behavior of a series of weakly ionic sugar-based surfactants, e.g., alky-β-D-gluco-pyranoside uronic acids, and of dodecyl-6-deoxy-6-amino-β-D-glucopyranoside, Figure 9.6, depend on solution pH, presumably because of changes in the charge density at the micellar interface [41].

Sugar-based zwitterionic surfactants have been synthesized by the following scheme: N-dodecyl-N,N-bis[3-aminopropyl]amine was synthesized by reacting the appropriate amine (dodecyl, hexadecyl, and octadecyl) with acrylonitrile in MeOH, followed by reduction of the cyano groups (CN → CH$_2$NH$_2$) with H$_2$/Raney nickel. N-alkyl-N,N-bis[3-(aldonamido)propyl]amines were synthesized by reacting the above-mentioned diamine with D-gluconic-δ-lactone; D-glucoheptonic-γ-lactone, or lactobionic acid [42]. These nonionic precursors were converted into zwitteri-onic surfactants by oxidation (N → N$^+$–O$^-$) with 30% H$_2$O$_2$ (wt%) solution in 2-propanol, and purified by recrystallization, Scheme 9.1 [43].

The syntheses of several sugar-based catanionic surfactants have been carried out. Thus the addition of RNH$_2$ to lactose in propan-2-ol, followed by reduction of the imine formed with NaBH$_4$ offered N-alkylamino-1-deoxylactitols (R = octyl; hexadecyl). Mixing equimolar amounts of these compounds with monocarboxylic acids, namely, octanoic; dodecanoic; 12(4-aminocoumarine) dodecanoic, or other acids, (perfluorooctyl- or perfluorodecyl)ethylphosphinic offered two-chain sur-factants, whereas mixing with 1,12-dodecane dicarboxylic acid offered two gemini catanionic glycolipids, Scheme 9.2 [44–46].

A bolaform amine has been synthesized by reacting bis(6-aminohexyl)amine with D-1,5-gluconolactone to produce bis(6-glucoamidohexyl)amine; the nucleo-philic addition to the C=O group of lactone is that by the less hindered primary nitrogen atoms. Reaction of this product with HO$_2$C(CH$_2$)$_n$CO$_2$H (n = 3, 5, and 6) produced the corresponding sugar-based catanionics, Scheme 9.3 [47].

A surfactant based on the lactobionate head-ion and a perflourinated hydrophobic chain (C$_8$F$_{17}$CH$_2$) has been synthesized by the ring opening of a perflourooctylated oxiran by 3-amino-propane-1,2-diol. The oxirane ring was

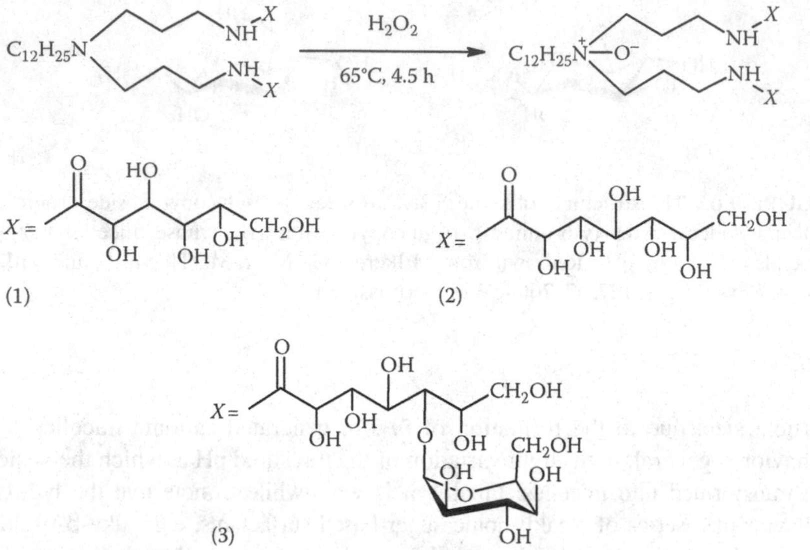

SCHEME 9.1   Synthesis of zwitterionic surfactants by oxidation of SBNS. Structures 1, 2, and 3 refer to N-dodecyl-N,N-bis[3-(D-gluconylamido)propyl]amine-N-oxide; N-dode-cyl-N,N-bis[3-(D-glucoheptonylamido)propyl]amine-N-oxide, and N-dodecyl-N,N-bis[3-(D-lactobinoylamido) propyl]amine-N-oxide, respectively. (Redrawn from Piasecki, A., Karczewski, S., and Syper, L., *J. Surf. Det.*, 4, 349, 2001. With permission.)

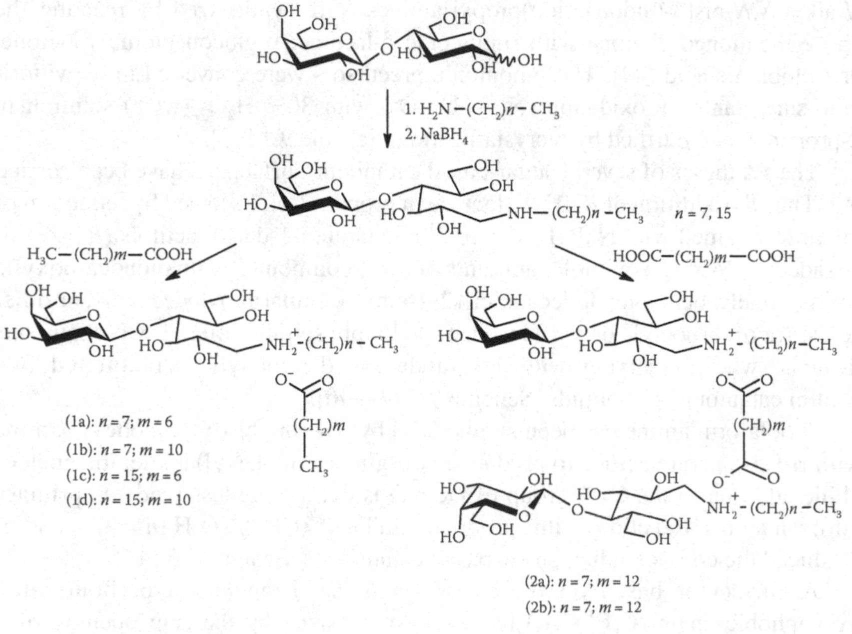

SCHEME 9.2   Synthesis of catanionic glycolipids. (Redrawn from Blanzat, M., Perez, E., Rico-Lattes, I., Prome, D., Prome, J.C., and Lattes, A., *Langmuir*, 15, 6163, 1999. With permission.)

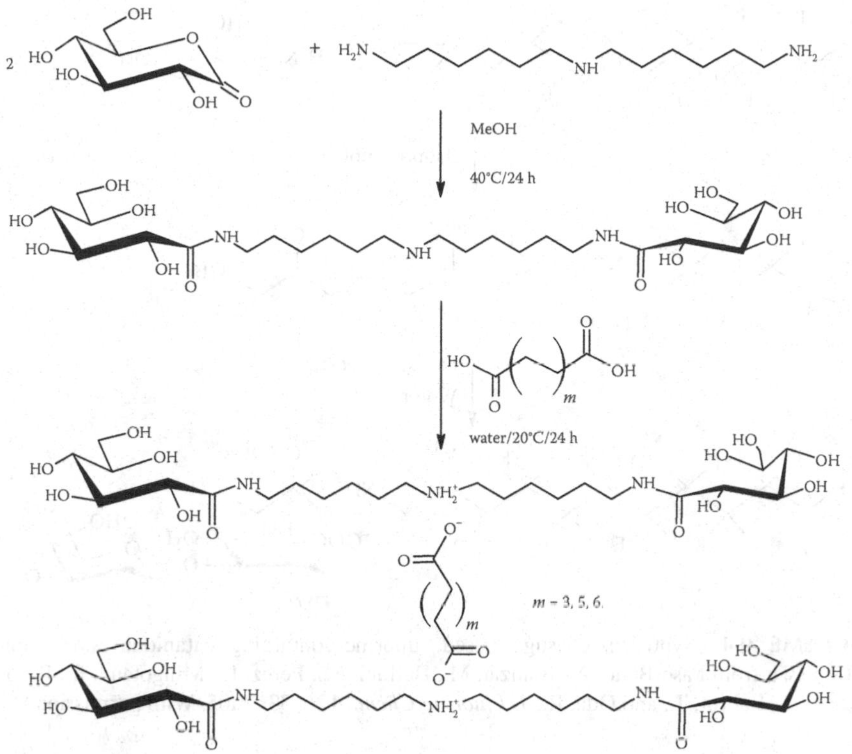

SCHEME 9.3 Synthesis of sugar-based bolaamphiphile and its conversion into catanionic surfactants. (Redrawn from Soussan, E., Pasc-Banu, A., Consola, S., Labrot, T., Perez, E., Blanzat, M., Oda, R., Vidal, C., and Rico-Lattes, I., *ChemPhysChem*, 6, 2492, 2005. With permission.)

regioselectively opened, induced by the electron-withdrawing effect of perflourooctyl group. No dialkylation occurred on the nitrogen of the aminodiol, probably due to steric hindrance, Scheme 9.4 [48].

9.2.3 SYNTHESES OF ANIONIC SBIS

The performance of ionic surfactants is related to their structure, including the length of the hydrophobic tail, the nature of the head-group and counterion, the number of head-groups present per molecule of surfactant, and the position (terminal or intermediate) of the ionic head-group in the molecule. Surfactants that carry multiple head-groups, or where the head-group does not occupy a terminal position act more as wetting agents and dispersants than as detergents. Sugars are poly-functional compounds, i.e., their direct functionalization is likely to produce a mixture of mono- and poly-functionalized surfactants. Therefore, the question of regioselective functionalization is important and will be focused upon, where data are available.

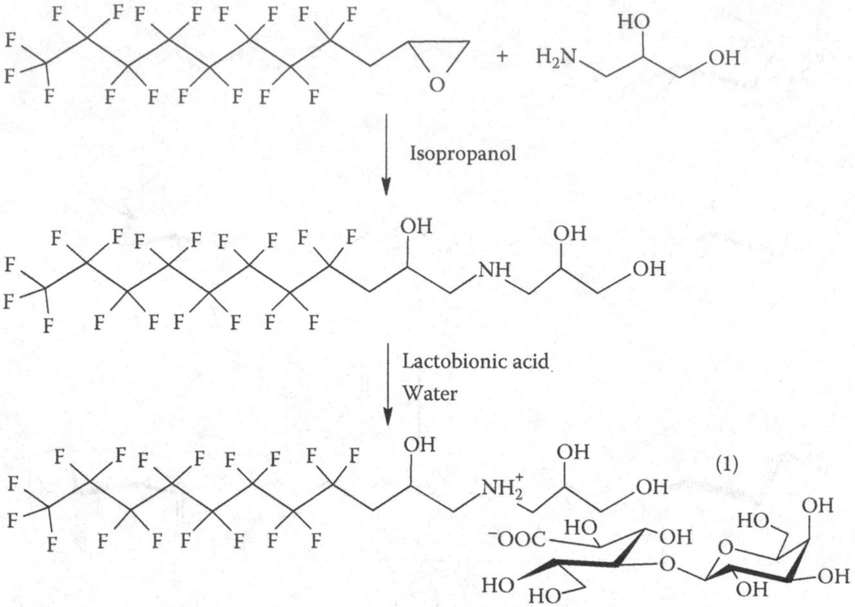

SCHEME 9.4 Synthesis of sugar-based, fluorine-containing catanionic surfactant. (Redrawn from Pasc-Banu, A., Blanzat, M., Belloni, M., Perez, E., Mingotaud, C., Rico-Lattes, I., Labrot, T., and Oda, R., *J. Fluorine Chem.*, 126, 33, 2005. With permission.)

9.2.3.1 Carboxylates

The simplest SBIS are derivatives of "onic" and "uronic" acids, e.g., gluconic acid glucuronic acid, respectively. In principle, the synthesis of the latter surfactants is straightforward because the ionic group can be introduced by direct oxidation of alkyl glucosides; glycosylation of O-unprotected uronic acid, or glycosylation of an appropriately O-protected uronic acid, all these approaches have been employed.

Direct oxidation of, e.g., the C_6-OH group of alkyl glucopyranosides is subject to complications due to competing cleavage of the glycosidic bond and subsequent difficulties in product purification. Nevertheless, oxidation of $C_{12}\alpha$GlucoPy by Adams' catalyst in buffered solution (NaHCO$_3$, pH = 7–8.5) offered sodium dodecyl-α-D-glucopyranosiduronate. Surprisingly, when the same reaction was applied to $C_{14}\alpha$MaltoPy only the 6'-OH was oxidized, leading to the monouronate derivative, Scheme 9.5 [49].

An alternative approach is to start from commercially available, or easily obtainable O-unprotected uronic acid; the synthetic challenge is to obtain tautomerically and anomerically pure glycosiduronic acid derivatives. Achieving the last two objectives is not trivial because of: tautomeric equilibria of some sugars in solution; *in situ* anomerization; competitive O-glycosydation, and esterification reactions [50,51]. An example of the latter problem is the (homogeneous) reaction

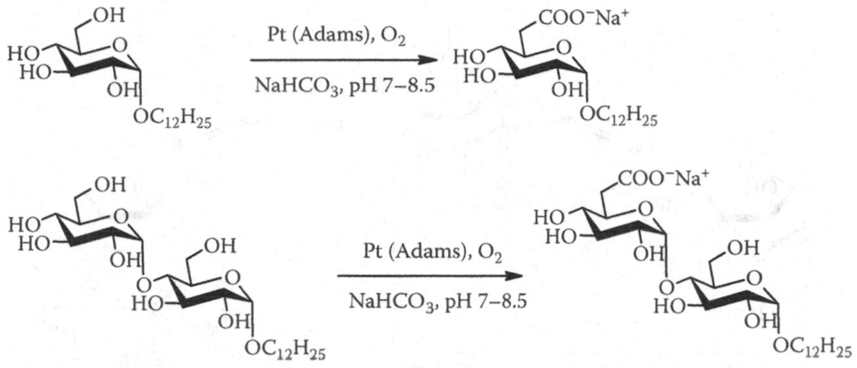

SCHEME 9.5 Synthesis of sodium dodecyl-α-D-glucopyranosiduronate by direct oxidation of O-unprotected $C_{12}\alpha$GlucoPy and $C_{12}\alpha$MaltoPy. (Redrawn from Boecker, T. and Thiem, J., *Tenside, Surf., Det.*, 26, 318, 1989. With permission.)

of D-galacturonic acid with methanol; this offers a mixture of methyl D-galacturonate and methyl (methyl galactopyranosiduronate); esterification proceeds faster than glycosylation. A reversal of chemoselectivity toward glycosidation was achieved, however, under heterogeneous reactions conditions, in the presence of Lewis acid catalysts (FeCl$_3$ or BF$_3$-OEt$_2$). Under these conditions, D-galacturonic acid reacted with ROH (R = octyl, decyl, and dodecyl, respectively) to give a mixture of products with predominance of α-pyranosidic and β-furanosidic compounds. The yields of the former compounds were increased to 65%, 61%, and 62%; α/β ratios of 24:1, 19:1, and 13:1, respectively, by ring expansion and anomerization of the latter ones [52].

D-Glucuronic acid and D-galacturonic acids were also treated with the same fatty alcohols, under heterogeneous reaction conditions, in the presence of FeCl$_3$ (3 equiv.) and CaCl$_2$ (2-equiv.), to give exclusively the corresponding alkyl-β-D-glucofuranosiduronic acids in 90% yield, for alkyl = decyl and dodecyl; and alkyl-β-D-galactofuranosiduronic acids in 60%, 70%, 65%, 65%, and 50% yield, for alkyl = octyl, decyl, dodecyl, tetradecyl, and 12-hydroxydodecyl, respectively. The preference for the kinetically favored furanosidic products, and the absence of esterification were explained on the bases of complex formation of the Ca^{2+} ion with HO-2 and HO-5 of the furanoid isomer, as shown in Scheme 9.6 [52,53].

Until recently, SBIS based on D-mannouronic acid were lacking because the monomer is not commercially available, and has not been separated from natural carbohydrates, e.g., alginate, a heteropolysaccharide composed of (1,4)-linked β-D-mannuronic acid and α-L-guluronic acid residues. This problem has been solved by using a sequence of reactions: controlled degradation of alginate, glycosylation, and hydrolysis, respectively. Thus sodium polymannuronate with DP of 16 was obtained from the hydrolysis of sodium alginate (from *Laminaria digitata*) by HCl at 95°C–100°C, followed by purification. Further hydrolysis of the latter

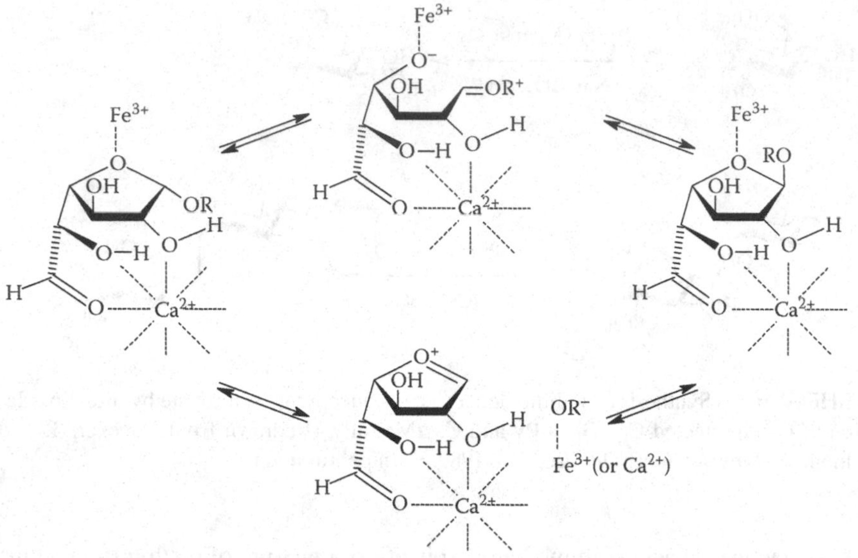

SCHEME 9.6 Complexation of Fe^{3+} and Ca^{2+} cations to the furanose ring. (Redrawn from Bertho, J.N., Ferrieres, V., and Plusquellec, D., *J. Chem. Soc., Chem. Commun.*, 1391, 1995. With permission.)

product with HCl, at pH of 3, followed by ultrafiltration yielded sodium oligoman-nuronate, with DP of 3. Butyl (butyl-α-D-mannopyranosiduronate) was obtained from the latter product by Fischer glycosylation and esterification with 1-butanol in the presence of MSA catalyst. The product with short-chain alkyl group was transformed into the corresponding R-α-D-mannopyranosiduronate (R = octyl, decyl, dodecyl, tetradecyl, and hexadecyl, respectively), by heating with an excess of the fatty alcohol, in presence of MSA catalyst. Alkaline hydrolysis of these R-α-D-mannopyranosiduronates, followed by purification and acidification gave the corresponding anionic, and nonionic surfactants, respectively, e.g., dodecyl-α-D-mannopyranosiduronate and dodecyl-α-D-mannopyranosiduronic acid, Scheme 9.7. A fact that may turn this scheme economically attractive is that it is possible to obtain the oligomannuronate in satisfactory yield, 12.5%, directly from brown sea-weeds whose cost is approximately 2% of that of commercial alginate [55].

Carboxylate sugar-based surfactants have been synthesized from O-unpro-tected sugars by another approach, condensation of lactose with $NH_2(CH_2)_{10}CO_2Na$ in aqueous methanol, followed by reduction of the imine formed with $NaBH_4$. The nitrogen atom present can be acylated by using *N*-alkanoyl-thiazolidine-2-thione (heptanoyl; nonanoyl; dodecanoyl; hexadecanoyl; and octadecanoyl) to give dou-ble-tail anionic surfactants, Scheme 9.8 [56].

More elaborate schemes were also employed for the synthesis of this class of SBIS. Thus D-glucorono-6,3-lactone [57] was converted into methyl 1,2,3,4-tetra-*O*-acetyl-β-D-glucopyranuronate by opening of the lactone ring with MeONa/MeOH,

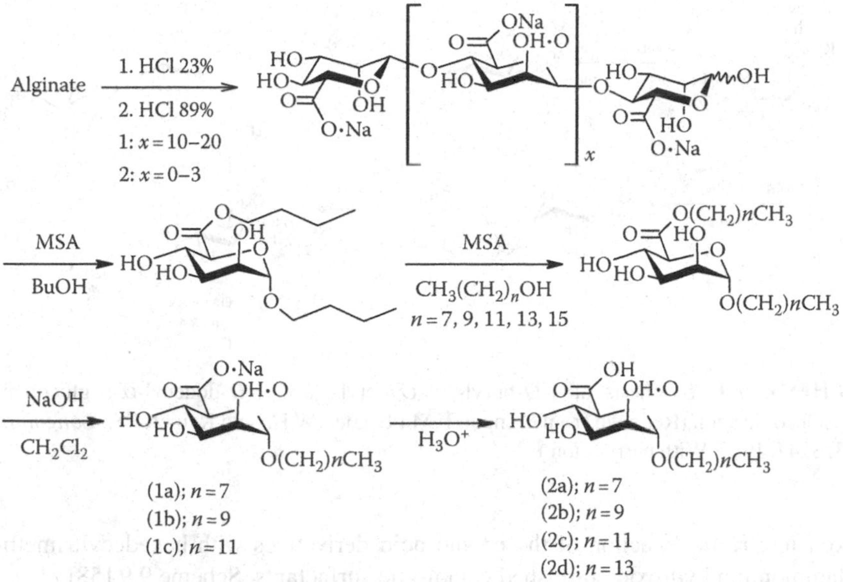

SCHEME 9.7 Stereoselective synthesis of D-mannuronate anionic and neutral sugar-based surfactants. (Redrawn from Roussel, M., Benvegnu, T., Lognone, V., Le Deit, H., Soutrel, I., Laurent, I., and Plusquellec, D., *Eur. J. Org. Chem.*, 3085, 2005. With permission.)

followed by acetylation of the hydroxyl groups with Ac$_2$O/pyridine. Several schemes were then employed for the glycosylation of the tetra-*O*-acetyl-glucopyranuronate. Reaction with alkyl bromides in the presence of Hg$_2$(CN)$_2$ or CF$_3$CO$_2$Ag catalysts failed to yield any product; reaction with aliphatic alcohols in CH$_2$Cl$_2$, in the presence of a Lewis acid catalyst (SnCl$_4$) gave low yields (36%, 30%, and 22%) of methyl 1-*O*-alkyl-2,3,4-tri-*O*-acetyl-β-D-glucopyranuronate (alkyl group = butyl, octyl, and dodecyl, respectively). The latter compounds were de-acetylated and converted into the corresponding uronic acid derivatives almost quantitatively by Na$_2$CO$_3$ in aqueous methanol, followed by treatment with cation

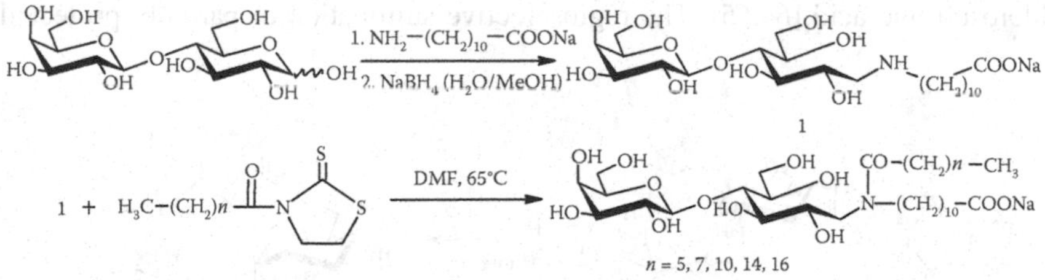

SCHEME 9.8 Synthesis of 1-(*N*-ω-sodium alkanoates)-*N'*-(acyl)-aminolactitols. (Redrawn from Rico-Lattes, I., Gouzy, M.F., Andre-Barres, C., Guidetti, B., and Lattes, A., *Biochimie*, 80, 483, 1998. With permission.)

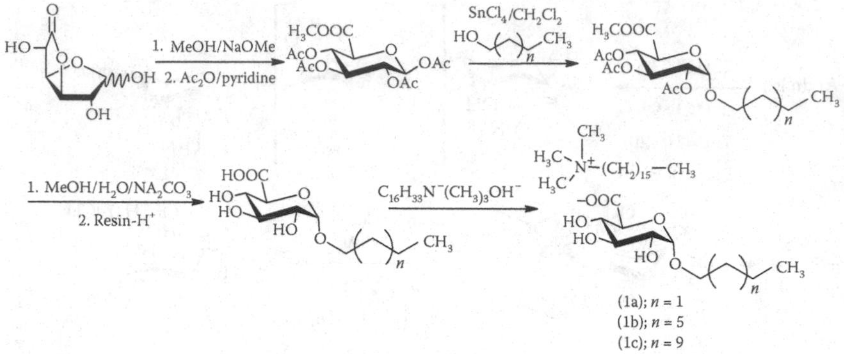

SCHEME 9.9 Synthesis of 1-*O*-butyl-, 1-*O*-octyl-, and 1-*O*-dodecyl-α-D-glucopyanosiduronic acid. (Redrawn from Menger, F.M., Binder, W.H., and Keiper, J.S., *Langmuir*, 13, 3247, 1997. With permission.)

exchange resin. Reaction of the uronic acid derivatives with hexadecyltrimethylammonium hydroxide furnished catanionic surfactants, Scheme 9.9 [58].

Sucrose *n*-octadecenylsuccinate was synthesized in a pioneering work, published on the synthesis of sucrose ethers and esters, and evaluation of their performance as surfactants. This product was obtained in 91% yield by the reaction of *n*-octadecenylsuccinic anhydride, with 10% excess sucrose in DMF, followed by purification, Scheme 9.10 [59].

9.2.3.2 Sulfates and Sulfonates

The direct regioselective synthesis of *O*-acyl, *O*-sulfosugar derivatives can be achieved by regioselective sulfonation of the sugar, followed by regioselective acylation or, alternatively, by regioselective acylation, followed by regioselective sulfonation. Introduction of the SO_3 group can be achieved by treating the sugar, or its derivative, with sulfonating agents, including complexes of SO_3/Lewis base, e.g., DMF, pyridine or trialkylamine. Other sulfonating agents include piperidine-*N*-sulfonic acid in a polar, aprotic solvent [60,61], sulfuric acid in the presence of *N,N'*-dicyclohexylcarbodiimide [62], or in acetic anhydride [63], and chlorosulfonic acid [64,65]. The regioselective sulfonation of partially protected

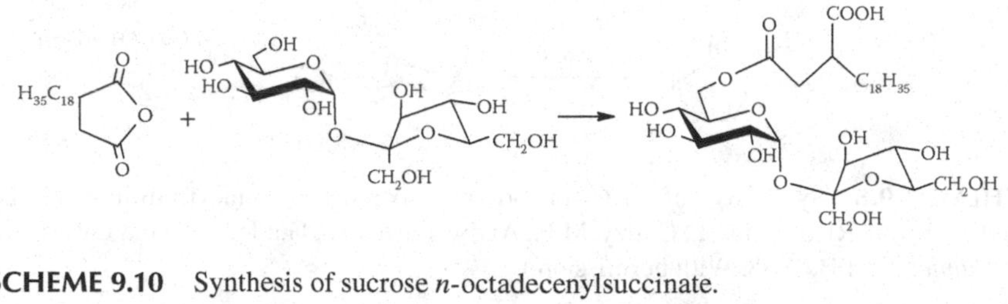

SCHEME 9.10 Synthesis of sucrose *n*-octadecenylsuccinate.

monosaccharides [66,67], and disaccharides [68], proceeds similarly to O-acylation, i.e., sulfonation of the primary OH group is preferred.

Several approaches are known for attaching the sulfonate group regioselectively. Thus, dibutylstannanediyl acetals have been employed for regioselective sulfonation of partially O-protected monosaccharides and disaccharides by using SO_3-triethylamine complex [69,70]. Another strategy is based on the nucleophilic opening of a cyclic sulfate, this method being also useful for the regioselective introduction of the nucleophilic group. This ring-opening is applicable to a wide variety of nucleophiles, including those of oxygen, nitrogen, and carbon [71–73].

The simplest SBIS carry head groups composed of three or four carbon atoms, i.e., derivatives of aldotriose (glyceraldehyde) and aldotetroses. To our knowledge, there is no information on surface active tetroses, and only few reports on derivative of glyceraldehyde. Sodium monolaurin sulfate was synthesized as follows: glycerol was esterified by reaction with equimolar amount of recrystallized dodecanoic acid. After purification (to 98.5% purity), the monolaurin was sulfated by reaction with $ClSO_3H$–NaCl mixture, and the product was purified by extraction with ethanol [74]. Fatty acid monoglyceride sulfates were prepared by sulfating purified monoglycerides with HSO_3Cl or SO_3/pyridine complex, followed by neutralization with NaOH [75,76]. Sulfated 1-monoglycerides of lauric, myristic, palmitic, stearic, and oleic acids were prepared by condensation of the fatty acids with isopropylene glycerol, hydrolysis with cold HCl, and sulfonation with a pyridine-SO_3 complex [77].

Sodium monoglyceride sulfate based on the acids from coconut oil is employed in detergent solid bars. The early commercial process consisted of transesterification of one mole of coconut oil with two moles of glycerol, in the presence of 18 moles of fuming sulfuric acid; excess of the latter is required to solubilize the very viscous product before neutralization. This process generates large amount of by-product (sodium sulfate) and, therefore, does not conform to the principles of green chemistry. A more interesting approach, applied to fatty acids of coconut and palm oils and their methyl esters, consists of reacting glycerol with oleum to give glycerol trisulfuric acid, followed by reaction with one mole of fatty acids, or their methyl esters, the product is extracted and neutralized, Scheme 9.11 [14].

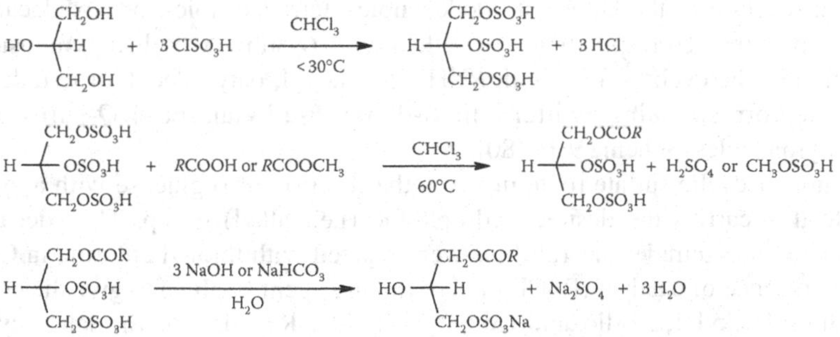

SCHEME 9.11 Synthesis of sodium monoglyceride sulfate. (Redrawn from Ahmed, F.U., *J. Am. Oil Chem. Soc.*, 67, 8, 1990. With permission.)

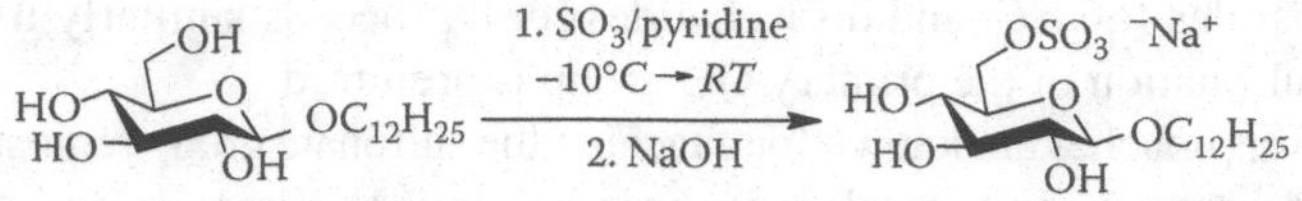

SCHEME 9.12　Synthesis of dodecyl-β-D-glucopyranoside-6-sulfate from O-unprotected C_{12}βGlucoPy. (Redrawn from Boecker, T. and Thiem, J., *Tenside, Surf., Det.*, 26, 318, 1989. With permission.)

Partially pure (24%–90%) monoglycerides from cottonseed, coconut, castor, vegetable tallow, and rice bran oils were sulfated either with concentrated H_2SO_4, or oleum to give sulfated monoglycerides. These were tested as emulsifiers, detergents, and dispersants for calcium soap. The authors claimed that neither high purity nor high monoglyceride content was a prerequisite for obtaining good sulfonated products [78], although it is known that high purity monoglycerides should be employed in order to avoid side reactions, e.g., addition of SO_3 to the double bond of oleic acid [79].

A straightforward strategy to obtain sugar-based anionic surfactants is to sulfonate alkyl glycosides, e.g., the reaction of C_{12}βGlucoPy with SO_3-pyridine complex at −10°C, followed by more agitation at room temperature and neutralization with NaOH gave sodium dodecyl-β-D-glucopyranoside-6-sulfate, Scheme 9.12 [49].

The same sulfonating reagent was reacted with C_{12}αGlucoPy, C_{12}βGlucoPy, and C_{14}βMaltoPy; the mono-sulfated products were purified (from the poly-sulfated ones) by HPLC, by employing methanol–water mixtures as eluents. All products obtained were characterized by ^1H, ^{13}C and 2-D NMR spectroscopy, and their specific rotation measured. In each case, several products were obtained, because the starting SBNS were O-unprotected, as shown in Scheme 9.13.

The previously mentioned (laborious) chromatographic purification of the mono-sulfated products may be avoided by carrying out regiospecific synthesis. A two-step synthesis was employed in order to obtain anionic and zwitterionic α-D-glucopyranoside-based surfactants. Reaction of α-D-glucopyranoside with thionyl chloride and pyridine in DMF/ethyl acetate gave the corresponding cyclic sulfite. ^1H NMR spectroscopy has indicated that the product is 4,6-cyclic sulfite, and that the configuration of the sulfoxide group is axial. Oxidation of this product with $NaIO_4$ in the presence of $RuCl_3$ catalyst gave the corresponding cyclic sulfate; reaction of the latter with dodecanoic-, tetradecanoic-, or hexadecanoic acid gave the corresponding 6-O-alkanoyl-4-O-sulfo-α-D-glucopyranosides. Reaction of the cyclic sulfate with RNH_2 (R = hexyl, octyl, decyl, or tetradecyl) gave the corresponding zwitterionic 6-deoxy-6-alkylamino-4-O-sulfo-α-D-glucopyranosides, Scheme 9.14 [80].

Another cyclic sulfate route involves the reaction of D-glucose with a cyclic sulfate that carries the desired hydrophobic (i.e., alkyl) group. Thus decane-, dodecane-, and tetradecane 1,2-diols were reacted with thionyl chloride in CCl_4, in the presence of catalyst ($RuCl_3$) and oxidizing agent ($NaIO_4$) to give the corresponding 4-alkyl-1,2,3-dioxathiolane-2,2-dioxide. Reaction of the latter cyclic sulfates with D-glucose and NaH in DMPU at room temperature gave the corresponding surfactants, e.g., sodium 1-(α- and β-D-glucopyranosyloxy)dodec-2-yl sulfate, α/β ratio ~1:3; 1:1 mixture of epimers. Acid-catalyzed hydrolysis of

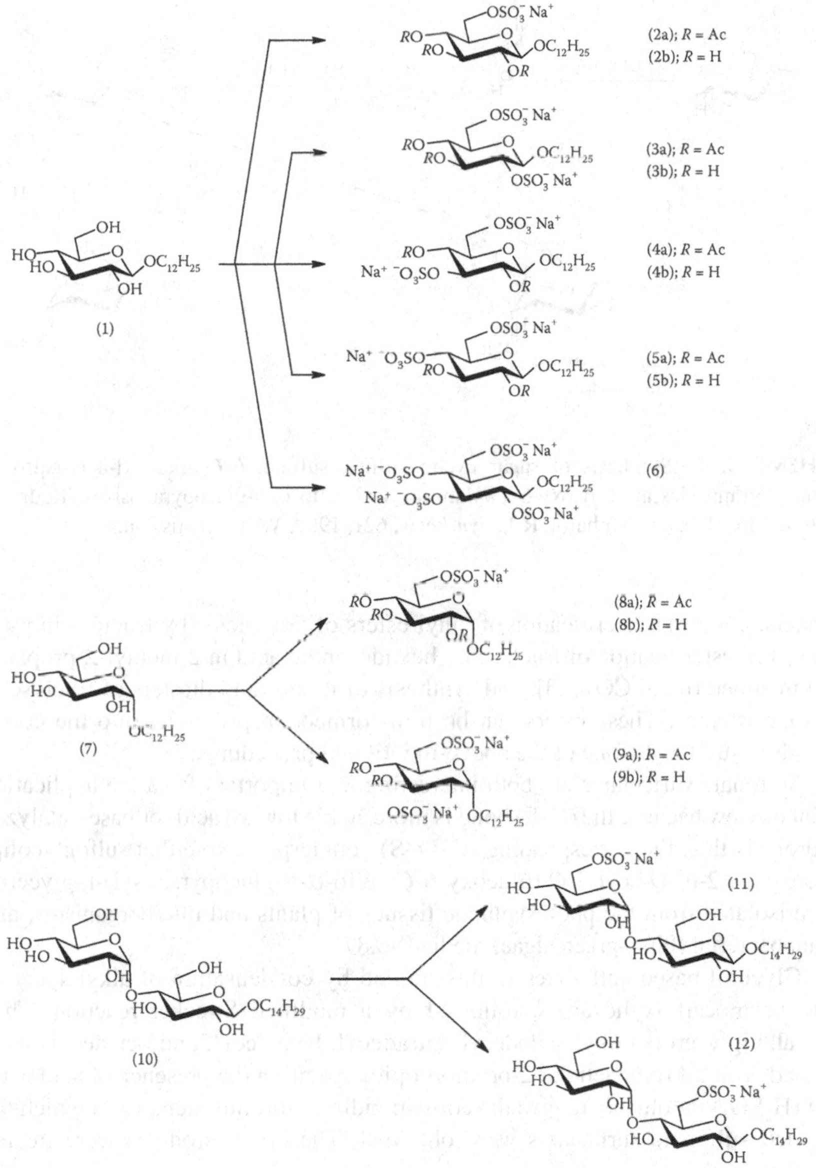

SCHEME 9.13 Synthesis of sugar-based sulfate surfactants by sulfonation and subsequent chromatographic purification of $C_{12}\beta$GlucoPy; $C_{12}\alpha$GlucoPy, and $C_{14}\beta$MaltoPy, respectively. (Redrawn from Boecker, T., Lindhorst, T.K., Thiem, J., and Vill, V., *Carbohydr. Res.*, 230, 245, 1992. With permission.)

these anionic surfactants gave the corresponding nonionic ones, 2-hydroxyalkyl-D-glucopyranosides, Scheme 9.15 [81].

To our knowledge, there are no reports on fructose-based ionic surfactants, although work has been done on lipase-catalyzed synthesis of fructose mono-and

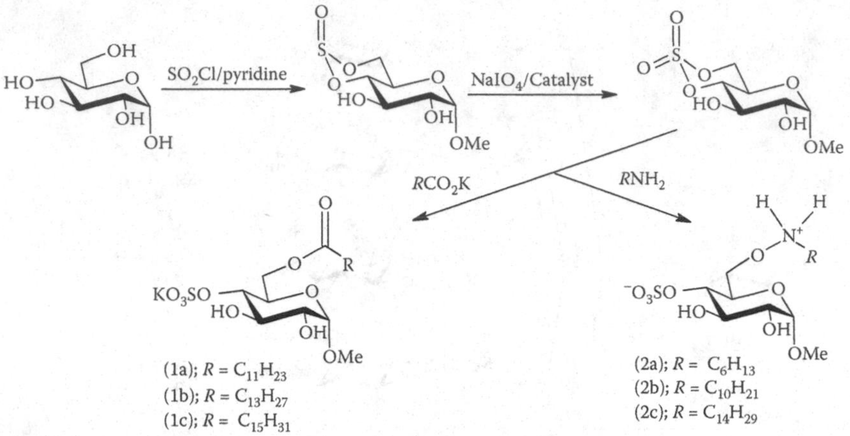

SCHEME 9.14 Synthesis of sugar cyclic sulfite, sulfate, 6-O-alkanoyl-4-O-sulfo-α-D-glucopyranosides, and 6-deoxy-6-alkylamino-4-O-sulfo-α-D-glucopyranosides. (Redrawn from Bazin, H.G. and Linhardt, R.J., *Synthesis*, 621, 1999. With permission.)

diesters, e.g., transesterification of vinyl esters of fatty acids by fructose in pyridine [82]; esterification of fructose by hexadecanoic acid in 2-methyl-2-propanol and in supercritical CO_2 [83]; and synthesis of mono- and diesters of fructose in acetone [84,85]. These esters can be transformed, in principle, into the corresponding sulfates by one of the above-mentioned procedures.

Sulfonate surfactants are both interesting and important from the application point of view because the (C–S) bond is more stable toward acid- or base-catalyzed hydrolysis than the corresponding (C–O–S) counterpart. Note that sulfoglycolipids, e.g., 1,2-di-O-acyl-3-O-(6-deoxy-6-C-sulfo-α-D-glucopyranosyl)-L-glycerol, were isolated from the photosynthetic tissues of plants and microorganisms, and cyanobacterial (blue–green algae) media [86,87].

Glycerol-based sulfonates were obtained by condensation of alkylglycerols with bromocarboxylic acids, followed by a modified Strecker reaction. Thus 1-O-alkylglycerols (alkyl = dodecyl, tetradecyl, hexadecyl, and octadecyl) were reacted with 2-bromoactic or 2-bromopropionic acid in the presence of acid catalyst (H_2SO_4) in toluene, to give the corresponding 2-bromoesters, from which the targeted sulfonate surfactants were obtained. The crude products were treated

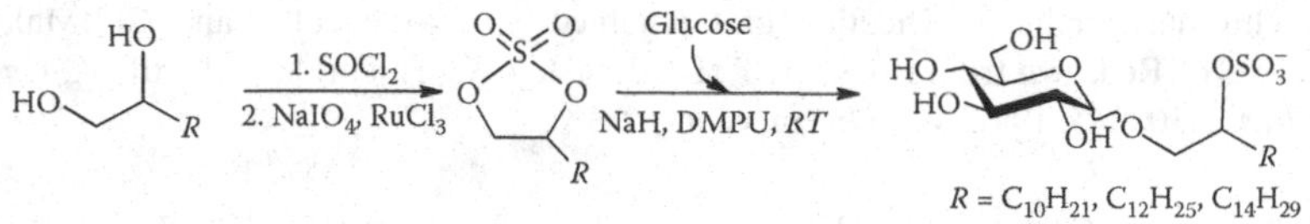

SCHEME 9.15 Synthesis of 4-alkyl-1,2,3-dioxathiolane-2,2-dioxide (-decyl, dodecyl tetradecyl). (Redrawn from Klotz, W. and Schmidt, R.R., *Synthesis*, 687, 1996. With permission.)

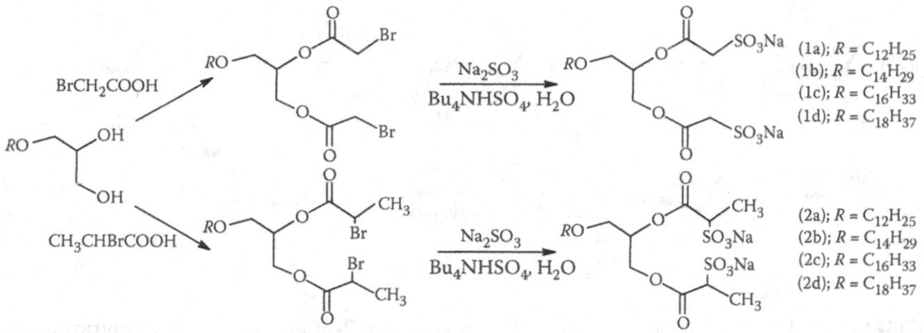

SCHEME 9.16 Synthesis of bis(sodium sulfonated ester) of 1-*O*-alkylglycerol (-dodecyl, -tetradecyl, -hexadecyl, -octadecyl). (Redrawn from Ono, D., Yamamura, S., Nakamura, M., and Takeda, T., *J. Surf. Det.*, 1, 201, 1998. With permission.)

with sodium sulfite and catalytic amounts of tetrabutylammonium bisulfate; the overall yields ranged from 60% to 65%, based on the 1-*O*-alkylglycerol employed, Scheme 9.16 [88].

Sugars with different functional groups, in particular 2-amino-D-glucose are interesting because it is possible to functionalize each group separately. For example, the large difference in nucleophilicity between the NH$_2$ and OH groups is employed to synthesize 2-acylamido-2-deoxy-D-glucopyranoses in good yields [89]. On the other hand, methyl 2-amino-2-deoxy- or methyl 2-acylamido-2-deoxy-D-glucopyranosides can be obtained by reacting the appropriate starting material (2-amino-D-glucose, or 2-acylamido-2-deoxy-D-glucopyranose) with methanol in the presence of mineral acid catalyst [18,90]. These approaches have been employed to synthesize both sulfates and sulfonates. Thus, 2-acylamido (octanamido; dodecanamido; hexadecanamido)-2,6-dideoxy-D-glucopyranose-6-sulphonates were synthesized by two alternative approaches. In the first, 2-amino-2,6-dideoxy-D-glucopyranose-6-sulphonic acid was N-acylated by the corresponding fatty acid chloride (octanoyl, dodecanoyl, hexadecanoyl) in the presence of NaHCO$_3$ in aqueous acetone. The precursor sulfonic acid was obtained by the following sequence: Acetylation of 2-aminoglucose, tosylation, oxidation in buffered solution, hydrolysis of the protecting acetyl groups, Scheme 9.17 [91].

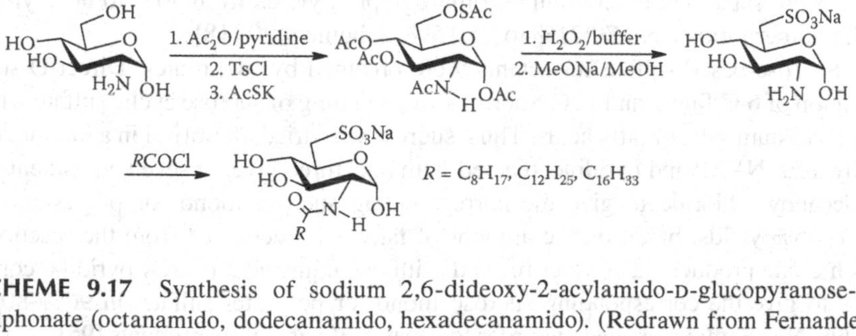

SCHEME 9.17 Synthesis of sodium 2,6-dideoxy-2-acylamido-D-glucopyranose-6-sulphonate (octanamido, dodecanamido, hexadecanamido). (Redrawn from Fernandez-Bolanos, J., Maya Castilla, I., and Fernandez-Bolanos Guzman, J., *Carbohydr. Res.*, 173, 33, 1988. With permission.)

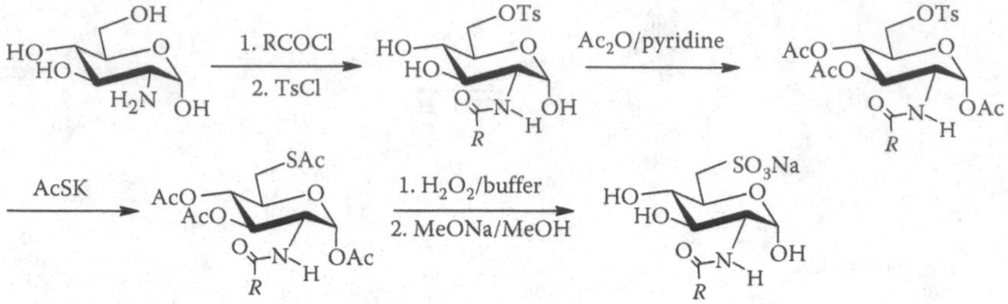

SCHEME 9.18 Synthesis of sodium 2,6-dideoxy-2-acylamido-D-glucopyranose-6-sulphonate (octanamido, dodecanamido, hexadecanamido). (Redrawn from Fernandez-Bolanos, J., Maya Castilla, I., and Fernandez-Bolanos Guzman, J., *Carbohydr. Res.*, 173, 33, 1988. With permission.)

In the second route, acylation of the aminosugar by the same fatty acid chlorides was carried out first, followed by tosylation, protection of the hydroxyl groups, nucleophilic substitution of the tosylate group, oxidation, and hydrolysis, Scheme 9.18 [91,92].

Finally, sodium 3-*O*-[6-deoxy-6-sulfo-β-D-glucopyranosyl)-1-*O*-linolenoyl-2-*O*-palmitoyl-*sn*-glycerol was obtained by the sequence of reactions shown below, Scheme 9.19 [93].

The compounds 2-acylamido-2-deoxy-D-glucopyranoses were prepared by reacting the appropriate acyl chlorides (octanoyl, dodecanoyl, and hexadecanoyl) with 2-amino-D-glucose under vigorous stirring, first at −10°C for 1 h, then at room temperature for two more hours. The solids obtained were washed with water until free of Cl⁻, dried, washed with anhydrous ethyl ether, recrystallized from absolute ethanol, and dried under reduced pressure. Mixtures of α and β anomers were obtained (72%:28%) because 2-amino-D-glucose undergoes mutarotation, similar to D-glucose [94]. The products were converted into the corresponding methyl 2-acylamido-2-deoxy-D-glucopyranosides by refluxing with anhydrous methanolic HCl (final [acid] = 1.5%) for optimized length of time. Sulfonation with SO_3-pyridine complex was carried out at −10°C, followed by neutralization, drying, and extraction with methanol. The monosulfated surfactants were separated by column chromatography; yields from 40% to 56% yield α:β anomer ratios from 79:21% to 85:15%, Scheme 9.20 [18].

Sucrose-based sulfate surfactants were obtained by two routes: Direct O-sulfonation of 6-C_nSucro and 1′-C_nSucro, or ring opening of sucrose cyclic sulfate with the potassium salts of fatty acids. Thus, sucrose was dried, dissolved in a mixture of anhydrous NMeP and pyridine, reacted with one, three, five, or seven equivalent of dodecanoyl chloride to give the corresponding sucrose mono- or polyesters in 95%–85% yields, based on the amount of fatty acid recovered from the reaction. Each crude product was further reacted with one equivalent of SO_3-pyridine complex to give the corresponding sucrose mono- or polyester sulfate, in 96%–80% yields. No attempt was made in order to separate the isomeric products [95].

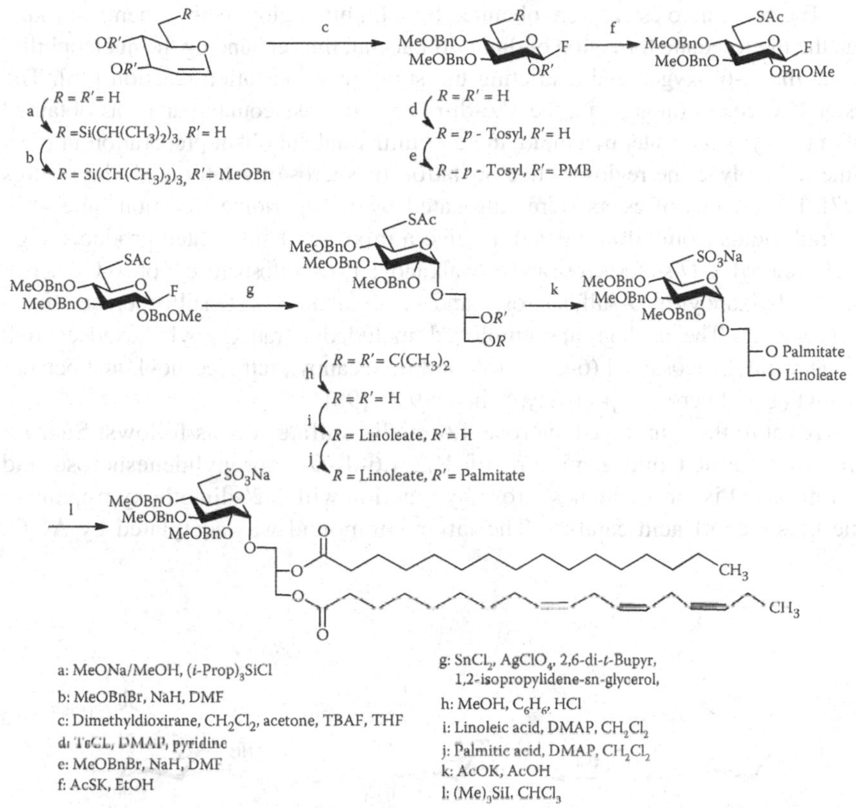

SCHEME 9.19 Synthesis of 3-*O*-[6-deoxy-6-sulfo-β-D-glucopyranosyl)-1-*O*-linolenoyl-2-*O*-palmitoyl-*sn*-glycerol. (Redrawn from Gordon, D.M. and Danishefsky, S.J., *J. Am. Chem. Soc.*, 114, 659, 1992. With permission.)

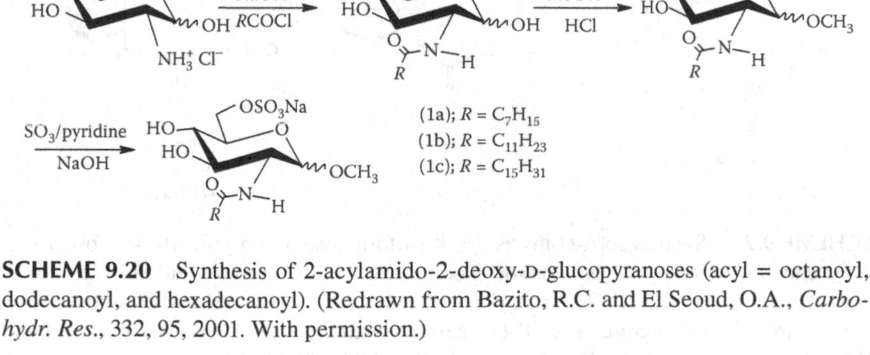

SCHEME 9.20 Synthesis of 2-acylamido-2-deoxy-D-glucopyranoses (acyl = octanoyl, dodecanoyl, and hexadecanoyl). (Redrawn from Bazito, R.C. and El Seoud, O.A., *Carbohydr. Res.*, 332, 95, 2001. With permission.)

The 6-C_nSucro esters were obtained by a highly regioselective chemical route, via the intermediate formation of dibutyltin acetal, thus enhancing the nucleophilicity at the C-6 oxygen and restricting the subsequent acylation reaction [96]. The ester 1′-C_nSucro (along with the 1′,6-di-O-acyl sucrose counterpart) was obtained by an enzymatic route, by employing a commercial subtilisin preparation in pyridine to catalyze the regioselective acylation of sucrose by fatty acid vinyl esters [97]. Both series of esters were sulfonated by SO_3/pyridine (reaction time 48 h; overall yields from 80% to 91%), to give a mixture of sulfonated products, e.g., 6-O-alkanoyl-4′-O-sulfosucrose and 6-O-alkanoyl-1′-O-sulfosucrose (from 6-C_nSucro), or 1′-O-alkanoyl-6′-O-sulfosucrose and 1′-O-alkanoyl-6-O-sulfosucrose (from 1′-C_nSucro). The acyl groups employed included tetradecanoyl, hexadecanoyl, octadecanoyl, eicosanoyl (6-C_nSucro), and dodecanoyl, tetradecanoyl, and octadecanoyl (1′-C_nSucro), respectively, Scheme 9.21 [98].

A route that employed sucrose 4,6-cyclic sulfate was as follows: Sucrose was transformed into a mixture of 1′,2:4,6-di-O-isopropylidenesucrose and 4,6-mono-O-isopropylidenesucrose by reaction with 2,2′-dimethoxypropane in the presence of acid catalyst. The latter compound was acetylated by Ac_2O;

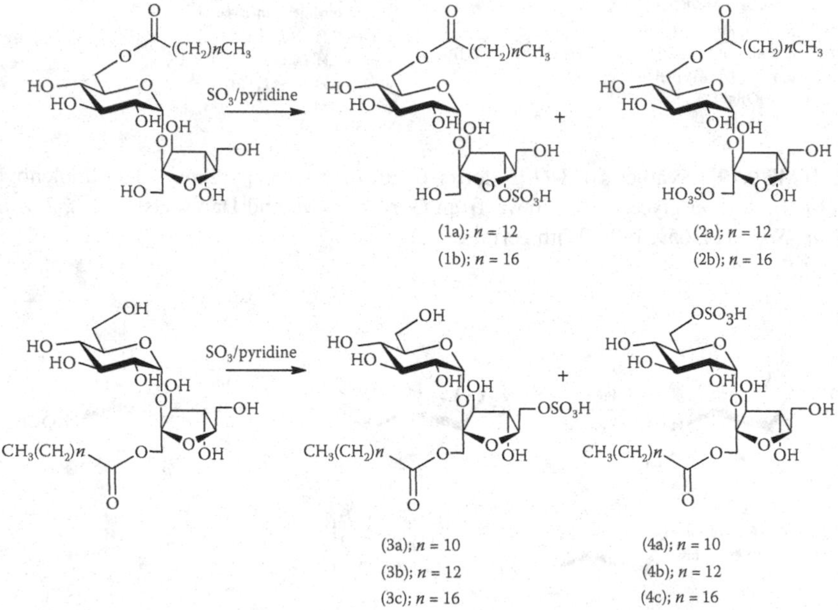

SCHEME 9.21　Synthesis of 6-O-myristoyl-4′-sulfosucrose, 6-O-myristoyl-1′-sulfosucrose, 6-O-stearoyl-4′-sulfosucrose, 6-O-stearoyl-1′-sulfosucrose, 1′-O-lauroyl-6′-sulfosucrose, 1′-O-lauroyl-6-sulfosucrose, 1′-O-myristoyl-6′-sulfosucrose, 1′-O-myristoyl-6-sulfosucrose, 1′-O-stearoyl-6′-sulfosucrose, and 1′-O-stearoyl-6-sulfosucrose. (Redrawn from Bazin, H.G., Polat, T., and Linhardt, R.J., *Carbohydr. Res.*, 309, 189, 1998. With permission.)

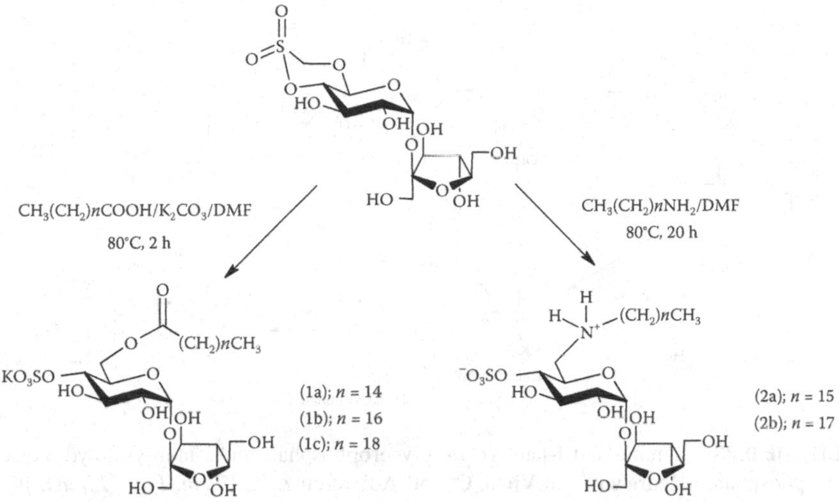

SCHEME 9.22 Synthesis of 6-O-palmitoyl-4-sulfosucrose, 6-O-stearoyl-4-sulfosucrose, 6-O-eicosanoyl-4-sulfosucrose, 6-O-hexadecylamino-4-sulfosucrose, and 6-O-octadecylamino-4-sulfosucrose. (Redrawn from Bazin, H.G., Polat, T., and Linhardt, R.J., *Carbohydr. Res.*, 309, 189, 1998. With permission.)

reacted with $SOCl_2$; the cyclic sulfite ring was oxidized to cyclic sulfate with $NaIO_4$ in the presence of $RuCl_3$ catalyst; de-acetylated by reaction with Et_3N/MeOH to give sucrose 4,6-cyclic sulfate. The required surfactants were synthesized by ring opening of the latter with fatty acids (hexadecanoic; octadecanoic; eicosanoic). Zwitterionic surfactants were obtained by ring opening of the same sulfate with hexadecylamine and octadecylamine. The surface activity properties of the sucrose monoesters obtained were determined and compared with those of commercially available ionic and nonionic surfactants, Scheme 9.22 [98].

9.2.3.3 Phosphates

Enzyme catalysis has been employed to synthesize phosphate surfactants based on glycerol and 1,3-dihydroxyacetone. Thus microbial phospholipase D catalyzes the transfer reaction of the dipalmitoyl-phosphatidyl residue of 1,2-dipalmitoyl-3-sn-phosphatidylcholine to dihydroxyacetone in a biphasic system, to yield 1,2-dipalmitoyl-3-sn-phosphatidyldihydroxyacetone [99]. The same enzyme catalyzes the transphosphatidylation reaction of secondary alcohols, and βGlucoPy, with natural phosphatidyl choline in water/ethyl acetate buffered emulsion system [100].

Transphosphatidylation of 1-monolauroyl-rac-glycerol or 1-monolauroyl-dihydroxyacetone with phosphatidylcholine was carried out efficiently in the

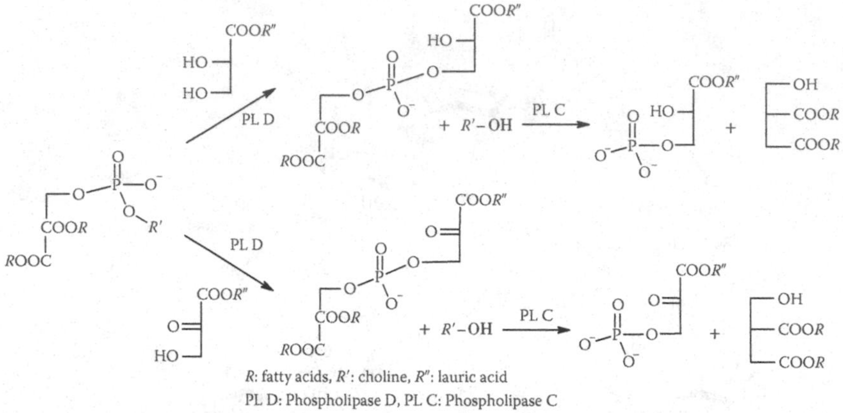

R: fatty acids, R': choline, R": lauric acid
PL D: Phospholipase D, PL C: Phospholipase C

SCHEME 9.23 Synthesis of 1-lauroyl-*rac*-glycerophosphate and 1-lauroyl-dihydroxyacetonephosphate. (Redrawn from Virto, C. and Adlercreutz, P., *Chem. Phys. Lipids*, 104, 175, 2000. With permission.)

presence of phospholipase D (from *Streptomyces* sp.) in a two-phase system (ether/aqueous buffer). Phospholipase C (from *Bacillus cereus*) was then employed in order to hydrolyze 1-lauroyl-phosphatidylglycerol or 1-lauroyl-phosphatidyldihydroxyacetone into the corresponding 1-lauroyl-*rac*-glycerophosphate and 1-lauroyl-dihydroxyacetonephosphate, respectively, Scheme 9.23 [101].

The compound 1,2,3,4-tetra-*O*-acetyl-D-glucopyranose-6-phosphate was prepared by acetylation of commercially available glucose phosphate. Reaction of the latter with hexadecanol in the presence of pyridine and trichloroacetonitrile, followed by chromatographic purification gave 1,2,3,4-tetra-*O*-acetyl-D-glucopyranosyl-hexadecyl phosphate; from which the surfactant 6-D-glucopyranosyl-hexadecyl phosphate was obtained by alkaline hydrolysis with MeONa/MeOH [102].

A series of alkyl- and partially fluorinated alkyl phosphate esters of glucose has been synthesized by several routs. Thus 1,2,3,4-tetra-*O*-acetyl-β-D-glucopyranoside was phosphorylated by the reaction with $ROPOCl_2$ ($R = C_6F_{13}C_2H_4-,C_8F_{17}C_2H_4-$, and $C_{10}H_{21}-$) in chloroform, in the presence of Et_3N. The phosphate diester produced was hydrolyzed to the corresponding monoester by reaction with triethylammonium hydrogencarbonate; treatment of the product with NaOMe/MeOH gave the required D-glucose-6-sodium alkyl- or partially fluorinated alkyl phosphate. Phosphorylation of the protected tetra-*O*-acetyl-β-D-glucopyranoside has been also carried out with alkyl phosphoroditriazolide, prepared *in situ* from the reaction of $C_6F_{13}C_2H_4O$-$POCl_2$ and 1*H*-,2,4-triazole in THF, in the presence of Et_3N. The workup for obtaining D-glucose-6-sodium tridecafluorooctyl phosphate was similar to that described above. The triazole-based reagent has been employed for the phosphorylation of 1,2:5,6-di-*O*-isopropylidene-α-D-glucofuranose; hydrolysis of the intermediate

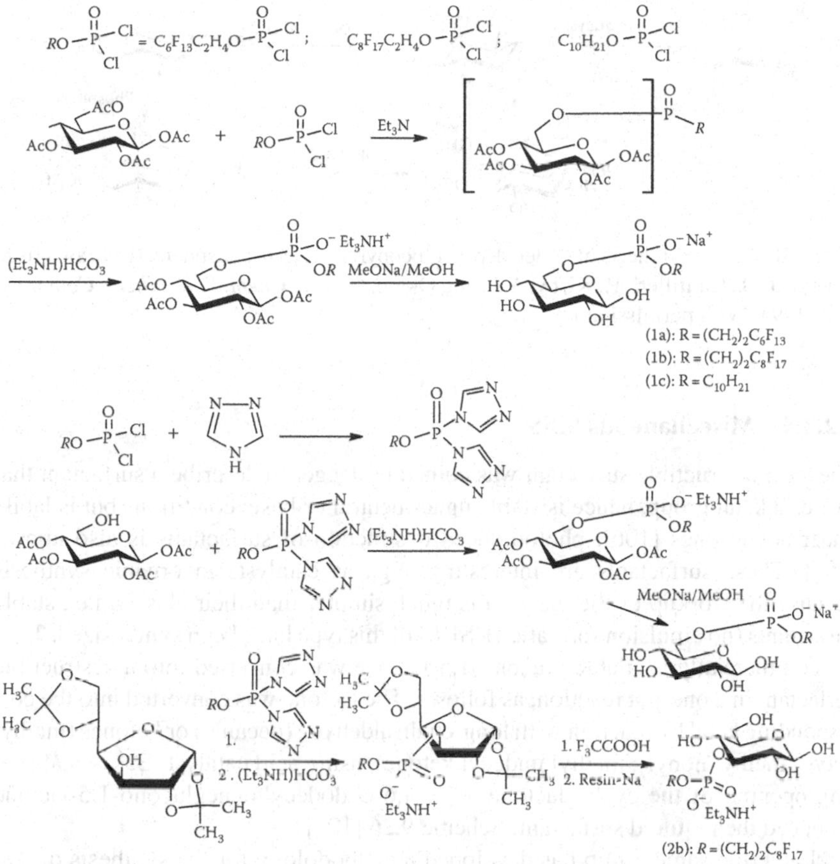

SCHEME 9.24 Synthesis of D-glucose-6-sodium tridecafluorooctyl-; heptadecafluoro-decyl- and decylphosphate, and D-glucose-3-sodium heptadecafluorodecyl phosphate. (Redrawn from Milius, A., Greiner, J., and Riess, J.G., *Carbohydr. Res.*, 229, 323, 1992. With permission.)

1,2:5,6-di-*O*-isopropylidene-α-D-glucofuranose 3-(triethylammonium heptadecaflu-orodecyl phosphate) to the corresponding (thermodynamic product) D-glucose 3-(sodium heptadecafluorodecyl phosphate) was carried out under acidic conditions, Scheme 9.24 [103].

Dodecyl-β-D-glucopyranoside-6-phosphate was obtained by the following sequence of reactions: C_{12}βGlucoPy was prepared by the reaction of 1-dodecanol with 1-bromo-1-deoxy-2,3,4,6-tetra-*O*-acetyl-β-D-glucopyranoside, followed by removal of the blocking (acetyl) groups [104]. Selective mono phosphorylation of the latter was obtained by the reaction with dibenzyl diisopropylamino phospho-ramidate. The reaction product was transformed in the SBIS, in quantitative yield, by catalytic hydrogenation on Pd/C/THF, Scheme 9.25 [105].

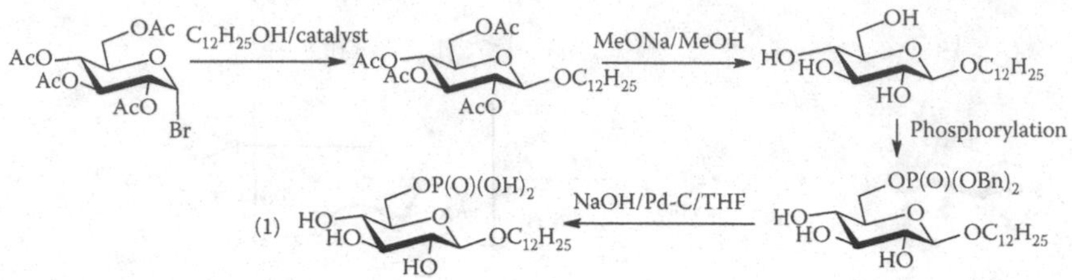

SCHEME 9.25 Synthesis of dodecyl-β-D-glucopyranoside-6-phosphate. (Redrawn from Jones, R.F.D., Camilleri, P., Kirby, A.J., and Okafo, G.N., *J. Chem. Soc., Chem. Commun.*, 1311, 1994. With permission.)

9.2.3.4 Miscellaneous SBIS

The term destructible surfactant was coined by Jaeger to describe a surfactant that carries a ketal group, hence is stable under neutral or basic conditions but is labile under acidic ones [106]; photoinduced destruction of surfactants is also known [107]. These surfactants are interesting, e.g., as catalysts in organic synthesis because the workup of the reaction is much simpler than their classic, i.e., stable surfactants (no emulsion formation). SBIS of this type have been synthesized. Thus the commercially available glucono-1,5-lactone was converted into a destructible surfactant in a one-pot reaction, as follows: The lactone was converted into the corresponding ketal by reaction with long-chain aldehyde (decanal) or ketones (methyl octyl-, methyl nonyl-; methyl undecyl ketone), using acid catalyst. Base-catalyzed ring opening of the cyclic lactone, e.g., 4,6-O-dodecylideneglucono-1,5-lactone produced the required surfactant, Scheme 9.26 [108].

Later, the same group has developed a methodology for the synthesis of two anionic and one cationic D-glucosamine-derived destructible surfactants: sodium methyl 4,6-O-alkylidene-2-(carboxylatomethylamino)-2-deoxy-D-glucopyranoside; methyl 4,6-O- alkylidene-2-deoxy-2-(trimethylamino)-D-glucopyranoside iodide; and sodium methyl 2-acetamide-4,6-O-alkylidene-3-O-[1-(caboxylato)ethyl]-2-deoxy-D-glucopyranoside.

The following are the examples of the synthetic steps: 1,1-dimethoxyalkane was obtained by the acid-catalyzed condensation of alkanal with methyl orthoformate; *N*-acetyl-D-glucosamine was obtained by acetylation of the corresponding

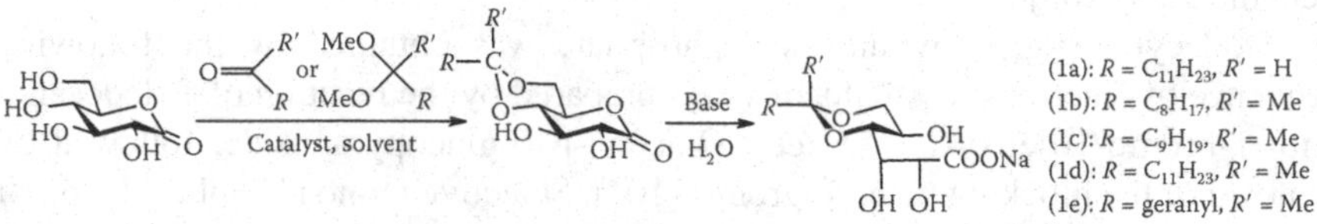

(1a): R = $C_{11}H_{23}$, R' = H
(1b): R = C_8H_{17}, R' = Me
(1c): R = C_9H_{19}, R' = Me
(1d): R = $C_{11}H_{23}$, R' = Me
(1e): R = geranyl, R' = Me

SCHEME 9.26 Synthesis of a destructible surfactant. (Redrawn from Kida, T., Masuyama, A., and Okahara, M., *Tetrahedron Lett.*, 31, 5939, 1990. With permission.)

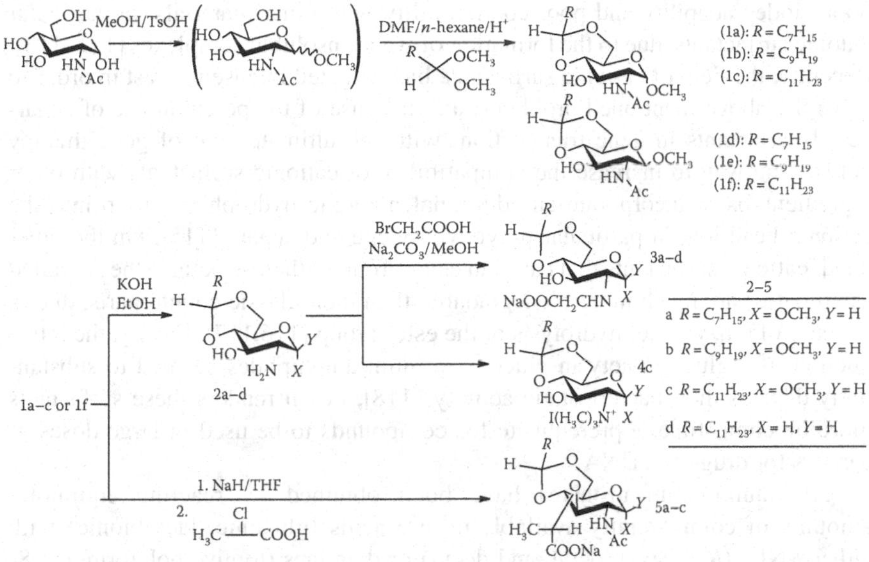

SCHEME 9.27 Synthesis of sodium methyl 4,6-*O*-alkylidene-2-(carboxylatomethylamino)-2-deoxy-D-glucopyranoside; methyl 4,6-*O*-alkylidene-2-deoxy-2-(trimethylamino)-D-glucopyranoside iodide; and sodium methyl 2-acetamide-4,6-*O*-alkylidene-3-*O*-[1-(caboxylato)ethyl]-2-deoxy-D-glucopyranoside. (Redrawn from Kida, T., Yurugi, K., Masuyama, A., Nakatsuji, Y., Ono, D., and Takeda, T., *J. Am. Oil Chem. Soc.*, 72, 773, 1995. With permission.)

amine with acetic anhydride; 2-acetamide-4,6-*O*-alkylidene-2-deoxy-D-glucopyranoside was obtained by the acid-catalyzed condensation of *N*-acetyl-D-glucosamine with the appropriate aliphatic aldehyde dimethyl acetal. Reaction of the condensation product with 2-chloropropionic acid gave an anionic surfactant, whereas its deacetylation followed by reaction with either bromoacetic acid or methyl iodide gave anionic-, and cationic surfactants, respectively, Scheme 9.27 [109].

Finally, novel amphiphiles were prepared under homogenous solution conditions by benzylation of pea, potato, and waxy maize starch followed by the introduction of carboxymethyl (anionic) or hydroxypropyltrimethylammonium (cationic) moieties. Their surface active properties depend on the DS values of the hydrophobic (benzyl) and hydrophilic (charged) groups, surface tensions of 36.8 and 48.8 mN/m were measured for the anionic and cationic derivative, respectively [110].

9.2.3.5 Cationic SBIS

Classic cationic surfactants, e.g., tetraalkylammonium halides and alkylpyridinium halides are employed, *inter alia*, as fabric softeners, hair conditioners, corrosion inhibitors, and antimicrobial agents. Unfortunately, most classic cationics show poor biodegradability and poor compatibility with other materials, in particular anionic surfactants, due to the formation of water insoluble complexes [111–113]. Recently, the field of cationic surfactants has attracted intense interest in order to solve the above-mentioned problems, and because of the potential use of sugar-based surfactants in gene transfection, with the ultimate goal of gene therapy [114]. One way to increase the compatibility of cationic surfactants with other ingredients is to incorporate an additional nonionic hydrophilic group into the cationic head-ion, in particular polyoxyethylene and sugars [115]. On the other hand, cationic surfactants that carry an ester group in their structure (the so-called esterquates) are much more biodegradable than their classic counterparts, due to the ease of (enzymatic) hydrolysis of the ester group [116,117]. Finally, the introduction of a gluco moiety in glucopyridinium amphiphiles seemed to substantially depress their bacteriostatic activity [118], i.e., it renders these surfactants more biocompatible, a prerequisite for compounds to be used in large doses as carriers for drugs and DNA.

Alkylammonium uronates have been obtained by reacting equimolar amounts of commercially available uronic acids (gluconic, lactobionic) with either RNH_2 (R = hexyl, octyl, and decyl) or diamines (to give bolaforms), 1,8-diaminooctane; 1,10-diaminodecane; 1,12-diaminododecane. Water was employed as a solvent, the reaction was carried out at room temperature, at pH 7; the products obtained were purified by freeze drying. The residual uronic acid in the product ranged from 0.9% (lactobionic acid) to 3.5% (gluconic acid), Figure 9.7 [119].

An early synthesis of a cationic sugar-based surfactant, 1-*O*-dodecyl-6-deoxy-6-dialkylammonium-D-glucopyranoside chloride, was carried out by reacting unprotected $C_{12}\beta$GlucoPy with tosyl chloride in the presence of triethylamine at 0°C, followed by nucleophilic substitution of the tosylate group with methanolic solution of dialkylamine, the product was neutralized with methanolic HCl; no synthetic details were given, Scheme 9.28 [49].

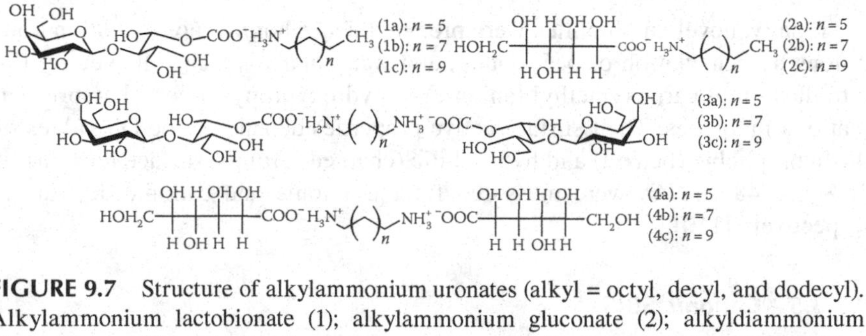

FIGURE 9.7 Structure of alkylammonium uronates (alkyl = octyl, decyl, and dodecyl). Alkylammonium lactobionate (1); alkylammonium gluconate (2); alkyldiammonium dilactobionate (3); and alkyldiammonium digluconate (4). (Redrawn from Moon, T.L., Blanzat, M., Labadie, L., Perez, E., and Rico-Lattes, I., *J. Dispersion Sci. Technol.*, 22, 167, 2001. With permission.)

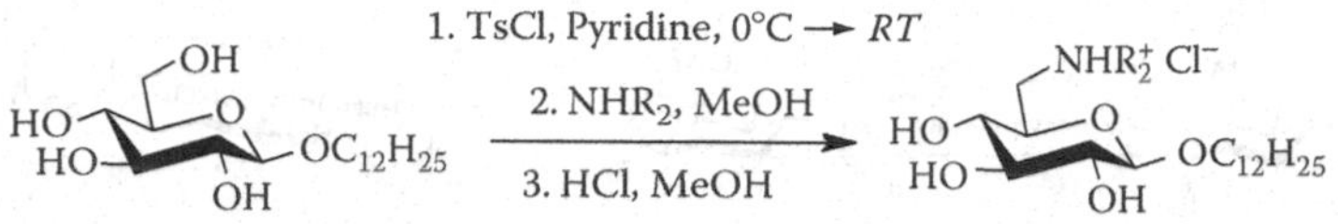

SCHEME 9.28 Synthesis of 1-*O*-dodecyl-6-deoxy-6-dialkylammonium-D-glucopyranoside chloride. (Redrawn from Boecker, T. and Thiem, *J.*, *Tenside, Surf., Det.*, 26, 318, 1989. With permission.)

A key intermediate, required for the synthesis of glucocationic surfactants is 2-bromoethyl-2,3,4,6-tetraacetyl-β-D-glucopyranoside. This was obtained in good yields (54% or 71%), with control of stereoselectivity (β-anomer) by reacting either 2,3,4,6-tetra-*O*-acetyl-α-D-glucopyranosyltrichloroacetimidate or penta-*O*-acetyl-β-glucopyranose with 2-bromoethanol in CH$_2$Cl$_2$, in the presence of BF$_3$. etherate, followed by anomerization by treatment with anhydrous FeCl$_3$/CH$_2$Cl$_2$. The product was then quaternized by treatment with *N*-alkyl-*N*,*N*-dimethylamine (alkyl = octyl, dodecyl, and hexadecyl) in ethanol, followed by deacetylation by treatment with MeONa/MeOH, Scheme 9.29 [120].

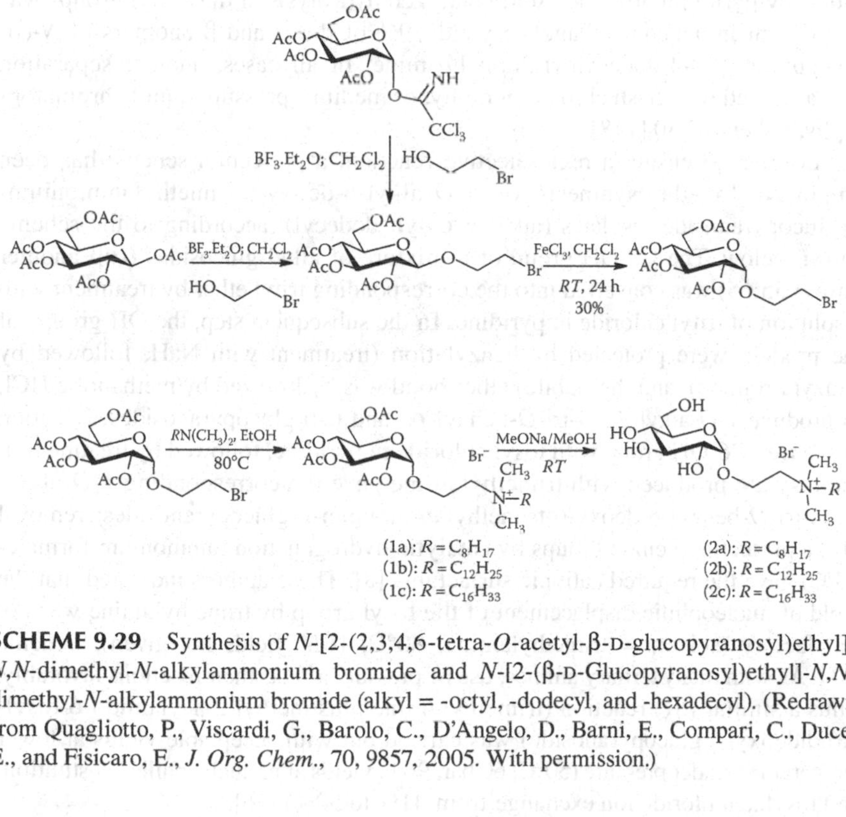

SCHEME 9.29 Synthesis of *N*-[2-(2,3,4,6-tetra-*O*-acetyl-β-D-glucopyranosyl)ethyl]-*N*,*N*-dimethyl-*N*-alkylammonium bromide and *N*-[2-(β-D-Glucopyranosyl)ethyl]-*N*,*N*-dimethyl-*N*-alkylammonium bromide (alkyl = -octyl, -dodecyl, and -hexadecyl). (Redrawn from Quagliotto, P., Viscardi, G., Barolo, C., D'Angelo, D., Barni, E., Compari, C., Duce, E., and Fisicaro, E., *J. Org. Chem.*, 70, 9857, 2005. With permission.)

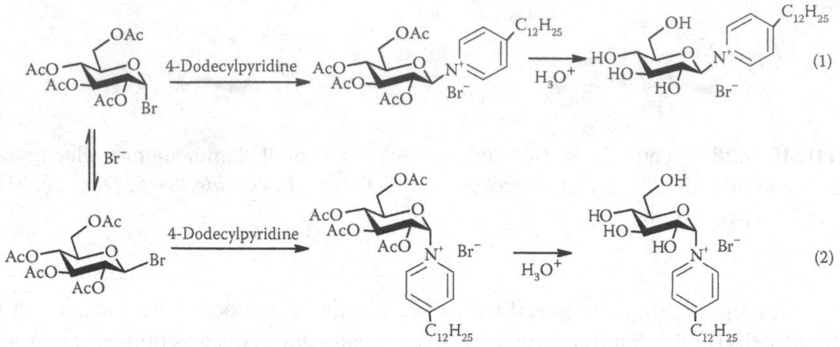

SCHEME 9.30 Synthesis of α and β anomers of N-(tetra-O-acetyl-D-glucopyranosyl)-4-dodecylpyridium bromide. (Redrawn from Viscardi, G., Quagliotto, P., Barolo, C., Savarino, P., Barni, E., and Fisicaro, E., *J. Org. Chem.*, 65, 8197, 2000. With permission.)

Dodecylpyridinium bromide-based surfactants that carries a sugar moiety attached to the heterocyclic ring have been synthesized. Thus the reaction of 4-dodecylpyridine with 2,3,4,6-tetra-O-acetyl-α-bromoglucose resulted in 71% yield of (20:80) of the α and β anomers of N-(tetra-O-acetyl-D-glucopyranosyl)-4-dodecylpyridium bromide. Acid-catalyzed hydrolysis of the acetyl groups was carried out in aqueous ethanol to yield 79% of the α and β anomers of N-(D-glucopyranosyl)-4-dodecylpyridium bromide. In all cases, anomer separation was achieved with flash chromatography, or medium-pressure liquid chromatography, Scheme 9.30 [118].

In order to ensure a regioselective reaction, a protection scheme has been employed for the synthesis of 1-O-alkyl-6-deoxy-6-trimethylammonium-D-glucopyranoside tosylates (alkyl = octyl; dodecyl), according to the scheme shown below. The C6-OH group of a mixture of alkyl glucosides (α:β anomer ratio = 35:65) was converted into the corresponding trityl ether by treatment with a solution of trityl chloride in pyridine. In the subsequent step, the OH groups of the product were protected by benzylation (treatment with NaH, followed by benzyl bromide), and the (labile) ether bond was hydrolyzed by methanolic HCl, to produce 1-O-alkyl-2,3,4-tri-O-benzyl-α- and β-D-glucopyranosides. Reaction of the free C6-OH group with tosyl chloride in pyridine, followed by treatment of the tosylate produced with trimethylamine gave the corresponding 1-O-alkyl-2,3,4-tri-O-benzyl-6-deoxy-6-trimethylammonium-D-glucopyranosides; removal of the protecting benzyl groups by catalytic hydrogenation (ammonium formate-Pd/C) gave the required cationic surfactants [8]. These authors indicated that the yield of nucleophilic displacement of the tosyl group by trimethylamine was not satisfactory (solvent = trimethylamine, 80°C, 16h; yields not given), Scheme 9.31. As with many tertiary amines, use of pressure affects their reactions favorably. Thus a similar (S_N) reaction (trimethylamine plus methyl 2-acylamido-6-tosyl-2,6-dideoxy-D-glucopyranoside) was carried out with acceptable yields at lower temperature, under pressure (50°C, 60 bar, 96 h; yields after nucleophilic substitution, and tosylate/chloride ion exchange from 41% to 50%) [16].

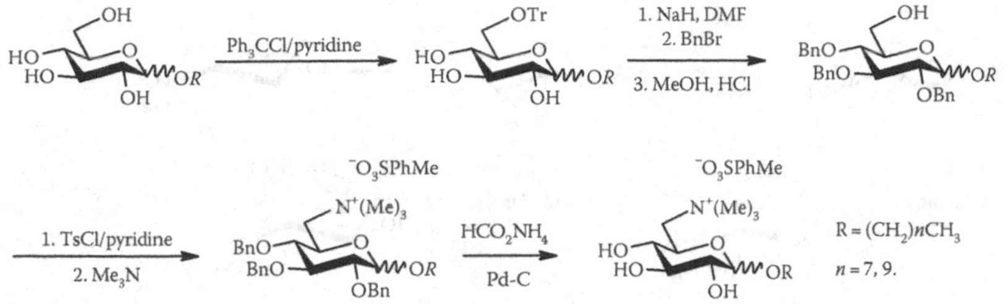

SCHEME 9.31 Synthesis of 1-*O*-octyl-6-deoxy-6-trimethylammonium-D-glucopyranoside tosylates and 1-*O*-dodecyl-6-deoxy-6-trimethylammonium-D-glucopyranoside tosylates. (Redrawn from Focher, B., Savelli, G., and Torri, G., *Chem. Phys. Lipids*, 53, 141, 1990. With permission.)

Two favorable structural features (presence of a sugar moiety and an ester group) have been incorporated in a novel series of sugar-based cationic surfactants. Acid-catalyzed condensation of α-D-glucose with 3-chloro-1-propanol produced a mixture of the α- and β-anomers of 3-chloropropyl-D-glucopyranoside, in 84% yield. Immobilized B-lipase from *Candida antartica* was employed to catalyze the esterification of fatty acids (decanoic, dodecanois, tetradecanoic, hexadecanois, and octadecanois) with the previous product. The cationic surfactants were obtained by the treatment of the 3-chloropropyl-6-*O*-alkanoyl-D-glucopyranosides with NaI in acetone, filtration of the NaCl formed, and treatment with trimethylamine at 0°C, Scheme 9.32 [121].

Another series of sugar-based cationic surfactants is that based on 2-aminoglucose. The publication of a recent review on the synthetic strategies to obtain (short-chain) *O*-glycosides of 2-amino-2-deoxysugars should encourage more work on the use of this starting material because, in principle, several of the schemes shown for short-chain compounds can be adapted in order to obtain sugar-based cationic surfactants [122]. The above-mentioned amino sugar was reacted with the appropriate acyl chloride (octanoyl, dodecanoyl, and hexadecanoyl)

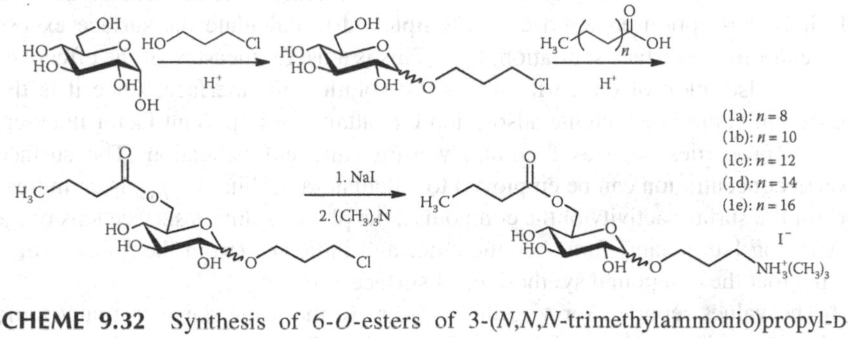

SCHEME 9.32 Synthesis of 6-*O*-esters of 3-(*N*,*N*,*N*-trimethylammonio)propyl-D-glucopyranoside. (6-*O*-acyl side chain are decanoyl, dodecanoyl, tetradecanoyl, hexadecanoyl, and octadecanoyl). (Redrawn from Kirk, O., Pedersen, F.D., and Fuglsang, C.C., *J. Surf. Det.*, 1, 37, 1998. With permission.)

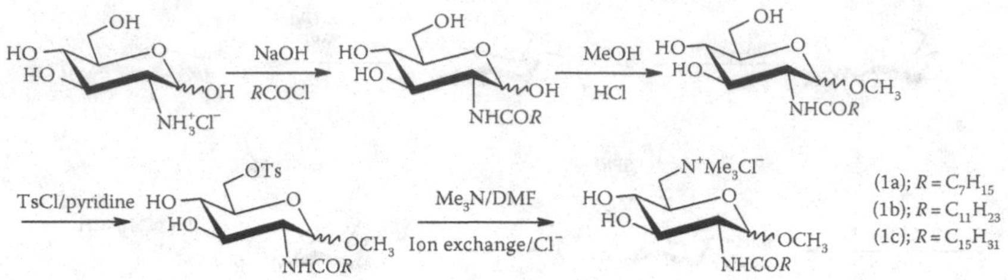

SCHEME 9.33 Synthesis of 2-acylamido-6-trimethylammonio-2,6-dideoxy-D-glucopy-ranoside chlorides (acyl = octyl, dodecyl, and hexadecyl). (Redrawn from Bazito, R.C. and El Seoud, O.A., *J. Surf. Det.*, 4, 395, 2001. With permission.)

to yield acylamido-D-glucoses. The products obtained were converted into pre-dominantly α-anomers (81% ± 2%) methyl 2-acylamido-2-deoxy-D-glucopyranosides by reaction with anhydrous methanolic HCl solution, followed by neutralization and purification. Reaction with tosyl chloride in pyridine, followed by reaction with trimethylamine/DMF under pressure (60 bar, 50°C, 96 h), and ion exchange on macro-porous resin in the chloride form furnished the surfactants methyl 2-acylamido-6-trimethylammonio-2,6-dideoxy-D-glucopyranoside chlorides; yields 41%, 50%, 44%; α:β anomer ratio 89:11, 84:16, and 74:26, for the C_8, C_{12}, and C_{16} derivatives, respectively, Scheme 9.33 [16].

9.3 PHYSICOCHEMICAL PROPERTIES OF AQUEOUS SOLUTIONS OF SBIS

Table 9.1 shows the reported solution properties of SBIS, as well those of a few of the precursor SBNS.

Because the properties shown in Table 9.1 are similar to those of solutions of classic ionic surfactants, the same experimental techniques are employed for both classes of surfactants, in particular, surface tension, conductivity, fluorescence, and light scattering; the same information about the aggregates formed can be calculated from the data obtained. Thus measurement of surface tension and use of Gibbs adsorption equation can be employed to calculate the surface excess concentration at surface saturation, Γ_{max}. This is a useful measure of the effective-ness of adsorption of the surfactant at the solution/air interface, since it is the maximum value to which the adsorption can attain, an important factor in deter-mining properties such as foaming, wetting, and emulsification. The surface excess concentration can be employed to calculate A_{min}. Finally, γ_{cmc} is an indica-tion of the surface activity of the compound; the plot of solution surface tension, γ, versus log [surfactant] furnishes the cmc, and indicates (from the absence of a "dip") that the compound synthesized is surface active pure.

The values reported for γ_{cmc} and Γ_{max} are in the same range of many ionic surfactants; whereas those of A_{min} are higher than those reported for simple ionic surfactants, because of the bulkiness of the sugar moiety [125,126]. Several A_{min}

TABLE 9.1
Solution Properties of SBIS and Some SBS

Entry	Scheme Number; Compound Number	Number of C in Hydrophobic Chain; or in the Two Chains	T, °C; cmc or cac, mmol/L	Other Information[a] γ_{cmc}, mN/m; Γ_{max}, mol/cm²; A_{min}, nm²; α_{mic}, N_{agg}	Ref.
			Cationic and Zwitterionic		
1	Scheme 9.1; (1)	12	25; 3.51	γ_{cmc} = ~35[c]; A_{min} = 0.35–0.45	[43]
2	Scheme 9.1; (2)	12	25; 3.15	γ_{cmc} = ~25[c]; A_{min} = 0.35–0.45	[43]
3	Scheme 9.1; (3)	12	25; 3.97	γ_{cmc} = ~31[c]; A_{min} = 0.35–0.45	[43]
4	Scheme 9.2; (1b)	7; 10	25; 2.1	γ_{cmc}, ~20[c]	[44]
5	Scheme 9.2; (1c)	15; 6	25; 0.17	γ_{cmc}, ~26[c]	[44]
6	Scheme 9.2; (1d)	15; 10	25; 0.01	γ_{cmc}, ~35[c]	[44]
7	Scheme 9.2; (2a)	7; 12	25; 2.5	γ_{cmc}, ~37[c]	[44]
8	Scheme 9.2; (2b)	15; 12	25; 0.01	γ_{cmc}, ~26[c]	[44]
9	Scheme 9.4; (1)	11	25; 1.0	γ_{cmc} = 20; A_{min} = 0.66	[48]
10	Scheme 9.9; (1a)	4; 16	NS[b]; 1.0	γ_{cmc}, ~38[c]	[58]
11	Scheme 9.9; (1b)	8; 16	NS[b]; 0.22	γ_{cmc}, ~29[c]	[58]
12	Scheme 9.9; (1c)	12; 16	NS[b]; 0.04	γ_{cmc}, ~22[c]	[58]
13	Scheme 9.22; (2a)	16	NS[b]; 0.014		[98]
14	Figure 9.7; (1a)	8	25; 34.6		[119]
15	Figure 9.7; (1b)	10	25; 5.89		[119]
16	Figure 9.7; (1c)	12	25; 1.78		[119]
17	Figure 9.7; (2b)	10	25; 15.8		[119]
18	Figure 9.7; (2c)	12	25; 5.0		[119]

(continued)

TABLE 9.1 (continued)
Solution Properties of SBIS and Some SBS

Entry	Scheme Number; Compound Number	Number of C in Hydrophobic Chain; or in the Two Chains	T, °C; cmc or cac, mmol/L	Other Information γ_{cmc} mN/m; Γ_{max} mol/cm²; A_{min} nm²; α_{mic}, N_{agg}	Ref.
			Anionic		
19		12	25; 5.2		[75]
20	Scheme 9.7; (1a)	8	1.62	$\gamma_{cmc} = 30.2$	[55]
21	Scheme 9.7; (1c)	12	0.28	$\gamma_{cmc} = 29.5$	[55]
22	Scheme 9.7; (2c)	12	0.13	$\gamma_{cmc} = 30.5$	[55]
23	Scheme 9.7; (2d)	14	0.16	$\gamma_{cmc} = 29.7$	[55]
24	Scheme 9.13; (1)	12	NS^b, 0.08	$\gamma_{cmc} = 36.0$	[17]
25	Scheme 9.13; (2b)	12	NS^b, 11.1	$\gamma_{cmc} = 34.5$	[17]
26	Scheme 9.13; (3b)	12	NS^b, 1.8	$\gamma_{cmc} = 60.0$	[17]
27	Scheme 9.13; (4b)	12	NS^b, 3.6	$\gamma_{cmc} = 40.5$	[17]
28	Scheme 9.13 (5a)	12	NS^b, 1.6	$\gamma_{cmc} = 62.0$	[17]
29	Scheme 9.13; (5b)	12	NS^b, 1.8	$\gamma_{cmc} = 50.0$	[17]
30	Scheme 9.13; (6)	12	NS^b, 13.2	$\gamma_{cmc} = 38.5$	[17]
31	Scheme 9.13; (7)	12	NS^b, 0.072	$\gamma_{cmc} = 28.0$	[17]
32	Scheme 9.13; (8b)	12	NS^b, 11.1	$\gamma_{cmc} = 34.5$	[17]
33	Scheme 9.13; (9a)	12	NS^b, 1.6	$\gamma_{cmc} = 62.0$	[17]
34	Scheme 9.13; (9b)	12	NS^b, 18.1	$\gamma_{cmc} = 42.0$	[17]
35	Scheme 9.13; (10)	14	NS^b, 0.015	$\gamma_{cmc} = 34.5$	[17]
36	Scheme 9.13; (11)	14	NS^b, 1.6	$\gamma_{cmc} = 48.5$	[17]
37	Scheme 9.14; (1a)	12	25; 0.43		[80]
38	Scheme 9.14; (1b)	14	25; 0.067		[80]
39	Scheme 9.14; (1c)	16	25; 0.034		[80]

#	Compound	n			Ref.
40	Scheme 9.16; (1a)	12	25: 0.90	$\gamma_{cmc} = 36.0$	[88]
41	Scheme 9.16; (1b)	14	25: 0.10	$\gamma_{cmc} = 36.5$	[88]
42	Scheme 9.16; (1c)	16	25: 0.052	$\gamma_{cmc} = 36.2$	[88]
43	Scheme 9.16; (1d)	18	25: 0.020	$\gamma_{cmc} = 37.3$	[88]
44	Scheme 9.16; (2a)	12	25: 1.0	$\gamma_{cmc} = 37.3$	[88]
45	Scheme 9.16; (2b)	14	25: 0.19	$\gamma_{cmc} = 37.0$	[88]
46	Scheme 9.16; (2c)	16	25: 0.064	$\gamma_{cmc} = 37.0$	[88]
47	Scheme 9.16; (2d)	18	25: 0.023	$\gamma_{cmc} = 38.0$	[88]
48	Scheme 9.20; (1a)	8	25, 24.4; 40; 24.9	$\gamma_{cmc} = 35.3; A_{min} = 0.94; \alpha_{mic} = 0.33; N_{agg} = 109$	[18]
49	Scheme 9.20; (1b)	12	25; 1.69; 40; 1.72	$\gamma_{cmc} = 34.7; A_{min} = 0.90; \alpha_{mic} = 0.24; N_{agg} = 125$	[18]
50	Scheme 9.20; (1c)	16	40, 0.012	$\gamma_{cmc} = 33.2; A_{min} = 0.92; \alpha_{mic} = 0.17; N_{agg} = 794$	[18]
51	Scheme 9.21; (1a)	14	NS^b: 0.053		[98]
52	Scheme 9.21; (1b)	18	NS^b: 0.0033		[98]
53	Scheme 9.21; (2a)	14	NS^b: 0.059		[98]
54	Scheme 9.21; (2b)	18	NS^b: 0.0038		[98]
55	Scheme 9.21; (3a)	12	NS^b: 0.065		[98]
56	Scheme 9.21; (3b)	14	NS^b: 0.048		[98]
57	Scheme 9.21; (4a)	12	NS^b: 0.071		[98]
58	Scheme 9.21; (4b)	14	NS^b: 0.052		[98]
59	Scheme 9.21; (3c)	18	NS^b: 0.0047		[98]
60	Scheme 9.21; (4c)	18	NS^b: 0.0056		[98]
61	Scheme 9.22; (1a)	16	NS^b: 0.048		[98]
62	Scheme 9.22; (1b)	18	NS^b: 0.011		[98]
63	Scheme 9.24; (1a)	8	20: 1.55		[124]
64	Scheme 9.24; (2b)	10	20; 1.68		[124]
65	Scheme 9.24; (1c)	10	20; 8.11		[124]
66	Scheme 9.25; (1)	12	NS^b: 0.25		[105]

(continued)

TABLE 9.1 (continued)
Solution Properties of SBIS and Some SBS

Entry	Scheme Number; Compound Number	Number of C in Hydrophobic Chain; or in the Two Chains	T, °C; cmc or cac, mmol/L	Other Information γ_{cmc} mN/m; Γ_{max} mol/cm²; A_{min} nm²; α_{mic} N_{agg}	Ref.
67	Scheme 9.27; (3a)	7	20; 1.9	$\gamma_{cmc} = 29$; $A_{min} = 0.51$	[109]
68	Scheme 9.27; (3b)	9	20; 0.58	$\gamma_{cmc} = 28$; $A_{min} = 0.5$	[109]
69	Scheme 9.27; (3c)	11	20; 0.022	$\gamma_{cmc} = 27$; $A_{min} = 0.5$	[109]
70	Scheme 9.27; (3d)	11	20; 0.045	$\gamma_{cmc} = 28$; $A_{min} = 0.51$	[109]
71	Scheme 9.27; (5a)	7	20; 2.0	$\gamma_{cmc} = 39$; $A_{min} = 0.67$	[109]
72	Scheme 9.27; (5b)	9	20; 0.36	$\gamma_{cmc} = 33$; $A_{min} = 0.66$	[109]
73	Scheme 9.27; (5c)	11	20; 0.086	$\gamma_{cmc} = 31$; $A_{min} = 0.66$	[109]
			Cationic		
74	Scheme 9.14; (2a)	6	25; 3.0	$\gamma_{cmc} = 32$; $A_{min} = 0.5$	[80]
75	Scheme 9.27; (4c)	11	20; 0.032		[109]
76	Scheme 9.29; (1a)	8	25; 29.2	$\gamma_{cmc} = 37.8$; $\Gamma_{max} = 1.06$; $A_{min} = 1.57$; $\alpha_{mic} = 0.56$	[120]
77	Scheme 9.29; (1b)	12	25; 5.48	$\gamma_{cmc} = 37.0$; $\Gamma_{max} = 0.90$; $A_{min} = 1.84$; $\alpha_{mic} = 0.53$	[120]

78	Scheme 9.29; (1c)	16	25; 1.25	$\gamma_{cmc} = 35.0; \Gamma_{max} = 1.11; A_{min} = 1.5; \alpha_{mic} = 0.44$	[120]
79	Scheme 9.29; (2b)	12	25; 21.3	$\gamma_{cmc} = 39.0; \Gamma_{max} = 1.28; A_{min} = 1.30; \alpha_{mic} = 0.29$	[120]
80	Scheme 9.29; (2c)	16	25; 1.42	$\gamma_{cmc} = 39.1; \Gamma_{max} = 1.32; A_{min} = 1.25; \alpha_{mic} = 0.30$	[120]
81	Scheme 9.30; (1)	12	25; 4.1	$\gamma_{cmc} = 38.3; \Gamma_{max} = 4.51; A_{min} = 0.37; \alpha_{mic} = 0.51$	[118]
82	Scheme 9.30; (2)	12	25; 5.2	$\gamma_{cmc} = 39.9; \Gamma_{max} = 3.31; A_{min} = 0.50; \alpha_{mic} = 0.27$	[118]
83	Scheme 9.32; (1a)	10	NS[b]; 0.69	$\gamma_{cmc} = 33.8$	[121]
84	Scheme 9.32; (1b)	12	NS[b]; 0.046	$\gamma_{cmc} = 29.6$	[121]
85	Scheme 9.32; (1c)	14	NS[b]; 0.057	$\gamma_{cmc} = 27.3$	[121]
86	Scheme 9.32; (1d)	16	NS[b]; 0.12	$\gamma_{cmc} = 33.5$	[121]
87	Scheme 9.32; (1e)	18	NS[b]; 0.15	$\gamma_{cmc} = 39.7$	[121]
88	Scheme 9.33; (1a)	8	25; 14.90	$\alpha_{mic} = 0.40; N_{agg} = 28$	[16]
89	Scheme 9.33; (1b)	12	25; 2.40	$\alpha_{mic} = 0.30; N_{agg} = 56$	[16]
90	Scheme 9.33; (1c)	16	25; 0.42	$\alpha_{mic} = 0.24; N_{agg} = 95$	[16]

[a] cmc; cac; γ_{cmc}; A_{min}; Γ_{max}; α_{mic}; N_{agg} refer to critical micelle concentration; critical aggregation concentration; surface tension at the cmc; surface excess concentration; minimum area/surfactant molecule at the solution/air interface, and aggregation number, respectively.

[b] NS, not specified.

[c] Values of γ_{cmc} determined by visual inspection of the plot of surface tension versus log [surfactant] are reported as "-".

are, however, smaller than the areas theoretically calculated for the sugar head-group. This indicates that the more probable arrangement for the (hydrophilic) sugar moiety is tilt, i.e., is not parallel to the micellar interface [42,120]. Based on A_{min}, and χ_{20} (the surfactant mol fraction required to reduce γ_{H_2O} by 20 mN/m) it is possible to calculate Gibbs free energy of adsorption, i.e., that due to the transfer of the surfactant molecule from bulk solution to the interface, ΔG°_{ads}, and to split the latter into contributions from the surfactant segments:

$$\Delta G^\circ_{ads} = \Delta G^\circ_{ads,CH_3} + N_{CH_2}\,\Delta G^\circ_{ads,CH_2} + \Delta G^\circ_{ads,head\text{-}group} \qquad (9.1)$$

These contributions are due to the terminal CH_3 group of the hydrophobic chain, $\Delta G^\circ_{ads,CH_3}$; the methylene groups of the alkyl chain, ($N_{CH_2}\,\Delta G^\circ_{ads,CH_2}$), and the head-group, $\Delta G^\circ_{ads,head\text{-}group}$. Equation 9.1 predicts a linear correlation between ΔG°_{ads} and N_{CH_2}, where the intercept includes contribution from the terminal methyl plus the head-group. Since $\Delta G^\circ_{ads,CH_3}$ is independent of the chain length of the surfactant, its contribution is constant in a homologous series. That is, differences of the intercepts essentially reflect the transfer of the head-group from bulk solution to the solution/air interface. A similar expression, Equation 9.2, can be written for the formation of the

$$\Delta G^\circ_{mic} = \Delta G^\circ_{mic,CH_3} + N_{CH_2}\,\Delta G^\circ_{mic,CH_2} + \Delta G^\circ_{mic,head\text{-}group} \qquad (9.2)$$

Where the free energies refer to the transfer from bulk solution to the micellar pseudo-phase [125,126]. Equations 9.1 and 9.2 have been applied to a series of cationic and anionic sugar-based surfactants. The following data were calculated for the adsorption of methyl 2-acylamido-2-deoxy-6-O-sulfo-D-glucopyranoside (all free energies in kJ/mol): $\Delta G^\circ_{ads,CH_2} = -3.5$; $\Delta G^\circ_{ads,(CH_3+head\text{-}group)} = -40$; $\Delta G^\circ_{mic,CH_2} = -3.4$; and $\Delta G^\circ_{mic,(CH_3+head\text{-}group)} = -12$. The corresponding values for the micelliza-tion of methyl 2-acylamido-6-trimethylammonio-2,6-dideoxy-D-glucopyrano-side chlorides are: $\Delta G^\circ_{mic,CH_2} = -2.4$; $\Delta G^\circ_{mic,(CH_3+head\text{-}group)} = -18.3$. The free energies of transfer of the methylene group from bulk solution to the micellar pseudo-phase are in the same range calculated for other surfactants, e.g., alkyl sulfates, −3.4; ethoxylated alkyl sulfates, −3.7; alkylbenzyldimethylammonium chlorides, −2.7; and alkyltrimethylammonium chlorides, −3.5. The micellization of the sugar-based head-groups is however much more favorable, as shown by the following values of $\Delta G^\circ_{mic,(CH_3+head\text{-}group)}$, that we have calculated for the classic surfactants: −2.8, 0.1, −3.3, and 9.1 for alkyl sulfates, ethoxylated alkyl sulfates, alkylbenzyldimethylam-moinum chlorides, and alkyltrimethylammonium chlorides, respectively. The more favorable $\Delta G^\circ_{mic,(CH_3+head\text{-}group)}$ the SBIS is probably due to head-group interactions in the interfacial region, either directly (via the NHCO and OH groups) or via a water intermediary [16,123]. The formation of direct or water-mediated hydrogen bonding between surfactant monomers that carry the amide group has been demonstrated by FTIR [127].

The conductivity data are usually employed for the determination of cmc, and α_{cmc}. Several of the α_{cmc} reported in Table 9.1 are high, in the range of 0.44–0.56; this implies extensive dissociation of the counterion. In part, this result is due to use of Frahm's method, i.e., by dividing the slopes of the straight lines above and below the cmc. Although this method is simple, it is only a useful approximation when N_{agg} is not available [128]. The reason is that the conductivity of the micelle (a macro-ion) is not taken into account; this leads to relatively high α_{mic} [127]. Evans equation avoids this overestimation of α_{mic} [129], the required N_{agg} may be experimentally determined or theoretically calculated. Therefore, the degrees of dissociation of the counterions of these SBIS are only slightly higher than those of the classical surfactants, because the head-groups of the former surfactants are bulkier. For the same hydrophobic moiety, α_{mic}, is higher for O-acetylated cationic than the corresponding nonacetylated ones [120], in agreement with the fact that crowding around the quaternary (N) causes charge screening, thus increases counterion dissociation [130].

Considering other properties, in particular those relevant to applications, several points are worth mentioning: (1) The Krafft temperatures of most of the surfactants synthesized are below room temperature, sometimes <0°C [43–88], i.e., their solutions are stable; (2) The foam height and foam stability are important properties of surfactants, these are usually evaluated by the Ross-Miles foam test. The height of foam formed by SBIS and the foam stability are comparable, or better than those of classic surfactants. For example, the foam volumes of sulfated monoglycerides at 0 time, and after 1 and 5 min are higher than that of SDS [88]; mixtures of N-dodecyl-N,N-bis[3-(D-gluconylamido)propyl]amine-N-oxide; N-dodecyl-N,N-bis[3-(D-glucoheptonylamido) propyl]amine-N-oxide, and N-dodecyl-N,N-bis[3-(D-lactobinoylamido) propyl] amine-N-oxide with SDS form high and stable foam [43], the foam formed by cationic 3(trimethylammonio)propyl-6-O-alkanoyl-D-glucopyranoside iodides is more stable than that formed by SDS [121]; (3) Introduction of an additional charge increases the cmc noticeably. e.g., in sulfated $C_{12}\alpha GlucoPy$, $C_{12}\beta GlucoPy$, and $C_{14}\beta MaltoPy$ the second sulfate group increases the cmc by approximately 100-fold as a result of electrostatic repulsion between the head-ions [17]. The cmc measured for several SBIS, e.g., the 6- and 1'-O-acylsucrose derivatives are more than an order of magnitude lower than that of the commercially prepared ionic surfactants, e.g. SDS and sodium tetradecylsulfate. As expected, the cmc, hence the surface activity, depend on the positions of the acyl, as well as the ionic group; it decreases with longer acyl chain [98]; (4) As stated above, introduction of a sugar moiety in the head-group increases the compatibility of cationic surfactants with other compounds, in particular anionic surfactants. Indeed, instant precipitation occurs when benzalkonium chloride (mixture of alkylbenzyldimethylammonium chlorides) solution (>125 ppm) is added to SDS solution (500 or 1000 ppm). On the other hand, no precipitation occurred when the same experiments were repeated in the presence of 1000 ppm of 3(trimethylammonio)propyl-6-O-dodecanoyl-D-glucopyranoside iodide. This noticeable increases in compatibility with anionic surfactants makes it possible to utilize their disinfectant effect in complex matrixes (i.e., formulations), thereby making the antimicrobial system much more robust [121]; (5) The stability of some chemo-cleavable (destructible) SBIS have been investigated. The stability of two sulfonated esters of 1-O-dodecylglycerol (with normal and branched sulfonate side chain) toward acid and base was determined. Complete decomposition by 1 mol/L DCl took 2 and 4 days, respectively; whereas it took 3 days and more than 1 week, respectively, for complete hydrolysis by 0.5 mol/L DCl. The same surfactants are hydrolyzed immediately by 0.5 mol/L NaOD, but more slowly by 0.01 mol/L NaOD, 45 min and 4.5 h, respectively [88]. All sodium methyl 2-acetamide-4,6-O-alkylidene-3-O-[1-(carboxylato)ethyl]-2-deoxy-α-D-glucopyranosides decomposed into non-surface-active products under acidic conditions. The experiments were carried out by using solutions of the surfactants, 25 mmol/L in 2% HCl. The time for complete decomposition showed dependence on the structure, ranging from approximately 20 to 110 h. Under these conditions, two surfactants did not decompose completely, even after 140 h [109]; and (6) The biodegradability of several SBIS has been evaluated, usually, by the BOD method; the results were compared with those of SDS and LAS. The biodegradability of sulfonated esters of 1-O-alkylglycerol is comparable, or higher than that of SDS, and much higher than that of LAS [88]; A BOD test was carried out and the results were compared with those of sodium dodecanoate; 2-acetamide-4,6-O-alkyldene-3-O-[1-(carboxylato)ethyl]-2-deoxy-α-D-glucopyranosides showed a comparable, or higher biodegradability than sodium dodecanoate [109].

9.4 APPLICATIONS OF SBIS

9.4.1 MEDICAL AND PHARMACEUTICAL PROPERTIES

9.4.1.1 Antibacterial Action

As expected, the cationic sugar-based surfactants have antibacterial properties; these are modified by the introduction of the sugar moiety. The antibacterial effect of the surfactants 3(trimethylammonio)propyl-6-O-alkanoyl-D-glucopyranoside iodides were tested against the growth of the bacteria *Micrococcus luteus* and *Vibrio alginolyticus*, and the fungus strain *Fusarium oxysporum* at the levels of 50 and 500 ppm, respectively. The efficiency increased as a function of increasing the surfactant hydrophobicity, i.e., its log P value. The antimicrobial potency of these surfactants, however, is considerably lower than that of the most frequently employed disinfectant, benzalkonium chloride; the latter inhibited both bacterial test strains at the level of 10 ppm, and the fungal test strain at 100 ppm. Even so, taking into account the poor compatibility of benzalkonium chloride with anionic surfactants, the antifungal effect of the SBIS is potentially useful [121].

9.4.1.2 Anti-HIV Activity

The anti-HIV activity (EC_{50}) and the toxicity (CC_{50}) of several sugar-based surfactants have been evaluated. The EC_{50} is the concentration of compound that saves 50% of infected cells; CC_{50} is the concentration leading to the death of 50% of healthy cells. Drugs with higher efficacy are associated with low EC_{50} and high CC_{50}, i.e., have a high selectivity index (SI) = CC_{50}/EC_{50}. Values of the latter ranged from 1 to 4.5, but the most lipophilic compound (a mixture of N-hexade-cylaminolactitol and dodecyldicarboxylic acid, log P = 8.4) showed SI > 200. Thus, polyvalent interactions between the virus and the galactosylceramide receptor are more effectively inhibited by gemini compounds bearing two galactose moieties than by compounds bearing a single sugar moiety [44]. It is interesting that all compounds tested exhibited antiviral activity below their critical aggregation concentration (cac), i.e., these sugar-based catanionics are active in their monomeric state. On the other hand, their cellular toxicity (given by CC_{50}) was principally associated with their aggregated state.

A series of polyanion inhibitors for HIV were synthesized by γ-irradiation-initiated polymerization of micellar solutions of SBIS carrying an unsaturated tail. Examples are sodium 1-O-(undec-10-enyl)-β-D-glucopyranoside-6-sulfate and sodium 1-O-(undec-10-enyl)-β-D-glucuronate. All compounds synthesized have modest anti-HIV activity (in CEM-4 cells); IC_{50} ranged from 0.79 to 20 μg/mL based on either the cytopathogenic effect assay or reverse transcriptase [131].

9.4.1.3 Vaccines

Combinations of hydrophobic, negatively charged sucrose fatty ester sulfates plus submicron emulsions of squalene-in-water are strong adjuvants for humoral and cell-mediated immunity; these formulations are promising adjuvants for vaccine containing poor immunogens. Thus the activity of sucrose derivatives toward a recombinant glycoprotein was determined in animal species. Compared to antigen alone, up to 3000-fold higher virus neutralizing antibody titers and 10-fold higher cellular responses against swine fever virus were observed in pigs after two immunizations with the sucrose derivatives combined with a squalene-in-water emulsion. The lipophilicity of the derivative was crucial; this explains the fact that sucrose esters containing one sulfate and seven decanoic or dodecanoic esters exerted the highest adjuvanticity. Derivatives without the sulfate group, with fewer esters groups, or with shorter chain (C6 and C8 acids) were less effective. Enhanced humoral and cell-mediated immune responses lasted for at least 24 weeks [95].

9.4.1.4 Dermatological Applications

A new kind of catanionic assembly was developed that associates a SBNS, dodecy-lactylamine, and a nonsteroidal anti-inflammatory drug (NSAID, general formula ArCH(R)-COOH), e.g., indomethacin (1-(4-chlorobenzoyl)-5-methoxy-2-methyl-1H-indole-3-acetic acid), ibuprofen (α-methyl-4-(2-methylpropyl)benzeneacetic acid) or ketoprofen (3-benzoyl-α-methylbenzeneacetic acid) [132]. The reactions between the two components lead to the formation of catanioinc surfactants; these form vesicles spontaneously, whose cac, γ_{cac}, hydrodynamic radii, and log P were found to be 0.5, 0.5, and 0.6 mmol/L; 31, 28, and 30 mN/m; 40, 90, and 20 nm; 4.68,

4.59, 4.15 for the catanionics of the SBNS with indomethacin, ibuprofen, and keto-profen, respectively. They were tested as potential NSAID delivery systems for der-matological application. All compounds were active against the arachidonic acid-induced ear mouse oedema. NSAIDs alone showed percentages of inhibition close to 45%, whereas the anti-inflammatory activity of the catanionic assemblies ranged from 48% to 87%. Indomethacin showed the highest activity, while ibupro-fen was the least active compound. These results showed that the NSAIDs maintain their therapeutic properties when associated with a biocompatible sugar-based sur-factant through electrostatic and hydrophobic interactions. Additionally, the topical administration of catanionic assemblies of indomethacin and ketoprofen decreased the oedema more efficiently than the nonassociated NSAID of the control groups (increase of 77% and 56% for indomethacin and ketoprofen, respectively).

The effect of administration of indomethacin as catanionic vesicle was com-pared with that of the free drug, employed as sodium salt, under the same condi-tions. The skin permeation of the free drug was twofold higher, due to the slower diffusion of its aggregate-containing counterpart. In summary, catanionic vesicles can be potentially useful for drug delivery. The above-discussed example shows that they improved the anti-inflammatory activity of the NSAID; induce a slower diffusion of the anti-inflammatory drug through the skin, probably leading to a prolonged time of residence in the targeted sites, and are clinically safe, thanks to the use of biocompatible surfactants and an aqueous vehicle [133].

9.4.1.5 Gene Therapy

Although viral vectors display a highly effective gene delivery and transfection efficacy *in vivo*, these systems suffer from several drawbacks, with regard to bio-hazard and safety. Consequently, research in the development of improved nonvi-ral vectors for therapeutic applications is justified. Cationic lipids have been exploited as nonviral DNA vectors (lipoplexes), and have been shown to give rise to good transfection efficiency *in vitro*, although they mostly failed to sustain this efficiency *in vivo*.

The application of pH-sensitive sugar-based gemini surfactants may provide a simpler and programmable alternative for *in vivo* gene delivery. In an aqueous environment at physiological pH, these compounds form bilayer vesicles, and undergo a lamellar-to-micellar phase transition in the endosomal pH range as a consequence of protonation. Likewise, lipoplexes made with these amphiphiles exhibit a lamellar morphology at physiological pH and a nonlamellar phase at acidic pH.

A single example is discussed here; a detailed account of these applications is given in Chapter 8. Five different sugar-based gemini surfactants were syn-thesized, all contained oleoyl tails. The head-group is either mannose or glu-cose, connected via an ethylene oxide or an aliphatic spacer. It was reasoned that the property of undergoing a pH-triggered structural change would convey colloidal stability to gemini lipoplexes before cellular uptake, while exerting destabilizing properties necessary for gene delivery only after internalization within (mildly acidic) endosomal compartments. In principle, this colloidal stability could be exploited *in vivo*, avoiding massive aggregation of these cationic lipids while in the circulation. The data are consistent with this reasoning in that effective transfection of these gemini surfactants was observed *in vitro*, two of them led to 70% of transfected cells with a good cell survival. The effect of the head-group, glucose versus mannose, or the effect of the spacer, C6-alkyl versus ethylene oxide, does not appear to modulate the level of transfection to a significant extent [134].

9.4.2 Supramolecular Assemblies and Dendrimers

Self-assembled nanostructures of amphiphiles held together by noncovalent interactions are currently the focus of interest. These can generate bilayers in water, the most common forms being planar membranes, rods, helices, ribbons, and tubules [135,136]. Variations of the head-group and the alkyl chain were found to affect the type of clustering and the morphology. These morphologies are formed by catanionic sugar surfactants containing a fluorescent coumarinic moiety, obtained simply by an acid–base reaction between, e.g., 12-(4-aminocoumarin)-dodecanoic acid and *N*-octylaminolactitol. Examination by electron microscopy showed a large variety of assembly morphologies: polydisperse population of vesicles with diameters generally ranging from 60 to 400 nm; helices with a diameter of 40 nm, a pitch ranging from 40 to 50 nm, and a strip width ranging from 20 to 40 nm; helix/tubule transition with a diameter of 40 nm and a length of 360 nm; tubules resulting from the contraction of helices, with a diameter of 40 nm, reaching 1 μm in length. The formation of these morphologies is interesting because catanionics are known to form only vesicles and lamellar phases; this may be due to a combination of the chirality of the sugar head-group and the stacking of the coumarinic part, in agreement with molecular modeling of the catanionic surfactant molecule [45].

The architectural features of dendrimers include their precise constitutions with high overall symmetries, their well-defined internal cavities, and their nanometer dimensions. Saccharides have been used to functionalize dendrimers, leading to amphiphilic dendrimers with polar sugar heads; these are suitable for molecular recognition and catalysis. Glucose-persubstituted poly(amidoamine) dendrimers have been synthesized from commercially available poly(amidoamine) dendrimers and D-glucono-1,5-lactone and their physicochemical properties in water were examined. The general structure of the products synthesized is $[(glucose-NH(C_2H_4)NHCO)_2]_m-N(C_2H_4)N-[(OC-NH(C_2H_4)NH-glucose)_2]_m$, where $m = 0, 1, 2$, and 3, corresponding to 4, 8, 16, and 32 surface groups, respectively. In water, these amphiphilic dendrimers are able to solubilizate lipophilic substances, e.g., pyrene and phenyl cyclohexylketone; the solute concentration increased almost linearly as a function of the dendrimer concentration. The ratio of pyrene (I_1/I_3) vibronic bands decreased as a function of increasing dendrimer concentration and increasing m, showing concomitant deeper penetration into the aggregate structure, where the local polarity is lower. The size distribution also depends on m, as shown by the following diameters, calculated from dynamic light scattering: 46–264, 100–464, 215–1000, and 464–2150 nm, for $m = 0$, 1, 2, and 3, respectively. The solubilized (pro-chiral) ketone (0.1 mmol/L in 10 mmol/L, for dendrimer with $m = 3$) was reduced with $NaBH_4$, to give 95% of the corresponding alcohol, enriched in the (S) isomer [137].

9.4.3 MISCELLANEOUS

An effective intravenously injectable oxygen carrier based on perflourdecalin has been obtained by using small amounts of sodium (3- or 6-D-glucosyl)[2-(perfluoroalkyl)ethyl] phosphates as surfactant, or as cosurfactant with egg-yolk phospholipids. The fluorinated surfactant, either alone or with the phospholipids, give smaller particles than those obtained by use of the phospholipid alone, both at the time of preparation and after 1 month at 40°C [124]

9.4.4 MICELLAR ELECTROKINETIC CAPILLARY CHROMATOGRAPHY

Under the influence of an electric field, the elctrophoretic migration of a negatively charged micelle, e.g., SDS, is in the direction of the anode. At neutral or alkaline pH electroosmotic flow, toward the cathode, is strong. In fact, under these conditions the electroosmotic mobility of the bulk solution is higher than the electrophoretic mobility of the micelles, leading to an overall migration of the micelles in the direction of the cathode. In the absence of micelles, all neutral molecules migrate with the same mobility, under the influence of the electroosmotic flow only. The presence of micelles allows resolution of neutral analytes due to differences in their distribution between bulk aqueous solution and the micellar pseudo-phase.

Using MECC and 0.01 mol/L solution of dodecyl-β-D-glucopyranoside-6-phosphate, eight neutral analytes—mesityl oxide; nitrobenzene; toluene; o-, m-, and p-xylene; naphthalene; 1-nitronaphthalene; butylbenzene; and halofantrine—were separated in approximately 7 min; the order of separation is closely related to their log P values, i.e., less hydrophobic analytes are eluted first. Racemic dansyl derivatives of valine, phenylalanine and tryptophan, were also separated according to their log P values [105]. MEGA-9 and MEGA-10, octanoyl sucrose, C_8βGlucoPy, C_8βMaltoPy were successfully employed in the separation of several classes of compounds under neutral and alkaline solution conditions. Examples of analytes include a series of alkyl phenyl ketons, $RCOC_6H_5$, $R = C_1–C_7$, dichloro- and trichlorophenoxyacetic acids and their butyl and isopropyl esters, commercial urea-based herbicides, and dansyl amino acids. The migration times of the analytes increased as a function of increasing the borate buffer concentration, and as a function of increasing the pH (3.5–10) at a constant [borate buffer]. Efficient separations were obtained between 20 and 30 min [19,34]. The monophosphate and monosulfate derivatives of C_{12}βGlucoPy have been successfully employed in the separation of racemic mixtures of five dansylated amino acids (57 cm column, phosphate-borate buffer, pH 8, 20°C) in less than 20 min. The same experimental conditions were employed for the separation of enantiomers of cromakalin and 1,1'-dinaphthyl-2,2'-diyl hydrogen phosphate; mephenytoin and hydroxymephenytoin; and 3,4-dimethyl-dioxo-2-phenylperhydro-1,4-oxazepine. Unlike naturally occurring surfactants, in particular cyclodextrins and bile salts, which are available in one stereoisomeric form, some sugar-based surfactants can be readily synthesized in both the D- and L-forms. This may be analytically desirable, especially when an accurate estimate of a low-level enantiomeric impurity is needed, but this follows the main antipode [138]. Two commercially available amphiphilic anionic aminosaccharide derivatives were investigated as chiral selector additives in MECC. Each surfactant has a glucamine backbone carrying three hydrocarbon chains and three carboxylic groups. Resolution of dansylated amino acids or a new quinoline-based antimicrobial agent was achieved [139].

9.5 CONCLUSIONS

Sugar-based ionic surfactants are interesting materials that can be obtained from a variety of mono- and disaccharides and their oligomers. They are efficient as surfactants, biodegradable, biocompatible, and have exciting potential applications in separation, pharmaceutical, and medical sciences. Equally important, their syntheses conform to the basic principles of green chemistry, in particular, they are obtained from renewable, practically unlimited feedstock; their manufacturing generates little waste, and involves a high degree of atom-economy. The amount of information on the physicochemical properties of solutions of these surfactants, a prerequisite for rational applications, is still very modest. Thus systematic studies on the structure/property relationship, their phase behavior, the nature of their interfacial regions, the mechanism of solubilization of polar and nonpolar substances and the average location of the solubilizates in the micellar pseudo-phases, the formation of oil-in-water (O/W) and water-in-oil (W/O) emulsions and microemulsions—just to mention some examples—are welcome and timely.

ACKNOWLEDGMENTS

We thank the State of São Paulo Research Foundation (FAPESP) for financial support (grant 2004/15400-5), and the National Council for Scientific and Technological Research (CNPq) for a predoctoral fellowship to P. D. Galgano and research productivity fellowship to O. A. El Seoud (grant 305547/2003-8). We thank Marcelo C. S. Bandoria for his help in adding the references to a data base.

ABBREVIATIONS AND SYMBOLS

Ac	acetyl group
Ac_2O	acetic anhydride
2-aminoglucose	2-amino-2-deoxy-D-glucopyranose
Bn	benzyl group
BOD	biological oxygen demand
Bz	benzoyl group.

$C_n\alpha GlucoPy$	n-alkyl-α-D-glucopyranoside, thus $C_8\alpha GlucoPy$ is n-octyl-α-D-glucopyranoside
$C_n\beta GlucoPy$	n-alkyl-β-D-glucopyranoside
$C_n\alpha MaltoPy$	n-alkyl-α-D-maltopyranoside, thus $C_8\alpha MaltoPy$ is n-octyl-α-D-maltopyranoside
$C_n\beta MaltoPy$	n-alkyl-β-D-maltopyranoside, thus $C_{12}\beta MaltoPy$ is n-dodecyl-β-D-maltopyranoside
DMF	N,N-dimethylformamide
DMAP	4-N,N-dimethylaminopyridine
DMPU	N,N'-dimethylhexahydropyrimidin-2-one
DP	average degree of polymerization of a polymer
EO	ethylene oxide unit ($-CH_2-CH_2-O$)
Et	ethyl group
$\log P$	measure of the lipophilicity of a compound, given by the partition of the compound between (mutually saturated) water and 1-octanol ($\log P = [\text{substance}]_{\text{octanol}}/[\text{substance}]_{\text{water}}$)
Me	methyl group
MECC	micellar electrokinetic capillary chromatography
MEGA-C_n	N-alkanoyl-N-alkyl-1-glucamine. Thus MEGA-12 is N-dodecanoyl-N-methyl-1-glucamine
MSA	methanesulfonic acid
NMeP	N-methylpyrrolidinone
NSAID	nonsteroidal anti-inflammatory drug
SBIS	sugar-based ionic surfactant
SBNS	sugar-based nonionic surfactant
Ts	4-methylphenylsulfonyl group
trityl	triphenylmethyl
X-C_nSucro	sucrose ester substituted at the position (X) of the glucose moiety thus 6-C_8Sucro is 6-O-n-octanoylsucrose
X'-C_nSucro	sucrose ester substituted at the position (X') of the fructose moiety thus 1'-C_{12}Sucro 1'-O-n-lauroylsucrose
α- or β-D-Glucose	α- or β-D-glucopyranoside

REFERENCES

1. Anastas, P. T. and Lankey, R. L., Sustainability through green chemistry and engineering, in *Advancing Sustainability through Green Chemistry and Engineering*, Anastas, P. T., and Williamson, T. C., (Eds.), Oxford University Press, New York, p. 1, 1998.
2. Tundo, P., Anastas, P., Black, D., Breen, J., Collins, T., Memoli, S., Miyamoto, J., Polyakoff, M., and Tumas, W., *Pure Appl. Chem.*, 2000, 72, 1207.
3. Gruesbeck, C. and Rase, H. F., *Ind. Eng. Chem. Prod. Res. Develop.*, 1972, *11*, 74.
4. Gupta, S., Chemistry, chemical and physical properties and raw materials, in *Soap Technology for the 1990's*, Spitz, L. (Ed.), *Am. Oil Chem. Soc.*, Champaign, p. 48, 1990.
5. Tsukada, K., Matsuda, M., Kadono, Y., Horio, M., and Matsuda, Y., CESIO, *Proceedings of the World Surfactants Congress*, Munich, 1, p. 190, 1984.

6. Allen, D. K. and Tao, B. Y., *J. Surf. Det.*, 1999, *2*, 383.
7. Hass, H. B., Early history of sucrose esters. *Sugar Esters Symposium*, Noyes Development Co., Park Ridge, 1, 1968.
8. Focher, B., Savelli, G., and Torri, G., *Chem. Phys. Lipids*, 1990, *53*, 141.
9. Hill, K., von Rybinski, W., and Stoll, G., *Alkyl Polyglycosides: Technology, Properties and Applications*, Wiley-VCH, Weinheim, 1997.
10. Laughlin, R. C., Fu, Y. C., Wireko, F. C., Scheibel, J. J., and Munyon, R. L., N-alkanoyl-N-alkyl-1-glycamines, in *Novel Surfactants, Surfactant Science Series*, Holmberg, K., (Ed.), 2nd ed., Marcel Dekker, New York, Vol. 114, p. 1, 2003.
11. Hauthal, H. G., *Chem. Unserer Zeit*, 1992, *26*, 293.
12. Mian, N., Anderson, C. E., and Kent, P. W., *Biochem. J.*, 1979, *181*, 387.
13. Fabry, B. and Schumacher, A., DE 3941061, 1991.
14. Ahmed, F. U., *J. Am. Oil Chem. Soc.*, 1990, *67*, 8.
15. Fabry, B., Weuthen, M., and Schumacher, A., DE 4006841, 1991.
16. Bazito, R. C. and El Seoud, O. A., *J. Surf. Det.*, 2001, *4*, 395.
17. Boecker, T., Lindhorst, T. K., Thiem, J., and Vill, V., *Carbohydr. Res.*, 1992, *230*, 245.
18. Bazito, R. C. and El Seoud, O. A., *Carbohydr. Res.*, 2001, *332*, 95.
19. Smith, J. T. and El Rassi, Z., *J. Chromatogr. A*, 1994, *685*, 131.
20. Johnsson, M. and Engberts, J. B. F. N., *J. Phys. Org. Chem.*, 2004, *17*, 934.
21. Binkley, R. W., *Modern Carbohydrate Chemistry*, Marcel Dekker, New York, 1998.
22. Robyt, J. F., *Essentials of Carbohydrate Chemistry*, Springer, Heidelberg, 1998.
23. Hao, J., Hoffmann, H., and Horbaschek, K., *Langmuir*, 2001, *17*, 4151.
24. Beber, R. C., Bunton, C., Savelli, G., and Nome, F., *Prog. Colloid Polym. Sci.*, 2004, *128*, 249.
25. Bunton, C. A., *J. Phys. Org. Chem.*, 2005, *18*, 115.
26. Marques, E. F., Regev, O., Khan, A., and Lindman, B., *Adv. Colloid Interface Sci.*, 2003, *100*, 83.
27. Tondre, C. and Caillet, C., *Adv. Colloid Interface Sci.*, 2001, *93*, 115.
28. Anderson, J. L., Eyring, E. M., and Whittaker, M. P., *J. Phys. Chem.*, 1964, *68*, 1128.
29. Greenhill-Hooper, M. J., Austerberry, M. S., and Render, C. M., *Proceedings of the 5th World Surfactants Congress*, Firenze, 683, 2000.
30. Smith, J. T. and El Rassi, Z., *J. Microcol. Sep.*, 1994, *6*, 127.
31. Smith, J. T., Nashabeh, W., and El Rassi, Z., *Anal. Chem.*, 1994, *66*, 1119.
32. Cai, J. and El Rassi, Z., *J. Chromatogr.*, 1992, *608*, 31.
33. Tegeler, T. and El Rassi, Z., *J. AOAC Int.*, 1999, *82*, 1542.
34. Mechref, Y., Smith, J. T., and El Rassi, Z., *J. Liq. Chromatogr.*, 1995, *18*, 3769.
35. Mechref, Y. and El Rassi, Z., *J. Chromatogr. A*, 1996, *724*, 285.
36. Menger, F. M. and Littau, C. A., *J. Am. Chem. Soc.*, 1991, *113*, 1451.
37. Baba, T., Zheng, L. Q., Minamikawa, H., and Hato, M., *J. Colloid Interface Sci.*, 2000, *223*, 235.
38. Zheng, L. Q., Shui, L. L., Shen, Q., Li, G. Z., Baba, T., Minamikawa, H., and Hato, M., *Colloids Surf., A*, 2002, *207*, 215.
39. Izatt, R. M., Rytting, J. H., Hansen, L. D., and Christensen, J. J., *J. Am. Chem. Soc.*, 1966, *88*, 2641.
40. Wasungu, L., Stuart, M. C. A., Scarzello, M., Engberts, J. B. F. N., and Hoekstra, D., *Biochim. Biophys. Acta: Biomembranes*, 2006, *1758*, 1677.
41. Milkereit, G., Morr, M., Thiem, J., and Vill, V., *Chem. Phys. Lipids*, 2004, *127*, 47.
42. Wilk, K. A., Syper, L., Burczyk, B., Sokolowski, A., and Domagalska, B. W., *J. Surf. Det.*, 2000, *3*, 185.
43. Piasecki, A., Karczewski, S., and Syper, L., *J. Surf. Det.*, 2001, *4*, 349.

44. Blanzat, M., Perez, E., Rico-Lattes, I., Prome, D., Prome, J. C., and Lattes, A., *Langmuir*, 1999, *15*, 6163.

45. Blanzat, M., Massip, S., Speziale, V., Perez, E., and Rico-Lattes, I., *Langmuir*, 2001, *17*, 3512.

46. Pasc-Banu, A., Stan, R., Blanzat, M., Perez, E., Rico-Lattes, I., Lattes, A., Labrot, T., and Oda, R., *Colloids Surf.*, A, 2004, *242*, 195.

47. Soussan, E., Pasc-Banu, A., Consola, S., Labrot, T., Perez, E., Blanzat, M., Oda, R., Vidal, C., and Rico-Lattes, I., *ChemPhysChem*, 2005, *6*, 2492.

48. Pasc-Banu, A., Blanzat, M., Belloni, M., Perez, E., Mingotaud, C., Rico-Lattes, I., Labrot, T., and Oda, R., *J. Fluorine Chem.*, 2005, *126*, 33.

49. Boecker, T. and Thiem, J., *Tenside, Surf., Det.*, 1989, *26*, 318.

50. Larsson, K. and Petersson, G., *Carbohydr. Res.*, 1974, *34*, 323.

51. Vlahov, J. and Snatzke, G., *Liebigs Ann. Chem.*, 1983, *570*.

52. Bertho, J. N., Ferrieres, V., and Plusquellec, D., *J. Chem. Soc., Chem. Commun.*, 1995, *1391*.

53. Ferrieres, V., Bertho, J. N., and Plusquellec, D., *Carbohydr. Res.*, 1998, *311*, 25.

54. Mabeau, S. and Kloareg, B., *J. Exp. Bot.*, 1987, *38*, 1573.

55. Roussel, M., Benvegnu, T., Lognone, V., Le Deit, H., Soutrel, I., Laurent, I., and Plusquellec, D., *Eur. J. Org. Chem.*, 2005, *3085*.

56. Rico-Lattes, I., Gouzy, M. F., Andre-Barres, C., Guidetti, B., and Lattes, A., *Biochimie*, 1998, *80*, 483.

57. Bollenback, G. N., Long, J. W., Benjamin, D. G., and Lindquist, J. A., *J. Am. Chem. Soc.*, 1955, *77*, 3310.

58. Menger, F. M., Binder, W. H., and Keiper, J. S., *Langmuir*, 1997, *13*, 3247.

59. Gaertner, V. R., *J. Am. Oil Chem. Soc.*, 1961, *38*, 410.

60. Gilbert, E. E., *Chem. Rev.*, 1962, *62*, 549.

61. Nagasawa, K., Harada, H., Hayashi, S., and Misawa, T., *Carbohydr. Res.*, 1972, *21*, 420.

62. Mumma, R. O., Hoiberg, C. P., and Simpson, R., *Carbohydr. Res.*, 1970, *14*, 119.

63. Wagenknecht, W., Nehls, I., Koetz, J., Philipp, B., and Ludwig, J., *Cellul. Chem. Tech.*, 1991, *25*, 343.

64. Wolfrom, M. L. and Han, T. M. S., *J. Am. Chem. Soc.*, 1959, *81*, 1764.

65. Hatanaka, K., Yoshida, T., Miyahara, S., Sato, T., Ono, F., Uryu, T., and Kuzuhara, H., *J. Med. Chem.*, 1987, *30*, 810.

66. Guiseley, K. B. and Ruoff, P. M., *J. Org. Chem.*, 1961, *26*, 1248.

67. Peat, S., Turvey, J. R., Clancy, M. J., and Williams, T. P., *J. Chem. Soc.*, 1960, *4761*.

68. Takiura, K. and Honda, S., *Yakugaku Zasshi*, 1967, *87*, 997; *CA* 68:20712.

69. Langston, S., Bernet, B., and Vasella, A., *Helv. Chim. Acta*, 1994, *77*, 2341.

70. Guilbert, B., Davis, N. J., Pearce, M., Aplin, R. T., and Flitsch, S. L., *Tetrahedron: Asymmetry*, 1994, *5*, 2163.

71. Tomalia, D. A. and Falk, J. C., *J. Heterocycl. Chem.*, 1972, *9*, 891.

72. Gao, Y. and Sharpless, K. B., *J. Am. Chem. Soc.*, 1988, *110*, 7538.

73. Polat, T. and Linhardt, R. J., *J. Surf. Det.*, 2001, *4*, 415.

74. Biswas, A. K. and Mukherji, B. K., *J. Phys. Chem.*, 1960, *64*, 1.

75. Biswas, A. K. and Mukherji, B. K., *J. Am. Oil Chem. Soc.*, 1960, *37*, 171.

76. Yamashita, K., Koen, K., and Nagaoka, J., JP 50070322, 1975.

77. Chamanlal, R., Gadgoli, H. G., and Kane, J. G., *J. Oil Technol. Assoc. India*, 1972, *4*, 41.

78. Ramayya, D. A., Kumar, Y. S. C., and Rao, S. D. T., *Indian Oil Soap J.*, 1966, *31*, 335.

79. Schwartz, A. M., Perry, J. W., and Berch, J., *Surface Active Agents and Detergents*, Vol. 2, Interscience Pub., New York, 1958.

80. Bazin, H. G. and Linhardt, R. J., *Synthesis*, 1999, *621*.
81. Klotz, W. and Schmidt, R. R., *Synthesis*, 1996, *687*.
82. Sin, Y. M., Cho, K. W., and Lee, T. H., *Biotechnol. Lett.*, 1998, *20*, 91.
83. Sabeder, S., Habulin, M., and Knez, Z., *Ind. Eng. Chem. Res.*, 2005, *44*, 9631.
84. Sekeroglu, G., Fadiloglu, S., and Ibanoglu, E., *J. Sci. Food. Agric.*, 2002, *82*, 1516.
85. Arcos, J. A., Bernabe, M., and Otero, C., *Enzyme Microb. Technol.*, 1998, *22*, 27.
86. Benson, A. A., The plant sulfolipid, in *Advances in Lipid Research*, Paoletti, R. and Kritchevsky, D., (Eds.), Academic Press, Vol. 1, p. 387, 1963.
87. Weislow, O. S., Kiser, R., Fine, D. L., Bader, J., Shoemaker, R. H., and Boyd, M. R., *J. Natl. Cancer Inst.*, 1989, *81*, 577.
88. Ono, D., Yamamura, S., Nakamura, M., and Takeda, T., *J. Surf. Det.*, 1998, *1*, 201.
89. Inoue, Y., Onodera, K., Kitaoka, S., and Hirano, S., *J. Am. Chem. Soc.*, 1956, *78*, 4722.
90. Conchie, J. and Levy, G. A., Methyl glycopyranosides by the Koenigs-Knorr method, in *Methods in Carbohydrate Chemistry*, Whistler, R. L. and Wolfrom, M. L., (Eds.), Academic Press, New York, p. 332, 1963.
91. Fernandez-Bolanos, J., Maya Castilla, I., and Fernandez-Bolanos Guzman, J., *Carbohydr. Res.*, 1988, *173*, 33.
92. Fernandez-Bolanos, J., Maya Castilla, I., and Fernandez-Bolanos Guzman, J., *An. Quim. Ser. C*, 1986, *82*, 200.
93. Gordon, D. M. and Danishefsky, S. J., *J. Am. Chem. Soc.*, 1992, *114*, 659.
94. Angyal, S. J., *Adv. Carbohydr. Chem. Biochem.*, 1984, *42*, 15.
95. Blom, A. G. and Hilgers, L. A. T., *Vaccine*, 2004, *23*, 743.
96. Vlahov, I. R., Vlahova, P. I., and Linhardt, R. J., *J. Carbohydr. Chem.*, 1997, *16*, 1.
97. Polat, T., Bazin, H. G., and Linhardt, R. J., *J. Carbohydr. Chem.*, 1997, *16*, 1319.
98. Bazin, H. G., Polat, T., and Linhardt, R. J., *Carbohydr. Res.*, 1998, *309*, 189.
99. Takami, M. and Suzuki, Y., *Biosci., Biotechnol., Biochem.*, 1994, *58*, 2136.
100. D'Arrigo, P., Piergianni, V., Pedrocchi-Fantoni, G., and Servi, S., *J. Chem. Soc., Chem. Commun.*, 1995, *2505*.
101. Virto, C. and Adlercreutz, P., *Chem. Phys. Lipids*, 2000, *104*, 175.
102. Neumann, J. M., Herve, M., Debouzy, J. C., Iglesias Guerra, F., Gouyette, C., Dupraz, B., and Tam, H. D., *J. Am. Chem. Soc.*, 1989, *111*, 4270.
103. Milius, A., Greiner, J., and Riess, J. G., *Carbohydr. Res.*, 1992, *229*, 323.
104. Havlinova, B., Kosik, M., Kovac, P., and Blazej, A., *Tenside Det.*, 1978, *15*, 72.
105. Jones, R. F. D., Camilleri, P., Kirby, A. J., and Okafo, G. N., *J. Chem. Soc., Chem. Commun.*, 1994, *1311*.
106. Jaeger, D. A. and Frey, M. R., *J. Org. Chem.*, 1982, *47*, 311.
107. Eastoe, J., *Prog. Colloid Polym. Sci.*, 2006, *133*, 106.
108. Kida, T., Masuyama, A., and Okahara, M., *Tetrahedron Lett.*, 1990, *31*, 5939.
109. Kida, T., Yurugi, K., Masuyama, A., Nakatsuji, Y., Ono, D., and Takeda, T., *J. Am. Oil Chem. Soc.*, 1995, *72*, 773.
110. Heinze, T., Rensing, S., and Koschella, A., *Starch/Staerke*, 2007, *59*, 199.
111. Richmond, J. M., in *Cationic Surfactants: Organic Chemistry*, Richmond, J. M. (Ed.), Marcel Dekker, New York, 1990.
112. Singer, E. J., Biological evaluation, in *Cationic Surfactants: Analytical and Biological Evaluation*, Cross, J. and Singer, E. J., Eds., Surfactant Science Series, Marcel Dekker, New York, Vol. 53, p. 29, 1994.
113. Boethling, R. S., *Water Res.*, 1984, *18*, 1061.
114. Kirby, A. J., Camilleri, P., Engberts, J. B. F. N., Feiters, M. C., Nolte, R. J. M., Soderman, O., Bergsma, M., Bell, P. C.; Fielden, M. L., Garcia Rodriguez, C. L., Guedat, P., Kremer, A., McGregor, C., Perrin, C., Ronsin, G., and van Eijk, M. C. P., *Angew. Chem., Int. Ed.*, 2003, *42*, 1448.

115. Holland, P. M. and Rubingh, D. N., Cationic surfactants in mixed surfactant systems, in *Cationic Surfactants: Physical Chemistry*, Rubingh, D. N. and Holland, P. M. Eds., Surfactant Science Series, Marcel Dekker, New York, Vol. 37, p. 141, 1990.
116. Overkempe, C., Annerling, A., van Ginkel, C. G., Thomas, P. C., Boltersdorf, D., and Speelman, J., Esterquats, in *Novel Surfactants, Surfactant Science Series*, Homberg, K. (Ed.), 2nd ed., Marcel Dekker, New York, Vol. 114, p. 347, 2003.
117. Puchta, R., Krings, P., and Sandkuehler, P., *Tenside, Surf. Det..*, 1993, *30*, 186.
118. Viscardi, G., Quagliotto, P., Barolo, C., Savarino, P., Barni, E., and Fisicaro, E., *J. Org. Chem.*, 2000, *65*, 8197.
119. Moon, T. L., Blanzat, M., Labadie, L., Perez, E., and Rico-Lattes, I., *J. Dispersion Sci. Technol.*, 2001, *22*, 167.
120. Quagliotto, P., Viscardi, G., Barolo, C., D'Angelo, D., Barni, E., Compari, C., Duce, E., and Fisicaro, E., *J. Org. Chem.*, 2005, *70*, 9857.
121. Kirk, O., Pedersen, F. D., and Fuglsang, C. C., *J. Surf. Det.*, 1998, *1*, 37.
122. Bongat, A. F. G. and Demchenko, A. V., *Carbohydr. Res.*, 2007, *342*, 374.
123. Bazito, R. C. and El Seoud, O. A., *Langmuir*, 2002, *18*, 4362.
124. Milius, A., Greiner, J., and Riess, J. G., *Colloids Surf.*, 1992, *63*, 281.
125. Rosen, M. J., *Surfactants and Interfacial Phenomena*, 2nd ed., Wiley, New York, 1989.
126. Hiemenz, P. C., and Rajagopalan, R., *Principles of Colloid and Surface Chemistry*, Marcel Dekker, New York, 1997.
127. Shimizu, S. and El Seoud, O. A., *Langmuir*, 2003, *19*, 238.
128. Frahm, J., Diekmann, S., and Haase, A., *Ber. Bunsenges Phys. Chem.*, 1980, *84*, 566.
129. Evans, H. C., *J. Chem. Soc.*, 1956, *579*.
130. Buckingham, S. A., Garvey, C. J., and Warr, G. G., *J. Phys. Chem.*, 1993, *97*, 10236.
131. Leydet, A., Jeantet-Segonds, C., Barthelemy, P., Boyer, B., and Roque, J. P., *Recl. Trav. Chim. Pays-Bas*, 1996, *115*, 421.
132. Latge, P., Rico, I., Garelli, R., and Lattes, A., *J. Dispersion Sci. Technol.*, 1991, *12*, 227.
133. Consola, S., Blanzat, M., Perez, E., Garrigues, J. C., Bordat, P., and Rico-Lattes, I., *Chem. Eur. J.*, 2007, *13*, 3039.
134. Wasungu, L., Scarzello, M., van Dam, G., Molema, G., Wagenaar, A., Engberts, J. B. F. N., and Hoekstra, D., *J. Mol. Med.*, 2006, *84*, 774.
135. Dill, K. A. and Flory, P. J., *Proc. Natl. Acad. Sci. USA*, 1980, *77*, 3115.
136. Menger, F. M., *Acc. Chem. Res.*, 1979, *12*, 111.
137. Schmitzer, A., Perez, E., Rico-Lattes, I., Lattes, A., and Rosca, S., *Langmuir*, 1999, *15*, 4397.
138. Tickle, D. C., Okafo, G. N., Camilleri, P., Jones, R. F. D., Kirby, A. J., *Anal. Chem.*, 1994, *66*, 4121.
139. Horimai, T., Arai, T., and Sato, Y., *J. Chromatogr. A*, 2000, *875*, 295.

10 Sugar-Based Surfactants with Isoprenoid-Type Hydrophobic Chains: Physicochemical and Biophysical Aspects

Masakatsu Hato, Hiroyuki Minamikawa, and Tadashi Kato

CONTENTS

10.1 INTRODUCTION

Sugar-based surfactants are becoming increasingly important from the ecological and industrial viewpoints. The rationale for the interest in these surfactants lies in the fact that they can be synthesized from renewable resources and are generally nontoxic and biodegradable [1]. Moreover, recent research interests are expanding toward the applications of the sugar-based surfactants in the fields of drug and gene delivery and of nanotechnology [2].

In aqueous media, a surfactant displays a specific temperature, T_K, (Krafft-eutectic temperature or a hydrated solid–liquid crystalline phase transition temperature) at which the surfactant undergoes a transition from a frozen to a fluid state to form a range of molecular assemblies such as a normal (or inverted) micelle, a lamellar liquid crystal, a hexagonal liquid crystal, a cubic liquid crystal, etc. [3]. These fluid phases are crucial to the functions of surfactant/water systems. At temperatures below T_K, the surfactant precipitates as a hydrated solid and is difficult to handle. Thus, to be usable as a surfactant, T_K must be lower than working temperatures, most preferably well below room temperatures [4].

Carbohydrates form relatively hard crystals with high melting points, since their principal cohesive forces are hydrogen bonds [5]. This also holds for the sugar-based surfactants; the inter-headgroup hydrogen bonds substantially contribute to the crystal stability of the sugar-based surfactants [6]. Thus, T_K values of conventional sugar-based surfactants are unusually high for nonionic surfactants [7]. This situation is clearly seen in Figure 10.1, where the values of T_K of typical sugar-based surfactants, n-alkyl $\alpha(\beta)$-D-glycosides, are plotted as functions of the type of sugar headgroup and the alkyl chain length [8–10]. It is noted that the type of sugar headgroup largely determines the values of T_K. As a rule of thumb, the galactose, fucose, and xylose headgroups give higher T_K values. On the other hand, the mannose or glucose headgroups give relatively low T_K values. Nevertheless, the T_K values of C$_{12}$-glucosides are already 55°C (1-O-dodecyl-α-D-glucoside) and 36°C (1-O-dodecyl-β-D-glucoside). The T_K value of 1-O-dodecyl-α-D-mannoside is as high as 43°C. As further extension of the alkyl chain will further increase the value of T_K, the alkyl chain lengths of the conventional sugar-based surfactants are generally limited in a range C8–C12.

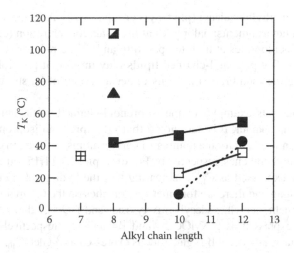

FIGURE 10.1 T_K, for alkyl α(β)-D-glycosides [8–10] as functions of the sugar headgroup and hydrocarbon chain length. (■): alkyl α-D-glucoside, (□): alkyl β-D-glucoside, (●): alkyl α-D-mannoside, (⊞): 1-*O*-heptyl-β-D-xyloside, (▲): 1-*O*-octyl-β-fucoside, (◪):1-*O*-octyl-α-galactoside.

This fact should not be underestimated. The aqueous phases available in the conventional sugar-based surfactant/water systems are mainly of the normal type [7,8,11], and inverted liquid crystalline phases can appear only at significantly high temperatures. For the same reason, majority of the synthetic analogues of natural membrane glycolipids can form liquid crystalline phases at temperatures significantly higher than room temperature [12]. This imposes a severe limitation in exploiting sugar-based surfactants in many technical applications. A new approach to depress T_K of the sugar-based surfactants is, therefore, necessary to fully realize their technical potentials.

Sugar-based surfactants with isoprenoid-type hydrophobic chains, hereafter referred to as isoprenoid-chained surfactants, are a new class of surfactants that largely overcome the high T_K problem inherent in the conventional sugar-based surfactants.

Isoprenoids are derivatives of terpenes and found in nearly every living creature. Biosynthetically produced from "isoprene unit," these compounds are acyclic or cyclic oligomers of C_5 chains with methyl branches. These are a large class of biological compounds such as geraniol and limonene (monoterpenes C_{10}), farnesol (sesquiterpenes C_{15}), and retinal and phytol (diterpenes C_{20}). Squalene (C_{30}) is a critical precursor to steroids and vitamin A is a C_{20} diterpene alcohol. In industry, the isoprenoids have been widely used as essential oil additives for commodity chemicals and cosmetics, but their usage as surfactant hydrophobic chains have been relatively limited. Important examples of isoprenoid-chained surfactants (lipids) are found in archaebacterial plasma membranes, where the polar lipids are all derived from a common basic core structure, glycerol di-(or tetra-)ether of

3,7,11,15-tetramethylhexadecyl (phytanyl) group [13,14]. Archaebacteria survive in exceptional environments: halophiles at high salt concentration (e.g., saturated brine), thermoacidophiles at high temperature and acidic pH, for example, 90°C and pH 2 [13]. The phytanyl-chained lipids play important physiological roles in maintaining the membrane functions under a variety of harsh environmental conditions [15].

Figure 10.2 lists examples of the isoprenoid-chained surfactants, which are described in this chapter. It is here noted that we express an isoprenoid chain as C_{p+q}, where p and q stand for the number of carbon atoms in the main chain and the number of methyl branches, respectively. For example, 3,7,11,15-tetramethylhexadecyl group is expressed as C_{16+4}, indicating that the hydrophobic chain is of 16 carbon atoms long and there are four methyl branches on the C_{16} main chain. Thus, 1-O-(3,7,11,15-tetramethylhexadecyl)-β-D-xyloside and 1-O-(3,7-dimethyloctyl)-β-D-glucoside are expressed as β-XylOC$_{16+4}$ and β-GlcOC$_{8+2}$, respectively. A straight-chained surfactant, 1-O-decyl-β-D-glucoside is expressed as β-GlcOC$_{10}$.

10.2 SYNTHESIS

Since the time of Emil Fisher [16,17], simple alkyl glycosides have been prepared by alcoholysis. For more complex alcohols, Königs–Knorr glycosylation [18] and its numerous modifications have been applied to preparation of well-defined sugar-based surfactants up to now. However, the fundamental issues in glycosylation have not changed [19]: which hydrophobic chain "acceptors" one should choose, how activated sugar part "donors" can be prepared, and what stereoselective glycosylation one should select as the most appropriate coupling reaction of individual donor and acceptor.

Figure 10.3 shows the synthetic strategy to the isoprenoid-chained sugar-based surfactants. As the glycosylation acceptors, the highly branched hydrophobic chains can be derived from a class of abundant natural products, isoprenoid alcohols such as geraniol, farnesol, and phytol. The isoprenoid alcohols are hydrogenated in the presence of palladium or platinum on activated carbon in a moderate to quantitative yield. In the absence of chain unsaturation, the obtained saturated alcohols are chemically more stable against autoxidation in comparison with unsaturated fatty alcohols such as oleyl alcohol. The double-chained derivatives, 1,3-di-O-alkylglycerol, can be prepared in the reaction of epichlorohydrin with 2 mole equivalent of the single-chained alcohol in the presence of sodium hydride.

To prepare activated forms of sugars "glycosylation donors" one introduces a leaving group to the sugar reducing terminus and protective groups to the other hydroxy groups. Acetyl protective group has been widely used, because one can efficiently attain β-selective glycosylation by the neighboring group effect [19] via the acetyl group at 2-position on the pyranose or furanose ring. The typical leaving groups used for donors are bromide, acetate, trichloroacetimidate, and so on. Among them, trichloroacetimidate [20,21] is suitable to oligosaccharide derivatives because of the selective and mild reaction conditions both at the preparation and glycosylation steps.

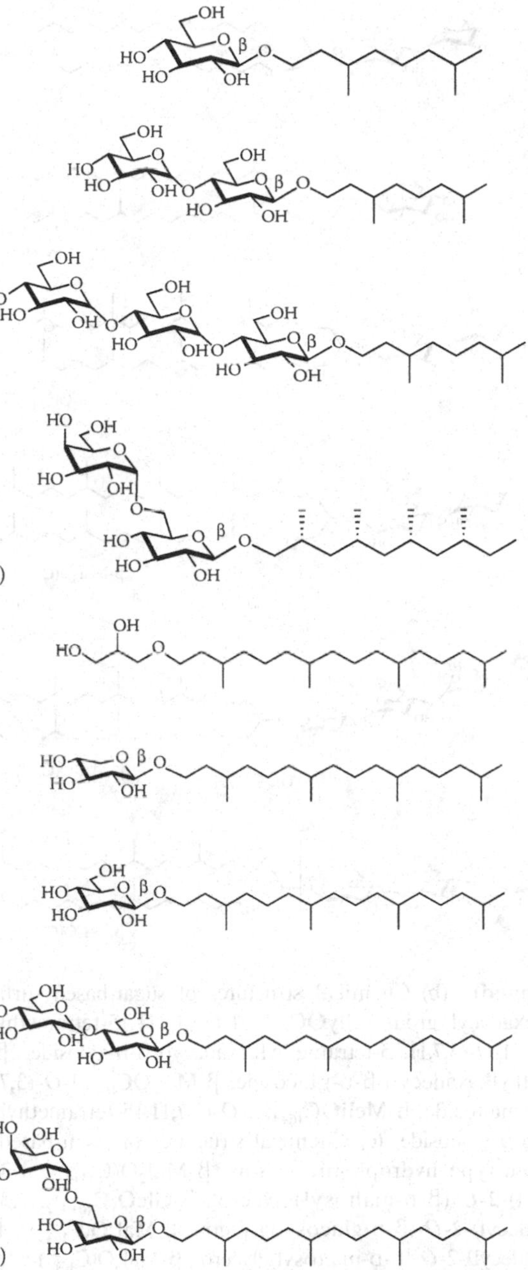

β-GlcOC_{8+2}

β-Mal_2OC_{8+2}

β-Mal_3OC_{8+2}

β-MeliOC_{10+4}

(a)

GlyOC_{16+4}

β-XylOC_{16+4}

β-GlcOC_{16+4}

β-Mal_2OC_{16+4}

β-MeliOC_{16+4}

(b)

FIGURE 10.2 (a) Chemical structures of sugar-based surfactants with 3,7-dimethyloctyl and (2*R*, 4*R*, 6*R*, 8*R*)-2, 4, 6, 8-tetramethyldecyl group. β-GlcOC_{8+2}: 1-*O*-(3,7–dimethyloctyl)-β-D-glucoside, β-Mal_2OC_{8+2}: 1-*O*-(3,7–dimethyloctyl)-β-D-maltoside, β-Mal_3OC_{8+2}: 1-*O*-(3, 7–dimethyloctyl)-β-D-maltotrioside, β-MeliOC_{10+4}: 1-*O*-[(2*R*, 4*R*, 6*R*, 8*R*) -2, 4, 6, 8-tetramethyldecyl]-6-*O*-(α-D-galactosyl)-β-D-glucoside.

(continued)

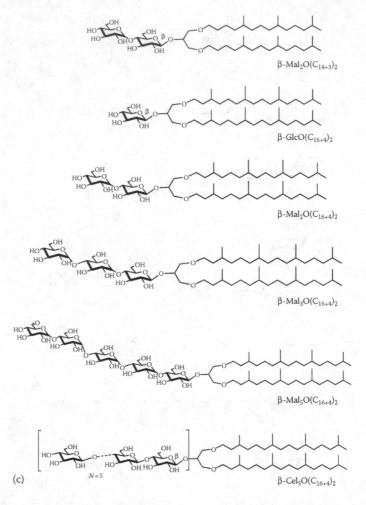

FIGURE 10.2 (continued) (b) Chemical structures of sugar-based surfactants with 3,7,11,15-tetramethylhexadecyl group. GlyOC$_{16+4}$: 1-O-(3,7,11,15-tetramethylhexadecyl) glycerol, β-XylOC$_{16+4}$: 1-O-(3,7,11,15-tetramethylhexadecyl)-β-D-xyloside, β-GlcOC$_{16+4}$: 1-O-(3,7,11,15-tetramethylhexadecyl)-β-D-glucoside, β-Mal$_2$OC$_{16+4}$:1-O-(3,7,11,15-tetramethylhexadecyl)-β-D-maltoside, β-MeliOC$_{16+4}$: 1-O-(3,7,11,15-tetramethylhexadecyl)-6-O-(α-D-galactosyl)-β-D-glucoside. (c) Chemical structures of sugar-based surfactants with double isoprenoid-type hydrophobic chains. β-Mal$_2$O(C$_{14+3}$)$_2$: 1,3-Di-O-(5,9,13-trimethyltetradecyl)-2-O-(β-D-maltosyl)glycerol, β-GlcO(C$_{16+4}$)$_2$: 1,3-Di-O-(3,7,11,15-tetramethylhexadecyl)-2-O-(β-D-glucosyl)glycerol, β-Mal$_2$O(C$_{16+4}$)$_2$: 1,3-Di-O-(3,7,11,15-tetramethylhexadecyl)-2-O-(β-D-maltosyl)glycerol, β-Mal$_3$O(C$_{16+4}$)$_2$: 1,3-Di-O-(3,7,11,15-tetramethylhexadecyl)-2-O-(β-D-maltotriosyl)glycerol, β-Mal$_5$O(C$_{16+4}$)$_2$: 1,3-Di-O-(3,7,11,15-tetramethylhexadecyl)-2-O-(β-D-maltopentaosyl)glycerol, β-Cel$_5$O(C$_{16+4}$)$_2$:1,3-Di-O-(3,7,11,15-tetramethylhexadecyl)-2-O-(β-D-cellopentaosyl)glycerol.

Glycosyl donors   Isoprenoid chains

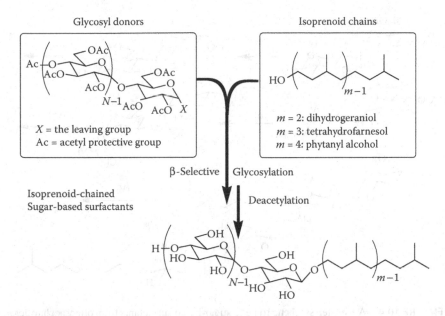

X = the leaving group
Ac = acetyl protective group

m = 2: dihydrogeraniol
m = 3: tetrahydrofarnesol
m = 4: phytanyl alcohol

β-Selective | Glycosylation

Isoprenoid-chained
Sugar-based surfactants

Deacetylation

FIGURE 10.3 Synthetic strategy to well-defined isoprenoid-chained sugar-based surfactants. X = the leaving group, N = the number of glucose residues, and m = the number of isoprenoid units.

Figure 10.4 shows a five-step synthetic route of isoprenoid-chained surfactants via trichloroacetimidate glycosylation, taking 1-O-β-(3,7-dimethyloctyl)-D-maltoside, β-Mal$_2$OC$_{8+2}$ as a typical example [22]. At the first step, a chromatographically purified oligosaccharide was acetylated with acetic anhydride in pyridine in the presence of a catalytic amount of 4-dimethylaminopyridine. At the second step, this peracetylated oligosaccharide was treated with 1.4 equivalent of hydrazine acetate in N,N-dimethylformamide (DMF) at 50°C for 2 h, leading to a selective deacetylation at the reducing terminus [23]. At the third step, this derivative was reacted with trichloroacetonitrile in dichloromethane in the presence of ~0.1 equivalent of cesium carbonate [24]. In the glycosylation coupling [20,21], the trichloroacetimidate was reacted with 3,7-dimethyl-1-octanol in one-to-one molar ratio in the presence of trimethylsilyl trifluoromethanesulfonate, followed by the conventional deacetylation by sodium methoxide. The overall yields of the reaction route were 50%–60%.

Provided the mono- or oligosaccharide possesses a reducing terminus, the present method allows us to couple a desired oligosaccharide β-glycosidically with a desired alkyl alcohol with a definite chain length. By employing this method, we are now able to synthesize desired series of chemically pure sugar-based surfactants. The alkyl furanosides with isoprenoid chains may be also attractive as a new class of sugar-based surfactants. For preparation of 1-O-alkyl-β-D-riboside

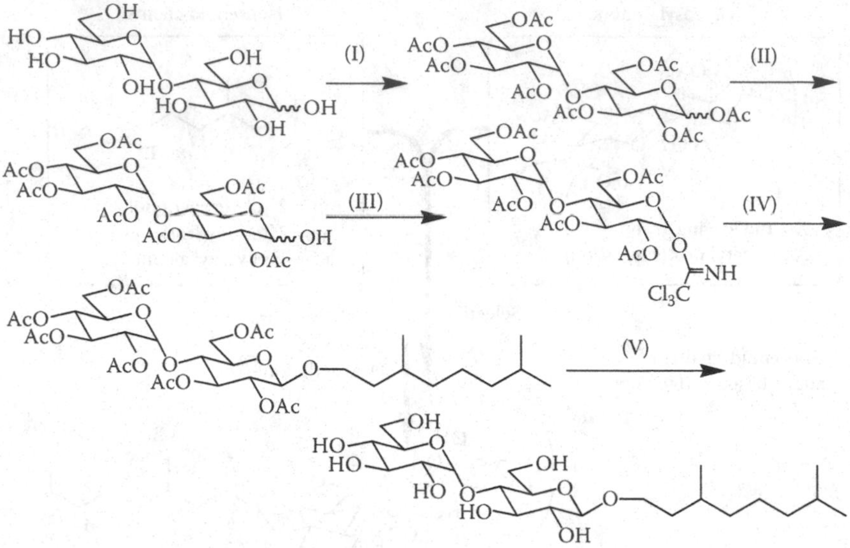

FIGURE 10.4 A five-step synthetic route to sugar-based surfactants from oligosaccharides. (I) Ac$_2$O, 4-dimethylaminopyridine/pyridine (r.t. > 3 h), (II) hydrazine acetate/DMF (50°C, 2 h). (III) CCl$_3$CN, Cs$_2$CO$_3$/CH$_2$Cl$_2$ (r.t. > 2 h), (IV) 3,7-dimethyl-1-octanol, TMSOTf, MS4A/CH$_2$Cl$_2$ (0°C, 6 h), (V) NaOMe/MeOH (0°C, 1 h).

possessing a primary alkyl chain, stannic chloride (tin(IV) tetrachloride) is the most effective Lewis acid at the glycosylation step; the alkyl furanoside was obtained via an orthoester intermediate in a good yield [25]. As a related polyol derivative, one can prepare O-alkylglycerols by O-alkylation of glycerol or a protected glycerol compound with an isoprenoid-chained bromide in the presence of a base.

10.3 PHYSICAL PROPERTIES

10.3.1 THERMAL PROPERTIES

Tables 10.1 through 10.3 summarize the values of T_K of the isoprenoid-chained surfactants. One important feature of the isoprenoid-chained surfactants is their low values of T_K. Particularly, the T_K values of surfactants with maltooligosaccharides (Mal$_N$) headgroups (Mal$_N$-surfactants), are generally below 0°C, even when the total number of carbon atoms in the hydrophobic group is as large as 40. We show DSC thermogram of the β-GlcOC$_{16+4}$/water system in Figure 10.5 [136] as a representative thermogram of Mal$_N$-surfactants [26]. The curve (A) shows the heating thermogram for a one-phase region of a lamellar, L$_\alpha$, phase (80.6 wt%), while curve (B) shows the thermogram for an L$_\alpha$ + W two-phase region just outside the one phase L$_\alpha$ region (77.4 wt%). No endothermic peak associated with the melting of hydrated solid β-GlcOC$_{16+4}$ was observed over the temperature range

TABLE 10.1

Phase Behavior and Structural Features of the Liquid Crystalline Phases of the Isoprenoid-Chained Surfactants with 3,7-Dimethyloctyl, and (2R, 4R, 6R, 8R)-2,4,6,8-Tetramethyldecyl Groups, and Straight-Chained Surfactants at 25°C

Surfactant	T_K (°C)	Phase Sequence	(n_w/n_L) (mol/mol)	d_{hc} (nm)	A (nm²)	Phase
β-GlcOC$_{8+2}$	<0	L$_1$–L$_{1''}$–L$_\alpha$	10.3	0.64	0.46	L$_\alpha$
β-Mal$_2$OC$_{8+2}$	<0	L$_1$–H$_I$–Q$_I$(Ia3d)–L$_\alpha$–L$_c$	16	1.15	0.51	H$_I$
			8.5	—	0.50	Q$_I$(Ia3d)
			3.0	0.64	0.46	L$_\alpha$
β-Mal$_3$OC$_{8+2}$	<0	L$_1$–H$_I$	19	1.15	0.52	H$_I$
β-MeliOC$_{10+4}$	~10	L$_1$–L$_{1''}$	—	—	1.0	Aggregate[a]
β-GlcOC$_8$ [9,11,75]	<0	L$_1$–H$_I$–Q$_I$(Ia3d)–L$_\alpha$–L$_c$	4.1	0.67	0.36	L$_\alpha$
β-GlcOC$_9$ [11,75]	—	L$_1$–(H$_I$)–Q$_I$–L$_\alpha$–L$_c$	3.9	0.75	0.38	L$_\alpha$
β-GlcOC$_{10}$ [8,11][b]	26	L$_1$–L$_{1''}$L$_\alpha$L$_c$	7.6	0.78	0.40	L$_\alpha$

[a] Bilayer/aggregate, a size of which is about 400 nm estimated from SANS measurement (=1 × 10^{-3} g/mL, ~20 times of the CMC at 50°C).

[b] Data at 30°C.

TABLE 10.2

Phase Behavior and Structural Features of the Liquid Crystalline Phases of the Isoprenoid-Chained Surfactants with 3,7,11,15-Tetramethylhexadecyl Group at 25°C

Surfactant	T_K (°C)	Phase Sequence	(n_w/n_L) (mol/mol)	d_{hc} (nm)	A (nm²)	Phase
GlyOC$_{16+4}$	<0	W–H$_{II}$–FI	4	0.98–1.33	0.37	H$_{II}$
β-XylOC$_{16+4}$	9	W–Q$_{II}$(Pn3m/Ia3d)–L$_\alpha$	13	1.3	0.41	Q$_{II}$(Pn3m)
			12	1.3	0.40	Q$_{II}$(Ia3d)
			5.2	1.35	0.42	L$_\alpha$
β-GlcOC$_{16+4}$	<0	W–L$_\alpha$–	8.6	1.3	0.49	L$_\alpha$
β-Mal$_2$OC$_{16+4}$	<0	W–L$_\alpha$–	13	1.1	0.50	L$_\alpha$
β-MeliOC$_{16+4}$	33	L$_1$–?	—	—	0.90	L$_1$[a]

[a] Cylindrical micelles in a diluted solution regime (c = 5×10^{-4} g/mL, ~25 times of the CMC).

TABLE 10.3

Phase Behavior and Structural Features of the Liquid Crystalline Phases of the Isoprenoid-Chained Surfactants with Double 3,7,11,15-Tetramethylhexadecyl and 5,9,13-Trimethyltetradecyl Groups at 25°C

Surfactant	T_K (°C)	Phase Sequence	(n_w/n_L) (mol/mol)	d_{hc}^a (nm)	A^a (nm²)	Phase
C-Mal$_2$O(C$_{14+3}$)$_2$	—	W–H$_{II}$–	—	—	—	H$_{II}$
C-GlcO(C$_{16+4}$)$_2^b$	<0	W–Q$_{II}$(Fd3m)–	4.4	0.9–1.9	0.48–0.51	Q$_{II}$(Fd3m)
C-Mal$_2$O(C$_{16+4}$)$_2$	<0	W–H$_{II}$–	20	1.2–1.7	0.75	H$_{II}$
C-Mal$_3$O(C$_{16+4}$)$_2$	<0	W–L$_\alpha$–	38	1.3	1.0	L$_\alpha$
C-Mal$_5$O(C$_{16+4}$)$_2$	<0	W–L$_\alpha$–	45	1.3	1.0	L$_\alpha$
C-Cel$_5$O(C$_{16+4}$)$_2$	135	Hydrated solid	—	—	—	—

[a] We assumed 1,3-di-O-3,7,11,15-tetramethylhexadecylglycerol as the hydrophobic part of the surfactants, so that three oxygen atoms are included in the hydrophobic part of the lipids.
[b] Data at 0°C.

from −100°C to 100°C. An endothermic peak at 0°C in the curve (B) is due to the melting of ice [136]. Essentially the similar results, that is, no transition associated with melting of hydrated solid lipid takes place above 0°C, were obtained for other Mal$_N$-surfactants, for example, β-GlcOC$_{8+2}$, β-Mal$_2$OC$_{16+4}$, and β-Mal$_N$O(C$_{16+4}$)$_2$/water systems with $N = 1$–5 [27,28], indicating that the values of T_K of the Mal$_N$-surfactants are below 0°C.

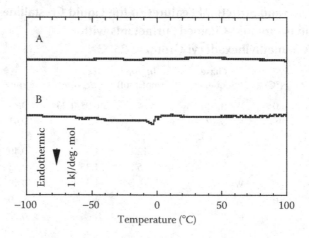

FIGURE 10.5 DSC heating thermograms of aqueous β-GlcOC$_{16+4}$. (A) 80.6 wt% lipid (a one-phase region of the L$_\alpha$ phase). (B) 77.4 wt% lipid (a water + L$_\alpha$ two-phase region). The heating rate was 1.0°C/min. The molar isobaric heat capacity of the surfactant is taken as the y axis in the thermogram. (Reproduced from Salkar, R.J., Minamikawa, H., and Hato, M., *Chem. Phys. Lipids*, 127, 65, 2004. With permission.)

The decisive headgroup effects on the values of T_K of the isoprenoid-chained surfactants are also evident (Tables 10.1 through 10.3). For example, the values of T_K of β-XylOC$_{16+4}$, β-MeliOC$_{16+4}$, and β-Cel$_5$O(C$_{16+4}$)$_2$, are significantly higher than those of the Mal$_N$-surfactants, that is, 9°C, 33°C, and 135°C, respectively [29–31]. The remarkable difference in the T_K values of β-Mal$_5$O(C$_{16+4}$)$_2$ (<0°C) and β-Cel$_5$O(C$_{16+4}$)$_2$ (135°C) can be explained in terms of the different conformations of the headgroups (Figures 6 and 7 in Ref. [32]). Mal$_N$ and the cellooligosaccharides (Cel$_N$), both made of N glucose residues, are different in their configuration of their glycosidic linkages, β-1,4-O-glycosidic bonds and α-1,4-O-glycosidic bonds for Cel$_N$ and Mal$_N$, respectively. For the Cel$_N$-headgroups, which possess the same repeating structure as cellulose, the inter-headgroup cohesive forces rapidly increase as N increases, thereby suppressing the hydration-induced melting point depression of β-Cel$_5$O(C$_{16+4}$)$_2$(4). A very high value of T_K (>160°C) for a β-Cel$_5$O(C$_{12}$)$_2$/water system is also due to the same mechanism [32]. For the Mal$_N$-headgroups, which possess the same repeating structure as amylose, the inter-headgroup cohesive forces are reduced as N increases, thereby lowering the values of T_K [32]. The higher value of T_K of 1-O-heptyl-β-D-xyloside than that of 1-O-heptyl-β-D-glucoside may be inferred from Figure 10.1.

In summary, the values of T_K of the isoprenoid-chained surfactants with rationally selected sugar headgroups, for example, the Mal$_N$-headgroups, can be depressed below 0°C, even when the total number of carbon atoms in the hydrophobic group is as large as 40. This is in marked contrast to those of straight-chained sugar-based surfactants. Due to their low T_K values, the sugar-based isoprenoid-chained surfactants significantly expand our freedom to control the molecular structures and hence the aqueous phase structures at low temperatures, which is discussed in the following sections.

10.3.2 Phase Behavior

In this section we discuss the phase behavior of the isoprenoid-chained surfactants with dihydrogeranyl (C$_{8+2}$) and phytanyl (C$_{16+4}$) groups. The dihydrogeranyl group, C$_{8+2}$, contains eight carbon atoms in its main chain with two methyl branches at 3 and 7 positions (total of 10 carbon atoms). The phytanyl (C$_{16+4}$) group contains 16 backbone carbon atoms and 4 methyl groups at 3, 7, 11, and 15 positions (total of 20 carbon atoms). The hydrophilic headgroups are systematically altered, for example, maltooligosaccharides (Mal$_N$), glucose (Glc = Mal$_1$), xylose (Xyl), and glycerol (Gly) in order of the decreasing size (molecular cross-section area) of the headgroups. The effects of the headgroup stereochemistry on the phase behavior are also discussed by comparing the N-dependent phase behavior of surfactants with the Mal$_N$- and Cel$_N$-headgroups. Somewhat unexpected phase behavior of C$_{16+4}$-chained surfactant with α-1–6 linked sugar (melibiose) headgroup as compared to that of α-1–4 linked sugar (maltose) headgroup is also discussed. The (partial) phase diagram of each surfactant/water system was determined by small-angle x-ray scattering (SAXS) and polarizing microscopy measurements.

10.3.2.1 Sugar-Based Surfactants with 3,7–Dimethyloctyl Group

10.3.2.1.1 β-GlcOC$_{8+2}$/Water System

Figure 10.6a shows a partial phase diagram of the β-GlcOC$_{8+2}$/water system examined over 0.1 to 97.3 wt% β-GlcOC$_{8+2}$ and at temperatures from 20°C to 135°C [26,74]. At 25°C, there are two isotropic regions, one in a dilute concentration regime below about 0.17 wt% β-GlcOC$_{8+2}$, $L_{1'}$, and the one in

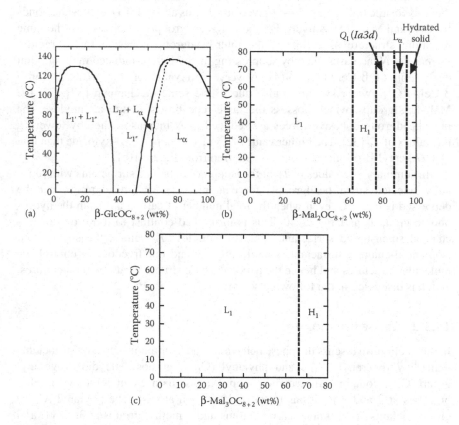

FIGURE 10.6 Partial phase diagrams of the isoprenoid-chained surfactants with 3,7- dimethyloctyl group. (a) Phase diagram of the β-GlcOC$_{8+2}$/water system (Reproduced from Salkar, R.J., Minamikawa, H., and Hato, M., *Chem. Phys. Lipids*, 127, 65, 2004. With permission; Yamashita, I., Kawabata, Y., Kato, T., Hato, M., and Minamikawa, H., *Colloid Surf. A*, 250, 485, 2004. With permission); (b) Partial phase diagram of the β-Mal$_2$OC$_{8+2}$/water system. The system was examined up to 96 wt% β-Mal$_2$OC$_{8+2}$ (monohydrate solid) (Reproduced from Minamikawa, H. and Hato, M., *Chem. Phys. Lipids*, 134, 151, 2005. With permission.); (c) Partial phase diagram of the β-Mal$_3$OC$_{8+2}$/water system. The system was examined up to 80 wt% β-Mal$_3$OC$_{8+2}$. (Reproduced from Minamikawa, H. and Hato, M., *Chem. Phys. Lipids*, 134, 151, 2005. With permission.)

a concentrated regime from 37 to 53 wt%, $L_{1''}$. The value of 0.17 wt% (~5 mM) β-GlcOC$_{8+2}$ is close to a concentration where the surface tension of aqueous β-GlcOC$_{8+2}$ first attains a minimum value (CMC) (Figure 10.10). Therefore, the $L_{1'}$ most probably consists of dilute micelles + molecularly dispersed β-GlcOC$_{8+2}$. Between 0.17 and 34 wt%, the system exhibits an $L_{1'}$ + $L_{1''}$ two-phase region. Above about 54 wt% β-GlcOC$_{8+2}$, the solution started to display birefringence and two SAXS diffraction peaks appeared with a d-spacing ratio of 1:0.5. This indicates the presence of a liquid crystalline phase, which is lamellar in origin. From 62 wt% to at least 97 wt%, the system forms an L_α one-phase region as evidenced from the d-spacing values versus concentration plot. Between 54 and 62 wt%, the L_α phase coexists with the $L_{1''}$ phase. The extent of the L_α phase goes on decreasing as temperature increases. Above 130°C, the L_α phase totally disappears giving way to an isotropic solution.

The β-GlcOC$_{8+2}$/water system displays an upper critical phenomenon of the two immiscible isotropic solutions. The $L_{1'}$+$L_{1''}$ two-phase region becomes narrower in concentration as the temperature increases. Above 130°C, the system is converted into an optically isotropic solution that consists of micelles and monodispersed surfactant. Thus, the β-GlcOC$_{8+2}$ is miscible with water in all proportions above 130°C. This behavior is in marked contrast to the generally observed "clouding phenomena" in nonionic surfactant/water systems including β-GlcOC$_{10}$ [11,33]. Similar upper critical phenomena have been observed for several surfactants whose aqueous solutions display immiscibility gaps, which lead to "cloud points." The examples are seen in the tetra- and penta-oxyethylene glycol decyl ether and also in the decyl and dodecyldimethylphosphine oxide [3,34,35]. The critical temperature of 130°C for the β-GlcOC$_{8+2}$/water system is significantly lower than those for the other surfactants with a C$_{10}$-chain, where the critical temperatures are close to 300°C.

10.3.2.1.2 β-Mal$_2$OC$_{8+2}$/Water and β-Mal$_3$OC$_{8+2}$/Water Systems

Figure 10.6b and c reports partial phase diagrams of β-Mal$_2$OC$_{8+2}$/water and β-Mal$_3$OC$_{8+2}$/water system, respectively [36]. The phase diagram of the β-Mal$_2$OC$_{8+2}$/water system is characterized by a large micellar solution region, L_1, followed by a series of normal type liquid crystalline phases, a normal hexagonal phase, H_I, a normal $Ia3d$ cubic phase, Q_I ($Ia3d$), and a lamellar phase, L_α, each separated by a nearly vertical phase boundary line. The β-Mal$_3$OC$_{8+2}$/water system is also characterized by the large L_1 and H_I phases. This is in marked contrast to the β-GlcOC$_{8+2}$/water system ($N = 1$), in which β-GlcOC$_{8+2}$ is practically insoluble in water. The β-Mal$_N$C$_{8+2}$/water system therefore shifts from water-insoluble to water-soluble system and from the L_α alone ($N = 1$) to the H_I dominating system as N increases from 1 to 2 to 3. It is also noted that the maximum amount of water molecules incorporated in the H_I phase is low and temperature insensitive, for example, about 16 moles of water per β-Mal$_2$OC$_{8+2}$ and about 20 moles of water per β-Mal$_3$OC$_{8+2}$, respectively. These values correspond to about two water molecules per OH group of the headgroup.

10.3.2.2 Sugar-Based Surfactants with 3,7,11, 15-Tetramethylhexadecyl Group

10.3.2.2.1 GlyOC$_{16+4}$/Water System
The partial phase diagram of the GlyOC$_{16+4}$/water system is reported in Figure 10.7a [37,38]. Below 65°C, the SAXS diffraction of the system is characterized

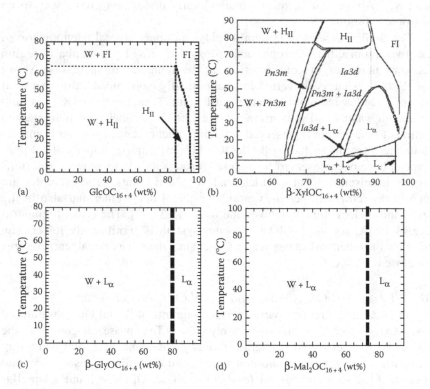

FIGURE 10.7 Partial phase diagrams of isoprenoid-chained surfactants with 3,7,11,15-tetramethyloctadecyl group. (a) GlyOC$_{16+4}$/water system (Reproduced from Hato, M., Minamikawa, H., Salkar, R.A., and Matsutani, S., *Progr. Colloid Polym. Sci.*, 123, 56, 2004. With permission), (b) β-XylOC$_{16+4}$/water system (Reproduced from Hato, M., Yamashita, I., Kato, T., and Abe, Y., *Langmuir*, 20, 11366, 2004. With permission), (c) β-GlcOC$_{16+4}$/water system (Reproduced from Hato, M., Minamikawa, H., Salkar, R.A., and Matsutani, S., *Progr. Colloid Polym. Sci.*, 123, 56, 2004. With permission), (d) β-Mal$_2$OC$_{16+4}$/water system (Reproduced from Hato, M., Minamikawa, H., Salkar, R.A., and Matsutani, S., *Progr. Colloid Polym. Sci.*, 123, 56, 2004. With permission).

by three diffraction peaks whose spacing ratio is $1:1/\sqrt{3}:1/2$ that is consistent with a two-dimensional hexagonal lattice. The value of the first-order diffraction at full hydration, $d_1 = 4.1_5$ nm (at 4°C), decreased only slightly as the temperature increases, about 0.01 nm/°C over a temperature range from 4°C to 60°C. By analogy with other lipid/water systems, the low hydration capacity of the mesophase indicates that the structure of this phase is of type H_{II}. On dilution below about 85 wt% surfactant, excess water, W, separates from the H_{II} phase to form a $W + H_{II}$ two-phase region rather than to form a micellar solution, L_1. At higher temperatures above about 65°C, the H_{II} phase transforms into a fluid isotropic phase, FI, which gives a broad SAXS peak at around 3.7 nm. Above about 96 wt% surfactant, the system no more forms a liquid crystalline phase. An $H_{II} + FI$ coexistence region is very narrow and could not be experimentally detected.

10.3.2.2.2 β-XylOC$_{16+4}$/Water System*

The partial phase diagram of the β-XylOC$_{16+4}$/water system (Figure 10.7b) exhibits rich phase behavior involving a crystalline L$_c$ phase, a wax-like solid, an L$_\alpha$ phase, an H_{II} phase, two inverted cubic phases of crystallographic space groups *Pn3m* and *Ia3d*, and a fluid isotropic phase, FI [29]. Anhydrous β-XylOC$_{16+4}$ is wax-like solid below about 43°C and did not adopt a crystalline state even after 4 month incubation at −10°C. Below 3 wt% water, a single broad peak characterizes the SAXS profile. Once the water content exceeds about 4 wt%, the β-XylOC$_{16+4}$/water system starts to form a crystalline lamellar phase L$_c$ below 9°C, and displays rich phase behavior as the temperature and the concentration varies. The L$_c$ phase consists of a mono-hydrate form of the surfactant, β-XylOC$_{16+4}$H$_2$O, which corresponds to about 96 wt% β-XylOC$_{16+4}$ as indicated by a vertical line in the phase diagram. Between 81 and 96 wt% β-XylOC$_{16+4}$, an L$_c$ + L$_\alpha$ phase coexistence region appears above 10°C, which is followed by a single L$_\alpha$ region that spans between approximately 96 and 83 wt% surfactant at 20°C. Above 20°C, the L$_\alpha$ region is triangular in shape with upper apex located at about 50°C and about 90 wt% surfactant. The *Ia3d* cubic phase, Q$_{II}$(*Ia3d*), is identified in a range from 68 to 82 wt% surfactant (30°C) and from 75 to presumably about 94 wt% (65°C). With more hydration, the *Ia3d* cubic phase gives way to a *Pn3m* cubic phase, Q$_{II}$(*Pn3m*). Compared with the *Ia3d* cubic phase, the *Pn3m* cubic phase region is considerably narrow. Above about 75°C, the cubic phases give way to the H_{II} phase, identified in a range from about 73 to 88 wt% surfactant (80°C). The H_{II} phase exists at least up to about 120°C, above which it melts into the fluid isotropic phase, FI (not indicated in the phase diagram). Finally it must be mentioned that the phase diagram of the β-XylOC$_{16+4}$/water system is similar to that of the monoolein/water [39], the alkyl β-D-glucosyl-*rac*-glycerol/water [40], and the phytantriol/water [41] systems, indicating that the phase behavior is largely determined by the average molecular shape rather than the detailed molecular structures.

---

* Reproduced from Hato, M., Yamashita, I., Kato, T., and Abe, Y. *Langmuir*, 20, 11366, 2004. With permission.

10.3.2.2.3 β-GlcOC$_{16+4}$/Water and β-Mal$_2$OC$_{16+4}$/ Water Systems

Figure 10.7c shows a phase diagram of β-GlcOC$_{16+4}$/water system, which was determined by the SAXS measurements up to 90 wt% β-GlcOC$_{16+4}$ [26,37]. The SAXS of the β-GlcOC$_{16+4}$/water system gives three diffractions with a d-spacing ratio of 1:1/2:1/3, corresponding to first-, second-, and third-order diffractions from a bilayer periodicity of an L$_\alpha$ phase. While the first d-spacing $d_1 = 4.3_8$ nm (at 25°C) is practically constant below 78 wt% lipid, the d_1 values decrease as the surfactant concentration is further increased. This indicates that a single-phase region of L$_\alpha$ starts above 78 wt% β-GlcOC$_{16+4}$, and extends at least up to 98 wt% surfactant that corresponds to β-GlcOC$_{16+4}$1/2H$_2$O and stable at least up to 100°C. Below 78 wt% surfactant, the system is in a W+ L$_\alpha$ two-phase regime. The phase boundary separating the biphasic and the one-phase L$_\alpha$ region is temperature insensitive, almost vertical to the concentration axis. The temperature effect on the d_1 values at full hydration is also small, about 0.001 nm/°C.

Figure 10.7d shows a phase diagram of the β-Mal$_2$OC$_{16+4}$/water system, the general feature of which is similar to the β-GlcOC$_{16+4}$/water system, indicating that addition of one glucose residue to the glucose headgroup does not appreciably affect the phase behavior [26,37]. The observed x-ray diffraction spacing in the ratio of 1:1/2:1/3 corresponds to an L$_\alpha$ phase. The value of the first-order diffraction line, $d_1 = 4.9_8$ nm (at 2°C) decreases only slightly with temperature: 0.002 nm/°C over the temperature range 2°C–70°C. On dilution below about 74 wt% surfactant, excess water separates from the L$_\alpha$ phase to form a W+ L$_\alpha$ two-phase region. The L$_\alpha$ phase extends at least up to 83 wt% surfactant (maximum concentration examined) and is stable at least up to 100°C. The major difference between the β-GlcOC$_{16+4}$/water and the β-Mal$_2$OC$_{16+4}$/water system lies in a maximum amount of hydration of the L$_\alpha$ phase: about 8 moles of water per β-GlcOC$_{16+4}$ and about 13 moles of water per β-Mal$_2$OC$_{16+4}$, respectively. It corresponds to about two water molecules per hydroxyl group of the headgroup, similar values as those found for the H$_I$ phase for the β-Mal$_2$OC$_{8+2}$/water and the β-Mal$_3$OC$_{8+2}$/water systems.

10.3.2.2.4 β-MeliOC$_{16+4}$/Water System

To conclude this section, we briefly describe the phase behavior of a 3,7,11,15-tetramethylhexadecyl-chained surfactant with α-1,6 linked sugar, melibiose, which consists of a galactose and glucose moiety, β-MeliOC$_{16+4}$. In contrast to β-Mal$_N$OC$_{16+4}$ ($T_K < 0$°C), a value of T_K of β-MeliOC$_{16+4}$ is as high as about 33°C [30]. Aqueous β-MeliOC$_{16+4}$ therefore gives a hydrated solid below 33°C. Above T_K, β-MeliOC$_{16+4}$ rapidly dissolves in water forming a clear transparent solution with a moderate foaming ability. This strongly suggests that β-MeliOC$_{16+4}$ is soluble in water by forming normal micelles at temperatures above its T_K. The SANS measurements for a 5×10^{-4} g/mL solution (~25 times of the CMC), corroborated the above observation; β-MeliOC$_{16+4}$ forms cylindrical micelles with the radius of about 1.9 nm and the length of about 29 nm (50°C) [30]. The formation of normal micelles of β-MeliOC$_{16+4}$ is in marked contrast to the phase behavior of the GlyOC$_{16+4}$/water, β-XylOC$_{16+4}$/

water, and β-Mal$_N$OC$_{16+4}$/water systems, where the surfactants forms an H$_{II}$, or a Q$_{II}$(Pn3m/Ia3d) or a L$_\alpha$ phase, representing another intriguing example that type of sugar headgroup has significant influences on the physical properties of sugar-based surfactants.

Finally it is interesting to note that β-MeliOC$_{10+4}$, which possesses the smaller hydrophobic chain than β-MeliOC$_{16+4}$, does not form normal micelles (Table 10.1). β-MeliOC$_{10+4}$ forms a two-phase solution of a β-MeliOC$_{10+4}$ rich and a poor phase above T_K (= ~15°C). The SANS measurements of the system indicates that the β-MeliOC$_{10+4}$ form disk-like large (bilayer) aggregates with the average radius of about 400 nm [30]. This somewhat puzzling result might be due to the reduced flexibility of the C$_{10+4}$ chain (an alternating –CH$_2$–CHCH$_3$– sequence) as compared to the C$_{16+4}$ chain [an alternating –(CH$_2$)$_3$– CHCH$_3$– sequence] [30].

10.3.2.3 Sugar-Based Surfactants with Double 3,7,11,15-Tetramethylhexadecyl Groups*

10.3.2.3.1 β-GlcO(C$_{16+4}$)$_2$/Water System
β-GlcO(C$_{16+4}$)$_2$/water system forms an *Fd3m* inverted micellar cubic phase (lattice constant: a_c = 14.8 nm, unit cell volume: V = 3.24 × 10^3 nm^3) over a temperature range at least from −20°C to 80°C, and is stable at least for 6 months at 0°C–25°C without any sign of decomposition of the lipid as examined by thin layer chromatography [28]. Maximum hydration of the *Fd3m* cubic phase, which was estimated by DSC measurement is about 4.4 moles of water per mole of β-GlcO(C$_{16+4}$)$_2$ (about 91 wt% at 0°C) [28]. The *Fd3m* structure consists of two types of discrete reverse micelles with different sizes; each unit cell consists of 8 larger ones and 16 smaller ones as shown in Figure 10.8a [42,43]. The interior of the micelles is filled by the polar headgroup of the surfactant and water, and the fluid hydrocarbon chains form a continuous three-dimensional matrix. In the present *Fd3m* cubic phase, the polar core radius of larger reverse micelle is 2.3 nm–1.83 nm, and the polar core radius of smaller reverse micelle is 1.46 nm–1.83 nm, respectively. The structure of the *Fd3m* cubic phase is consistent with the lipid molecular dimension [28].

10.3.2.3.2 β-Mal$_2$O(C$_{16+4}$)$_2$/Water System
A partial phase diagram of the β-Mal$_2$O(C$_{16+4}$)$_2$/water system is shown in Figure 10.8b [27]. The system is characterized by three SAXS peaks whose spacing ratio is 1:1/√3:1/2, consistent with a two-dimensional hexagonal lattice of type H$_{II}$. The value of the first-order diffraction, d_1 = 6.4 nm, increases only slightly with temperature: 0.002–0.006 nm/°C over the temperature range, 25°C–65°C. A one-phase region of the H$_{II}$ phase starts above 74 wt% β-Mal$_2$O(C$_{16+4}$)$_2$ and extends at least up to 82 wt% β-Mal$_2$O(C$_{16+4}$)$_2$. Below 74 wt%, excess water W (dilute aqueous

---

* Reproduced from Minamikawa, H. and Hato, M., Langmuir, 13, 2564, 1997; Minamikawa, H. and Hato, M., *Langmuir*, 14, 4503, 1998. With permission.

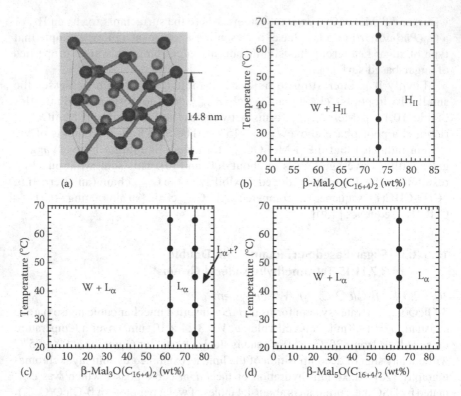

FIGURE 10.8 Partial phase diagrams of β-Mal$_N$O(C$_{16+4}$)$_2$/water systems. (a) Schematic representation of the structure of the cubic phase of space group *Fd3m* in the β-GlcO (C$_{16+4}$)$_2$/water system. The structure consists of eight globules of symmetry $\overline{4}3m$ (black), centered at position a, and 16 globules (grey) of symmetry $\overline{3}m$, centered at position d. (b) Partial phase diagram of β-Mal$_2$O(C$_{16+4}$)$_2$/water system examined up to 82 wt% β-Mal$_2$O(C$_{16+4}$)$_2$. (c) Partial phase diagram of β-Mal$_3$O(C$_{16+4}$)$_2$/water system examined up to 82 wt% surfactant. (d) Partial phase diagram of β-Mal$_5$O(C$_{16+4}$)$_2$/water examined up to 82 wt% surfactant. (Reproduced from Minamikawa, H. and Hato, M., *Langmuir*, 13, 2564, 1997. With permission.)

lipid) separates from the H$_{II}$ phase to form a two-phase region of W + H$_{II}$. The phase boundary is nearly independent of temperature, so the boundary line is practically parallel to the temperature axis.

10.3.2.3.3   β-Mal$_3$O(C$_{16+4}$)$_2$/Water and β-Mal$_5$O(C$_{16+4}$)$_2$/Water Systems

A partial phase diagram of the β-Mal$_3$O(C$_{16+4}$)$_2$/water system is shown in Figure 10.8c. A one-phase region of an L$_\alpha$ phase appears in the concentration range 62–75 wt% β-Mal$_3$O(C$_{16+4}$)$_2$ (water/lipid ratio, n_w/n_L, from 38 to 21). Below 62 wt%, excess water (dilute aqueous lipid) separates from the L$_\alpha$ phase to form a two-phase region of the L$_\alpha$ and dilute aqueous lipid, W. Above about 75 wt% β-Mal$_3$O(C$_{16+4}$)$_2$, a new peak at 5.1 nm at 25°C (5.2 nm at 65°C) is superposed to the previous set of diffraction peaks for the L_α phase. The values of the diffraction peaks for the L_α phase are identical to those obtained below 75 wt% β-Mal$_3$O(C$_{16+4}$)$_2$. This indicates that a second liquid crystalline phase coexists with the L_α phase above 75 wt% β-Mal$_3$O(C$_{16+4}$)$_2$. As only single diffraction peak was obtained from the second phase, we could not unequivocally define its phase structure [27].

A partial phase diagram of the β-Mal$_5$O(C$_{16+4}$)$_2$/water system shown in Figure 10.8d is similar to that of the β-Mal$_3$O(C$_{16+4}$)$_2$/water system [27]. A major difference is seen in that the L_α phase of the β-Mal$_5$O(C$_{16+4}$)$_2$/water system spans over a wider concentration range from 64 ± 3 wt% lipid ($n_w/n_L = 45$) to at least 82 wt% lipid ($n_w/n_L = 18$). Below 64 wt% lipid, a W + L_α two-phase region appears. The phase boundaries are also practically parallel to the temperature axis. As expected, owing to the bulkier Mal$_5$-headgroup, the polar layer thickness of the L_α phase is larger for the β-Mal$_5$O(C$_{16+4}$)$_2$/water system than the β-Mal$_3$O(C$_{16+4}$)$_2$/water system, 3.3 nm and 4.7 nm, respectively.

10.3.2.4 Molecular Correlation

Tables 10.1 through 10.3 summarize the aqueous phase behavior together with the structural parameters of the phases found in the isoprenoid-chained surfactant/water systems. (n_w/n_L), d_{hc}, and A denote the maximum amount of water swelled in the liquid crystalline phase, the hydrophobic chain length of the surfactant, and the average molecular cross-section area at the polar/apolar interface, respectively.

10.3.2.4.1 Molecular Structure–Aqueous Phase Behavior Relationship
The aqueous phase structures of the isoprenoid-chained surfactants are well correlated with the surfactant molecular structure. Given a hydrophobic group, for example, the spontaneous curvature of the Mal$_N$-surfactants consistently decrease as N increases, for example, from L_α to H$_I$ for β-Mal$_N$OC$_{8+2}$/water systems, and from an inverted micellar cubic phase Q$_{II}$ (*Fd3m*) to H$_{II}$ to L$_\alpha$ for β-Mal$_N$O(C$_{16+4}$)$_2$/water systems. The sign of the curvature is taken to be positive when the surfactant headgroup surface bends toward the water. This trend is attributable to the preferred "helical" conformation of the Mal$_N$-headgroups, which is similar to the "flexible helix" conformation of 1,4-α-D-glucan, amylose [44]. Due to the helical conformation, the cross-section area of the Mal$_N$-headgroups increases as N increases ([32,45], and Figure 6 in Ref. [32]), thereby shifting the spontaneous curvature of the surfactant monolayer toward negative direction as N increases, that is, increasing tendency for the surfactant monolayer to curve toward apolar region. Increasing cross-section area as N increases is observed; the values of fully compressed area of β-Mal$_N$O(C$_{16+4}$)$_2$ monolayer at the air/water interface are 0.70, 0.73, and 0.82 nm^2/molecule for $N = 1$, 3, and 5, respectively [46].

Essentially the same mechanism holds for the surfactants with a single 3,7,11,15-tetramethylhexadecyl group (Figure 10.7); the H$_{II}$ phase for the smallest glycerol-headgroup, and the L_α phase for the glucose (=Mal_1) and Mal_2-headgroups. Xylose differs from glucose in that the hydroxymethylene group, CH_2-OH, at carbon C5 of glucose is replaced by a hydrogen atom. Thus, the cross-section area of the xylose-headgroup would be intermediate between the glucose- and the glycerol-headgroups. The β-XylOC$_{16+4}$/water system in fact gives phase behavior intermediate between that of GlyOC$_{16+4}$, and β-GlcOC$_{16+4}$ [29,38].

It must, however, be noted that the observed systematic N-dependent phase behavior observed for the Mal_N-surfactants does not always hold for surfactants with other sugar-headgroups. As discussed in the Section 10.3.1, the Cel$_N$-headgroup prefers an "extended conformation" (similar to the "extended ribbon" conformation in 1,4-β-D-glucan cellulose) [44], where the cross-section area of the headgroups remains nearly constant irrespective of N ([32,45], and Figure 7 in Ref. [32]). Increased N in β-Cel$_N$O(C$_{16+4}$)$_2$ leads to tight inter-headgroup associations, thereby stabilizing the crystal phase rather than shifting the spontaneous curvature toward negative direction [32,45]. The same conformation-dependent phase behavior was observed for β-Cel$_N$O(C$_{12}$)$_2$/water and β-Mal$_N$O(C$_{12}$)$_2$/water systems [32,76]. The drastic differences in the phase behavior observed for the β-MeliOC$_{16+4}$/water and the β-Mal$_N$OC$_{16+4}$/water systems are attributable to the significantly large A value (0.9 nm^2) of MeliOC$_{16+4}$ as compared with 0.4–0.5 nm^2 for the other C$_{16+4}$ chained-surfactants (Table 10.2). The larger A value of MeliOC$_{16+4}$ most presumably arises from the three rotational freedoms of the (1→6) linked melibiose-headgroup as compared to the two rotational freedoms of the (1→4) linked Mal$_N$-headgroups [44]. These results again demonstrate a crucial contribution of the headgroup conformations on the physical properties of the sugar-based surfactants.

10.3.2.4.2   Hydration of the Liquid Crystalline Phases

As already noted in Section 10.3.2.3.3, the maximum amount of water molecules incorporated in the liquid crystalline phases (n_w/n_L) is relatively low and temperature insensitive (the phase boundaries are nearly parallel to the temperature axis). Let us look at the (n_w/n_L) values for the liquid crystalline phases in equilibrium with W or L_1 phase. The values for the H$_1$ phase of β-Mal$_2$OC$_{8+2}$ and β-Mal$_3$OC$_{8+2}$ are 10 and 16 water molecules per surfactant, corresponding to about two water molecules per OH group in the headgroup (Table 10.1). The similar hydration levels have also been observed for an H$_{II}$ and an L_α phase of 3,7,11,15-tetramethylhexadecyl-chained surfactant/water systems (Table 10.2). The hydration level of β-Mal$_N$O(C$_{16+4}$)$_2$ is slightly higher, about 3–4 water molecules per OH group, most presumably due to a larger molecular cross-section area A (Table 10.3). This suggests that apart from strongly bound primary hydrated water molecules, only a few extra water molecules can exist in a "hydration shell" of the sugar headgroups. The similar low hydration levels are also observed for other sugar-based surfactants such as 1-O-octyl-β-D-glucoside [47], monoglucosyldiglyceride [48], and digalactosyldiacylglycerol DGDG [49], indicating that the low levels of hydration are common for nonionic sugar-based surfactants [37].

Very small hydration amount (about one water molecule per OH group) of the $Q_{II}(Fd3m)$ phase in the β-GlcO(C_{16+4})$_2$/water system is due to nearly close packed β-GlcO(C_{16+4})$_2$ molecules in the $Fd3m$ cubic phase; the value of $A(=0.48-0.51\,nm^2)$ is close to that of the closely packed phytanyl chain, $0.48\,nm^2$. Larger hydration of the $Pn3m$ cubic phase of the β-XylOC$_{16+4}$/water system as compared to other C_{16+4} chained surfactants, is due to its bicontinuous structures of hyperbolic shape [50].

10.3.2.4.3   Salient Features of the Isoprenoid Chains
10.3.2.4.3.1   Static Aspects
As expected from the highly branched structures, the values of molecular cross-section area at the polar/apolar interface, A, of the isoprenoid-chained surfactants are generally larger than those of the straight-chained counterparts, for example, the A values for the L_α phase of β-Mal$_N$O(C_{16+4})$_2$ are about $1.0\,nm^2$, while values from 0.65 to $0.70\,nm^2$ were reported for the straight-chained lipids such as DGDG and phospholipids [51–53]. The larger A values of β-Mal$_N$OC$_{8+2}$ are also evident as compared to those of the straight-chained counterparts (Table 10.1).

The isoprenoid chain lengths in the L_α phase (d_{hc}), on the other hand, do not differ significantly from that of straight-chained counterparts (Table 10.1). A similar result was also reported for the A and d_{hc} values for the L_α phase of 1,2-di-(3,7,11,15-tetramethylhexadecanoyl)-sn-glycero-3-phosphocoline (DPhPC) and the straight-chain counterpart, 1,2-dipalmitoyl-sn-glycero-3-phosphocholine (DPPC). While the cross-section area of DPhPC, $0.76\,nm^2$, is significantly larger than $0.629\,nm^2$ of DPPC, the peak-to-peak distance of the electron density profile of the L_α phase of $3.8\,nm$ for DPhPC does not differ significantly from the value of $3.96\,nm$ for DPPC [54,55]. The larger A values of the isoprenoid-chained surfactants, appear largely attributable to the larger molecular volume of the isoprenoid chains.

The larger values of A have some consequences on the spontaneous curvature of the lipid monolayers as compared with straight-chain counterparts. The phase behavior has been reported for the plant glycolipids, monogalactosyl diglycerides (MGDGs), and DGDGs [53,56], whose hydrophobic tails are composed of straight fatty chains (mainly C_{16} and C_{18}) with different degree of unsaturation. While MGDG prefers an H_{II} phase, DGDG prefers an L_α phase over a temperature range from $-10°C$ to $80°C$, regardless of the degree of chain unsaturation. Essentially the same behavior was found in mono- and diglucosyl diglycerides from mycoplasma membrane [48], and synthetic galacto- and glucolipids [57–60]; as a rule of thumb, while monogalacto- and monoglucolipids prefer an H_{II} phase, diga-lacto- and diglucolipids prefer an L_α phase. Thus, the transition from H_{II}-prone to L_α-prone lipids occurs between $N = 1$ and $N = 2$ for the double straight-chained lipids. For the β-Mal$_N$O(C_{16+4})$_2$/water systems, on the other hand, the transition occurs between $N = 2$ and $N = 3$.

The recent molecular dynamics (MD) simulations of bilayers of DPhPC and DPPC have revealed detailed microscopic images of the phytanoyl chain; the chain branching reduces segmental order of the chain, which is closely related to a high gauche probability at the dihedrals in the vicinity of $tert$-carbons. This leads to the chain bending at the branched segments, reduced probability of parallel orientation of two chains, and increased probability of a chain caught between the chains of the neighboring surfactants [61].

Finally, we discuss a possible difference in the structures of the lamellar phase in β-GlcOC$_{8+2}$/water system from those in β-GlcOC$_{10}$ which has the same number of carbon atoms in a hydrophobic chain as β-GlcOC$_{8+2}$ and in β-GlcOC$_8$ which has the same extended length of a hydrophobic chain as β-GlcOC$_{8+2}$ [74].

When the bilayer sheets of constant thickness are stacked, the concentration dependence of the repeat distance d follows the swelling law

$$d = \frac{2d_{hc}}{\varphi_{hc}} \tag{10.1}$$

where $2d_{hc}$ and φ_{hc} are the thickness and volume fraction of the hydrophobic layer, respectively. According to Equation 10.1, a plot of log d versus log φ_{hc} gives a straight line with a slope −1. However, the observed slope is flatter than −1, which may result from the fact that the half-thickness d_{hc} is not constant, or that bilayers have water-filled defect [66–71]. Here, the results based on the latter interpretation are shown.

Hyde [72,73] has reported that the repeat distance for the lyotropic phase follows the generalized swelling law

$$d \propto \phi_{hc}^{-s} \tag{10.2}$$

where s is equal to the so-called packing parameter. According to his theory, $s = 1/2 \sim 2/3$ for mesh type structures (corresponding to bilayer sheets with water filled defects). So we have determined s values from the slope of the log d − log ϕ_{hc} plot. Figure 10.9a shows temperature dependence of the exponent s thus obtained [74]. This figure demonstrates that s is between 0.5 and 1 and increases in the order β-GlcOC$_8$ < β-GlcOC$_{10}$ < β-GlcOC$_{8+2}$, suggesting that the fraction of water-filled defects increases in the order β-GlcOC$_{8+2}$ < β-GlcOC$_{10}$ < β-GlcOC$_8$. Such an interpretation can be confirmed by the SAXS patterns for the three systems shown in Figure 10.9b [74]; the SAXS patterns for the β-GlcOC$_{10}$ and β-GlcOC$_8$ systems have broad components with a peak at about $q = 1.6\,nm^{-1}$, whereas there is no such component for the β-GlcOC$_{8+2}$ system. From the chemical structure of the surfactants used, we can infer that the packing parameter increases in the order β-GlcOC$_8$ < β-GlcOC$_{10}$ < β-GlcOC$_{8+2}$, which is consistent with the result that s increases in the same order.

10.3.2.4.3.2 Dynamic Aspects
The MD simulations of bilayers of DPhPC and DPPC have revealed distinct features of the dynamic properties of the phytanoyl chain as compared to the palmitoyl chain [61–63]. The rate of *trans*-gauche isomerization of the dihedral angles of the phytanoyl chain in DPhPC showed a significantly longer relaxation time profile throughout the chain; the relaxation time of 200–300 ps observed for the middle part

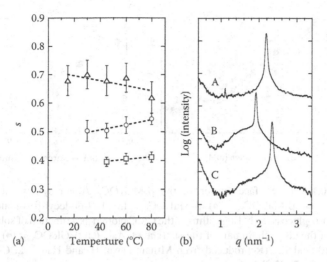

FIGURE 10.9 (a) Temperature dependence of the exponent s in Equation 10.2 for β-GlcOC$_{8+2}$/water (triangles), β-GlcOC$_{10}$/water (circles), and β-GlcOC$_8$/water (squares) systems. (b) Typical SAXS patterns for β-GlcOC$_{8+2}$/water (A), β-GlcOC$_{10}$/water (B), and β-GlcOC$_8$/water (C) systems for $\phi_{hc} = 0.45$ at 45°C (ϕ_{hc}: volume fraction of the hydrophobic layer). The concentration of surfactant is 80, 80, and 89 wt% for (A), (B), and (C), respectively. (Reproduced from Yamashita, I., Kawabata, Y., Kato, T., Hato, M., and Minamikawa, H., *Colloids Surf. A*, 250, 485, 2004. With permission.)

of the DPhPC chain is about five times that of DPPC. The peaks in the relaxation time profile in DPhPC occur at the *tert*-carbons, which are also observed in the ^{13}C NMR relaxation time profile [64]. The slower torsional motion of the phytanoyl chain leads to longer relaxation time of the wobbling motion of the hydrophobic chain, 3.3 ns for DPhPC and 1.7 ns for DPPC. Ten nanoseconds MD simulations demonstrate a smaller two-dimensional diffusion coefficient D of DPhPC than that of DPPC, which approximately agrees with those estimated by time-resolved fluorescence spectroscopy measurements [65]. The rotational relaxation time for DPhPC also appears longer than that of DPPC [63]. The slow chain dynamics appear to contribute to lower permeability of ions, water, and nonionic solutes through the phytanoyl-chained bilayers, which are discussed in Sections 10.4.2 and 10.4.3.

10.3.2.5   Micellar Solutions

10.3.2.5.1   Interfacial Behavior
Figure 10.10a reports the surface tension versus concentration curves for aqueous solutions of β-GlcOC$_{8+2}$ and β-Mal$_N$OC$_{8+2}$ together with those for β-GlcOC$_8$ and β-Mal$_2$OC$_{12}$ for comparison [36]. Table 10.4 lists the critical micelle concentration, CMC, the surface tension at the CMC, γ_{CMC}, the surface excess at the CMC, Γ_{CMC}, and the molecular cross-section area at the CMC, $A_{CMC,}$

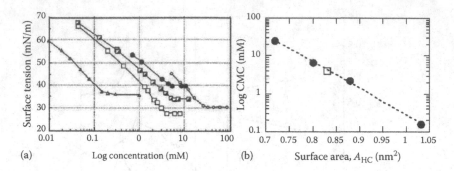

(a)   Log concentration (mM)   (b)   Surface area, A_{HC} (nm²)

FIGURE 10.10 (a) Surface tension of the β-Mal$_N$OC$_{8+2}$/water systems at 25°C. ($\square$) β-GlcOC$_{8+2}$, ($\blacksquare$) β-Mal$_2$OC$_{8+2}$, ($\bullet$) β-Mal$_3$OC$_{8+2}$, ($\triangle$) 1-O-dodecyl-β-D-maltoside, ($\bigcirc$) 1-O-octyl-β-D-glucoside. (b) Logarithm of the CMC plotted as a function of solvent accessible area of the alkyl chain part of the surfactant, A_{HC}. ($\square$) β-GlcOC$_{8+2}$; ($\bullet$) β-GlcOC$_n$ (n = 8, 9, 10, and 12). (Reproduced from Minamikawa, H. and Hato, M., *Chem. Phys. Lipids*, 134, 151, 2005. With permission.)

calculated from the relation $A_{CMC} = \dfrac{10^{18}}{N_A \Gamma_{CMC}}$ for aqueous solutions of β-GlcOC$_{8+2}$, β-Mal$_2$OC$_{8+2}$, and β-Mal$_3$OC$_{8+2}$ [36]. The data for β-GlcOC$_n$ (n = 8, 9, and 10), and β-Mal$_2$OC$_{10}$ are also included for comparison [8,77–79]. Γ_{CMC} was estimated from Gibbs adsorption isotherm,

$$\lim_{C \to CMC} \frac{\partial \gamma}{\partial \ln C}\left(\frac{\partial \ln C}{\partial \ln a}\right)_T = -RT\Gamma_{CMC} \tag{10.3}$$

TABLE 10.4

CMC, and Interfacial Properties of β-Mal$_N$OC$_{8+2}$/Water (N = 1, 2, 3) and β-GlcOC$_n$ (n = 8, 9, 10) Systems at 25°C

Surfactant	CMC (mM)	γ_{CMC} (mN/m)	Γ_{CMC} (μmol/m²)	A_{CMC} (nm²/molecule)	References
β-GlcOC$_{8+2}$	4.0[a]	27.7	4.4	0.38	[36]
β-Mal$_2$OC$_{8+2}$	5.3	33.7	3.5	0.47	[36]
β-Mal$_3$OC$_{8+2}$	5.0	39.6	3.1	0.52	[36]
β-GlcOC$_8$	25	30.1	3.9	0.42	[77]
β-GlcOC$_9$	6.5	—	—	—	[79]
β-GlcOC$_{10}$	2.2	27.7	3.5, 4.05	0.47, 0.41	[8,77,78]
β-Mal$_2$OC$_{10}$	2.0	35.6	3.0	0.56	[78]

Source: Reproduced from Minamikawa, H. and Hato, M., *Chem. Phys. Lipids*, 134, 151, 2005. With permission.

[a] A solution became slightly turbid near the CMC.

under the assumption that $\frac{\partial \ln C}{\partial \ln a} = 1$ below the CMC. C and a denote the surfactant concentration and the activity, respectively.

Table 10.4 demonstrates that the number of glucose residues in the headgroup, N, only marginally influences the CMC of β-Mal$_N$C$_{8+2}$; the CMC values lie in the range of 4–5 mM regardless of N. The CMC of nonionic sugar-based surfactants may therefore be expressed as

$$\ln \text{CMC} = \frac{\Delta G^\circ_{\text{micelle}}}{kT} = \frac{\omega A_{\text{HC}}}{kT} + \text{constant} \tag{10.4}$$

where

$\Delta G^\circ_{\text{micelle}}$ is the free energy of micellization

A_{HC} is the solvent (water) accessible surface area of the hydrocarbon chain

ω is the free energy difference (counted per unit area of the alkyl chain) passing from the bulk of the aqueous solution to the micelle [77]

k is Boltzmann's constant

T is temperature (K)

In Figure 10.10b, ln CMC is plotted against A_{HC} for β-GlcOC$_{8+2}$ and β-GlcOC$_n$ ($n = 8, 9, 10$) where A_{HC} is estimated by COSMO program (AM1, MOPAC Pro in Chem 3D Pro 3.5, Cambridge Soft), subtracting the OH group area from the areas of the corresponding alkyl alcohols [36]. The figure indicates a linear relation between ln CMC and A_{HC} regardless of the branching of the alkyl chain. The CMC of β-GlcOC$_{8+2}$ corresponds to that of alkyl β-D-glucoside with an alky chain length of 9.4 carbon atoms. Contrary to the CMC, the values of A_{CMC} and Γ_{CMC} consistently increase as N increases. This can again be interpreted in terms of a helical conformation of Mal$_N$-headgroups, where the cross-section area increases as N increases.

10.3.2.5.2 Micellar Size and Shape

Micellar size and shape are sometimes discussed in terms of the packing parameter [80], V/Al, where V is the volume of the hydrophobic chain, A is the optimum headgroup area per molecule, and l is the critical chain length. Because the molecular weights of C$_{8+2}$ and C$_{10}$ chains are the same, the V's for these two surfactants are nearly equal. On the other hand, l of the C$_{8+2}$ chain (1.16 nm) is shorter than that of the C$_{10}$ chain (1.42 nm) (note that l may be replaced by the extended length of the hydrophobic chain in the case of micelles). In addition, the surface area A of β-GlcOC$_{8+2}$ does not differ significantly from that of β-GlcOC$_{10}$, because they are dominated by the headgroup size. Thus it can be expected that the packing parameter for β-GlcOC$_{8+2}$ is larger than that of β-GlcOC$_{10}$, suggesting that the β-GlcOC$_{8+2}$ micelles are more elongated than the β-GlcOC$_{10}$ micelles.

In the β-GlcOC$_{8+2}$/water system, however, phase separation occurs below 34 wt% of surfactant at room temperature as shown in Figure 10.6a. Such a phase separation is also observed in the β-GlcOC$_{10}$/water system below 12 wt% of surfactant at room temperature [11]. Although this phase behavior also suggests that the β-GlcOC$_{8+2}$

micelles are more elongated than the β-GlcOC$_{10}$, we cannot compare the micellar size for these systems directly in the single-phase regions because their concentrations are too high to obtain quantitative information due to strong inter-micellar interactions.

In this section, therefore, we compare micellar properties of β-Mal$_2$OC$_{8+2}$/ water and β-Mal$_2$OC$_{10}$/water systems and also mixed surfactant systems of β-GlcOC$_{8+2}$/β-Mal$_2$OC$_{8+2}$/water (abbreviated as β-Glc/β-Mal$_2$OC$_{8+2}$, hereafter) and β-GlcOC$_{10}$/β-Mal$_2$OC$_{10}$/water (abbreviated as β-Glc/β-Mal$_2$OC$_{10}$, hereafter). To determine the micellar size, we have measured surfactant self-diffusion coefficients by using pulsed-gradient spin echo NMR (PGSE-NMR) measurements.

10.3.2.5.2.1 β-Mal$_2$OC$_{8+2}$/Water and β-Mal$_2$OC$_{10}$/Water Systems

The PGSE-NMR measures surfactant self-diffusion coefficient D. Assuming a two-site exchange model, the micellar diffusion coefficient D_{mic} can be obtained from the observed D. Table 10.5 presents hydrodynamic radii R_H of β-Mal$_2$OC$_{8+2}$ and β-Mal$_2$OC$_{10}$ calculated from D_{mic} and the Stokes–Einstein equation [81]. Table 10.5 also includes hydrodynamic radii of β-Mal$_2$OC$_8$ and

TABLE 10.5

Comparison of Hydrodynamic Radii of β-Mal$_2$OC$_{8+2}$ and β-Mal$_2$OC$_{10}$ Micelles with Those of β-Mal$_2$OC$_8$ and β-Mal$_2$OC$_{12}$ Calculated from Published Data at 25°C

		PGSE [81]			DLS [82]		SAXS/SANS [83]				
Surfactant	l/nm	C/mM	R_H/nm		C/mM	R_H/nm	C/mM	R_{maj}/ nm	e	t/nm	R_H/ nm[a]
β-Mal$_2$OC$_{8+2}$	1.16	22.7	2.5 ± 1.1								
		46.1	2.6 ± 0.5[b]								
		115.1	2.86 ± 0.25[b]								
β-Mal$_2$OC$_8$	1.16				30	2.59	92	2.37	1	1.22	2.4
					50	2.88					
β-Mal$_2$OC$_{10}$	1.42	22.6	2.68 ± 0.16								
		45.5	2.73 ± 0.09[b]								
		115.1	3.14 ± 0.12[b]								
β-Mal$_2$OC$_{12}$	1.67				9	3.48	100	3.44	0.59	0.62	3.0
					37	3.47					
					92	3.64					

Source: Kato, T., Kawabata, Y., Fujii, M., Kato, T., Hato, M., and Minamikawa, H., *J. Colloid Interface Sci.*, 312, 122, 2007. With permission.

[a] Calculated from the major radius (R_{maj}) and the axial ratio (e) by using the equation
$$R_H = \frac{R_{maj}\,(1-e^2)^{1/2}}{\tan^{-1}(e^{-2}-1)^{1/2}}.$$

[b] At 30°C.

β-Mal$_2$OC$_{12}$ calculated from published data on dynamic light scattering (DLS), small-angle neutron scattering (SANS), and small-angle x-ray scattering (SAXS) [82,83].

The R_H of β-Mal$_2$OC$_{10}$ micelles in the present study is nearly equal to or slightly larger than that of β-Mal$_2$OC$_8$ and smaller than that of β-Mal$_2$OC$_{12}$. It should be noted, however, that the hydrodynamic radius obtained from DLS is affected by inter-micellar interactions more strongly than that obtained from PGSE. Also, the R_H value calculated from scattering data should be smaller than that obtained from the diffusion coefficient because the latter includes hydration layers.

Table 10.5 shows that R_H's of β-Mal$_2$OC$_{8+2}$ and β-Mal$_2$OC$_{10}$ micelles are around 3 nm and depend on the concentration only slightly [81]. From the studies of He et al. (SAXS and SANS) and Durpuy et al. (SANS), the shell thickness of micelles is reported to be 1.22 nm and 0.62 nm, for β-Mal$_2$OC$_8$ and β-Mal$_2$OC$_{12}$, respectively [83], which can be regarded as the effective length of the maltose headgroup. On the other hand, the extended lengths of C$_{8+2}$ and C$_{10}$ chains are 1.16 nm and 1.42 nm, respectively. So the total lengths of β-Mal$_2$OC$_{8+2}$ and β-Mal$_2$OC$_{10}$ molecules are estimated to be 1.8–2.3 nm and 2.0–2.6 nm, respectively. Because the observed R_H values are nearly equal to or slightly larger than the molecular length thus estimated, micelles are expected to be spherical or slightly elongated. Table 10.5 also demonstrates that the micellar size of β-Mal$_2$OC$_{8+2}$ is almost equal to that of β-Mal$_2$OC$_{10}$. This may be due to the fact that the difference in the conformation of the alkyl group is hidden by the large hydrophilic group.

10.3.2.5.2.2 β-GlcOC$_{8+2}$ β-Mal$_2$OC$_{8+2}$/Water and β-GlcOC$_{10}$/β-Mal$_2$OC$_{10}$/
Water Systems

In this section, we use the mixing ratio expressed by the mole fraction of β-GlcOC$_{8+2}$ or β-GlcOC$_{10}$ in the total mixed solute defined as

$$X_G \equiv \frac{n_G}{n_M + n_G} \tag{10.5}$$

where n_G and n_M are the molar concentrations of β-GlcOC$_{8+2}$ (β-GlcOC$_{10}$) and β-Mal$_2$OC$_{8+2}$ (β-Mal$_2$OC$_{10}$), respectively, for the β-Glc/β-Mal$_2$OC$_{8+2}$ (β-Glc/β-Mal$_2$OC$_{10}$) system. In these systems, the mixed micelles having the same mixing ratio can be regarded as the micelles having the same hydrophilic group. As described previously, the CMC for the β-GlcOC$_{8+2}$ and β-Mal$_2$OC$_{8+2}$ are larger than those for β-GlcOC$_{10}$ and β-Mal$_2$OC$_{10}$. The same holds true for the mixed surfactant systems; the CMC of the mixed surfactant β-Glc/β-Mal$_2$OC$_{8+2}$ is larger than that of β-Glc/β-Mal$_2$OC$_{10}$.

Figure 10.11 shows the hydrodynamic radius of the mixed micelles as a function of X_G where the total surfactant concentration is kept constant at 2 wt% (it should be noted that the total surfactant concentration in molar unit varies from 41.4 to 59.4 mM as X_G increases from 0 to 0.9 [81]). One sees that the hydrodynamic

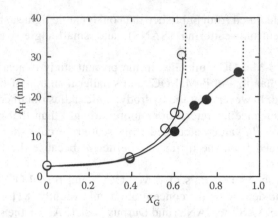

FIGURE 10.11 Hydrodynamic radii of the mixed micelles for β-GlcOC$_{8+2}$/β-Mal$_2$OC$_{8+2}$/water (○) and β-GlcOC$_{10}$/β-Mal$_2$OC$_{10}$/water (●) systems versus mole fraction of β-GlcOC$_{8+2}$ or β-GlcOC$_{10}$ in the total mixed solute X_G (see Equation 10.5) at 25°C. The total surfactant concentration is kept constant to 2 wt%. (Reproduced from Kato, T., Kawabata, Y., Fujii, M., Kato, T., Hato, M., and Minamikawa, H., *J. Colloid Interface Sci.*, 312, 122, 2007. With permission.)

radii for both systems begin to increase with X_G above $X_G \cong 0.4$. The R_H of β-Glc/β-Mal$_2$OC$_{8+2}$ micelles increases more rapidly than that of β-Glc/β-Mal$_2$OC$_{10}$ micelles, and then phase separation occurs at $X_G \cong 0.65$. On the other hand, the R_H of β-Glc/β-Mal$_2$OC$_{10}$ micelles continues to increase until the phase separation occurs at $X_G \cong 0.92$.

To elucidate the effects of alkyl chain conformation on micellar size in more detail, we have measured the self-diffusion coefficient as a function of total concentration keeping the mixing ratio constant. Figure 10.12 shows the hydrodynamic radii as a function of micellar concentration ($C - \text{CMC}$) for $X_G = 0.6$ at 25°C. One sees that micelles grow rapidly with increasing concentration and that R_H of β-Glc/β-Mal$_2$OC$_{8+2}$ micelles is much larger than that of β-Glc/β-Mal$_2$OC$_{10}$ micelles at higher concentrations [81].

Contrary to the pure surfactant systems, the hydrodynamic radii for the mixed systems at $X_G = 0.6$ substantially increase with increasing concentration. It is interesting that the R_H of β-Glc/β-Mal$_2$OC$_{8+2}$ is much larger than that of β-Glc/β-Mal$_2$OC$_{10}$ at higher concentrations in spite of the fact that the CMC of the former is larger than the latter.

Table 10.6 shows the surface area per molecule (A_0) calculated from the volume (V) and the extended length (l) of the hydrophobic chain for spherical and cylindrical (with infinite length) micelles by using the following relations [80,84]:

$$A_0 \text{ (sphere)} = \frac{3V}{l} \tag{10.6a}$$

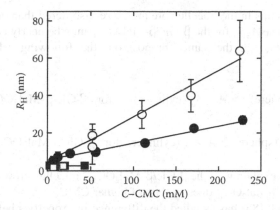

FIGURE 10.12 Hydrodynamic radii of mixed micelles for β-GlcOC$_{8+2}$/β-Mal$_2$OC$_{8+2}$/ water (○) and β-GlcOC$_{10}$/β-Mal$_2$OC$_{10}$/water (●) systems versus micellar concentration $(C - CMC)$ for $X_G = 0.6$ at 25°C. The data for β-Mal$_2$OC$_{8+2}$/water (open squares) and β-Mal$_2$OC$_{10}$/water (closed squares) systems are also shown for comparison. (Reproduced from Kato, T., Kawabata, Y., Fujii, M., Kato, T., Hato, M., and Minamikawa, H., *J. Colloid Interface Sci.*, 312, 122, 2007. With permission.)

$$A_0 \ (\text{cylinder}) = \frac{2V}{l} \tag{10.6b}$$

In these calculations, we used the relations $V = 0.274 + 0.269\ n_c$ $(n_c = 10)$ and $l = 0.15 + 0.1265\ n_c$ $(n_c = 8$ and $10)$. Surface area per molecule at air/water interface for a mixed surfactant system (A_{ST}) may be estimated from the relation

$$A_{ST} = A_{ST}(\text{Glc})\,X_G + A_{ST}(\text{Mal}_2)(1 - X_G)\,(X_G = 0.6) \tag{10.7}$$

where $A_{ST}(\text{Glc})$ and $A_{ST}(\text{Mal}_2)$ are the surface areas for β-GlcOC$_{8+2}$ and β-Mal$_2$OC$_{8+2}$, respectively. From the surface tension data [36], A_{ST} can be calculated to be about $0.42\,\text{nm}^2$. The surface area in micelles (A) should be larger than A_{ST} because the

---

TABLE 10.6
Surface Area Per Molecule Calculated from the Volume and Extended Length of the Hydrocarbon Chain for Spherical and Cylindrical Micelles

Hydrocarbon	V/nm^3	l/nm	A_0/nm^2 (Sphere)	A_0/nm^2 (Cylinder)
C$_{8+2}$	0.297	1.16	0.77	0.51
C$_{10}$	0.297	1.42	0.63	0.42

hydrocarbon chains in the micelle core are more disordered than at the air/water interface. Because A_{ST}'s for the β-Glc/β-Mal$_2$OC$_{10}$ may be nearly equal to that of β-Glc/β-Mal$_2$OC$_{8+2}$ at the same composition, the following relation can be expected.

$$A_0(\text{sphere}) \gg A \approx A_0 \text{ (cylinder)} \qquad \text{for } \beta\text{-Glc}/\beta\text{-Mal}_2\text{OC}_{8+2} \qquad (10.8)$$

$$A_0(\text{sphere}) \geq A > A_0(\text{cylinder}) \quad \text{for } \beta\text{-Glc}/\beta\text{-Mal}_2\text{OC}_{10} \qquad (10.9)$$

These relations explain why the β-Glc/β-Mal$_2$OC$_{8+2}$ micelles grows more rapidly than the β-Glc/β-Mal$_2$OC$_{10}$ does in spite of higher CMC.

Nilsson et al. [85] have studied the difference in properties between straight and branched hydrocarbon chains by comparing the phase behaviors and surfactant self-diffusion coefficients for 2-ethylhexyl β-D-glucoside (β-GlcOC$_2$C$_6$)/ water and β-GlcOC$_8$/water systems. They have also reported that the CMC of β-GlcOC$_2$C$_6$ is larger than that of β-GlcOC$_8$ and that the axial ratio of micelles of β-GlcOC$_2$C$_6$ (obtained from the micellar diffusion coefficient) is slightly larger than that of β-GlcOC$_8$ although these micelles are much smaller than the micelles of β-Glc/β-Mal$_2$OC$_{8+2}$ or β-Glc/β-Mal$_2$OC$_{10}$ at $X_G = 0$.

Söderman and coworkers [11,33,85–87] have studied phase behavior and micelle structures in aqueous systems of β-GlcOC$_n$ ($n = 8$, 9, 10) by using PGSE, SAXS, and cryo-TEM. They have found that a mixed system of β-GlcOC$_9$/ β-GlcOC$_{10}$/water exhibits a closed-loop liquid/liquid miscibility gap at a certain mixing ratio and that networks of bicontinuous micelles have been suggested to exist throughout the micellar regions. In our systems, contribution of such networks is not important at at least $X_G < 0.6$. It should be noted however that our conclusion is not affected by the existence of entanglement or network formation because contribution of them becomes larger as the micellar size increases.

10.4 BIOPHYSICAL PROPERTIES

The fact that the values of T_K can be depressed below 0°C has made the sugar-based isoprenoid-chained surfactants considerably easier to handle, opening up a new opportunity to prepare vesicles and/or black lipid membranes (BLMs) from them. Vesicles are not only useful for *in vitro* studies of membrane physical and chemical properties but also indispensable to a range of technical applications such as drug delivery systems (DDS). BLMs have been widely employed in studies of membrane-associated processes such as ion permeation studies via ion carriers, ion channels, or ion pumps [88]. The lipids generally used for these purposes have been naturally occurring lipids, for example, soybean phospholipids (SBPL) or egg yolk phosphatidylcholine (EPC). The BLMs formed from these lipids, however, are not very stable to oxidation, hydrolysis, or mechanical stress.

To demonstrate unique features of the vesicles of sugar-based surfactants (lipids), we start this section by a discussion on kinetic stability of the β-Mal$_3$O(C$_{16+4}$)$_2$ vesicles. The section is then followed by a second topic on ion permeability of BLM [46] and vesicle membranes [65] of β-Mal$_3$O(C$_{16+4}$)$_2$. Finally, functional reconstitution of membrane proteins into β-Mal$_3$O(C$_{16+4}$)$_2$ vesicles and BLM is discussed. Detailed procedures for the preparation of the vesicles and the BLM of the sugar-based surfactants have been described elsewhere [46,65,89,90].

10.4.1 KINETIC STABILITY OF β-MAL$_3$O(C$_{16+4}$)$_2$ VESICLES

It is well known that apparently neutral phosphatidylcholine (PC) vesicles are generally stable against changes in pH and ionic strength, for example, no aggregation of DPPC vesicles in a liquid-crystalline state occurs even in such concentrated salt solutions as $1\,M$ MgSO$_4$ or $0.2\,M$ LaCl$_3$ [91]. Since classical Derjaguin-Landau-Verwey-Overbeek (DLVO) theory cannot explain the observed PC vesicle stability, the modified DLVO theory with an additional short range repulsive force called as hydration force has been put forward to account for this stability [91,92]. Contrary to PCs, changes in electrolyte concentration have significant effects on aqueous behavior of nonionic sugar-based surfactants both in bulk solutions [93–97] and at the air–water interface [98–100]. Though molecular mechanisms underlying these phenomena are not fully understood, mechanisms based on ion-induced changes in interfacial hydration (or interfacial water structures) have often been assumed for explaining these phenomena. For example, Webb et al. [94,95] and Fragata et al. [96] have investigated a salt-induced aggregation of vesicles of chloroplast thylakoid galactolipid, DGDG and MGDG. Webb et al., found that DGDG vesicles exhibit reversible aggregation upon addition of salts and interpreted the phenomena in terms of salt-induced change in hydration of the bilayer surfaces that leads to a reduction of hydration repulsion. Fragata et al. later proposed a mechanism that the initial step in DGDG aggregation is an ion-induced decrease in interfacial polarity [96].

The unilamellar vesicles of β-Mal$_3$O(C$_{16+4}$)$_2$ also exhibit pH (or NaCl)-dependent aggregation–disaggregation in a reversible manner [90]. In this and the following sections, we show that the kinetic stability of the β-Mal$_3$O(C$_{16+4}$)$_2$ vesicles is largely controlled by the double layer forces, rather than by interfacial water-related phenomena such as changes in the hydration forces.

10.4.1.1 pH-Dependent Reversible Aggregation of β-Mal$_3$O(C$_{16+4}$)$_2$ Vesicles

Figure 10.13a reports the optical density of the β-Mal$_3$O(C$_{16+4}$)$_2$ vesicle suspension (A_{600}) as functions of the solution pH and NaCl concentration. When a concentrated HCl solution was added into the vesicles suspended in pure water ($A_{600} \approx 0.02$, pH6), the value of A_{600} increased rapidly at around pH 4.5 and reached a plateau value of $A_{600} \approx 0.45$ below pH 4 (the curve A in

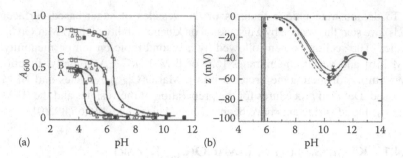

FIGURE 10.13 (a) Variations of optical density at 600 nm (A_{600}) of β-Mal$_3$O(C$_{16+4}$)$_2$ vesicle suspensions as functions of pH and NaCl concentration at 25°C. (A) 0 mM NaCl, (B) 1 mM NaCl, (C) 5 mM NaCl, (D) 10 mM NaCl; (open square) first addition of HCl, (circle) first addition of NaOH, (triangle) second addition of HCl, (closed square) second addition of NaOH. The curves are to guide the reader's eye. (b) ζ-Potentials of β-Mal$_3$O(C$_{16+4}$)$_2$ vesicles as a function of pH in 1 mM NaCl at 25°C. (●) measured ζ-potential derived from the Henry equation. (▲) graphically estimated ζ-potential derived from the O'Brien–White numerical calculation (From O'Brien, R.W. and White, L.R., *J. Chem. Soc. Faraday Trans. 2*, 74, 1607, 1978). (○) measured ζ-potential derived from the Ohshima–Healy–White equation (From Ohshima, H., Healy, T.W., and White, L.R., *J. Chem. Soc. Faraday Trans. 2*, 79, 1613, 1983). The bar indicates the standard deviation. Solid and dashed lines are calculated simulation curves on the basis of the Gouy–Chapman expression for a 1:1 type electrolyte solution [90]. The shaded bar indicates the aggregation threshold pH. (Reproduced from Baba, T., Zheng, L., Minamikawa, H., and Hato, M., *J. Colloid Interface Sci.*, 223, 235, 2000. With permission.)

Figure 10.13a). When a concentrated NaOH solution was then added to the turbid vesicle suspension ($A_{600} \approx 0.45$, pH 3.5), the value of A_{600} started to decrease at about pH 4.7 and reverted to the original value of $A_{600} \approx 0.02$ above pH 5.5 and remained practically constant as pH further increased (the curve A). Second titration cycle with HCl and NaOH followed the same curve as that of the first titration cycle. Essentially the same reversible A_{600}–pH profiles were observed for the vesicles suspended in the varying NaCl concentrations. We noted that a threshold value of pH, (pH)$_{th}$, below which A_{600} starts to steeply increase, shifts to higher values as the NaCl concentration increases, for example, 4.7, 5.7, 6.2, and 7.2 for 0 (distilled water), 1, 5, and 10 mM NaCl, respectively. Table 10.7 summarizes the accompanied changes in the vesicles diameter during the alternate addition of HCl and NaOH to the vesicle suspensions. It is clear that the vesicle diameter changes reversibly from a monomer value (~110 nm) to a significantly large value of >1500 nm in response to the pH cycles. The data clearly indicate that the pH-dependent aggregation–disaggregation process is reversible, and membrane fusion scarcely occurred during the pH cycles.

TABLE 10.7

Effects of NaCl Concentration on the Aggregation Threshold pH, $(pH)_{th}$, and the Vesicle Aggregation Reversibility Observed from Diameter of β-Mal$_3$O(C$_{16+4}$)$_2$ Vesicle at 25°C

C_{NaCl} (mM)	Aggregation Threshold pH	pH[a]	Ionic Strength[a] (mM)	Diameter (nm)
0.0	4.7	~6	(~10^{-3})	112 ± 20
		3.6	0.25	>1500
		9.1	0.26	113 ± 20
		3.2	0.89	>2000
		11.2	2.5	119 ± 30
1.0	5.7	~6	1.0	108 ± 15
		3.5	1.35	>2000
		9.1	1.37	108 ± 20
		3.7	1.55	>1500
		11.3	3.45	110 ± 30
10	7.2	3.5	10.4	>1500
		9.4	10.4	117 ± 30
		3.6	10.6	>2000
		9.1	12.5	120 ± 30

Source: Reproduced from Baba, T., Zheng, L., Minamikawa, H., and Hato, M., *J. Colloid Interface Sci.*, 223, 235, 2000. With permission.

[a] Given by the alternate addition (two cycles) of HCl and NaOH to a vesicle suspension.

10.4.1.2 ζ-Potential of β-Mal$_3$O(C$_{16+4}$)$_2$ Vesicles

As a typical example, pH-dependent ζ-potentials of the β-Mal$_3$O(C$_{16+4}$)$_2$ vesicles in 1 mM NaCl is shown in Figure 10.13b [90]. Two points must be mentioned. First, the β-Mal$_3$O(C$_{16+4}$)$_2$ vesicles are negatively charged in a pH range from 1.6 to 13. In the neutral to alkaline pH region, the negative value of the ζ-potential increases as pH increased, going through a maximum at pH $\approx$ 10, and then decreases as pH further increases. In the acidic region below pH 4, the ζ-potential decreases to about -3 mV at pH 3.6 and reaches nearly 0 mV at pH 1.6; the charge reversal does not occur at least up to pH 1.6. Second, the aggregation or dispersion of the vesicles correlates well with the magnitude of the ζ-potential as shown in Table 10.8 [90]. When the ζ-potential is above -10 to -15 mV, the vesicles remain as a monomer form and the aggregation is not a major event. Once the ζ-potential is lowered below about -5 mV, the rapid aggregation of the vesicles starts taking place. A critical ζ-potential value for the β-Mal$_3$O(C$_{16+4}$)$_2$ vesicles, aggregation, therefore, is around -10 mV.

TABLE 10.8

Correlation between ζ-Potential and pH-Dependent Aggregation of β-Mal$_3$O(C$_{16+4}$)$_2$ Vesicles

C_{NaCl} (mM)	pH	ζ-Potential (mV)	V_{tot}^{max} (kT)	Diameter (nm)	A_{600}
0.0	4	−5	≤1, 2.5[a]	>1500[b]	0.47[b]
(pure water)	6	−20	14	112 ± 20[c]	0.02[c]
	11	−60	130	119 ± 30[c]	0.02[c]
1.0	4	−8	1.5	>1500[b]	0.50[b]
	8	−20	12	—	0.02[c]
10	6	−5	≈0	—	0.75[b]
	9	−15	4	120 ± 30[c]	0.05[c]

Source: Reproduced from Baba, T., Zheng, L., Minamikawa, H., and Hato, M., *J. Colloid Interface Sci.*, 223, 235, 2000. With permission.

Note: V_{tot}^{max} are the values of maximum of the total potential $V_{tot}(d)$ estimated from Equations 10.10 through 10.12 under a potential constant condition with A_{eff} = 5×10^{-21} J. We assumed that ζ-potential is equal to φ_0.

[a] Values of maximum of the total potential $V_{tot}(d)$ estimated from Equations 10.10 through 10.12 under a charge constant condition.

[b] Aggregated vesicles.

[c] Dispersed vesicles.

10.4.1.3 Double Layer Forces Control the Kinetic Stability β-Mal$_3$O(C$_{16+4}$)$_2$ Vesicles

The good correlation between the magnitude of the ζ-potential and the vesicle aggregation phenomenon has prompted us to estimate the potential energy of interaction, $V_{tot}(d)$, between two vesicles. Following the classical DLVO theory, $V_{tot}(d)$ between two vesicles of radius a may be given by the sum of the energy of repulsive double layer forces, $V_{dl}(d)$, and the energy of attractive van der Waals forces, $V_{vdw}(d)$[101]

$$V_{tot}(d) = V_{dl}(d) + V_{vdw}(d) \tag{10.10}$$

where d is the inter-particle separation.

Under conditions of potential constant or charge constant conditions, $V_{dl}(d)$ may be expressed by [102,103]

$$V_{dl}(d) = 2\pi\varepsilon\varepsilon_0 a\varphi_0^2 \ln[1 + \exp(-\kappa d)] \text{ (potential constant)} \tag{10.11a}$$

$$V_{dl}(d) = -2\pi\varepsilon\varepsilon_0 a\varphi_0^2 \ln[1 - \exp(-\kappa d)] \text{ (charge constant)} \tag{10.11b}$$

where φ_0 is the surface potential of the vesicles. $V_{vdw}(d)$ may be expressed by [104]

$$V_{vdw}(d) = -(A_{eff}/12)(H_{11} + H_{22} - 2H_{12})$$ (10.12)

where, $H(x, y) = y/(x^2 + xy + x) + y/(x^2 + xy + x + y) + 2\ln[(x^2 + xy + x)/(x^2 + xy + x + y)]$

$$\begin{cases} x = (d + 2l)/2(a - l), y = 1 \quad \text{for } H_{11} \\ x = d/2a, y = 1 \quad \text{for } H_{22} \\ x = (d + l)/2(a - l), y = a/(a - l) \quad \text{for } H_{12} \end{cases}$$

A_{eff} denotes the effective Hamaker constant of the vesicle bilayer in water, and l ($= 5.8$ nm) is the hydrated bilayer thickness [27]. We take $A_{eff} \approx 5 \times 10^{-21}$ J for the β-Mal$_3$O(C$_{16+4}$)$_2$ vesicles in water [105].

The maximum values of the total potential, V_{tot}^{max} calculated as a function of φ_0 (we assume that the ζ-potential is equal to φ_0), are also listed in Table 10.8 [90]. When the surface potential of the vesicles, φ_0 is -10 mV, V_{tot}^{max} is at most $1-3kT$ at $I = 0-1$ mM, and close to $0kT$ at $I = 10$ mM, predicting the vesicle aggregation. Once the value of φ_0 exceeds -15 mV, V_{tot}^{max} becomes well above kT, indicating stable vesicle dispersion. $\varphi_0 \approx -10$ mV, therefore, represents a borderline condition between the vesicle dispersion and aggregation, in accord with the conclusion in the Section 10.4.1.2. We noted that the similar correlation also holds for the NaCl-dependent vesicle aggregation as shown in Table 10.9. Above discussion indicates that the kinetic stability of β-Mal$_3$O(C$_{16+4}$)$_2$ vesicles can be well described within the framework of the classical DLVO theory without invoking any additional short-range repulsive hydration force.

Finally it must be mentioned that the pH-dependent ζ-potentials can be described by the Gouy–Chapman theory with an assumption that the interfacial

TABLE 10.9
Correlation between ζ-Potential and NaCl-Dependent Aggregation of β-Mal$_3$O(C$_{16+4}$)$_2$ Vesicles

C_{NaCl} (mM)	pH	ζ-Potential (mV)	V_{tot}^{max} (kT)	Diameter (nm)	A_{600}
0.0 (pure water)	6	−20	14	112 ± 20	0.02[a]
1.0	6	−13	4–5	108 ± 15	0.05[a]
10	6	−5	≈0	>1500	0.75[b]

Source: Reproduced from Baba, T., Zheng, L., Minamikawa, H., and Hato, M., *J. Colloid Interface Sci.*, 223, 235, 2000. With permission.

Note: V_{tot}^{max} are the values of maximum of the total potential $V_{tot}(d)$ estimated from Equations 10.10 through 10.12 under potential constant condition with $A_{eff} = 5 \times 10^{-21}$ J. We assumed that ζ-potential is equal to φ_0.

[a] Dispersed vesicles.

[b] Aggregated vesicles.

charges arise from adsorption of OH⁻ ions at the vesicle–water interface and dissociation of the sugar headgroup in a higher pH regime above pH 10 [90].

10.4.1.4   Inter-Sugar-Based Surfactant Membrane Interactions*

To get a deeper insight into the mechanism and forces that control the phenomena, we here discuss results of direct force measurements between β-Mal$_3$O(C$_{16+4}$)$_2$ layers in aqueous solutions of different pH conditions.

It is well established that at long distances, lipid bilayers are attracted by van der Waals forces, $F_{vdw}(d)$, while at short distances, they are repelled by structural and solvation (hydration) forces, $F_S(d)$, that are associated with the intervening water and lipid headgroups [106–108]. For charged lipids, double layer forces, $F_{dl}(d)$, are superposed on the core interactions, $F_S(d)$ and $F_{vdw}(d)$. If the lipid layers are unsupported, the repulsion may be further enhanced by the thermal undulations of the lipid membranes, $F_u(d)$ [109,110]. Thus, at a specific surface separation, d, the total force of interaction, $F(d)$, may be given approximately by

$$F(d) \approx F_S(d) + F_{vdw}(d) + F_{dl}(d) + F_u(d) \tag{10.13}$$

As the influence of pH on the vesicle aggregation is best compared at pH 4 (the rapid aggregation) and at pH 9.8 (the stable dispersion), we examined how these two pH conditions affect the strength of the adhesion forces and a range of hydration forces between the β-Mal$_3$O(C$_{16+4}$)$_2$ layers. If ion-induced changes in the interfacial hydration such as dehydration of the β-Mal$_3$O(C$_{16+4}$)$_2$ headgroup really occurred to cause the vesicle aggregation, one would expect the increased strength of adhesion (or a decreased range of the hydration forces) at pH 4 as compared to that at pH 9.8.

We prepared β-Mal$_3$O(C$_{16+4}$)$_2$ layer-coated mica surfaces by the Langmuir Blodgett technique [111]. The first layer was a solid monolayer of dipalmitoylphosphatidylethanolamine (DPPE) deposited on mica surface. A second β-Mal$_3$O(C$_{16+4}$)$_2$ monolayer was then deposited onto the DPPE coated mica surfaces at a molecular cross-section area of 1.0 nm²/lipid that is equal to an equilibrium molecular area in the lamellar phase [27]. The distance zero ($D = 0$) was defined as a distance where the adhesive DPPE/DPPE contact takes place in air.

10.4.1.4.1   Forces between β-Mal$_3$O(C$_{16+4}$)$_2$ /DPPE/Mica Surfaces
An acidic condition, pH 4. A typical example of force versus distance curve between β-Mal$_3$O(C$_{16+4}$)$_2$ /DPPE/mica surfaces measured at pH 4 (0.1 mM HCl) is shown in Figure 10.14a. There was practically no measurable force between the β-Mal$_3$O(C$_{16+4}$)$_2$ layers at separations D beyond about 9 nm. At smaller separations, the gradient of the van der Waals forces exceeded the spring constant, and the surfaces jumped into contact at 5.7 ± 0.1 nm. With further compression, hydration forces dominated the interactions and the distance

---

* Reproduced from Korchowiec, B.M., Baba, T., Minamikawa, H., and Hato, M., *Langmuir*, 17, 1853, 2001. With permission.

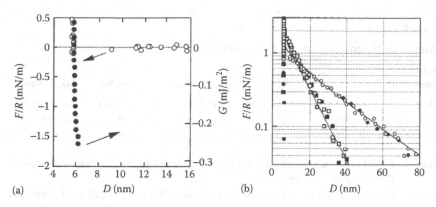

FIGURE 10.14 (a) Forces between β-Mal$_3$O(C$_{16+4}$)$_2$ layers in 0.1 mM HCl (pH 4). The jump to contact is indicated by the inward directed arrow and the position where the surfaces were pulled apart is indicated by the outward directed arrow. The open and filled circles represent an approach and a separation measurement, respectively. Practically no double-layer force was observed. The right-hand ordinate gives the interaction energy per unit area, G, for two parallel layers of the same material calculated according to the Derjaguin approximation $F(d)/R = 2\pi G(d)$ where R is the surface radius [116]. (b) Forces measured between β-Mal$_3$O(C$_{16+4}$)$_2$ layers at pH 9.8 in distilled water (○,●) and in a 1 mM NaCl solution at pH 9.7 (□,■). The open and filled symbols represent an approach and a separation measurement, respectively. (Reproduced from Korchowiec, B.M., Baba, T., Minamikawa, H., and Hato, M., *Langmuir*, 17, 1853, 2001. With permission.)

of a closest β-Mal$_3$O(C$_{16+4}$)$_2$ layer approach did not change appreciably even when the surfaces were compressed up to 4 mN/m. On separation, the surfaces jumped apart from $D_0 = 6.1 \pm 0.2$ nm. The adhesion minimum (the adhesion force) determined from the pull-off force was -1.7 ± 0.1 mN/m. It is also noted that the adhesion minimum (-1.6 ± 0.1 mN/m) and its position, D_0 (=6.1 ± 0.2 nm), in an aqueous 0.1 mM NaCl (pH ~ 5.7) were practically indistinguishable from those at pH 4.

An alkaline condition, pH 9.8. Since the negative values of ζ-potential of β-Mal$_3$O(C$_{16+4}$)$_2$ vesicles become appreciable in alkaline conditions above pH 8 (the Section 10.4.1.2), one may expect a double-layer force between the β-Mal$_3$O(C$_{16+4}$)$_2$ layers at pH 9.8. Figure 10.14b shows the force curves measured between β-Mal$_3$O(C$_{16+4}$)$_2$ layers in the alkaline conditions (NaOH in distilled water and in an aqueous 1 mM NaCl), where the long-range repulsions are observed. At pH 9.8 in distilled water (○,●), the force increases exponentially with a decay length of about 22 nm that can be compared with the Debye length of 21 nm expected from the ionic strength of the solution. As surface separation decreased, a force maximum was reached at about 7 nm, below which the surfaces experienced an attraction and jumped into contact at about 5.7 nm below which the hydration forces dominated the interactions. On separation, the surfaces jumped apart at 6.0 ± 0.2 nm with an adhesion minimum of −0.4 ± 0.1 mN/m. When 1 mM

TABLE 10.10

Adhesion Parameters at pH 4 and pH 9.8

Solution Condition	A_{600}	D_0 (nm)	d_l^{max} (nm)	F_0/R (mN/m)	G_0 (mJ/m^2)
pH 4	0.47[a]	6.1 ± 0.2	6.0 ± 0.1	−1.7 ± 0.1	−0.27 ± 0.01
pH 9.8[b]	0.02[c]	6.0 ± 0.2	6.0 ± 0.1	−1.8[d]	−0.29[d]
pH 9.7[e]	0.02[c]	6.0 ± 0.2	6.0 ± 0.1	−1.7[d]	−0.27[d]

Source: Reproduced from Korchowiec, B.M., Baba, T., Minamikawa, H., and Hato, M., *Langmuir*, 17, 1853, 2001. With permission.

Notes: A_{600}, values of the optical density of a vesicle suspension at 600 nm; D_0 equilibrium layer thickness, a sum of thickness of two β-Mal$_3$O(C$_{16+4}$)$_2$ monolayers and that of the intervening water layer at the adhesion minimum (d_0), that is, $D_0 = 2l + d_0$; d_l^{max}, lamellar repeat spacing of an L_α phase at maximal swelling.

F_0/R and G_0 (= $F_0/2\pi R$) denote adhesion force and adhesion energy, respectively, where R is the surface radius.

[a] Aggregated vesicles.

[b] In distilled water.

[c] Dispersed vesicles.

[d] Value that was estimated by subtracting a possible contribution of the double-layer forces from the measured total force.

[e] In 1 mM NaCl.

NaCl was added to a solution of pH 9.7 (□, ■), the range of the repulsion decreased as expected with a decay length of about 8.8 nm that is again favorably compared with the expected Debye length of 8.7 nm. The separation occurred at 6.0 ± 0.2 nm with an adhesion minimum of + 0.3 ± 0.1 mN/m. For both cases, the distance of the closest β-Mal$_3$O(C$_{16+4}$)$_2$ layer approach did not change on further compression up to 4 mN/m. The adhesion parameters of β-Mal$_3$O(C$_{16+4}$)$_2$ obtained from the force measurements are summarized in Table 10.10.

Surface potential of the β-Mal$_3$O(C$_{16+4}$)$_2$ layers at infinite separation, $|\phi_S^\infty|$, estimated by fitting the theoretical double-layer force curve [112–115] to the tail end of the experimental force curve is summarized in Table 10.11. The estimated $|\phi_S^\infty|$ values are favorably compared with the corresponding values of ζ-potential of the β-Mal$_3$O(C$_{16+4}$)$_2$ vesicles: −50 ± 10 mV for pH 9.7 in 1 mM NaCl and −60 ± 10 mV for pH 9.8 in distilled water. Since a ζ-potential of β-Mal$_3$O(C$_{16+4}$)$_2$ vesicle at pH 4 is 0 to approximately −5 mV, the expected double-layer force is too weak to be measured, in consistent with the force curve at pH 4 where practically no double-layer force was observed (Figure 10.14a). Hence, the pH-dependent force behavior observed appears well correlated to the pH-dependence of the ζ-potentials of the vesicles.

TABLE 10.11

Double Layer Parameters for β-Mal$_3$O(C$_{16+4}$)$_2$ at Infinite Separation at 22.5°C

Solution	$|\phi_S^\infty|^a$ (mV)	ζ-Potential[b] (mV)	Surface Charge (C/m^2)	Fraction Adsorbed, α	Adsorption Constant, K_a (M^{-1})
pH 4[c]	—	0 to −5	—	—	—
pH 9.8[d]	60	−60 ± 10	2.4×10^{-3}	0.015	2.5×10^3
pH 9.7[e]	45	−50 ± 10	4.1×10^{-3}	0.026	2.5×10^3

Source: Reproduced from Korchowiec, B.M., Baba, T., Minamikawa, H., and Hato, M., *Langmuir*, 17, 1853, 2001. With permission.

[a] Values calculated for an OHP position at a hypothetical headgroup/water interface [105].

[b] Values measured for β-Mal$_3$O(C$_{16+4}$)$_2$ vesicles of 100 nm in diameter [90].

[c] In 10^{-4} M HCl.

[d] NaOH in distilled water.

[e] NaOH in 1 mM NaCl.

10.4.1.4.2 pH Effects on the Adhesion Forces and the Hydration Forces

Straightforward comparison of the forces measured at pH 9.8 with those at pH 4 may be made after subtraction of expected double-layer forces from the experimental forces at pH 9.8. To do this, we assume that the adsorption of OH⁻ ions at the lipid surface can be expressed by

$$[OH^-]_S + [L]_S \overset{K_a}{\leftrightarrow} [LOH^-]_S \qquad (10.14)$$

The surface adsorption constant, K_a, can be described, as a first approximation, by a mass action law relating the OH⁻ concentration at the vesicle surface, $[OH^-]_S$, to the surface density of bound and unbound sites, $[LOH^-]_S$ and $[L]_S$, respectively,

$$K_a = \frac{[LOH^-]_S}{[OH^-]_S[L]_S} = \frac{\alpha e^{-e\phi_S^\infty/kT}}{(1-\alpha)[OH^-]_B}. \qquad (10.15)$$

where

$[OH^-]_B$ (=6.3×10^{-5} M) is the bulk OH⁻ concentration

$\alpha = [LOH^-]_S / \{[LOH^-]_S + [L]_S\}$ is a fraction of surfactant headgroups that have bound an OH⁻ ion

K_a can be calculated by the measurable parameters, $[OH^-]_B$, ϕ_S^∞, and α, which are listed in Table 10.11. The theoretical double-layer forces at pH 9.8 (in distilled water and in 1 mM NaCl) were then computed by solving the Poisson–Boltzmann equation for each value of $[OH^-]_B$, and K_a, [112–115] and are shown by the solid lines in Figure 10.14b.

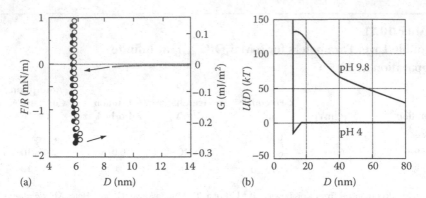

(a)   D (nm)   (b)   D (nm)

FIGURE 10.15 (a) Comparison of a hydration-dominated part of a force curve at pH 4 (○) and at pH 9.8 (●). The force curve at pH 9.8 was inferred after subtraction of the expected double-layer force shown in Figure 10.14b. The right-hand ordinate gives the interaction energy per unit area, G, for two parallel layers of the same material calculated according to the Derjaguin approximation, $F(d)/R = 2\pi G(d)$ where R is the surface radius [116]. (b) Schematic of pH effects on the interaction energy $U(D)$ between two spherical β-Mal$_3$O(C$_{16+4}$)$_2$ vesicles of radius 55 nm. (Reproduced from Korchowiec, B.M., Baba, T., Minamikawa, H., and Hato, M., *Langmuir*, 17, 1853, 2001. With permission.)

After subtracting the theoretical double-layer force, a hydration part of the force curve at pH 9.8 (●) is compared with that at pH 4 (○) in Figure 10.15a. Although this approach must be used with caution, in view of the uncertain validity of a macroscopic theory at these small distances, the results in Figure 10.15a show that the force curves at pH 9.8 and at pH 4 are practically indistinguishable. This means that the pH has negligible effect on a range of the hydration forces. Moreover, the adhesion minima exist always at the same separation $D_0 = 6.0$–6.1 (± 0.2) nm, and the values of adhesion forces (or energy) between the β-Mal$_3$O(C$_{16+4}$)$_2$ layers, after the double-layer force subtraction, are practically the same, -1.7 to -1.8 mN/m ($G_0 = 0.27$–0.29 mJ/m^2), irrespective of the solution pH (Table 10.10).

To summarize the preceding results, it can be concluded that the changes in pH, while affecting the double-layer forces, affect neither the magnitude of the adhesion forces nor the range of the hydration forces between the β-Mal$_3$O(C$_{16+4}$)$_2$ layers. As the headgroup hydration, preventing anhydrous contact, is an important factor affecting the strength of adhesion between the β-Mal$_3$O(C$_{16+4}$)$_2$ layers, the present results indicate that the often assumed "water structure" based mechanisms such as ion-induced dehydration do not occur in the present case. The lack of any appreciable effects of OH$^-$ adsorption on both the hydration forces and the adhesion forces may be understandable, considering that the fraction of lipids binding OH$^-$ ions was only 2%–3% (Table 10.11).

10.4.1.4.3　Forces that Control the β-Mal₃O(C₁₆₊₄)₂ Vesicle Aggregation

What is the major force that controls the pH-dependent β-Mal$_3$O(C$_{16+4}$)$_2$ vesicle aggregations? We propose that it is the repulsive double-layer forces that control the vesicle aggregation. In a simplest approximation, we modeled the vesicles as hard spheres. By the Derjaguin approximation [116], the interaction energy $U(D)$ between two spherical vesicles of radii, r (=55 nm for the vesicles used in the aggregation experiments [89]), may be estimated from the force versus distance curve measured by two crossed cylindrical geometry, $F(D)$, using,

$$U(D) = \left(\frac{r}{2R}\right) \int_{\infty}^{D} F(D)\mathrm{d}D \qquad (10.16)$$

where R is the radius of two crossed cylinders. Figure 10.15b shows a representation of $U(D)$ versus D curves at pH 4 and pH 9.8 (NaOH in distilled water), respectively. At pH 4, there was practically no potential barrier that prevents the aggregation, and the potential minimum is about $-16kT$. This is consistent with the kinetics of the aggregation at pH 4, where it occurs almost instantaneously. At pH 9.8, on the other hand, the potential barrier is as large as $\sim 130\,kT$ and a difference between the potential minimum and maximum was about $2-3\,kT$ consistent with the stable vesicle dispersion. Assuming that bending modulus of β-Mal$_3$O(C$_{16+4}$)$_2$ layer has similar magnitude as those for DGDG and phospholipids, $k_C \approx 10\,kT$ [117,118], the magnitude of the double-layer forces is roughly an order of magnitude greater than that expected from the undulations.

All the results discussed in this and previous sections indicate that the pH-dependent aggregation of the β-Mal$_3$O(C$_{16+4}$)$_2$ vesicles is controlled largely by the double-layer forces originating from most presumably pH-dependent adsorption of OH$^-$ onto the vesicle surface and the dissociation of hydroxyl groups in the headgroups at strong alkaline conditions (pH $\gg$ 10) [119]. Thus, in near neutral pH (pH 5.5–7), the β-Mal$_3$O(C$_{16+4}$)$_2$ vesicles are stabilized largely by the double-layer forces with the surface potentials of an order of -10 mV. Once the electrostatic barrier is suppressed by a small decrease in pH (or a slight increase in electrolyte concentration), the strong attractive potential sets in and the rapid aggregations take place. We have also noted that β-Mal$_3$O(C$_{16+4}$)$_2$ vesicles aggregate in the presence of millimolar order of electrolytes such as NaCl, and CaCl$_2$ just as for the case of DGDG. Recently, similar DLVO-based mechanism has been proposed to explain reversible vesicle flocculation of sugar-based gemini surfactant [97].

10.4.2　PERMEABILITY AND STABILITY OF BLM OF β-MAL$_N$O(C$_{16+4}$)$_2$

We prepared BLMs by the folding method [120] in the aperture (100–250 μm in diameter) located in a thin Teflon sheet (25 μm thick), which separates two Teflon chambers filled with 100 mM KCl unbuffered or 100–250 mM KCl/10 mM Hepes-Tris (pH 7.4).

Among β-Mal$_N$O(C$_{16+4}$)$_2$ (N = 1, 2, 3, 5), β-Mal$_3$O(C$_{16+4}$)$_2$ forms most stable BLM. Table 10.12 summarizes electric properties of the β-Mal$_3$O(C$_{16+4}$)$_2$ BLM

TABLE 10.12

Electric Properties of Planar Bilayer Membranes in 0.1M KCl at pH 6 and at 23°C

Surfactant	Conductance (nS/cm²)	Capacitance (μF/cm²)	Thickness (nm)	Rupture Potential (mV)
β-Mal₃O(C₁₆₊₄)₂	2.8	0.6–0.7	2.7–3.1	344 ± 63
SBPL	12	0.6–0.7	—	293 ± 44
EPC [135]	—	0.6	—	280 ± 70
DPhPC	3.2	0.6–0.7	—	411 ± 42

Source: Reproduced from Baba, T., Toshima, Y., Minamikawa, H., Hato, M., Suzuki, K., and Kamo, N., *Biochim. Biophys. Acta*, 1421, 91, 1999. With permission.

Note: SBPL, soybean phospholipids; EPC, egg yolk phosphatidylcholine; DPhPC, 1,2-diphytanoyl-*sn*-glycero-3-phosphocholine.

together with those of various phospholipids BLM [46]. The thickness of the β-Mal$_3$O(C$_{16+4}$)$_2$ BLM estimated from the capacitance, is 2.9 ± 0.2 nm, which is favorably compared with the hydrophobic region thickness of the β-Mal$_3$O(C$_{16+4}$)$_2$ L$_\alpha$ phase, 2.6 nm (twice the d_{hc} value in Table 10.3), confirming the existence of solvent-free single bilayer membrane in the aperture. The β-Mal$_3$O(C$_{16+4}$)$_2$ BLM at the applied voltage of 0 mV was, in the best preparation case, stable for at least 3 days, whereas that of SBPL was usually stable only for less than 8 h under the same experimental conditions [46]. The higher value of rupture potentials of β-Mal$_3$O(C$_{16+4}$)$_2$ membrane as compared with those of SBPL or EPC also indicates the better stability of the planar membranes thereof. The conductance values of various BLMs listed in Table 10.12 indicate that β-Mal$_3$O(C$_{16+4}$)$_2$ BLM gives a lower (about 1/4) conductance value than that of SBPL. The BLM from a phytanoyl chained-lipid, 1,2-diphytanoyl-*sn*-glycero-3-phosphocholine (DPhPC) also exhibit low conductance, which is comparable to that of β-Mal$_3$O(C$_{16+4}$)$_2$.

In contrast to β-Mal$_3$O(C$_{16+4}$)$_2$, which forms a stable BLM, we were unable to form stable BLMs from other surfactants examined, that is, β-GlcO(C$_{16+4}$)$_2$, β-Mal$_2$O(C$_{16+4}$)$_2$, and β-Mal$_5$O(C$_{16+4}$)$_2$. This may be ascribed to the spontaneous curvature of the surfactant monolayer. As discussed in Section 10.3.2.3, β-GlcO(C$_{16+4}$)$_2$ and β-Mal$_2$O(C$_{16+4}$)$_2$ form non-bilayer phases, a Q$_{II}$(*Fd3m*) and an H$_{II}$ phase, respectively. This means that the spontaneous curvatures of these surfactant monolayers are both positive, which is unfavorable to form a planar bilayer membrane of ~100 μm size. The observation that BLM of synthetic 1,2-di-*O*-phytanyl-3-*O*-glucosylglycerol (DPhGG), the molecular structure of which is close to that of β-GlcO(C$_{16+4}$)$_2$, is unstable [121] may also be attributable to unfavorable positive spontaneous curvature of the DPhGG monolayer. β-Mal$_5$O(C$_{16+4}$)$_2$ that forms an L$_\alpha$ phase also failed to form a stable BLM. The pressure–area isotherm at the water/air interface of the β-Mal$_5$O(C$_{16+4}$)$_2$ mono-layer is more (~10%) expanded than that of β-Mal$_3$O(C$_{16+4}$)$_2$ [46]. As inferred from this result, the spontaneous curvature of β-Mal$_5$O(C$_{16+4}$)$_2$ monolayer is most presumably too negative for the stable BLM formation. The above observations strongly suggest that the use of surfactants with optimum spontaneous curvatures is essential to form stable BLMs.

10.4.3   PROTON PERMEABILITY OF β-MAL$_3$O(C$_{16+4}$)$_2$ VESICLE MEMBRANE

Rates of proton permeation across the β-Mal$_3$O(C$_{16+4}$)$_2$ vesicle membranes were estimated by measuring rates of change in fluorescence intensity of pyranine at pH 7 and 25°C [65]. The net proton/hydroxyl ion permeability coefficient at pH 7.0 ($P_{H/OH}$) through β-Mal$_3$O(C$_{16+4}$)$_2$ and a synthetic archaeal phospholipids DPhPC were 4.5 ± 1.4 and 3.6 ± 1.1 cm/s, respectively. On the other hand, the $P_{H/OH}$ values for vesicles of straight-chained lipids, EPC, and plant galactolipids, DGDG, were 17 ± 3 and > 23cm/s, respectively; about 4–5 times larger than those for the vesicles of the phytanyl or pytanoyl-chained lipids.

The low permeability feature of the C$_{16+4}$ chained lipids β-Mal$_3$O(C$_{16+4}$)$_2$ both in a form of BLM and vesicle is in accord with reported facts that vesicle membranes composed of archaeal lipids exhibit much lower permeability against solutes such as H$^+$ or hydrophilic dyes [122–125]. The recent experimental and MD simulations indicated that the reduction of chain dynamics rather than the lower hydration accounts for the lower permeability of the highly branched chains [62,63,65].

10.4.4   RECONSTITUTION OF MEMBRANE PROTEINS

We first describe a trial to assess potentials of the β-Mal$_3$O(C$_{16+4}$)$_2$ BLM to the functional reconstitution of membrane proteins, a channel protein derived from octopus retina microvilli vesicles [46].

For the reconstitution of octopus retina channels, the fusion method [88] was employed. After the formation of the BLM, the microvilli vesicles, which were solubilized with 2% 1-O-octyl-β-D-glucoside (β-GlcOC$_8$), were added to a gently stirred aqueous phase in one of the chamber (a grounded side). To enhance the fusion between the BLM and the microvilli vesicles, osmotic gradient and exter-nal voltage were applied. For the β-Mal$_3$O(C$_{16+4}$)$_2$ alone BLM, the fusion with vesicles did not take place. In contrast to this, incorporation of ion channels into SBPL bilayer proceeded successfully. SBPL contains a certain amount of nega-tively charged lipids, which may promote the fusion with the microvilli vesicles. We, therefore, employed a BLM, consisting of 5 mol% of a negatively charged glycolipid, sulfoquinovosyl diacylglycerol (SQDG), and 95 mol% β-Mal$_3$O(C$_{16+4}$)$_2$. For the SQDG-doped membrane, current fluctuations due to incorporated channels were observed as shown in Figure 10.16a [46]. The current observed can be favorably compared with the reported fact that the ion channels of microvilli vesicles show specific conductance ranging from 10 to 20pS [126].

We finally show the usefulness of the β-Mal$_3$O(C$_{16+4}$)$_2$ vesicles as a matrix for functional reconstitution of membrane proteins, taking photosystem II complex

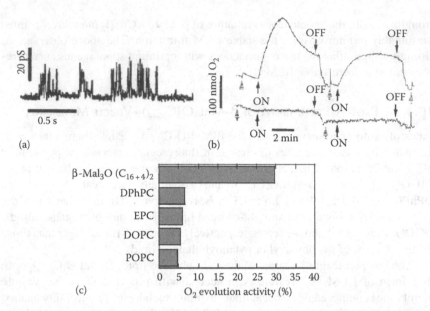

FIGURE 10.16 (a) Typical current fluctuation of the ion channels of octopus retina microvilli vesicles in BLM of β-Mal$_3$(OC$_{16+4}$)$_2$/SQDG (95:5, mol/mol) at 23°C. (Reproduced from Baba, T., Toshima, Y., Minamikawa, H., Hato, M., Suzuki, K., and Kamo, N., *Biochim. Biophys. Acta*, 1421, 91, 1999. With permission.) (b) Light-induced oxygen evolution of reconstituted PS II at 40°C. Effects of incorporated amount of PS II in β-Mal$_3$O(C$_{16+4}$)$_2$/SQDG (9:1, mol/mol) vesicles. 10μg Chl$a$/mL, [β-Mal$_3$O(C$_{16+4}$)$_2$] = 0.5 mM (upper trace); 2μg Chl$a$/mL, [β-Mal$_3$O(C$_{16+4}$)$_2$] = 0.5 mM (lower trace). Volume of reaction mixture = 2 mL. The dashed arrows indicate the addition of 100 mM DCBQ (8 μL each). The arrows labeled by ON; saturated red light (λ > 610 nm) was applied. Those labeled by OFF; light was removed. (Reproduced from Baba, T., Minamiakwa, H., Hato, M., Motoki, A., Hirano, M., Zhou, D., and Kawasaki, K., *Bochim. Biophys. Res. Commun.*, 265, 734, 1999. With permission.) (c) Effects of lipids on the oxygen evolution activity of PS II incorporated into vesicle at 40°C. The 100% activity of O$_2$ evolution was defined as the one measured in the Mes/Hepes/betaine buffer just before initiating reconstitution procedure, 1900 ± 50μmol O$_2$/mg Chl$a$·h. All the vesicles contain 10 mol% of SQDG. (Reproduced from Baba, T., Minamiakwa, H., Hato, M., Motoki, A., Hirano, M., Zhou, D., and Kawasaki, K., *Bochim. Biophys. Res. Commun.*, 265, 734, 1999. With permission.)

(PS II) from a thermophilic cyanobacterium *Synechococcus elongatus* grown at 55°C as an example [89]. PS II is one of the most important protein complexes involved in oxygenic photosynthesis and performs a series of photochemical reactions common in cyanobacteria, algae, and higher plants and is composed of most presumably more than 15 subunits [127]. PS II is relatively labile; it readily loses its water splitting activity owing to loss of the Mn-cluster from the complexes [128].

Thylakoid membranes of *S. elongatus* are composed of four major glycerolipids: 56 mol% MGDG, 23 mol% DGDG, 15 mol% SQDG, and 6 mol%

phosphatidylglycerol (PG). As to the hydrocarbon chain composition, about 74 mol% of the hydrophobic chains are C_{16}-chains: 56 mol% of saturated palmitic acid ($C_{16:0}$), and 18 mol% of monounsaturated palmitoleic acid ($C_{16:1}$). Thus, β-Mal$_3$O(C$_{16+4}$)$_2$ appears suitable in terms of the hydrophobic chain length to fulfill the hydrophobic matching. As SQDG is the major anionic glycolipid constituent of the *S. elongatus* thylakoid membranes and is considered essential to the PS II activity, we employed a mixture of β-Mal$_3$O(C$_{16+4}$)$_2$/SQDG (9:1, mol/mol) as a reconstitution matrix. The presence of negatively charged SQDG in β-Mal$_3$O(C$_{16+4}$)$_2$ vesicles is also essential to obtain stable vesicle suspensions in the Mes-Hepes/betaine buffer. In the absence of SQDG, β-Mal$_3$O(C$_{16+4}$)$_2$ vesicles readily aggregated to form precipitates (Section 10.4.1). The reconstitution of PS II was performed as follows. A mixture of lipid and SQDG (9:1, mol/mol) was solubilized with β-GlcOC$_8$ in a Mes-Hepes/betaine buffer (1 mM Mes/1 mM Hepes/10 mM NaCl/5 mM MgCl$_2$/1 M betaine, pH 7.0). PS II (solubilized with β-GlcOC$_8$) was then added to the solubilized β-Mal$_3$O(C$_{16+4}$)$_2$ to give a final PS II concentration of 8 µg Chl$a$ /mL of the Mes-Hepes/betaine buffer containing 2 mM β-Mal$_3$O(C$_{16+4}$)$_2$, 0.22 mM SQDG, and 40 mM β-GlcOC$_8$. The solubilized β-Mal$_3$O(C$_{16+4}$)$_2$/SQDG/ β-GlcOC$_8$/PS II mixture was briefly sonicated and then dialyzed against 100-fold volume of the Mes-Hepes/betaine buffer containing Bio-Beads SM-2 at 4°C for 12 h under dark [88]. The reconstitution trials of PS II into a range of phospholipid vesicles, for example, EPC, palmitoyloleoyl-PC (POPC), and dioleoyl-PC (DOPC), and DPhPC were also performed by the same procedures mentioned above. Freeze-fracture electron microscopy (FFEM) images of PS II reconstituted vesicle and an ultracentrifugation technique indicated that more than 90% of PS II used were incorporated into the β-Mal$_3$O(C$_{16+4}$)$_2$ vesicles in a transmembraneous manner. It is also noted that essentially the same results were obtained for the PSII/EPC vesicles [89].

Oxygen-evolving activity of reconstituted PS II. Typical examples of light-induced oxygen evolution of reconstituted PS II/β-Mal$_3$O(C$_{16+4}$)$_2$ vesicles with different PS II concentrations are shown in Figure 10.16b [89]. In this experiment, 100 mM 2,6-dichloro-p-benzoquinone (DCBQ) as an electron acceptor was added just before each on–off cycle. When the actinic light ($\lambda > 610$ nm) at saturated intensity was removed, air bubbling was immediately started to remove evolved oxygen, quickly re-equilibrating the oxygen level in the reaction mixture. For the first on–off cycle of the upper trace (10 µg Chl$a$/mL), the oxygen evolution exhibits a maximum followed by a spontaneous decrease of the oxygen level even though the actinic light is continuously applied. The spontaneous decrease in oxygen concentration was due to the bubble formation of supersaturated oxygen. As the oxygen-evolving activity was sufficiently high, the oxygen bubbles appeared at some stages of the reaction, eventually leading to the decrease of the oxygen level in the reaction mixture. The succeeding on–off cycles gave a response about 60% of that of the preceding cycle, indicating that partial inactivation of PS II occurred during the exposure of saturated actinic light. Nevertheless, at least six on–off cycles could be repeated until total extinction of the PS II activity was reached (data not shown). With a reduced incorporated amount of PS II ($2\,\mu g$ Chl$a$/mL), the oxygen evolution was significantly reduced as expected (a lower trace).

Figure 10.16c [89] compares the oxygen-evolving activity of the PS II/β-Mal$_3$O(C$_{16+4}$)$_2$ system with those of several PS II/PC systems, where PS II reconstitution into PC vesicles was performed using the same experimental procedures as in β-Mal$_3$O(C$_{16+4}$)$_2$. We define 100% of oxygen-evolving activity as the one measured for PS II in the Mes-Hepes/betaine buffer just before initiating the reconstitution procedures. Note that the incorporated amount of PS II and the FFEM images of PS II/EPC system was practically the same as those of PS II/β-Mal$_3$O(C$_{16+4}$)$_2$ system. Nevertheless, PS II reconstituted in β-Mal$_3$O(C$_{16+4}$)$_2$ showed 5–6-fold higher activity than that in the PC-vesicles examined. As the activity of PS II/DPhPC system was comparable to that of the PS II/EPC system, phytanoyl chain does not seem a major factor for the higher activity of PS II/β-Mal$_3$O(C$_{16+4}$)$_2$ system. Bearing in mind that the glycolipids (MGDG and DGDG) are the major lipid components of the thylakoid membrane, the sugar moieties may have some consequences to support the oxygen-evolving activities of PS II, though a definite conclusion awaits further studies. The present results reveal that well-designed synthetic sugar-based surfactants are effective for the functional reconstitution of labile membrane protein complexes, such as PS II.

10.5 SUMMARY AND FUTURE PERSPECTIVES

In this chapter we have described the recent progress in the understanding of the physical and biophysical properties of sugar-based isoprenoid-chained surfactants/water systems. Though our knowledge of them is still in its infancy, several features of isoprenoid-chained surfactant/water systems have emerged.

1. The sugar-based isoprenoid-chained surfactants can afford a greater control of aqueous phase structures at low temperatures.

    Provided the sugar headgroups are rationally selected, the T_K values of the isoprenoid-chained surfactants can be depressed below 0°C even when the total number of carbon atoms in the hydrophobic group is as large as 40. This is in marked contrast to the straight-chained surfactant counterparts, where the values of T_K rapidly increase as the chain length increases. The sugar-based isoprenoid-chained surfactants, therefore, can afford a greater control of aqueous phase structures at low temperatures; being able to cover practically a full range of lyotropic phases from normal micelles to an inverted micellar cubic phase. They are particularly useful in preparing vesicles or BLM, and a range of inverted liquid crystalline phases at low temperatures.

2. Solute permeability through the isoprenoid chained surfactants membranes is significantly lower than that of the straight-chained surfactant membranes.

    The reduced chain dynamics of the isoprenoid chains contributes to the low solute permeability of the isoprenoid-chained surfactant membranes and most presumably to the enhanced stability of the bilayer membranes. This feature is preferable for many technical applications.

3. Rational combinations of the sugar headgroups and the hydrophobic groups are crucial to control the physical properties of the sugar-based surfactant/water systems.

   Physical properties of the sugar-based surfactant/water systems depend critically on the inter-headgroup interactions, which depend on the type of the sugar residues, the type of the glycosidic linkage, the overall conformation of the headgroup, nonbonding interactions between groups in adjacent residues, sugar–water interactions, etc. Explicit accounts of the headgroup interactions, therefore, are crucial to design surfactants and to understand a particular system concerned. Among the oligosaccharide headgroups so far examined, the maltooligosaccharides serve as the convenient and useful headgroups both for controlling the aqueous phase structures and depressing the values of T_K [129]. However, the types of sugar headgroups and the hydrophobic groups so far investigated have been still limited. More systematic survey of combinations of diverse sugar headgroups and hydrophobic groups deserves attention and will surely expand the scope of the isoprenoid-chained surfactants [130].

4. "Sugar-surfaces" behave as "hard-surfaces" interacting via a short-range attractive potential.

   The forces (or potentials) between nonionic sugar-based surfactant membranes in aqueous media have been directly measured for synthetic and naturally occurring oligosaccharide-based surfactants ($N = 1$–3) [105,131–134]. Despite the diverse headgroups and the lipid structures examined, the measured potential profiles have one feature in common; the "sugar-surfaces" behave, in a simplest approximation, as "hard-surfaces" interacting via a short-range attractive potential as seen in Figure 10.15a. This underlies their somewhat unexpected behavior such as the ion and pH sensitive nature of the vesicles, which is very different from that of phospholipid vesicles.

5. The sugar-based isoprenoid-chained surfactants may be useful in many technical applications.

   The ease of low temperature control of the aqueous phase structures, the low solute permeability, and the success of the functional reconstitution of membrane proteins may envisage the sugar-based isoprenoid-chained surfactants as attractive materials in a variety of technical fields, for example, DDS, solubilization/crystallization of membrane proteins, and new building blocks for nano-materials. In addition, their hydrophobic chain will help their adsorption onto solid substrates, and the carbohydrate headgroups can serve for specific interactions with lectins, toxins, and other proteins. The molecular recognition functionality of the carbohydrates may be suitable for high-throughput bioassay on quartz crystal microbalance (QCM) measurement, surface plasmon resonance (SPR) spectroscopy, and so on [2].

ACKNOWLEDGMENTS

We acknowledge the fruitful cooperation over the years with coworkers in Tsukuba and elsewhere. The work described here has received financial support from several sources, of which the most important are Research and Development Projects of Industrial Science and Technology Frontier Program supported by AIST, and International Joint Research Program supported by NEDO.

REFERENCES

1. Hill, K., von Rybinski, W., and Stoll, G., *Alkyl Polyglycosides*. VCH, Weinheim, 1997.
2. Kitamoto, D., Toma, K., and Hato, M., in *Handbook of Nanostructured Biomaterials and Their Applications in Nano-biotechnology*, Vol.1., Nalwa, S. H., Ed., American Scientific Publishers, Stevenson Ranch, 2005, Chapter 6.
3. Laughlin, R. G., in *The Aqueous Phase Behavior of Surfactants*. Academic Press, London, 1994, Chapters 3 and 5.
4. Shinoda, K., in *Solution and Solubility*, 3rd ed., Maruzen, Tokyo, 1991, Chapter 1.
5. Jeffrey, G. A., *Acc. Chem. Res.*, 1986, *19*, 168.
6. Abe, Y., Harata, K., Fujiwara, M., and Ohbu, K., *J. Chem. Soc. Perkin Trans.*, 1998, 2, 177.
7. Hato, M., *Curr. Opin. Colloid Interface Sci.*, 2001, *6*, 268.
8. Boyd, B. J., Drummond, G. J., Krodkiewska, I., and Grieser, F., *Langmuir*, 2000, *16*, 7359.
9. Sakya, P., Seddon, J. M., and Vill, V., *Liq. Cryst.*, 1997, *23*, 409.
10. Shinoyama, H., Gama, Y., Nakahara, H., Ishigami, Y., and Yasui, T., *Bull. Chem. Soc. Jpn.*, 1991, *64*, 29.
11. Nilsson, F., Söderman, O., Hansson, P., and Johansson, I., *Langmuir*, 1998, *14*, 4050.
12. Koynova, R. and Caffrey, M., *Chem. Phys. Lipids*, 1994, *69*, 181.
13. Kates, M., in *Handbook of Lipid Research 6, Glycolipids, Phosphoglycolipids, and Sulfoglycolipids*, Kates, M., Ed., Plenum Press, New York and London, 1990, pp. 1–122.
14. Smith, P. F., in *Microbial Lipids*, Vol. 1, Ratledge, C. and Wilkinson, S. G., Eds., Academic Press, New York, 1988, pp. 489–525.
15. Langworthy, T. A. and Pond, J. L., in *Thermophiles: General, Molecular, and Applied Microbiology*, Brock, T. D., Ed., John Wiley & Sons, New York, 1986, pp. 107–135.
16. Fischer, E. and Helperich, B., *Justus Liebigs Ann. Chem.*, 1911, *383*, 68.
17. Fischer, E., *Ber. Dtsh. Chem. Ges.*, 1893, *26*, 2400.
18. Königs, W. and Knorr, E., *Ber.*, 1901, *34*, 957.
19. Veeneman, G. H., in *Carbohydrate Chemistry*, Boon G.-J., Ed., Blackie Academic & Professional, London, 1998, Chapter 4.
20. Schmidt, R. R. and Kläger, R., *Angew. Chem., Int. Ed. Engl.*, 1985, *24*, 65.
21. Schmidt, R. R., *Angew. Chem., Int. Ed., Engl.*, 1986, *25*, 212.
22. Minamikawa, H., Murakami, T., and Hato, M., *Chem. Phys. Lipids*, 1994, *72*, 111.
23. Exoffier, G., Gagnaire, D. G., and Utille, J.-P., *Carbohydr. Res.*, 1975, *39*, 368.
24. Urban, F. J., Moore, B. S., and Breitenbach, R., *Tetrahedron Lett.*, 1990, *31*, 4421.
25. Hanessian, S. and Banoub, J., *Carbohydr. Res.*, 1997, *59*, 261.

26. Salkar, R. J., Minamikawa, H., and Hato, M., *Chem. Phys. Lipids*, 2004, *127*, 65.
27. Minamikawa, H. and Hato, M., *Langmuir*, 1997, *13*, 2564.
28. Minamikawa, H. and Hato, M., *Langmuir*, 1998, *14*, 4503.
29. Hato, M., Yamashita, I., Kato, K., and Abe, Y., *Langmuir*, 2004, *20*, 11366.
30. Milkereit, G., Garamus, V. M., Yamashita, J., Hato, M., Morr, M., and Vill, V., *J. Phys. Chem. B.*, 2005, *109*, 1599.
31. Hato, M., Seguer, J. B., and Minamikawa, H., *Stud. Surf. Sci. Catal.*, 2001, *132*, 725.
32. Hato, M. and Minamikawa, H., *Langmuir*, 1996, *12*, 1658.
33. Nilsson, F., Söderman, O., and Reimer, J., *Langmuir*, 1998, *14*, 6396.
34. Lang, J. C. and Morgan, R. D., *J. Chem. Phys.*, 1980, *73*, 5849.
35. Kahlweit, M. and Strey, R., *Angew. Chem. Int. Ed. Engl*, 1985, *24*, 654.
36. Minamikawa, H. and Hato, M., *Chem. Phys. Lipids*, 2005, *134*, 151.
37. Hato, M., Minamikawa, H., Salkar, R. A., and Matsutani, S., *Progr. Colloid Polym. Sci.*, 2004, *127*, 65.
38. Hato, M., Minamikawa, H., Salkar, R. A., and Matsutani, S., *Langmuir*, 2002, *18*, 3425.
39. Hyde, S. T., Andersson, S., Ericsson, B., and Larsson, K., *Z. Kristallogr.*, 1984, *168*, 213.
40. Turner, D. C., Wang, Z.-G., Gruner, S. M., Mannock, D. A., and McElhaney, N., *J. Phys. II France*, 1992, *2*, 2039.
41. Barauskas, J. and Landh, T., *Langmuir*, 2003, *19*, 9562.
42. Seddon, J. M., *Biochemistry*, 1990, *29*, 7997.
43. Seddon, J. M., Bartle, E. A., and Mingins, J., *J. Phys. Condens. Matter*, 1990, *2*, SA285.
44. Kennedy, J. F., in *Carbohydrate Chemistry*, Kennedy, J. F., Ed., Clarendon Press, Oxford, 1988, Chapter 1.
45. Tamada, K., Minamikawa, H., Hato, M., and Miyano, K., *Langmuir*, 1996, *12*, 1666.
46. Baba, T., Toshima, Y., Minamikawa,H., Hato, M., Suzuki, K., and Kamo, N., *Biochim. Biophys. Acta*, 1999, *1421*, 91.
47. Nilsson, F., Söderman, O., and Johansson, I., *Langmuir*, 1996, *12*, 902.
48. Wieslander, A., Ulmius, J., Lindblom, G., and Fontell, K., *Biochim. Biophys. Acta*, 1978, *512*, 241.
49. McDaniel, R. V., *Biochim. Biophys. Acta*, 1998, *940*, 158.
50. Larsson, K., in *Lipids-Molecular Organization, Physical Functions and Technical Applications*, The Oily Press, Dundee, 1994, Chapter 3.
51. Chapman, R. M., Williams, B. D., and Ladbrooke, B. D., *Chem. Phys. Lipids.*, 1967, *1*, 445.
52. Gruner, S. M., Tate, M. W., Kirk, G. L., So, P. T. C., Turner, D. C., Keane, D.T., Tilcock, C. P. S., and Cullis, P. R., *Biochemistry*, 1988, *27*, 2853.
53. Shipley, G. G., Green, J. P., and Nichols, B. W., *Biochim. Biophys. Acta*, 1973, *311*, 53.
54. We, Y., He, K., Ludtke, S. J., and Huang, H. W., *Biophys. J.*, 1995, *68*, 2361.
55. Nagel, J. F., Zhang, R., Tristram-Nagle, S., Sun, W., Petrache, H. L., and Suter, R. M., *Biophys. J.*, 1995, *68*, 2361.
56. Gounaris, K., Mannock, D. A., Sen, A., Brain, A. P. R., Williams, W. P., and Quinn, J. P., *Biochim. Biophys. Acta*, 1983, *732*, 229.
57. Mannock, D. A., Lewis, R. N. A. H., Sen, A., and McElhaney, R. N., *Biochemistry*, 1988, *27*, 6852.
58. Hinz, H. J., Kuttenreich, H., Meyer, R., Renner, M., and Fründ, R., *Biochemistry*, 1991, *30*, 5125.

59. Endo, T., Inoue, K., and Nojima, S., *J. Biochem. (Tokyo)*, 1982, *92*, 953.
60. Iwamoto, K., Sunamoto, J., Inoue, K., Endo, T., and Nojima, S., *Biochim. Biophys. Acta*, 1982, *691*, 44.
61. Shinoda, W., Mikami, M., Baba, T., and Hato, M., *J. Phys. Chem. B*, 2003, *107*, 14030.
62. Shinoda, W., Mikami, M., Baba, T., and Hato, M., *J. Phys. Chem. B*, 2004, *108*, 9346.
63. Shinoda, W., Mikami, M., Baba, T., and Hato, M., *Chem. Phys. Lett.*, 2004, *390*, 35.
64. Degani, H., Danon, A., and Caplan, S. R., *Biochemistry*, 1980, *19*, 1626.
65. Baba, T., Minamikawa, H., Hato, M., and Handa, T., *Biophys. J.*, 2001, *81*, 3377.
66. Holmes, M. C., and Charvolin, J., *J. Phys. Chem.*, 1984, *88*, 810.
67. Kekicheff, P., Cabane, B., and Rawiso, M., *J. Physique, Lett.*, 1984, *45*, L-813.
68. Funari, S. S., Holmes, M. C., and Tiddy, G. J. T., *J. Phys. Chem.*, 1994, *98*, 3015.
69. Fairhurst, C. E., Holmes, M. C., and Leaver, M. S., *Langmuir*, 1997, *17*, 4964.
70. Gustafsson, J., Oradd, G., Lindblom, G., Olsson, U., and Almgren, M., *Langmuir*, 1997, *13*, 852.
71. Minewaki, K., Kato, T., Yoshida, H., Imai, M., and Ito, K., *Langmuir*, 2001, *17*, 1864.
72. Hyde, S. T., *Colloque de Physique*, 1990, *51*, C7.
73. Hyde, S. T., *Colloids Surf. A*, 1995, *103*, 227.
74. Yamashita, I., Kawabata, Y., Kato, T., Hato, M., and Minamikawa, H., *Colloids Surf. A*, 2004, *250*, 485.
75. Dörfler, H.-D. and Göpfert, A., *J. Disp. Sci. Technol.*, 1999, *20*, 35–58.
76. Hato, M., Minamikawa, H., and Seguer, J. B., *J. Phys. Chem. B.*, 1998, *102*, 11035.
77. Shinoda, K., Yamaguchi, T., and Hori, R., *Bull. Chem. Soc. Jpn.*, 1961, *34*, 237.
78. Aveyard, R., Binks, B. P., Chen, J., and Fletcher, I., *Langmuir*, 1998, *14*, 4699.
79. De Grip, W. J. and Bovee-Geurts, P. H. M., *Chem. Phys. Lipids*, 1979, *23*, 321.
80. Israelachvili, J., in *Intermolecular and Surface Forces*, 2nd ed., Academic Press, New York, 1992, Chapter 17.
81. Kato, T., Kawabata, Y., Fujii, M., Kato, T., Hato, M., and Minamikawa, H., *J. Colloid Interface Sci.*, 2007, *312*, 122.
82. Focher, B., Savelli, G., Torri, G., Vecchio, G., McKenzie, D. C., Nicoli, D. F., and Bunton, C. A., *Chem. Phys. Lett.*, 1989, *158*, 491.
83. He, L., Garamus, V. M., Funari, S. S., Malfois, M., Willumeit, R., and Niemeyer, B., *J. Phys. Chem. B.*, 2002, *106*, 7596; Durpuy, C., Auvray, X., Petipas, C., Rico-Lattes, I., and Lattes, A., *Langmuir*, 1997, *13*, 3965.
84. Ben-Shaul, A. and Gelbart, W. M., in *Micelles, Membranes, Microemulsions, and Mono-layers*, Gelbart, W. M., Ben-Shaul, A., and Roux, D., Eds., Springer, New York, 1994, p. 1.
85. Nilsson, F., Söderman, O., and Johansson, I., *J. Colloid Interface Sci.*, 1998, *203*, 131.
86. Whiddon, C., Söderman, O., and Hansson, P., *Langmuir*, 2002, *18*, 4610.
87. Whiddon, C., Reimer, J., and Söderman, O., *Langmuir*, 2004, *20*, 2172.
88. Miller, C., *Ion Channel Reconstitution*, Plenum, New York, 1986.
89. Baba, T., Minamiakwa, H., Hato, M., Motoki, A., Hirano, M., Zhou, D., and Kawasaki, K., *Biochim. Biophys. Res. Commun.*, 1999, *265*, 734.
90. Baba, T., Zheng, L.-Q., Minamikawa, H., and Hato, M., *J. Colloid Interface Sci.*, 2000, *223*, 235.
91. Minami, H., Inoue, T., and Shimozawa, R., *Langmuir*, 1996, *12*, 3574.
92. Gamon, B. L., Virden, J. W., and Berg, J. C., *J. Colloid Interface Sci.*, 1989, *132*, 125.
93. Balzer, D., *Langmuir*, 1993, *9*, 3375.

94. Webb, M. S., Tilcock, C. P. S., and Green, B. R., *Biochim. Biophys. Acta*, 1988, *938*, 323.
95. Webb, M. S. and Green, B. R., *Biochim. Biophys. Acta*, 1990, *1030*, 231.
96. Fragata, M., Menikh, A., and Robert, S., *J. Phys. Chem.*, 1993, *97*, 13920.
97. Johnsson, M., Wagenaar, A., and Engberts, J. B. F. N., *J. Am. Chem. Soc.*, 2003, *125*, 757.
98. Zhang, L., Somasundaran, P., and Maltesh, C., *Langmuir*, 1996, *12*, 2371.
99. Tomoasia-Cotisel, M., Zsakó, J., Chifu, E., and Quinn, P. J., *Chem. Phys. Lipids*, 1983, *34*, 55.
100. Johnston, D. S., Coppard, E., and Chapman, D., *Biochim. Biophys. Acta*, 1985, *815*, 325.
101. Verwey, E. J. and Overbeek, J. Th. G., *Theory of Stability of Lyophobic Colloid*, Elsevier, Amsterdam, 1948.
102. Hogg, R., Healy, T. W., and Fuerstenau, D. W., *Trans. Faraday Soc.*, 1966, *62*, 1638.
103. Wiese, G. and Healy, T. W., *Trans. Faraday Soc.*, 1970, *66*, 490.
104. Vold, M. J., *J. Colloid Sci.*, 1961, *16*, 1.
105. Korchowiec, B. M., Baba, T., Minamikawa, H., and Hato, H., *Langmuir*, 2001, *17*, 1853.
106. Rand, R. P. and Parsegian, V. A., *Biochim. Biophys. Acta*, 1989, *988*, 351.
107. Israelachvili, J. N. and Wennerström, H., *Nature*, 1996, *379*, 219.
108. McIntosh, T. J. and Simon, S. A., *Biochemistry*, 1993, *32*, 8374.
109. Helfrich, W., *Z. Naturforsch.*, 1978, *33a*, 305.
110. Evans, E. A. and Parsegian, V. A., *Proc. Natl. Acad. Sci. U. S. A.*, 1986, *83*, 7132.
111. Blodgett, K. B. and Langmuir, I., *Phys. Rev.*, 1937, *51*, 964.
112. Ninham, B. W. and Parsegian, A. D., *J. Theor. Biol.*, 1971, *31*, 405.
113. Ohshima, H., *Colloid Polym. Sci.*, 1976, *254*, 484.
114. Ohshima, H. and Mitsusi, T., *J. Colloid Interface Sci.*, 1978, *63*, 525.
115. Chan, D. Y., Pashley, R. M., and White, L. R., *J. Colloid Interface Sci.*, 1980, *77*, 283.
116. Derjaguin, B. V., *Kolloid-Z.*, 1934, *69*, 155.
117. Lorenzen, S., Servuss, R. M., and Helfrich, W., *Biophys. J.*, 1986, *50*, 565.
118. Evans, E. and Rawicz, W., *Phys. Rev. Lett.*, 1990, *64*, 2094.
119. Doppert, H. L. and Staverman, A. J., *J. Polym. Sci. A-1*, 1966, *4*, 2367.
120. Montal, M. and Muller, P., *Proc. Natl. Acad. Sci., U. S. A.*, 1972, *69*, 3561.
121. Stern, J., Freisleben, H.-J., Janku, S., and Ring, K., *Biochim. Biophys. Acta*, 1992, *1128*, 227.
122. Gambacorta, A., Gliozzi, A., and Rosa, M. De., *J. Microbial. Biotechnol.*, 1995, *11*, 115.
123. Rosa, M. De, *Thin Solid Films*, 1996, *284/285*, 13.
124. Elferink, M. G. L., De Wit, J. G., Driessen, A. J. M., and Konings, W. N., *Biochim. Biophys. Acta*, 1994, *1193*, 247.
125. Dannenmuller, O., Arakawa, K., Eguchi, T., Kakinuma, K., Blanc, S., Albrecht, A.-M., Schumtz, M., Nakatani, Y., and Ourisson, G., *Chem. Eur. J.*, 2000, *6*, 645.
126. Hirata, H., Ohno, K., and Tsuda, M., *Seibutsubutsuri (Biophysics)*, 1986, *26*, S108.
127. Ikeuchi, M., *Bot. Mag. Tokyo*, 1992, *105*, 327.
128. Nash, D., Miyao, M., and Murata, N., *Biochim. Biophys. Acta*, 1985, *807*, 127.
129. Hato, M., Minamikawa, H., Tamada, K., Baba, T., and Tanabe, Y., *Adv. Colloid Interface Sci.*, 1999, *80*, 233.
130. Yamashita, J., Shiono, M., and Hato, M., *J. Phys. Chem. B*, 2008, in press.
131. Waltermo, Å., Manev, E., Pugh, R., and Claesson, P. M., *J. Disp. Sci. Technol.*, 1994, *15*, 273.
132. Person, C. M. and Claesson, P. M., *Langmuir*, 2000, *16*, 10227.
133. Ricoul, F., Dubois, M., Belloni, L., Zemb, T., André-Barrés, C., and Rico-Lattes, I., *Langmuir*, 1993, *14*, 2645.
134. McIntosh, T. J., *Curr. Opin. Struct. Biol.*, 2000, *10*, 481, and references cited therein.
135. Robello, M. and Gliozzi, A., *Biochim. Biophys. Acta*, 1989, *982*, 173.

11 Micellar Properties and Molecular Interactions in Binary Surfactant Systems Containing a Sugar-Based Surfactant

Cristóbal Carnero Ruiz

CONTENTS

11.1 INTRODUCTION

The study of interfacial and bulk properties of solutions composed of mixtures of surfactants has become in recent years a topic of increasing interest in the field of self-assembly of amphiphiles. Due to the appearance of interactions of a different nature between the component surfactants, it is often observed that the mixed surfactant systems present a very different behavior from those formed by single surfactants, and this fact has very interesting implications of both applied and theoretical character. Therefore, in order to characterize these mixed systems, many efforts have been carried out by different workers since the early 1980s. In this regard, the reader is referred to references [1–3]. These three excellent monographs contain valuable information covering both fundamental and applied aspects of this subject.

The advantages of mixed surfactant systems in many technical applications, over those constituted by a single surfactant, are numerous [4–7]. Mixed micelles composed of ionic and nonionic surfactants display an expanded colloidal stability when compared with the pure nonionic micelle. Furthermore, the size of an ionic micelle, which usually forms small globular aggregates at a low surfactant concentration, may be increased upon addition of a nonionic surfactant. The enhancement of these two properties, stability and size, would enhance the capability of incorporating different solutes in the micellar phase, which has relevant implications in many applications of micellar solutions [8]. For instance, pure cationic surfactants are poor detergents since they neutralize the negative charges on fibers or solutes, but it has been shown that this property can be improved by using a cationic–nonionic mixture [4]. In addition, there are some important properties of great interest in several applications of surfactants that are determined by the micellar composition. A remarkable example refers to the microenvironmental properties, which play a decisive role in areas such as micellar catalysis. It is well known, for instance, that the local polarity or micropolarity can modify not only the velocity but also the mechanism of the reaction [9].

On the other hand, mixed surfactant systems have also been the subject of considerable attention from a theoretical point of view. Note that, in order to interpret the different experimental results, it is particularly interesting to get appropriate mixing thermodynamic models. The most significant aspects on theoretical studies of mixed micellization have been recently reviewed [10]. In this respect, most noteworthy are the efforts made by Blankschtein's group [11–15] and by Nagarajan [16,17] in order to provide a theory that is not only well founded, from a thermodynamic point of view, but also capable of predicting the more relevant properties of mixed surfactant systems.

Because polyoxyethylene type surfactants are widely employed in both biochemical research and other technical applications, the mixtures of these surfactants with common ionic surfactants, such as alkyl sulfates, alkyl benzene sulfonates, or alkyltrimethylammonium salts, have been thoroughly investigated. However, mixed systems involving less common surfactants, including sugar-based ones, have been much less studied [4]. Among sugar-based surfactants, alkyl polyglucosides (APGs) are quickly gaining acceptance in applications ranging from manual dishwashing detergents to all-purpose cleaners and laundry detergents. These surfactants show remarkable physicochemical properties, which often differ clearly from those of other nonionic surfactants. For example, it has been found that APGs show synergism with different types of surfactants, including the three primary surfactant linear alkyl benzene sulfonate (LAS), secondary alkane sulfonate (SAS), fatty alcohol sulfate (FAS), and also, although less pronounced, with fatty alcohol ether sulfates (FAES). However, other ethoxylated nonionic surfactants do not show any synergism with FAES [18]. In addition, APGs are widely used in dishwashing detergents in both conventional and concentrated forms. In the case of conventional dishwashing detergents, the partial or total replacement of FAES by APGs leads to an increase in performance, while the use of APGs is advantageous in concentrated detergents because, due to the synergic behavior with anionic surfactants, highly effective products with relatively low surfactant content can be prepared [19].

In recent years, considerable efforts have been made to elucidate the behavior in solution and different structural aspects in mixed systems composed of a sugar-based surfactant, particularly those belonging to the groups of APGs and fatty acid glucamides (FAGs), and both ionic and nonionic conventional surfactants. This chapter deals with work in the field of mixed surfactant systems, involving a sugar-based surfactant, from 1997 to date. The outline of the chapter is as follows. First of all, the main aspects of the thermodynamic models most frequently used to describe the behavior of mixed systems of surfactants are summarized in Section 11.2. Sections 11.3 and 11.4 are devoted to the revision of experimental results in mixed systems composed of APGs and FAGs with other surfactants, respectively. In these sections, the interpretation of the results according to the viewpoints of the original authors is presented. In Section 11.5, some conclusions and several suggestions for future research are given. Finally, we summarize the list of symbols and abbreviations for the surfactants used in this chapter.

11.2 PARAMETERS DESCRIBING A MIXED SURFACTANT SYSTEM

Several thermodynamic models to describe the behavior of a mixed surfactant system have been proposed [20]. Among them, those based on the pseudo-phase-separation approach are by far the most widely used. According to this model, the micelles are considered as a macroscopic phase in equilibrium with a solution containing the corresponding monomers, so that the condition of thermodynamic equilibrium between phases applies. When two surfactants are mixed together both ideal and nonideal behaviors are possible. The treatment for ideal mixed micelles was developed by Clint [21] and can be summarized as follows. Let us call μ_i the chemical potential of the ith surfactant monomer in the bulk of a mixed surfactant system, which is given by

$$\mu_i = \mu_i^0 + RT \ln C_i^m \tag{11.1}$$

where
μ_i^0 is its standard chemical potential
C_i^m is the concentration of the monomeric surfactant i in the bulk

In the mixed micelle, the chemical potential of component i can be expressed as

$$\mu_i^M = \mu_i^0 + RT \ln C_i + RT \ln x_i \tag{11.2}$$

where C_i is the critical micelle concentration (CMC) of the pure component i and x_i the mole fraction of surfactant i in the mixed micelles, respectively. By applying the condition of micellization phase equilibrium ($\mu_i = \mu_i^M$), the monomer concentration can be written as

$$C_i^m = x_i C_i = \alpha_i C^* \tag{11.3}$$

where
C^* is the mixed CMC
α_i is the mole fraction of surfactant i in the bulk

In the case of a binary mixture, the mixed CMC, C^*, can be expressed as

$$\frac{1}{C^*} = \sum_{i=1}^{2} \frac{\alpha_i}{C_i} \tag{11.4}$$

This same treatment can be applied to the case of nonideal mixtures in which the concentrations are to be replaced by the corresponding activities, a_i, which are related to the molar fraction of the ith component by $a_i = f_i x_i$. Therefore, the chemical potential of the component i in the mixed micelle can be now written as

$$\mu_i^M = \mu_i^0 + RT \ln C_i + RT \ln f_i x_i \tag{11.5}$$

In this way, the above treatment yields the following expression for the monomer concentration:

$$C_i^m = x_i f_i C_i = \alpha_i C^* \tag{11.6}$$

and the mixed CMC, C^*, is given by

$$\frac{1}{C^*} = \sum_{i=1}^{2} \frac{\alpha_i}{f_i C_i} \tag{11.7}$$

By using the regular solution theory (RST), Rubingh [1,22] introduced the following equation for the activity coefficients, f_i,

$$f_i = \exp\left[\beta^M \left(1-x_i\right)^2\right] \tag{11.8}$$

where β^M is an interaction parameter characterizing the interactions between the two surfactants in the mixed micelle. This parameter is defined as

$$\beta^M = \frac{N_A \left(W_{11} + W_{22} - 2W_{12}\right)}{RT} \tag{11.9}$$

where

N_A is Avogadro's number

W_{ij} are the pairwise interaction energies between monomeric species in the mixed micelle

Note that β^M is an indication not only of the degree of interaction between the surfactants but also accounts for the deviation from ideality. A negative value of β^M implies an attractive interaction; the more negative the value of β^M, the greater the attraction. From a physical point of view, the β^M parameter can be interpreted in terms of an energetic parameter that represents the excess Gibbs free energy of mixing. This interpretation is correct if, according to RST, the excess entropy of mixing equals zero.

Appropriately combining Equations 11.6 through 11.8, the following equations can be derived:

$$\frac{x_1^2 \ln\left(\alpha_1 C^*/x_1 C_1\right)}{\left(1-x_1\right)^2 \ln\left[\left(1-\alpha_1\right)C^*/\left(1-x_1\right)C_2\right]} = 1 \tag{11.10}$$

$$\beta^M = \frac{\ln\left(\alpha_1 C^*/x_1 C_1\right)}{\left(1-x_1\right)^2} \tag{11.11}$$

Equation 11.10 relates the mole fraction of surfactant 1 in the mixed micelle (x_1) with the CMC of the binary system, with the mole fraction of surfactant 1 in the solution (α_1), and with the CMC of pure surfactants (C_1 and C_2). This equation can be solved iteratively for x_1 and then the interaction parameter β^M can be determined from Equation 11.11.

It should be mentioned that the validity of RST to describe nonideal mixing in mixed surfactant micelles has been questioned [5]. On the one hand, while RST assumes the excess entropy of mixing to be zero, as mentioned above, calorimetric measurements have shown that this magnitude is nonzero in some mixed surfactant systems. On the other, if this approach accurately describes the nonideal mixing behavior, the β^M parameter should be constant for any composition. However, β^M values obtained in many binary surfactant systems show considerable variations with the solution composition [14]. It has been established that some variability in β^M may be due to experimental error in measurements of the mixture CMCs [5]. Nevertheless, large changes of β^M with the mixture composition could indicate that RST may not be appropriate to describe the nonideal mixing behavior. Despite these limitations, RST is a very used and convenient method for analyzing experimental CMC measurements of mixed micellar systems because the β^M parameter quantitatively indicates the extent of nonideality in a single number that can be easily compared among different pairs of surfactants [14].

Besides the micellar formation process, surfactants undergo another important physical process that has much interest from a technological point of view. They adsorb onto interfaces, thereby lowering their surface tension. Therefore, surface interaction and adsorption of the surfactant mixture components is another phenomenon that has received considerable attention from many researchers in this field. To describe the adsorption process at the air–liquid interface of a binary surfactant system, Rosen and Hua [23] extended the regular solution treatment of Rubingh for mixed surfactant micelles to the adsorption at the air–liquid interface. The aforementioned authors proposed two equations, with a structure similar to Equations 11.10 and 11.11, respectively, in terms of the interfacial mole fraction of each surfactant (x_i^σ) at the adsorbed interfacial monolayer, and an interfacial molecular interaction parameter (β^σ), which accounts for the deviation from ideality in the mixed monolayer formation at the air–liquid interface. The relevant equations for this approach are the following:

$$\frac{\left(x_1^\sigma\right)^2 \ln\left(\alpha_1 C_{mix} / x_1^\sigma C_1^0\right)}{\left(1 - x_1^\sigma\right)^2 \ln\left[\left(1 - \alpha_1\right)C_{mix} / \left(1 - x_1^\sigma\right)C_2^0\right]} = 1 \tag{11.12}$$

$$\beta^\sigma = \frac{\ln\left(\alpha_1 C_{mix} / x_1^\sigma C_1^0\right)}{\left(1 - x_1^0\right)^2} \tag{11.13}$$

where C_{mix}, C_1^0, and C_2^0 are the concentrations of the mixture and pure surfactants 1 and 2, respectively, required to produce a given surface tension value [24].

When a certain property of the mixed system improves in comparison to that attained by either of the pure surfactants, it is said that the system presents synergism. On the basis of the nonideal model, the conditions for the occurrence of synergism in both efficiency and effectiveness in the reduction of surface tension and mixed micelle formation have been derived mathematically [24,25]. For example, with regards to mixed micelle formation, synergism is present when the CMC of any mixture of two surfactants is smaller than that of either individual surfactant. In this case, the two conditions that a mixture must obey in order to exhibit synergism are [25] (1) an attractive interaction of the individual components, $\beta^M < 0$, and (2) $|\beta^M| > |\ln (C_1/C_2)|$. Mixed systems that fulfill these two conditions show a minimum in CMC at a certain composition. Hua and Rosen [25] found that this composition is the point where the mole fraction of surfactant 1 in the mixed micelle, x_1, equals its mole fraction in the bulk, α^*, which is given by

$$\alpha^* = \frac{\ln(C_1/C_2) + \beta^M}{2\beta^M} \quad (11.14)$$

The minimum CMC value of the mixture at this point, C^*_{min}, is

$$C^*_{min} = C_1 \exp\left\{\beta^M\left[\frac{\beta^M - \ln(C_1/C_2)}{2\beta^M}\right]^2\right\} \quad (11.15)$$

Later on, Maeda [26] proposed a new approach for mixed micelles involving ionic species, which is applicable to systems with moderately high ionic strength where the short range of electrostatic interaction is no longer negligible. In the formulation of Maeda, which is based on the phase separation model, thermodynamic stability is described by the standard free energy change due to the micellization process, ΔG_{mic}, given as a function of the mole fraction of the ionic component in the mixed micelle, x_2, by

$$\frac{\Delta G_{mic}}{RT} = B_0 + B_1 x_2 + B_2 x_2^2 \quad (11.16)$$

Here B_0 is an independent term related to the CMC of the nonionic component, expressed in the mole fraction scale as

$$B_0 = \ln C_1 \quad (11.17)$$

The parameter B_1 is related to the standard free energy change upon replacement of a nonionic monomer in the nonionic pure micelle with an ionic monomer. The last coefficient, B_2, is equivalent to β^M in RST, specifically

$$B_2 = -\beta^M \quad (11.18)$$

Finally, the parameters B_1 and B_2 are related to the CMC values of the pure surfactants via

$$\ln\left(\frac{C_2}{C_1}\right) = B_1 + B_2 \quad (11.19)$$

Some authors consider that the treatments based on RST, concerning mixed adsorbed films and mixed micelles, are incomplete from a thermodynamic point of view because they ignore the participation of the solvent and do not consider the dissociation of ionic surfactants [27]. Moreover, the fact that both the mixed adsorbed film and mixed micelles of ionic surfactants are accompanied by the electric double layer makes it difficult to suppose that RST is applicable to these systems [28,29]. Motomura et al. [29] showed that the composition of surfactants in the mixed adsorbed interfacial monolayer can be estimated directly from surface tension measurements without introducing additional assumptions. What is more, considering the mixed micelles as a macroscopic bulk phase, and assuming that intramicellar thermodynamic quantities are given by the excess thermodynamic functions similar to those used for the mixed adsorption in the air–liquid interface, the aforementioned authors could evaluate the composition of the mixed micelles. The relevant equations resulting from the so-called Motomura theory, for the composition of mixed monolayers and mixed micelles, are the following:

$$X_2^S = X_2 - \left(\frac{X_1 X_2}{c}\right)\left(\frac{\partial c}{\partial X_2}\right)_{T,p,c} \tag{11.20}$$

$$X_2^M = X_2 - \left(\frac{X_1 X_2}{C^*}\right)\left(\frac{\partial C^*}{\partial X_2}\right)_{T,p} \tag{11.21}$$

where
X_1 and X_2 are the mole fraction of surfactants 1 and 2 in the bulk solution
X_2^S and X_2^M are the mole fractions of surfactant 2 in the mixed monolayer and the mixed micelle, respectively
c is the total concentration of mixed solution at a given value of surface tension
C^* is the critical micelle concentration

The total concentration and mole fractions of surfactants in the bulk are defined as follows:

$$c = v_1 c_1 + v_2 c_2 \tag{11.22}$$

$$X_1 = \frac{v_1 c_1}{c}, \quad X_2 = \frac{v_2 c_2}{c} \tag{11.23}$$

where v_1 and v_2, and c_1 and c_2 are the number of ions dissociated by the surfactants and their corresponding concentrations, respectively.

From a quantitative point of view, the extent of the nonideality of mixing in the adsorbed monolayer and micelles are expressed by the excess free energy of adsorption ($\Delta G^{S,EXC}$) and micelle formation ($\Delta G^{M,EXC}$), respectively, which can be determined by applying the following equations [28,30]:

$$\Delta G^{S,EXC} = RT \sum_{i=1}^{2} X_i^S \ln f_i^S \tag{11.24}$$

$$\Delta G^{M,EXC} = RT \sum_{i=1}^{2} X_i^M \ln f_i^M \tag{11.25}$$

where f_i^S and f_i^M are the activity coefficients of surfactant i in the adsorbed film and the micelle, respectively, evaluated by the following equations:

$$f_i^S = \frac{c\, X_i}{c_i^0\, X_i^S} \qquad (11.26)$$

$$f_i^M = \frac{C * X_i}{C_i\, X_i^M} \qquad (11.27)$$

where
    c_i^0 is the concentration of the single surfactant i at a given σ
    C_i is the CMC of the pure surfactant i

11.3 MIXED SYSTEMS INVOLVING ALKYL POLYGLUCOSIDES

A number of studies have been conducted on mixed micellar systems composed of some APG and another surfactant. Most of the papers analyze combinations with ionic surfactants, but binary mixtures with zwitterionic and nonionic surfactants have also been reported. Table 11.1 presents a list of mixtures of APGs and other surfactants with an indication of the interaction parameters (β^M and β^σ) characterizing the mixed system when available. With the purpose of structuring this section we distinguish between those studies aiming at the characterization of an individual mixed system and those focusing on the effect of the head-group nature and of the hydrophobic chain length.

11.3.1 MIXTURES WITH ANIONIC SURFACTANTS

Among the APG surfactants, n-octyl-β-D-glucoside (β-C_8G_1) has probably been the one most widely investigated, and some studies on the mixed micellization and adsorption in the air–liquid interface involving this surfactant have recently been reported [31–34]. Of course, the anionic surfactant mainly considered to examine the mixing behavior of APG surfactants has been sodium dodecyl sulfate (SDS).

Motivated by the fact that the addition of β-C_8G_1 achieved the recovery of the native properties of proteins previously denatured by the presence of SDS, Kameyama et al. [31] studied the effect of salt concentration on micelle formation in the mixed system constituted by β-C_8G_1 and SDS. By using surface tension measurements, they determined the CMC values of the mixed systems in the whole composition range at fixed NaCl concentrations (20, 75, and 150 mM). These CMC data were acceptably described by RST and a negative deviation from ideal behavior was found in all the cases, indicating that the formation of β-C_8G_1/SDS micelles is favorable. However, the value of the interaction parameter β^M showed small variations with the presence of salt (Table 11.1). In general, the results were discussed on the basis of the effect of salt concentration on the CMC of the pure and mixed systems. In this way, the change in the CMC of the mixed systems in the presence of salt was analyzed with the Corrin–Harkins

TABLE 11.1
Interaction Parameters of Surfactant Mixtures Involving APGs at 25°C

Mixtures	Systems	Medium	β^M	β^σ	References
Nonionic–anionic	β-C_8G_1/SDS	20 mM NaCl	−2.5	—	[31]
	β-C_8G_1/SDS	75 mM NaCl	−2.1	—	[31]
	β-C_8G_1/SDS	150 mM NaCl	−2.3	—	[31]
	β-SC_8G_1/SDS	100 mM NaCl	−1.52	—	[36]
	β-$C_{10}G_1$/SDS	Water	−2.3 (−2.4)	—	[37,39]
	β-$C_{10}G_1$/SDS	10 mM NaCl	−2.7	—	[39]
	β-$C_{10}G_1$/SDS	50 mM NaCl	−2.2	—	[39]
	β-$C_{10}G_1$/SDS	300 mM NaCl	−1.7	—	[39]
	β-$C_{10}G_2$/SDS	Water	−3.3	—	[37]
	β-$C_{12}G_1$/SDS	Water	−1.8	—	[37]
	β-$C_{12}G_2$/SDS	Water	−2.7 (−4.1)	—	[37,47]
	β-$C_{10}G_2$/DESS	100 mM NaCl (pH = 5.7)	−1.2	−1.5	[45]
	β-$C_{10}G_1$/DESS	100 mM NaCl (pH = 5.7)	−1.4	−1.8	[45]
	β-$C_{12}G_2$/DESS	100 mM NaCl (pH = 5.7)	−1.3	−1.4	[45]
	β-C_8G_1/NaOl	5 mM NaOH	−3.2	—	[55]
	β-$C_{10}G_1$/NaOl	5 mM NaOH	−2.8	—	[55]
	β-$C_{10}G_2$/DBS	10 mM NaCl	−2.1	—	[38]
Nonionic–cationic	β-C_8G_1/DTAB	Water	−0.73 (−2.4)	—	[32,56]
	β-C_8G_1/TTAB	Water	−1.37 (−0.8)	—	[33,56]
	β-C_8G_1/CTAB	Water	−2.5		[56]
	β-$C_{10}G_1$/DTAB	Water	−4.1	—	[37]
	β-$C_{10}G_1$/DeTAB	100 mM NaCl (pH = 9.0)	−1.2	−1.2	[44]
	β-$C_{10}G_2$/DeTAB	100 mM NaCl (pH = 9.0)	−0.3	−0.3	[44]
	β-$C_{12}G_2$/DTAC	100 mM NaCl (pH = 5.7)	−0.76	−1.0	[44]
	β-$C_{12}G_2$/DTAC	100 mM NaCl (pH = 9.0)	−1.5	−1.9	[44]
	β-$C_{12}G_2$/TTAB	100 mM NaCl (pH = 9.0)	−1.3	−1.8	[44]
	β-$C_{12}G_1$/DTAB	Water	—	—	[49,50]
	β-$C_{10}G_1$/CTPPB	Water	−2.22	—	[52]
	β-$C_{10}G_1$/CTBPB	Water	−2.15	—	[52]
	β-$C_{10}G_1$/CTAB	Water	−0.28	—	[52]
	β-$C_{12}G_1$/CTPPB	Water	−0.13	—	[52]
	β-$C_{12}G_1$/CTBPB	Water	−0.13	—	[52]

TABLE 11.1 (continued)
Interaction Parameters of Surfactant Mixtures Involving APGs at 25°C

Mixtures	Systems	Medium	β^M	β^σ	References
	β-C$_{12}$G$_1$/CTAB	Water	−1.28	—	[52]
	β-C$_{10}$G$_2$/(C$_{10}$N)$_2$	100 mM NaCl (pH = 9.0)	−1.9	−2.7	[44]
	β-C$_{10}$G$_2$/(C$_{10}$N)$_2$O	100 mM NaCl (pH = 9.0)	−1.7	−2.3	[44]
	β-C$_{10}$G$_2$/(C$_{10}$N)$_2$ OH	100 mM NaCl (pH = 9.0)	−1.4	−2.9	[44]
	β-C$_{10}$G$_2$/(C$_{10}$N)$_2$ (OH)$_2$	100 mM NaCl (pH = 9.0)	−1.7	−2.0	[44]
	β-C$_{10}$G$_1$/(C$_{10}$N)$_2$	100 mM NaCl (pH = 9.0)	−1.9	−4.0	[44]
	β-C$_{10}$G$_1$/(C$_{10}$N)$_2$O	100 mM NaCl (pH = 9.0)	−1.5	−3.3	[44]
	β-C$_{10}$G$_1$/(C$_{10}$N)$_2$ OH	100 mM NaCl (pH = 9.0)	−1.2	−4.2	[44]
	β-C$_{10}$G$_1$/(C$_{10}$N)$_2$ (OH)$_2$	100 mM NaCl (pH = 9.0)	−1.4	−3.1	[44]
	β-C$_8$G$_1$/10-2-10	Water	−2.2	—	[56]
	β-C$_8$G$_1$/12-2-12	Water	−2.9	—	[56]
	β-C$_8$G$_1$/14-2-14	Water	−5.7	—	[56]
	β-C$_8$G$_1$/16-2-16	Water	—	—	[56]
	β-C$_8$G$_1$/12-0-8	Water	−1.8	—	[56]
	β-C$_8$G$_1$/12-0-10	Water	−4.6	—	[56]
	β-C$_8$G$_1$/12-0-12	Water	—	—	[56]
	β-C$_8$G$_1$/12-016	Water	—	—	[56]
Nonionic–zwitterionic	β-C$_{12}$G$_2$/DBMG	100 mM NaCl (pH = 5.7)	−1.1	−1.7	[45]
	β-C$_8$G$_1$/DPS	Water	−2.1	—	[56]
	β-C$_8$G$_1$/TPS	Water	−4.0	—	[56]
	β-C$_8$G$_1$/HPS	Water	—	—	[56]
	β-C$_{12}$G$_1$/DSB	Water	—	—	[51]
	β-C$_{12}$G$_2$/DSB	Water	—	—	[51]
	β-C$_{12}$G$_2$/DDAB	Water	−0.01		[47]
Nonionic–nonionic	β-C$_{10}$G$_1$/C$_{12}$E$_7$	Water	−0.04	—	[37]
	β-C$_{10}$G$_1$/β-C$_{10}$G$_2$	Water (100 mM NaCl)	−2.0 (−0.3)	(−0.2)	[37,44]
	β-C$_{12}$G$_2$/C$_{12}$E$_7$	100 mM NaCl	−0.05	−0.7	[44]
	β-C$_{12}$G$_2$/C$_{12}$E$_5$	pH = 6.5	−0.05	—	[48]
	β-C$_8$G$_1$/C$_{12}$E$_4$	Water[a]	−5.1	—	[46]
	β-C$_{10}$G$_2$/C$_{10}$E$_8$	10 mM NaCl	−0.3	−0.5	[38]

[a] $T = 30°C$.

equation, whose plots showed a fairly linear relationship for all mixed micellar compositions. This behavior seems to indicate that counterion binding plays a decisive role in the mixed micelle formation. Nevertheless, because of the change in CMC of nonionic surfactants with salt concentration is mainly due to a "salting-out" effect, particularly in regions where the mole fraction of the ionic species is low, the effect of salt concentration on the CMC of the mixed β-C_8G_1/SDS systems was attributed to two contributions, the counterion binding and the salting-out effect.

n-Octyl-β-D-thioglucoside (β-SC_8G_1) is a related APG surfactant that differs from β-C_8G_1 only in that the hydrophilic group is linked by a thioether to the hydrophobic chain, this structural peculiarity providing β-SC_8G_1 with solution properties substantially different from those of β-C_8G_1 [35]. Therefore, a different behavior for the β-SC_8G_1/SDS system compared with β-C_8G_1/SDS should be expected. Recently, the mixed system formed by β-SC_8G_1 and SDS in the presence of 0.1 M NaCl has been characterized [36]. This investigation was carried out by using the fluorescence probe technique and includes not only the study of the variation in the CMC with the solution composition, but also the change in the micellar aggregation number and in the microenvironmental properties with the content of the ionic component. Figure 11.1a shows the CMC values of the β-SC_8G_1/SDS mixed systems, as obtained by the pyrene 1:3 ratio method, as a function of the mole fraction of SDS in the bulk solution. Data in Figure 11.1a were analyzed by RST, which yield an interaction parameter of −1.52, indicating a rather less attractive interaction for β-SC_8G_1/SDS than that observed for β-C_8G_1/SDS in similar conditions [31]. This attractive interaction was ascribed to two contributions, namely, the electrostatic stabilization—by which the intercalation of the nonionic component shields the repulsive interaction between the negatively charged head groups of SDS—and the existence of an attractive interaction ion-dipole, which could be significant in the present mixed system due to the high charge density of the sulfate group of SDS. One can also observe in Figure 11.1a that the addition of small amounts of ionic surfactant considerably reduces the CMC of the mixed system. This effect, which has been observed in other mixed systems involving APG surfactants [37,38], has been justified by the reduction in the steric hindrance between the glycoside head groups of the nonionic micelle as a result of the incorporation of the ionic surfactant to form the mixed micelle [37]. Figure 11.1b shows the variation in micelle composition with that of the solution. From this figure, one observes that the content of the ionic component in the mixed micelle is higher than that expected from an ideal behavior and, in addition, that small amounts of the ionic component produce mixed micelles with a considerable amount of this component.

The effect of the micellar composition on the size of the β-SC_8G_1/SDS micelles, through the change in the mean micellar aggregation number (N_{agg}), was also investigated [36]. Table 11.2 lists the N_{agg} values obtained by the static quenching method, as well as the contribution of each component calculated by using the micellar composition values as determined by RST. From data in Table 11.2,

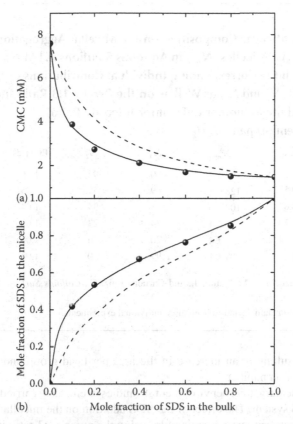

FIGURE 11.1 Data for the mixture of n-octyl-β-D-thioglucoside (β-SC$_8$G$_1$) with sodium dodecyl sulfate (SDS). (a) CMC of the β-SC$_8$G$_1$/SDS system as a function of the mole fraction of SDS in the solution, and (b) Micellar composition versus the bulk composition. Symbols represent the experimental values, the dashed and solid lines are the phase separation model predictions according to an ideal behavior and RST ($\beta^M = -1.52$), respectively. (From Hierrezuelo, J.M., Aguiar, J., and Carnero Ruiz, C., *Colloids Surf. A*, 264, 29, 2005. With permission.)

one observes that the aggregation number increases, reaching a maximum value at $\alpha_{SDS} = 0.2$, and then remains practically constant. A similar tendency has been previously observed for the C$_{12}$E$_6$/SDS system in 0.1 M NaCl [15]. The trend observed in Table 11.2 for the mixed aggregation number was rationalized on the basis of two opposite effects: the electrostatic repulsions and the steric interactions between the head groups of the surfactants forming the mixed micelle. When the participation of SDS is low, the contribution of the electrostatic repulsions is not significant, so a micellar growth is observed. However, as the ionic content increases, the favorable steric effect is overcome by the increasing electrostatic

TABLE 11.2

Effect of the Micellar Composition on the Micellar Aggregation Numbers of SDS/β-SC$_8$G$_1$ Micelles, N_{agg}, in Aqueous Solutions 0.1 M NaCl at 25°C, Together with the Corresponding Individual Contributions of Both Components, N_1 and N_2, as Well as on the Pyrene 1:3 Ratio Index, Py (1:3), and the Monomer to Excimer Intensity Ratio of 1,3-dipyrenilpropane, I_M/I_E

α_{SDS}	x_{SDS}	N_{agg}^a	N_1	N_2	Py (1:3)	I_M/I_E
0	0	92 ± 1	0	92	1.06	4.44
0.2	0.54	128 ± 1	69	59	1.05	2.23
0.4	0.67	102 ± 3	68	34	1.04	1.15
0.6	0.76	97 ± 1	74	23	1.05	1.29
0.8	0.85	106 ± 4	70	16	1.06	1.79
1	1	99 ± 1	99	0	1.08	2.25

Source: Hierrezuelo, J.M., Aguiar, J., and Carnero Ruiz, C., *Colloids Surf. A*, 264, 29, 2005. With permission.

[a] Mean value ± standard deviation from three individual experiments.

repulsions, resulting in an increase in the area per head group and, therefore, a lower aggregation number.

Complementary microenvironmental studies were also carried out for the β-SC$_8$G$_1$/SDS system. The effect of the SDS addition on the micellar micropolarity and microviscosity was examined by using the pyrene 1:3 ratio index and the intramolecular excimer formation of 1,3-dipyrenilpropane (P3P), respectively. Representative results of these experiments are listed in Table 11.2. It is observed that the pyrene 1:3 ratio variations are hardly significant and can be correlated with the initial increase and subsequent reduction in the aggregation number. However, the behavior of the monomer to excimer intensity ratio, I_M/I_E, reflects a considerable initial reduction in the microviscosity as the SDS content increases, down to a minimum value at $\alpha_{SDS} = 0.4$, increasing then for higher values of α_{SDS}. The initial decrease in micellar microviscosity was attributed to the increasing participation of the ionic component in the mixed micelle, which produces an increment of the electrostatic interactions between the charged head groups, causing the formation of micelles with a looser structure. The subsequent increase of I_M/I_E is probably due to a better packing as a result of more favorable steric interactions.

The mixed system composed of n-decyl-β-D-glucoside (β-C$_{10}$G$_1$) and SDS has also been the subject of study by several researchers [37,39,40]. Bergström et al. [39] have studied the mixed micellization of the β-C$_{10}$G$_1$/SDS system at different concentrations of added NaCl. The main objective of this work was the introduction of a new model to evaluate synergistic effects from experimentally determined

CMC values of surfactant mixtures. The aforementioned authors demonstrated that the proposed model, based on the Poisson–Boltzmann mean field theory for planar, cylindrical, and spherical geometries, respectively, showed a better agreement with data than the conventional RST. They found that RST systematically underestimates synergism at low fractions and overestimates it at high fractions of the ionic component. With regard to the behavior of the mixed system, it was found that the synergism increases with added salt up 10 mM, followed by a decrease upon further addition of salt.

This same mixed system was also studied from a structural point of view. By using small-angle neutron scattering (SANS) and static light scattering (SLS) measurements, Bergström et al. [40] investigated the size and structure of mixed β-$C_{10}G_1$/SDS micelles formed in the absence of salt and at 10 and 100 mM NaCl. It was found that, at low salt concentrations (0 and 10 mM NaCl), prolate ellipsoidal micelles are formed, which grow substantially in length as the salt concentration is raised to 100 mM. At this salt concentration, different structures were observed depending on the system composition. For instance, at surfactant ratios of [SDS]/ [β-$C_{10}G_1$] = 3:1 and 1:1 the micelles were rigid rods. However, when the nonionic component was in excess ([SDS]/[β-$C_{10}G_1$] = 1:3) very long flexible wormlike micelles were observed.

11.3.2 Mixtures with Cationic Surfactants

A few papers on the mixing behavior of some APG with cationic surfactants have been published [32–34,41]. In addition, other mixed systems formed by APGs and both common and gemini cationic surfactants have been investigated in the context of wider studies aiming at the influence of the head group or the chain length. These systems will be discussed in the corresponding section.

Aicart and coworkers have reported, in two different papers, the micellar properties of two mixed systems composed of β-C_8G_1 and two quaternary ammonium salts: decyltrimethylammonium bromide (DTAB) [32] and tetradecyltrimethylammonium bromide (TTAB) [33]. Recently, the system β-C_8G_1/DTAB has been revisited by Matsubara et al. [34].

The work by Aicart's group was developed on the basis of conductometric, ultrasonic, densitometric, and static fluorescence measurements. By means of these techniques, they determined, in the whole composition range, the CMC, the dissociation degree, the aggregation number, the hydration numbers, the isentropic compressibility, the apparent and partial molar properties of the mixed micelles, and the change of partial molar quantities due to the mixed micellization process. In addition, the behavior of both mixed systems was analyzed by using several descriptive and predictive theoretical models.

The thermodynamic studies carried out on both systems allowed for the determination, among others quantities, of the difference between the hydration number in the micellar phase and the monomeric one, informing about the change in the number of hydration molecules surrounding surfactant molecules due to the micellization process. In this respect, an interesting observation was that a considerable percentage, higher in the case of the β-C_8G_1/DTAB system, of water molecules surrounding the monomers is lost when the micelles are formed. This fact was interpreted in the sense that not only the hydrophobic chain but also the polar head group of the surfactant is dehydrated through the micellization process.

As previously mentioned, the experimental CMC values were used to analyze the behavior of both mixed systems by using different theoretical models. However, we will comment here only the results obtained by the more frequently used models of Rubingh (RST) and Motomura (M), which are presented in Table 11.3. Data in Table 11.3 indicate that the CMCs of the mixed systems are similar to those of the cationic component, particularly in the case of the β-C_8G_1/DTAB system. Only in the β-C_8G_1 rich region, at molar fractions above 0.8, a significant

TABLE 11.3
CMC and Micellar Composition (Experimental Values, $(x_1)_{exp}$, Motomura Model, $(x_1)_M$, and RST Treatment, $(x_1)_{RST}$), Activity Coefficients, f_1 and f_2, and Interaction Parameter, β^M, for the Mixed Systems β-C_8G_1/DTAB and β-C_8G_1/TTAB as Determined by RST, Together with the Mixed Micellar Aggregation Number, N_{agg}, and Partial Contribution of the Nonionic Component $(N_{agg})_1$

Mixed System	α_1	CMC (mM)	$(x_1)_{exp}$	$(x_1)_M$	$(x_1)_{RST}$	f_1	f_2	β^M	N_{agg}	$(N_{agg})_1$
β-C_8G_1/	0	15.19	0	0	0				58	0
DTAB	0.1995	15.36	0.191	0.187	0.177	0.69	0.98	−0.55	56	11
	0.4036	15.71	0.409	0.386	0.336	0.75	0.93	−0.66	56	23
	0.5024	15.88	0.508	0.451	0.410	0.77	0.88	−0.75	59	30
	0.6019	16.40	0.603	0.505	0.483	0.81	0.83	−0.79	60	36
	0.7826	17.97	0.766	0.620	0.631	0.88	0.70	−0.91	62	48
	1	25.24	1	1	1			$(\beta^M)_{av} =$	80	80
								−0.73		
β-C_8G_1/	0	3.63	0	0	0				62	0
TTAB	0.3000	4.25	0.306	0.106	0.152	0.33	0.97	−1.5	66	20
	0.4974	5.17	0.502	0.195	0.230	0.44	0.93	−1.4	68	34
	0.7497	7.51	0.747	0.370	0.375	0.60	0.83	−1.3	79	59
	0.8491	9.59	0.841	0.441	0.468	0.69	0.75	−1.3	77	65
	0.9206	12.57	—	0.650	0.578	0.79	0.65	−1.3	—	—
	1	25.24	1	1	1			$(\beta^M)_{av} =$	80	80
								−1.37		

Sources: del Burgo, P., Junquera, E., and Aicart, E., *Langmuir*, 20, 1587, 2004; Lainez, A., del Burgo, P., Junquera, E., and Aicart, E., *Langmuir*, 20, 5745, 2004.

change in the CMC is observed, this effect being more pronounced in the case of the β-C_8G_1/TTAB system. It seems that this fact can be attributed to the greater difference of hydrophobicity between the pure components in the later case. In addition, it is to be noted that, although both systems show a negative deviation from the ideal behavior, $\beta^M < 0$, the condition of $|\beta^M| > |\ln (C_1/C_2)|$ is not fulfilled by these systems, so they do not exhibit synergistic behavior.

A noticeable aspect of the work of Aicart and coworkers is the application of a new procedure previously proposed by them to determine experimentally the micellar composition of the mixed system. These experimental values, $(x_1)_{exp}$, together with those obtained by RST and Motomura treatment are also listed in Table 11.3. One observes that, while the values of x_1 given by RST and Motomura theory are similar, the experimental values are systematically higher, and similar to the bulk composition. In the authors' opinion, this fact reflects the poor predictions given by RST and Motomura theory.

Table 11.3 also lists the mean aggregation numbers, N_{agg}, obtained for both pure and mixed micelles by the steady-state fluorescence quenching method, as well as the partial aggregation number of the nonionic component estimated by assuming the experimental values of the micellar composition, $(N_{agg})_1$. It can be seen that the aggregation numbers for the β-C_8G_1/DTAB system remain close to the value of pure DTAB almost in the whole composition range, and clear changes are only observed in the very rich β-C_8G_1 region. In contrast, for the β-C_8G_1/TTAB system a monotonic increase in N_{agg} is observed, as the participation of the nonionic component increases. The authors concluded that the mixed micelles formed are always spherical, except in the close vicinity of pure β-C_8G_1.

With the purpose of examining the effect of the bulkiness of the head group and hydrogen bond formation between β-C_8G_1 molecules on the molecular packing in the mixed adsorbed film and micelle, Matsubara et al. [34] carried out surface tension measurements of aqueous solutions of β-C_8G_1/DTAB mixtures at 298.15 K under atmospheric pressure by the drop volume method. They treated their experimental results on the basis of Motomura theory. Thus, the excess free energy of adsorption and micelle formation as a function of the composition of the adsorbed film and micelles, respectively, were evaluated. Figures 11.2 and 11.3 show the corresponding results. From data in Figure 11.2 one observes that the excess free energy of adsorption values are negative, indicating that the mutual interaction between β-C_8G_1 and DTAB in the adsorbed film is stronger than that between molecules of the same species. In addition, Figure 11.2 shows that the absolute value of the excess free energy of adsorption increases with decreasing surface tension, which is related to the more favorable packing of the surfactant molecules in the adsorbed film. Figure 11.3 shows the excess free energy of micelle formation, as determined by Equation 11.25, together with the excess free energy in the adsorbed film at the CMC. It was established that the energetic superiority of the micelle over the adsorbed film suggests that the hydrogen bond formation between β-C_8G_1 molecules is less effective in the pure β-C_8G_1 micelle, due to its curved geometry, than in the pure β-C_8G_1 adsorbed film. It is therefore believed that the intercalation of DTAB monomers between

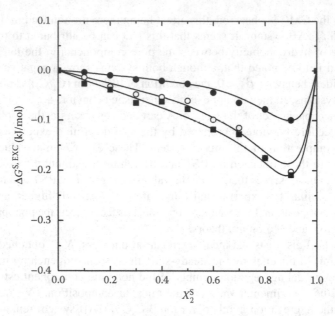

FIGURE 11.2 Excess Gibbs energy of adsorption versus composition of adsorbed film plot for the β-C_8G_1/DTAB mixed system at 41 mN/m (■), 45 mN/m (○), and 50 mN/m (●). (From Matsubara, H., Obata, H., Matsuda, T., and Aratono, M., *Colloids Surf. A*, 315, 183, 2008. With permission.)

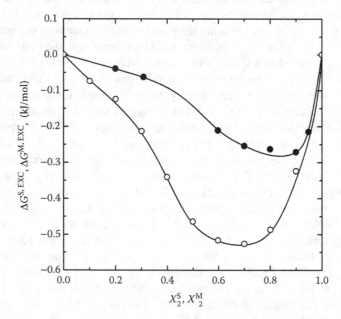

FIGURE 11.3 Excess Gibbs energy of micelle formation versus composition of micelle plot for the β-C_8G_1/DTAB mixed system (○). Black circles show the excess Gibbs of adsorption extrapolated to CMC. (From Matsubara, H., Obata, H., Matsuda, T., and Aratono, M., *Colloids Surf. A*, 315, 183, 2008. With permission.)

β-C_8G_1 molecules can stabilize the mixed micelle by the ion–dipole interaction instead of the hydrogen bond between β-C_8G_1 molecules. The main conclusion drawn from this study is that although the interaction between β-C_8G_1 and DTAB is suitable to form effectively packed aggregates both in the adsorbed film and in the micelle, the mechanism controlling molecular packing in each system is different. Although hydrogen bond formation is considered the main contribution to determine the physical properties of the adsorbed film, the ion–dipole interaction between the head groups of different components is more effective in the curved micellar geometry, being the main cause of the attractive interaction stabilizing the mixed micelle [34].

Recently, by using nuclear magnetic resonance (NMR) spectroscopy, Yang et al. [41] have investigated mixed micelles composed of the gemini cationic surfactant octane-1,8 bis(dodecyl dimethyl ammonium chloride) (12-8-12) and n-dodecyl-β-D-maltoside (β-$C_{12}G_2$). Several NMR techniques, providing information at a molecular level, were used to gain new insights into the interactions and the relative arrangement of surfactant molecules in the mixed aggregates. It was found that the surfactant interactions between both surfactants depend on the molar ratio in which the surfactants are mixed. Whereas at low molar ratios of β-$C_{12}G_2$/12-8-12 (1:1–1:3) the intermolecular interactions increase gradually, at higher molar ratios (1:4–1:8) the intermolecular interactions decrease and the intramolecular interaction of proton–proton dipolar coupling becomes more dominant. The short distance found between the methylene group next to the ammonium head of 12-8-12 and the penultimate sugar ring of β-$C_{12}G_2$ suggested that the primary hydrophilic interactions occurring in the mixed micelles may be produced in this site. Lastly, it was observed that the size of the mixed micelles, at a fixed total surfactant concentration, is largely determined by the presence of the nonionic component.

11.3.3 Mixtures with Nonionic Surfactants

There are not many articles concerning micellization studies of APGs with nonionic surfactants. Some of these investigations are focused on aspects such as structure and phase behavior rather than on the study of the molecular interactions in the mixed micelle. Representative examples of these studies are the papers of Söderman and coworkers [42,43]. These authors have investigated the phase behavior in mixtures of n-decyl-β-D-glucopyranoside (β-$C_{10}G_1$) and n-nonyl-β-D-glucopyranoside (β-C_9G_1), as well as the effect of salts on the consolute boundaries and microstructure of the micellar regions surrounding the closed-miscibility gap of mixed systems.

Zhang and Somasundaran [44] have studied micellar growth in micelles of β-$C_{12}G_2$ and nonyl phenol ethoxylated decyl ether (NP-10), and in their 1:1 molar ratio mixture. They found that pure and mixed micelles are asymmetrical in shape at the CMC, but micellar growth was observed at concentrations immediately above the CMC. In addition, their results from both sedimentation velocity and sedimentation equilibrium suggest a coexistence of two types of micelles in pure

NP-10 solutions and in its mixture with β-$C_{12}G_2$, while only one micellar species is present in pure β-$C_{12}G_2$ solutions.

Analysis of the molecular interactions of a few nonionic surfactant pairs has been reported [37,38,45–47]. Table 11.1 lists some values for the interaction parameters (β^M and β^σ) in mixtures of nonionic surfactants. The data show that in most cases the almost zero value for the interaction parameters in these mixtures indicates a nearly ideal behavior, as expected for this kind of combination. There are some exceptions which deserve some additional comments. For example, the mixed system composed by β-$C_{10}G_1$ and n-decyl-β-D-maltoside (β-$C_{10}G_2$) has been investigated by Sierra and Svensson [37] and by Rosen et al. [45]. The formers found a considerable deviation from the ideal behavior for the β-$C_{10}G_1$/β-$C_{10}G_2$ system ($\beta^M = -2.0$), similar to the value they found for the β-$C_{10}G_1$/SDS system. These authors indicated that this favorable interaction cannot be explained by means of electrostatic interactions, because it would result in a net repulsive interaction between both components. Therefore, they considered the possibility of a structurally suitable packing, allowing for the formation of hydrogen bonds between the glysosidic units of β-$C_{10}G_1$ and β-$C_{10}G_2$, which would be more favorable than the hydrogen bonds of the glucose moieties with the surrounding water molecules. By contrast, Rosen et al. [45] claimed that the high β^M value reported by Sierra and Svensson for the β-$C_{10}G_1$/β-$C_{10}G_2$ system is due to the value of 3.23 mM for the CMC of β-$C_{10}G_2$ found by these authors, which differs from the CMC value close to 2.0 mM in 0.1 M NaCl obtained by them.

As part of a comparative study of the micellar regions of two mixed β-C_8G_1 systems, one with a nonionic surfactant (tetraethylene glycol monododecyl ether, $C_{12}E_4$) and another with a water-soluble polymer (polyethylene glycol 20,000), the β-C_8G_1/$C_{12}E_4$ system has been characterized by Sanz et al. [47]. The CMC data at different compositions, obtained by surface tension measurements, were fitted via Rubingh's treatment. It was found that the value of β^M varied between -4.1 and -6.0 for the mixtures investigated, giving an average value of -5.1. This value implies a significant attractive interaction between components in the mixed micelle, comparable with that observed for mixtures of nonionic surfactants of the polyoxyethylene type with anionic surfactants with sulfate or sulfonate head groups. The aforementioned authors concluded that the sugar head group behaves similarly in that respect to an anionic sulfate head group, and attributed this behavior to a similarly large polarizability of the β-C_8G_1 head group that can interact strongly with the ethylene oxide head groups of $C_{12}E_4$ [47].

11.3.4 EFFECT OF THE POLAR HEAD GROUP

The influence of the head group nature on the micellization and surface behavior of mixtures of APGs with other surfactants has been addressed by a number of researchers [37,38,48–53]. In this section, we summarize the relevant research on this subject.

Hines et al. [48] carried out a comparative study on the micellization and adsorption in the air–liquid interface of mixtures of β-$C_{12}G_2$ with SDS and the zwitterionic surfactant n-dodecyl-N,N'-dimethylamino betaine (DDAB). In this case, as the hydrophobic chain is a dodecyl for all three surfactants considered, one can analyze the possible nonideality of the mixed systems in terms of interactions between the head groups. From surface tension measurements, the CMC values through the whole composition range were obtained for the two mixed systems (β-C$_{12}$G$_2$/SDS and β-C$_{12}$G$_2$/DDAB). The experimental data were treated according to RST and the interaction parameters in micelles, β^M, and at the surface, β^σ, were determined. In addition, they used neutron reflection measurements to obtain directly the composition at the interface. In this way, the composition of the interface was obtained from experiment either via the Gibbs equation or, more directly, from neutron reflection measurements. Figure 11.4 shows the CMC

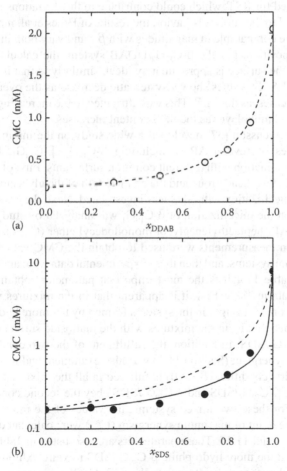

(a) x_{DDAB}

(b) x_{SDS}

FIGURE 11.4 CMC as a function of the bulk composition for the systems: (a) β-C12G2/DDAB and (b) β-C12G2/SDS. Symbols are experimental data, the dashed lines represent the phase separation model according to an ideal behavior, whereas the solid line is the best fit to the experimental data according to RST. (From data in Hines, J.D., Thomas, R.K., Garret, P.R., Rennie, G.K., and Penfold, J., *J. Phys. Chem. B*, 101, 9215, 1997).

experimental values as a function of the bulk composition for the two studied systems. One observes (Figure 11.4a) that β-$C_{12}G_2$ mixes ideally with DDAB. This behavior was explained on the basis of two aspects: (1) the absence of electrostatic interactions due to the electrically neutral character of DDAB at a pH of 7 and (2) the absence of steric interactions due to the similarity between the head group sections of each component. On the other hand, data in Figure 11.4b indicates that in the β-$C_{12}G_2$/SDS system there is a considerable negative deviation from ideal mixing, due to attractive interactions between the two surfactants. It is also noted in Figure 11.4b that the CMC experimental values for this system (β-$C_{12}G_2$/SDS) are not well described by RST. The authors considered that the charge on the SDS monomers will tend to ensure that they are as widely spaced as possible, a situation that does not correspond with the random mixing that is supposedly required for RST, which could contribute to the deviations in β^M.

With regard to the surface behavior, the results of Hines et al. indicate that the values of β^σ are comparable in magnitudes with β^M and vary both in surface pressure and composition. For the β-$C_{12}G_2$/DDAB system, the calculations suggest that mixing at the surface is approximately ideal, similarly to that in the micelles. For β-$C_{12}G_2$/SDS, β^σ values show a systematic decrease as the overall SDS concentration increases, as does β^M. This was attributed to the increasing electrostatic repulsions in the monolayer as the SDS content increases.

Sierra and Svensson [37] developed a wide study on the mixing behavior of binary mixtures of several APGs, including β-$C_{10}G_1$, β-$C_{12}G_1$, β-$C_{10}G_2$, and β-$C_{12}G_2$, in combination with different common surfactants. First of all, the effect of the nonionic polar head group and the chain length of the glycosidic surfactants in the mixed micellization with SDS was investigated. Next, the effect of the head group nature on the micellization of β-$C_{10}G_1$ was analyzed by studying the mixtures with DTAB, heptaethyleneglycol monododecyl ether ($C_{12}E_7$), and β-$C_{10}G_2$. Surface tension measurements were used to obtain the CMC values of pure and mixed surfactant systems, and then these experimental data were analyzed according to RST. Table 11.4 lists the most important parameters obtained from this study. From data in Table 11.4, it is apparent that in the mixtures with SDS the interaction becomes stronger in the systems formed by the more hydrophilic nonionic surfactants, that is, in the mixtures with the maltoside surfactants, β-$C_{10}G_2$/SDS and β-$C_{12}G_2$/SDS. In addition, the fulfillment of the two conditions for the occurrence of synergism (Section 11.2) was also examined, and it was found that although the first condition ($\beta^M < 0$) is fulfilled in all the mixtures, only the two mixed systems β-$C_{10}G_1$/SDS and β-$C_{10}G_2$/SDS obey the second condition ($|\beta^M| > |\ln (C_1/C_2)|$). For these two mixed systems, the values of the mole fraction (α^*) and CMC at the point of maximum synergism (C_{min}^*) were calculated according to Equations 11.14 and 11.15. These parameters are also listed in Table 11.4, and it can be seen that the more hydrophilic β-$C_{10}G_2$/SDS mixture exhibits a maximum synergism at a lower mole fraction than the β-$C_{10}G_1$/SDS system.

In relation to the head group nature of the cosurfactant (surfactant 2), one observes in Table 11.4 that micelle formation of β-$C_{10}G_1$ appears to be more favorable with DTAB than with SDS. This result was interpreted in the sense that

TABLE 11.4
Interaction Parameters for the Mixed Micelle Formation (β^M), Ratio between the CMC of Each of the Pure Surfactants in the Mixture ($|\ln(C_1/C_2)|$), Molar Fraction of the Nonionic Glycosidic Surfactant (α^*), and CMC of the Mixture (C^*_{min}) at the Point of the Maximum Synergism for Different Surfactant Mixtures

Surf. 1	Surf. 2	β^M	$|\ln(C_1/C_2)|$	α^*	C^*_{min} (mM)
β-$C_{10}G_1$	SDS	−2.3	1.1	0.74	1.70
β-$C_{10}G_2$	SDS	−3.3	0.7	0.60	1.93
β-$C_{12}G_1$	SDS	−1.8	4.0		
β-$C_{12}G_2$	SDS	−2.7	3.7		
β-$C_{10}G_1$	DTAB	−4.1	2.0	0.74	1.50
β-$C_{10}G_1$	$C_{12}E_7$	−0.04			
β-$C_{10}G_1$	β-$C_{12}G_2$	−2.0	0.5	0.63	1.52

Source: Sierra, M.L. and Svensson, M., *Langmuir*, 15, 2301, 1999. With permission.

β-$C_{10}G_1$ has a certain anionic character, resulting in a stronger interaction with DTAB than with SDS. Another aspect to be considered is that the trimethylammonium head group is much bulkier than the sulfate head group, and it is possible that the favorable interaction is related not only to screening of the electrostatic repulsions of the cationic head groups but also to a decrease in the steric hindrance [37]. The results obtained in this study on the mixing behavior of mixtures of nonionic surfactants, β-$C_{10}G_1/C_{12}E_7$ and β-$C_{10}G_1/\beta$-$C_{10}G_2$, have previously been commented.

Liljekvist and Kronberg [38] carried out a comparative study between the mixed micelle behavior of an APG (β-$C_{10}G_2$) and an anionic surfactant (dodecyl benzenesulfonate or DBS), and that of an ethoxylated surfactant (octaethyleneglycol monodecyl ether or $C_{10}E_8$) with the same anionic surfactant, for which the CMC of the single surfactants are very similar. The CMC experimental results were analyzed according to RST, and indicated that the nonionic–anionic mixed systems (β-$C_{10}G_2$/DBS and $C_{10}E_8$/DBS) show synergistic effects as regards both mixed micelle formation and surface tension reduction. It was found that in the mixture with $C_{10}E_8$, these synergistic effects are more pronounced than in the mixture containing the APG surfactant. This behavior was attributed to the fact that while the head group of β-$C_{10}G_2$ has a slightly anionic character, that of $C_{10}E_8$ has a cationic character, which causes stronger net attractive interactions between the head groups of $C_{10}E_8$ and DBS. This view was supported by the fact that the β-$C_{10}G_2$/$C_{10}E_8$ mixture deviates slightly from ideal behavior, indicating a net attraction between the two nonionic surfactants, consistent with the occurrence of charges slightly negative and positive in character for β-$C_{10}G_2$ and $C_{10}E_8$, respectively.

TABLE 11.5

CMC, Micellar Composition, x_1, and Interaction Parameter, β^M, for the Mixtures of β-$C_{12}G_2$ with SDS, DTAB and $C_{12}E_5$ in Pure Water at 25°C

| α_1 | β-$C_{12}G_2$/SDS | | | β-$C_{12}G_2$/DTAB | | | β-$C_{12}G_2$/$C_{12}E_5$ | | |
	CMC (mM)	x_1	β^M	CMC (mM)	x_1	β^M	CMC (mM)	x_1	β^M
0	8	0		15	0		0.065	0	
0.012				6.7	0.51	−0.43			
0.10				1.5	0.83	−0.98			
0.25	0.44	0.73	−3.25	0.64	0.90	−1.36	0.078	0.11	0.01
0.50	0.26	0.80	−3.77	0.35	0.95	−1.54	0.096	0.26	0.03
0.75	0.20	0.87	−4.00				0.013	0.52	0.11
1	0.18	1		0.18	1		0.18	1	

Source: Zhang, R., Zhang, L., and Somasundaran, P., *J. Colloid Interface Sci.*, 278, 453, 2004.

Somasundaran and coworkers [49] reported the interactions in mixtures of SDS, DTAB, and pentaethyleneglycol monododecyl ether ($C_{12}E_5$) with β-$C_{12}G_2$. In this case, all the selected surfactants had a 12-carbon chain and different head groups. Thus all changes in the interactions are due to the differences in head groups. Table 11.5 lists the values of CMC, micellar composition, and interaction parameters, estimated from the RST treatment of the surface tension data, for the β-$C_{12}G_2$/SDS, β-$C_{12}G_2$/DTAB, and β-$C_{12}G_2$/$C_{12}E_5$ mixtures at different bulk compositions. From data in Table 11.5, it should first be noted that in the mixtures with ionic surfactants, the interaction parameter β^M becomes more negative as the content of β-$C_{12}G_2$ increases. This was attributed to the fact that the presence of the nonionic component causes a decrease in the surface charge density of the micelles, so that the mixed micelles of ionic and nonionic surfactants are more stable than the micelles containing only the ionic surfactant. Therefore, the higher the content of β-$C_{12}G_2$, the lower the surface charge density, and so the stronger the interaction. In relation to the nature of the interactions involved, the ion–dipole interactions between the head groups seem to play a decisive role in ionic–nonionic mixed micelles. Moreover, the steric interactions, due to the different sizes of the different surfactant head groups, must also be considered. Both partial charge neutralization and reduction of steric repulsion favor mixed micellization [49]. One also observes (Table 11.5) that the interaction parameters for the β-$C_{12}G_2$/DTAB system are less negative than those of the β-$C_{12}G_2$/SDS system, indicating weaker interactions between β-$C_{12}G_2$ and DTAB. This was ascribed to the fact that the nitrogen atom in the hydrophilic group of DTAB is screened by three methyl groups that hinder formation of ion–dipole interaction. The low values of the interaction parameters found in the case of the β-$C_{12}G_2$/$C_{12}E_5$ system seem reasonable since both surfactants are nonionic. Among the possible driving forces for interaction in this mixture, dipole–dipole, dipole-induce–dipole, and London dispersion type must be considered. These interactions are much weaker than the ion–dipole ones, which are mainly responsible for the stability of the ionic–nonionic mixed micelles. Finally, Somasundaran and coworkers [49] indicated that, although to simplify the RST treatment only the contribution of electrostatic interactions is considered, steric interactions between the surfactant hydrophilic heads and packing restrictions of the hydrophobic groups in the core can be expected to contribute to the stability of mixed micelles.

Wydro et al. [50–52] have published a series of three articles about the surfaces and micellar properties of mixtures composed of β-$C_{12}G_2$ and β-$C_{12}G_1$ with ionic and zwitterionic surfactants. In the first of these papers, the binary cationic–nonionic mixed systems β-$C_{12}G_2$/DTAB and β-$C_{12}G_1$/DTAB were investigated by surface tension measurements and subsequent analysis of the experimental data by applying the Motomura theory [50]. The aim of this study was to examine the influence of the size of the polar group on the surface and micellar properties of the mixed systems. In this case, two nonionic sugar-based surfactants (β-$C_{12}G_2$ and β-$C_{12}G_1$) with different polar group sizes and the same hydrocarbon chain length were combined with the same cationic surfactant (DTAB). It was found that both mixtures behave nonideally in both the mixed monolayers and micelles. The values of the excess free energy of adsorption and micelle formation were found to be more negative for β-$C_{12}G_1$/DTAB as compared to β-$C_{12}G_2$/DTAB, suggesting stronger attraction between the surfactants in the former system than the latter. This was explained by considering the smaller area per adsorbed molecule occupied by β-$C_{12}G_1$ as compared to β-$C_{12}G_2$, causing the distance between DTAB and β-$C_{12}G_1$ molecules to be smaller than between DTAB and β-$C_{12}G_2$ molecules, and therefore van der Waals interactions are stronger for β-$C_{12}G_1$/DTAB. However, it was observed that whereas the size of the head group affects the composition of mixed monolayers, such dependency was not observed in the case of the mixed micelles, which was attributed to the different geometry of the air–water interface in comparison to the micellar surface.

The second paper reported by Wydro et al. [51] deals with a new theoretical model for the description of the experimental surface tension isotherm of surfactant systems. In this case, the model was applied to the mixtures β-$C_{12}G_2$/SDS and β-$C_{12}G_1$/DTAB, and the results were compared with those obtained with the Motomura theory. Since the main objective of this work was to test the proposed theoretical model rather than the characterization of the mixed systems, we will only mention that, according to the authors, the advantage of the new model resides in the fact that it allows for correlation of its parameters with some molecular properties of the system studied.

In the third paper of the series, Wydro [52] studied the surface and micellar properties of the mixtures constituted by β-$C_{12}G_2$ and β-$C_{12}G_1$ with the zwitterionic surfactant dodedcylsulfobetaine (DSB). In this way, he examined the influence of the size of the hydrophilic group of the sugar-based surfactants on the mixed adsorbed film and on micellar formation. The adsorption isotherms of the pure and mixed surfactants were obtained at 25°C by using surface tension measurements, and the experimental data were analyzed according to the Motomura theory. This allowed for the determination of the composition of the adsorbed films and micelles, as well as the interactions between the surfactant components, which were quantitatively described by the excess free energy of mixing in monolayers, $\Delta G^{S,EXC}$, and micelles, $\Delta G^{M,EXC}$. Figures 11.5 and 11.6 show the values of $\Delta G^{S,EXC}$ and $\Delta G^{M,EXC}$ plotted as functions of the mixed adsorbed film composition at a given surface tension and the mixed micelle composition, respectively, for the two studied systems. From these figures, it is seen that $\Delta G^{S,EXC}$ and $\Delta G^{M,EXC}$ are negative within the whole range of the composition of the mixed monolayers and micelles, respectively, indicating that in mixed aggregates, the

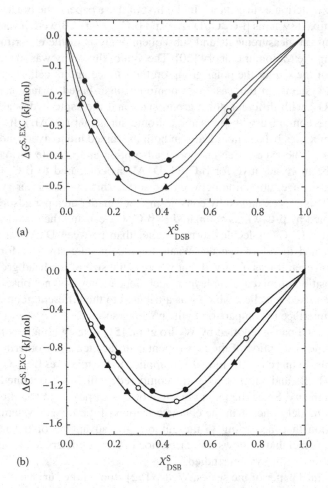

FIGURE 11.5 The excess free energy of mixing ($\Delta G^{S,\,EXC}$) versus composition of mixed adsorbed film ($(X^S)_{DSB}$) plots for the mixtures: (a) β-$C_{12}G_2$/DSB, and (b) β-$C_{12}G_1$/DSB. (▲) 45 mN/m, (○) 50 mN/m, and (●) 55 mN/m. (From Wydro, P., *J. Colloid Interface Sci.*, 316, 107, 2007. With permission.)

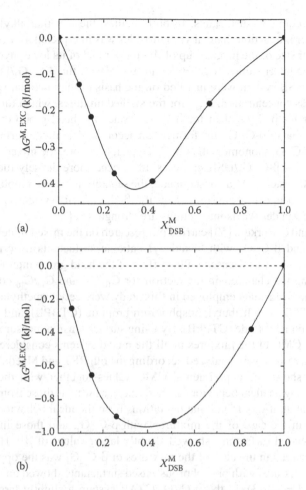

FIGURE 11.6 The excess free energy of mixing ($\Delta G^{M,\,EXC}$) versus composition of mixed micelle ($X_{DSB}{}^{M}$) plots for the mixtures: (a) β-$C_{12}G_2$/DSB, and (b) β-$C_{12}G_1$/DSB. (From Wydro, P., *J. Colloid Interface Sci.*, 316, 107, 2007. With permission.)

interactions between molecules are more attractive than those in the respective pure monolayers and micelles. This behavior was attributed to the fact that the intercalation of monomers of the nonionic component prevents the electrostatic repulsions between the monomers of the zwitterionic component and, at the same time, the steric interactions between the bulky head groups of the nonionic component are reduced by the presence of DSB. In addition, by comparing Figures 11.5 and 11.6, it can also be noticed that the values of $\Delta G^{S,EXC}$ are more negative for the β-$C_{12}G_1$/ DSB system than for the β-$C_{12}G_2$/DSB one, suggesting that DSB interacts more strongly with β-$C_{12}G_1$ than with β-$C_{12}G_2$. This result is in good agreement with those obtained by Rosen and Sulthana [46] in a wide study on the interaction of different alkyl glycosides, including β-$C_{10}G_1$, β-$C_{12}G_1$, β-$C_{10}G_2$, and β-$C_{12}G_2$, with several common surfactants. In both studies, the fact that alkyl glucosides show stronger interaction than alkyl maltosides was interpreted as a consequence of the smaller size of the head group of the former and of its lower hydrophilicity. Moreover, the lower values of $\Delta G^{S,EXC}$ and $\Delta G^{M,EXC}$ for the β-$C_{12}G_1$/DSB than for the β-$C_{12}G_2$/DSB system were justified on the basis of the maximal values of the surface excess concentration, Γ_{max}, for the studied mixtures, which turns out to be much higher for β-$C_{12}G_1$ than for β-$C_{12}G_2$. Since the head group of β-$C_{12}G_1$ is smaller than that of β-$C_{12}G_2$, the former can accommodate in the surface a greater number of β-$C_{12}G_1$ monomers than β-$C_{12}G_2$ and, therefore, the mixed monolayers and micelles for β-$C_{12}G_1$/DSB mixtures are packed more densely than those for β-$C_{12}G_2$/DSB systems. As a consequence, the separation between molecules in the mixed film and micelles constituted by the β-$C_{12}G_1$/DSB system is shorter and therefore the van der Waals interactions are stronger [52].

Palepu and coworkers [53] carried out research on the mixed micellar systems of β-$C_{10}G_1$ and β-$C_{12}G_1$ with hexadecyl cationic surfactants differing in head groups. This study was focused on the effect of the head group interaction nature while keeping the chain/chain interaction for C_{10}/C_{16} and C_{12}/C_{16} constant. The three cationic surfactants employed in this study were cetyltrimethylammnonium bromide (CTAB), cetyltributylphosphonium bromide (CTBPB), and cetyltriphenylphosphonium bromide (CTPPB). By using surface tension measurements, they obtained the CMC of the mixtures in all the mixed systems considered, and their experimental results were analyzed according to both RST and Maeda's approach. Figure 11.7 shows the experimental CMC values together with the calculated values assuming an ideal behavior (dashed lines), as well as those from RST treatment. In general, it was found that deviations from the ideal behavior were more pronounced in the case of the mixtures with β-$C_{10}G_1$, and those involving the phosphorus-based surfactants showed slightly larger values of β^M. However, the behavior observed in the case of the mixtures of β-$C_{12}G_1$ was the opposite, since whereas the mixtures with phosphorous based surfactants showed an almost ideal behavior ($\beta^M = -0.13$), in the β-$C_{12}G_1$/CTAB system a slightly larger negative value of $\beta^M(-1.28)$ was observed. In the opinion of the authors, the difference between the head groups of these cationic surfactants could reside in the distribution of their electric charge. According to previous molecular mechanics studies [54], whereas the negative charge in the phosphonium bromide surfactants is located on the phosphorus atom, in CTAB the charge is distributed among neighboring atoms, resulting in a net negative charge on the nitrogen atom. Again, in this study, the favorable effects produced by the reduction in the ionic head group repulsions and also in the steric hindrance of the nonionic head group were invoked to explain the greater stability of the mixed micelles.

Li et al. [45] have studied the interactions of a series of cationic gemini surfactants with several sugar-based surfactants at pH 9 in 0.1 M NaCl at 25°C. In this work, to determine whether cationic gemini surfactants showed stronger interaction with sugar-based surfactants than the comparable cationic conventional surfactant (DeTAB and DTAB), surface tension measurements were made on a series of diquaternary gemini surfactants and their mixtures with β-$C_{10}G_2$,

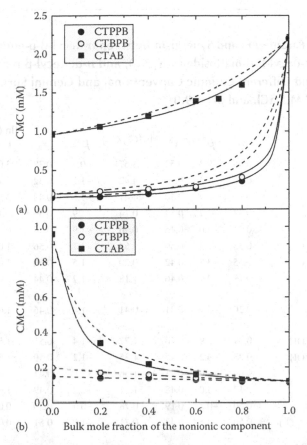

FIGURE 11.7 CMC versus mole fraction in the bulk of the nonionic component for the mixtures of $\beta\text{-}C_{10}G_1$ (a) and $\beta\text{-}C_{12}G_1$ (b) with different hexadecyl cationic surfactants. Symbols are experimental data, dashed and solid lines are the phase separation model predictions according to an ideal behavior and RST, respectively. (From Rodgers, C., Moulins, J., and Palepu, R., *Tenside Surf. Det.*, 43, 310, 2006.)

$\beta\text{-}C_{10}G_1$, and $\beta\text{-}C_{12}G_2$. The five cationic gemini surfactants investigated were $C_{12}H_{25}(CH_3)_2N^+\text{-}(CH_2)_4\text{-}N^+(CH_3)_2C_{12}H_{25}]\cdot 2Br^-((C_{12}N)_2)$ and $[C_{10}H_{21}(CH_3)_2N^+\text{- spacer -}N^+(CH_3)_2\,C_{10}H_{21}]\cdot 2Br^-$ with four different spacers: $-(CH_2)_4-((C_{10}N)_2)$, $-CH_2CH(OH)CH(OH)CH_2-((C_{10}N)_2(OH)_2)$, $-CH_2CH_2OCH_2CH_2-((C_{10}N)_2O)$, and $-CH_2CH(OH)CH_2-((C_{10}N)_2OH)$.

Table 11.6 lists the interaction parameters in both the adsorption in the air–liquid interface and the micellar formation of the all mixed systems studied by Li et al. [45] as well as those showing the occurrence of synergism in the corresponding mixtures. First of all, in relation to adsorption in the air–liquid interface, one can observe that each of the cationic gemini surfactants, when mixed with either $\beta\text{-}C_{10}G_2$ or $\beta\text{-}C_{10}G_1$, shows a stronger interaction than the comparable cationic

TABLE 11.6

Interaction Parameters and Synergism between n-decyl-β-D-glucoside (β-$C_{10}G_1$), n-decyl-β-D-maltoside (β-$C_{10}G_2$), and n-dodecyl-β-D-maltoside (β-$C_{12}G_2$) and Different Cationic Conventional and Gemini Surfactants at 25°C, 0.1 M NaCl, and pH = 9.0

Systems	α_2	β^σ	x_2^g	$\left\lvert \ln\left(C_1^0/C_2^0\right)\right\rvert$	β^M	x_2	$\lvert \ln(C_1/C_2)\rvert$	$\beta^\sigma - \beta^M$
β-$C_{10}G_2$/DeTAB	0.96	−0.3	0.52	3.67	−0.3	0.38	3.09	0
β-$C_{10}G_1$/DeTAB	0.96	−1.2	0.52	4.42	−1.2	0.32	3.10	0
β-$C_{12}G_2$/DTAB	0.97	−1.9	0.40	4.41	−1.5	0.49	3.56	−0.4
β-$C_{10}G_2$/$(C_{10}N)_2$	0.57	−2.7	0.52	0.14	−1.9	0.56	0.18	−0.8
β-$C_{10}G_1$/$(C_{10}N)_2$	0.57	−4.0	0.45	0.96	−1.9	0.56	0.15	−2.1
β-$C_{10}G_2$/$(C_{10}N)_2O$	0.35	−2.3	0.50	0.58	−1.7	0.56	1.04	−0.6
β-$C_{10}G_1$/$(C_{10}N)_2O$	0.35	−3.3	0.42	0.20	−1.5	0.56	1.03	−1.8
β-$C_{10}G_2$/$(C_{10}N)_2$ (OH)$_2$	0.18	−2.0	0.46	1.18	−1.7	0.44	0.90	−0.3
β-$C_{10}G_1$/$(C_{10}N)_2$ (OH)$_2$	0.20	−3.1	0.41	0.41	−1.4	0.46	1.07	−1.7
β-$C_{10}G_2$/$(C_{10}N)_2OH$	0.79	−2.9	0.56	1.77	−1.4	0.55	1.59	−1.5
β-$C_{10}G_1$/$(C_{10}N)_2OH$	0.83	−4.2	0.45	1.35	−1.2	0.50	1.58	−3.0
β-$C_{12}G_2$/$(C_{10}N)_2$	0.09	−3.0	0.41	1.36	−2.2	0.41	1.63	−0.7
β-$C_{12}G_2$/$(C_{10}N)_2$	0.17	−3.0	0.48	1.36	−2.2	0.49	1.63	−0.7
β-$C_{10}G_2$/β-$C_{10}G_1$	0.50	−0.2	0.19	0.78	−0.3	0.50	0.01	0.1
(1:1 β-$C_{10}G_2$/β-$C_{10}G_1$)/ $(C_{10}N)_2$	0.50	−3.5	0.43	0.71	−1.9	0.51	0.04	−1.6

Source: Li, F., Rosen, M.J., and Sulthana, S.B., *Langmuir*, 17, 1037, 2001. With permission.

Note: α_2 is the molar fraction in the bulk of the cationic surfactant; for β-$C_{10}G_2$/β-$C_{10}G_1$ is the molar fraction of either surfactant.

conventional surfactant, as indicated by its more negative β^σ value. This fact was attributed to the fact that the gemini has two positive groups interacting with the weakly negatively charged maltoside or glucoside at pH = 9. In addition, the interaction with β-$C_{10}G_1$ of both the conventional and the gemini surfactants is stronger than with β-$C_{10}G_2$. This was justified on the basis of the larger hydrophilic head group of β-$C_{10}G_2$ than that of β-$C_{10}G_1$ and, hence, a lower negative charge density and consequently a weaker interaction with cationic surfactants at the air–liquid interface. Finally, one can observe that the mixtures with β-$C_{10}G_1$, compared to those with β-$C_{10}G_2$, have a greater preference to be adsorbed at the air–liquid interface rather than to form mixed micelles, which is probably due to the smaller $A_{\min}$ (cross-sectional area of the surfactant at the air–liquid interface) value in the former mixtures compared to the latter.

As regards mixed micellar formation, data in Table 11.6 indicate that all the cationic gemini surfactants with the exception of the β-$C_{10}G_1$/$(C_{10}N)_2OH$ system, when mixed with either β-$C_{10}G_2$ or β-$C_{10}G_1$ show more negative β^M values, indicating stronger interaction in the mixed micelles. Moreover, it is observed that the interactions of the gemini surfactants with the alkyl maltosides or alkyl glucosides in the mixed micelles are weaker than in the mixed monolayer at the aqueous air–liquid interfaces. This fact is believed to be due the greater difficulty of incorporating the two hydrophobic groups of the gemini into the convex mixed micelles compared to that of accommodating them at the planar interface [55]. On the other hand, due to the curved hydrophilic surface of micelles, the larger hydrophilic group of maltoside, as compared to that of glucoside, appears not to have as great an influence on the interaction in the mixed micelle as it does at the planar air–liquid interface. This is supported by the fact that the interaction parameter β^M of the mixtures involving alkyl maltosides is not much different from those with alkyl glucosides.

From data in Table 11.6, it is also possible to extract conclusions about the spacer effect on β^σ and β^M values. With the exception of the β-$C_{10}G_1$/$(C_{10}N)_2OH$ system, showing the strongest interaction at the air–liquid interface and the weakest interaction at the mixed micellar formation, it is seen that the C_{10} gemini surfactant with the smallest spacer (smallest A_{min}) shows the strongest interaction at the planar air–liquid interface with C_{10} alkyl glucoside or C_{10} alkyl maltoside and the weakest interaction with them in mixed micelle formation. In addition, a more hydrophilic spacer in the gemini surfactants weakens the interaction with the sugar-based surfactants for both mixed monolayers and mixed micelle formation.

11.3.5 Effect of the Hydrophobic Chain Length

Another interesting aspect investigated by different researchers is the influence of the hydrophobic chain length, either of the nonionic sugar-based surfactant or of the cosurfactant, on the interactions controlling mixed micellar formation.

As mentioned, Sierra and Svensson [37] have studied the effect of the nonionic chain length of some glycosidic surfactants in the mixed micellization with SDS (see Table 11.4). If we compare the data in Table 11.1 for the mixtures β-$C_{10}G_1$/SDS and β-$C_{12}G_1$/SDS, and for β-$C_{10}G_2$/SDS and β-$C_{12}G_2$/SDS, one observes that a decrease in the chain length of the nonionic surfactant leads to a stronger interaction (β^M more negative) in micelle formation. These results are in good agreement with those obtained by Kakehashi et al. [56] for the mixtures of β-C_8G_1 and β-$C_{10}G_1$ with sodium oleate (NaOl). These authors found that the β^M value (-3.2) for β-C_8G_1/NaOl was more negative than that for β-$C_{10}G_1$/NaOl (-2.8). According to RST, the molecular interactions responsible for the stabilization of the mixed micelles are electrostatic in nature. Moreover, from RST it is not possible to obtain information on the presence of short-range attractive interactions between the components in the mixed micelle, as these interactions are hidden by the contribution from the electrostatic ones. Therefore, in order to rationalize the mixing behavior, the aforementioned authors discussed the interaction between surfactants in terms of the parameter B_1 introduced by Maeda [26]. As previously established (Section 11.2), the parameter B_1 accounts for the standard free energy change when a nonionic monomer, in the pure nonionic micelle, is replaced by an ionic one. So, it does not contain the electrostatic free energy contribution but does contain information on the short-range attractive interactions between the ionic and nonionic species. In ionic/nonionic systems with CMC values of the same order of magnitude, it is often observed that the hydrocarbon chain of the ionic species is longer than that of the nonionic one. For these systems, the hydrocarbon tail dissimilarity results in a negative value of B_1. In the present case, the B_1 value for the β-C_8G_1/NaOl mixed system was −7.0 and that for β-$C_{10}G_1$/NaOl was −4.1, indicating that a decreasing interaction of nonelectrostatic nature occurs in the mixed system as the alkyl nonionic chain length of the glycosidic surfactant increases. Note that this result is consistent with the interpretation given by Maeda's approach, in the sense that the short-range attractive interactions are due to the hydrocarbon tail dissimilarity of the ionic and nonionic components. In summary, a decrease in the hydrocarbon chain length of the glycoside surfactant resulted in a stronger interaction with NaOl from both $β^M$ and B_1 values.

Recently, Bakshi and Kaur [57] have carried out a fluorescence study on mixed micellization of a number of different cosurfactants with β-C_8G_1 and β-$C_{12}G_1$. In this study, the authors evaluated different micellization parameters, including CMC, micropolarity, and micellar aggregation numbers. The cosurfactants chosen were monomeric cationics (MC) (DTAB, TTAB, and CTAB), zwitterionics (ZI) (3-(N,N-dimethyldodecylammonio)propane sulfonate or DPS, 3-(N,N-dimethyltetradecylammonio)propane sulfonate or TPS, and 3-(N,N-dimethylhexadecylammonio)propane sulfonate or HPS), dimeric cationics (DC) such as dimethylene-bis(alkyldimethylammonium bromide) (m-2-m, where m is 10, 12, 14, and 16), and, finally, unequal twin-tail cationics (TC) such as dodecylalkyldimethylammonium bromide (m-0-n, where m is 12, and n is 8, 10, 12, and 16). In this way, the appropriate selection of cosurfactants permitted the analysis of the influence of hydrophobic tail changes on the micellar properties of the mixed systems.

First of all, the CMC experimental values obtained for all the binary mixtures were treated using RST treatment. In this way, both the micelle composition (x_1) and the RST interaction parameter of mixed micellization ($β^M$) were evaluated for the whole series of mixtures. Figure 11.8 shows the average interaction parameter of micellization ($β_{average}$), in the case of mixtures with β-C_8G_1, as a function of the number of carbon atoms (n_C) in the hydrophobic tail of a series of cosurfactants in a particular category of binary mixtures. The n_C values for DC and TC were taken for the single carbon tail in the former case, but for the variable carbon tail in the latter. Data in Figure 11.8 indicate that the $β_{average}$ for all the mixtures with β-C_8G_1 essentially decreases with n_C, thus making $β_{average}$ more negative and, thus, showing stronger interactions. It is important to point out that the only difference between the cosurfactants MC and TC is that, having identical head groups, they have single and twin hydrophobic tails, respectively. Therefore, as far as the head group nature is concerned, the same interactions with β-C_8G_1

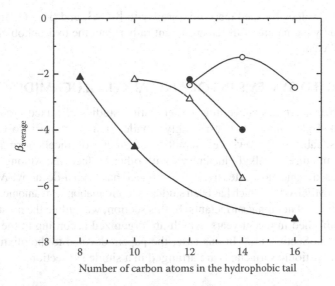

FIGURE 11.8 Plots of the average interaction parameter of micellization ($\beta_{average}$) versus the number of carbon atoms in the hydrophobic chain of cosurfactant for mixtures of β-C_8G_1 with: ($\circ$) Monomeric cationics (MC); DTAB, TTAB, and CTAB, ($\bullet$) ZI; DPS, TPS, and HPS, ($\triangle$)DC; m-2-m, where m is 10, 12, 14, and 16, and ($\blacktriangle$)TC; m-0-n, where m is 12, and n is 8, 10, 12, and 16. (From Bakshi, M.S. and Kaur, G., *J. Colloid Interface Sci.*, 289, 551, 2005. With permission.)

should be expected. Consequently, the observed differences in $\beta_{average}$ values can be attributed to the respective hydrophobicity of the cosurfactant. Therefore, from data in Figure 11.8, it could be concluded that hydrophobic interactions have a significant influence on the variation of $\beta_{average}$. Surprisingly, for the mixtures with β-$C_{12}G_1$ (data not shown), no more clear behavior is observed. The $\beta_{average}$ value mainly increases with n_C and becomes positive at a higher hydrophobicity of the cosurfactant in all the binary mixtures except in the TC/β-$C_{12}G_1$ ones. In this case, the trend is not very clear, because $\beta_{average}$ is positive for the TC with one of the twin-tails C_8, while negative with C_{16}. In the authors' opinion, the origin of the antagonism with increasing hydrophobicity of the cosurfactant, observed in most cases, indicates the incompatibility of the longer hydrophobic tails with that of β-$C_{12}G_1$ in the micellar core, which might be due to the packing geometry constraints.

Another interesting micellar parameter determined by Bakshi and Kaur [57] was the micellar aggregation number (N_{agg}). It was observed that N_{agg} shows a similar tendency in mixtures containing MC and ZI irrespectively of the presence of β-C_8G_1 or β-$C_{12}G_1$, increasing N_{agg} with n_C in both cases. At the same time, an opposite trend was found for the mixtures of both nonionic surfactants with DC and TC. This behavior, together with the fact that the trends of N_{agg} with n_C cannot be completely related to the interactions in the corresponding mixtures, suggests that the N_{agg} values for the binary mixtures with β-C_8G_1 and β-$C_{12}G_1$ are mainly influenced by the nature of the cosurfactant rather than the hydrophobicity of the sugar surfactant.

11.4 MIXED SYSTEMS INVOLVING ALKYL GLUCAMIDES

The number of articles concerning micellization studies of mixed systems containing alkyl glucamides is considerably smaller than of those involving alkyl glucosides. Table 11.7 shows the interaction parameter of micelle formation, β^M, in several mixtures of alkyl glucamides with other surfactants. Among the alkyl glucamides, the one most extensively investigated has been n-decanoyl-N-methyl-glucamide (MEGA-10), which has been studied in combination with anionic, cationic, zwitterionic, and nonionic surfactants. In this section, we outline the most relevant articles published in recent years, which are organized according to the criterion established in Section 11.3. In this case, the papers devoted to the mixtures with anionic and cationic surfactants are arranged in a single subsection.

TABLE 11.7
Interaction Parameters of Mixed Micelles in Surfactant Mixtures Involving FAGs at 30°C

Mixture	Systems	Medium	β^M	References
Nonionic–anionic	MEGA-10/NaDC	pH = 10	0.15	[57]
	MEGA-10/ (α-SMy·Me)	Water	−2.1	[63]
	MEGA-10/SDS	0.1 M NaCl	−2.08	[64]
	MEGA-10/SDS	Water	−2.13	[69]
	$G_{10}Glu_2$/LiFOS	Water (at 40°C)	−5.0	[71]
	$G_{12}Glu_2$/LiFOS	Water (at 40°C)	−0.2	[71]
	DBNMG/SDS	Water	—	[59]
	TBNMG/SDS	Water	—	[62]
Nonionic–cationic	MEGA-10/DTAB	Water	−0.21 (−0.652)	[74,75]
	MEGA-10/TTAB	Water	−0.63 (−0.408)	[74,75]
	MEGA-10/CTAB	Water	−1.36 (−1.52)	[74,75]
	MEGA-10/CTPPB	Water (at 24°C)	−3.91	[66]
Nonionic– zwitterionic	MEGA-8/CHAPS	pH = 7.2	−2.1	[73]
	MEGA-9/CHAPS	pH = 7.2	−1.8	[73]
	MEGA-10/CHAPS	pH = 7.2	−1.1	[73]
Nonionic–nonionic	MEGA-10/$C_{12}E_{10}$	Water	−6.5	[68]
	MEGA-10/$C_{16}E_{10}$	Water	−6.7	[68]
	MEGA-10/$C_{12}E_8$	Water	0	[69]
	MEGA-8/MEGA-9	Water	0	[67]

11.4.1 Mixtures with Ionic Surfactants

Yunomiya et al. [58] have reported a study describing the mixed micelle formation of a bile salt, sodium deoxycholate (NaDC), with MEGA-10 as a function of composition in the surfactant mixture. In this study, the CMC of the mixed system in the entire composition range was determined by means of surface tension measurements, and the experimental data were analyzed by both RST and Motomura theory. Determined CMC values showed a slight positive deviation from ideal mixing, which were larger at the composition range of 0.2–0.4. From RST treatment, an average interaction parameter of 0.15 was obtained, but it was observed that the theoretical curve of the CMC against bulk composition, simulated with $\beta^M = 0.15$, does not agree well with experimental values. This poor agreement was attributed to the great difference in the chemical structure of both surfactants. On the other hand, when the experimental data were analyzed with Motomura theory, the presence of an azeotrope (around $\alpha_{MEGA-10} = 0.2$) in the plots of the CMC as a function of both micelle and bulk composition was observed. This fact seems to indicate that the aggregation state is separated in two parts; below and above the azeotrope, inducing, therefore, differences in the properties of mixed micelles in both ranges. In addition, the calculated micellar composition curve was almost coincident with the CMC curve, indicating that the composition of micelle forming species was practically the same as that of singly dispersed species.

Considerable efforts have been made by Griffiths et al. to characterize the aggregation behavior and structures of the mixed micelles formed in solutions of SDS and the sugar-based surfactant dodecylmalono-bis-N-methylglucamide (DBNMG) [59–62]. In this research, experimental techniques such as surface tension, fluorescence, small-angle neutron scattering (SANS), NMR, and electron paramagnetic resonance (EPR) were used. First of all, the micellar formation of the DBNMG/SDS system was studied determining the mixed CMC values by both fluorescence and surface tension measurements. All CMC values were found to lie below ideal mixing behavior. This negative (attractive) deviation was reflected in the negative β^M values obtained by the RST treatment. However, it was observed that this interaction parameter is strongly dependent on system composition, being very negative at very low SDS mole fractions and becoming less negative, but still indicating favorable interaction, at increasing SDS content. This behavior was interpreted in the sense that the reduction in steric interactions between the bulky head groups of DBNMG on introduction of SDS molecules is greater than the reduction in electrostatic repulsions between the sulfate head group in a SDS micelle on adding DBNMG [60].

The average aggregation number of the DBNMG/SDS mixed system, as obtained from fluorescence quenching studies, was found to be largely invariant with composition (ranging from 63 for pure SDS to 70 for 1:1 molar ratio of the mixture). From SANS measurements, it was confirmed that the mixed micelles are of a size and shape very similar to those of the two pure surfactants. Moreover, it was observed that the mixed system is insensitive to the presence of 0.1 M NaCl, since upon addition of this salt concentration the CMC and the micelle size and shape resulted unchanged. This behavior was justified on the basis of the sufficient separation that the head group of the nonionic component introduces between the sulfate head groups of SDS, in such a way that any changes in ionic strength have no effect on the electrostatic interactions [60].

Because of these peculiar properties, the mixed system constituted by DBNMG and SDS can be considered a particularly suitable model for studies of nonionic/anionic micelles and, therefore, it has later been employed in structural studies focused on checking the effect of the insertion of DBNMG into the hydration of SDS micelles [61] or in the characterization of this mixed system as a variable reaction medium [62]. In the first paper [61], the spin–probe technique was used to analyze the change in hydration of SDS micelles upon addition of the nonionic component for SDS-rich compositions. In this region, it was observed that the hydration decreased linearly with the number of inserted DBNMG monomers, which was interpreted as being due to the expulsion of water molecules from the polar shell surrounding the hydrocarbon core of the SDS micelle when one DBNMG monomer replaces one SDS monomer in the anionic micelle. In the second article [62], by using experimental techniques such as time-resolved fluorescence quenching (TRFQ), EPR, and SANS, a complete description of the DBNMG/SDS mixed micelles as a variable reaction medium was provided. The research focused on the following aspects: structural characterization of the mixed micelle, including micellar aggregation numbers and thickness of the polar shell; micellar hydration; and microviscosity. To summarize, we briefly mention the main observations from this study. The mixed micelles turned out to be similar in size and shape to the two pure surfactant micelles. Essentially, replacing an SDS monomer in the micelle by a DBNMG monomer had little effect on the structure of the micelle, except to displace a volume of solvating water comparable to the difference in the respective head group volumes. Furthermore, the addition of DBNMG substantially increases the local viscosity of the head group region.

It should be mentioned that an additional study was performed by the same group on the related mixed system composed of SDS and tetradecylmalono-bis-N-methylglucamide (TBNMG) [63]. As regards interactions in the mixed micelles, this system shows a peculiar behavior, since β^M ranges from positive for $\alpha_{SDS} \leq 0.6$ to negative at higher values of α_{SDS}. From a structural point of view, it was found that shell thickness decreased monotonically with the participation of SDS in the micelle, but the core radius and the aggregation number turned out to be invariant with the composition. At a high SDS content in the mixed micelle, the smaller sulfate head group resulted in a thinner head group but with a significantly greater volume fraction of water. Consequently, microviscosity was found to be much lower at high x_{SDS}.

Okano et al. [64] have reported on the micellization process and the adsorbed film formation of the binary mixed system composed of MEGA-10 and the sodium salt of α-sulfonatomyristic acid methyl ester (α-SMy·Me). On the basis of surface tension measurements, the CMCs and the surface excess for the mixed systems in the whole composition range were determined at 30°C. In order to analyze the mixed micellization process, the CMC experimental data were treated according to RST and an average interaction parameter, $(\beta^M)_{avg}$, of -2.1 was determined. This negative value was considered as a sign of interaction acting especially between the head groups of both surfactants, which is considerably greater than the hydrophobic interaction occurring between the hydrophobic chains. The micellar composition curve, obtained by plotting the CMC as a function of the mole fraction of MEGA-10 in micelles, indicated the existence of an azeotrope formed at a mole fraction in the bulk of this component around 0.4. In addition, at this mole fraction, the highest stabilization in free energy was afforded for the mixed system. The aforementioned authors derived a treatment, also based on RST, to estimate the composition of the adsorbed film phase formed at the CMC. By using the corresponding equations, they constructed a phase diagram containing the singly dispersed phase curve (CMC vs. $\alpha_{MEGA-10}$) and the adsorbed film phase curve (CMC vs. the composition of the adsorbed film). From this last diagram an azeotrope was found to be formed by the 1:1 mixture ($\alpha_{MEGA-10} = 0.5$), which was interpreted in the sense that the composition in micelles differs from that in the adsorbed film.

Hierrezuelo et al. [65] have recently investigated the micellar properties of the mixed system constituted by MEGA-10 and SDS in 0.1 M NaCl. By using the fluorescence probe technique, aspects such as CMC, mean aggregation number, and microenvironmental properties of the mixed micelles were determined. The CMC experimental data were treated by two mixing thermodynamic models within the framework of the pseudophase separation approach, including the conventional RST and a recent treatment proposed by Maeda [66]. For this system, RST yields an interaction parameter of -2.08, a value practically identical to that found for the MEGA-10/α-SMy·Me systems [64]. However, Maeda's treatment [65] provided a wider perspective on this system, because it revealed that short-range attractive interactions play an important role in the stabilization of the mixed system. The mean aggregation number was found to increase initially with the participation of the ionic component, remain roughly constant, and then decrease slightly for mixtures with a high content of SDS. This behavior was analyzed on the basis of two opposite effects: the repulsive interactions between the head group of the ionic surfactant, which prevent micellar increase, and the favorable steric interactions occurring when this component replaces the nonionic component in the mixed micelle. Finally, the microenvironmental studies are consistent with a decrease in the micellar hydration and with the formation of less permeated and ordered micelles as the presence of the ionic component increases.

Recently, Sehgal et al. [67] have reported on the interfacial and micellar properties of the binary mixture of MEGA-10 and CTPPB in water at 24°C. In this case, a negative deviation from ideality was found ($\beta^M = -3.91$), but a considerable contribution due to short-range attractive interactions was also observed, as estimated by the Maeda parameter ($B_1 = -7.51$) [26]. On the contrary to the previous case, the aggregation number decreased with the participation of the ionic surfactant in the mixed micelle, and the results of the microenvironmental properties indicated the formation of more rigid and ordered micelles.

11.4.2 Mixtures with Nonionic Surfactants

To the best of our knowledge, only four nonionic/nonionic mixed systems involving alkyl glucamides have been studied in the last 10 years [68–70]; and often some of this research was a part of a more extensive study.

Kawaizumi et al. [68] measured some volumetric properties of mixtures composed of MEGA-8 and MEGA-9 in an aqueous medium. They observed that not all the properties of the mixed system exhibited ideal behavior. For example, it was found that the MEGA-8/MEGA-9 mixture behaves ideally with regard to the partial molar volume but not so for the partial molar adiabatic compressibility. This observation was interpreted on the basis of the arrangement of the surfactant monomers within the mixed micelle. Two alternative situations were proposed: (1) an irregular micellar surface, and (2) a smooth surface. In the former, the head groups of the surfactant with either shorter or larger tail, depending on its participation in the mixed micelle, should contribute to the increase of the solvation layer, resulting in a negative deviation of the compressibility behavior for the mixed system. According to the second situation, if the surface is smooth, one must assume that the change in compressibility is due to the effect produced in the micellar core as a result of the entanglement of the shorter and longer hydrocarbon tails in this micellar region.

An extensive study on the interfacial and thermodynamic properties of the mixed systems composed of MEGA-10 and two different polyoxyethylene (10) alkyl ethers ($C_{12}E_{10}$ and $C_{16}E_{10}$) has been carried out by Sulthana et al. [69]. These authors employed surface tension, steady-state fluorescence, and ^1H NMR measurements. CMC data of pure and mixed surfactants as obtained by surface tension measurements in a temperature range (from 30°C to 45°C) resulted to be lower than those expected for ideal behavior. In fact, for both MEGA-10/$C_{12}E_{10}$ and MEGA-10/$C_{16}E_{10}$ mixed systems, negative β^M values were obtained. These β^M values were found to vary depending on the composition of the mixture, ranging from −8.7 to −5.1 for MEGA-10/$C_{12}E_{10}$ at 30°C, and from −7.3 to −5.4 for MEGA-10/$C_{16}E_{10}$ at the same temperature. These unexpected results were confirmed by proton NMR spectroscopy and are indicative of considerable attractive interactions between the surfactant components, higher than those observed in the case of mixtures of MEGA-10 with ionic surfactants. In order to explain the interactions responsible for micellar stability in these mixed systems, four effects were invoked: (1) the interaction between the head groups of the surfactants through hydration, (2) a small repulsive interaction between the oxonium ion of polyoxyethylene and the slightly positive nitrogen atom of the MEGA-10, (3) incorporation of MEGA-10 molecules lowering the steric repulsions between the large oxyethylene head groups, and (4) decreased hydration of the hydrophilic moiety of MEGA-10 due to the addition of $C_{12}E_{10}$ or $C_{16}E_{10}$. On the other hand, by using the steady-state fluorescence technique, micellar aggregation number and micropolarity were estimated. It was observed that while the number of aggregations were found to be lower for the pure components compared to the mixtures at all composition ratios, the micropolarity was, in general, lower than that expected for an ideal mixed micellar system. This fact was interpreted in the sense that the structure of the mixed micelles is apparently rigid enough for the penetration of water molecules from the bulk into the mixed micelle to be considerably reduced.

11.4.3 EFFECT OF THE POLAR HEAD GROUP

With the aim of examining the role of the hydrophilic group structure of the cosurfactant in the properties of the mixed micelles, Hierrezuelo et al. [70] carried out a study on the mixtures of MEGA-10 with three different surfactants, including SDS, DTAB, and $C_{12}E_8$, all of them having the same hydrophobic chain. With regard to the interactions involved in these mixed systems, it was observed that while the MEGA-10/$C_{12}E_8$ system behaves ideally ($\beta^M = 0$), the MEGA-10/DTAB mixture shows a slight negative deviation from ideal behavior ($\beta^M = -0.21$), and in the MEGA-10/SDS system a moderate interaction occurs ($\beta^M = -2.13$). According to previous research carried out by Somasundaram and coworkers [49], the fact that SDS exerts a more pronounced effect on the mixed micelle can be attributed to two factors: (1) the interaction of ion–dipole between the head groups of the surfactants, and (2) the ability of the hydrophilic groups of MEGA-10 to envelope the small sulfate group of SDS. In addition, data about the effect of the participation of the cosurfactant in the microenvironmental properties support this view, in the sense that SDS is also the cosurfactant producing the most pronounced effect on the micellar microstructure.

Akisada et al. [71] have carried out a circular dichroism study of the interaction between MEGA-10 with a number of surface active agents, including ionic and nonionic species. These authors claimed that this experimental technique is a useful tool to explore the interaction between an optically active carbonyl and a hydrophilic group in aqueous solutions; therefore they focused their study on this aspect, attaining some interesting conclusions. In relation to the charge effect of the hydrophilic group, they observed that the interaction between the carbonyl group of MEGA-10 and the hydrophilic group of the cosurfactant does not depend on the sign of its charge, but rather on the charge density in the micelle. In addition, the carbonyl group of MEGA-10 neither interacts with nonionic surface active agents nor with the catanionic surfactant in which the anion and cation are decanesulfonate and decyltrimethylammonium, respectively. However, it does interact with ionic surfactants.

11.4.4 EFFECT OF THE HYDROPHOBIC CHAIN LENGTH

The influence of the hydrophobic chain length on the properties of mixed systems involving alkyl glucamides has been investigated by several groups [72–76]. By using surface tensiometry, NMR, and light scattering, Arai et al. [72] have studied the micellar properties of two lactose-based surfactants (n-decyllactobionamide or $C_{10}Glu_2$ and n-dodecyllactobionamide or $C_{12}Glu_2$) with the anionic surfactant lithium perfluorooctanesulfonate (LiFOS). As regards the interaction parameters estimated from the modified RST, they found significant differences between both mixed systems, while the mixture of $C_{10}Glu_2$/LiFOS shows a large negative deviation ($\beta^M = -5.0$), for the $C_{12}Glu_2$/LiFOS system, the interaction parameter was found to be much smaller ($\beta^M = -0.2$). Since both pairs have the same hydrophilic heads, they attributed this difference to the interactions of the hydrophobic tails. It was also shown that these saccharide derivatives interact in a different way with LiFOS as compared to the corresponding ethoxylates. On the other hand, the aggregation number and micellar size for both mixed micelles were determined by light scattering measurements, and it was observed that these properties are essentially controlled by the participation of the ionic component in both mixed systems. Lastly, to examine the effects of the micellar composition on micellar compactness, studies on the solubility of decaflurobyphenyl in the mixed micelles were carried out. An enhancement was observed in the amount of solubilization of this substance at a C_nGlu_2 mole fraction of 0.4, which was correlated with the compactness of surfactants in the mixed micelles, this fact being supported by the NMR and light scattering measurements.

Besides the investigations commented in Section 4.1 about the properties of the mixed systems formed by DBNMG and TBNMG with SDS, Griffiths and coworkers have evaluated the influence of the hydrocarbon chain length in an additional paper [73]. In this study, the main objective was to analyze the behavior of a polymer/surfactant mixture when the polymer (gelatin) is "selective," that is, the interaction is only observed with one of the two surfactants, in this case SDS. As to the interactions occurring in these mixed micelles, we briefly mention that the CMCs of n-alkyl malono-bis-N-methylglucamides (with $n = 10$, 12, and 14) were measured in binary mixtures with the anionic SDS. For the binary mixtures, the $n = 10$ and 12 compounds showed a synergistic interaction with SDS. However, the $n = 14$ compound was strongly antagonistic at low SDS mole fractions but synergistic at high SDS mole fractions. Furthermore, for the DBNMG/SDS mixture no growth or change in micelle shape was observed. However, when the alkyl lengths of the nonionic and ionic component are different, that is, for the DeBNMG/SDS and TBNMG/SDS mixtures, the SANS data suggest a significant change in the shape of the micelle.

Ko et al. [74] have studied the micellization and adsorbed film formation of three combinations of mixed surfactant systems composed of the zwitterionic surfactant 3-[(3-cholamidopropyl) dimethylammonio] propanosulfonic acid (CHAPS) with different n-alkyl-N-methylglucamides (MEGA-n, with $n = 8$, 9, and 10). By means of surface tension measurements, the surface activity and the CMC values of the mixtures in the whole composition range were determined. On the basis of RST, the relations between compositions of the singly dispersed phase, the micellar phase and the adsorbed film were estimated, and then the interaction parameters in micelles (β^M) and in the adsorbed film phase (β^σ) were obtained. With relation to micelle formation, it was found that all the mixed systems show a negative deviation from ideal mixing, producing the averaged β^M values of -2.1 for MEGA-8/CHAPS, -1.8 for MEGA-9/CHAPS, and -1.1 for MEGA-10/CHAPS, that is, the increasing order is in quite strong contrast with the hydrophobic chain length of the nonionic component. In addition, the three mixed systems were found to show explicit synergism in micelle formation. Nevertheless, although negative deviations from ideality were observed for the adsorbed film formation, producing averaged β^σ values of -5.0 for MEGA-8/CHAPS, -2.0 for MEGA-9/CHAPS, and -1.5 for MEGA-10/CHAPS, the term of efficiency for evaluating the extent of synergism is not satisfied except for the MEGA-8/CHAPS system. On the other hand, the RST treatment of the experimental data (CMC and the total surfactant concentration at a determined surface tension versus bulk composition) showed that between bulk composition and micellar composition, an azeotropic phenomenon was observed for the three mixed systems. However, this phenomenon was not observed for the adsorbed film formation in the case of MEGA-10/CHAPS, which was ascribed to the weaker interaction between the hydrophilic groups caused by steric hindrance when compared with the other mixed systems.

Hierrezuelo et al. [75] have reported research based on mixed binary systems comprising MEGA-10 and three different n-alkyltrimethylammonium bromides ($n = 12$ (DTAB), 14 (TTAB), and 16 (CTAB)) using conductance and fluorescence probe techniques. A year later, Ray et al. [76] published another experimental study on the same mixed systems. The focus of both studies was essentially the same, that is, to examine the effect of the hydrocarbon chain length of the ionic component on the properties of the mixed system. In both studies, CMCs, micellar aggregation numbers, and micropolarity of pure and mixed systems were determined. However, whereas Hierrezuelo et al. [75] estimated the micellar microviscosity, Ray et al. [76] reported additional results on interfacial properties and thermodynamics of the micellar formation. In both cases, the CMC experimental values were analyzed in the light of the theories of Rubingh (RST) and Maeda. In Figure 11.9a are plotted the CMC experimental values reported by Hierrezuelo et al. [75] together with the predictions for both RST (solid lines) and ideal behavior (dashed lines), and Figure 11.9b shows the micellar composition as a function of bulk composition as obtained from RST for the three systems. The interaction parameters β^M reported for these mixed systems by the two groups, as well as the characteristic parameters of Maeda's approach, are listed in Table 11.8 for comparison. It can be seen that the β^M values found by Hierrezuelo et al. [75] increase monotonically with the hydrophobic chain length of the ionic component, indicating an increasing attractive interaction between both surfactants in the mixed micelle. In contrast, this tendency is not observed in the β^M values reported by Ray et al. [76]. However, although the parameter B_1 values of each study were not in good agreement, one can observe the same trend in both cases, suggesting that increasing interaction of nonelectrostatic nature (short-range interactions) occurs in the mixed system as the length of the alkyl chain of the cosurfactant increases. According to Maeda [26] this result is quite reasonable, since the origin of the short-range attractive interactions resides in the hydrocarbon chain dissimilarity of the ionic and nonionic components. In relation to the thermodynamics of micellization, Ray et al. [76] found that the Gibbs free energy of micellization according to Maeda's approach is appreciably lower than that obtained from the conventional pseudophase model. They claimed that the

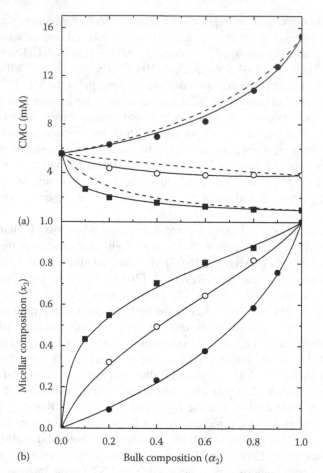

FIGURE 11.9 (a) Experimental (scatter) and predicted (lines) CMC values for the mixtures of MEGA-10 with different alkyltrimethylammonium bromides as a function of the mole fraction of the solution. (b) Plots of the micellar composition as a function of the molar fraction of the solution for the same systems. The dashed lines represent the phase separation model according to an ideal behavior, whereas the solid line is the best fit to the experimental data according to RST for (•) MEGA-10/DTAB, $\beta_{12} = -0.21$, (○) MEGA-10/TTAB, $\beta_{12} = -0.63$, and (■) MEGA-10/CTAB, $\beta_{12} = -1.36$. (From Hierrezuelo, J.M., Aguiar, J., and Carnero Ruiz, C., *J. Colloid Interface Sci.*, 294, 449, 2006. With permission.)

discrepancy arises for the neglect of counterion condensation in Maeda's treatment, which was considered to be a weakness of this formalism.

As can be seen in Figure 11.10, the micellar aggregation numbers obtained by Hierrezuelo et al. [75] indicated two different tendencies; whereas the systems with DTAB and TTAB (Figure 11.10a) show a negative deviation from ideality, that with CTAB (Figure 11.10b) deviates positively. This fact was ascribed to the higher participation of the ionic component in the case of CTAB (see Figure 11.9b),

TABLE 11.8

Interaction Parameters for the Mixtures of MEGA-10 with Different *n*-alkyltrimethylammonium Bromides (*n* = 12 (DTAB), 14 (TTAB), and 16 (CTAB)) according to the Theories of Rubingh (RST) and Maeda

System	B_0	B_1	B_2 (or $-\beta^M$)
MEGA-10/DTAB	−9.20 (−9.17)	0.799 (0.362)	0.21 (0.652)
MEGA-10/TTAB	−9.20 (−9.17)	−1.001 (−0.797)	0.63 (0.408)
MEGA-10/CTAB	−9.20 (−9.17)	−3.089 (−3.29)	1.36 (1.52)

Sources: Hierrezuelo, J.M., Aguiar, J., and Carnero Ruiz, C., *J. Colloid Interface Sci.*, 294, 449, 2006; Ray, G.B., Chakraborty, I., Ghosh, S., and Moulik, S.P., *J. Colloid Interface Sci.*, 307, 543, 2007.

Note: Within parentheses are the values of Ray et al. [76].

producing therefore a predominance of the repulsive electrostatic interactions in this mixed system. The observations of Ray et al. [76] in relation to the micellar aggregation numbers of the mixed systems were rather focused on the effect of the cosurfactant on the micellar shape as revealed by the so-called packing parameter. They concluded that for pure nonionic micelles, and mixed micelles of MEGA-10/CTAB with a high content of the nonionic component, the micellar shapes were nonspherical (prolate or oblate), but the remaining micellar systems have spherical shape.

11.5 CONCLUSIONS AND OUTLOOK

In the past few years, considerable progress has been made in the characterization of the interactions responsible for the micellar stability and surface activity of binary surfactant mixtures involving sugar-based surfactants. Particularly, the number of studies addressing the combinations of APGs with other conventional surfactants have been substantially more numerous that those of alkyl glucamides. However, the absence of studies concerning other kinds of sugar-based surfactants is conspicuous.

Research carried out to date indicate that both hydrophobic and electrostatic interactions, as well as better packing of the head groups in the mixed systems, must be considered in order to explain the behavior of alkyl glucosides and alkyl glucamides in surfactant mixtures. In general, the interaction parameters reported in mixtures with ionic surfactants resemble those previously published in mixtures of conventional nonionic polyoxyethylene surfactants. From the studies resumed in this chapter, it seems clear that a decrease in the alkyl chain length of both alkyl glucosides and alkyl glucamides results in a stronger interaction with other surfactants [37,74]. However, the effect produced by an increase in the number of glucose units in the polar group seems to depend on the nature of the

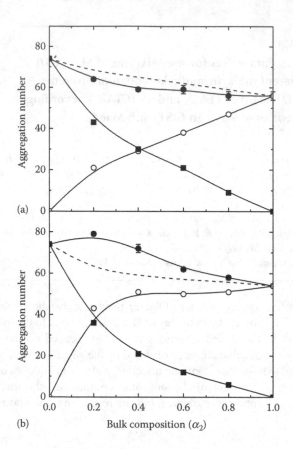

FIGURE 11.10 Micellar aggregation numbers as a function of the molar fraction of the ionic component in the solution (α_2): (a) MEGA-10/TTAB, (b) MEGA-10/CTAB. (●) Experimental values, (○) partial contribution of the ionic component, and (■) partial contribution of the nonionic component. The dashed lines represent the phase separation model according to an ideal behavior. (From Hierrezuelo, J.M., Aguiar, J., and Carnero Ruiz, C., *J. Colloid Interface Sci.*, 294, 449, 2006. With permission.)

cosurfactant considered. For example, while β-$C_{10}G_2$ shows a stronger interaction than β-$C_{10}G_1$ when mixed with SDS [37], DSB interacts more strongly with β-$C_{12}G_1$ than with β-$C_{12}G_2$ [52]. Although often the mixtures of sugar-based surfactants with polyoxyethylene ones present an almost ideal behavior, some reports indicate the occurrence of considerable interactions between both surfactants. Since the nature of the interactions involved in these mixtures is not well characterized, it seems necessary to look into these mixed systems more thoroughly.

Most of the advances in sugar-based mixed surfactant systems refer to the concentration range close to CMC, that is, to the dilute region. Because of the peculiar behavior of single sugar-based surfactants in the high concentration range, especially of the APGs, it should be interesting to achieve a significant development in the characterization of these mixed systems in the concentrated regime, which would allow us to establish suitable structure–property relationships, useful for current and future applications. Note that some specific micellar aggregates, such as, for example, wormlike micelles, apart from their traditional use as fracturing fluids for oil recovery are finding an increasing number of applications in a range of fields, including novel uses in pharmaceutical sciences and biomedical applications [77]. Finally, the possible vesicle formation in complex systems composed of mixtures of alkyl glucosides with other conventional surfactants is another important aspect to be investigated. The knowledge of the conditions under which vesicles could be formed is decisive to understand and optimize some biotechnological applications of these systems, particularly those involved in the processes of extraction and solubilization of lipids [78].

ACKNOWLEDGMENT

The author is grateful to the Spanish Education and Science Ministry (Project CTQ2005-04513) for financial support.

GLOSSARY OF NOTATION

SYMBOLS

Regular Solution Theory

μ_i	chemical potential of the ith surfactant monomer in the bulk
μ_i^0	standard chemical potential
μ_i^M	chemical potential of the ith component in the mixed micelle
C_i^m	concentration of the monomeric surfactant i in the bulk
C_i	critical micelle concentration of the pure component i
x_i	mole fraction of the surfactant i in the mixed micelle
α_i	mole fraction of the surfactant i in the bulk
C^*	critical micelle concentration of the mixed system
a_i	activity of the component i in the mixed micelle
f_i	activity coefficient of the component i in the mixed micelle
R	gas constant
T	absolute temperature
β^M	interaction parameter between the surfactants in the mixed micelle
β^σ	interaction parameter between the surfactants in the interface air–liquid
C_{mix}	concentration of the mixture required to produce a given surface tension value
C_i^0	concentration of the pure surfactant i required to produce a given surface tension value
α^*	mole fraction in the bulk in which the mole fraction of the surfactant i in the mixed micelle equals its mole fraction in the bulk
C_{min}^*	minimum critical micelle concentration for α^*
B_0	independent term related to the critical micelle concentration of the nonionic component in Maeda's model

B_1 second parameter in Maeda's model related to the standard free
             energy change upon replacement of a nonionic monomer in the
             nonionic pure micelle with an ionic one
B_2 third parameter in Maeda's model equivalent to β^M ($B_2 = -\beta^M$)

Motomura Theory

X_i^S mole fraction of the surfactant i in the mixed monolayer
X_i^M mole fraction of the surfactant i in the mixed micelle
X_i mole fraction of the surfactant i in the bulk solution
C total concentration of mixed solution at a given surface tension value
c_i^0 concentration of single surfactant i at a given surface tension value
C_i critical micelle concentration of the pure surfactant i
ν_i number of ions dissociated by the surfactant i
c_i concentration of the ionic species dissociated
$\Delta G^{M,EXC}$ excess free energy of micelle formation
$\Delta G^{S,EXC}$ excess free energy of adsorption
f_i^S activity coefficient of the surfactant i in the adsorbed film
f_i^M activity coefficient of the surfactant i in the micelle

ABBREVIATIONS OF SURFACTANTS

SDS	sodium dodecyl sulfate
α-SMy·Me	α-sulfonatomyristic acid methyl ester
β-$C_{10}G_2$	n-decyl-β-D-maltoside
β-$C_{12}G_2$	n-dodecyl-β-D-maltoside
β-C_8G_1	n-octyl-β-D-glucoside
β-SC_8G_1	n-octyl-β-D-thioglucoside
β-C_9G_1	n-nonyl-β-D-glucoside
β-$C_{10}G_1$	n-decyl-β-D-glucoside
β-$C_{12}G_1$	n-dodecyl-β-D-glucoside
DBS	dodecyl benzenesulfonate
$C_{10}E_8$	octaethyleneglycol monodecyl ether
$C_{12}E_4$	tetraethyleneglycol monododecyl ether
$C_{12}E_5$	pentaethyleneglycol monododecyl ether
$C_{12}E_7$	heptaethyleneglycol monododecyl ether
$C_{12}E_8$	octaethyleneglycol monododecyl ether
$C_{12}E_{10}$	decaethyleneglycol monododecyl ether
$C_{16}E_{10}$	decaethyleneglycol monohexadecyl ether (Brij 56)
NP-10	nonyl phenol ethoxylated decyl ether
DDAB	n-dodecyl-N,N'-dimethylamino betaine
DSB	dodecylsulfobetaina
DeTAB	decyltrimethylammonium bromide
DTAB	dodecyltrimethylammonium bromide
TTAB	tetradecyltrimethylammonium bromide

CTAB	cetyltrimethylammnonium bromide
CTBPB	cetyltributylphosphonium bromide
CTPPB	cetyltriphenylphosphonium bromide
DeTAC	decyltrimethylammonium chloride
DTAC	dodecyltrimethylammonium chloride
$C_{10}Glu_2$	n-decyllactobionamide
$C_{12}Glu_2$	n-dodecyllactobionamide
SDC	sodium deoxycholate
DeBNMG	decylmalono-bis-N-methylglucamide
DBNMG	dodecylmalono-bis-N-methylglucamide
TBNMG	tetradecylmalono-bis-N-methylglucamide
MEGA-8	n-octanoyl-N-methylglucamide
MEGA-9	n-nonanoyl-N-methylglucamide
MEGA-10	n-decanoyl-N-methylglucamide
CHAPS	3-[(3-cholamidopropyl) dimethylammonio] propanosulfonic acid
DPS	3-(N,N-dimethyldodecylammonio) propane sulfonate
TPS	3-(N,N-dimethyltetradecylammonio) propane sulfonate
HTS	3-(N,N-dimethylhexadecylammonio) propane sulfonate
m-2-m	dimethylene-bis(alkyldimethylammonium bromide (where m is 10, 12, 14, and 16)
m-2-n	dodecylalkyldimethylammonium bromide (where m is 12 and n is 8, 10, 12, and 16)
DESS	sodium dodecylethoxy sulphate
DBMG	dodecyl-N-benzyl-N-methylglycine
NaOl	olic acid sodium salt

Gemini surfactants with formula: $[C_{10}H_{21}(CH_3)_2N^+-spacer-N^+(CH_3)_2C_{10}H_{21}]\cdot 2Br^-$

$(C_{12}N)_2$	$[C_{12}H_{25}(CH_3)_2N^+—(CH_2)_4—N^+(CH_3)_2C_{12}H_{25}]\cdot 2Br^-$
$(C_{10}N)_2$	spacer $\rightarrow$ — $(CH_2)_4$ —
$(C_{10}N)_2(OH)_2$	spacer $\rightarrow$ — $CH_2CH(OH)CH(OH)CH_2$ —
$(C_{10}N)_2O$	spacer $\rightarrow$ — $CH_2CH_2OCH_2CH_2$ —
$(C_{10}N)_2OH$	spacer $\rightarrow$ — $CH_2CH(OH)CH_2$ —

REFERENCES

1. Holland, P.M. and Rubingh, D.N., *Mixed Surfactant Systems. ACS Symposium Series*, vol. 501, American Chemical Society, Washington, DC, 1992.
2. Ogino, K. and Abe, M., *Mixed Surfactant Systems. Surfactant Science Series*, vol. 46, Marcel Dekker, New York, 1993.
3. Abe, M. and Scamehorn, J.F., *Mixed Surfactant Systems. Surfactant Science Series*, vol. 124, Marcel Dekker, New York, 2005.
4. Hill, R.M., in *Mixed Surfactant Systems*, Ogino, K. and Abe, M. (Eds.), Marcel Dekker, New York, 1993, chap. 11.
5. Hoffmann, H. and Pössnecker, G., *Langmuir*, 1994, *10*, 381.
6. Desai, T.R. and Dixit, S.G., *J. Colloid Interface Sci.*, 1996, *177*, 471.
7. Kronberg, B., *Curr. Opin. Colloid Interface Sci.*, 1997, *2*, 456.
8. de Oliveira, H.P.M. and Gehlen, M.H., *Langmuir*, 2002, *18*, 3792.

9. Abe, M. and Ogino, K., in *Mixed Surfactant Systems*, Ogino, K. and Abe, M. (Eds.), Marcel Dekker, New York, 1993, chap. 1.
10. Hines, J.D., *Curr Opin Colloid Interface Sci.*, 2001, *6*, 350.
11. Puvvada, S. and Blankschtein, D., *J. Phys. Chem.*, 1992, *96*, 5576.
12. Puvvada, S. and Blankschtein, D., *J. Phys. Chem.*, 1996, *96*, 5579.
13. Sarmoria, C., Puvvada, S., and Blankschtein, D., *Langmuir*, 1992, *8*, 2690.
14. Shiloach, A. and Blankschtein, D., *Langmuir*, 1998, *14*, 1618.
15. Shiloach, A. and Blankschtein, D., *Langmuir*, 1998, *14*, 7166.
16. Nagarajan, R., *Langmuir*, 1985, *1*, 331.
17. Nagarajan, R., *Adv. Colloid Interface Sci.*, 1986, *26*, 205.
18. von Rybinsky, W. and Hill, K., *Angew. Chem. Int. Ed.*, 1998, *37*, 1238.
19. Gunjikar, J.P., Ware, A.M., and Momin, S.A., *J. Disp. Sci. Technol.*, 2006, *27*, 265.
20. Nishikido, N., in *Mixed Surfactant Systems*, Ogino, K. and Abe, M. (Eds.), Marcel Dekker, New York, 1993, chap. 2.
21. Clint, J.H., *J. Chem. Soc. Faraday I*, 1975, *71*, 1327.
22. Holland, P.M., *Adv. Colloid Interface Sci.*, 1986, *26*, 111.
23. Rosen, M.J. and Hua, X.Y., *J. Colloid Interface Sci.*, 1982, *86*, 164.
24. Rosen, M.J., *Surfactants and Interfacial Phenomena*, 2nd ed., Wiley, New York, 1989, chap. 11.
25. Hua, X.Y. and Rosen, M.J., *J. Colloid Interface Sci.*, 1982, *90*, 212.
26. Maeda, H., *J. Colloid Interface Sci.*, 1995, *172*, 98.
27. Lucassen-Reynders, E.H., *J. Colloid Interface Sci.*, 1982, *85*, 178.
28. Motomura, K., Ando, N., Matsuki, H., and Aratono, M., *J. Colloid Interface Sci.*, 1990, *139*, 188.
29. Motomura, K. and Aratono, M., in *Mixed Surfactant Systems*, Ogino, K. and Abe, M. (Eds.), Marcel Dekker, New York, 1993, chap. 4, and references herein.
30. Iyota, H., Todoroki, N., Ikeda, N., Motomura, K., Otha, A., and Aratono, M., *J. Colloid Interface Sci.*, 1999, *216*, 41.
31. Kameyama, K., Muroya, A., and Takagi, T., *J. Colloid Interface Sci.*, 1997, *196*, 48.
32. del Burgo, P., Junquera, E., and Aicart, E., *Langmuir*, 2004, *20*, 1587.
33. Lainez, A., del Burgo, P., Junquera, E., and Aicart, E., *Langmuir*, 2004, *20*, 5745.
34. Matsubara, H., Obata, H., Matsuda, T., and Aratono, M., *Colloids Surf. A*, 2008, *315*, 183.
35. Saito, S. and Tsuchiya, T., *Biochem. J.*, 1984, *222*, 829.
36. Hierrezuelo, J.M., Aguiar, J., and Carnero Ruiz, C., *Colloids Surf. A*, 2005, *264*, 29.
37. Sierra, M.L. and Svensson, M., *Langmuir*, 1999, *15*, 2301.
38. Liljekvist, P. and Kronberg, B., *J. Colloid Interface Sci.*, 2000, *222*, 159.
39. Bergström, M., Jonsson, P., Persson, M., and Eriksson, J.C., *Langmuir*, 2003, *19*, 10719.
40. Bergström, L.M., Bastardo, L.A., and Garamus, V.M., *J. Phys. Chem. B*, 2005, *109*, 12387.
41. Yang, Q., Zhou, Q., and Somasundaran, P., *Colloids Surf. A*, 2007, *305*, 22.
42. Nilsson, F., Söderman, O., and Reiner, J., *Langmuir*, 1998, *14*, 6396.
43. Whiddon, C., Söderman, O., and Hansson, P., *Langmuir*, 2002, *18*, 4610.
44. Zang, R. and Somasundaran, P., *Langmuir*, 2004, *20*, 8552.
45. Li, F., Rosen, M.J., and Sulthana, S.B., *Langmuir*, 2001, *17*, 1037.
46. Rosen, M.J. and Sulthana, S.B., *J. Colloid Interface Sci.*, 2001, *239*, 528.

47. Sanz, M.A., Granizo, N., Gradzielski, M., Rodrigo, M.M., and Valiente, M., *Colloid. Polym. Sci.*, 2005, *283*, 646.
48. Hines, J.D., Thomas, R.K., Garret, P.R., Rennie, G.K., and Penfold, J., *J. Phys. Chem. B*, 1997, *101*, 9215.
49. Zhang, R., Zhang, L., and Somasundaran, P., *J. Colloid Interface Sci.*, 2004, *278*, 453.
50. Wydro, P. and Paluch, M., *Colloids Surf. A*, 2004, *245*, 75.
51. Jarek, E., Wydro, P., Warszynsky, P., Paluch, M., *J. Colloid Interface Sci.*, 2006, *293*, 194.
52. Wydro, P., *J. Colloid Interface Sci.*, 2007, *316*, 107.
53. Rodgers, C., Moulins, J., and Palepu, R., *Tenside Surf. Det.*, 2006, *43*, 310.
54. Mohareb, M.M., Ghosh, K.K., Orlova, G., and Palepu, R.M., *J. Phys. Org. Chem.*, 2006, *19*, 281.
55. Liu, L. and Rosen, M.J., *J. Colloid Interface Sci.*, 1991, *179*, 454.
56. Kakehashi, R., Shizuma, M., Yamamura, S., and Takeda, T., *J. Colloid Interface Sci.*, 2004, *279*, 253.
57. Bakshi, M.S. and Kaur, G., *J. Colloid Interface Sci.*, 2005, *289*, 551.
58. Yunomiya, Y., Kunitake, T., Tanaka, T., Sugihara, G., and Nakashima, T., *J. Colloid Interface Sci.*, 1998, *208*, 1.
59. Griffiths, P.C., Stilbs, P., Paulsen, K., Howe, A.M., and Pitt, A.R., *J. Phys. Chem. B*, 1997, *101*, 915.
60. Griffiths, P.C., Whatton, M.L., Abbott, R.J., Kwun, W., Pitt, A.R., Howe, A.M., King, S.M., and Heenan, R.K., *J. Colloid Interface Sci.*, 1999, *215*, 114.
61. Bales, B.L., Howe, A.M., Pitt, A.R., Roe, J.A., and Griffiths, P.C., *J. Phys. Chem. B*, 2000, *104*, 264.
62. Bales, B.L., Ranganathan, R., and Griffiths, P.C., *J. Phys. Chem. B*, 2001, *105*, 7465.
63. Griffiths, P.C., Cheung, A.Y.F., Finney, G.J., Farley, C., Pitt, A.R., Howe, A.M., King, S.M., Heenan, R.K., and Bales, B.L., *Langmuir*, 2002, *18*, 1065.
64. Okano, T., Tamura, T., Abe, Y., Tsuchida, T., Lee, S., and Sugihara, G., *Langmuir*, 2000, *16*, 1508.
65. Hierrezuelo, J.M., Aguiar, J., and Carnero Ruiz, C., *Langmuir*, 2004, *20*, 10419.
66. Maeda, H., *J. Phys. Chem. B*, 2004, *108*, 6043.
67. Sehgal, P., Kosaka, O., and Doe, H., *Colloid Polym. Sci.*, 2008, *286*, 275.
68. Kawaizumi, F., Kuzuhara, T., and Nomura, H., *Langmuir*, 1998, *14*, 3749.
69. Sulthana, S.B., Rao, P.V.C., Bhat, S.G.T., Nakano, T.Y., Sugihara, G., and Rakshit, A.K., *Langmuir*, 2000, *16*, 980.
70. Hierrezuelo, J.M., Aguiar, J., and Carnero Ruiz, C., *Mol. Phys.*, 2005, *103*, 3299.
71. Akisada, H., Kuwahara, J., Kunisaki, M., Nishikawa, K., Akagi, S., Wada, M., Kuwata, A., and Iwamoto, S., *Colloid Polym. Sci.*, 2004, *283*, 169.
72. Arai, T., Takasugi, K., and Esumi, K., *J. Colloid Interface Sci.*, 1998, *197*, 94.
73. Griffiths, P.C., Roe, J.A., Jenkins, R.L., Reeve, J., and Cheung, A.Y.F., *Langmuir*, 2000, *16*, 9983.
74. Ko, J.-S., Oh, S.-W., Kim, K.-W., Nakashima, N., Nagadome, S., and Sugihara, G., *Colloids Surf. B*, 2005, *45*, 90.
75. Hierrezuelo, J.M., Aguiar, J., and Carnero Ruiz, C., *J. Colloid Interface Sci.*, 2006, *294*, 449.
76. Ray, G.B., Chakraborty, I., Ghosh, S., and Moulik, S.P., *J. Colloid Interface Sci.*, 2007, *307*, 543.
77. Dreiss, C.A., *Soft Matter*, 2007, *3*, 956.
78. Stubenrauch, C., *Curr. Opin. Colloid Interface Sci.*, 2001, *6*, 160.

12 Microemulsions Stabilized by Sugar Surfactants

Cosima Stubenrauch and Thomas Sottmann

CONTENTS

12.1 INTRODUCTION

Microemulsions are thermodynamically stable and macroscopically isotropic mixtures of at least three components, namely water, oil, and surfactant, where the surfactant forms an extended film separating water and oil on a nanoscale. The curvature of this surfactant film depends on various parameters such as the temperature, the salt, or the cosurfactant concentration. With these parameters a phase inversion can be induced, i.e. that an oil-in-water (o/w)-droplet microemulsion can be inverted into a water-in-oil (w/o)-droplet microemulsion via a bicontinuous microemulsion. In the special case of ternary water–oil–surfactant systems where the surfactant is a nonionic n-alkyl polyglycol ether (C_iE_j) the phase inversion can be induced simply by changing the temperature as is illustrated by the test tubes in Figure 12.1 (reviewed in Ref. [1]). Figure 12.1 represents a schematic $T(\gamma)$-section through the phase prism of such a ternary system. Starting with equal volumes of water and oil and measuring the phase diagram as a function of the temperature T and the surfactant mass fraction γ, one obtains the well-known "fish"-shape phase diagram [2]. At $\gamma < \gamma_0$ no microemulsion is formed as the surfactant is needed to saturate the solvents [3]. At $\gamma > \gamma_0$ and low T an o/w-microemulsion coexists with an excess oil phase ($\underline{2}$), while at high T a w/o-microemulsion coexists with an excess water phase ($\overline{2}$). At intermediate T and $\gamma < \tilde{\gamma}$ a microemulsion coexists with an excess water and an excess oil phase (3). The microstructure of this middle-phase microemulsion is bicontinuous around the mean temperature $\tilde{T}$, while it consists of a w/o network structure next to the 3–$\overline{2}$; and of an o/w network structure next to the $\underline{2}$–3 phase boundary. With increasing γ, more water and oil are solubilized until a 1-phase microemulsion appears (1).

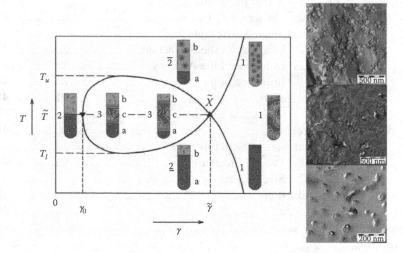

FIGURE 12.1   (Left) Schematic $T(\gamma)$-section of the system H_2O–oil–nonionic surfactant at constant oil/(water + oil) volume fraction $\phi = 0.50$. The phase inversion is illustrated by the test tubes. The $\tilde{X}$-point indicates the minimum amount of surfactant that is needed to solubilize water and oil. (right) Freeze fracture electron microscopy images of a w/o (top), a bicontinuous (middle), and an o/w microemulsion (bottom). (The FFEM images are taken from Belkoura, L., Stubenrauch, C., and Strey, R., *Langmuir*, 20, 4391, 2004. With permission.)

The $\tilde{X}$-point indicates the minimum amount of surfactant $(\tilde{\gamma})$ that is needed to solubilize water and oil at $\tilde{T}$ at which a bicontinuous microemulsion is formed.

It is important to realize that the microstructure has to be determined independently and cannot be extracted from phase studies [1,4]. Useful complementary techniques are (1) transmission electron microscopy, TEM, (2) NMR self-diffusometry, (3) electric conductivity, (4) scattering techniques in general, and Small angle neutron scattering, SANS, in particular. The combination of these techniques allows us to distinguish between bicontinuous and discrete droplet structures (TEM, NMR, electric conductivity) as well as to determine the domain size ξ (SANS) from which the mean curvature H of the surfactant monolayer can be calculated. All these parameters are interlinked as is clearly seen in Figures 12.1 and 12.2. The freeze fracture electron microscopy (FFEM) images (Figure 12.1, right) show a truly bicontinuous structure at $T = \tilde{T}$, which is supported by similar self-diffusion coefficients D of water and oil (Figure 12.2, top). Note that often the relative self-diffusion coefficients D/D_0 are plotted versus the temperature, with D_0 being the self-diffusion coefficient of the pure solvents at the respective temperature. In this case the values of water and oil are equal at $T = \tilde{T}$. In addition to equal D/D_0 values one finds a maximum for the domain size ξ (Figure 12.2, middle) and a mean curvature H of zero (Figure 12.2, bottom) at $T = \tilde{T}$. We will come back to this point in Section 12.3. Due to the enormous knowledge gained over the last 20 years we are now able to directly correlate microstructure and phase behavior, i.e. measuring the phase diagram is often sufficient to obtain information about the microstructure. Although we dedicated Section 12.2 to the phase behavior and Section 12.3 to the microstructure, we will use this knowledge and discuss some of the results presented in Section 12.2 in terms of microstructures and vice versa.

Let us come back to the focus of this chapter, namely microemulsions stabilized by sugar surfactants. The temperature sensitivity of microemulsions stabilized by C_iE_j surfactants can be a disadvantage in certain fields of application. It is at this point that microemulsions stabilized with sugar surfactants come into play as their properties are usually temperature insensitive. The sugar surfactants we are referring to are alkyl polyglucosides, C_nG_m, with n representing the number of C-atoms in the hydrophobic chain and m representing the degree of glycosidation. Although alkyl glucosides have been known for nearly 50 years [5], the formation of microemulsions with sugar surfactants has been addressed only recently. First experimental studies were made using technical grade C_nG_m surfactants [6–14], before systematic studies with pure C_nG_m surfactants were reported [15–28]. The weak temperature dependence of the head group hydration of C_nG_m makes it difficult to induce the phase inversion of a water–oil–alkyl glucoside system by temperature variation. In this case, the method of choice is to mix the sugar surfactant with a long-chain alcohol [6,15,17,27] or a C_iE_j surfactant [19]. In particular the combination of alkyl glucosides and natural alcohols and oils, respectively, is important for formulations of nontoxic microemulsions [7,8] and thus for applications in pharmaceutical products and cosmetics [29–31]. In these systems, the composition of the mixed interfacial film has to be

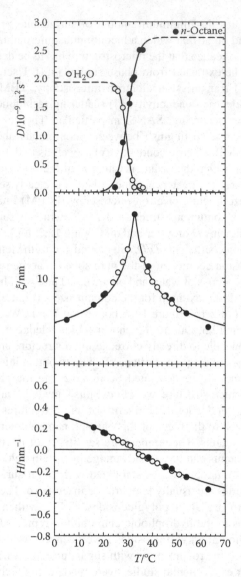

FIGURE 12.2   Self-diffusion coefficients D of water and oil (top), characteristic domain size ξ (middle), and mean curvature H of the interfacial film (bottom) in the one phase region of the system H_2O–n-octane–$C_{12}E_5$ as a function of temperature T. At $T_m = 32.6°C$ the microemulsion is bicontinuous, which is illustrated by similar D-values for water and oil, a maximum of ξ, and $H = 0$. Values for D have been determined via NMR diffusometry, while ξ and H have been determined via SANS measurements (filled symbols) and calculated form the composition, respectively. The solid lines in the $D(T)$-plot only guide the eyes, while the solid lines in the $\xi(T)$- and $H(T)$-plots are calculated according to empirical descriptions (see Ref. [1] for further details). (Middle and bottom figures are redrawn with data from Sottmann, T. and Strey, R., in *Soft Colloids V—Fundamentals in Interface and Colloid Science*, Elsevier, Amsterdam, 2005.)

regarded as the parameter which determines the properties of the microemulsion [17,25,32].*

Two very detailed reviews about the basic concepts of microemulsions have been published only recently [1,33]. Although we will explain some general concepts of microemulsions for the sake of clarity, the reader is referred to Refs. [1,33] for details. The general properties of sugar surfactants have also been summarized several times in the last 10 years [29,34–36]. Although these reviews include the ability of sugar surfactants to form microemulsions, microemulsions stabilized by sugar surfactants have been discussed neither extensively nor completely and the present chapter aims to fill this gap. We will hence review published results to which unpublished data will be added, which will, in turn, allow us to draw general conclusions. The focus is on microemulsions stabilized by alkyl polyglycosides although there are some studies in which other types of sugar surfactants were used [37–40].

12.2 PHASE BEHAVIOR

12.2.1 BACKGROUND

As mentioned above, the phase behavior of microemulsions has been reviewed in detail only recently [1,33]. Thus we restrict ourselves to some general concepts, which are needed to understand the phase diagrams shown below.

12.2.1.1 Ternary Systems

We start with the general phase behavior of a ternary water (A)–oil (B)–nonionic surfactant (C) system, which was briefly addressed in Section 12.1. A ternary system is described by four thermodynamic variables, namely T and p as well as two composition variables. As the effect of the pressure is weak compared to that of the temperature [41,42], p is usually kept constant so that the phase behavior can be represented in an upright phase prism with the Gibbs triangle A-B-C as the base and T as the ordinate. Each point in the phase prism is unambiguously defined by T and two composition variables. Regarding the latter it has been proven useful to choose the volume fraction of the oil in the mixture of water and oil

$$\phi = \frac{V_\text{B}}{V_\text{A} + V_\text{B}} \tag{12.1}$$

and the surfactant mass fraction in the mixture of all three components

$$\gamma = \frac{m_\text{C}}{m_\text{A} + m_\text{B} + m_\text{C}}. \tag{12.2}$$

---

* Note that the composition of the mixed interfacial film is a tuning parameter for both temperature-insensitive $C_n G_m$ [17,25] and temperature-sensitive $C_i E_j$ microemulsions [32], while the temperature influences only the latter.

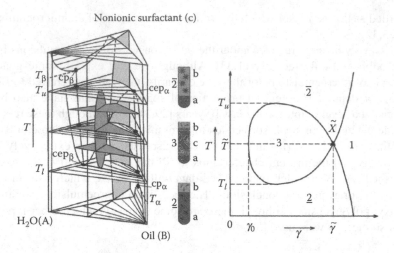

FIGURE 12.3 (Left) Phase prism of the system H_2O–oil–nonionic surfactant showing the temperature dependency of the phase behavior [2]. The phase inversion is illustrated by the test tubes on the right hand side of the prism. (Right) A convenient way to study the occurring phases is to perform a section at a constant oil/(water + oil) ratio ϕ as a function of the temperature T and the surfactant mass fraction γ [70]. This section results in phase boundaries resembling the shape of a fish. Note that the three-phase region in the phase prism is shown in dark gray and the one-phase region in light gray.

The phase prism of such a ternary system is shown in Figure 12.3 (left). T_l and T_u are the lower and the upper temperatures of the three-phase body vanishing in the limit of the lower and upper critical tie-lines drawn as thick solid lines. The critical endpoints cep_α and cep_β are connected by the trajectory of the middle phase originating from the lower critical point cp_β of the binary water-surfactant system and the upper critical point cp_α of the oil-surfactant system. A section through the phase prism at equal volumes of water and oil, i.e. at $\phi = 0.5$ (right), leads to the same $T(\gamma)$-section as the one shown in Figure 12.1. This section has been redrawn for the sake of clarity and for illustrating the difference between ternary and quaternary systems.

12.2.1.2 Quaternary Systems

Talking about microemulsions stabilized with alkyl glucosides most of the time a cosurfactant is involved. Thus we have to find a way of presenting and studying the phase behavior of a quaternary water (A)–oil (B)–surfactant (C)–cosurfactant (D) system. Most commonly these systems are studied at constant pressure and temperature so that the phase behavior can be represented in a phase tetrahedron (instead of a phase prism) with the ternary system as base and the cosurfactant placed on top [32,43–45]. In this case three composition variables are needed to describe each single point in the tetrahedron, for example the oil/(oil + water) volume fraction ϕ (Equation 12.1), the mass fraction of surfactant, and cosurfactant in the total mixture

$$\gamma = \frac{m_C + m_D}{m_A + m_B + m_C + m_D} \quad (12.3)$$

and the mass fraction of cosurfactant in the surfactant plus cosurfactant mixture

$$\delta = \frac{m_D}{m_C + m_D} \quad (12.4)$$

The phase tetrahedron of such a quaternary system is shown schematically in Figure 12.4 (left). As was the case for the ternary system, a section through the phase tetrahedron at equal volumes of water and oil, i.e. at $\phi = 0.5$, also leads to a "fish"-shape phase diagram, as can be seen in Figure 12.4 (right). Although this section is the correct representation of the phase behavior of a quaternary system at constant p, T, and ϕ, a more commonly used representation is a rectangular $\delta(\gamma)$-plot (Figure 12.5) as it can be directly compared with the $T(\gamma)$-plot of a ternary system. While the abscissa represents the surfactant mass fraction γ in both cases, the ordinate represents the temperature T and the mass fraction of cosurfactant δ, respectively. Comparing Figure 12.5 with Figure 12.3 (right) one clearly sees the similarities and differences between a $T(\gamma)$-plot and a $\delta(\gamma)$-plot. In both cases an $\tilde{X}$-point is found, which indicates the minimum amount of surfactant ($\tilde{\gamma}$) that is needed to solubilize water and oil at a certain temperature $\tilde{T}$ or amount of cosurfactant $\tilde{\delta}$. Moreover the monomeric surfactant solubility γ_0 can be extracted from the "head" of the fish shape phase diagrams which is defined by the coordinates $(\gamma_0, \tilde{T})$ and (γ_0, δ_0), respectively. On the other hand, a $\delta(\gamma)$-plot results in a distorted three-phase region. This distortion is caused by the different monomeric

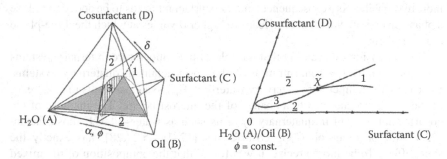

FIGURE 12.4 (Left) Schematic phase tetrahedron of a quaternary water–oil–surfactant–cosurfactant system at constant temperature T. (Right) A convenient way to study the occurring phases is to perform a section at a constant oil/(water + oil) ratio ϕ [46]. Again this section results in phase boundaries resembling the shape of a fish. A phase inversion is induced by adding a hydrophobic cosurfactant.

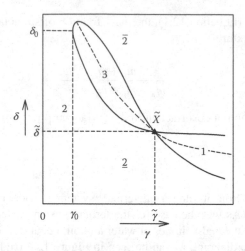

FIGURE 12.5 Schematic view of a rectangular $\delta(\gamma)$-section through the phase tetrahedron of the quaternary system H_2O–oil–surfactant–cosurfactant at constant oil/(water + oil) ratio ϕ and temperature T. The phase boundaries shift to higher δ-values with decreasing γ because of the different monomeric solubilities of surfactant and cosurfactant in water and oil, respectively. See text for further details.

solubilities of surfactant and cosurfactant in water and oil, respectively. Let us assume that the monomeric solubility of the hydrophobic cosurfactant in water and of the hydrophilic surfactant in oil, respectively, can be neglected, which is indeed the case for water–n-alkane–C_nG_m–C_iE_0 systems. Here, C_iE_0 is a medium or long chain alcohol, i.e. $i \geq 4$. If we also assume that the monomeric solubility of the cosurfactant in oil is higher than that of the surfactant in water, a phase diagram similar to the one seen in Figure 12.5 is obtained [17,32,46,47]. Diluting the system with water and oil (i.e. decreasing γ) leads to an extraction of the hydrophobic cosurfactant from the surfactant film, thus rendering the film more and more hydrophilic. As a consequence, more cosurfactant needs to be added to induce a phase inversion, which, in turn, leads to higher δ-values at which the three-phase region is formed.*

In conclusion one can say that the role of the temperature in ternary systems is taken over by the concentration of the cosurfactant in quaternary systems. Increasing the temperature in ternary water–oil–C_iE_j systems results in a decrease of the curvature and thus to a change of the microstructure. The increase of the cosurfactant amount in quaternary systems such as water–oil–C_iE_j–cosurfactant [32,43–45] or water–oil–C_nG_m–cosurfactant [17–19,25,27,28][†] has exactly the same effect. To be more precise, it was found that the composition of the mixed

---

* Note that the phase diagram is distorted downwards if the monomeric solubility of the cosurfactant in oil is lower than that of the surfactant in water [44,45].

† Note that the temperature-insensitivity of ternary water–oil–C_nG_m systems requires the addition of a cosurfactant to form microemulsions in the first place.

surfactant film (surfactant + cosurfactant) controls the behavior of quaternary systems in the same way as the temperature does in ternary systems (see Ref. [1] and references therein). In other words, it is not the total mass fraction of cosurfactant in the surfactant mixture (δ) but rather the mass fraction of cosurfactant in the mixed surfactant film (δ_i) that corresponds to the temperature. Plotting δ_i on the y-axis indeed allows us to compare quantitatively $T(\gamma)$-sections with $\delta_i(\gamma)$-sections. In the following, all experimentally determined phase diagrams for pure and technical surfactants will be presented in triangular $W_D(\gamma)$-sections. However, we will come back to the rectangular presentation in Section 12.2.4 as it clearly illustrates the general pattern of microemulsion phase diagrams as well as the difference between δ and δ_i.

12.2.2 Pure Sugar Surfactants

12.2.2.1 Ternary Systems

As mentioned above, microemulsions stabilized by sugar surfactants usually contain a cosurfactant. Without a cosurfactant only o/w-microemulsions are formed, which is mainly due to the fact that the high hydrophilicity of sugar surfactants cannot be changed significantly by temperature variation. The situation changes if polar oils are used instead of the classical n-alkanes. In this case ternary water–polar oil–C_nG_m systems indeed form o/w-, w/o-, and bicontinuous microemulsions and the phase inversion can be induced via temperature variation [20–23,26] as is the case for classical ternary water–n-alkane–C_iE_j systems. In Figure 12.6 two $T(\gamma)$-sections of ternary systems at constant oil/(water + oil) ratios $\phi = 0.50$ are shown, namely H_2O–1,2-dichloroethane–β-C_9G_1 [26] and H_2O–$C_{2.25}OC_2OC^*_{2.25}$–β-C_8G_1 [21]. In the former case the determination of the lower phase boundary ($\overline{2}$–3) was not possible due to the high Krafft temperature. Moreover, the density of the chlorinated oil is larger than the density of water. Thus the state where an o/w microemulsion coexists with an oil excess phase corresponds to $\overline{2}$, while the coexistence of a w/o-microemulsion and a water excess phase corresponds to 2. Regarding the $C_{2.25}OC_2OC_{2.25}$ system it has to be noted that this system is near the tricritical point and rather an ideal mixture than a structured microemulsion (curvature model nonapplicable). The phase inversion is obtained because the sugar surfactant is soluble in water at low and soluble in the hydrophilic oil at high temperatures. Increasing the chain length of the surfactant, i.e. using C_9G_1 and $C_{10}G_1$ leads to microstructured systems [23]. Apart from these pecularities the resulting phase diagrams look exactly the same as those obtained for ternary water–n-alkane–C_iE_j systems and can thus be discussed in the same way: an increase in temperature changes the solubility and obviously the curvature of the surfactant film. However, as already mentioned, the head group of

---

* Fractional carbon numbers for the alkyl ethylene glycol ethers were made by mixing $C_2OC_2OC_2$ and $C_4OC_2OC_4$.

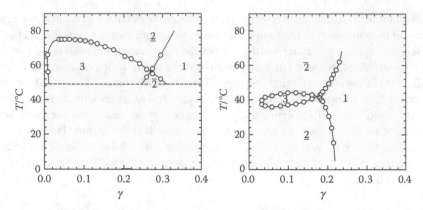

FIGURE 12.6 (Left) $T(\gamma)$-section of the ternary system H_2O–1,2-dichloroethane–β-C_9G_1 at constant oil/(water + oil) ratio $\phi = 0.50$. The determination of the phase diagram is limited by the Krafft temperature, which is at $T = 49°C$. (Right) $T(\gamma)$-section of the ternary system H_2O–$C_{2.25}OC_2OC_{2.25}$–β-C_8G_1 at constant oil/(water + oil) ratio $\phi = 0.50$. In this system the Krafft temperature lies below $0°C$. (Redrawn from (right) Ryan, L.D. and Kaler, E.W., *J. Phys. Chem. B*, 102, 7549, 1998; (left) Egger, H., Sottmann, T., Strey, R., Valero, C., and Berkessel, A., *Tenside Surf. Det.*, 39, 17, 2002.)

C_nG_m surfactants does not dehydrate. Thus the change of the curvature must be induced by the hydrophobic part, which increases in size with increasing temperature due to the increasing number of chain conformations and the increasing penetration of oil molecules into the surfactant layer. As the penetration is directly correlated to the solubility of the surfactant in the respective oil phase the following picture emerges: the better the solubility of a C_nG_m surfactant in an oil phase (i.e. the more polar the oil), the larger the oil penetration and thus the larger the ability of the system to change the curvature. At high enough oil solubility, a phase inversion in ternary water–oil–C_nG_m systems can indeed be induced by changing the temperature as is seen in Figure 12.6. However, the large majority of studies have been carried out with unpolar oils (either *n*-alkanes or cyclohexane) so that in the following only quaternary systems containing water, an unpolar oil, a sugar surfactant (pure or technical), and a cosurfactant are discussed.

12.2.2.2 Quaternary Systems

In Figures 12.7 through 12.9 various phase diagrams of the quaternary systems water–*n*-alkane–β-C_nG_m–alcohol are shown to illustrate the influence of the surfactant (Figure 12.7), of the alcohol (Figure 12.8), and of the solvent (Figures 12.7 through 12.9), respectively. The coordinates of the $\tilde{X}$-points, i.e. $\tilde{\gamma}$ and $\tilde{\delta}$, for all systems are listed in Table 12.1.

12.2.2.2.1 Influence of the Surfactant
One of the central questions of microemulsion formulation has always been the quest for high efficiency, i.e. finding microemulsion systems in which the amount

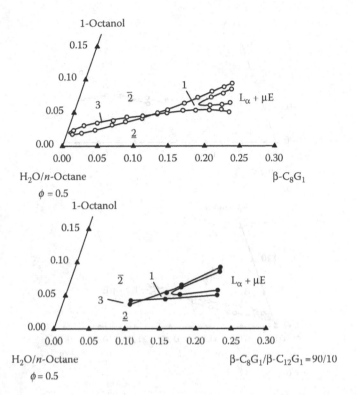

FIGURE 12.7 Sections through the phase tetrahedron of the systems H_2O–n-octane–β-C_nG_m–1-octanol at a constant oil/(water + oil) ratio of $\phi = 0.5$ and $T = 25°C$ with β-$C_nG_m =$ β-C_8G_1 (top) and β-C_8G_1/β-$C_{12}G_1 = 90/10$ (bottom), respectively. Increasing the (average) hydrophobic chain length of the surfactant leads to a decrease of the stability range of the bicontinuous one-phase microemulsion due to the increased extension of the lamellar phase (L_α). (Top figure redrawn with data from Kluge, K., PhD Thesis, University of Cologne, 2000 (ISBN 3-89722-457-7).)

of surfactant necessary to solubilize oil in water or vice versa is minimized. One way to increase the efficiency is to lengthen the hydrophobic alkyl chain of the surfactant. However, for low molecular weight surfactants a gain in efficiency is accompanied by the formation of a lamellar (L_α) phase [1,48]. As can be seen in Figure 12.7, the same general trend is observed for sugar surfactants. The upper figure shows the phase diagram of the system H_2O–n-octane–β-C_8G_1–1-octanol and it is clearly seen that the extension of the one-phase bicontinuous microemulsion (1) is limited by the adjacent two-phase region (L_α + μE), in which a microemulsion coexists with an L_α phase. Replacing 10 wt% of β-C_8G_1 by β-$C_{12}G_1$ already leads to a significant change, although the average chain length of the surfactant has been increased only by 0.4 carbon atoms (the mixture corresponds to an average molecular structure of β-$C_{8.4}G_1$). While such a small increase of the hydrophobic chain increases the efficiency only slightly (as is seen in Table 12.1 a small

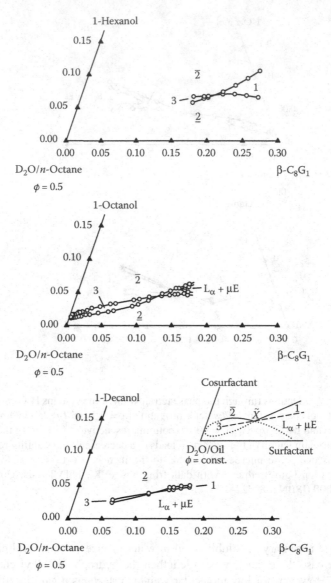

FIGURE 12.8 Sections through the phase tetrahedron of the systems D$_2$O–n-octane–β-C$_8$G$_1$–1-hexanol (top), 1-octanol (middle), 1-decanol (bottom) at a constant oil/(water + oil) ratio of $\phi = 0.5$ and $T = 25°C$. Increasing the hydrophobic chain length of the cosurfactant from 1-hexanol to 1-octanol shifts the $\tilde{X}$-Point to lower surfactant concentrations, i.e. the efficiency increases, and narrows the extension of the one-phase region. In the case of 1-decanol the L$_\alpha$ phase dominates the lower part of the phase diagram which makes the determination of the $\tilde{X}$-point impossible (see inlet). Inlet (bottom): Schematic drawing of the effect an L$_\alpha$ phase has on the phase diagram. (Middle figure redrawn with data from Kluge, K., Sottmann, T., Stubenrauch, C., and Strey, R., *Tenside Surf. Det.*, 38, 30, 2001.)

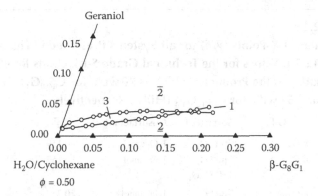

FIGURE 12.9 Section through the phase tetrahedron of the system H_2O–cyclohexane–β-C_8G_1–geraniol at a constant oil/(water + oil) ratio of $\phi = 0.5$ and $T = 25°C$. (Figure redrawn with data from Stubenrauch, C., Paeplow, B., and Findenegg, G.H., *Langmuir*, 13, 3652, 1997).

decrease in both $\tilde{\gamma}$ and $\tilde{\delta}$ is observed), the extension of the ($L_\alpha + \mu E$)-phase region increases remarkably at the expense of the one-phase microemulsion. Thus increasing the average chain length even further does not lead to an increased efficiency but to the formation of a broad L_α phase which totally suppresses the one-phase microemulsion at this oil to water plus oil ratio. For example, replacing β-C_8G_1 by β-$C_{10}G_1$ results in a phase diagram in which the $\tilde{X}$-point cannot be determined at $\phi = 0.5$ due to the broad L_α phase (data not shown). However, this only holds true for pure surfactants. It will be shown in Section 12.2.3 that a one-phase microemulsion can be observed at $\phi = 0.5$ if long chain technical surfactants are used.

12.2.2.2.2 Influence of the Alcohol

Looking at Figure 12.8 and Table 12.1 one sees that an exchange of 1-hexanol by 1-octanol leads to a drastic increase of the efficiency, i.e. to a much lower $\tilde{\gamma}$-value. The low efficiency of the system containing 1-hexanol is due to the high water solubility of 1-hexanol, which is 6.0 g in 1 L at 25°C compared to 0.54 g in 1 L for 1-octanol [49]. In other words, a lot of 1-hexanol is needed to saturate the aqueous phase. The exchange of 1-octanol by 1-decanol changes the phase behavior even more significantly. As it is shown in the inlet of Figure 12.8, the L_α phase dominates the lower part of the phase diagram. Thus the phase sequence $L_\alpha + \mu E$, $1, \bar{2}$ is found with increasing amount of 1-decanol and a determination of the $\tilde{X}$-point becomes impossible. Discussing the influence of an alcohol on the phase behavior in general, one has to consider its twofold role: the alcohol partitions between the bulk phases and the surfactant film. Thus, on the one hand, the alcohol acts as a co-solvent making the oil phase more hydrophilic and the aqueous phase more hydrophobic, respectively. On the other hand, the alcohol mixes into the surfactant film making it increasingly hydrophobic. Although the spontaneous curvature

TABLE 12.1

Coordinates of $\tilde{X}$-Points ($\tilde{\gamma},\tilde{\delta}$) for all Systems Presented in Figures 12.7 through 12.13. Values for the Technical Grade Surfactants Refer to the Active Matter of the Products, Which is 70 wt% for $C_{8/10}G_{1.7}$ (APG225, Henkel) and 51 wt% for $C_{10/12}G_{1.4}$ (Hüls), Respectively

Water	Oil	Surfactant	Cosurfactant	$\tilde{\gamma}$	$\tilde{\delta}$
H_2O	n-Octane	β-C_8G_1	1-Octanol	0.159	0.296
H_2O	n-Octane	β-C_8G_1 / β-$C_{12}G_1$	1-Octanol	0.157	0.274
D_2O	n-Octane	β-C_8G_1	1-Hexanol	0.258	0.267
D_2O	n-Octane	β-C_8G_1	1-Octanol	0.160	0.275
H_2O	n-Octane	β-C_8G_1	1-Hexanol	~0.26	>0.27[a]
H_2O	Cyclohexane	β-C_8G_1	1-Hexanol	~0.20	<0.18[b]
H_2O	Cyclohexane	β-C_8G_1	Geraniol	0.20	0.18[c]
H_2O	Cyclohexane	$C_{8/10}G_{1.7}$	Geraniol	0.11	0.51
H_2O	Cyclohexane	$C_{10/12}G_{1.4}$	Geraniol	0.045	0.28
H_2O	Cyclohexane	$C_{8/10}G_{1.7}$	1-Hexanol	0.12	0.36
H_2O	Cyclohexane	$C_{8/10}G_{1.7}$	Geraniol	0.11	0.51
H_2O	Cyclohexane	$C_{8/10}G_{1.7}$	Geraniol/Nerol	0.11	0.53
H_2O	Cyclohexane	$C_{8/10}G_{1.7}$	Nerol	0.15	0.59
H_2O	n-Decane	$C_{8/10}G_{1.7}$	1-Hexanol	0.19	0.39
H_2O	n-Decane	$C_{8/10}G_{1.7}$	Geraniol	0.16	0.49[d]
H_2O	n-Decane	$C_{8/10}G_{1.7}$	1-Hexanol	0.19	0.39
H_2O	Cyclohexane	$C_{8/10}G_{1.7}$	1-Hexanol	0.12	0.36

Note: Some of the data are shown twice for the sake of clarity.

[a] Although this phase diagram has not been measured, we know from the system D_2O–n-octane–β-C_8G_1–1-octanol that an exchange of D_2O by H_2O does not effect the $\tilde{\gamma}$-value and increases only slightly the $\tilde{\delta}$-value. Replacing D_2O in the system D_2O–n-octane–β-C_8G_1–1-hexanol by H_2O leads to the listed values.

[b] Although this phase diagram has not been measured, we know from the system H_2O–cyclohexane–$C_{8/10}G_{1.7}$–geraniol that an exchange of geraniol by 1-hexanol changes only slightly the $\tilde{\gamma}$-value, while it reduces significantly the $\tilde{\delta}$-value. Replacing geraniol in the system H_2O–cyclohexane–β-C_8G_1–geraniol by 1-hexanol leads to the listed values.

[c] In Ref. [17] a $\tilde{\delta}$-value of 0.22 is mentioned, which is a misprint.

[d] Phase diagram is not shown; data are extracted from Ref. [11].

of the surfactant film is lowered by both effects, the latter is predominant, since the OH-group of the alcohol is small compared to the head group of the "main" surfactant. By increasing the concentration of the alcohol, one enriches the film in alcohol and thus decreases its curvature, which, in turn, induces a phase inversion from o/w- to o/w-droplet microemulsions. Thus it is the volume fraction of the alcohol in the surfactant film $\delta_{V,i}$

$$\delta_{V,i} = \frac{V_{D,i}}{V_{C,i} + V_{D,i}} \tag{12.5}$$

rather than the overall mass fractions δ (Equation 12.4) that tune the curvature in a quaternary system. The volume of "interfacial" surfactant and cosurfactant

$$\gamma_{V,i} = \frac{V_{C,i} + V_{D,i}}{V_A + V_B + V_C + V_D} \tag{12.6}$$

determines the amphiphilicity of the surfactant/cosurfactant mixture. Regarding the influence of an alcohol on the phase behavior in general, and on the $\tilde{X}$-point in particular, it is known that both $\tilde{\gamma}_{V,i}$ and $\tilde{\delta}_{V,i}$ decrease with increasing chain length of the alcohol. Unfortunately, it is very time-consuming to determine these values so that for most of the studied systems this information is not available. Furthermore, increasing either the chain length of the surfactant or cosurfactant leads to the formation of an extended lamellar phase which can be regarded as a consequence of the increasing amphiphilicity of the surfactant/ cosurfactant mixture. This lamellar phase can become so dominant that the determination of the $\tilde{X}$-point is not possible. For the three systems shown in Figure 12.8 it is only for the system with 1-octanol that $\gamma_{V,i}$- and $\delta_{V,i}$-values have been determined (see Section 12.2.4).

12.2.2.2.3 Influence of the Solvent

Comparing Figures 12.7 (top) and 12.8 (middle), one sees the effect of replacing H_2O by D_2O. While the coordinates of the $\tilde{X}$-point are not affected significantly (Table 12.1), the extension of the L_α phase is much broader in the D_2O system, which, in turn, leads to a very small region of the one-phase microemulsion. The same observation, namely an increased stability of the L_α phase in the presence of D_2O, has also been made for binary water–C_iE_j systems and has been explained qualitatively with a higher hydrophobicity of the surfactant in D_2O compared to H_2O [50]. Although the H_2O–D_2O exchange is very interesting from a scientific point of view, it is the influence of the oil phase on the phase behavior that has to be answered as far as applications are concerned. The general trend is known from extensive studies of ternary water–oil–C_iE_j systems and can be summarized as follows: the less hydrophobic the oil is, the smaller the extension of the three-phase region, the $\tilde{X}$-point of which shifts to both lower $\tilde{T}$- and $\tilde{\gamma}$-values [51,52]. A ranking of some oils in terms of hydrophobicity leads to: n-dodecane > n-octane > cyclohexane > toluene. Unfortunately, a similar systematic study has not been carried out with pure sugar surfactants. However, three studies with technical grade sugar surfactants are available in which the hydrophobicity of the oil was changed [11,14,15] (see Section 12.2.3). Thus based on the knowledge gained for the ternary systems and the technical grade sugar surfactants we can deduce the following general trends. In Figure 12.9 the phase diagram of the system H_2O–cyclohexane–β-C_8G_1–geraniol* is shown. Note that compared to the system shown in Figure 12.7 (top) not only the oil but also the alcohol is different. However, we will see in Section 12.2.3 that replacing geraniol by 1-hexanol leads to similar $\tilde{\gamma}$- and lower $\tilde{\delta}$ -values. Thus if we replaced geraniol by 1-hexanol in the present system, we would expect to obtain a $\tilde{\gamma}$-value of ~0.2 and a $\tilde{\delta}$ -value < 0.18. With this assumption we can now compare the systems H_2O–cyclohexane–β-C_8G_1–1-hexanol and H_2O–n-octane–β-C_8G_1–1-hexanol. Looking at Table 12.1, one clearly sees that the latter is less efficient, as is expected for a more hydrophobic oil. In conclusion, one can certainly say that the hydrophobicity of the oil influences quaternary water–oil–C_nG_m–alcohol systems in the same way as ternary water–oil–C_iE_j systems: the less hydrophobic the oil is, the smaller the extension of the three-phase region, the $\tilde{X}$-point of which shifts to both lower $\tilde{\gamma}$ - and $\tilde{\delta}$ -values. We will see in the following section that this is indeed the case for technical surfactants.

12.2.3 TECHNICAL SUGAR SURFACTANTS

In applications, technical grade sugar surfactants have been used for the formulation of microemulsions mainly for economic reasons. These surfactants are complex mixtures of C_nG_m homologues consisting of species with different degrees of glycosidation, chain length, and stereoisomerism. Typical chain lengths are n = 8/10, 10/12, and 12/14, while the head group sizes vary from m = 1.3–1.7. Very often technical sugar surfactants are supplied as aqueous solutions, i.e. that only parts of the product are "active matter," which has to be taken into account if one compares the efficiencies of different products. In order to classify technical sugar surfactants regarding their efficiency, Kahlweit et al. studied both pure and technical surfactants. Using the short chain cosurfactant 1-pentanol they found that the efficiencies of pure β-$C_{10}G_1$ (Sigma) and pure β-$C_{12}G_1$ (Sigma) are about the same as those of the inexpensive technical grade products $C_{10/12}G_{1.3}$ (Hüls, 50 wt% active matter) and $C_{12/14}G_{1.3}$ (Hüls, 50 wt% active matter), respectively [15]. Thus the slightly higher carbon number of the technical product is compensated by its slightly higher degree of glycosidation. It is important to mention that in [15] the symbols γ and δ have a different meaning than in the present study and in all other studies. Denoting the symbols used in [15] with γ^* and w_D (weight fraction of the cosurfactant), respectively, it holds

$$\gamma^* = \frac{m_C}{m_A + m_B + m_C} = \frac{\gamma(1-\delta)}{1-\delta\gamma} \tag{12.7}$$

---

* Geraniol is a monoterpene (trans-3,7-dimethyl-2,6-octadien-1-ol) the structure of which is shown in Figure 12.12.

and

$$w_{\text{D}} = \frac{m_{\text{D}}}{m_{\text{A}} + m_{\text{B}} + m_{\text{C}} + m_{\text{D}}} = \delta\gamma \qquad (12.8)$$

Thus we recalculated all values given in Ref. [15] in order to compare these results with those obtained by others [6,11,14]. In the following, we present phase diagrams that were measured by Stubenrauch et al. [11 and unpublished results] and compare them with literature values focussing on the efficiency of the systems, i.e. on the coordinates of the $\widetilde{X}$-points ($\widetilde{\gamma},\widetilde{\delta}$). In Figures 12.10, 12.11, and 12.13 various phase diagrams of the systems water–oil–technical C_nG_m–alcohol are shown to illustrate the influence of the surfactant (Figure 12.10), of the alcohol (Figure 12.11), and of the oil (Figures 12.13), respectively. The technical product

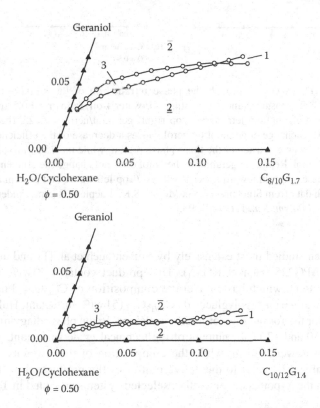

FIGURE 12.10 Sections through the phase tetrahedron of the system H₂O–cyclohexane–technical alkyl polyglucoside (APG)–geraniol at a constant oil/(water + oil) ratio of $\phi = 0.5$ and $T = 25°C$. APG225 = $C_{8/10}G_{1.7}$ (top), $C_{10/12}G_{1.4}$ (bottom). Increasing the hydrophobic chain length leads to a significant increase of the efficiency. Note that for technical APGs a shift of the phase diagram to lower amounts of the cosurfactant with decreasing surfactant concentration is observed. (Top figure redrawn with data from Stubenrauch, C., Mehta, S.K., Paeplow, B., and Findenegg, G.H., *Progr. Colloid Polym. Sci.*, 111, 92, 1998.)

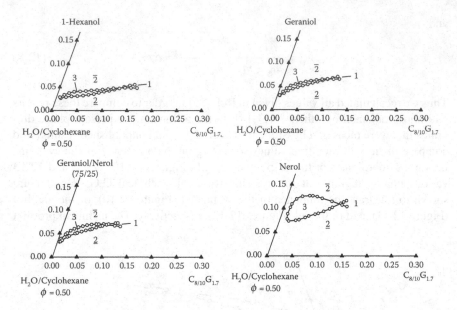

FIGURE 12.11 Sections through the phase tetrahedron of the systems H_2O–cyclohexane–APG 225–cosurfactant at a constant oil/(water + oil) ratio of $\phi = 0.5$ and $T = 25°C$ (top left) geraniol, (bottom left) nerol, (top right) geraniol/nerol = 75/25, (bottom right) 1-hexanol. An exchange of geraniol by nerol causes a decrease of the efficiency which is accompanied by an extension of the three phase region, while the 75/25 mixture behaves like pure geraniol. Replacing geraniol by hexanol does not change the efficiency but only shifts the phase diagram toward lower δ-values. (Top left figure and bottom right figure redrawn with data from Stubenrauch, C., Mehta, S.K., Paeplow, B., and Findenegg, G.H., *Progr. Colloid Polym. Sci.*, 111, 92, 1998.)

that has been studied most extensively by Stubenrauch et al. [11 and unpublished results] is APG225 (Henkel KGaA). This product contains 70 wt% surfactant (=active matter), which has an average composition of $C_{8/10}G_{1.7}$. Furthermore, the technical grade alkyl polyglucoside $C_{10/12}G_{1.4}$ (51 wt% surfactant, Hüls AG) has been used for the formulation of microemulsions. In the phase diagrams shown in Figures 12.10 and 12.11 the amount of the technical grade surfactant is the total mass that was weighted in, while the composition of the $\widetilde{X}$-points which are listed in Table 12.1 refer to the active matter of the products. For the sake of comparison the $\widetilde{X}$-points of some other selected systems are listed in Table 12.2.

12.2.3.1 Influence of the Surfactant

Not surprisingly, the same general trend as for the pure surfactants is observed, namely an increase of the efficiency with increasing hydrophobic chain. As is seen in Figure 12.10, replacing $C_{8/10}G_{1.7}$ by $C_{10/12}G_{1.4}$ leads to a significant shift of the phase diagram toward lower surfactant and cosurfactant (= geraniol[5]) concentrations.

TABLE 12.2
Selection of $\tilde{X}$-Points Coordinates $(\tilde{\gamma}, \tilde{\delta})$ for Microemulsions Stabilized by Technical Grade Sugar Surfactants

Water	Oil	Surfactant	Cosurfactant	$\tilde{\gamma}$	$\tilde{\delta}$	Reference
H_2O	n-Octane	$C_{8/10}G_{1.3}$ (50 wt%, DCI[a])	1-Butanol	0.32	0.45	[14]
H_2O	n-Octane	$C_{12/14}G_{1.3}$ (50 wt%, Hüls)	1-Butanol	0.20	0.32	[15]
H_2O	n-Octane	$C_{12/14}G_{1.4}$ (50 wt%, DCI[a])	1-Butanol	0.22	0.34	[14]
H_2O	n-Decane	$C_{8/10}G_{1.7}$ (70 wt%, Henkel)	1-Hexanol	0.19	0.39	[11]
H_2O	n-Decane	$C_{12/14}G_{1.3}$ (50 wt%, Hüls)	1-Hexanol	0.033	0.49	[15]
H_2O	n-Decane	$C_{12/14}G_{1.3}$ (50 wt%, Hüls)	1-Butanol	0.27	0.36	[15]
H_2O	n-Decane	$C_{12/14}G_{1.3}$ (50 wt%, Hüls)	1-Pentanol	0.14	0.29	[15]
H_2O	n-Decane	$C_{12/14}G_{1.3}$ (50 wt%, Hüls)	1-Hexanol	0.033	0.49	[15]
H_2O	Cyclohexane	$C_{10/12}G_{1.4}$ (70 wt%, Henkel)	Geraniol	0.045	0.28	[11]
H_2O	Cyclohexane	$C_{12}G_{1.8}$ (100 wt%, Kao)	2-Ethyl hexyl glycerol ether	0.04	0.27	[6]
H_2O	n-Dodecane	$C_{12/14}G_{1.3}$ (50 wt%, Hüls)	1-Pentanol	0.20	0.29	[15]
H_2O	n-Decane	$C_{12/14}G_{1.3}$ (50 wt%, Hüls)	1-Pentanol	0.14	0.29	[15]
H_2O	n-Octane	$C_{12/14}G_{1.3}$ (50 wt%, Hüls)	1-Pentanol	0.09	0.28	[15]

Source: Fukuda, K., Söderman, O., Lindman, B., and Shinoda, K., *Langmuir*, 9, 2921, 1993; Stubenrauch, C., Mehta, S.K., Paeplow, B., and Findenegg, G.H., *Progr. Colloid Polym. Sci.*, 111, 92, 1998; Chai, J., Li, G., Zhang, G., Lu, J., and Wang, Z., *Colloids Surfaces A:Physicochem. Eng. Aspects*, 231, 173, 2003; Kahlweit, M., Busse, G., and Faulhaber, B., *Langmuir*, 11, 3382, 1995.

Note: Values for the technical grade surfactants refer to the active matter, which is indicated for each product. As in Ref. [15] a different nomenclature is used, the respective $\tilde{\gamma}$- and $\tilde{\delta}$-values were recalculated to compare them with the results of other authors (see text for further details). Note that some of the data are shown twice and that some data of Table 12.1 are included for the sake of clarity.

[a] Daily Chemical Industry, China.

What is particularly remarkable is the decrease of the $\tilde{\delta}$-value (Table 12.1), which reflects the high hydrophilicity of $C_{8/10}G_{1.7}$. For $C_{8/10}G_{1.7}$ an enormous amount of alcohol is needed to compensate for the large head group and thus to induce a phase inversion. These results show that two effects always have to be considered if one discusses the influence of the surfactant structure on the location of the $\tilde{X}$-point, namely an increase in efficiency (decrease of $\tilde{\gamma}$-value) with increasing hydrophobic chain and a decrease of the alcohol content (decrease of $\tilde{\delta}$-value) with decreasing head group. These two effects are also clearly seen if one compares the pure system H_2O–cyclohexane–β-C_8G_1–geraniol with H_2O–cyclohexane–$C_{8/10}G_{1.7}$–geraniol (see Table 12.1). The advantage of using technical grade sugar surfactants is not only their lower cost but also the fact that efficient microemulsions can be formulated in most cases without facing the problem of a broadly extended L_α-phase. Comparing Figures 12.7 and 12.10 one clearly sees the difference: although the technical surfactant is much more efficient an L_α phase was not observed in the studied concentration range. This observation is not surprising either as surfactant mixtures usually suppress the formation of highly ordered structures because it is difficult to form densely packed mono- and bilayers with surfactants of different sizes.

12.2.3.2 Influence of the Alcohol

In Figure 12.11 the influence of the alcohol on the phase behavior of the quaternary systems water–cyclohexane–$C_{8/10}G_{1.7}$–alcohol is seen. Three different alcohols and one alcohol mixture were used, namely 1-hexanol, geraniol (*trans*-3,7-dimethyl-2,6-octadien-1-ol), nerol (*cis*-3,7-dimethyl-2,6-octadien-1-ol), and a 75:25 mixture of geraniol:nerol. The structures of the two monoterpenes are shown in Figure 12.12. Trying to describe the trends seen in Figure 12.11, one faces the same problem as discussed in relation to Figure 12.8, namely that the $\tilde{\gamma}_{V,i}$- and $\tilde{\delta}_{V,i}$-values are needed to be able to quantitatively compare the different systems. As these values are not available we can only give qualitative explanations for the different $\tilde{X}$-points (Table 12.1) as was done for the phase diagrams shown in Figure 12.8.

Comparing 1-hexanol with geraniol one expects the latter system to be more efficient as geraniol has a longer and more voluminous hydrophobic chain. However, this is not reflected in the $\tilde{\gamma}$-values, which are nearly equal for the two systems. One reason for the larger efficiency at the same $\tilde{\gamma}$-value can be a larger total monomeric solubility of geraniol. Recalling that 1-hexanol has a high water

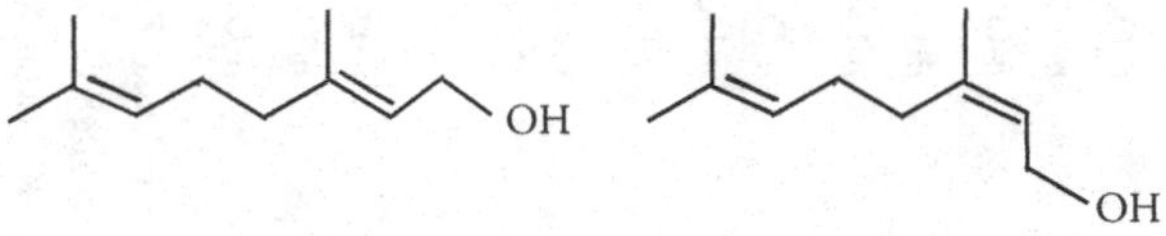

FIGURE 12.12 Molecular structures of two monoterpenes, namely geraniol = *trans*-3,7-dimethyl-2,6-octadien-1-ol (left) and nerol = *cis*-3,7-dimethyl-2,6-octadien-1-ol (right).

solubility and an oil solubility that can by no means be neglected, the oil solubility of geraniol must be high, which is also reflected in the huge $\tilde{\delta}$-value.* Speculative as it may be, subtracting the monomeric solubilities of both alcohol and surfactant from the total amount of alcohol and surfactant, respectively, should result in the following trends for the interfacial parameters; $\tilde{\gamma}_{v,i}$ (1-hexanol) > $\tilde{\gamma}_{v,i}$ (geraniol) and $\tilde{\delta}_{v,i}$ (1-hexanol) > $\tilde{\delta}_{v,i}$ (geraniol).

An exchange of geraniol by its *cis*-isomer nerol illustrates the influence of the molecular conformation. A comparison of the two-phase diagrams leads to two striking observations. Firstly, both the $\tilde{\gamma}$-value and the $\tilde{\delta}$-value increase slightly, which indicates a lower efficiency of the $C_{8/10}G_{1.7}$ + nerol mixture. This decrease in efficiency can be explained by the molecular structure: the *cis*-isomer has a shorter effective chain length due to the *cis*-substituted double bond. On the other hand the hydrophobic part of the *cis*-isomer has a larger volume which should have the opposite effect on the average curvature of the surfactant layer. Secondly, the three-phase region of the nerol containing system has a much broader extension compared to the geraniol containing system, which is another indication of a less efficient system: the broader the three-phase region, the less efficient the surfactant + cosurfactant mixture. Last but not least, the 75:25 mixture of geraniol:nerol behaves as expected if one compares this phase diagram with those of the single alcohols. The extension of the three-phase region is a little bit broader than that of the geraniol system, while an influence on the location of the $\tilde{X}$-point is not visible. Again slight differences of the water and oil solubilities of the two isomers could be an explanation for canceling out the effects different alcohols have on the location of the $\tilde{X}$-point. In conclusion, one can say that it is not necessarily the total number of C-atoms in the hydrophobic part of the alcohol that matters but the effective chain length. Comparing the two isomers geraniol and nerol, one finds that the latter has a shorter effective chain length and thus the respective surfactant + *cis*-isomer mixture is less efficient than the corresponding surfactant + *trans*-isomer mixture. A more detailed analysis of the role the hydrophobic volume of the alcohol plays would only be possible if the $\tilde{\gamma}_{v,i}$- and $\tilde{\delta}_{v,i}$-values were known.

A clear correlation between the chain length of the alcohol and the efficiency of the surfactant + alcohol mixture was observed for two technical surfactants, namely $C_{12/14}G_{1.3}$ from Hüls (Table 12.2) and $C_{8/10}G_{1.3}$ from DCI [14]. The values for microemulsions containing $C_{12/14}G_{1.3}$ clearly show that the efficiency increases with increasing chain length of the alcohol. Again the $\tilde{\delta}$-values do not follow a clear trend due to the different solubilities of the alcohols in water and oil, respectively. The same trend has been reported for the $\tilde{\gamma}$-value of $C_{8/10}G_{1.3}$ + alcohol mixtures, while the corresponding $\tilde{\delta}$-values do not change with increasing chain length as was the case for the pure β-C_8G_1 (see Table 12.1 and Figure 12.8). We conclude this section with a last example which illustrates the complexity but

---

* The same general trend is observed if *n*-decane instead of cyclohexane is used as oil phase (see Table 12.1).

also the beauty of tuning the efficiency of microemulsions by changing the alcohol and/or the surfactant. As can be seen in Table 12.2, the mixture $C_{10/12}G_{1.4}$ + geraniol has the same efficiency as $C_{12}G_{1.8}$ + 2-ethyl hexyl glycerol ether. In other words, the possibility of playing around with three components (oil, surfactant, cosurfactant) opens up a wide variety of combinations that can be used to optimize a system. It is the knowledge about the influence of each of these components that allows us nowadays to formulate systems with desired properties.

12.2.3.3 Influence of the Oil

In Figure 12.13 the influence of the oil on the phase diagrams is shown. Note that such a direct comparison was not possible for the pure surfactants as the respective phase diagrams are not available. Thus we extrapolated the behavior from known

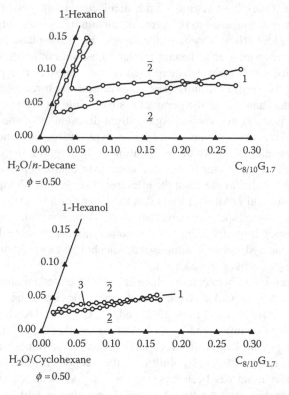

FIGURE 12.13 Section through the phase tetrahedron of the systems H_2O–oil–APG 225–1-hexanol at a constant oil/(water + oil) ratio of $\phi = 0.5$ and $T = 25°C$. (top) n-decane, (bottom) cyclohexane. Replacing n-decane by the more hydrophilic cyclohexane leads to an increased efficiency and a reduction of the three-phase region. In the former case an extension of the three phase region to very high cosurfactant concentrations δ at low surfactant mass fractions γ was observed. (Redrawn with data from Stubenrauch, C., Mehta, S.K., Paeplow, B., and Findenegg, G.H., *Progr. Colloid Polym. Sci.*, 111, 92, 1998.)

phase diagrams and concluded that the pure sugar surfactants behave exactly like C_iE_j surfactants regarding the influence of the oil, i.e. that the efficiency of the surfactant + cosurfactant mixture increases with decreasing hydrophobicity of the oil. Figure 12.13 is simply to be seen as the experimental proof of this "extrapolation." Replacing n-decane by cyclohexane leads to a much lower $\tilde{\gamma}$-value and to a much smaller three-phase region. The $\tilde{\delta}$-value, however, is not affected very much, which again can be explained with the different monomeric solubilities of 1-hexanol in the two oils. As 1-hexanol is much better soluble in cyclohexane, the $\tilde{\delta}$-value decreases only slightly. Note that for the pure C_nG_m we compared n-octane with cyclohexane. As the difference in hydrophobicities is smaller than that between n-decane and cyclohexane, the decrease of the $\tilde{\delta}$-value is more pronounced. A more detailed study was carried out with $C_{12/14}G_{1.3}$ from Hüls (Table 12.2) and $C_{12/14}G_{1.4}$ from DCI [14] where the chain length of the alkane was varied systematically. In both studies the behavior known from C_iE_j surfactants was found, namely a significant increase of the efficiency with decreasing oil chain length. While in the former system the $\tilde{\delta}$-values do not change with decreasing alkane chain length, they decrease in the latter.

12.2.4 GENERAL PATTERN

In the foregoing sections we discussed the twofold role of the alcohol, namely as co-solvent and cosurfactant. The situation becomes even more complex if one considers that the main surfactant also partitions, though to a smaller extent, between the interface and the bulk phases. We concluded that it is the composition of the interface rather than the total composition which is the relevant tuning parameter if more than one surface-active species is present in the system. However, it is not easy to determine the interfacial composition and to our knowledge there are only two studies with microemulsions stabilized by sugar surfactants in which the interfacial composition was indeed quantified [17,25]. Comparing these results with those obtained for the quaternary system H_2O–n-octane–C_8E_5–1-octanol [32] one sees that the general pattern, which will be explained in the following, is the same [27].

The high solubility of hydrophobic surface active compounds in the oil phase results in a distortion of the three-phase body, as was discussed qualitatively in relation to Figure 12.5. Moreover it was shown that the composition of the interfacial film remains constant along the trajectory of balanced microemulsions (see hatched line in Figure 12.5) across the entire three-phase region [17]. These results imply that the distortion of the three-phase body in the δ, γ-plane is caused by the different monomeric solubilities of the surfactant and the alcohol in water and oil. In other words, if water and oil are presaturated with surfactant and cosurfactant one obtains an undistorted three-phase body which is symmetrical with respect to the balanced interfacial composition. This has indeed been observed for H_2O–n-octane–C_8E_5–1-octanol [32] and H_2O–cyclohexane–β-C_8G_1–geraniol [17]. As the different solubilities of surfactant and alcohol in water and oil have been taken into account by presaturating the solvents, are the alcohol can now be considered solely as cosurfactant which is incorporated into the interfacial film and thus changes its spontaneous curvature. Increasing proportions of alcohol in the interfacial film alter the curvature of the surfactant film from positive to negative and thus induce a phase transition from $\underline{2}$ (o/w-droplet microemulsion) to $\overline{2}$ (w/o-droplet microemulsion). A slightly different approach was used to get the phase diagrams shown in Figure 12.14. In this case no presaturated solvents were used but the monomeric solubilities of surfactant and cosurfactant were determined independently via density measurements [25,27]. These solubilities were than subtracted from the total amount of surfactant and cosurfactant which leads to the undistorted phase diagram seen in Figure 12.14.

In conclusion, one can say that the knowledge of the monomeric solubilities allows us to plot $\delta_{V,i}$ versus the total surfactant concentration γ_V (Figure 12.14).* In contrast to the $\delta_V(\gamma_V)$-plot which shows a phase diagram with a distorted three phase region (closed symbols), the $\delta_{V,i}(\gamma_V)$-plot results in a symmetric horizontal phase diagram (open symbols). As the latter diagram is similar to a

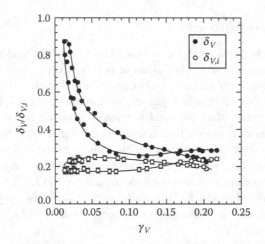

FIGURE 12.14 Rectangular $\delta_V(\gamma_V)$- and $\delta_{V,i}(\gamma_V)$-sections through the phase tetrahedron of the system H_2O–n-octane–β-C_8G_1–1-octanol at a constant oil/(water + oil) ratio of $\phi = 0.5$ and $T = 25°C$. The filled symbols represent the phase diagram as a function of the overall volume fractions δ_V and γ_V. Because of the high monomeric solubility of the cosurfactant in the oil the phase diagram is distorted toward high δ_V-values at low γ_V-values. However, when the volume fraction of the cosurfactant in the internal surface $\delta_{V,i}$ is plotted versus γ_V (white symbols) an undistorted phase diagram is observed (see text for details). (Redrawn with data from Kluge, K., Sottmann, T., Stubenrauch, C., and Strey, R., *Tenside Surf. Det.*, 38, 30, 2001.)

---

* Some authors use mass fractions (δ, γ), while others prefer volume fractions (δ_V, γ_V). The difference is indicated by the subscript "V".

$T(\gamma)$-diagram of ternary water–oil–C_iE_j–surfactant systems the composition of the interface $\delta_{V,i}$ and the temperature T are apparently equivalent parameters with respect to the phase behavior. Experimentally, a horizontal fish can be obtained by presaturating the solvents with surfactant and alcohol according to their monomeric solubilities.

12.3 MICROSTRUCTURE

12.3.1 BACKGROUND

The most striking feature of microemulsions is their complex microstructure. Although macroscopically homogeneous, they are heterogeneous on a nanoscale. Surfactant molecules form an interfacial film, the nature and properties of which are essential for the formation of microemulsions. 50 years ago, Winsor [53] and Schulman [54] suggested that microemulsions consist of spherical droplets (w/o or o/w) with one exception. They believed that the middle phase of a three-phase system (microemulsion plus excess water and oil phases) has a layered, lamellar structure. In 1970 Shinoda [55] and in 1976 Scriven [56] suggested a bicontinuous structure of the surfactant-rich middle phase, which was proven 12 years later with the help of NMR self-diffusion measurements [57,58] and the direct visualization by FFEM [57,59]. Further studies of the microstructure by NMR self-diffusion, TEM, and scattering techniques (SAXS and SANS) led to the result that not only droplets and bicontinuous structures are formed but also wormlike structures as well as sample spanning networks. It has been realized that the main parameter determining the microstructure is the mean curvature H of the interfacial film. Thus, controlling the curvature is the ultimate goal for choosing any desired structure.

In Section 12.2 we discussed the influence of the solvent, the surfactant, and the alcohol on the phase behavior. As was mentioned in the introduction, phase behavior and microstructure have to be studied separately in the first place. However, due to the enormous knowledge gained over the last 20 years we are now able to directly correlate microstructure and phase behavior. Having said this we can conclude that all parameters influencing the phase behavior automatically influence the microstructure, i.e. the curvature H of the interfacial film. Thus there are countless tuning parameters for the curvature. The easiest example of a tuning parameter is the temperature T. In ternary water–oil–C_iE_j systems the microstructure can just be controlled via the temperature as is illustrated schematically in Figure 12.15 (left). An increase in temperature decreases the curvature of the surfactant film, which is mainly caused by the dehydration of the ethylene oxide units. The higher the temperature the smaller the average area per head group and thus the more negative the curvature (curved toward water). Expressed in terms of microstructures one can say that an increase of temperature transforms an o/w ($H > 0$) into a w/o droplet ($H < 0$) microemulsion via a bicontinuous structure of zero mean curvature ($H = 0$). FFEM images are provided in Figure 12.15 (middle) to visualize the structural change [60]. However, as already

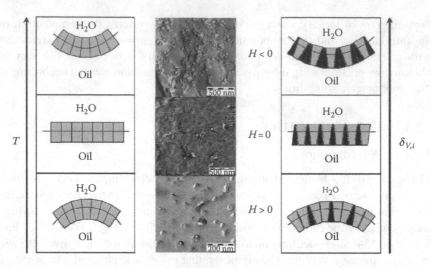

FIGURE 12.15 Schematic drawing of the mean curvature H of the interfacial film. In a ternary system H_2O–oil–nonionic surfactant the mean curvature mainly depends on the temperature T, while in a quaternary system H_2O–oil–surfactant–cosurfactant (at T = constant) it is the composition of the interfacial film $\delta_{V,i}$ that controls the curvature. Thus a phase inversion can be induced either by changing T or $\delta_{V,i}$. Freeze Fracture Electron microscopy images of the structures occurring at different values of H are shown in the middle (see also Figure 12.1).

mentioned, the head group of C_nG_m surfactants does not dehydrate with increasing temperature so that it is difficult to tune ternary water–n-alkane–C_nG_m systems through the phase inversion by temperature variation (with the exception of systems with hydrophilic oils as was discussed in Section 12.2.2). The method of choice is to mix the hydrophilic sugar surfactant with a hydrophobic surfactant or just a long-chain alcohol, a so-called cosurfactant. One of the most important results is that in the resulting quaternary systems the composition of the mixed interfacial film $\delta_{V,i}$ takes over the role of the temperature in ternary water–n-alkane–C_iE_j systems, which is schematically illustrated in Figure 12.15 (right). The curvature and thus the microstructure of the interfacial film changes continuously with increasing $\delta_{V,i}$ as it does with increasing T. This can be understood if one considers that the head group area of the alcohol is smaller than that of the sugar surfactant. Thus the increasing alcohol fraction causes a decrease of the average head group area (as was the case for the dehydration of the C_iE_j surfactants), which, in turn, leads to a decrease of the mean curvature.

To sum up, one can say that the general behavior of ternary and quaternary microemulsions is equal if the appropriate tuning parameters are used. In the case of ternary temperature-sensitive systems the temperature dominates the properties, whereas the composition of the interfacial film has to be discussed in the case of temperature-insensitive quaternary systems, or, in more general terms, in the case of quaternary systems under isothermal conditions. A scaling description has been derived for ternary water–oil–C_iE_j systems [61–63], which was later extended to quaternary water–oil–C_nG_m–alcohol systems [25,27]. It has been found that the same scaling description can be used for both types of microemulsions if the tuning parameter of the curvature is known and transferred to the corresponding reduced tuning parameter. In the following, we will discuss the type and size of microstructures observed for microemulsions stabilized by sugar surfactants. Although with FFEM both type and size of the structure could in principal be determined (see Figure 12.15) complementary techniques are required due to the following reasons. Firstly, the access to high resolution TEMs is limited and even having access to TEM facilities, preparing the samples and interpreting the respective images require a lot of expertise. Artifacts due to wrong sample preparation [60] and/or misinterpretation of images (see e.g. introduction of [64]) are the main problems. Furthermore, more quantitative informations about frequently occurring distances, i.e. the size of the structures, can be obtained from scattering techniques. Thus complementing TEM images with results from NMR and scattering techniques has been proven to be the best combination for determining the structures of microemulsions [1,4].

12.3.2 Type and Size of Structure

12.3.2.1 Type of Structure

The main question in relation to the type of structure is whether the structure is discrete or bicontinuous. The most commonly used technique to answer this question is NMR diffusometry via which the self-diffusion coefficients of the various components of the system can be measured. Knowing the self-diffusion coefficients of the two solvents, i.e. of water and oil, one can easily discriminate between droplet and bicontinuous microemulsions [6,18,28,57,58,65]. Having determined the self-diffusion coefficients of water and oil in a microemulsion the connectivity of the solvents and therefore the shape of the microstructure can be deduced by plotting the relative self-diffusion coefficients D/D_0 (D_0 is the self-diffusion coefficient of the pure solvent) versus the tuning parameter of the system, which is usually the temperature T or the composition if the interfacial film $\delta_{V,i}$. For the three limiting cases of o/w-droplets, w/o-droplets, and bicontinuous structures one obtains the following relations:

- o/w-droplets: D/D_0 (H_2O) $\gg D/D_0$ (oil)
- w/o-droplets: D/D_0 (H_2O) $\ll D/D_0$ (oil)
- bicontinuous: D/D_0 (H_2O) $\sim D/D_0$ (oil)

The first two relations hold as the droplets are smaller than the distance a molecule diffuses in an NMR experiment. Therefore the experiment is not sensitive to the molecular displacement within the droplets, but only to the translation of the entire droplet. To our knowledge four studies exist in which the microstructure of a quaternary microemulsion stabilized by a sugar surfactant has been investigated

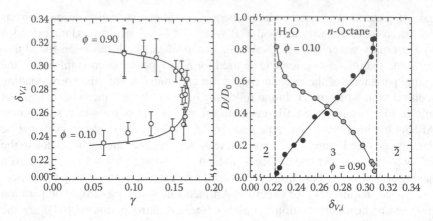

FIGURE 12.16 $\widetilde{X}$-points and relative self-diffusion coefficients of the system H_2O–n-octane–β-C_8G_1–1-octanol. (left) $\widetilde{X}$-points at oil/(water + oil) volume fractions from $\phi = 0.1$ to $\phi = 0.9$ and $T = 25°C$. Given are the values of γ and $\delta_{V,i}$ at the respective $\widetilde{X}$-point. (right) Relative self-diffusion coefficients of water (H_2O/D_2O = 10/90) and n-octane at different oil/(water + oil) volume fractions ϕ as a function of the composition of the interfacial layer $\delta_{V,i}$. Note that around $\phi = 0.5$ the microemulsion is bicontinuous and thus the relative self-diffusion coefficients of water and oil are nearly equal. At very low and very high ϕ's the relative diffusion coefficient of one component is small, which points to a rather discrete structure of the microemulsion. (Figures are redrawn with data from Reimer, J., Söderman, O., Sottmann, T., Kluge, K., and Strey, R., *Langmuir*, 19, 10692, 2003.)

via NMR diffusometry [6,18,28,66]. The aim of all studies was to monitor the conversion of o/w-droplets into w/o-droplets via a bicontinuous microemulsion. Although different sections through the phase prism were chosen by the authors the results are qualitatively the same. An example is discussed in the following.

In Figure 12.16 (left), the γ- and $\delta_{V,i}$-values at the $\widetilde{X}$-points of the system H_2O–n-octane–β-C_8G_1–1-octanol are shown for various oil/(water + oil) volume fractions ϕ. Connecting the $\widetilde{X}$-points leads to the so-called trajectory of the middle phase (see Ref. [28] for details). In Figure 12.16 (right) the relative self-diffusion coefficients D/D_0 of water and n-octane* are plotted versus the relevant tuning parameter, namely $\delta_{V,i}$. Note that low $\delta_{V,i}$-values correspond to low ϕ-values and high $\delta_{V,i}$-values to high ϕ-values, respectively. Figure 12.16 clearly shows the variation of the relative self-diffusion coefficients for both water and n-octane. At low $\delta_{V,i}$-values the water self-diffusion is rapid, while that of oil is slow. At high $\delta_{V,i}$-values the opposite behavior is observed. Finally, at intermediate $\delta_{V,i}$-values the relative self-diffusion coefficients are similar and even equal at $\delta_{V,i,m} = 0.27$. At the limits of the three-phase region (dashed lines) the relative self-diffusion coefficient of the discrete phase (droplets) asymptotically

---

* The self-diffusion coefficients of the pure solvents at $T = 25°C$ are $D_0(H_2O) = 1.9 \times 10^{-9}$ m² s⁻¹ and $D_0(n\text{-octane}) = 2.5 \times 10^{-9}$ m² s⁻¹.

approaches 0, whereas that of the continuous phase approaches 1. To sum up, one can say that the data given in Figure 12.16 (right) imply a change in the structure from o/w-droplets over a bicontinuous structure to w/o-droplets with increasing $\delta_{V,i}$. This observation can be rationalized by a change of the mean curvature H with increasing fraction of 1-octanol in the interfacial film (see discussion in relation to Figure 12.15). The balanced state of the microemulsion is defined as the point where the relative self-diffusion coefficients of water and oil are equal. Thus at $\delta_{V,i,m} = 0.27$ the microemulsion is truly bicontinuous, i.e. the interfacial film between the water and the oil domains is sample-spanning and has a mean curvature of $H = 0$.

12.3.2.2   Size of Structure

For a complete characterization of the microstructure we need both the type and size of the structure. To determine the latter usually small angle neutron scattering (SANS) experiments are carried out (reviewed in Ref. [1]). A detailed analysis of SANS curves leads not only to the characteristic length scales ξ of the structure but also to the mean curvature H of the interfacial film and the average area per surfactant molecule a_c, respectively. In the following the variation of ξ as a function of $\delta_{V,i}$ will be described, while the change of H with $\delta_{V,i}$ will be addressed in Section 12.3.3. For the three limiting cases of o/w-droplets, w/o-droplets, and bicontinuous structures the characteristic length scale ξ corresponds to:

* o/w-droplets: ξ = radius r of oil droplets
* w/o-droplets: ξ = radius r of water droplets
* bicontinuous: ξ = "mean diameter" of oil and water domains, respectively.

In the latter case people also use the term "periodicity of the structure," which is $d = 2\xi$ [67]. To our knowledge only three studies exist in which the microstructure of a quaternary microemulsion stabilized by a sugar surfactant has been investigated via SANS [25,28,66]. All three studies deal with the system H_2O–n-octane–β-C_8G_1–1-octanol in which the protonated solvents have been exchanged by deuterated solvents in order to measure in bulk and film contrast, respectively. The most important results of these studies will be discussed in the following.

In order to quantify the variation of the microstructure in the system H_2O–n-octane–β-C_8G_1–1-octanol SANS-measurements were carried out as a function of the composition of the interfacial film $\delta_{V,i}$. Samples were prepared in the droplet region near the emulsification failure boundary at the oil- and water-rich side of the system in film contrast (both water and oil are deuterated). Additionally, at $\phi = 0.50$ a sample in the one-phase region next to the $\tilde{X}$-point was prepared in bulk contrast (deuterated water and protonated oil). The experimentally observed scattering intensity of a droplet microemulsion is given by

$$I(q) = N \ P(q) \ S(q)$$

(12.9)

where

    N is the number density of the aggregates
    $P(q)$ is the particle form factor
    $S(q)$ is the interparticle form factor

Since all samples studied had a fairly low volume fraction of dispersed aggregates, the value of $S(q)$ has only a minor influence on the outcome of the fitting procedure and therefore $S(q)$ was set to unity. The $P(q)$ used to fit the data was that of polydisperse spherical shells with diffuse boundaries [68]. The inlets on the left (o/w-droplets) and the right (w/o-droplets) hand side of Figure 12.17 demonstrate

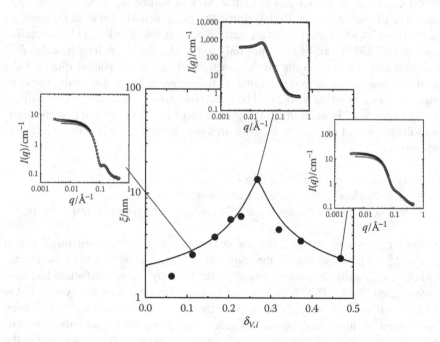

FIGURE 12.17 SANS curves and characteristic length scales ξ of the system H_2O–n-octane–β-C_8G_1–1-octanol along the trajectory of the middle phase as a function of the interfacial composition $\delta_{V,i}$ at $T = 25°C$. Three scattering curves and the corresponding fits (straight lines) are shown. (Left) o/w-droplet microemulsion at $\delta_{V,i} = 0.11$ (= low oil content, $\phi = 0.014$); (middle) bicontinuous microemulsion at $\delta_{V,i} = 0.27$ (= equal volumes of water and oil, $\phi = 0.5$) which is reflected in a characteristic correlation peak; (right) w/o-droplet microemulsion at $\delta_{V,i} = 0.47$ (= high oil content, $\phi = 0.979$). The scattering curves of the droplet structures were measured in film contrast and fitted with Equation 12.9. The scattering curve for the bicontinuous structure was measured in bulk contrast and fitted with the Fourier transformation of Equation 12.11. The solid line in the $\xi(\delta_{V,i})$-plot is calculated according to Equation 12.11. (From Kluge, K., Sottmann, T., Stubenrauch, C., and Strey, R., *Tenside Surf. Det.*, 38, 30, 2001.)

that the used particle form factor of polydisperse shells with diffuse boundaries describes the scattering curves very well. However, the deviations in the low q-range could be an indication of slightly deformed, i.e. elongated droplets. Furthermore, the polydispersity of the aggregates (reflected by an unclear fine structure at large q) on the oil-rich side is larger than on the water-rich side, where the first minimum at large q is still clearly visible. In bicontinuous microemulsions, the characteristic length scale of the structure can be determined from the scattering peak at low q-values. Following Teubner and Strey [69] the peak can be described by the Fourier transformation of the correlation function.

$$\Gamma(r) = \frac{d}{2\pi r} e^{-r/\xi_{TS}} \sin\left[\frac{2\pi r}{d_{TS}}\right] \tag{12.10}$$

where
ξ_{TS} is the correlation length
d_{TS} is the periodicity of the structure

It can be seen in the inlet in the middle of Figure 12.17 that the scattering peak is well described by this model.

Having analyzed the SANS data in the droplet and the bicontinuous regime, one can plot the characteristic length scale ξ versus the composition of the interface $\delta_{V,i}$. As is seen in Figure 12.17, with increasing $\delta_{V,i}$ the characteristic length scale ξ runs through a maximum at $\xi = 13.6$ nm in the middle of the three-phase region. A similar behavior is found for C_iE_j surfactants with increasing temperature [1,70]. It is not only the shape of the curves that is similar but also their empirical description. For C_iE_j an expression for the temperature dependence of ξ has been developed empirically (reviewed in Ref. [1]), namely

$$\xi = \frac{3}{\frac{1}{2}\left|c_1 + c_2\right| + \sqrt{\frac{1}{2}\left(c_1^2 + c_2^2\right)} + \sqrt[4]{\frac{1}{2}\left(c_1^4 + c_2^4\right)}} \tag{12.11}$$

where c_1 and c_2 are the two principal curvatures of the structure ($c_1 = c_2$ for droplets; $c_1 = -c_2$ for a bicontinuous structure). Taking the expressions for the principal curvatures and replacing the temperatures by the corresponding $\delta_{V,i}$ values one obtains

$$c_1 = c\frac{\delta_{V,i,u} - \delta_{V,i}}{1 + c\left|v_C/a_C\left(\delta_{V,i,u} - \delta_{V,i}\right)\right|}$$

and

$$c_2 = c\frac{\delta_{V,i,l} - \delta_{V,i}}{1 + c\left|v_C/a_C\left(\delta_{V,i,l} - \delta_{V,i}\right)\right|} \tag{12.12}$$

where
c is the composition dependent coefficient of the mean curvature H
$\delta_{V,i,u}$ and $\delta_{V,i,l}$ are the upper and lower limits of the three-phase region

v_C is the average volume of the surfactant molecules in the interfacial film
a_C is the average head group area of the surfactant molecules in the interfacial film

Describing the experimentally observed dependence of ξ on $\delta_{V,i}$ with Equation 12.11 one obtains a very good agreement as is seen in Figure 12.17. In other words, the variation of the length scales with the composition of the interfacial film is described quantitatively both in the droplet and the bicontinuous regimes by an empirical expression, which has been derived from the dependence of the principal curvatures on $\delta_{V,i}$. Thus the composition of the interface $\delta_{V,i}$ in quaternary systems and the temperature T in ternary systems are equivalent parameters not only with respect to the phase behavior but also with respect to the variation of the shape and size of the microstructure.

12.3.3 GENERAL PATTERN

The extensive studies of the microstructure of the two model systems H_2O–n-octane–$C_{12}E_5$ and H_2O–n-octane–β-C_8G_1–1-octanol, respectively, allow us to draw some general conclusions. The general pattern becomes obvious if one compares Figures 12.2 and 12.18. In the former the self-diffusion coefficients D of the solvents, the characteristic length scales ξ, and the mean curvatures H are plotted versus the temperature T, while in the latter case these parameters are plotted versus the composition of the interfacial layer $\delta_{V,i}$. For the sake of completeness we will briefly explain how the mean curvature H of the amphiphilic film can be obtained (see Ref. [1] for more details). Knowing the length scale ξ and the shape of the microstructure, the variation of the mean curvature H can be evaluated quantitatively. While for spherical droplets ($r_1 = r_2 = r$) the mean curvature H is given by $H = 1/r = 1/\xi$, we know from TEM that in the middle of the three-phase body the structure of the amphiphilic film is saddle-shaped, i.e. $H = 0$. Note that by definition the curvatures are positive if the interfacial film is curved toward oil and negative if it is curved toward water. Thus the data points shown in Figures 12.2 and 12.18 have been calculated from the respective ξ-values. The temperature dependence of H (solid line in Figure 12.2) is given by

$$H = c\frac{T_m - T}{1 + c\left|v_C/a_C\left(T_m - T\right)\right|} \tag{12.13}$$

Accordingly we can write for the dependence of H on the interfacial composition (solid line in Figure 12.18)

$$H = c\frac{\delta_{V,i,m} - \delta_{V,i}}{1 + c\left|v_C/a_C\left(\delta_{V,i,m} - \delta_{V,i}\right)\right|} \tag{12.14}$$

Looking at Figures 12.2 and 12.18 one sees that H of both systems decreases steadily with increasing T and $\delta_{V,i}$, respectively, and changes sign at T_m and $\delta_{V,i,m}$. In addition, same general trends are observed for the variation of the length scales

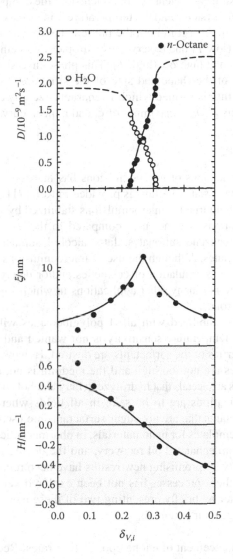

FIGURE 12.18 Self-diffusion coefficients D of water ($H_2O/D_2O = 10/90$) and oil (top), characteristic domain sizes ξ (middle), and mean curvatures H of the interfacial film (bottom) in the one phase region of the system H_2O–n-octane–β-C_8G_1–1-octanol as a function of composition of the mixed interface $\delta_{V,i}$. At $\delta_{V,i} = 0.27$ the microemulsion is bicontinuous, which is illustrated by equal D-values of water and oil, a maximum of ξ, and $H = 0$. Values for D have been determined via NMR diffusometry, while ξ and H have been determined via SANS measurements. The solid lines in the $D(\delta_{V,i})$-plot only guide the eyes, while the solid lines in the $\xi(\delta_{V,i})$- and $H(\delta_{V,i})$-plot are calculated according to Equations 12.11 and 12.14, respectively.

ξ and the self-diffusion coefficients D. In conclusion, the composition of the interface $\delta_{V,i}$ in quaternary systems and the temperature T in ternary systems are equivalent parameters with respect to both the phase behavior and the microstructure. Starting at low $\delta_{V,i}$ (low T) one observes o/w droplet microemulsions, while w/o droplets are observed at high $\delta_{V,i}$ (high T). This phase inversion takes place via a continuous change of the shape and size of the structure. At $\delta_{V,i,m}$ (T_m) a truly bicontinuous structure is formed which is characterized by equal relative self-diffusion coefficients D/D_0, a maximum of ξ, and a mean curvature H of zero.

12.4   OUTLOOK

The number of applications of microemulsions has increased in the past and an overview of the current state of affairs is provided in Ref. [71]. However, this general trend does not hold true for microemulsions stabilized by sugar surfactants as their use in applications is rather new compared to the use of the well-known "workhorses" alkylbenzene sulfonates, fatty alcohol sulfates, alkyl polyglycol ethers, and ether sulfates. Although the use of microemulsions stabilized by alkyl polyglucosides for the formulation of cosmetics and/or detergents is well-established [29], there are not many other applications in which sugar surfactants currently play a major role.

Microemulsions stabilized with alkyl polyglucosides will be attractive for applications where temperature sensitivity is not wanted and the favorable eco-toxicological properties of the surfactants are desired. However, they can only be used if temperatures are not too high and the medium is not too acidic because alkyl polyglucosides are acetals that hydrolyze to fatty alcohol and glucose. Potential "large-scale" applications are to be seen in all fields where microemulsions stabilized by surfactants that are not sugar surfactants are used nowadays, e.g. as reaction medium, templates for nanomaterials, in pharmaceutical applications, for soil decontamination, enhanced oil recovery, and the degreasing of leather [71]. On a "small scale," very promising new results have been reported but the potential for scaling up these processes has not been explored yet. We would like to conclude this chapter by briefly presenting two of these new results and leave it to the reader to judge their potential.

- Dramatic enhancement of enone epoxidation rates: Recently epoxidation reactions of the α,β-unsaturated enone *trans*-chalcone in different nonionic microemulsions have been studied, in the presence as well as in the absence of a phase-transfer agent (PTA) [72]. The obtained reaction profiles were compared with those for the corresponding surfactant-free two-phase systems for which a time constant of $\tau = 114\,min$ was found in the presence of PAT. The epoxidation of *trans*-chalcone in microemulsions stabilized by C_iE_j surfactants was much faster, namely $\tau = 66\,min$ with PAT and $\tau = 77\,min$ without PAT. Using n-octyl β-D-glucopyranoside (β-C_8G_1) as surfactant one obtains a conversion twice as fast as that in the C_iE_j microemulsion, namely $\tau = 35\,min$

without PAT. However, the presence of PTA did not accelerate the reaction further, as is reflected in a T-value of 33 min. The epoxidation of vitamin K_3, the second model system, was even more accelerated. The reaction was, by a factor of approximately 35, faster in the microemulsion ($\tau = 1.44$ min) than in the corresponding two-phase system ($\tau = 57$ min).

- Polymerization in microemulsion glasses: Bicontinuous sugar-based microemulsions containing the liquid monomer (DVB) were used for the synthesis of poly-DVB with a bicontinuous structure [73]. The base microemulsion consists of water–DVB–technical grade C_8G_1 and $C_{12}G_1$ (APG 225 and APG 600 from Henkel). The aqueous phase was replaced by sugar and the microemulsion was then dehydrated to the solid glass state without phase separation. Photopolymerization of the liquid DVB within the solid microemulsion glass template also proceeded without phase separation. Comparing SANS spectra of the microemulsion glasses before and after polymerization one sees that there is hardly any change in the microstructure. Following the polymerization, the sugar template was dissolved and poly-DVB membranes with ~25 nm pores were obtained.

ACKNOWLEDGMENTS

We thank Michael Klostermann for the hours he spent drawing the figures and Prof. Reinhard Strey for his continuous support both scientifically and personally.

REFERENCES

1. Sottmann, T. and Strey, R., in *Soft Colloids V–Fundamentals in Interface and Colloid Science*, Lyklema, J. (Ed.), Elsevier, Amsterdam, 2005, Chapter 5.
2. Kahlweit, M. and Strey, R., *Angew. Chem. Int. Ed. Engl.*, 1985, *24*, 654.
3. Burauer, S., Sachert, T., Sottmann, T., and Strey, R., *Phys. Chem. Chem. Phys.*, 1999, *1*, 4299.
4. Kahlweit, M., Strey, R., Haase, D., Kunieda, H., Schmeling, T., Faulhaber, B., Borkovec, M., Eicke, H.-F., Busse, G., Eggers, F., Funck, T., Richman, H., Magid, L., Södermann, O., Stilbs, P., Winkler, J., Dittrich, A., and Jahn, W., *J. Colloid Interf. Sci.*, 1987, *118*, 436.
5. Shinoda, K., Yamanaka, T., and Kinoshita, K., *J. Phys. Chem.*, 1959, 63, 648.
6. Fukuda, K., Söderman, O., Lindman, B., and Shinoda, K., *Langmuir*, 1993, *9*, 2921.
7. Kahlweit, M., Busse, G., and Faulhaber, B., *Langmuir*, 1996, *12*, 861.
8. Kahlweit, M., Busse, G., and Faulhaber, B., *Langmuir*, 1997, *13*, 5249.
9. von Rybinski, W. and Hill, K., *Angew. Chem. Int. Ed.*, 1998, *37*, 1328.
10. von Rybinski, W., Guckenbiehl, B., and Tesmann, H., *Colloids Surfaces A: Physicochem. Eng. Aspects*, 1998, *142*, 333.
11. Stubenrauch, C., Mehta, S.K., Paeplow, B., and Findenegg, G.H., *Progr. Colloid Polym. Sci.*, 1998, *111*, 92.
12. Comelles, F., *J. Dispersion Sci. Technol.*, 1999, *20*, 491.

13. Mönch, G. and Ilgenfritz, G., *Colloid Polym. Sci.*, 2000, *278*, 687.
14. Chai, J., Li, G., Zhang, G., Lu, J., and Wang, Z., *Colloids Surfaces A:Physicochem. Eng. Aspects*, 2003, *231*, 173.
15. Kahlweit, M., Busse, G., and Faulhaber, B., *Langmuir*, 1995, *11*, 3382.
16. Stubenrauch, C., Kutschmann, E.-M., Paeplow, B., and Findenegg, G.H., *Tenside Surf. Det.*, 1996, *33*, 237.
17. Stubenrauch, C., Paeplow, B., and Findenegg, G.H., *Langmuir*, 1997, *13*, 3652.
18. Stubenrauch, C. and Findenegg, G.H., *Langmuir*, 1998, *14*, 6005.
19. Ryan, L.D., Schubert, K.-V., and Kaler, E.W., *Langmuir*, 1997, *13*, 1510.
20. Ryan, L.D. and Kaler, E.W., *Langmuir*, 1997, *13*, 5222.
21. Ryan, L.D. and Kaler, E.W., *J. Phys. Chem. B*, 1998, *102*, 7549.
22. Ryan, L.D. and Kaler, E.W., *J. Colloid Interface Sci.*, 1999, *210*, 251.
23. Ryan, L.D. and Kaler, E.W., *Langmuir*, 1999, *15*, 92.
24. Häntzschel, D., Enders, S., Kahl, H., and Quitzsch, K., *PCCP*, 1999, *1*, 5703.
25. Kluge, K., Sottmann, T., Stubenrauch, C., and Strey, R., *Tenside Surf. Det.*, 2001, *38*, 30.
26. Egger, H., Sottmann, T., Strey, R., Valero, C., and Berkessel, A., *Tenside Surf. Det.*, 2002, *39*, 17.
27. Sottmann, T., Kluge, K., Strey, R., Reimer, J., and Söderman, O., *Langmuir*, 2002, *18*, 3058.
28. Reimer, J., Söderman, O., Sottmann, T., Kluge, K., and Strey, R., *Langmuir*, 2003, *19*, 10692.
29. *Alkyl polyglycosides*, Hill, K., von Rybinski, W., and Stoll, G. (Eds.), VCH, Weinheim, 1997.
30. von Rybinski, W., Hloucha, M., and Johansson, I., in *Microemulsions: Background, New Concepts, Applications, Perspectives*, Stubenrauch, C. (Ed.), Wiley-Blackwell, Oxford, to be published 2008, Chapter 8.
31. Patravale, V.B. and Date, A., in *Microemulsions: Background, New Concepts, Applications, Perspectives*, Stubenrauch, C. (Ed.), Wiley-Blackwell, Oxford, to be published 2008, Chapter 9.
32. Penders, M.H.G.M. and Strey, R., *J. Phys. Chem.*, 1995, *99*, 10313.
33. Sottmann, T. and Stubenrauch, C., in *Microemulsions: Background, New Concepts, Applications, Perspectives*, Stubenrauch, C. (Ed.), Wiley-Blackwell, Oxford, to be published 2008, Chapter 1.
34. Söderman, O. and Johansson, I., *Curr. Opin. Colloid Interface Sci.*, 2000, *4*, 391.
35. *Nonionic surfactants: Alkyl polyglucosides*, Balzer, D., and Luders, H. (Eds.), Marcel Dekker, New York, 2000.
36. Stubenrauch, C., *Current Opinion in Colloid & Interface Sci.*, 2001, *6*, 160.
37. Bastogne, F. and David, C., *Colloids Surfaces A: Physicochem. Eng. Aspects*, 1998, *139*, 311.
38. Bastogne, F., Nagy, B.J., and David, C., *Colloids Surfaces A: Physicochem. Eng. Aspects*, 1999, *148*, 245.
39. Bastogne, F. and David, C., *J. Photochem. Photobiol. A: Chem.*, 2000, *136*, 93.
40. Bolzinger-Thevenin, M.A., Grossiord, J.L., and Poelman, M.C., *Langmuir*, 1999, *15*, 2307.
41. Sassen, C.L., Cassielles, A.G., De Loos, T.W., and De Swaan-Arons, J., *Fluid Phase Equilibria*, 1992, *72*, 173.
42. Schneider, G.M., *Pure Appl. Chem.*, 1983, *55*, 479.
43. Strey, R., *J. Phys. Chem.*, 1991, *96*, 4537.

44. Kunieda, H. and Sato, Y., in *Organized Solutions*, Friberg, S. and Lindman, B., (Eds.), Marcel Dekker, New York, 1992, p. 67.
45. Yamaguchi, S. and Kunieda, H., *Langmuir*, 1997, *13*, 6995.
46. Kunieda, H. and Shinoda, K., *J. Colloid Interface Sci.*, 1985, *107*, 107.
47. Kunieda, H. and Ishikawa, N., *J. Colloid Interface Sci.*, 1985, *107*, 122.
48. Kahlweit, M., Strey, R., and Firman, P., *J. Phys. Chem.*, 1986, *90*, 671.
49. *Handbook of Chemistry and Physics*, 82nd Edition, Lide, D.R., (Ed.), CRC Press, Boca Raton, 2001.
50. Stubenrauch, C., Burauer, S., Strey, R., and Schmidt, C., *Liquid Crystals*, 2004, *31*, 39.
51. Kahlweit, M., Lessner, E., and Strey, R., *J. Phys. Chem.*, 1983, *87*, 5032.
52. Kahlweit, M., Strey, R., Firman, P., Haase, D., Jen, J., and Schomäcker, R., *Langmuir*, 1988, *4*, 499.
53. Winsor, P.A., *Solvent Properties of Amphiphilic Compounds*, Butterworth & Co Ltd., London, 1954.
54. Bowcott, J.E. and Schulman, J.H., *Z. Elektrochem.*, 1955, *59*, 283.
55. Saito, H. and Shinoda, K., *J. Colloid Interface Sci.*, 1970, *32*, 647.
56. Scriven, L.E., *Nature*, 1976, *263*, 123.
57. Bodet, J.F., Bellare, J.R., Davis, H.T., Scriven, L.E., and Miller, W.G., *J. Phys. Chem.*, 1988, *92*, 1898.
58. Lindman, B., Shinoda, K., Olsson, U., Anderson, D., Karlström, G., and Wennerström, H., *Colloids Surfaces*, 1989, *38*, 205.
59. Jahn, W. and Strey, R., *J. Phys. Chem.*, 1988, *92*, 2294.
60. Belkoura, L., Stubenrauch, C., and Strey, R., *Langmuir*, 2004, *20*, 4391.
61. Leitao, H., Somoza, A.M., Telo da Gama, M.M., Sottmann, T., and Strey, R., *J. Chem. Phys.*, 1996, *105*, 2875.
62. Sottmann, T. and Strey, R., *J. Phys. Condens. Matter*, 1996, *8*, A39.
63. Sottmann, T. and Strey, R., *J. Chem. Phys.*, 1997, *106*, 8606.
64. Burauer, S., Belkoura, L., Stubenrauch, C., and Strey, R., *Colloids Surfaces A: Physicochem. Eng. Aspects*, 2003, *228*, 159.
65. Lindman, B. and Olsson, U., *Ber. Bunsenges. Phys. Chem.*, 1996, *100*, 344.
66. Kluge, K., Der schlüssel zum verständnis von mikroemulsionen aus zuckertensiden: Die interne grenzfläche, PhD Thesis, University of Cologne, 2000 (ISBN 3-89722-457-7).
67. Sottmann, T., Strey, R., and Chen, S.H., *J. Chem. Phys.*, 1997, *106*, 6483.
68. Gradzielski, M., Langevin, D., Magid, L., and Strey, R., *J. Phys. Chem.*, 1995, *99*, 13232.
69. Teubner, M. and Strey, R., *J. Chem. Phys.*, 1987, *87*, 3195.
70. Strey, R., *Colloid Polym. Sci.*, 1994, *272*, 1005.
71. *Microemulsions: Background, New Concepts, Applications, Perspectives*, C. Stubenrauch (Ed.), Wiley-Blackwell, Oxford, to be published in 2008.
72. Wielpütz, T., Sottmann, T., Strey, R., Schmidt, F., and Berkessel, A., *Chem.- Eur. J.*, 2006, *12*, 7565.
73. Gao, F., Ho, C.C., and Co, C.C., *Macromolecules*, 2006, *39*, 9467.

13 Self-Organization of the Ternary Dialkyldimethylammonium Bromide/ OBG/Water Systems: Mixed Nanoaggregates, Vesicles, and Micelles

Elena Junquera and Emilio Aicart

CONTENTS

13.1   INTRODUCTION

It is well known that mixed vesicles or liposomes are formed in water solution, when at least one of the surfactants of a mixture has a double hydrophobic chain length [1,2]. Additionally, the presence of a second surfactant with a single chain drives, within the appropriate range of concentration and composition, to the formation of a wide variety of structures (mixed vesicles and micelles) [1–4]. These mixed colloidal systems are of great interest because they often improve the solubilization of hydrophobic moieties, and in many times show a high synergism, which reports an extraordinary change in the properties with respect to the pure surfactant systems [3,4]. Nowadays, the study of mixed colloidal systems represents an important field, from a technological point of view due to their applications as biological delivery systems (drugs, enzymes, DNA, etc.), in food industry, cosmetics, oil recovery, detergency, etc. [1–5]. In addition, these systems constitute simplified model systems to mimic biological membranes, the single chain surfactant acting as a purifying solubilizer agent [3–5].

The objective of this chapter is to report an experimental and theoretical physicochemical characterization of this kind of mixed supramolecular aggregates in water solution, focused in those systems composed by the dialkyl dimethylammonium bromides (di-C_nDMAB), with $n = 10$, 12, and a nonionic sugar surfactant, the n-octyl-β-D-glucopyranoside, (OBG), of great interest in biochemistry. Special attention is paid to the transitions between the supramolecular assemblies formed at "highly diluted," "very diluted," and "moderately diluted" concentration domains within the whole range of composition (mole fraction). The complexity of these mixed systems with surfactants containing one and two hydrophobic chains, and the wide variety of aggregates formed as a function of the composition and/or concentration, strongly suggest the necessity of carrying out studies by using different experimental methods, as well as theoretical models.

13.2   GENERAL OVERVIEW

Surfactants are amphiphilic molecules with a hydrophylic polar head and a hydrophobic region, generally one or two hydrocarbon chains. Surfactants can be nonionic (if the polar head has not charge) or ionic (cationic or anionic) when it is charged. Due to this particular structure, surfactants associate in water solution at a certain concentration, named critical aggregation concentration, CAC, thus forming supramolecular aggregates [6–10].

When surfactants have a single chain, the aggregates formed at the critical micelle concentration, CMC, are named micelles. Micellar solutions are characterized, in addition to the CMC, by the mean aggregation number of monomers, N, and, if the surfactant is ionic, by the dissociation degree of micelles, β [6–10]. It is assumed that at concentrations higher than CMC, the number of aggregates increases, but the size and shape of the initial spherical micelles remains roughly constant, at least up to the formation of new micelle structures, i.e., a transition from spherical to globular micelles [6–8,10]. If the surfactant is ionic, counterions are partially associated to the micelle surface compensating in part to the electrostatic repulsion between the ionic polar heads of the surfactants [6,8,10]. When two or more surfactants with a single chain are in solution, mixed micelles form at concentrations over the mixed critical micelle concentration, CMC* [4]. Interaction between similar but not equal hydrocarbon chains is possible because the hydrophobic effect, responsible of the surfactant association, is not restricted to identical hydrocarbon chains. In many mixed surfactant systems a synergistic effect, normally attributed to nonideal mixing effects of the aggregates, appears when the CMC* of the mixture is lower than any of the CMC of the pure surfactants [4]. This aspect is of relevant interest in the industry, because it permits to obtain mixed aggregates with lower surfactant concentrations and in many cases with more enhanced properties than the pure ones. Mixed micelle systems are characterized not only by the total parameters of the solution as CMC*, N^*, and β, but also through the partial parameters, such as the partial composition of both the monomeric, CMC_i^*, and micellar phases, X_i^{agg}, as well as the partial aggregation numbers, N_i^* [4,11].

Vesicles and liposomes are colloidal closed nanostructures constituted by one or multiple bilayers of 4–6 nm thick, where at least one of the surfactants of the mixture has a double chain [1,2,12]. In aqueous solution, the polar head points to the bulk while the hydrocarbon chains form the hydrophobic core of the membrane. Due to this particular morphology, these aggregates can encapsulate polar molecules inside the polar cavity and, furthermore, they can also solubilize hydrophobic moieties into the bilayer, offering a variety of practical applications [1–3,8,12]. Vesicles are characterized through the critical vesicle concentration, CVC, its lamellarity, size polydispersity, and micropolarity of the different vesicle microdomains. Depending on the number of bilayers, vesicles can be: unilamellar, with an unique bilayer, and multilamellar, when they contain two or more bilayers [3]. The size of the vesicles can vary from 20–100 nm of diameter for the small unilamellar vesicles (SUV), to 100–500 nm for the large unilamellar vesicles (LUV) [1,2,9,12]. Depending on its composition, vesicles are divided in two groups: (1) liposomes, composed by one or several natural phospholipidic surfactants, together with other components as cholesterol, proteins, etc., and (2) nonphospholipidic vesicles. Liposomes are metastable aggregation states, with high kinetic stability but thermodynamically nonstable [13,14]. Its preparation needs high energy and its size, morphology, and stability depends very much on the method used [12,13,15–17]. Nonphospholipid dialkyl chain surfactants form, at CVC, spontaneously and stable vesicles, the size and polydispersity being stable for longer periods of time [15,17–20] Vesicles are usually prepared dispersing an appropriate amount of surfactant in water, the formation of uniform and monodisperse unilamellar vesicles being favored by sonication [2,9,18,19]. They can be also prepared by methods similar to liposomes, but in this case their properties are independent on the method of choice [13]. Mixed vesicles may form in aqueous solution not only in systems with a double chain surfactant but also in systems with a double chain surfactant mixed with a single chain surfactant

[2,3,14,19,21–27], and also in mixtures of cationic and anionic surfactants, named catanionic vesicles [13,28–30]. Vesicle systems have increased their interest, not only because they have a higher thermodynamic stability than liposomes but also because their size, charge, and permeability can be conveniently fit changing concentration or chain length of the surfactants. The thermodynamic stability of the pure and mixed spontaneous vesicles has been explained by a model proposed by Safran et al., which is based on the elastic curvature theory [31–34]. According with this model, vesicles are stable if the spontaneous curvature of inner and outer monolayers of the bilayer is equal in magnitude but opposite in sign. In mixed vesicles this feature is easy to understand if one assumes an asymmetric composition of the two monolayers. In the case of pure vesicles, asymmetry of the monolayers only can be possible if the number of surfactant molecules is higher in the outer monolayer than that in the inner one.

Mixed vesicles are widely used as model systems of biological membranes for studying its purification, solubilization, and reconstitution [22,26,35–41]. These studies are based on the capability of single chain surfactants to solubilize poor soluble aqueous materials by forming mixed micelles [2]. Thus, the addition of a single chain surfactant to a vesicle aqueous solution breaks the vesicles by solubilizing the double chain surfactant within the mixed micelles [2,12]. This solubilization process has been described by Lichtenberg et al., using a three-stage model [42,43]. This model defines a stage I, where the single chain surfactant, according to an equilibrium distribution, partitions between the bulk and the vesicle bilayer. This stage ends, at a certain single chain surfactant concentration, when the bilayer saturates. It defines a stage II, where the addition of more single chain surfactant produces the partial destruction of vesicles by forming mixed micelles coexisting with mixed vesicles, and a stage III, where all double chain surfactant of the vesicles are solubilized into the mixed micelles. The three-stage model is in agreement with the presence of the next four domains well documented in the literature [25,42,43]: (1) the prevesicle domain, where only the presence of monomers is expected; (2) the mixed vesicles domain, at total surfactant concentration above the CVC*; (3) a region where the mixed micelles start to form and coexist with mixed vesicles, above the CMC*; and (4) the mixed micelles domain (the mixed vesicles are disrupted and solubilized by the mixed micelles), above the total mixed critical micelle concentration, CMC^{*}_{tot}. Nevertheless, although the model explains quite well the solubilization process of the vesicles, the mechanism is not yet clear. It is known that the structure of the aggregates is strongly dependent on temperature, composition, and concentration of the mixed system [12].

In the literature, some studies have been reported to analyze the formation of liposomes or vesicles in the presence of a single-chain surfactant, but only few of them include a sugar surfactant [24,25,44–47]. In any case, most of these works were mainly focused on the study of the vesicle-to-micelle transition, i.e., on the medium concentration range. However, rigorous studies in the very diluted region, or even highly diluted region, just when the formation of the aggregates takes place, has been recently reported by our research group [24,25,44]. In addition, the formation of mixed nanoaggregates at a total surfactant concentration lower than CVC*, on the highly diluted concentration range, has been recently demonstrated, for the first time [24,48]. Based on these new findings, a new model, that includes the three-stage model of Lichtenberg et al., has also been proposed [49].

13.3 SYSTEMS OBJECT OF STUDY

In this chapter we present a review of the knowledge of the mixed aggregation behavior of ternary systems constituted by a cationic double-chain surfactant of the dialkyldimethylammonium bromide series (di-C_nDMAB, with n = 10, 12), and a nonionic single-chain sugar surfactant, as the OBG. All the surfactants have one or two hydrophobic chains of comparable length but differ on the charge of their polar head (see Scheme 13.1).

Quaternary alkylammonium salts are essential in many biological processes due to their function in several cellular processes and/or physiological actions [50]. Since last century, these salts have been used as antiseptics, disinfectants and, at moderate concentration, they act as fungicides, bactericides, and antivirals [50–53]. In addition, these compounds are used as emulsionant agents in the cosmetics, detergency, and pharmacological industries [50–52]. Alkyltrimethyl ammonium bromides (C_mTAB, with m = 10–18) are cationic surfactants with a single hydrocarbon tail that form spherical micelles in aqueous solution. The CMC aggregation number, N, and dissociation degree, β, of these micelles have been widely analyzed [54–62]. Dialkyldimethylammonium bromides (di-C_nDAB) are less studied double chain cationic surfactants that, depending on its concentration, aggregate in water solution forming flat bilayers (lamelles) or thermodynamically stable spherical vesicles [63]. Phase diagrams of di-C_nDAB/H_2O (n = 10–18) binary systems, where vesicles are present, are widely studied in the dilute region [64–71]. It has been reported that CVC decreases, and the size of the vesicles increases as long as the hydrocarbon chain is longer [71]. This behavior is similar to the micelle solutions of single chain surfactants, but the CVC values of di-C_nDMAB in water are about 100 times lower than the CMC of C_mTAB of similar length [6,7,9].

On the other hand, alkylglycosides are surfactants consisting of one or two hydrocarbon tails linked to a sugar residue through a monoester bond. The most

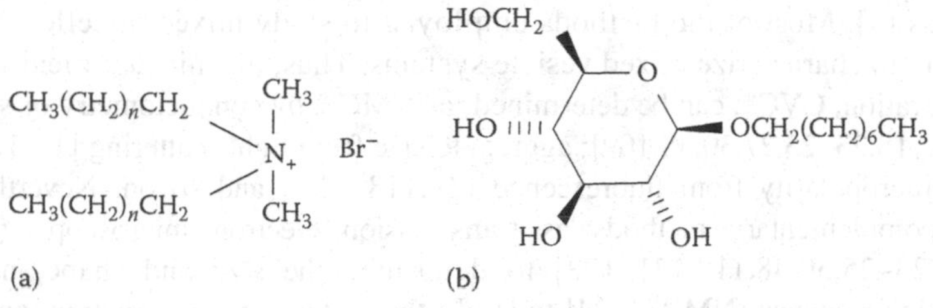

$CH_3(CH_2)_nCH_2$

$CH_3(CH_2)_nCH_2$

N_+ Br^-

CH_3

CH_3

(a)

$HOCH_2$

HO ''''

HO OH

O

$OCH_2(CH_2)_6CH_3$

(b)

SCHEME 13.1 (a) di-C_nDAB, with n = 8 and (b) OBG surfactant molecules.

important sugar surfactants are those where this linkage involves the anomeric C, giving rise to two stereoisomers (α or β) with different physical properties. Alkylglycosides are nontoxic, nonirritant, and biodegradable nonionic surfactants widely used on the crystallization and solubilization of membrane proteins [72–76]. These characteristics make them of great interest in the alimentary, pharmacological, and cosmetic industries [72,74,77–86]. Due to the absence of charge, alkylglycosides do not decrease the enzymatic activity. In addition, as cellular membrane is glicosilated, these compounds play an important role in the membrane stability and in the antigen–antibody response [87,88], which increases their biochemistry and pharmacological applications [88]. Due to its biochemistry applications, n-octyl-β-D-glucopyranoside is the more studied alkylglycoside. Shinoda [89] was the first to report the CMC of alkylglycosides in water from surface tension. The value of the CMC for the OBG, obtained from surface tension, ultrasonic velocity, isoentropic compressibility [89–91] is lower than that of typical ionic surfactants as C_8TAB, but higher than that of nonionic surfactants of similar hydrophobic length as octylpolyoxyethers, C_8E_n. In protein extraction, OBG is preferred to other nonionic surfactants as Brij 36-T and the TX-100 because it provides a better dialysis process. In addition, OBG does not present the cloud-point phenomena [89,92,93]. The aggregation number of OBG has been studied from several methods concluding that it depends strongly on the surfactant concentration [91,94], observing spherical micelles just passing by the CMC, and globular or ellipsoidal ones at higher concentrations [90,95].

13.4   METHODS OF STUDY

13.4.1   Experimental Characterization

The experimental characterization of mixed micelle systems is carried out using several experimental methods. Surface tension [96–99], conductometry [100–103], density, and ultrasonic velocity [91] are used to determine the total CMC*; conductometry or ion-selective electrode to estimate the dissociation degree of ionic micelles [101,104,105]; fluorescence or light scattering to obtain the total aggregation number, N^* [82–84,101,106–108]; surface tension to analyze the surface properties [60,109]; and viscosimetry to study rheological properties [110–113]. Additional information, as thermodynamic parameters of the micellization process, can be determined from density or ultrasonic velocity [101,114], and calorimetric methods [4]. Most of the methods employed to study mixed micelle systems are valid to characterize mixed vesicle systems. Thus, the mixed critical vesicle concentration, CVC*, can be determined, as CMC*, by conductometry or surface tension [18,23–25,27,30,48,109]; aggregate size from light scattering [19,27,115–117]; micropolarity from fluorescence [28,118–120], and so on. Nevertheless, other complementary methods as transmission electron microscopy (TEM) [18,19,23–25,30,48,115,121–123] to determine the size and shape, nuclear magnetic resonance (NMR) [124] to study the surfactant conformation, and zeta potential [18,23–25,48,109,125] to determine the surface charge of the aggregates are also needed. In this chapter, experimental results obtained from electrochemical and spectroscopic studies together with those obtained with TEM are reported. These data show information about the multiple aggregation processes and the characteristics of the aggregates.

13.4.1.1 Electrochemistry Characterization: Conductometry and Zeta Potential

Conductometry has been used to characterize the mixed systems studied in the chapter. In addition, when vesicles are present, zeta potential is also used. Electrical conductivity of an electrolytic solution is due to the ion mobility, which depends on the charge, size, and number of ionic species that are present in solution. Specific conductivity, $\kappa (= A\, 1/R)$, can be determined from the electrical resistance of the solution, R, and the cell constant, A, which only depends on the cell geometry [126,127]. Since the mobility of ions is different if surfactants are as monomers or associated, conductometry can help in colloidal solutions' characterization to determine parameters such as CMC of micelles and its dissociation degree, β, and CVC of vesicles. Nevertheless, this technique only can be used if at least one of the surfactant forming aggregates is ionic.

Capillary electrophoresis is the relative motion of a charged particle in an electrolyte solution in the presence of an external electric field applied parallel to the particle–solution interface [128–131]. The electrophoretic velocity, v_e, is related to the electric field through the electrophoretic mobility, μ_e, by

$$v_e - \mu_e E \tag{13.1}$$

In the frontier between the mobile and immobile phases an electrokinetic potential, named zeta potential, ζ, is generated, which is assumed to be equal to the surface potential [128,130–132]. This zeta potential is related to the electrophoretic mobility by the Henry equation [128,130–134]:

$$\zeta = \frac{3\eta}{2\varepsilon_0 \varepsilon_r f(\kappa_D a)} \mu_e \tag{13.2}$$

where
ε_0 is the permittivity of the vacuum
ε_r is the relative permittivity of the medium
η is the viscosity of water

The Henry function, $f(\kappa_D a)$, which depends on the particle shape (κ_D being the reciprocal Debye length and a, the particle size), includes the polarization deformation of the electrical interface, and for spherical particles can be well estimated (relative errors less than 1%) using the Ohshima approximation [128,130]:

$$f(\kappa_D a) = 1 + \frac{1}{2\left[1 + \dfrac{2.5}{\kappa_D a(1 + 2e^{-\kappa_D a})}\right]^3} \tag{13.3}$$

To apply Equation 13.3, size particles must be known. This information can be taken from TEM or light scattering. For this reason, many researchers very often use other approximations with less rigor, and thus driving to erroneous results as it has been fully discussed [25,130,131].

The surface charge density enclosed by the shear plane, σ_ζ, can be calculated from zeta potential, assuming a Gouy–Chapman double layer, by using the Loeb equation [135]:

$$\sigma_\zeta = \frac{2\varepsilon_0 \varepsilon_r \kappa_D k_B T}{ze}\left[\sinh\left(\frac{ze\zeta}{2k_B T}\right) + \frac{2}{\kappa_D a}\tanh\left(\frac{ze\zeta}{4k_B T}\right)\right] \qquad (13.4)$$

where
  e is the elemental charge
  z is the valence of the ion
  k_B is the Boltzmann constant
  T is the absolute temperature

This equation, which includes a correction term that takes into account the curvature of the vesicle, is not normally used in the literature since it needs information about the size of the particle. It decreases the error on the calculation of σ_ζ from 20% to 5% [130].

13.4.1.2 Microscopic Characterization: TEM and Cryo-TEM

Electron microscopy is based on the interaction between matter and electrons. This interaction origins an extensive type of signals, because one part of the electrons is transmitted, other part is scattered, and another one provokes interactions that produce different phenomena as light emission, Auger, x-ray, etc. [136]. All these signals can be used to get information about the matter under study, and, depending on the signal, one can get various types of electronic microscopes, such as scanned, transmission, atomic force, tunnel, and so on. In particular, TEM employs the electrons transmission/dispersion to produces images, the electrons diffraction to get information of the crystalline structure, and the x-ray emission to know the sample composition. Sample must be very thin (100–250 nm) to improve the quality of the images [132,136]. Nowadays, novel methods have been developed to prepare samples containing supramolecular aggregates, and to determine their electronic density, size, and shape [132]. Among others, the negative stained method [137], based on re-covering the sample with a heavy metal, and cryo-TEM [138,139], which consists of an ultrafast cryogenization of the sample with liquid ethanol, are widely used to elucidate the structure, size, shape, and morphology of colloidal aggregates.

13.4.1.3 Spectroscopic Characterization: Fluorescence Spectroscopy

Fluorescent probes are extensively used to get information about the structure, interactions, and dynamics of colloidal aggregates [118,140]. Those studies are based on the specific sensitivity of the photophysical properties of many fluorophores to its microenvironment; the energies of the fundamental and excited states of the probe are affected by the solvent and the characteristics of the solubilization site where the probe is housed within the aggregate. Thus, the emission length and the quantum fluorescence depend on the probe microenvironment [140]. In micellar systems, fluorescence spectroscopy may inform about the aggregation number, N, of the micelles [99]. It is based on the quenching of the emission intensity, I, of a fluorescent probe (usually pyrene) solubilized in the micelle by the addition of a quencher, Q. The slope of the line corresponding to ln I versus $[Q]$ is related to N [11,99,141]. In addition, the ratio I_I/I_{III} inform about the micropolarity of the microenvironment where the fluorescent probe is housed into the micelle (I_I and I_{III} are the intensities of the first and third emission peaks of pyrene fluorescence spectra) [11,91,99,141].

In vesicle systems, most of the fluorescent probes contain an aromatic ring joint to a donor or acceptor substituent, usually in para-position [118]. The results shown in this chapter are obtained using two fluorescent probes; one anionic, potassium 2-(p-toluidino-naphthalene-6-sulfonate (TNS), and other nonionic, 6-propionyl-2-dimethylaminonaphthalene (PRODAN). As can be seen in the Scheme 13.2, both probes have a naphthalene ring linked to an acceptor group and to a donor one, which is p-toluidino (in the TNS) and dimethylamino (in the PRODAN).

Both probes slightly emit in water but exhibit an intense fluorescence when they associate to a macromolecule or a membrane [118,119,140]. In the case of positively charged surfaces, as the di-C$_n$DMAB/OBG system, anionic TNS is not repelled and surfactant–TNS interactions allow the probe to be located in a more water-restricted environment, thus enhancing the fluorescence quantum yield. On the other hand, the nonionic character of PRODAN may supply information either of ionic or nonionic aggregates. It is well known [140] that excitation of a probe yields to the formation of an excited locally state (LE) (via $\pi \to \pi^*$ electronic transition). Nevertheless, in the last decade there is great controversy regarding the existence of other excited states and the mechanisms by which these states form and deactivate. Several researchers point to the formation of an intramolecular charge-transfer excited state, with a coplanar orientation of the rings (ICT state) [142–147], while others propose an intramolecular charge-transfer excited state, with a twist around the NH moiety (TICT state) to achieve a perpendicular

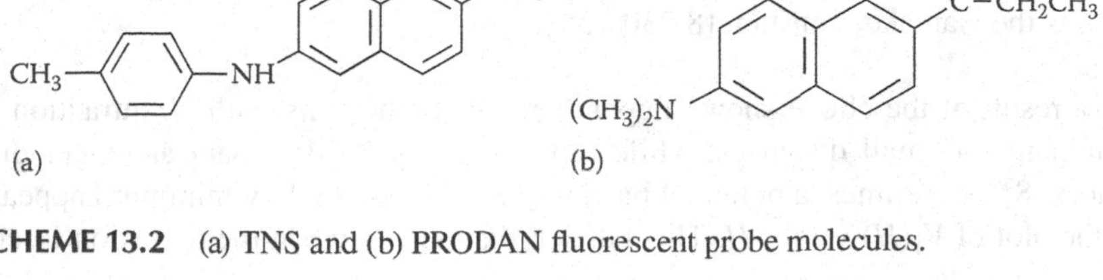

(a)   (b)

SCHEME 13.2   (a) TNS and (b) PRODAN fluorescent probe molecules.

configuration between the naphthalene ring and phenyl ring (TNS) or dimethyl-amino moiety (PRODAN) [148–151]. Transference of nitrogen lone pair to the sulfonate (TNS) or acetyl substituents (PRODAN) of the naphthyl ring, is stabilized in polar media [118,140]. This photophysical scheme has been questioned in supramolecular assemblies [28,119,152–154]. Nowadays [18,25,44,48,118], it is believed that the emission band of this type of fluorescent probes consist of several peaks or emission components, at different wavelengths, attributed to several microenvironments where the probe may be housed, each one having distinct hydrophobicity, microviscosity, rigidity and solvation characteristics. Accordingly, the overall emission of the probe can be deconvoluted into the optimum number of reproducible overlapping curves, each one assignable to $\pi \rightarrow \pi^*$ emissions corresponding to the probe housed within different microenvironments. In addition, polarity of those microdomains can be estimated by comparing the wavelength of the emission peaks with those reported for the probe in different solvents of known dielectric constant [118,155].

13.4.2 THEORETICAL CHARACTERIZATION ANALYSIS

13.4.2.1 DLVO Theory

The Derjaugin, Landau, Verwey, and Overbeek (DLVO) was the first theory aimed to quantitatively describe the stability of colloidal suspensions [156–158]. Essentially, this theory states that the interaction among two colloidal particles results from the balance between attractive dispersion forces (London or van der Waals forces), and repulsive electrostatic forces (due to the overlapping of electric interfaces). It assumes that the interacting particles are big, compared with the electrolyte ions, which are considered as immobile in solution at a certain distance, and evaluates this interaction as a function of the distance, as briefly resumed next.

The stability of colloidal aggregates depends on the total interaction energy, V_T, which, for spherical aggregates of equal radius, a, at a distance H, is expressed as [8,128,130,131]

$$V_T(H) = V_R(H) + V_A(H) = \frac{64\pi an^\circ k_B T}{\kappa_D^2}\Gamma^2 \exp(-\kappa_D H) - \frac{Aa}{12H} \quad (13.5)$$

where
$V_R(H)$ and $V_A(H)$ are the interaction energies due to electrostatic repulsion and attractive dispersion, respectively
n° is the number of ions per volume
Γ is a constant depending on the potential energy and temperature
A is the Hamaker constant [8,130,159]

As a result of the above shown dependency of both terms with H, attraction is dominant for small distances, while repulsion may be dominant at longer distances [8]. Sometimes, a potential barrier at H and a secondary minimum appears in the plot of $V_T(H)$ versus H. This barrier, that occurs when the repulsion forces are higher than the attractive ones, is important when analyzing the colloidal stability. Although attractive energy depends on the particle Hamaker constant and cannot be modified, the repulsive energy that depends on the Stern layer potential or on κ_D^{-1}, can be altered by, for example, changing the ionic strength. Furthermore, if the barrier energy is low, and at a certain electrolyte concentration, known as critical coagulation concentration, CCC, becomes less than zero, the particles will coagulate.

For many years, DLVO theory has constituted a solid basis to describe the colloidal stability, but it cannot satisfactory explain several behaviors, mainly, in concentrated colloidal suspensions, in systems with high surface potential, ψ_0, or in the case of big colloidal particles. For this reason, several models came up with the aim of overcoming its main failures, among which we emphasize in this chapter the models proposed by Inoue [160] and Sogami [161].

13.4.2.2 Inoue's Model

Inoue et al. [160] combine the DLVO electrostatic potential with an expression derived by Hogg et al. [162], to build the electrostatic repulsive term, $V_R(H)$, which for two spherical colloidal particles results in

$$V_R(H) = \frac{\varepsilon a \psi_0^2}{2} \ln\left[1 + \exp(-\kappa_D H)\right]$$ (13.6)

where
  ε is the dielectric constant of the medium
  ψ_0 is the surface potential of the particle

On the other hand, this model uses the following expression, derived by Vold [163], to build the attractive interaction potential for spherical particles:

$$V_A(H) = -\frac{A_{232}}{12}\left(F_{11} + F_{22} - 2F_{12}\right)$$ (13.7)

where
  A_{232} is the effective constant of Hamaker
  F_{ij} is given by

$$F_{ij} = \frac{y}{x^2 + xy + x} + \frac{y}{x^2 + xy + x + y} + 2\ln\left(\frac{x^2 + xy + x}{x^2 + xy + x + y}\right)$$ (13.8)

where x and y are related with H, a, and the interface thickness, d_{dc}, through the following relations:

$$\text{For } F_{11} : y = 1 \quad x = \frac{H + 2d_{dc}}{2(a - d_{dc})}$$ (13.9)

$$\text{For } F_{22}: y = 1 \quad x = \frac{H}{2a} \quad (13.10)$$

$$\text{For } F_{12}: y = \frac{a}{a - d_{dc}} \quad x = \frac{H + d_{dc}}{2(a - d_{dc})} \quad (13.11)$$

Combining Equations 13.6 and 13.7, the resulting total interaction potential proposed by Inoue's model is

$$V_T(H) = \frac{\varepsilon a \psi_0^2}{2} \ln\left[1 + \exp(-\kappa_D H)\right] - \frac{A_{232}}{12}\left(F_{11} + F_{22} - 2F_{12}\right) \quad (13.12)$$

This model has been mainly used for checking the colloidal stability of a wide variety of systems, mostly at high ionic strength [160].

13.4.2.3 Sogami's Model

This model proposes that not only a medium-range repulsion term but also a long-range weak attractive interaction term should be included to estimate the overall electrostatic interaction between the particles [161,164]. This attractive interaction normally drives to a secondary minimum in the total potential at a distance H_{min}, which is outside of the attractive van der Waals range, this secondary minimum being responsible for the flocculation phenomena. Briefly, Sogami's model calculates the total electrostatic energy, E_T, as a function of the position of the particles, by integrating over the total volume, V, as follows:

$$E_T = \frac{1}{2}\sum_l Z_l e \int \psi(r)\rho_l(r)dV + \frac{1}{2}\int \psi(r)\left[-\frac{\varepsilon\kappa_D^2}{4\pi}\phi(r)\right]dV \quad (13.13)$$

where the first term represents the electrostatic energy of all particles in the field of potential $\psi(r)$, and the second term is the electrostatic energy referred only to the counterions. This last term is responsible of the attractive electrostatic force among the particles. Sogami et al. obtain free Gibbs energy, G, from the total electrostatic energy, E_T, and the Helmholtz free energy, F, by the following expression:

$$G = G^0 + \frac{1}{2}\sum_{m \neq l} U_{ml}^G + \sum_l V_l^G \quad (13.14)$$

where G^0 is the free Gibbs energy in the limit of $e^2 = 0$. The adiabatic potential, U_{ml}^G, that only depends on the relative configuration and orientation of two particles l and m, and the adiabatic potential of a particle n due to the ions around it, V_l^G, are obtained by

$$U_{ml}^G = \frac{(Ze)^2}{\varepsilon}\left[\frac{\sinh(\kappa_D a)}{\kappa_D a}\right]^2 \left[1 + \kappa_D a \coth(\kappa_D a) - 1/2\kappa_D R_c\right]\frac{\exp(-\kappa_D R_c)}{R_c} \quad (13.15)$$

$$V_l^G = \frac{(Ze)^2}{4\varepsilon a}\left[\frac{1-\exp(-2\kappa_D a)}{2\kappa_D a}+\exp(-2\kappa_D a)\right]+\frac{(Ze)^2}{2\varepsilon a}\left[\frac{1-\exp(-2\kappa_D a)}{2\kappa_D a}\right] \quad (13.16)$$

where

R_c $(= H + 2a)$ is the distance between the centers of the two particles

Ze is the charge density of the colloidal particle

The U^G_{ml} potential, which reflexes the effective attraction among the particles at long distances, is the Sogami's potential, from now on V_T. It shows a minimum at

$$R_{c,min} = (H+2a)_{min} = \frac{C+1+\left[(C+1)(C+3)\right]^{1/2}}{\kappa_D} \quad (13.17)$$

where C is defined by

$$C = \kappa_D a \coth(\kappa_D a) \quad (13.18)$$

This model, which predicts that ionic particles repulse at short distances and attract at longer distances, has been mainly applied to analyze the colloidal stability of latex particles [161].

13.4.2.4 Further Developments

Sogami's model only considers the electrostatic interaction at medium-to-long distances. For that reason, the predictions of the Sogami's model are far away from those of DLVO and Inoue's models. The Sogami's model was developed to analyze particle interactions within a range of distances that are longer than those used in the above-mentioned theoretical colloidal models. We have recently proposed [18] a more general model that takes into account the interaction between particles (i.e., vesicles) at short and medium-to-long distances. It combines the electrostatic potential at medium–long distances proposed by Sogami (Equation 13.15) with the total potential at any distance proposed by DLVO (Equation 13.5) or Inoue's (Equation 13.12) model in order to estimate the total interaction energy between vesicles. Thus, the total interaction, V_T (H), between two spherical particles (i.e., vesicles) of equal radius and at a distance H can be obtained considering a DLVO–Sogami potential (Equation 13.19), or an Inoue–Sogami potential (Equation 13.20), respectively, as follows:

$$V_T(H) = \frac{(Ze)^2}{\varepsilon}\left[\frac{\sinh(\kappa_D a)}{\kappa_D a}\right]^2\left[1+\kappa_D a\coth(\kappa_D a)-1/2\kappa_D(H+2a)\right]$$

$$\times \frac{\exp\left[-\kappa_D(H+2a)\right]}{(H+2a)}+\frac{64\pi a n^0 k_B T}{\kappa_D^2}\Gamma^2\exp(-\kappa_D H)-\frac{Aa}{12H} \quad (13.19)$$

or

$$V_T(H) = \frac{(Ze)^2}{\varepsilon}\left[\frac{\sinh(\kappa_D a)}{\kappa_D a}\right]^2 \left[1 + \kappa_D a \coth(\kappa_D a) - 1/2\kappa_D(H + 2a)\right]$$

$$\times \frac{\exp[-\kappa_D(H + 2a)]}{(H + 2a)} + \frac{\varepsilon a \psi_0^2}{2}\ln[1 + \exp(-\kappa_D H)] - \frac{A_{232}}{12}(F_{11} + F_{22} - 2F_{12})$$

$$(13.20)$$

13.5 RESULTS AND DISCUSSION

Aqueous solutions of the ionic/nonionic ternary systems di-C_nDMAB/OBG, (n = 10, 12) have been characterized through conductivity, ζ-potential, TEM, and cryo-TEM measurements. In several reported mixed systems, mixed nanoaggregates were postulated in a concentration range between the mono-mer region and the CMC* (in micellar systems) [120,165–168], or the CVC* (in vesicle systems) [18,23,24,28,48]. With the aim of clarifying the forma-tion of different nanoaggregates in mixed systems containing the OBG sugar surfactant, the conductivity study has been done over three different concen-tration ranges: (1) the herein called "highly diluted" concentration range, i.e., $0 < [S]_{tot} < 1$ mM, approximately; (2) the herein called "very diluted" concen-tration range, i.e., $0 < [S]_{tot} < 10$ mM; and (3) the so-called "moderately diluted" concentration range, i.e., $0 < [S]_{tot} < 30$ mM. We must note here that most of the studies reported in the literature are focused in the moderately diluted region [19,22,27,169]. It is obvious that in studies covering wide ranges of concentration, specific details of the highly diluted region are lost. For this reason, we have divided the usual concentration range (moderately diluted) in the three above-mentioned concentration regions. Special attention is paid to the highly diluted and very diluted regions, scarcely studied in the literature [18,21,23–25,44,48]. Determination of the concentration domains among the three regions has been done from conductivity as a function of total surfactant concentration, $[S]_{tot}$, at several constant molar fractions of the dialkyl surfactant, α_1.

Figures 13.1 through 13.4 show, as examples, the specific conductivity, κ, as a function of $[S]_{tot}$, at several constant values of α_1, in the three regions for the di-C_{12}DMAB/OBG system and for the di-C_{10}DMAB/OBG system in the very diluted region. These kinds of studies can give information about the formation of nanoaggregates (Figure 13.1), vesicles (Figures 13.2 and 13.4), and micelles (Figure 13.3). Since OBG is a nonionic surfactant, the conductivity reflects only small changes due to the counterions, and ionic di-C_nDMA$^+$ monomers and/or aggregates. It is remarkable the extremely low κ values of Figure 13.1, and even Figure 13.2, compared with those reported for other vesicle systems [19,27] or micelle systems [11,101,141,170]. It is attributed not only to the low mobility of the double-chain surfactant and the larger vesicles compared to single-chain

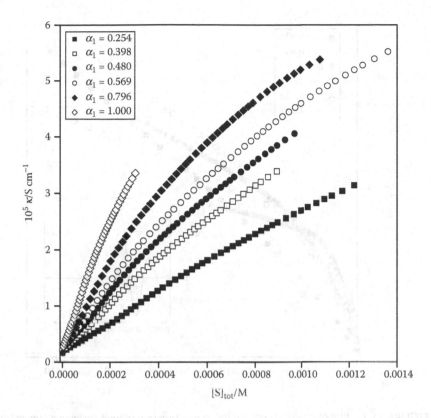

FIGURE 13.1 Plot of specific conductivity, κ, as a function of total surfactant concentration, $[S]_{tot}$ in the highly diluted range, at 298.15 K, at various fixed values of the molar fraction, α_1, for the mixed system di-C_{12}DMAB/OBG. (From Junquera, E., del Burgo, P., Arranz, R., Llorca, O., and Aicart, E., *Langmuir*, 21, 1795, 2005. With permission.)

surfactant and the micelles but also to the extremely low range of concentration studied. Conductivity curves of di-C_{12}DMAB (Figure 13.1) and di-C_{10}DMAB (Figure 13.4) in water show only one break, which indicates the formation of a unique type of supramolecular aggregates. In the case of mixed systems, in the di-C_{12}DMAB/OBG system, one break can be observed in the curves at the lowest concentration region (Figure 13.1), while in the very diluted (Figure 13.2) and in moderately diluted (Figure 13.3) regions, two breaks can be seen, which indicates the presence of three types of mixed nanoaggregates. In the case of the di-C_{10}DMAB/OBG system, a unique break in κ in the very diluted region (Figure 13.4) can be seen. These breaks in conductivity permit to determine the CACs. The first break (Figure 13.1) has been assigned to the mixed critical nanoaggregate concentration, CPVC*, and the second (Figures 13.2 and 13.4) to the CVC*. These critical values have been calculated from the concentration at which the third derivative of the experimental property is equal to zero ($\partial^3\kappa/\partial c^3 = 0$, see the

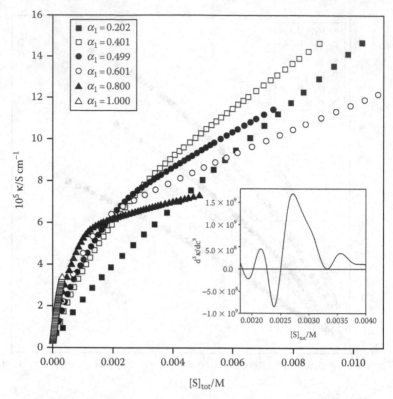

FIGURE 13.2   Plot of specific conductivity, κ, as a function of total surfactant concentration, $[S]_{tot}$ in the very diluted range, at 298.15 K, at constant molar fraction, α_1, for the mixed system di-C_{12}DMAB/OBG. The inset at the bottom shows the third derivative of conductivity with respect to concentration for $\alpha_1 = 0.499$. (From Junquera, E., del Burgo, P., Arranz, R., Llorca, O., and Aicart, E., *Langmuir,* 21, 1795, 2005. With permission.)

inset at the bottom of Figure 13.2, as an example). Table 13.1 reports CPVC* and CVC* values plotted in Figure 13.5. The values of CMC for OBG and the CVC for pure di-C_nDMAB are in agreement with most literature data [27,91,141,170–172]. In the di-C_{12}DMAB/OBG system, since CVC and CMC values for pure surfactants (0.165 and 25.24 mM, respectively) are separated enough in concentration units, the mixed critical prevesicle concentration, CPVC*, can be determined (Figure 13.1), which is not possible in the di-C_{10}DMAB/OBG system, whose CVC and CMC values for pure surfactants (1.325 and 25.24 mM, respectively) are much closer. These features, also previously observed in cationic–cationic systems (di-C_nDMAB/C_mTAB) [18,21,23,48], point to the fact that the presence of CPVC* depends on the difference between the CVC and the CMC of pure surfactants and it does not depend on the ionic or nonionic character of the surfactants.

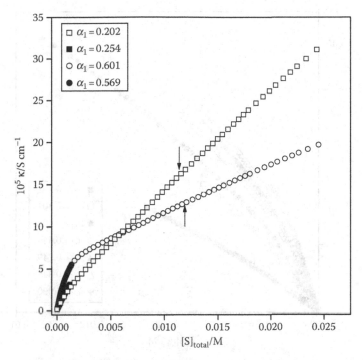

FIGURE 13.3 Plot of specific conductivity, κ, as a function of total surfactant concentration, $[S]_{tot}$ in both the very diluted range (solid symbols) and the moderately diluted range (open symbols), at 298.15 K, at two fixed values of the molar fraction, α_1, for the mixed system di-C_{12}DMAB/OBG. Arrows indicate the CMC* estimated from the third derivative method. (From Junquera, E., del Burgo, P., Arranz, R., Llorca, O., and Aicart, E., *Langmuir*, 21, 1795, 2005. With permission.)

In Table 13.1 and Figure 13.5, both critical concentrations, CPVC* and CVC*, decrease with α_1. In other words, as long as the total content on double-chain surfactant increases, either prevesicle nanoaggregates or vesicles tend to form at lower concentrations; at the limit ($\alpha_1 = 1$), CVC* tends to the CVC for pure di-C_nDMAB. This behavior is similar to the one found in mixed micelles forming systems consisting of single-chain surfactants [11,101,170] and in some mixed-vesicle systems of single/double chain surfactants [18,21], but contrasts with the trend found in other mixed vesicle systems [18,23,48]. Furthermore, it is worth noting that CPVC* values of the di-C_{12}DMAB/OBG system reported herein and those reported by the other kind of mixed systems [18,23,48] are comparable, while CVC* values are one order of magnitude higher in the present case. Given that the double-chain surfactant is similar in those systems, it seems that the presence of the nonionic OBG surfactant stabilizes the nanoaggregates within a wider range of concentration.

The vesicle concentration domain is limited by CVC*, on the monomers-to-mixed vesicle transition, and by CMC*$_{tot}$, the concentration at which the mixed

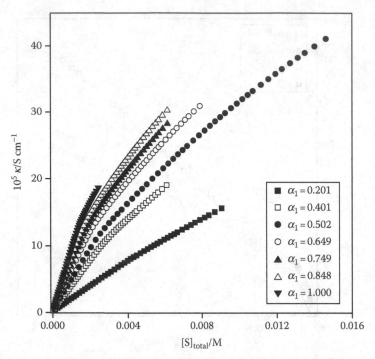

FIGURE 13.4 Specific conductivity, κ, as a function of total surfactant concentration, $[S]_{tot}$, in the very diluted range, at 298.15 K, at constant molar fraction, α_1, for the mixed system di-C_{10}DMAB/OBG. (Modified from Junquera, E., del Burgo, P., Boskovic, J., and Aicart, E., *Langmuir*, 21, 7143, 2005. With permission.)

TABLE 13.1
Values of Experimental Molar Fraction, α_1, Mixed Critical Prevesicle Concentration, CPVC*, and Mixed Critical Vesicles Concentration, CVC*, Obtained from Experimental Conductivity Data, for the Mixed Systems di-C_nDMAB/OBG

α_1	CVC* (mM)	α_1	CPVC* (mM)
	di-C_{10}DMAB/OBG		
0.201	4.210		
0.401	2.807		
0.502	2.426		
0.649	1.885		
0.749	1.735		
0.848	1.511		
1	1.325		

TABLE 13.1 (continued)
Values of Experimental Molar Fraction, α_1, Mixed Critical Prevesicle Concentration, CPVC*, and Mixed Critical Vesicles Concentration, CVC*, Obtained from Experimental Conductivity Data, for the Mixed Systems di-C$_n$DMAB/OBG

α_1	CVC* (mM)	α_1	CPVC* (mM)
	di-C$_{12}$DMAB/OBG		
0.202	4.856	0.254	0.468
0.401	4.079	0.398	0.407
0.499	2.501	0.480	0.375
0.601	1.117	0.569	0.266
0.800	1.045	0.796	0.215
1	0.165	1	—

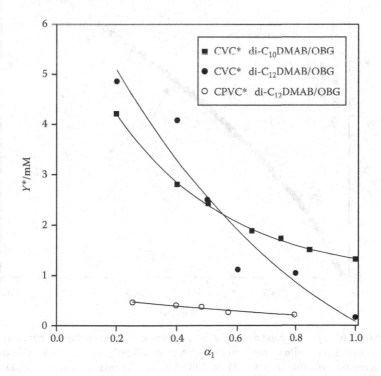

FIGURE 13.5 CVC* and CPVC* as a function of the molar fraction, α_1, for the mixed systems di-C$_n$DMAB/OBG.

vesicles are totally solubilized forming mixed micelles. According to the three-stage model [42,43], prior to this total mixed vesicle-to-mixed micelle transition, there must exist a concentration, named CMC*, at which mixed micelles start to form, and coexist with mixed vesicles up to CMC^*_{tot}. For this reason, the formation of mixed micelles of di-C_nDMAB and OBG surfactants can be studied by measuring conductivities of several mole fractions covering a wider concentration range, i.e., the moderately diluted region ($0 < [S]_{tot} < 30\,mM$), which allow to estimate the mentioned CMC*. The change in κ at this CMC* (vesicles-to-micelles) is expected to be small and more difficult to detect than CVC* (monomers-to-vesicles). Figures 13.3 and 13.6 show, as examples, the conductivity in this moderately diluted region, together with κ values measured in the very diluted region, at two molar fractions (around $\alpha_1 = 0.2$ and 0.6) for the di-C_{12}DMAB/OBG system (Figure 13.3) and at $\alpha_1 = 0.5$ for the di-C_{10}DMAB/OBG system (Figure 13.6), respectively.

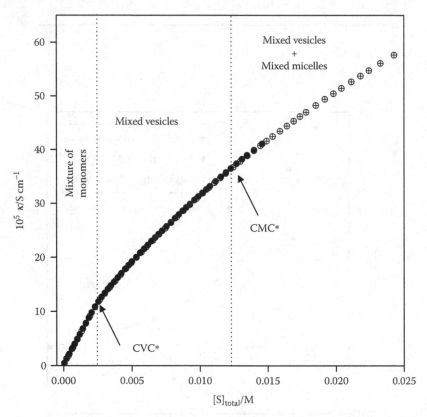

FIGURE 13.6 Specific conductivity, κ, as a function of total surfactant concentration, $[S]_{tot}$, in both the very diluted range (solid symbols, $\alpha_1 = 0.502$), and the moderately diluted range (crossed symbols, $\alpha_1 = 0.501$), at 298.15 K, for the mixed system di-C_{10}DMAB/OBG. (From Junquera, E., del Burgo, P., Boskovic, J., and Aicart, E., *Langmuir*, 21, 7143, 2005. With permission.)

The agreement between the experimental data done in both ranges is remarkable, which confirms the reproducibility of the conductivity technique. Although very small, a break on κ can be seen at the CMC*. The presence of the double-chain surfactant, irrespective of its length, approximately halves the CMC* with respect to the value for pure OBG (CMC = 25.24 mM) [91]. This fact means that the solubilization of the cationic–nonionic vesicles (vesicle-to-micelle transition) is clearly favored, reinforcing one of the well-known applications of OBG on the solubilization and reconstitution processes of membranes. Furthermore, Table 13.1 and Figures 13.3 and 13.6 show that as the content of double-chain surfactant increases, the total concentration of the surfactants, $[S]_{tot}$, that is needed to start the solubilization of vesicles, increases as well. This feature indicates that the three-stage model of Lichtenberg is also valid in studies where $[S]_{tot}$ changes but the molar fraction of the mixed system remains constant.

The TEM experiments have been done to confirm and assign the conductivity changes to structural transitions between specific supramolecular nanoaggregates. These results have permitted to visualize the different type of aggregates potentially present in solution and to determine its size, shape, and morphology. Figure 13.7 shows, as an example, how cryo-TEM micrographs confirm the presence of vesicles on samples of di-C_{10}DMAB in water [25]. Pure di-C_{10}DMAB vesicles are spherical and unilamellar, with an average diameter of about 36 nm, quite similar to pure di-C_{12}DMAB vesicles [23]. It seems that two C atoms per hydrocarbon surfactant tail do not affect the overall size and shape of the aggregates. In mixed systems, cryo-TEM experiments have been run on samples at different molar fractions, α_1: (1) on a concentration range where vesicles is supposed to be present, i.e., samples with CVC* < $[S]_{tot}$ < CMC*; and (2) on the nanoaggregates concentration domain, i.e., CPVC* < $[S]_{tot}$ < CVC*. Figure 13.8 shows the images of the vitrified samples of the di-C_{12}DMAB/OBG system in the very diluted region. Vesicles are clearly observed in the micrographs; they are essentially spherical, the average diameter being about 100 nm, although a certain contribution of pseudo-ellipsoidal vesicles can be considered. A similar cryo-TEM study for the mixed di-C_{10}DMAB/OBG system in the very diluted region (Figure 13.9) and in the

FIGURE 13.7 Cryo-TEM micrographs of a di-C_{10}DMAB solution containing pure vesicles ($\alpha_1 = 1$) at $[S]_{tot} = 2.935$ mM. (From Junquera, E., del Burgo, P., Boskovic, J., and Aicart, E., *Langmuir*, 21, 7143, 2005. With permission.)

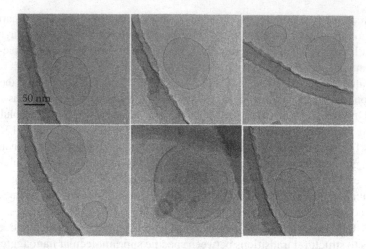

FIGURE 13.8 Cryo-TEM micrographs of a di-C_{12}DMAB/OBG solution containing mixed vesicles ($\alpha_1 = 0.503$) at $[S]_{tot} = 2.677$ mM. (From Junquera, E., del Burgo, P., Arranz, R., Llorca, O., and Aicart, E., *Langmuir*, 21, 1795, 2005. With permission.)

moderately diluted region (Figure 13.10) also confirm the presence of mixed vesicles. They appear to be also spherical and mainly unilamellar, although a certain contribution of vesicles with onionskin structures is also noticeable. The averaged diameters are around 90 nm at $\alpha_1 = 0.501$ and $[S]_{tot} = 8.98$ mM, and around 103 nm at $\alpha_1 = 0.848$, $[S]_{tot} = 2.992$ mM, comparable in magnitude with those of di-C_{12}DMAB/OBG mixed vesicles [25]. However, when cryo-TEM experiments have been done on samples of the di-C_{12}DMAB/OBG mixed system at such $[S]_{tot}$, that CPVC* < $[S]_{tot}$ < CVC*, colloidal aggregates have been not observed. In order to clarify this point, another microscopy technique, such as TEM of negatively stained samples that permits to study smaller colloidal aggregates has been used. Figure 13.11 shows one of the micrographs of the TEM experiments for a solution of di-C_{12}DMAB/OBG system, within the nanoaggregates domain.

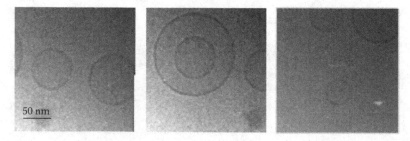

FIGURE 13.9 Cryo-TEM micrographs of a di-C_{10}DMAB/OBG solution containing mixed vesicles at $\alpha_1 = 0.501$ and $[S]_{tot} = 8.980$ mM. (From Junquera, E., del Burgo, P., Boskovic, J., and Aicart, E., *Langmuir*, 21, 7143, 2005. With permission.)

FIGURE 13.10 Cryo-TEM micrographs of a di-C_{10}DMAB/OBG solution containing mixed vesicles at $\alpha_1 = 0.848$ and $[S]_{tot} = 2.992$ mM. (From Junquera, E., del Burgo, P., Boskovic, J., and Aicart, E., *Langmuir*, 21, 7143, 2005. With permission.)

The presence of aggregates with an average diameter around 17 nm can be observed, confirming the existence of aggregation phenomena in the prevesicle region of mixed systems containing a dialkyldimethylammonium and a nonionic sugar surfactant [25]. The incapacity of cryo-TEM to detect these nanoaggregates has nothing to do with the resolution of the technique, but with the structure of the aggregates, which does not form a lipidic bilayer. It is remarkable that, in both systems, most of the vesicles are spherical and unilamellar, but with sizes clearly higher than those reported for the vesicles of pure double-chain surfactant. However, a clear pattern of the effect of α_1 on the sizes of mixed vesicles is not

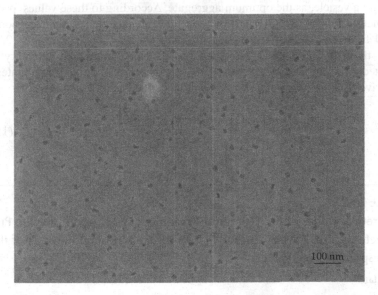

FIGURE 13.11 TEM micrograph of a di-C_{12}DMAB/OBG solution containing mixed nanoaggregates at $\alpha_1 = 0.566$ and initial $[S]_{tot} = 0.791$ mM. (From Junquera, E., del Burgo, P., Arranz, R., Llorca, O., and Aicart, E., *Langmuir*, 21, 1795, 2005. With permission.)

observed. According to the model proposed by Safran et al. [31,32], as long as the total molar fraction α_1 decreases, the content on single-chain surfactant should increase in the inner layer and even more in the outer layer. This feature will drive to a lower inner curvature and, thus, to a bigger vesicle. Nevertheless, only from the cryo-TEM results, it seems quite speculative to infer whether OBG goes preferentially to the outer or inner layer.

It is well known that total free energy of the aggregation process results from two main factors, i.e., the repulsive or attractive electrostatic interactions between the head polar groups, and the attractive energy, coming from the hydrocarbon chain packing. In the cationic/nonionic systems, both effects (repulsive electrostatic interactions between the cationic quaternary ammonium head groups of di-C_nDMA$^+$ shielded by nonionic OBG head, and molecular packing contributions coming from the hydrocarbon tails) must be considered. The experimental information can be checked together with the one obtained from the packing parameter, P ($= v/a_o l_c$), proposed by Israelachvili [8], which relates the molecular information of a surfactant molecule. The aggregates will be spherical micelles for $P < 1/3$, globular micelles for $1/3 < P < 1/2$, vesicles or flexible bilayers for $1/2 < P < 1$, and plane bilayers for $P \approx 1$. The volume, v and the critical chain length, l_c, of the hydrophobic tail, obtained according to the Tanford's model [10,174,175]; the optimal headgroup area, a_o, estimated from the aggregation number of the single-chain surfactant assuming spherical micelles [7]; and P are reported in Table 13.2. It seems that the decrease on the length of the tail of the double-chain surfactant does not affect to the packing parameter that is compatible with a vesicle, as the optimum aggregate. According to these values, packing parameters for di-C_nDMAB are in total agreement with the vesicle geometry found from cryo-TEM experiments for pure double-chain surfactants [23,25], while those for pure OBG correspond to its globular micelle geometry, also reported from fluorescence results [91]. In the case of mixed aggregates, an effective packing parameter, $P_{\text{effective}}$, can be defined as follows [24]:

$$P_{\text{effective}} = \left(\frac{v}{a_o l_c} \right)_{\text{effective}} = \frac{\sum_i v_i X_i^{\text{agg}}}{\sum_i a_{o,i} X_i^{\text{agg}} \sum_i l_{c,i} X_i^{\text{agg}}} \qquad (13.21)$$

TABLE 13.2
Theoretical Values of the Structural Parameters, v, a_o, and l_c, for the Pure Surfactants, di-C_nDMAB and OBG, and the Packing Parameter, P, of the Corresponding Aggregates

Surfactant	v (Å^3)	a_o (Å^2)	l_c (Å)	P
di-C_{10}DMAB	593.4	75	14.15	0.56
di-C_{12}DMAB	700.4	75	16.68	0.56
OBG	242.6	60	11.62	0.35

where X_i^{agg} is the molar fraction in the aggregate. Effective packing parameter, calculated with Equation 13.21, shows a gradual decrease from 0.56 of the pure di-C_nDMAB vesicles to 0.35 of the pure OBG globular micelles, as would be expected. The presence of only mixed vesicles or only mixed micelles or coexistence of both depends very much on both α_1 and $[S]_{tot}$, because the decrease on $P_{effective}$ is gradual as the vesicle-to-micelle transition occurs; for example, for the di-C_{10}DMAB/OBG system at $\alpha_1 = 0.5$, $P_{effective}$ results 0.54 at $[S]_{tot} = 15\,mM$, 0.51 at $[S]_{tot} = 26\,mM$, 0.49 at $[S]_{tot} = 36\,mM$, and 0.48 at $[S]_{tot} = 46\,mM$. Considering that CVC* = 2.426 mM and CMC* = 12.3 mM at $\alpha_1 = 0.5$, and keeping in mind the range of P for globular micelles ($1/3 < P < 1/2$) and for vesicles ($1/2 < P < 1$), the resulting $P_{effective}$ indicate that there exists a concentration range, above CMC*, within mixed vesicles and mixed micelles must coexist prior to the total disruption and solubilization of the vesicles at higher concentrations, as proposed in Section 13.2. In addition, it is worth noting the gradual change of $P_{effective}$, as $[S]_{tot}$ increases, justifying the gradual change on the specific conductivity found in this region of coexistence of both types of aggregates (see Figure 13.6). However, at higher α_1, for example at $\alpha_1 = 0.85$, the resulting $P_{effective}$ are all higher than 0.5, independent of $[S]_{tot}$ and even at high $[S]_{tot}$ concentrations (e.g., $P_{effective} = 0.55$ at $[S]_{tot} = 40\,mM$ and even $P_{effective} = 0.54$ at $[S]_{tot} = 100\,mM$), which means that even at high α_1, mixed vesicles are present. For the other system, di-C_{12}DMAB/OBG, at $\alpha_1 = 0.2$ (CMC* = 11.4 mM), $P_{effective}$ goes from 0.53 at $[S]_{tot} = 12\,mM$ to 0.47 at $[S]_{tot} = 15\,mM$. It means that vesicles should coexist with micelles in the close vicinity of CMC*, but they are solubilized at higher concentrations. Furthermore, a cryo-TEM experiment has been run in the micelles domain, for a solution of di-C_{10}DMAB/OBG at $\alpha_1 = 0.502$ and $[S]_{tot} = 38\,mM$, well above CMC*. Mixed vesicles could not be visualized in any of the micrographs, confirming that at such concentration all the vesicles have been solubilized by the mixed micelles, in concordance with the theoretical predictions ($P_{effective} = 0.49$).

Combining conductometry, TEM, and cryo-TEM results, three critical aggregation concentrations can be unambiguously assigned, as can be resumed in Scheme 13.3: (1) the CPVC*, unique break in the conductivity observed in the highly dilution region or the first observed in the very diluted region, which corresponds to the formation of mixed prevesicle nanoaggregates, (2) the CVC*, the second break in the very diluted region or the first in the moderately dilute region, which corresponds to the initial formation of mixed vesicles, and (3) the CMC*, the second break in the moderately dilute region, which corresponds to the initial formation of mixed micelles which coexist with mixed vesicles. All these features reveal, up to moderate concentrations, the presence of the five concentration domains: (1) mixture of monomers ($[S]_{tot} <$ CPVC*), (2) prevesicle nanoaggregates (CPVC* $< [S]_{tot} <$ CVC*), (3) mixed vesicles (CVC* $< [S]_{tot} <$ CMC*), (4) mixed vesicles together with mixed micelles (CMC* $< [S]_{tot} <$ CMC*$_{tot}$), and (5) mixed micelles ($[S]_{tot} >$ CMC*$_{tot}$), this domains being very much dependent on α_1.

The charged surface of pure and mixed vesicles can also be studied by electrophoretic mobility, μ_E, which has been measured for solutions of di-C_nDMAB/ OBG systems at several molar fractions α_1 and at two or three different $[S]_{tot}$ per

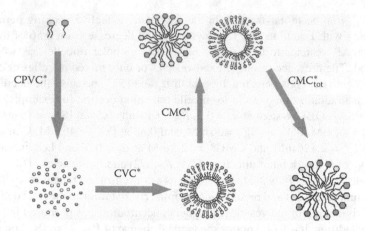

SCHEME 13.3 Extended model proposed from the three-stage model of Lithengberg et al.

molar fraction. These $[S]_{tot}$ concentrations have been chosen in such a manner that nanoaggregates or vesicles are present in solution. ζ-Potential has been calculated from electrophoretic mobility using Equation 13.2. The Henry function, $f(\kappa_D\ a)$, which depends on the particle size has been calculated using Equation 13.3, where a, the particle radius, is taken from TEM and cryo-TEM results. Figure 13.12

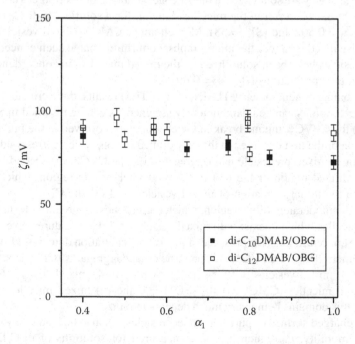

FIGURE 13.12 Plot of ζ-potential as a function of the molar fraction, α_1, for various $[S]_{tot}$ concentrations of the mixed systems di-C_nDMAB/OBG, containing vesicles.

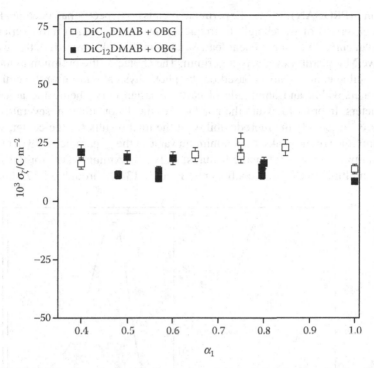

FIGURE 13.13 Surface charge density, σ_ζ, as a function of molar fraction, α_1, for various $[S]_{tot}$ concentrations of the mixed systems di-C_nDMAB/OBG, containing vesicles. (From Junquera, E., del Burgo, P., Boskovic, J., and Aicart, E., *Langmuir*, 21, 7143, 2005. With permission.)

shows ζ-potential values as a function of the molar fraction, α_1, for solutions containing mixed prevesicle nanoaggregates or vesicles, this ζ-potential being almost constant, around (85 ± 5) mV. It is worth remarking that the prevesicle nano-aggregates, with sizes around 10 times smaller than those of mixed vesicles and around twice smaller than pure vesicles, show similar ζ-potential values to those presented by vesicles. The independency of the ζ-potential with both $[S]_{tot}$ and α_1 contrasts with the changes observed on conductivity and cryo-TEM experiments.

Figure 13.13 shows the surface charge density, σ_ζ, calculated with Equation 13.4, as a function of α_1, for di-C_nDMAB/OBG vesicles [24,25]. An average value of $(17 \pm 5) \times 10^{-3}$ C m^{-2}, almost constant with α_1, has been obtained. A comparison of these σ_ζ values with those reported for the cationic–cationic systems di-C_nDMAB/C_mEDMAB [21,23,48] points to that nonionic OBG molecules that shield the electrostatic repulsion between the cationic quaternary ammonium heads, thus stabilizing the aggregates, increasing the counterion dissociation degree and, accordingly, ζ-potential and σ_ζ.

The structure and morphology of di-C_nDMAB/OBG mixed aggregates has been analyzed by measuring the fluorescent emission of the anionic (TNS) and nonionic (PRODAN) probes. Experimental emission spectra has been analyzed, after conversion of wavelength to frequency, by deconvolution into overlapping Gaussian curves with a nonlinear least-squares multipeaks procedure that uses an iterative Marquardt–Levenberg algorithm. The choice of the optimum number of reproducible components is based on the photophysical data of probe emission, the center, width, and amplitude of each Gaussian curve being the adjustable parameters. In order to assure the goodness of the deconvolution, several control tests are chosen: (1) the reproducibility of the final results for the center, width, and amplitude of the peaks; (2) a minimum value in the χ^2 parameter; (3) a random residual plot with no systematic features; and (4) maximum value for the square of the multiple correlation coefficients. Figures 13.14 through 13.17 show the

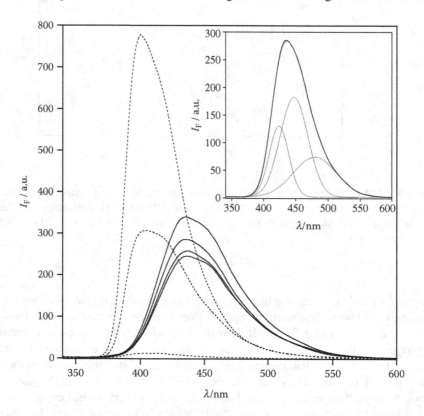

FIGURE 13.14 Emission fluorescence spectra of TNS ([TNS] = 2.07 µM) immersed on aqueous solutions of di-C_{10}DMAB/OBG at: (i) prevesicle concentrations (dash lines, in increasing intensity order), $\alpha_1 = 1$, $[S]_{tot} = 0.070$ mM; $\alpha_1 = 1$, $[S]_{tot} = 0.696$ mM; and $\alpha_1 = 0.502$, $[S]_{tot} = 0.698$ mM; and (ii) mixed vesicle concentrations (solid lines, in increasing intensity order), $\alpha_1 = 0.502$, $[S]_{tot} = 5.970$ mM; $\alpha_1 = 0.649$, $[S]_{tot} = 5.417$ mM; $\alpha_1 = 1$, $[S]_{tot} = 3.113$ mM; and $\alpha_1 = 0.502$, $[S]_{tot} = 8.132$ mM. The inset on the top shows the deconvolution of the emission fluorescence band of TNS immersed on a vesicle solution of di-C_{10}DMAB ($\alpha_1 = 1$, $[S]_{tot} = 3.113$ mM) on three Gaussian components. (From Junquera, E., del Burgo, P., Boskovic, J., and Aicart, E., *Langmuir*, 21, 7143, 2005. With permission.)

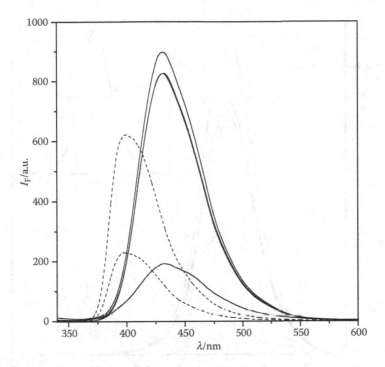

FIGURE 13.15 Emission fluorescence spectra of TNS ([TNS] = 2.07 μM) immersed on aqueous solutions of di-C_{12}DMAB/OBG at: (i) prevesicle concentrations (dash lines, in increasing intensity order), $\alpha_1 = 1$, $[S]_{tot} = 0.067$ mM; and $\alpha_1 = 0.499$, $[S]_{tot} = 0.070$ mM; and (ii) mixed vesicle concentrations (solid lines, in increasing intensity order), $\alpha_1 = 1$, $[S]_{tot} = 0.306$ mM; $\alpha_1 = 0.601$, $[S]_{tot}$ 3.076 mM; $\alpha_1 = 0.499$, $[S]_{tot} = 7.308$ mM; and $\alpha_1 = 0.499$, $[S]_{tot} = 4.461$ mM. (From del Burgo, P., Aicart, E., and Junquera, E., *Appl. Spectrosc.*, 60, 1307, 2006. With permission.)

emission spectra of TNS and PRODAN for a series of di-C_nDMAB/OBG/Probe aqueous solutions at different molar fractions, α_1, and total surfactant concentration, $[S]_{tot}$. These concentrations are chosen to have either monomers, prevesicles, or vesicles, according to the CPVC* and CVC* values obtained from conductivity experiments. Irrespective of the probe used, two groups of peaks are clearly observed in all the figures, those corresponding to monomers or prevesicle solutions (dashed lines), and those corresponding to vesicle solutions (solid lines). In Figures 13.14 and 13.15, the bands of TNS emission of prevesicle solutions (dashed lines) are blue shifted with respect to the TNS emission bands of vesicle solutions (solid lines), while in Figures 13.16 and 13.17, the bands of the PRODAN emission prevesicle solutions (dashed lines) are red shifted with respect to the TNS emission bands of vesicle solutions (solid lines). The insets on the top in Figures 13.14 and 13.16 show, as examples, the deconvolution of TNS and PRODAN emissions. Tables 13.3 and 13.4 resume the fit parameters for all the

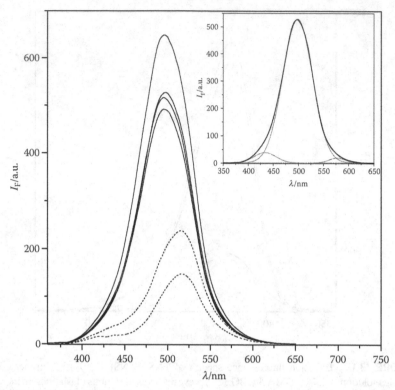

FIGURE 13.16 Emission fluorescence spectra of PRODAN ([PRODAN] = 5.02 μM) immersed on aqueous solutions of di-C_{10}DMAB/OBG at: (i) prevesicle concentrations (dash lines, in increasing intensity order), $\alpha_1 = 0.502$, $[S]_{tot} = 0.701$ mM; $\alpha_1 = 1$, $[S]_{tot} = 0.700$ mM; and (ii) mixed vesicle concentrations (solid lines, in increasing intensity order), $\alpha_1 = 0.649$, $[S]_{tot} = 5.455$ mM; $\alpha_1 = 0.502$, $[S]_{tot} = 8.134$ mM; $\alpha_1 = 0.502$, $[S]_{tot} = 5.986$ mM; and $\alpha_1 = 1$, $[S]_{tot} = 3.122$ mM. The inset on the top shows the deconvolution of the emission fluorescence band of PRODAN immersed on a vesicle solution of di-C_{10}DMAB ($\alpha_1 = 0.502$, $[S]_{tot} = 5.986$ mM) on three Gaussian components. (From Junquera, E., del Burgo, P., Boskovic, J., and Aicart, E., *Langmuir*, 21, 7143, 2005. With permission.)

experimental spectra of TNS and PRODAN, respectively. The mathematical treatment points to the presence of three microenvironments, in both prevesicle and vesicle solutions. Note that λ_i remains roughly constant within each group, irrespective of (1) the molar fraction, α_1; (2) the monomer, prevesicle, and vesicle concentration and; (3) the length of the double chain hydrocarbon tail of the surfactant. Assignment of the bands to the microenvironments is based on polarity criteria [28,119,152–154]; thus, as more hydrophobic is the microenvironment where the probe is housed more blue shifted (shorter wavelength) the band will appear. In the presence of vesicles, the probe may be solubilized in the hydrophobic bilayer or in the vesicle surface. Prevesicle nanoaggregates also offer distinct sites

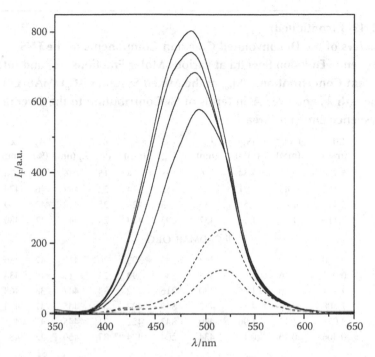

FIGURE 13.17 Emission fluorescence spectra of PRODAN ([PRODAN] = 5.02 μM) immersed on aqueous solutions of di-C_{12}DMAB/OBG at: (i) prevesicle concentrations (dash lines, in increasing intensity order), $\alpha_1 = 0.499$, $[S]_{tot} = 0.070$ mM; and $\alpha_1 = 1$, $[S]_{tot} = 0.067$ mM; and (ii) mixed vesicle concentrations (solid lines, in increasing intensity order), $\alpha_1 = 1$, $[S]_{tot} = 0.326$ mM; $\alpha_1 = 0.499$, $[S]_{tot} = 4.488$ mM; $\alpha_1 = 0.499$, $[S]_{tot} = 7.327$ mM; and $\alpha_1 = 0.601$, $[S]_{tot} = 3.111$ mM. (From del Burgo, P., Aicart, E., and Junquera, E., *Appl. Spectrosc.*, 60, 1307, 2006. With permission.)

TABLE 13.3
Parameters of the Deconvoluted Gaussian Components of the TNS Fluorescence Emission Spectra at Various Molar Fractions, α_1, and Total Surfactant Concentrations, $[S]_{tot}$ of the Mixed System di-C_nDMAB/OBG: Wavelength λ_i, and Area A_i in Terms of % Contribution to the Overall Fluorescence Emission Area

α_1	$[S]_{tot}$ (mM)	CVC* (mM)	$[S]_v$ (mM)	λ_{max} (nm)	I_{max}	λ_1 (nm)	A_1 (%)	λ_2 (nm)	A_2 (%)	λ_3 (nm)	A_3 (%)
				di-C_{10}DMAB/OBG							
0.502	0.698	2.43	—	396	803	395	22	412	46	437	32
1	0.070	1.33	—	416	12	397	18	415	49	445	33
1	0.696	1.33	—	402	314	396	17	415	43	444	40

(continued)

TABLE 13.3 (continued)

Parameters of the Deconvoluted Gaussian Components of the TNS Fluorescence Emission Spectra at Various Molar Fractions, α_1, and Total Surfactant Concentrations, $[S]_{tot}$ of the Mixed System di-C_nDMAB/OBG: Wavelength $\lambda_{i'}$ and Area A_i in Terms of % Contribution to the Overall Fluorescence Emission Area

α_1	$[S]_{tot}$ (mM)	CVC* (mM)	$[S]_v$ (mM)	λ_{max} (nm)	I_{max}	λ_1 (nm)	A_1 (%)	λ_2 (nm)	A_2 (%)	λ_3 (nm)	A_3 (%)
0.502	5.970	2.43	3.54	436	258	424	15	446	49	476	35
0.502	8.132	2.43	5.71	433	352	424	22	447	46	479	32
0.649	5.417	1.89	3.53	435	263	424	25	448	43	479	32
1	3.113	1.33	1.79	437	293	424	24	447	47	480	29
di-C_{12}DMAB/OBG											
0.499	0.070	2.50	—	397	635	393	20	410	47	436	33
1	0.067	0.16	—	398	236	393	22	411	48	438	30
0.499	4.461	2.50	1.96	430	916	422	34	447	38	468	27
0.499	7.308	2.50	4.81	432	845	422	30	445	47	474	23
0.601	3.076	1.12	1.96	430	843	422	31	445	47	475	22
1	0.306	0.16	0.15	434	204	424	37	450	43	487	19

Note: CVC* and $[S]_v$ are the mixed critical vesicle concentration and the surfactant concentration on vesicle form.

TABLE 13.4

Parameters of the Deconvoluted Gaussian Components of the PRODAN Fluorescence Emission Spectra at Various Molar Fractions, α_1, and Total Surfactant Concentrations, $[S]_{tot}$ of the Mixed System di-C_nDMAB/OBG: Wavelength $\lambda_{i'}$ and Area A_i in Terms of % Contribution to the Overall Fluorescence Emission Area

α_1	$[S]_{tot}$ (mM)	CVC* (mM)	$[S]_v$ (mM)	λ_{max} (nm)	I_{max}	λ_1 (nm)	A_1 (%)	λ_2 (nm)	A_2 (%)	λ_3 (nm)	A_3 (%)
di-C_{10}DMAB/OBG											
0.502	0.701	2.43	—	516	152	460	10	515	85	576	5
1	0.700	1.33	—	515	247	460	20	515	75	576	5
0.502	5.986	2.43	3.56	495	538	434	5	499	93	577	2
0.502	8.134	2.43	5.71	493	526	432	5	498	93	579	2
0.649	5.455	1.89	3.57	498	502	433	5	498	94	578	1
1	3.123	1.33	1.80	495	665	440	7	499	91	579	2

TABLE 13.4 (continued)

Parameters of the Deconvoluted Gaussian Components of the PRODAN Fluorescence Emission Spectra at Various Molar Fractions, α_1, and Total Surfactant Concentrations, $[S]_{tot}$ of the Mixed System di-C_nDMAB/OBG: Wavelength λ_i, and Area A_i in Terms of % Contribution to the Overall Fluorescence Emission Area

α_1	$[S]_{tot}$ (mM)	CVC* (mM)	$[S]_v$ (mM)	λ_{max} (nm)	I_{max}	λ_1 (nm)	A_1 (%)	λ_2 (nm)	A_2 (%)	λ_3 (nm)	A_3 (%)
				di-C_{12}DMAB/OBG							
0.499	0.070	2.50	—	518	129	438	9	515	86	576	5
1	0.067	0.16	—	519	250	456	12	515	83	576	5
0.499	4.488	2.50	1.99	492	697	471	60	505	39	577	1
0.499	7.327	2.50	4.83	482	764	462	49	501	49	576	2
0.601	3.111	1.12	1.99	483	819	464	54	503	45	575	1
1	0.326	0.16	0.17	493	590	480	64	510	34	578	2

Note: CVC* and $[S]_v$ are the mixed critical vesicle concentration and the surfactant concentration on vesicle form.

of variable polarity, inside of the nanoaggregate and as a labile cluster where, due to a certain water penetration, the probe experiments a less apolar microenvironment. Last kind of associations is well documented in the literature [18,24,48,118,120] and permit to explain how fluorescent probes that weakly emit in water present a notorious emission in the presence of surfactants at concentration below its CVC or CMC, where only monomers should be present. Furthermore, irrespective of the type of aggregates, there is a third microenvironment present in solution—the bulk. It is expected that the emission band for this third environment is red shifted because it is supposed to be the more polar one. Effectively, this happens for the prevesicle and vesicle solutions of PRODAN, but anionic TNS experiments a higher polarity on the positive charged vesicle surface than that in the bulk and shows a different pattern. This only can be due to the anionic character of TNS interacting with the positive charged surface of the vesicles. Based on all the mentioned premises, the assignment done by the different bands is detailed as follows.

The Gaussian components of TNS emission spectra of the prevesicle solutions (Table 13.3, rows 1–3 for di-C_{10}DMAB/OBG and rows 1–2 for di-C_{12}DMAB/OBG) are centered at around $\lambda_1 = (394 \pm 4)$, $\lambda_2 = (412 \pm 3)$, and $\lambda_3 = (440 \pm 5)$ nm, while the peak components of spectra of the vesicle solutions (Table 13.3, last four rows for di-C_{10}DMAB/OBG and last four rows for di-C_{12}DMAB/OBG) are centered at around $\lambda_1 = (423 \pm 4)$, $\lambda_2 = (445 \pm 5)$, and $\lambda_3 = (477 \pm 10)$ nm. These peaks can be assigned to the next microenvironments: (1) the 445 nm emission, the unique peak in common on vesicle and prevesicle solutions, is assigned to

TNS in the bulk; in the presence of vesicles, (2) the emission at around 423 nm is attributed to TNS inside the bilayer, because as it is the most blue shifted emission it is expected to be the more hydrophobic region; and (3) the emission at around 477 nm, the most red shifted one, is assumed to come from TNS in the vesicle surface. This indicates that TNS experiences a lower energy transition when it is housed in the surface, than that one found in the bulk, the charged head of the surfactant yielding a higher polar microenvironment than that in the bulk. Furthermore, as it has been reported that TNS fluoresces in aqueous surfactant solutions below CVC or CMC [18,25,44,48,120], assignment of the other two peaks on the spectra for prevesicle solutions is as follows: (1) the emission at around 394 nm is attributed to TNS housed within clusters formed by the di-C_{10}DMA$^+$ or di-C_{12}DMA$^+$ and OBG monomers. As these clusters possess a rigid hydrophobic cavity, the band is blue shifted with respect to that found in the bulk; and (2) the peak at around 412 nm may be attributed to the formation of ion–pair complexes or to the presence of certain microdomains in the solution, that must be more loose than the hydrophobic cavity, since this microenvironment should be more polar than that rigid hydrophobic cavity, but less polar than the bulk. This assignment is supported by the literature [18,25,44,48,120,167], where it is reported that the association of a fluorescent probe with monomers or clusters of surfactant monomers in aqueous media could be the responsible for the emission observed at concentrations below the CMC. It is also remarkable (see Table 13.3) that in the presence of vesicles I_{max} remains basically constant at different α_1 and constant $[S]_v$ (= $[S]_{tot}$ – CVC*), confirming that vesicle concentration mainly governs the intensity of the peak. On the other hand, it can be observed as well that in the absence of vesicles, I_{max} decreases with α_1, at constant $[S]_{tot}$, which indicates that the sugar surfactant has a participation in the microenvironments by creating, together with the dialkylsurfactant, a more hydrophobic region than the one formed by di-C_nDMA$^+$ alone. In any case, the micropolarity of the microdomains below CVC* is lower than that one found within the vesicles, which produces a blue shift of the TNS spectra with respect to the spectra obtained in the presence of vesicles, in agreement with previous results for TNS in surfactant solutions of a single tail surfactant (C_{16}TAB) [120].

The Gaussian components of the PRODAN emission spectra obtained for prevesicle solutions (Table 13.4, rows 1 and 2 for di-C_{10}DMAB/OBG and rows 1 and 2 for di-C_{12}DMAB/OBG) are centered at around λ_1 = (450 ± 10), λ_2 = (515 ± 1), and λ_3 = (576 ± 2) nm, while those for vesicle solutions (Table 13.4, last four rows for di-C_{10}DMAB/OBG and last four rows for di-C_{12}DMAB/OBG) are centered at around λ_1 = (460 ± 20), λ_2 = (508 ± 10), and λ_3 = (576 ± 5) nm. The peaks of PRODAN are assigned as follows: (1) the emission at 576 nm, the unique common peak on vesicle and prevesicle solutions, is assigned to PRODAN in the bulk; in the presence of vesicles, (2) the emission at 460 nm is attributed to PRODAN inside the hydrophobic bilayer; and (3) the emission at around 508 nm is assumed to come from PRODAN in the vesicle surface, this band blue being shifted with respect to that from the bulk. In prevesicle solutions, the assignment of the other two peaks is similar to that proposed for TNS: (1) the emission at around 450 nm is also attributed to PRODAN immersed within clusters formed by the di-C_nDMA$^+$ and OBG monomers; and (2) the peak at around 515 nm is attributed to the loose microdomains, since the formation of ion-pair complexes is not possible in this case, given the nonionic character of PRODAN.

Several features can be deduced from Table 13.4: (1) at constant α_1, I_{max} remains roughly constant with increasing $[S]_v$ in the vesicles domain, which can be explained if the probe is almost totally immersed in the vesicle; (2) at constant $[S]_v$, I_{max} decreases with α_1; (3) in the absence of vesicles, I_{max} is lower than that in its presence, irrespective of α_1, while at constant $[S]_{tot}$ (below CVC*), I_{max} increases with α_1, which indicate that, in contrast with what it is found for TNS, in the case of PRODAN, the OBG is not mainly responsible for the hydrophobicity of the microdomains. In conclusion, the micropolarity that PRODAN experiments in the microdomains below CVC* is slightly higher than that felt within the vesicles, this fact produces a red shift of the spectra compared with those obtained in the presence of vesicles.

Table 13.3 (TNS) and Table 13.4 (PRODAN) also resume the areas A_i of the deconvoluted bands for each emission peak, in terms of percentage contribution to the overall fluorescence emission. Since quantum yield of fluorescence is usually much more intense in hydrophobic than in hydrophilic environments, a larger spectral area of a peak that is blue-shifted with respect to other is not necessarily correlated with a higher probe concentration. But if the peak with the larger spectral area is red-shifted, a correlation can be established between this intensity (or area) and a higher probe concentration in this more polar microenvironment. In this case, the quantum yield is expected to be lower, so a larger area necessarily has to do with a higher probe concentration. In that sense, in view of the results reported in Tables 13.3 and 13.4, it is possible to say that both probes prefer to be solubilized within the vesicles, because $(A_1 + A_3) > A_2$ for TNS and $(A_1 + A_2) > A_3$ for PRODAN. Nevertheless, TNS remains on the bulk in remarkable proportion $(A_2 \approx 45\%)$, while that of PRODAN in the bulk is negligible $(A_3 \approx 2\%)$. This fact can be explained by the nonionic character of PRODAN, which favor the solubilization in hydrophobic media better than the anionic TNS probe does. Furthermore, both TNS and PRODAN prefer the vesicle surface to the bilayer for the system di-C_{10}DMAB/OBG $(A_3 > A_1$ for TNS and $A_2 > A_1$ for PRODAN). In contrast, nothing conclusive can be said about the preferred solubilization site of both probes in the case of di-C_{12}DMAB/OBG vesicles $(A_1 > A_3$ for TNS and $A_1 > A_2$ for PRODAN). These results could be indicating that length of the double-chain surfactant and, accordingly, the thickness of the hydrophobic bilayer are important factors with respect to the solubilization site of the probe.

A comparison of λ_i of the bands reported in Tables 13.3 and 13.4 with the λ_{max} values found in the literature for these probes in a series of pure solvents of known dielectric constant [118,155] points to several conclusions with respect to the polarity of the colloidal microenvironments herein studied: (1) the results obtained for TNS indicate that the hydrophobic core of the studied vesicles $(\lambda_1 = 423$ nm) where TNS is housed presents a polarity similar to that of butanol $(\varepsilon = 17)$, while that of the vesicle surface $(\lambda_3 = 477$ nm) resembles an ethylene glycol environment ($\varepsilon = 42$); and (2) the results obtained from PRODAN spectra reveal an hydrophobic core ($\lambda_1 = 460$ nm) where PRODAN is housed similar to that of dimethylformamide ($\varepsilon = 37$), and a palisade ($\lambda_2 = 505$ nm) with a polarity similar to that of methanol ($\varepsilon = 33$). As the vesicles for both systems are very similar, except that the di-C_{12}DMAB/OBG bilayer is about 0.5 nm thicker than that of di-C_{10}DMAB/OBG, the fluorescence results point to the fact that the non-ionic PRODAN is located deeper in the hydrophobic core of the vesicles than the anionic TNS, and, furthermore, it selects zones of vesicle surface richer on non-ionic OBG than TNS does.

From a theoretical point of view, we have applied to the di-C_nDMAB/OBG ($n = 10$, 12) systems the original DLVO theory, and also the Inoue and Sogami models and their combinations thereof, which try to improve DLVO predictions. For this purpose, calculations have been made by using the equations resumed in the Section 13.4.2, where all the magnitudes are defined with the exception of ψ_0, which has been substituted by the experimental zeta potential, ζ, as usually done in this type of calculations [128,131]. Figure 13.18 shows as an example a

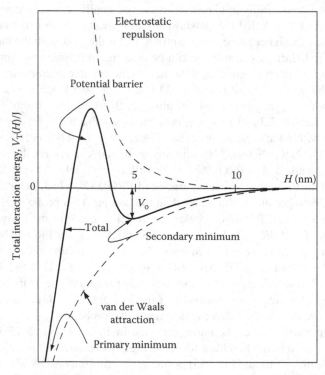

FIGURE 13.18 Simplified diagram showing the primary and secondary minimum and the potential barrier according to a typical DLVO interaction potential. (Adapted from Israelachvili, P., *Intermolecular and Surfaces Forces*, Academic Press Inc., London, 1992. With permission.)

TABLE 13.5

Values of the Interaction Energy and Distance at the Energy Barrier, $V_{T,max}$ and H_{max}, and/or at the Secondary Minimum, $V_{T,min}$ and H_{min}, of the DLVO Theory, the Theoretical Models of Inoue and Sogami, and the Combinations thereof, Applied to di-C_nDMAB/OBG Vesicles at Different Molar Fractions, α_1, of the System

		DLVO		Inoue		Sogami		DLVO–Sogami		Inoue–Sogami	
		10^{19}		10^{21}		10^{22}		10^{19}		10^{21}	
α_1	$[S]_{tot}$ (mM)	$V_{T,max}$ (J)	H_{max} (nm)	$V_{T,max}$ (J)	H_{max} (nm)	$V_{T,min}$ (J)	H_{min} (nm)	$V_{T,max}$ (J)	H_{max} (nm)	$V_{T,max}$ (J)	H_{max} (nm)
					di-C_{10}DMAB/OBG						
0.401	6.076	9.8	1.0	33	4.1	−7.5	18	10	1.0	46	3.3
0.649	7.890	9.4	0.9	19	4.1	−6.9	13	9.6	0.8	26	3.4
0.749	9.788	9.7	0.8	17	3.8	−7.7	11	9.9	0.8	23	3.2
0.749	6.116	10	0.9	22	4.1	−7.3	13	10	0.9	30	6.4
0.848	9.772	9.0	0.8	9.8	4.3	−6.6	10	9.2	0.8	13	3.6
1	2.427	3.6	1.0	9.3	4.5	−2.4	18	3.7	1.0	15	3.4
					di-C_{12}DMAB/OBG						
0.401	8.859	10	0.9	33	3.8	−8.6	15	11	0.9	46	3.1
0.480	1.55	13	1.2	68	4.6	−11	33	14	1.2	102	3.5
0.500	7.327	11	0.9	29	4.1	−8.3	15	11	0.9	40	3.3
0.569	2.191	14	1.1	58	4.4	−11	25	14	1.1	84	3.5
0.569	1.361	14	1.3	60	4.9	−10	33	14	1.2	88	3.8
0.601	4.978	13	0.9	40	4.0	−10	16	13	0.9	56	3.2
0.796	1.729	17	1.1	77	4.2	−14	24	18	1.1	110	3.3
0.796	1.078	18	1.2	82	4.6	−14	31	18	1.2	118	3.6
0.800	4.889	13	0.9	26	4.6	−8.9	14	14	0.9	36	3.7
1	0.305	4.7	1.6	24	5.2	—	—	5.1	1.5	55	3.0

simplified diagram showing the primary and secondary minimum, and the potential barrier according to a typical DLVO interaction potential and Table 13.5 resumes all the theoretical parameters calculated for the systems herein reported. As can be deduced from the table, both DLVO and Inoue's models yield similar trends for pure di-C_nDMAB vesicles, with a primary minimum at very short distances ($H \to 0$) and an energy barrier, but without a secondary minimum. The existence of an energy barrier confirms the stability of the system, avoiding particle coagulation, in agreement with experimental evidences. Note (see Table 13.5, $\alpha_1 = 1$, last row for each system) that the energy barrier is higher ($V_{T,max} \approx 10^{-19}$ J) and appears at shorter distances (H_{max} around 1 nm) when applying DLVO theory than in the case of Inoue's model ($V_{T,max} \approx 10^{-21}$ J and H_{max} around 5 nm). However, Sogami's model, developed for medium-to-long distances, predicts total energy values much less than the other models and an energy minimum at longer distances ($H_{min} \approx 18$ nm) that would be pointing to a flocculation regime in the medium distance range. In the case of mixed di-C_nDMAB/OBG vesicles, several general conclusions regarding DLVO and Inoue models, valid at any distance H, can be remarked:

1. An energy barrier at H_{max} is predicted at around 1 nm with the DLVO theory and at around 4–5 nm with the Inoue's model for both systems at any composition. Both distance values are much lower than vesicles size. Furthermore, no secondary minimum appears, revealing the total absence of flocculation phenomena.
2. The energy barrier (higher again with DLVO) seems unaffected by the molar fraction, α_1, for DLVO and the contrary is true for Inoue. On the other hand, Sogami's model predicts a minimum for both systems, irrespective of the molar fraction, α_1, at longer distances ($H_{min} \approx 15$ nm for di-C_{10}DMAB/OBG and 15–30 nm for di-C_{12}DMAB/OBG). This minimum is characterized by a very low stabilization energy, much lower than that one found for latex dispersions [161].

Additionally, the distance at which the secondary minimum, H_{min}, of the vesicles theoretically studied in this chapter appears is one order of magnitude lower than that one of the above-mentioned latex dispersions. Figure 13.19a and b shows, as an example, the repulsive, V_R, attractive, V_A, and total interaction energies, V_T, as a function of the distance, H, obtained by the DLVO theory and the Inoue's model, while Figure 13.19c shows the total interaction energy obtained with the three theories, DLVO, Inoue, and Sogami, for pure di-C_{10}DMAB vesicles. Similar plots have been drawn for mixed di-C_nDMAB/OBG vesicles. Table 13.5 also reports the results obtained when applying both extended models (DLVO–Sogami potential and Inoue–Sogami potential) to the vesicles previously studied with DLVO, Inoue, and Sogami models, at several molar fractions. From this table, it can be deduced that the inclusion of Sogami's electrostatic potential at medium-to-long distances in the DLVO theory does not almost modify $V_{T,max}$, or H_{max} values and, accordingly, all the corresponding conclusions commented for the DLVO theory remain unaffected. It is due to the fact that $V_{T,Sogami}$ is less than 1% of the DLVO potential. The minimum predicted by the Sogami model at medium distances disappears when using the extended DLVO–Sogami model. However, when the results obtained with Inoue's model are compared with those yielded by Inoue–Sogami's model, appreciable differences can be observed: $V_{T,max}$ are 30% higher with this extended model. The position of the energy barrier, H_{max}, is also affected, being shifted around 30% toward shorter distances (3–4 nm). With respect to the secondary minimum, found at medium distances with the Sogami's model, it is noticeable that it also disappears with the Inoue–Sogami model.

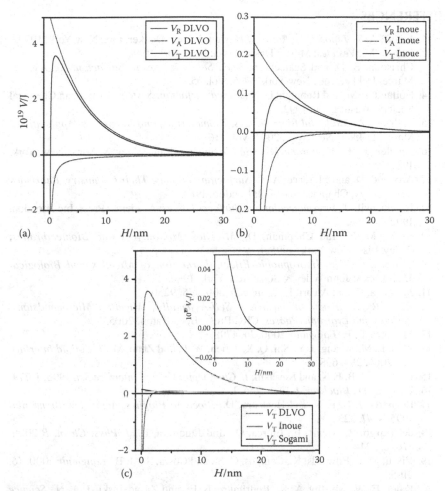

FIGURE 13.19 Plot of the total potential, V_T, and the attractive, V_A, and repulsive, V_R, contributions as a function of interparticle distance, H, obtained from: (a) DLVO theory and (b) Inoue's model; (c) Comparison of the results for V_T versus H, obtained from DLVO, Inoue, and Sogami models. The inset shows a zoom view of $V_{T,Sogami}$ versus H, showing the secondary minimum predicted by Sogami's model.

It can be concluded that, generally and qualitatively speaking, the models proposed by Inoue, Sogami, or even the extended DLVO–Sogami, and Inoue–Sogami models, do not significantly improve the theoretical predictions of diluted vesicle systems with respect to those of the classical DLVO theory. However, it is remarkable that the extended models analyzed in this chapter have the advantage of being applicable in a wide range of interparticle distance, which allows for the characterization of diluted or even medium concentrated vesicle systems.

REFERENCES

1. Janoff, A. S., *Liposomes: Rational Design*; Marcel Dekker, Inc.: New York, 1999.
2. Rosoff, M., *Vesicles*; Marcel Dekker, Inc.: New York, 1996.
3. Christian, S. D. and Scamehorn, J. F., *Solubilization in Surfactant Aggregates*; Marcel Dekker, Inc.: New York, 1995; Vol. 55.
4. Holland, P. M. and Rubingh, D. N., *Mixed Surfactant Systems*; American Chemical Society: Washington, 1992.
5. Elaissari, A., *Colloidal Biomolecules, Biomaterials, and Biomedical Applications*; Marcel Dekker: New York, 2004; Vol. 116.
6. Fendler, J. H., *Membrane Mimetic Chemistry*; John Wiley & Sons: New York, 1982.
7. Attwood, D. and Florence, A. T., *Surfactant Systems: Their Chemistry, Pharmacy and Biology*; Chapman and Hall: London, 1983.
8. Israelachvili, J., *Intermolecular and Surfaces Forces*; Academic Press, Inc.: London, 1992.
9. Jones, M. N. and Chapman, D., *Micelles, Monolayers and Biomembranes*; Wiley-Liss: New York, 1995.
10. Tanford, C., *The Hydrophobic Effect: Formation of Micelles and Biological Membranes*; John Wiley & Sons: New York, 1980.
11. Junquera, E. and Aicart, E., *Langmuir* 2002, *18*, 9250.
12. Zana, R., *Dynamics of Surfactant Self-Assemblies. Micelles, Microemulsions, Vesicles and Lyotropic Phases*; CRC Press: Boca Raton, 2005.
13. Marques, E. F., *Langmuir* 2000, *16*, 4798.
14. Zhai, L. M., Zhang, J. Y., Shi, Q. X., Chen, W. J., and Zhao, M., *J. Colloid Interface Sci.* 2005, *284*, 698.
15. Engberts, J. B. F. N. and Kevelam, J., *Curr. Opin. Colloid Interface Sci.* 1996, *1*, 779.
16. Lasic, D. D., *Biochem. J.* 1988, *256*, 1.
17. Engberts, J. B. F. N. and Hoekstra, D., *Biochim. Biophys. Acta: Rev. Biomembr.* 1995, *1241*, 323.
18. del Burgo, P., Aicart, E., Llorca, O., and Junquera, E., *J. Phys. Chem. B* 2006, *110*, 23524.
19. Viseu, M. I., Edwards, K., Campos, C. S., and Costa, S. M. B., *Langmuir* 2000, *16*, 2105.
20. Kaler, E. W., Murthy, A. K., Rodriguez, B. E., and Zasadzinski, J. A. N., *Science* 1989, *245*, 1371.
21. del Burgo, P., Aicart, E., and Junquera, E., *Colloids Surf. A: Physicochem. Eng. Aspects* 2007, *292*, 165.
22. Fontana, A., De Maria, P., Siani, G., and Robinson, B. H., *Colloids Surf. B: Biointerfaces* 2003, *32*, 365.
23. Junquera, E., Arranz, R., and Aicart, E., *Langmuir* 2004, *20*, 6619.
24. Junquera, E., del Burgo, P., Arranz, R., Llorca, O., and Aicart, E., *Langmuir* 2005, *21*, 1795.
25. Junquera, E., del Burgo, P., Boskovic, J., and Aicart, E., *Langmuir* 2005, *21*, 7143.
26. Ollivon, M., Lesieur, S., Grabielle-Madelmont, C., and Paternostre, M., *Biochim. Biophys. Acta* 2000, *1508*, 34.
27. Viseu, M. I., Velazquez, M. M., Campos, C. S., Garcia-Mateos, I., and Costa, S. M. B., *Langmuir* 2000, *16*, 4882.
28. Karukstis, K. K., Zielemiuk, C. A., and Fox, M. J., *Langmuir* 2003, *19*, 10054.
29. Almgren, M. and Rangelov, S., *Langmuir* 2004, *20*, 6611.

30. Yatcilla, M. T., Herrington, K. L., Brasher, L. L., and Kaler, E. W., *J. Phys. Chem.* 1996, *100*, 5874.
31. Safran, A., Pincus, P., and Andelman, D., *Science* 1990, *248*, 354.
32. Safran, A., Pincus, P. A., Andelman, D., and MacKintosh, F. C., *Phys. Rev. A* 1991, *43*, 1071.
33. Helfrich, W., *Zeit. Naturf. C. J. Biosci.* 1973, *C 28*, 693.
34. Helfrich, W., *J. Phys.* 1986, *47*, 321.
35. Deo, N. and Somasundaran, P., *Colloids Surf. A: Physicochem. Eng. Aspects* 2001, *186*, 33.
36. Deo, N. and Somasundaran, P., *Langmuir* 2003, *19*, 2007.
37. Egermayer, M. and Piculell, L., *J. Phys. Chem. B* 2003, *107*, 14147.
38. Hildebrand, A., Neubert, R., Garidel, P., and Blume, A., *Langmuir* 2002, *18*, 2836.
39. Hildebrand, A., Beyer, K., Neubert, R., Garidel, P., and Blume, A., *Colloids Surf. B: Biointerfaces* 2003, *32*, 335.
40. Kawasaki, H., Imahayashi, R., Tanaka, S., Almgren, M., Karlsson, G., and Maeda, H., *J. Phys. Chem. B* 2003, *107*, 8661.
41. Majhi, P. R. and Blume, A., *J. Phys. Chem. B* 2002, *106*, 10753.
42. Lichtenberg, D., Robson, R. J., and Dennis, E. A., *Biochim. Biophys. Acta* 1983, *737*, 285.
43. Lichtenberg, D., *Biochim. Biophys. Acta* 1985, *821*, 470.
44. del Burgo, P., Aicart, E., and Junquera, E., *Appl. Spectrosc.* 2006, *60*, 1307.
45. Lopez, O., Cocera, M., Coderch, L., Parra, J. L., Barsukov, L., and de la Maza, A., *J. Phys. Chem. B* 2001, *105*, 9879.
46. Ribosa, I., Sanchez-Leal, J., Comelles, F., and Garcia, M. T., *J. Colloid Interface Sci.* 1997, *187*, 443.
47. Vacklin, H. P., Tiberg, F., and Thomas, R. K., *Biochim. Biophys. Acta* 2005, *1668*, 17.
48. Aicart, E., del Burgo, P., Llorca, O., and Junquera, E., *Langmuir* 2006, *22*, 4027.
49. del Burgo, P., Ph. D. Thesis, Universidad Complutense, Madrid, Spain, 2007.
50. Jungermann, E., *Cationic Surfactants*; Marcel Dekker: New York, 1970.
51. Argy, G., Bricout, F., d'Hermies, F., and Cheymol, A., *C. R. Acad. Sci. Ser. III. Sci. Vie. Life Sci.* 1999, *322*, 863.
52. Budavari, S., *The Merck Index*; Merck & Co., Inc.: Whitehouse Station, New Jersey, 1996.
53. Arévalo, J. M., Arribas, J. L., Hernández, M. J., Lizán, M., and Herruzo, R., *Medicina Preventiva* 2001, *7*, 17.
54. Aswal, V. K. and Goyal, P. S., *Chem. Phys. Lett.* 2002, *364*, 44.
55. Czerniawski, M., *Rocz. Chem.* 1966, *40*, 1935.
56. Klevens, H. B., *J. Phys. Colloid Chem.* 1948, *52*, 130.
57. Lianos, P. and Zana, R., *J. Colloid Interface Sci.* 1982, *88*, 594.
58. Lu, J. R., Simister, E. A., Thomas, R. K., and Penfold, J., *J. Phys. Chem.* 1993, *97*, 13907.
59. Zana, R., *J. Colloid Interface Sci.* 1980, *78*, 330.
60. Moulik, S. P., Haque, M. E., Jana, P. K., and Das, A. R., *J. Phys. Chem.* 1996, *100*, 701.
61. Zana, R. and Levy, H., *Colloids Surf. A: Physicochem. Eng. Aspects* 1997, *127*, 229.
62. Bales, B. L. and Zana, R., *J. Phys. Chem. B* 2002, *106*, 1926.
63. Wiacek, A. E., Chibowski, E., and Wilk, K., *Colloids Surf. A: Physicochem. Eng. Aspects* 2001, *193*, 51.
64. Blandamer, M. J., Briggs, B., Cullis, P. M., Irlam, K. D., Kirby, S. D., and Engberts, J., *J. Mol. Liq.* 1998, *75*, 181.

65. Clancy, S. F., Steiger, P. H., Tanner, D. A., Thies, M., and Paradies, H. H., *J. Phys. Chem.* 1994, *98*, 11143.
66. Dubois, M., Gulikkrzywicki, T., and Cabane, B., *Langmuir* 1993, *9*, 673.
67. Fontell, K., Ceglie, A., Lindman, B., and Ninham, B., *Acta Chemica Scandinavica Series A: Phys. Inorg. Chem.* 1986, *40*, 247.
68. Marques, E., Khan, A., Miguel, M. D., and Lindman, B., *J. Phys. Chem.* 1993, *97*, 4729.
69. Marques, E. F., Regev, O., Khan, A., Miguel, M. D., and Lindman, B., *J. Phys. Chem. B* 1999, *103*, 8353.
70. Svitova, T., Smirnova, Y., and Yakubov, G., *Colloids Surf. A: Physicochem. Eng. Aspects* 1995, *101*, 251.
71. Svitova, T. F., Smirnova, Y. P., Pisarev, S. A., and Berezina, N. A., *Colloids Surf. A: Physicochem. Eng. Aspects* 1995, *98*, 107.
72. Baron, C. and Thompson, T. E., *Biochim. Biophys. Acta* 1975, *382*, 276.
73. Korsaric, N., *Biosurfactants: Production. Properties. Applications*; Marcel Dekker, Inc.: New York, 1993; Vol. 48.
74. Stubbs, G. W., Gilbert, S. H., and Litman, B. J., *Biochim. Biophys. Acta* 1976, *425*, 45.
75. Garavito, R. M. and Rosenbush, J. P., *J. Cell. Biol.* 1980, *86*, 327.
76. Michel, H. and Oesterhelt, D., *Proc. Nat. Acad. Sci., U S A* 1980, *77*, 1283.
77. Cabezas, J. A., Reglero, A., Garcia, J. I., Rodrigo, M., Martinez, V., Cabezas, M., Fitt, H., and Hueso, P., *Jano* 1989, *37*, 43.
78. Ceccarelli, B., Aporti, and Finesso, M. *Adv. Exp. Med. Biol.* 1976, *71*, 275.
79. Grip, W. J. and Bovee-Geurts, P. H. M., *Chem. Phys. Lipids* 1979, *23*, 321.
80. Hakomori, S., *Invest. Ciencia* 1986, *118*, 14.
81. Carnero-Ruiz, C., Molina-Bolivar, J. A., Aguiar, J., and Peula-Garcia, J. M., *Colloids Surf. A: Physicochem. Eng. Aspects* 2004, *249*, 35.
82. Hierrezuelo, J. M., Aguiar, J., and Carnero-Ruiz, C., *Langmuir* 2004, *20*, 10419.
83. Hierrezuelo, J. M., Aguiar, J., and Ruiz, C. C., *Colloids Surf. A: Physicochem. Eng. Aspects* 2005, *264*, 29.
84. Hierrezuelo, J. M., Aguiar, J., and Ruiz, C. C., *J. Colloid Interface Sci.* 2006, *294*, 449.
85. Molina-Bolivar, J. A., Aguiar, J., Peula-Garcia, J. M., and Carnero-Ruiz, C., *J. Phys. Chem. B* 2004, *108*, 12813.
86. Molina-Bolivar, J. A., Hierrezuelo, J. M., and Ruiz, C. C., *J. Phys. Chem. B* 2006, *110*, 12089.
87. Hinz, H. J., Six, L., Ruess, K. P., and Lieflander, M., *Biochemistry* 1985, *24*, 806.
88. Crowe, L. M. M., Mouradian, R., Crowe, J. H., Jackson, S. A., and Womersley, C., *Biochim. Biophys. Acta* 1984, *769*, 141.
89. Shinoda, K., Yamaguchi, T., and Hori, R., *Bull. Chem. Soc. Jpn.* 1961, *34*, 237.
90. Kameyama, K. and Takagi, T., *J. Colloid Interface Sci.* 1990, *137*, 1.
91. Pastor, O., Junquera, E., and Aicart, E., *Langmuir* 1998, *14*, 2950.
92. Jeffrey, G. A., *Acc. Chem. Res.* 1986, *19*, 168.
93. von Rybinski, W., *Curr. Opin. Colloid Interface Sci.* 1996, *1*, 587.
94. La Mesa, C., Bonincontro, A., and Sesta, B., *Colloid Polym. Sci.* 1993, *271*, 1165.
95. Roxhy, R. W. and Mills, B. P., *J. Phys. Chem.* 1990, *94*, 456.
96. Rosen, M. J. and Sulthana, S. B., *J. Colloid Interface Sci.* 2001, *239*, 528.
97. Sulthana, S. B., Rao, P. V. C., Bhat, S. G. T., Nakano, T. Y., Sugihara, G., and Rakshit, A. K., *Langmuir* 2000, *16*, 980.
98. Yunomiya, Y., Kunitake, T., Tanaka, T., Sugihara, G., and Nakashima, T., *J. Colloid Interface Sci.* 1998, *208*, 1.

99. Zana, R., *Surfactant Solutions: New Methods of Investigation*; Marcel Dekker Inc.: New York, 1987.

100. Rhode, A. and Sackman, E., *J. Colloid Interface Sci.* 1979, *23*, 900.

101. Junquera, E., Ortega, F., and Aicart, E., *Langmuir* 2003, *19*, 4923.

102. Leaist, D. G. and MacEwan, K., *J. Phys. Chem. B* 2001, *105*, 690.

103. Valiente, M., Thunig, C., Munkert, U., Lenz, U., and Hoffmann, H., *J. Colloid Interface Sci.* 1993, *160*, 39.

104. Bakshi, M. S., *J. Colloid Interface Sci.* 2000, *227*, 78.

105. Bakshi, M. S. and Kaur, G., *J. Mol. Liq.* 2000, *88*, 15.

106. Ghosh, S., *J. Colloid Interface Sci.* 2001, *244*, 128.

107. Gorski, N., Gradzielski, M., and Hoffmann, H., *Langmuir* 1994, *10*, 2594.

108. Sugihara, G., Era, Y., Funatsu, M., Kunitake, T., Lee, S., and Sasaki, Y., *J. Colloid Interface Sci.* 1997, *187*, 435.

109. Tomasic, V., Stefanic, I., and Filipovic-Vincekovic, N., *Colloid Polym. Sci.* 1999, *277*, 153.

110. Jobe, D. J. and Reinsborough, V. C., *Aust. J. Chem.* 1984, *37*, 303.

111. Ishikawa, M., Matsumura, K., Esumi, K., and Meguro, K., *J. Colloid Interface Sci.* 1991, *141*, 10.

112. Valiente, M. and Rodenas, E., *J. Phys. Chem.* 1991, *95*, 3368.

113. Jobe, D. J. and Reinsborough, V. C., *Can. J. Chem.* 1984, *62*, 280.

114. Yamanaka, M. and Kaneshina, S., *J. Solution Chem.* 1990, *19*, 729.

115. Lim, W. H. and Lawrence, M. J., *Phys. Chem. Chem. Phys.* 2004, *6*, 1380.

116. Schonfelder, E. and Hoffmann, H., *Ber. Bunsenges. Phys. Chem.* 1994, *98*, 842.

117. Yue, B. H., Huang, C. Y., Nieh, M. P., Glinka, C. J., and Katsaras, J., *J. Phys. Chem. B* 2005, *109*, 609.

118. Karukstis, K. K., *Encapsulation of Fluorophores in Multiple Microenvironments in Surfactant-Based Supramolecular Assemblies*; Academic Press: London, 2001; Vol. 3: *Nanostructured Materials, Micelles and Colloids*.

119. Karukstis, K. K., McCormack, S. A., McQueen, T. M., and Goto, K. F., *Langmuir* 2004, *20*, 64.

120. Niu, S., Gopidas, K. R., Turro, N. J., and Gabor, G., *Langmuir* 1992, *8*, 1271.

121. Sun, C. Q., Hanasaka, A., Kashiwagi, H., and Ueno, M., *Biochim. Biophys. Acta* 2000, *1467*, 18.

122. Tsuchiya, K., Nakanishi, H., Sakai, H., and Abe, M., *Langmuir* 2004, *20*, 2117.

123. Villeneuve, M., Kaneshina, S., Imae, T., and Aratono, M., *Langmuir* 1999, *15*, 2029.

124. Pluckthum, A., DeBony, J., Fanni, T., and Dennis, E. A., *Biochim. Biophys. Acta* 1986, *856*, 144.

125. Carrion, F. J., de la Maza, A., and Parra, J. L., *J. Colloid Interface Sci.* 1994, *164*, 78.

126. Smedley, S. I., *The Interpretation of Ionic Conductivity in Liquids*; Plenum Press: New York, 1980.

127. Junquera, E. and Aicart, E., *Rev. Sci. Instrum.* 1994, *65*, 2672.

128. Delgado, A. V., *Interfacial Electrokinetics and Electrophoresis*; Marcel Dekker, Inc.: New York, 2002; Vol. 106.

129. Hunter, R. J., *Zeta Potential in Colloids Science. Principles and Applications*; Academic Press: London, 1981.

130. Ohshima, H. and Furusawa, K., *Electrical Phenomena at Interfaces. Fundamentals, Measurements, and Applications*; Marcel Dekker, Inc.: New York, 1998.

131. Hunter, R. J., *Foundations of Colloid Science*; Oxford Science Publications: Oxford, 1995.

132. Evans, D. F. and Wennerström, H., *The Colloidal Domain: Where Physics, Chemistry, Biology and Technology Meet*; Wiley-VCH: New York, 1999.

133. Henry, D. C., *Proc. Roy. Soc. Lond., Ser. A* 1931, *133*, 106.
134. Evans, H. C., *J. Chem. Soc.* 1956, 579.
135. Loeb, A. L., Overbeek, J. T. G., and Wiersema, P. H., *The Electrical Double Layer around a Spherical Colloid Particle*; MIT Press: Cambridge, MA, 1961.
136. Williams, D. B., *Transmission Electron Microscopy*; Plenum Press: New York, 1996; Vol. Basic I.
137. Kunitake, T. and Okahata, Y., *J. Am. Chem. Soc.* 1980, *102*, 549.
138. Dubochet, J., Adrian, M., Chang, J. J., Homo, J. C., Lepault, J., McDowall, A. W., and Schultz, P., *Q. Rev. Biophys.* 1988, *21*, 129.
139. Llorca, O., McCormack, E., Hynes, G., Grantham, J., Cordell, J., Carrascosa, J. L., Willison, K. R., Fernández, J. J., and Valpuesta, J. M., *Nature* 1999, *402*, 693.
140. Lakowicz, J. R., *Principles of Fluorescence Spectroscopy*; Kluwer Academic/Plenum Publishers: New York, 1999.
141. del Burgo, P., Junquera, E., and Aicart, E., *Langmuir* 2004, *20*, 1587.
142. Lobo, B. C. and Abelt, C. J., *J. Phys. Chem. A* 2003, *107*, 10938.
143. Il'ichev, Y. V., Kuhnle, W., and Zachariasse, K. A., *J. Phys. Chem. A* 1998, *102*, 5670.
144. Kawski, A., *Z. Naturforsch.* 1999, *54*, 379.
145. Zachariasse, K. A., van der Haar, T., Hebecker, A., Leinhos, U., and Kiihnle, W., *Pure Appl. Chem.* 1993, *65*, 1745.
146. Zachariasse, K. A., Grobys, M., van der Haar, T., Hebecker, A., Il'ichev, Y. V., Morawski, O., Riicker, I., and Kiihnle, W., *J. Photochem. Photobiol. A* 1997, *105*, 373.
147. Zachariasse, K. A., Druzhinin, S. I., Bosch, W., and Machinek, R., *J. Am. Chem. Soc.* 2004, *126*, 1705.
148. Parusel, A. B., Nowak, W., Grimme, S., and Köhler, G., *J. Phys. Chem.* 1998, *102*, 7149.
149. Grabowski, Z. R., Rotkiewicz, K., Siemiarexuk, A., Cowley, D. J., and Baumann, W., *Nouv. J. Chim.* 1979, *3*, 443.
150. Rettig, W., Paeplow, B., Herbst, H., Mullen, K., Desvergne, J. P., and Bouas-Laurent, H., *New J. Chem.* 1999, *23*, 453.
151. Jödicke, C. J. and Lüthi, H. P., *J. Am. Chem. Soc.* 2002, *125*, 252.
152. Asakawa, T., Mouri, M., Miyagishi, S., and Nishida, M., *Langmuir* 1989, *5*, 343.
153. Mandal, D., Pal, S. K., Datta, A., and Bhattacharyya, K., *Anal. Sci.* 1998, *14*, 199.
154. Manoj, K. M., Jayakumar, R., and Rakshit, S. K., *Langmuir* 1996, *12*, 4068.
155. Lide, D. R., *CRC Handbook of Chemistry and Physics*; CRC Press: Boca Raton, FL, 2004.
156. Derjaguin, B. V., *Kolloid Z.* 1934, *69*, 155.
157. Derjaguin, B. V. and Landau, L., *Zh. Eksp. Teoret. Fiz.* 1945, *15*, 663.
158. Verwey, E. J. W. and Overbeek, J. T. G., *Theory of the Stability of Lyophobic Colloids*; Elsevier: Amsterdam, 1948.
159. Hamaker, H. C., *Physica* 1937, *4*, 1058.
160. Minami, H., Inoue, T., and Shimozawa, R., *J. Colloid Interface Sci.* 1993, *158*, 460.
161. Sogami, I. and Ise, N., *J. Chem. Phys.* 1984, *81*, 6320.
162. Hogg, R., Healy, T. W., and Fuerstenau, D. W., *Trans. Faraday Soc.* 1966, *62*, 1638.
163. Vold, M. J., *J. Colloid Sci.* 1961, *19*, 1.
164. Sogami, I., *Phys. Lett. A* 1983, *96*, 199.
165. Aicart, E., Jobe, D. J., Skalski, B. D., and Verrall, R. E., *J. Phys. Chem.* 1992, *96*, 2348.

166. Jobe, D. J., Verrall, R. E., Skalski, B., and Aicart, E., *J. Phys. Chem.* 1992, *96*, 6811.
167. Karukstis, K. K., Savin, D. A., Loftus, C. T., and D'Angelo, N. D., *J. Colloid Interface Sci.* 1998, *203*, 157.
168. Verrall, R. E., Jobe, D. J., and Aicart, E., *J. Mol. Liq.* 1995, *65/66*, 195.
169. Bai, G. Y., Wang, Y., Wang, Y., Han, B., and Yan, H., *Langmuir* 2001, *17*, 3522.
170. del Burgo, P., Aicart, E., and Junquera, E., *Colloids Surf. A: Physicochem. Eng. Aspects* 2007, *292*, 165.
171. Lainez, A., del Burgo, P., Junquera, E., and Aicart, E., *Langmuir* 2004, *20*, 5745.
172. Treiner, C. and Makayssi, A., *Langmuir* 1992, *8*, 794.
173. Matsumoto, H., Hieuchi, T., and Horie, K., *Colloid Polym. Sci.* 1989, *267*, 71.
174. Tanford, C., *J. Phys. Chem.* 1972, *76*, 3020.
175. Tanford, C., *J. Phys. Chem.* 1974, *78*, 2469.

14 Comparative Study of the Interaction between Nonionic Surfactants and Bovine Serum Albumin: Sugar-Based vs. Ethoxylated Surfactants

Cristóbal Carnero Ruiz, José M. Hierrezuelo, José M. Peula-García, and Juan Aguiar

CONTENTS

14.1 INTRODUCTION

The interactions between proteins and low molecular weight surfactants have broad relevance in many biotechnological disciplines. These interactions play a decisive role in applications such as drug delivery, cosmetics, and solubilizing processes used for the extraction, isolation and purification of biological macromolecules to name but a few. Moreover, since the fundamental principles underlying the formation of protein–micellar complexes are common to other assemblies such as reverse micelles, bilayers, liposomes, and biological membranes [1,2], an understanding of these interactions can be used to our advantage in the selection of the surfactant [3,4]. Therefore, the study of the interactions responsible for the binding of surfactants to proteins has been the subject of extensive research for many years using a number of different experimental techniques [4–7].

Proteins are copolymers built up of amino acids containing polar and nonpolar groups, which have a prescribed sequence. Globular proteins are compact macromolecules with a well-defined structure on different levels, which, unlike other polymers, are monodisperse and essentially lack conformational freedom [8]. Serum albumins are the most abundant proteins in blood plasma and have the function of incorporating and transporting lipids such as fats, cholesterol, and derivatives, into the lymph or bloodstream. Bovine serum albumin (BSA) is an important globular protein frequently used for research purposes and as a reference in clinical analyses and biochemistry research due to its stability, water solubility, and versatile binding capability [9,10]. Ionic surfactants interact more effectively with proteins than nonionic surfactants. However, it is well known that the charged group of the anionic surfactants interacts electrostatically with the oppositely charged group of the amino acids causing the denaturation of the protein [11]. This effect has been attributed to the ability of these surfactants to associate micelle-like aggregates on the surface of the protein promoting unfolding and, consequently, protein structure alterations [12]. For this reason, many studies have focused on the binding of sodium dodecyl sulfate (SDS) and other anionic surfactants, the studies involving cationic and nonionic surfactants being relatively scarcer [10–27].

Because of their favorable physicochemical properties, nonionic surfactants are extensively used in many fields of technology and research [28]. From the point of view of protein–surfactant interactions, the nonionic surfactants are generally less effective than the ionic ones. However, there are applications that require preservation of protein functionality and the use of nonionic surfactants is preferred in these cases. The nature of the interactions responsible for the binding of ethoxylated nonionic surfactants to proteins is being characterized. These investigations indicate that the hydrophobic moiety of surfactants can bind to the nonpolar amino acids, whereas the hydrophilic ethyleneoxide chain can interact with the peptide bond and with one or more polar amino acids residues, probably by electrostatic interactions and hydrogen bonding [28,29].

Sugar-based surfactants are attracting increasing interest due to advantages with regard to performance, health of consumers, and environmental compatibility as compared to the more common ethoxylated nonionic surfactants [30]. In addition, these surfactants share properties such as high solubilizing power, high critical micelle concentration (CMC), no denaturation of proteins, high solubility in water, and stability. These properties are essential for applications in the biomembrane field, implying solubilization and purification of biopolymers from their natural matrixes by selective solubilization [31]. Octyl-β-D-glucopyranoside (OG) is probably the most popular sugar-based surfactant, belonging to the alkyl polyglucosides (APG) group, and has been widely used in biomembrane research and reconstitution processes of biological membranes [32,33]. Octyl-β-D-thioglucopyranoside (OTG) is a related APG surfactant that differs from OG only in how the hydrophilic group is linked by a thioether to the hydrophobic chain (Figure 14.1). Its CMC has been determined to be around 8 or 9 mM [34,35] (a low value in comparison to 25 mM for OG [35]) suggesting a more hydrophobic character for the thio-glycosylated surfactant. In addition, OTG shows a similar power of solubilization with a higher stability and lower cost [31]. Nevertheless, although OTG has scarcely been used in the biomembrane field, recent investigations [31,32,36] showing the advantageous properties of OTG over OG suggest a promising role for the former in that area. On the other hand, N-decanoyl-N-methylglucamide (MEGA-10) (Figure 14.1) is another important sugar surfactant, belonging to the group of the fatty acid glucamides. It has been used as a membrane protein solubilizer since it was synthesized by Hildreth. Since this

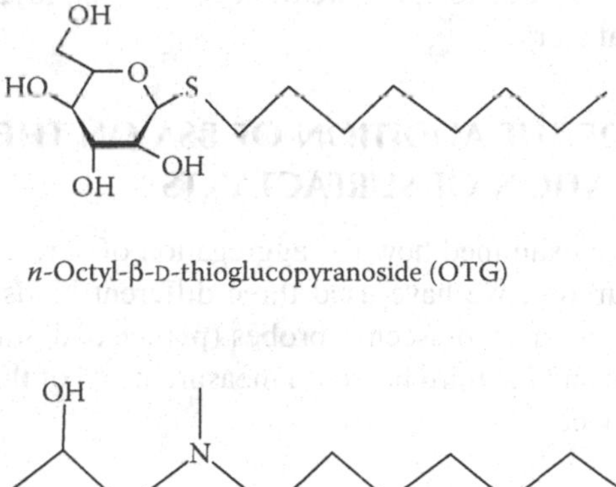

n-Octyl-β-D-thioglucopyranoside (OTG)

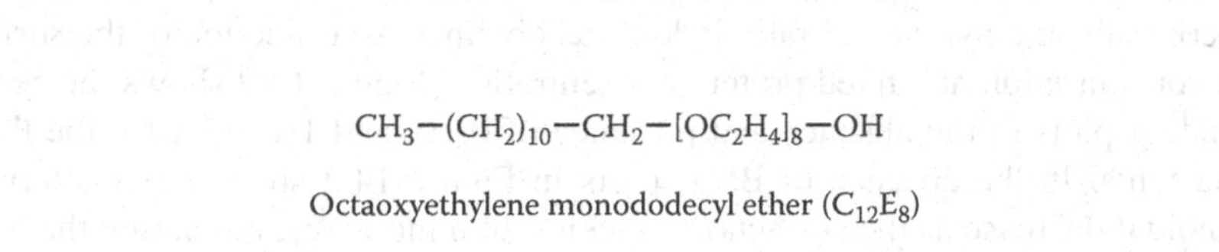

N-Decanoyl-N-methylglucamide (MEGA-10)

$$CH_3-(CH_2)_{10}-CH_2-[OC_2H_4]_8-OH$$

Octaoxyethylene monododecyl ether ($C_{12}E_8$)

FIGURE 14.1   Molecular structures of the surfactants investigated in the present study.

surfactant seems to maintain protein function [37], as can be seen in Figure 14.1, the main difference between MEGA-10 and OTG lies in the hydrophilic moiety structure. The linear head-group of MEGA-10 and amino substitution may contribute to differences not only in steric packing constraints but also in hydrogen bonding possibilities in comparison with OTG [38]. Therefore, it seems interesting to examine the binding behavior of both surfactants to a model protein such as BSA, as this kind of study could provide valuable information about the nature of the interactions involved in the process [39–41]. From a practical point of view, it is also important to analyze this behavior in comparison with that of conventional ethoxylated nonionic surfactants. With this purpose, we have chosen the ethoxylated nonionic surfactant octaoxyethylene monododecyl ether ($C_{12}E_8$) (Figure 14.1), which is also commonly used in many processes of the membrane protein field, including solubilization and crystallization, and in reconstitution experiments [3].

This chapter presents a comparative study of the binding processes between the three surfactants OTG, MEGA-10 and $C_{12}E_8$, and BSA using the techniques of tensiometry, steady-state fluorescence, and dynamic light scattering (DLS). Some experimental data concerning other nonionic surfactants are also included. We will examine, in particular, two aspects: the nature of the binding between the protein and surfactants, and the structure of the protein–surfactant complexes. The present study has been carried out in a glycine-HCl buffer of ionic strength 0.1 M in NaCl at pH 5.75. Moreover, whereas the work temperature used in the case of OTG and $C_{12}E_8$ was 25°C, a temperature of 30°C was employed for MEGA-10 in all the experimental work.

14.2 EFFECT OF THE ADDITION OF BSA ON THE AGGREGATION OF SURFACTANTS

First of all, we have examined how the aggregation of surfactants is affected by BSA. With this purpose, we have used three different kinds of studies, two of them based on the use of fluorescence probes (pyrene and 8-anilinonaphthalene-1-sulfonate [ANS]) and the third based on measurements of the surface tension in the air–liquid interface.

14.2.1 PYRENE 1:3 RATIO INDEX

The aggregation process of the surfactants was followed by the well-documented pyrene 1:3 ratio method [42] in both the presence and absence of protein. In these experiments, the pyrene 1:3 ratio index was obtained as a function of the surfactant concentration at a fixed protein concentration. Figure 14.2 shows the corresponding plots in the absence and presence of 0.1% and 1% BSA for the three surfactants. In the absence of BSA, plots in Figure 14.2 show a characteristic sigmoidal decrease as the surfactant concentration increases, indicating the transition from the pre- to postmicellar region. Below the CMC, the pyrene 1:3 ratio value corresponds to a polar aqueous environment. As the surfactant concentration increases the pyrene 1:3 ratio decreases rapidly, indicating that pyrene is

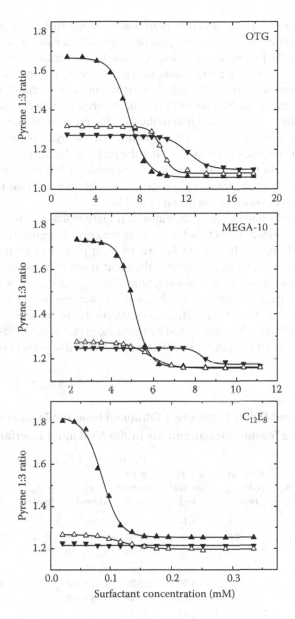

FIGURE 14.2 Plots of pyrene 1:3 ratio versus the total surfactant concentration at different BSA concentrations: (▲) 0, (△) 0.1%, (▼) 1.0%.

sensing a more hydrophobic environment. Above the CMC, the pyrene 1:3 ratio reaches a roughly constant value because of the incorporation of the probe into the hydrophobic region of the micelles. By using the data treatment described previously [43], the CMC values obtained by the pure surfactants in buffered solutions were 7.1 mM for OTG, 5.1 mM for MEGA-10, and 0.086 mM for $C_{12}E_8$.

In the presence of protein, it is observed that when the surfactant concentration is low and micelles are not formed, pyrene is bound to a hydrophobic site of the protein, where the pyrene 1:3 ratio index is much lower than in water but not so low as in the micellar microenvironment. Nevertheless, as soon as the micelles are formed, pyrene prefers the micellar hydrophobic environment. It is also observed that, except in the case of OTG, the differences between the pyrene 1:3 ratio index bound to protein and that in the micellar media are rather small. In the case of $C_{12}E_8$, it is observed that the polarities of the hydrophobic sites of the protein and the micelles are close, so that the probe molecule probably migrates from the protein to the micelle, and it is impossible to determine the surfactant concentration where the transition occurs. It should be noted that this behavior was observed previously by Vasilescu et al. [44].

The free surfactant concentration at which aggregation begins to occur in the presence of a polymer is often referred to as the critical aggregation concentration (CAC). However, the data in Figure 14.2 are plotted in terms of the total surfactant concentration and, therefore, the point at which transition occurs is an apparent critical aggregation concentration $(CAC)_{app}$. In the cases of OTG and MEGA-10, the data in Figure 14.2 indicate that in the presence of BSA, the transition is shifted to higher surfactant concentrations. In the case of OTG, the $(CAC)_{app}$ values were also determined by using the previous data treatment (Table 14.1). The fact that the pyrene 1:3 ratio curves are shifted to higher concentrations

TABLE 14.1
CMC Values and Other Parameters Obtained from the Treatment Based on the Surface Tension Measurements in the Air–Liquid Interface

Surfactant	BSA (w/v)	CMC or $(CAC)_{app}^a$ (mM)	$a \times 10^3$ (mmol/ m^2)	$(\Gamma_p{}^0 \cdot f) \times 10^3$ (mmol/ m^2)	K_S (L/mol)	$(C_S)_{sat}$ (mM)	$(C_S^f)_{sat}$ (mM)	v_{sat}
OTG[b]	0	7.2 (7.1)	3.74	—	—	—	—	—
	0.1	8.5 (9.8)	4.08	2.58	8.69×10^3	8.41	7.23	78.7
	1	10.1 (12.1)	4.45	2.68	5.98×10^3	10.12	7.23	19.3
MEGA-10[c]	0	4.4 (5.1)	3.41	—	—	—	—	—
	0.1	5.1	3.57	2.29	1.91×10^4	5.07	4.38	45.8
	1	6.8	4.10	2.44	1.08×10^4	6.77	4.35	16.1
$C_{12}E_8$[b]	0	0.075 (0.086)	2.63	—	—	—	—	—
	0.1	0.130	3.58	2.43	8.70×10^5	0.131	0.075	3.7

Source: Hierrezuelo, J.M., PhD Thesis, Málaga University, Málaga, Spain, 2007.
[a] Within parenthesis CMC values as determined by the pyrene 1:3 ratio method.
[b] $T = 298\,K$.
[c] $T = 303\,K$.

of surfactant indicates that part of the surfactant is no longer available for forming micelles. In other words, the shift to the right of the curves in Figure 14.2 reveals partitioning of the surfactant into new sites (probably the hydrophobic patches of the protein) indicating interaction of the surfactant with these hydrophobic sites [39–41].

14.2.2 FLUORESCENCE OF ANS

8-anilinonaphthalene-1-sulfonate (ANS) is a hydrophobic probe widely utilized for examining the nonpolar character of proteins, membranes, and other micro-heterogeneous systems [45]. ANS fluorescence is extremely sensitive to changes in the probe microenvironment. This probe shows an emission maximum at 515 nm in water with an emission quantum yield of 0.004. In comparison, when solubilized in 1% BSA, the emission maximum is blue shifted to 472 nm and its emission quantum yield is considerably enhanced ($\approx$0.7) [23,44]. This is a much higher value than that in micelles ($\approx$0.2) [46]. Due to its peculiar photophysical characteristics, ANS has previously been used as a probe to examine the protein–surfactant interaction in several systems [24,46].

Figure 14.3 shows the effect of the surfactant concentration on the relative fluorescence intensity of ANS bound to BSA for the three systems studied. From this figure it is observed that, from a certain surfactant concentration the fluorescence of ANS decreases with increasing surfactant concentration. It is believed

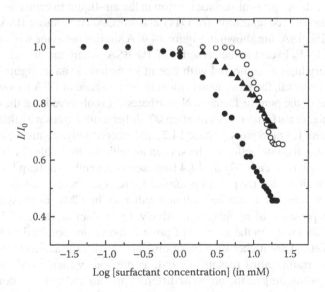

FIGURE 14.3 Effect of the surfactant concentration on the relative fluorescence intensity of ANS (20 μM) solubilized in BSA (30 μM). I and I_0 are the fluorescence intensity of ANS in the presence and the absence of surfactant, respectively. ($\bullet$) $C_{12}E_8$, ($\blacktriangle$) OTG, and (O) MEGA-10.

that this effect occurs mainly because the surfactant monomers displace some ANS molecules previously bound to the protein [46], which are probably transferred either to the micellar clusters adsorbed on the protein surface or even to the free micelles of surfactant. Therefore, the break point of the plots in Figure 14.3 should provide the surfactant concentration required for aggregation. These break points occur at higher concentrations than the corresponding CMCs in the case of $C_{12}E_8$ (≈ 0.4 mM) and MEGA-10 (≈ 5.0 mM), but lower than in the case of OTG (≈ 2.0 mM). Vasilescu et al. [44] have previously pointed out that the threshold concentration required for surfactant aggregation as determined by fluorescence methods can be overestimated owing to the contribution of the fluorescence emitted by the probe from different hydrophobic domains. Therefore, it seems reasonable to take the results in Figure 14.3 only from a qualitative point of view. It is clear, however, that the data in Figure 14.3 indicate a certain interaction between protein and surfactants, with OTG being the species with a higher affinity to BSA.

14.2.3 SURFACE TENSION MEASUREMENTS

The effect of protein addition on the aggregation process of the surfactants was investigated by using surface tension measurements, a very different experimental technique from the two employed previously. This technique has been widely used by many workers to obtain information on both ionic [12,22,25,47] and nonionic surfactants [48–51].

Representative plots of surface tension in the air–liquid interface as a function of the surfactant concentration for OTG and MEGA-10, without BSA and with 0.1% and 1% BSA, are shown in Figure 14.4. A similar behavior was observed in the case of $C_{12}E_8$ except in the presence of 1% BSA, where the protein concentration was very high as compared with that of surfactant. Data in Figure 14.4 indicate that, in general, the tensiometric plots in the presence of BSA follow a similar trend to that of the pure surfactants. Nevertheless, it is observed that the curves are shifted to higher surfactant concentrations. This behavior agrees with that observed for the pyrene 1:3 ratio plots (Figure 14.2) and, accordingly, it must be interpreted as evidence of the interaction of the surfactant with the available sites of the protein. It can also be seen in Figure 14.4 that each plot exhibits a sharp break, showing the transition from the pre- to postmicellar region. These break points, which are listed in Table 14.1, can be assigned either to the CMC or $(CAC)_{app}$, in the absence or presence of protein, respectively. From data in Table 14.1 it can be observed that, except in the absence of protein, the values obtained by the pyrene 1:3 ratio method are higher than those obtained by the surface tension measurements. This result seems reasonable and, as previously mentioned, is probably due to the partitioning of the probe in different protein hydrophobic domains.

Another interesting aspect of Figure 14.4 is that in the postmicellar region all plots show constant surface tension values that are roughly independent of the protein concentration. This fact suggests that in the postmicellar region, the pure surfactant seems to control the surface tension in the air–liquid interface and,

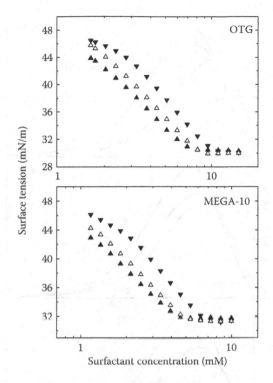

FIGURE 14.4 Surface tension of surfactant solutions versus the total surfactant concentration at 298 K for OTG and at 303 K for MEGA-10 in varying BSA concentrations: (▲) 0, (△) 0.1%, (▼) 1.0%.

therefore, it can be assumed that the protein–surfactant complexes are surface inactive species. Similar observations have previously been reported by a number of authors for several protein–nonionic surfactant systems [48–51].

14.3 EFFECT OF SURFACTANTS ON THE FLUORESCENCE OF BSA

The fluorescent behavior of BSA is mainly characterized by the presence of two tryptophan (Trp) residues (W^{135} and W^{214}), which contribute greatly to its intrinsic fluorescence. It has been established that the Trp residue located at position 135 (W^{135}) is buried into a hydrophobic pocket and lies near the surface of the albumin molecule in the second helix of the first domain, whereas that located at position 214 resides in the hydrophobic cavity of the IIA subdomain, corresponding to the so-called Sudlow I binding region [18,21]. Interaction of BSA with surfactants is usually accompanied by changes in fluorescence intensity and shifts in the emission maximum that are mainly attributed to changes in the position or orientation of the Trp residues, altering their exposure to the solvent and producing modifications

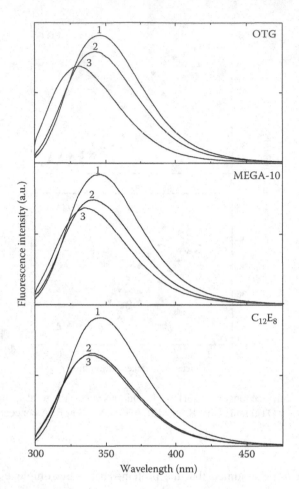

FIGURE 14.5 Effect of the surfactant addition on the fluorescence spectra of BSA (30 μM): (1) native BSA, (2) with a surfactant concentration 2 mM, (3) with a surfactant concentration 20 mM.

of the quantum yield. Therefore, fluorescence studies can be used to monitor changes on the tertiary structure of the protein induced by the surfactant binding [18]. Fluorescence excited at 295 nm was used to monitor the changes in the tertiary structure of BSA induced by the interaction with OTG, MEGA-10, and $C_{12}E_8$. Figure 14.5 shows representative fluorescence emission spectra of native BSA and BSA mixed with the three surfactants at different concentrations (2 and 20 mM). In this figure, it is observed that in the absence of surfactant a band appears, centered at 345 nm, which is clearly associated with native BSA. At low surfactant concentration (2 mM), it is seen that both OTG and MEGA-10 produce a decrease in the Trp fluorescence accompanied by a slight blue shift of the emission maximum (343 nm). This spectrum corresponds to protein–surfactant complexes, indicating interaction between protein and surfactant, but where the binding only involves localized conformational changes that do not affect the global structure of the protein. At low surfactant concentration, these changes for $C_{12}E_8$ are somewhat stronger due to the relatively higher ratio than that of BSA. Note that 2 mM is below the CMC of OTG and MEGA-10, but corresponds to a supra-micellar concentration in the case of $C_{12}E_8$. At high surfactant concentration (20 mM), quenching and blue shift are more pronounced for OTG than for MEGA-10. In the case of MEGA-10, the emission band has a maximum at 335 nm, whereas that of OTG is located at 331 nm. It is important to note that the effect produced by both surfactants, and in particular OTG, is similar to that observed by Tabak and coworkers with different ionic surfactants [18,21], but is more pronounced than that observed recently in the case of $C_{12}E_8$ and other systems involving conventional ethoxylated nonionic surfactants [29,52]. According to Tabak and coworkers [18,21], the effect produced by OTG and MEGA-10 can be ascribed to a protein partial denaturation induced by the surfactant binding. However, it is remarkable that the effect produced by OTG and MEGA-10 on BSA structure is less deleterious than that exerted by a typical ionic surfactant such as SDS, but is more pronounced than that caused by conventional ethoxylated nonionic surfactants.

14.4 BINDING OF SURFACTANTS TO BSA

The character of the interactions between proteins and surfactants can be better understood by analyzing the binding process of the surfactant to the protein by determining the so-called binding curve or binding isotherm. A binding isotherm shows the average number of surfactant molecules bound per protein molecule, v, as a function of the logarithm of the free surfactant concentration [5,53]. In general, the binding isotherm of a surfactant, for example SDS, onto a protein is characterized by the existence of four regions (Figure 14.6). In the initial region, at lowest surfactant concentration, there is some binding to specific high-energy

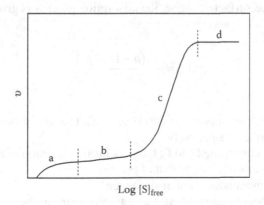

FIGURE 14.6 Schematic representation of a protein–surfactant binding isotherm, showing a plot of the number of surfactant molecules bound per protein molecule, v, as a function of the logarithm of the free surfactant concentration, $[S]_{free}$. (Adapted from Jones, M.J., *Biochem. J.*, 151, 109, 1975.)

sites on the protein. Either a plateau or a slowly rising section follows this region and a noncooperative binding occurs. Beyond, a third region is observed where a massive increase in binding due to cooperative interactions takes place. Finally, the binding curve shows a plateau indicating that further binding of the surfactant does not occur on the protein, and free micelles are formed as excess surfactant is added. This is the so-called saturation region [11,53].

The amount of surfactant bound to the protein is usually obtained by means of equilibrium dialysis assays. However, in many cases, this technique can present serious problems. For example, some nonionic surfactants such as Triton X-100 do not penetrate the dialysis membrane fast enough for dialysis experiments [48]. Therefore, to obtain information on the binding between protein and surfactant it is important to utilize alternative methods. In this chapter we have used two different methods, which are described below.

14.4.1 Method Based on Surface Tension Measurements

Binding of surfactants to BSA was initially quantified by application of the method proposed by Nishikido et al. [48], which is based on surface tension measurements. This method has been successfully applied to a number of hydrophilic protein–nonionic surfactant systems, including $C_{12}E_6$/BSA and $C_{12}E_6$/lysozyme [48], and Triton X-100/BSA [49]. Nishiyama and Maeda [50] applied a simplified version of that approach to examine the binding between a number of ethoxylated nonionic surfactants and lysozyme. The method proposed by Nishikido et al. [48] assumes that the protein–surfactant complexes are surface inactive. This assumption is supported by the experimental observation that the surface tension values above the break point in the tensiometric plots are, within the experimental error, identical regardless of the protein concentration (see Figure 14.4). This is probably due to the fact that the hydrophobic areas of protein and surfactant are practically hidden by the formation of the complex [48]. In the case of the surfactant alone, the adsorption process is given by the Langmuir equation

$$\Gamma_S = \frac{K_S \left(a - \Gamma_P f\right) C_S^f}{1 + K_S C_S^f} \tag{14.1}$$

where

Γ_S and Γ_P are the amounts of adsorbed surfactant and protein at the air–liquid interface, respectively

f is the factor converting Γ_P to Γ_S basis, defined as the ratio of the maximum adsorption of surfactant to that of protein

a is the maximum adsorption of surfactant

C_S^f is the concentration of surfactant monomer or the free surfactant concentration

K_S is the Langmuir constant for adsorption of the surfactant at the air–liquid interface

Since the Langmuir equation applies to the adsorption of protein at low concentrations, one may write for the protein adsorption

$$\Gamma_P f = \frac{K_P \left(a - \Gamma_S\right) C_P^f}{1 + K_P C_P^f} \tag{14.2}$$

and also

$$\Gamma_P^0 f = \frac{K_P a C_P^0}{1 + K_P C_P^0} \tag{14.3}$$

where

the superscript "0" denotes the quantity for the adsorption in the aqueous solutions of the protein alone

the subscript "P" refers to protein

If the protein concentration is low enough, so that $K_P C_P^f \approx K_P C_P^0 \ll 1$, it can be shown by using Equations 14.2 and 14.3 that

$$\Gamma_P f = \Gamma_P^0 f + K_P \left(\Gamma_S - a\right)\left(C_P^0 - C_P^f\right) - K_P C_P^0 \Gamma_S \tag{14.4}$$

Nishikido et al. [48] established that, in the case of the adsorption of nonionic surfactant alone, the value of Γ_S is nearly equal to that of a. Therefore, in a certain protein concentration range, the magnitude of the second term is expected to be much less than those of the other terms in Equation 14.4, hence, this equation can be expressed as

$$\Gamma_P f \approx \Gamma_P^0 f - K_P C_P^0 \Gamma_S = \Gamma_P^0 f - \frac{\Gamma_P^0 f}{a}\Gamma_S \tag{14.5}$$

This equation indicates that Γ_P decreases linearly with Γ_S. By introducing Equation 14.5 into Equation 14.1, we obtain

$$\Gamma_S = \frac{K_S a \left(a - \Gamma_P^0 f\right) C_S^f}{a + K_S \left(a - \Gamma_P^0 f\right) C_S^f} \tag{14.6}$$

In the case of low concentrations of protein and surfactant, the Gibbs adsorption equation is

$$\left(\frac{\partial \gamma}{\partial \mu_S}\right)_{T,\mu_P} \approx \left(\frac{\partial \gamma}{\partial \ln C_S^f}\right)_{T,C_P} = -RT\,\Gamma_S \tag{14.7}$$

where

γ is the surface tension of an aqueous solution containing protein and surfactant

μ refers to the chemical potential

Introducing the value of Γ_S, given by Equation 14.6, and integrating, we obtain

$$\gamma = -aRT \ln\left[a+K_S\left(a-\Gamma_P^0 f\right)C_S^f\right]+\gamma_P+aRT \ln a \tag{14.8}$$

where γ_P is the surface tension of the solution containing protein but no surfactant. In the case of a pure surfactant solution, the surface tension is given by

$$\gamma_S = -aRT \ln\left(1+K_S C_S^0\right)+\gamma_{\text{water}} \tag{14.9}$$

where

C_S^0 is the concentration of pure surfactant
γ_{water} is the surface tension of pure water

If one assumes that at the surfactant concentration where the surface tension begins to stay constant, surfactant monomers begin to form micelles by themselves and that the concentration of monomers is identical to the CMC of the surfactant alone, CMC^0, and since $\gamma_{\text{CMC}} = \gamma_{\text{CMC}^0}$, as previously stated (Figure 14.4), the following relations can be obtained from Equations 14.8 and 14.9:

$$\gamma_{\text{CMC}} = -aRT \ln\left(1+K_S \text{CMC}^0\right)+\gamma_{\text{water}} \tag{14.10}$$

$$\Gamma_P^0 f = \frac{a\left(1+K_S \text{CMC}^0\right)\left[1-\exp\left(-\dfrac{\gamma_{\text{water}}-\gamma_P}{aRT}\right)\right]}{K_S \text{CMC}^0} \tag{14.11}$$

In this way, by combining the surface tension data in Equations 14.8, 14.10, and 14.11, the average number of bound surfactant monomers per protein molecule, υ, can be obtained through the equation

$$\upsilon = \frac{C_S-C_S^f}{C_P} \tag{14.12}$$

where C_S is the total surfactant concentration. The procedure consists of obtaining a from the slope of the adsorption isotherm in the air–liquid interface, in the premicellar region, according to the Gibbs adsorption equation. Then, K_S may be determined by the values of a, CMC^0, and γ_{water} in Equation 14.10. After obtaining $\Gamma_P^0 f$ from Equation 14.11, C_S^f can finally be calculated by applying Equation 14.8. Once C_S^f is determined, the average number of bound surfactant monomers per protein molecule, υ, can be estimated from Equation 14.12.

As an example, the adsorption isotherm of $C_{12}E_8$ in the presence of 0.1% BSA and the corresponding plot of the free surfactant concentration against the total surfactant concentration, as obtained by applying the above treatment, are shown in Figure 14.7. The data in Figure 14.7 indicate that over a wide range of surfactant

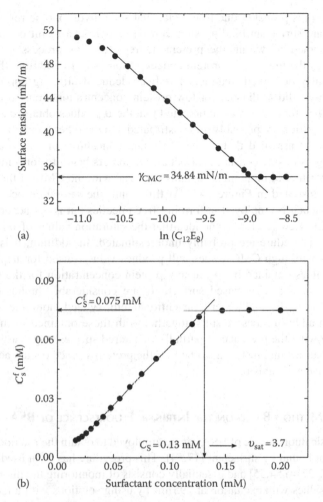

FIGURE 14.7 (a) Surface tension of $C_{12}E_8$ versus the total surfactant concentration, and (b) free surfactant concentration, C_S^f, versus the total surfactant concentration at 298 K and in the presence of 0.1% BSA.

concentration, the concentration of free surfactant, C_S^f, increases linearly with the total surfactant concentration until a saturation value, $(C_S^0)_{sat}$, is reached. This can be determined by the intersection of the two straight lines, as also shown in Figure 14.7. The saturation average number of bound surfactant monomers per protein molecule, v_{sat}, can be estimated. Table 14.1 lists the parameters that have been obtained for the three surfactants studied using the procedure based on surface tension measurements. Data in Table 14.1 indicate that in all cases the constant K_S and the maximum number of bound surfactants per protein molecule are greater for smaller protein concentration, in good agreement with the observations of Nishikido et al. for the $C_{12}E_6$/BSA system [48]. The decrease of v_{sat} with protein concentration is probably due to the fact that some hydrophobic regions of proteins that are surfactant-binding sites can be hidden as a result of the protein–protein interactions within the protein aggregates. This process is effectively enhanced by the increase in protein concentration [48]. In addition, the fact that the previously described treatment has been deduced utilizing approximations that are only valid in the case of low protein concentrations must be taken into account. Therefore, it may be anticipated that the v_{sat} values obtained at high protein concentration are probably underestimated. On the other hand, the aforementioned authors assumed that at the surfactant concentration where the surface tension begins to stay constant, surfactant monomers begin to form micelles and the concentration of monomers remains identical with the CMC of the surfactant alone, as illustrated in Figure 14.7. At this point, the saturation occurs and the added surfactant is employed in forming free micelles but is not adsorbed on the protein. This view reinforces the idea that the saturation values of v reported by the present procedure are probably underestimated. In addition, it is clear that surfactants with high CMC values will produce values raised for v_{sat}. Nevertheless, the values obtained by v_{sat} at low protein concentration for the two mixed systems involving sugar-based surfactants are considerably higher than those obtained for $C_{12}E_8$ and other conventional ethoxylated nonionic surfactants [48,49]. In addition, these results, together with those obtained in the previous section showing the presence of partially denatured species, suggest that the formation of surfactant clusters adsorbed on the protein surface occurs according to a cooperative mechanism.

14.4.2 METHOD BASED ON THE INTRINSIC FLUORESCENCE OF BSA

The intrinsic fluorescence of BSA can be employed to obtain the fraction of protein occupied or bound by surfactant [17,54]. This procedure has been used by several authors [17,27,46,54,55] and essentially consists of monitoring the fluorescence of the Trp residues with excitation at 295 nm by using solutions with a fixed protein concentration and increasing surfactant concentrations. The fractional change of protein fluorescence due to surfactant binding, θ, can be estimated by [17,54]

$$\theta = \frac{I_{obs} - I_f}{I_{min} - I_f}$$ (14.13)

where
I_{obs} is the fluorescence intensity at a certain surfactant concentration
I_f is the fluorescence intensity in the absence of surfactant
I_{min} is the fluorescence intensity under saturation binding condition

It must be remarked that this procedure does not measure the average number of surfactant molecules per protein molecule, v, but rather the fractional occupation of surfactant binding sites [54]. Thus, if a protein molecule has n_0 available binding sites for a determined surfactant and at a certain stage surfactant molecules bind to n sites, then the fraction of protein occupied by surfactant is $\theta = n/n_0$ [17]. In other words, in the absence of surfactant or when no surfactant is bound to the protein, I_{obs} equals I_f, and $\theta = 0$. When the protein is occupied by the maximum number of surfactant molecules and saturation is reached, $I_{obs} = I_{min}$, and $\theta = 1$. In this way, by following the change of the intrinsic fluorescence intensity of BSA upon surfactant addition, one can examine how the protein is bound by the surfactant through the parameter θ. Figure 14.8a shows the intrinsic fluorescence intensity of BSA as a function of the surfactant concentration for all the studied systems. In general, it is observed that binding of surfactant is accompanied by quenching of the intrinsic protein fluorescence until a roughly constant minimum value is reached, indicating saturation of the binding process between surfactant and protein. Figure 14.8b shows the binding curve obtained by plotting the fractional occupation

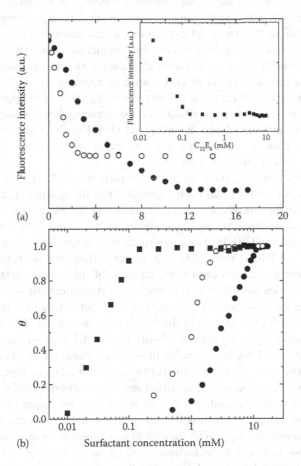

FIGURE 14.8 (a) Plots of the intrinsic fluorescence of BSA (30 μM) as a function of the total surfactant concentration. (b) Binding curves showing the fraction of a protein molecule bound by surfactant (θ) as a function of log of the total surfactant concentration: (■) $C_{12}E_8$, (○) MEGA-10, and (●) OTG.

of protein against the total surfactant concentration. Data in Figure 14.8b show a similar behavior for the three surfactants and can be interpreted in the light of the conceptual scheme proposed by Jones [5,53] discussed previously and displayed graphically in Figure 14.6. Since the parameters θ and v have different meanings and, moreover, we have plotted θ against the total surfactant concentration, the binding curves in Figure 14.8b do not allow a direct comparison with the binding isotherm model proposed by Jones (Figure 14.6). However, from data in Figure 14.8 it is possible to draw some conclusions about how the protein binding sites are occupied by the surfactant. The main difference between the binding curves in Figure 14.8b is observed at low surfactant concentration. Note that in the case of the sugar surfactants, the initial rise of the fractional occupation is more gradual than that of $C_{12}E_8$. This behavior could be related to a major contribution of specific interactions (electrostatic and hydrogen bonding) in the case of OTG and MEGA-10. For these two surfactants it is also observed that, starting from a certain concentration, the fractional occupation, θ, increases abruptly in a narrow concentration range. Taking into account the concentration range where this binding occurs, this behavior seems to reflect the massive adsorption of surfactant onto the protein surface as a result of cooperative mechanism. Note that this strong binding of the surfactant occurs at slightly below the CMC of the pure surfactant, in agreement with results on manifold other surfactant-polymer systems in which either the surfactant or polymer is uncharged [56]. Finally, a region is observed where the fractional occupation reaches a plateau ($\theta = 1$), which evidently corresponds to the saturation binding region.

On the other hand, in the case of the interactions involved in the binding of small ligands to albumin, it has been established that the hydrophobic interactions may be stronger than the electrostatic ones. However, the electrostatic interactions are also significant because uncharged hydrophobic chains have lower binding affinity to the protein [18]. Previous reports on interactions between nonionic surfactants and BSA have concluded that the binding observed reflects the condensation of nonionic surfactants onto nonpolar areas of the protein surface [5]. This conclusion has been corroborated by calorimetric measurements showing that the binding of nonionic surfactants is associated with rather small enthalpy changes comparable with those involved in the micellar formation [57,58]. Nielsen et al. [59] have studied (by using high sensitivity isothermal titration calorimetry) the interaction between BSA and a number of surfactants with C12 acyl chains. These included the two nonionic surfactants hepta- and penta(ethylene glycol) monododecyl ether ($C_{12}E_7$ and $C_{12}E_5$). The aforementioned authors found that for these nonionic surfactants, the binding enthalpy is considerably larger than for any of the anionic surfactants, suggesting that electrostatic interactions contribute favorably to the binding enthalpy. With the aim of explaining these observations, these authors proposed a possible mechanism based on the existence of hydrogen bonding between the ethylene glycol chain and the protein. However, they could not reconcile this contribution with a large positive heat capacity change associated with the binding process. In this sense, it must be mentioned that Singh and

Kishore [29] have recently found that polar interactions play a significant role in the binding between Triton X-100 and BSA.

With regard to the role of the electrostatic interactions in the binding process of the surfactants considered in the present investigation, it is important to mention that according to previous studies, the hydrophilic group of OTG behaves more similarly to that of certain homologous series of ionic surfactants, including alkyl sulfates and alkyltrimethylammonium bromides, than to the nonionic ones [34]. In fact, the existence of a small but significant charge in micelles formed by APG surfactants has been previously established [60]. According to this view, a relatively larger contribution of the electrostatic interactions in the case of OTG, as compared with the other two surfactants, should be expected. This could serve as an explanation of the observed higher affinity of OTG for BSA, as well as of the fact that this surfactant causes larger conformational changes, as deduced from the observed alterations in the intrinsic fluorescence of BSA. In addition, the different binding behaviors observed for the sugar surfactants and the ethoxylated surfactants may, at least in part, be explained by two aspects: the fact that the former have larger CMC values, delaying the competition between the binding processes and the micellar formation, and the higher capability of sugar-based surfactants to form hydrogen bonds between the hydroxyl groups of their headgroups and the protein.

14.5 DETERMINATION OF AGGREGATION NUMBERS

The analysis of the binding process between surfactant and protein does not afford any information about the structure of the surfactant–protein complexes. However, an estimation of the mean aggregation numbers of the self-assembled structures present in the mixed system could provide valuable data from a structural point of view. In addition, if the binding process between the surfactants and BSA is dominated by a cooperative mechanism, as suggested by the previous findings, the formation of clusters of surfactants adsorbed onto the protein surface should be expected. In order to examine this point, we have carried out a determination of the aggregation numbers of surfactant in the absence and in the presence of protein, by using the well-established fluorescence static quenching method [61]. Vasilescu et al. [44] have applied both the static and time-resolved methods for the determination of aggregation numbers of micellar aggregates formed in several protein–surfactant systems. These authors have discussed the applicability of both methods to these systems thoroughly. The problem lies in the possibility that molecules of probe, quencher, and surfactant adsorb on the protein before the cooperative, aggregative adsorption. Moreover, it is further probable that molecules adsorbed to specific sites on the protein are transferred into the micelle-like aggregates. Since it was not possible to differentiate between these situations they calculated the aggregation numbers as if all surfactant, probe, and quencher were micellized. Under this assumption, they obtained comparable results by the application of both static and dynamic methods [44].

Static quenching experiments were carried out by using pyrene as a probe and cetylpyridinium chloride (CPyC) as a quencher, since this pair ensures the fulfillment of the appropriate requirements for the application of this method [62]. The results of the quenching experiments were analyzed by using the following equation:

$$\ln\left(\frac{I_0}{I}\right) = \frac{N_{agg}}{[S] - CMC}[Q] \tag{14.14}$$

where

I_0 and I are the fluorescence intensities in the absence and presence of quencher, respectively

N_{agg} is the mean aggregation number

[S] is the total surfactant concentration

[Q] is the quencher concentration

The fluorescence intensities at 383 nm were plotted according to Equation 14.14 (Figure 14.9), and an acceptable linear relationship ($r > 0.99$) was observed in all

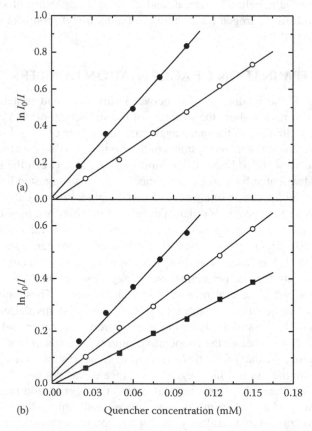

FIGURE 14.9 Quenching plots of pyrene fluorescence in (a) pure surfactant solutions, and (b) in the presence of 1% BSA: (●) OTG, (○) MEGA-10, (■) $C_{12}E_8$.

TABLE 14.2

Mean Aggregation Number Values, N_{agg}, for the OTG/BSA, MEGA-10/BSA, and $C_{12}E_8$/BSA Systems, as Determined by the Steady-State Fluorescence Quenching Method

Surfactant	[BSA] (% w/v)	N_{agg}^a
OTG	0	107 ± 1
	1	44 ± 1
MEGA-10	0	74 ± 1
	1	46 ± 1
$C_{12}E_8$	0	(94)
	1	52 ± 1 (64)

Source: Hierrezuelo, J.M., PhD Thesis, Málaga University, Málaga, Spain, 2007.

[a] Within parentheses N_{agg} values obtained by Vasilescu et al. [44] through the dynamic quenching method, at pH = 7.0 and ionic strength 0.2 M NaCl.

the cases. From plots in Figure 14.9, the N_{agg} values listed in Table 14.2 were obtained. Firstly, it should be noted that, while the N_{agg} value obtained for OTG in the absence of protein is lower than the value previously determined by using static light scattering measurements at the same ionic strength [63], the value obtained in the case of MEGA-10 agrees well with literature values [64,65]. Data in Table 14.2 indicate that the N_{agg} values in the presence of 1% BSA are considerably reduced. A similar behavior was observed previously by Vasilescu et al. [44] in a number of protein–surfactant systems, including those constituted by the nonionic surfactant $C_{12}E_8$ and the proteins BSA and lysozyme. It is important to mention that in studies dealing with polymer–surfactant systems, a decrease in the aggregation number is generally reported compared to the value of the surfactant water system. This change is considered to be a strong indication of a polymer–surfactant interaction [66]. The N_{agg} values that we have obtained in the presence of 1% BSA represent the average value of two types of aggregates: the free micelles of surfactant and the micelle-like aggregates adsorbed on the surface protein by hydrophobic interactions. Note that this result is consistent with the formation of a necklace structure for the protein–surfactant complexes [16] where the micellar clusters, which are smaller than the free micelles, would be adsorbed on the protein surface.

14.6 RESONANCE ENERGY TRANSFER STUDIES

As mentioned above, there are signs that the binding mechanism between surfactant and protein is concomitant with possible conformational changes in the protein structure as a result of the interactions involved in the process. This fact has previously been suggested by other authors [44]. We have carried out a study of resonance energy transfer (RET) from Trp residues of BSA to ANS as an exogenous acceptor with the purpose of examining this point. RET processes have been widely used to obtain structural and dynamic information on several macromolecular assemblies, including macromolecule–ligand complexes, polymer–surfactant systems, and protein–surfactant systems [46,67–69].

The rate constant for RET, k_{ET}, occurring by dipolar interactions between a donor and acceptor, is given by Förster's theory [70]:

$$k_{ET} = \frac{1}{\tau_D} \left(\frac{R_0}{R} \right)^6 \tag{14.15}$$

where

τ_D is the donor lifetime in the absence of acceptor
R is the donor–acceptor distance
R_0 is the Förster distance, a characteristic parameter in Förster's model related to the spectral properties of chromophores by

$$R_0^6 = 8.8 \times 10^{-25} K^2 n^{-4} \Phi_D J \ (\text{cm}^6) \tag{14.16}$$

where

K^2 is a factor describing the relative orientation in space of the transition dipoles of the donor and acceptor
n is the refractive index of the medium
Φ_D is the quantum yield for the donor in the absence of an acceptor
J is the spectral overlap integral, which represents the degree of spectral overlap between donor emission and acceptor absorption

This spectral overlap integral is given by

$$J = \frac{\int_0^\infty I_D(\lambda) \varepsilon_A(\lambda) \lambda^4 \mathrm{d}\lambda}{\int_0^\infty I_D(\lambda) \mathrm{d}\lambda} \ \left(\text{M}^{-1} \ \text{cm}^3 \right) \tag{14.17}$$

where

$I_D(\lambda)$ is the fluorescence intensity of the donor, expressed in arbitrary units
$\varepsilon_A(\lambda)$ is the molar coefficient of the acceptor in M^{-1}/cm
λ is the wavelength in cm

From Equation 14.15 it is obvious that the strong dependence of the RET process on the distance R makes RET studies especially useful for estimating changes in the separation distance between donor and acceptor. Likewise, it is also evident that an efficient energy transfer process requires a significant overlap between donor emission and acceptor absorption spectra. Since the emission

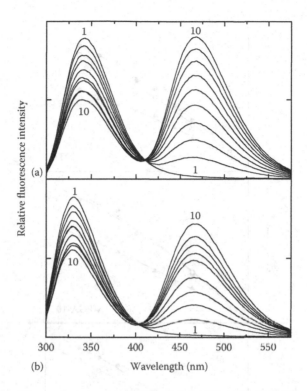

FIGURE 14.10 Fluorescence spectra of BSA solutions (30 μM) in (a) 2 mM, and (b) 20 mM OTG and varying concentrations of ANS: $1 \rightarrow 0$ mM and $10 \rightarrow 0.15$ mM ($\lambda_{exc} = 295$ nm).

spectrum of Trp shows a good overlap with the absorption spectrum of ANS, this donor–acceptor pair has been previously employed in the study of systems involving proteins [67], including protein–surfactant systems [46].

Figure 14.10 shows the combined spectra of the Trp–ANS pair in mixed solutions of OTG and BSA. These spectra were recorded by using fixed concentrations of surfactant (2 mM in Figure 14.10a and 20 mM in Figure 14.10b) and protein (30 μM), and increasing ANS concentrations. From data in Figure 14.10, it can be seen that as the acceptor concentration increases, the quenching of Trp is accompanied by enhancement of the acceptor fluorescence, showing an efficient energy transfer between Trp and ANS in the mixed systems. Similar results were obtained for the systems involving MEGA-10 and $C_{12}E_8$. Figure 14.11 shows the effect of the surfactant concentration on the efficiency of the RET process as followed by the sensitized acceptor fluorescence. From this figure, it is observed that the efficiency in the energy transfer between Trp and ANS depends on the surfactant concentration. However, it is to be noted that the behavior observed in Figure 14.11 can be due to two effects: (1) a real reduction in the efficiency of the RET process due to a certain alteration in the protein structure, which would produce the decrease in the spectral overlap between donor emission and acceptor absorption or even a

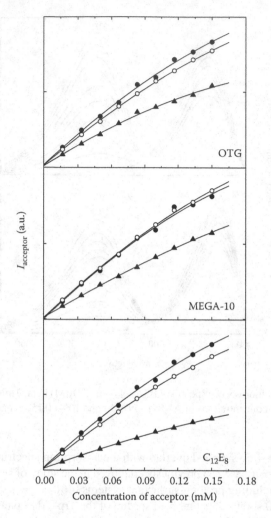

FIGURE 14.11 Enhancement of ANS emission as a function of the acceptor concentration at different surfactant concentrations: (●) 0 mM, (○) 2 mM, (▲) 20 mM.

larger separation between donor and acceptor; and (2) the possible displacement of some acceptor molecules previously bound to the protein, induced by the surfactant binding, which would probably be transferred either to the micellar clusters adsorbed on the protein surface or to the free micelles of surfactant. This second possibility can be analyzed by considering the effect of the surfactant concentration on the ANS fluorescence shown in Figure 14.3. From this figure it can be observed that, except in the case of $C_{12}E_8$, the presence of a surfactant of concentration 2 mM does not produce reduction in the fluorescence of ANS, indicating that this probe does not change its localization. However, at high surfactant concentration (20 mM) data in Figure 14.3 clearly show that some bound ANS molecules are displaced from the protein sites and are being transferred to different environments where the fluorescence quantum yield is considerably reduced. Therefore, it seems evident that whereas data in Figure 14.11 at 2 mM OTG and MEGA-10 reflect a real energy transfer process, those at 20 mM, and also data corresponding to $C_{12}E_8$, are the result of two contributions: a real RET process and the displacement of a number of ANS molecules from the protein to micellar-like aggregates adsorbed onto the protein or even to the free micelles formed by the surfactants. Consequently, it is not possible to draw conclusions about the effect of the surfactant on the protein structure from energy transfer data obtained at high surfactant concentration.

Figure 14.11 indicates that at low surfactant concentration (2 mM), the RET process is practically unaffected in the presence of MEGA-10. The OTG binding induces a slight reduction in that process. Since the rate constant for RET is strongly dependent upon the donor–acceptor distance, it may be anticipated that in the presence of 2 mM MEGA-10, the binding of surfactant to protein occurs without alterations in the protein structure. However, the same concentration of OTG produces an effect on the protein conformational structure, causing a larger separation between donor and acceptor.

14.7 DYNAMIC LIGHT SCATTERING STUDIES

The DLS technique has been proved to be a powerful tool for the investigation of the structure and dynamics of protein–surfactant systems [71–74]. The usefulness of this technique resides in two fundamental aspects. First, DLS provides data about the mobility and diffusion coefficients of macromolecules, which are of great conceptual and practical importance. Second, under certain circumstances those data can be directly interpreted in terms of the structure of scattering species. We have employed DLS for obtaining additional information on the protein–surfactant binding process and, in particular, to investigate the size and structures of the surfactant–protein complexes formed.

14.7.1 THEORETICAL BACKGROUND

The DLS technique measures fluctuations in the intensity of scattered light that arise from the thermal motion of particles in the medium. In practice, in a DLS experiment a digital correlator obtains the intensity autocorrelation function $g_I(t)$. This is related to the field autocorrelation function $g_E(t)$ for a Gaussian scattering process through the Siegert relation [75]

$$g_I = 1 + C |g_E(t)|^2 \qquad (14.18)$$

where
    t is the delay time
    C is the Siegert constant, an experimental fitting parameter of the measuring device

For a dilute dispersion of noninteracting macromolecules, $g_E(t)$ decays exponentially according to

$$g_E(t) = \exp\left(-\frac{t}{\tau}\right) \tag{14.19}$$

where τ is the relaxation time of the scattered field correlation function, which is related to the so-called apparent diffusion coefficient of the scattering species, D, by

$$\tau = \frac{1}{Dq^2} \tag{14.20}$$

where the scattering vector, q, is related to the dispersion angle, α, by $q = (4\pi n_0/\lambda)$ $\sin(\alpha/2)$, with n_0 being the refractive index of the solvent and λ the wavelength of light in vacuum.

In the case of a sample with N different scattering species, with relaxation times τ_i and diffusion coefficients D_i, the autocorrelation functions of the field and intensity are given, respectively, by [75]

$$g_E = \sum_{i=1}^{N} a_i \exp\left(-\frac{t}{\tau_i}\right) \tag{14.21}$$

and

$$g_I(t) = A + \left[\sum_{i=1}^{N} a_i \exp\left(-\frac{t}{\tau_i}\right)\right]^2 \tag{14.22}$$

In the case of two characteristic decay times, that is, of two different sizes of scattering species, Equation 14.22 can be expressed as

$$g_I(t) = A + \left[a_1 \exp\left(-\frac{t}{\tau_1}\right) + a_2 \exp\left(-\frac{t}{\tau_2}\right)\right]^2 \tag{14.23}$$

where

a_1 and a_2 are expansion coefficients known as amplitudes, representing the relative contributions of each particle size

τ_1 and τ_2 are the relaxation times of the respective species

It has been established that these relaxation times must differ by at least a factor of two in order to obtain statistically significant regression [76]. Equation 14.20 allows for the determination of the diffusion coefficients. The hydrodynamic radii, R_H, can be determined using the Stokes–Einstein relationship:

$$R_H = \frac{k_B T}{6\pi\eta_0 D} \tag{14.24}$$

where

η_0 is the solvent viscosity
k_B is the Boltzmann constant
T is the absolute temperature of the dispersion

14.7.2 DLS ON PROTEIN–SURFACTANT SYSTEMS

The DLS measurements were carried out on solutions with a fixed protein concentration and increasing surfactant concentrations ranging from the pre- to postmicellar region. In addition, single systems constituted only by BSA and pure surfactants at concentrations well above the CMC were examined. The aim of the present study is twofold. On the one hand, if the size of the protein increases in the premicellar region, it seems reasonable to attribute this change to the interaction between surfactant and protein. On the other hand, in the postmicellar range, the presence of two scattering species (protein–surfactant complexes and free micelles) should be expected.

First of all, the apparent hydrodynamic radius, R_H, of each system as a function of the total surfactant concentration assuming a unimodal distribution of aggregates has been obtained. Figure 14.12 shows the results obtained for the three systems. In fact, this analysis only provides relevant results in the premicellar region. Note that as soon as the free micelles are formed, as their concentration is much higher than that of the protein these species will control the light scattering process. In the case of the $C_{12}E_8$/BSA system, the behavior shown in Figure 14.12

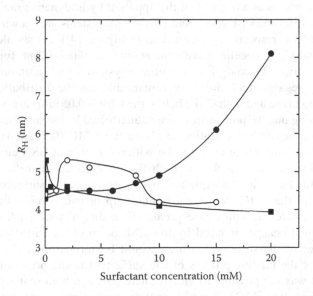

FIGURE 14.12 Effect of the surfactant concentration on the hydrodynamic radius of the surfactant–protein systems containing a fixed protein concentration (30 μM): (■) $C_{12}E_8$, (○) MEGA-10, (●) OTG.

is identical to that previously reported [74,77]. Because of the low CMC of $C_{12}E_8$, the determination of the hydrodynamic radius of the $C_{12}E_8$/BSA complex is difficult [74]. However, at low surfactant concentration, a small increase of R_H is observed, which can be attributed to surfactant binding. At high surfactant concentration, the observed R_H corresponds to the free $C_{12}E_8$ micelles, which dominate the relaxation distribution. The behavior for the MEGA-10/BSA system (Figure 14.12) is very similar. In this case, as the CMC of MEGA-10 is much higher, the trend of the hydrodynamic radius of the MEGA-10/BSA complex is easily observed. At low surfactant concentration it is shown that the hydrodynamic radius increases with the surfactant concentration, indicating the growth of the MEGA-10/BSA complex as a result of the surfactant binding. Above a certain surfactant concentration, the hydrodynamic radius decreases until a constant value is reached, corresponding to that of the free micelles of MEGA-10. Finally, for the OTG/BSA system, the behavior observed in Figure 14.12 is apparently different. In the premicellar region, a slight increase of the hydrodynamic radius with the surfactant concentration is observed, and this becomes considerable when the concentration overcomes the CMC. This tendency reflects the same behavior previously discussed for the systems involving $C_{12}E_8$ and MEGA-10. The difference resides in the fact that the size of OTG micelles is strongly dependent on its concentration, as previously reported by several authors [63,78]. Therefore, at the postmicellar region when the OTG free micelles are formed, the hydrodynamic radius increases with the surfactant concentration.

These behaviors are better illustrated by analyzing the distribution of scattered intensities as a function of the apparent hydrodynamic radius. Representative plots for the $C_{12}E_8$/BSA and OTG/BSA systems are shown in Figures 14.13 and 14.14, respectively. From data in Figure 14.13, it should be noted that BSA and $C_{12}E_8$ micelles have similar sizes (around 4.3 nm for BSA and 3.8 nm for $C_{12}E_8$). The analysis of the mixed system reveals a unimodal distribution of aggregates in all cases. It is remarkable that the distribution patterns of the protein alone and with $C_{12}E_8$ below the CMC (0.05 mM) are nearly identical, indicating that the protein structure is unaffected by the surfactant binding. When the surfactant concentration is above the CMC (0.5 and 10 mM), the unimodal distribution is observed to be wider than that of BSA alone, and its size is slightly larger than that of both the pure protein and the free $C_{12}E_8$ micelles. This fact could be explained by assuming that at surfactant concentrations above the CMC the system is made up of two species: the protein–surfactant complexes, with a size greater than that of pure BSA due to the conformational changes induced by the adsorption of the surfactant, and the free micelles of the surfactant, which would be formed once the saturation condition for the binding process of the surfactant to the protein is reached. However, it was not possible to discriminate between both scattering species by using Equation 14.23 because protein–surfactant complexes and free micelles of $C_{12}E_8$ present very similar relaxation times. A similar situation was observed in the case of the mixed systems involving MEGA-10.

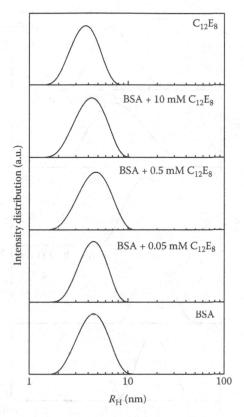

FIGURE 14.13 Hydrodynamic radii distribution from DLS measurements carried out with solutions of constant protein concentration, 30 µM, and increasing surfactant concentrations for the BSA/$C_{12}E_8$ system.

Figure 14.14 shows several plots of the distribution of scattered intensities as a function of the apparent hydrodynamic radius, R_H, for the OTG/BSA system. In this figure we can note the following aspects: (1) both pure and mixed systems present a unimodal distribution of aggregates which increase in size when the surfactant rises above the CMC, (2) the free micelles of OTG and the protein alone have sizes that are very different (10.0 nm for OTG and 4.3 nm for BSA), (3) the distribution patterns of the protein alone and with 2 mM OTG are largely identical, suggesting that the protein structure is practically unaffected by the surfactant binding. However, in the presence of 6 mM OTG (which is still below the CMC), the distribution pattern is wider than that of BSA and its maximum is shifted to higher values of R_H. This fact can be attributed to some conformational alteration in the protein structure due to the cooperative binding between surfactant and protein, and (4) when the surfactant concentration is well above the CMC (20 mM), the peak of the distribution corresponds to that of the free OTG micelles formed at the same surfactant concentration, but its bandwidth is greater. As previously

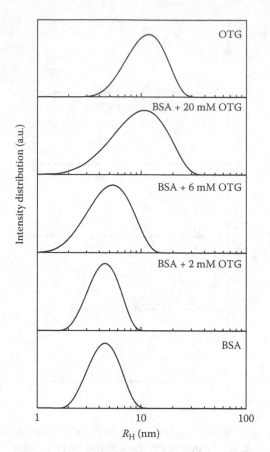

FIGURE 14.14 Hydrodynamic radii distribution from DLS measurements carried out with solutions of constant protein concentration, $30\,\mu M$, and increasing surfactant concentrations for the BSA/OTG system.

discussed in the case of the mixed system involving $C_{12}E_8$, this fact could be the result of the superposition of two dispersant species: the protein–surfactant complexes and the free micelles of the surfactant.

Since the sizes and, as a result, the relaxation times of the species in Figure 14.14 are very different, we have attempted to discriminate between the protein–surfactant complexes and the free micelles of the surfactant by using a treatment based on the analysis of the corresponding autocorrelation functions according to the following strategy. On the basis of the observations above, a model similar to that schematically represented in Figure 14.15 mimicking the interactions between OTG and BSA can be conjectured. According to this model, two different experimental situations can be considered. At low surfactant concentration, that is, as long as the free surfactant concentration is below the CMC of pure surfactant in aqueous medium, the surfactant binds to BSA through a noncooperative mechanism

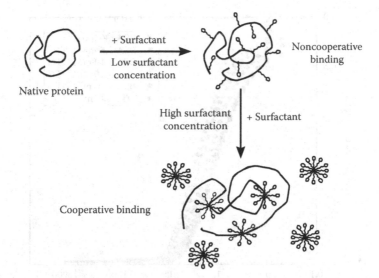

FIGURE 14.15 Proposed binding model for the interaction between OTG and BSA. At low surfactant concentration, below CMC, monomers of surfactant are adsorbed on the protein by a noncooperative mechanism. In the presence of an excess of surfactant, above CMC, a cooperative mechanism operates, producing surfactant–protein complexes coexisting with the free micelles of the surfactant.

mainly controlled by hydrophobic interactions. The hydrophobic chains of OTG would be adsorbed on the protein surface with their head-groups directed toward the bulk water and the protein acting as a nucleus on which OTG monomers condense below the CMC. In this concentration range, the binding process would probably lead to some minor conformational change in the protein structure and, as a result, to a slight increase in the size of the complex, but where no free micelles would be present. When the surfactant concentration becomes high enough, a cooperative mechanism operates, leading to the formation of micelle-like structures adsorbed onto the protein surface and to the formation of free micelles of the surfactant. At this stage, a partial protein denaturation should be expected, and saturated protein–surfactant complexes and free micelles will coexist in the system.

In order to examine the previous hypothesis, a treatment of the autocorrelation functions as obtained by the DLS measurements was performed. Figure 14.16 shows a typical set of intensity autocorrelation functions for the OTG/BSA system at surfactant concentrations below and above the CMC. On the basis of Equation 14.23, and according to Hitscherich et al. [76], the appropriate number of exponentials required to fit an autocorrelation function was determined by the relative weight of the two exponential terms a_1 and a_2 of Equation 14.23. The criterion employed was the following: as long as $a_2 \leq 0.1 \, a_1$ we use a single exponential fit,

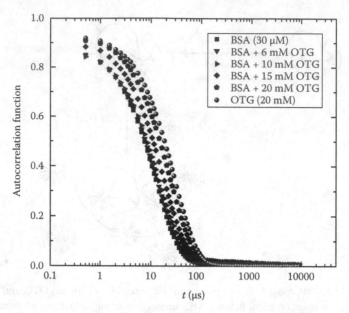

FIGURE 14.16 Normalized autocorrelation functions for BSA (30 μM) at different OTG concentrations, together with that corresponding to the pure OTG at supra-micellar concentration (20 mM).

but when $a_2 > 0.1 \, a_1$ the autocorrelation functions were fitted to a double exponential. The fit parameters (relaxation times) that we have obtained, together with the corresponding diffusion coefficients and hydrodynamic radii, are listed in Table 14.3. From data in Table 14.3, it can be observed that, below the CMC of the surfactant (≤10 mM) only single scattering species with relaxation times ranging between 27 and 30 μs exist. However, when the surfactant concentration is well above the CMC, two relaxation times appear. In the premicellar region the relaxation time increases slightly with surfactant concentration. This increase is, of course, reflected in the behavior of the corresponding hydrodynamic radius R_1, which increases from a value of 4.6 nm for the protein alone to 5.3 nm for the protein–surfactant complex. This growth (of around 15%) suggests a partial protein denaturation induced by the surfactant binding, in accordance with our previous findings, and in particular those changes affecting the emission spectrum of the intrinsic fluorescence of BSA (see Figure 14.5). Above 8 mM OTG the value of R_1 remains largely constant. However, above 10 mM a second species arises whose relaxation time (and hydrodynamic radius) seems to correspond to that of the OTG-free micelles. When saturation takes place, which happens near the CMC, two types of structures are observed to coexist in the medium; surfactant–protein complexes and free micelles of the surfactant. On the other hand, the values of R_H as obtained by the mono-exponential analysis compare well with the R_1 values in the premicellar region. However, the apparent increase in R_H observed in the postmicellar region actually reflects the superposition of the two coexisting

TABLE 14.3
Relaxation Times (τ_i), Diffusion Coefficients (D_i), and Hydrodynamic Radii (R_i) Obtained from the Fit of the Autocorrelations Functions according to Equation 14.14 by Using Either a Single or a Double Exponential Fitting (see text), Together with the Values of the Hydrodynamic Radii (R_H) and the Polydispersity Indexes (PDI) Determined by Considering an Only Population of Scattering Species

[OTG] (mM)	τ_1 (μs)	τ_2 (μs)	$D_1 \times 10^{11}$ (m²/s)	$D_2 \times 10^{11}$ (m²/s)	R_1 (nm)	R_2 (nm)	R_H (nm)	PDI
0	26.96		5.32		4.6		4.3	0.205
2	28.40		5.05		4.9		4.5	0.254
4	28.54		5.03		4.9		4.5	0.249
6	28.62		5.01		4.9		4.5	0.232
8	30.68		4.68		5.2		4.7	0.269
10	31.26		4.59		5.3		4.9	0.310
15	29.96	70.30	4.79	2.04	5.1	12.0	6.1	0.214
20	30.32	62.72	4.73	2.29	5.2	10.7	8.1	0.184
OTG		60.58		2.37		10.4	10.0	0.103

Source: Hierrezuelo, J.M., PhD Thesis, Málaga University, Málaga, Spain, 2007.

species. While the size of the surfactant–protein complexes remains constant, that of the surfactant-free micelles increases with the surfactant concentration. In this context, the behavior of the polydispersity index (PDI) (see Table 14.3) is also significant. Note that, in the premicellar region, PDI increases with the surfactant concentration, as expected. Nevertheless, above 10 mM OTG the PDI decreases until it reaches a value close to 0.1. This reduction reflects the fact that the higher molecular-weight component (i.e., the free micelles) is predominant in the medium and, hence, dominates the PDI.

14.8 CONCLUDING REMARKS

We have carried out a comparative study of the interactions of BSA with two sugar-based surfactants, OTG and MEGA-10, and the ethoxylated surfactant $C_{12}E_8$. From the results obtained in this research, the following conclusions can be drawn:

1. The binding of the surfactants to BSA has been demonstrated both by pyrene 1:3 ratio index measurements and by surface tension data for the air–liquid interface. Although the studies based on the pyrene 1:3 ratio method were, in general, not conclusive from a quantitative point of view, those derived from the adsorption isotherm in the air–liquid interface

allowed for the determination of the main characteristic parameters of the binding process.

2. The analysis of the intrinsic fluorescence of BSA upon surfactant addition showed, in the case of OTG, a behavior consistent with a partial protein denaturation induced by surfactant binding. However, there is no evidence that MEGA-10 and $C_{12}E_8$ cause major structural alterations on binding to BSA.

3. Our binding data indicate that the affinity of the surfactant toward BSA correlates well with the CMC value of the surfactant, with OTG being shown to have a greater tendency to bind on the protein. It was found that, at low surfactant concentration an initial noncooperative binding occurs. However, at a surfactant concentration slightly below the corresponding CMC, a massive adsorption of surfactant is produced, which is consistent with a mechanism cooperative in nature.

4. In all of the mixed systems, a reduction in the mean aggregation numbers was observed in comparison to the values for the surfactants in the absence of protein. These results were interpreted as evidence of the formation of micelle-like clusters, which are smaller than the free micelles, adsorbed on the protein surface.

5. From the protein binding sites, the RET results were only conclusive at low surfactant concentration. In this concentration range it was observed that the efficiency of the RET process between Trp residues and ANS is reduced in the presence of OTG. This has been interpreted as evidence that the binding of the surfactant produces some conformational modification in the protein structure.

6. For all of the mixed systems in the premicellar range, a unimodal distribution of aggregates, where the hydrodynamic radius increases with surfactant concentration, was observed. However, in the postmicellar region, the DLS results were only conclusive for the OTG/BSA system. In this case, it was possible to discriminate between two scattering species: protein–surfactant complexes and OTG-free micelles. From these results, together with previous binding data, a conceptual scheme based on the so-called necklace and bead model has been proposed.

ACKNOWLEDGMENT

This work has been financially supported by the Spanish Education and Science Ministry (Project CTQ2005–04513).

REFERENCES

1. Israelachvili, J.N., Marcelja, S., and Horn, R.G., *Q. Rev. Biophys.*, 1980, *13*, 121.
2. Tanford, C., in *The Hydrophobic Effect: Formation of Micelles and Biological Membranes*, 2nd ed., Wiley, New York, 1980, Chap. 14.
3. le Marie, M., Champeil, P., and Møller, J.V., *Biochim. Biophys. Acta*, 2000, *1508*, 86.
4. Garavito, R.M. and Ferguson-Miller, S., *J. Biol. Chem.*, 2001, *276*, 32403.

5. Jones, M.N., *Chem. Soc. Rev.*, 1992, *21*, 127.
6. Ananthapadmanabhan, K.P., in *Interactions of Surfactants with Polymers and Proteins*, Goddard, E.D. and Ananthapadmanabhan, K.P., (Eds.), CRC Press Inc., London, 1993, Chap. 8.
7. La Mesa, C., *J. Colloid Interface Sci.*, 2005, *286*, 148.
8. Holmberg, K., Jönsson, B., Kronber, B., and Lindman, B., in *Surfactants and Polymers in Aqueous Solutions*, 2nd ed., Wiley, London, 2003, Chap. 14.
9. Peters, T. Jr., in *All about Albumin Biochemistry, Genetics, and Medical Applications*. Academic Press, San Diego, C.A. 1996, Chap. 2.
10. Schweitzer, B., Felippe, A.C., Dal Bó, A., Minatti, E., and Zanette, D., *Macromol. Symp.*, 2005, *229*, 208.
11. Turro, N.J., Lei, X.-G., Ananthapadmanabhan, K.P., and Aronson, M., *Langmuir*, 1995, *11*, 2525.
12. Santos, S.F., Zanette, D., Fischer, H., and Itri, R., *J. Colloid Interface Sci.*, 2003, *262*, 400.
13. Tanner, R.E., Herpigny, B., Chen, S.-H., and Rha, C.K., *J. Chem. Phys.*, 1982, *76*, 3866.
14. Chen, S.-H. and Teixeira, J., *Phys. Rev. Lett.*, 1986, *57*, 2583.
15. Guo, X.-H. and Chen, S.-H., *Phys. Rev. Lett.*, 1990, *64*, 2579.
16. Guo, X.-H. and Chen, S.-H., *Chem. Phys.*, 1990, *149*, 129.
17. Das, R., Guha, D., Mitra, S., Kar, S., Lahiri, S., and Mukherjee, S., *J. Phys. Chem. A*, 1997, *101*, 4042.
18. Gelamo, E.L. and Tabak, M., *Spectrochim. Acta Part A*, 2000, *56*, 2255.
19. Valstar, A., Vasilescu, M., Vigouroux, C., Stilbs, P., and Almgren, M., *Langmuir*, 2001, *17*, 3208.
20. Stenstam, A., Khan, A., and Wennerström, H., *Langmuir*, 2001, *17*, 7513.
21. Gelano, E.L., Silva, C.H.T.P., Imasato, H., and Tabak, M., *Biochim. Biophys. Acta*, 2002, *1594*, 84.
22. Ghosh, S. and Banerjee, A., *Biomacromolecules*, 2002, *3*, 9.
23. Deo, N., Jockusch, S., Turro, N.J., and Somasundaram, P., *Langmuir*, 2003, *19*, 5083.
24. Hazra, P., Chakrabarty, D., Chakraborty, A., and Sarkar, N., *Biochem. Bioph. Res. Q.*, 2004, *314*, 543.
25. Shweitzer, B., Zanette, D., and Itri, R., *J. Colloid Interface Sci.*, 2004, *277*, 285.
26. Sabín, J., Prieto, G., González-Pérez, A., Ruso, J.M., and Sarmiento, F., *Biomacromolecules*, 2006, *7*, 176.
27. De, S., Das, S., and Girigoswami, A., *Colloids Surf. B*, 2007, *54*, 74.
28. Cserháti, T., *Environ. Health Perspect.*, 1995, *103*, 358.
29. Singh, K.S. and Kishore, N., *J. Phys. Chem. B*, 2006, *110*, 9728.
30. Hill, K. and Rhode, O., *Fett/Lipid*, 1999, *101*, 25.
31. Saito, S. and Tsuchiya, T., *Biochem. J.*, 1984, *222*, 829.
32. Wenk, M.R. and Seelig, J., *Biophys. J.*, 1997, *73*, 2565.
33. Pastor, O., Junquera, E., and Aicart, E., *Langmuir*, 1998, *14*, 2950.
34. Molina-Bolívar, J.A., Aguiar, J., Peula-García, J.M., and Carnero Ruiz, C., *J. Phys. Chem. B*, 2004, *108*, 12813.
35. Frindi, M., Michels, B., and Zana, R., *J. Phys. Chem.*, 1992, *96*, 8137.
36. Chami, M., Pehau-Arnaudet, G., Lambert, O., Rank, J.-L., Lèvy, D., and Rigaud, J.-L., *J. Struct. Biol.*, 2001, *133*, 64.
37. Hildreth, J.E.K., *Biochem. J.*, 1982, *207*, 363.
38. Walter, A., Suchy, S.E., and Vinson, P.K., *Biochim. Biophys. Acta*, 1990, *1029*, 67.
39. Hierrezuelo, J.M., Agregación y estructuras en sistemas supramoleculares formados en disoluciones acuosas de tensioactivos de base azucarda, PhD thesis, Málaga University, Málaga, Spain, 2007.

40. Carnero Ruiz, C., Hierrezuelo, J.M., Aguiar, J., and Peula-García, J.M., *Biomacromolecules*, 2007, *8*, 2497.
41. Carnero Ruiz, C., Hierrezuelo, J.M., Peula-García, J.M., and Aguiar, J., *The Open Macromolecules Journal*, 2008, *2*, 6.
42. Kalyanasundaram, K. and Thomas, J.K., *J. Am. Chem. Soc.*, 1977, *99*, 2039.
43. Aguiar, J., Carpena, P., Molina-Bolívar, J.M., and Carnero Ruiz, C., *J. Colloid Interface Sci.*, 2003, *258*, 116.
44. Vasilescu, M., Angelescu, D., Almgren, M., and Valstar, A., *Langmuir*, 1999, *15*, 2635.
45. Slavik, J., *Biochim. Biophys. Acta*, 1982, *694*, 1.
46. De, S., Girigoswami, A., and Das, S., *J. Colloid Interface Sci.*, 2005, *285*, 562.
47. Lu, R.-C., Cao, A.-N., Lai, L.H., Zhu, B.-Y., Zhao, G.-X., and Xiao, J.-X., *Colloids Surf. B*, 2005, *41*, 139.
48. Nishikido, N., Takahara, T., Kobayashi, H., and Tanaka, M., *Bull. Chem. Soc. Jpn.*, 1982, *55*, 3085.
49. Tribout, M., Paredes, S., González-Mañas, J.M., and Goñi, F.M., *J. Biochem. Biophys. Methods*, 1991, *22*, 129.
50. Nishiyama, H. and Maeda, H., *Biophys. Chem.*, 1992, *44*, 199.
51. Zadymova, N.M., Yampol'skaya, G.P., and Filatova, L.Y., *Colloid J.*, 2006, *68*, 162.
52. Liu, J., Xu, G.-Y., and Yu, L., *J. Disper. Sci. Technol.*, 2006, *27*, 835.
53. Jones, M.J., *Biochem. J.*, 1975, *151*, 109.
54. Andreu, J.M. and Muñoz, J.M., *Biochemistry*, 1986, *25*, 5220.
55. Pi, Y., Shang, Y., Peng, C., Liu, H., Hu, Y., and Jiang, J., *Biopolymers*, 2006, *83*, 243.
56. Sjögren, H., Ericsson, C.A., Evenäs, J., and Ulvenlund, S., *Biophys. J.*, 2005, *89*, 4219.
57. Córdoba, J., Reboiras, M.D., and Jones, M.N., *Int. J. Biol. Macromol.*, 1988, *10*, 270.
58. Sukow, W.W., Sandberg, H.E., Lewis, E.A., Eatough, D.J., and Hansen, L.D., *Biochemistry*, 1980, *19*, 912.
59. Nielsen, A.D., Borch, K., and Westh, P., *Biochim. Biophys. Acta*, 2000, *1479*, 321.
60. Balzer, D., *Langmuir*, 1993, *9*, 3375.
61. Turro, N.J. and Yekta, A., *J. Am. Chem. Soc.*, 1978, *100*, 5951.
62. Zana, R., in *Surfactant Solutions: New Methods of Investigation*, Zana, R., (Ed.), Marcell Dekker, Inc., New York, 1987, Chap. 5.
63. Molina-Bolívar, J.A., Hierrezuelo, J.M., and Carnero Ruiz, C., *J. Phys. Chem. B*, 2006, *110*, 12089.
64. Hierrezuelo, J.M., Aguiar, J., and Carnero Ruiz, C., *Langmuir*, 2004, *20*, 10419.
65. Molina-Bolívar, J.A., Hierrezuelo, J.M., and Carnero Ruiz, C., *J. Colloid Interface Sci.*, 2007, *313*, 656.
66. Lindman, B. and Thalberg, K., in *Interactions of Surfactants with Polymers and Proteins*. Goddard, E.D. and Ananthapadmanabhan, K.P. (Eds.), CRC press, London, 1993, Chap. 5.
67. Wu, P. and Brand, L., *Anal. Biochem.*, 1994, *218*, 1.
68. Hayakawa, K., Nakano, T., Satake, I., and Kwak, J.C.T., *Langmuir*, 1996, *12*, 269.
69. Lakowicz, J.R., in *Principles of Fluorescence Spectroscopy*, 3rd ed., Springer, New York, 2006, Chap. 13.
70. Förster, T., *Z. Naturforsch.*, 1949, *4A*, 321.
71. Tanner, R.E., Herpigny, B., Chen, S.-H., and Rha, C.K., *J. Chem. Phys.*, 1982, *76*, 3866.
72. Gimel, J.C. and Brown, W., *J. Chem. Phys.*, 1996, *104*, 8112.

73. Saxena, A., Antony, T., and Bohidar, H.B., *J. Phys. Chem. B*, 1998, *102*, 5063.
74. Valstar, A., Almgren, M., Brown, W., and Vasilescu, M., *Langmuir*, 2000, *16*, 922.
75. Berne, B.J. and Pecora, R., in *Dynamic Light Scattering. With Applications to Chemistry, Biology, and Physics*, Dover Publications, New York, 2000, Chap. 4.
76. Hitscherich, C., Aseyev, V., Wiencek, J., and Loll, P.J., *Acta Cryst.*, 2001, *D57*, 1020.
77. Valstar, A., Protein-surfactant interactions, PhD thesis. Uppsala University, Uppsala, Sweden, 2000.
78. Diab, C., Winnik, F.M., and Tribet, C., *Langmuir*, 2007, *23*, 3025.

15 Applications of Alkyl Glucosides in the Solubilization of Liposomes and Cell Membranes

Olga López, Mercedes Cócera, Luisa Coderch,
Lucyanna Barbosa-Barros, José L. Parra,
and Alfonso de la Maza

CONTENTS

15.1   INTRODUCTION

The study the membrane proteins within specialized membrane microdomains, such as lipids enriched in sphingolipids, glycerophospholipids, and cholesterol [1–3] presents some specific problems. These domains, also called surfactant-resistant membranes (DRMs), have been shown to play key roles in cell signaling and protein sorting. Historically, DRMs have been detected by their resistance to solubilization by cold Triton X-100. However, it has been shown that the characteristics of these DRMs are dependant upon the surfactants used in their isolation. For example, Schuck et al. showed that the amounts and types of proteins and lipids associated with DRMs varied dramatically when different surfactants were used to isolate the membrane domain [1]. Another important aspect in membrane protein research is that the surfactant used must be able to maintain a target membrane in a soluble, native and functional form in the absence of the membrane. A surfactant-solubilized state is usually required for protein purification, since biological membranes must be disrupted with surfactants to separate individual proteins. Furthermore, many biochemical and biophysical techniques, require solubilized and monodisperse protein. To minimize the potential problems of protein denaturation and aggregation, one of the central issues is to find conditions where the physical environment of the protein was minimally perturbed despite the dissolution of the original membrane. In fact, most membrane proteins can be readily maintained in a functional, active form while in their native membranes, but may be highly prone to loss of function and aggregation in the surfactant-solubilized state. Thus, a surfactant that is able to closely mimic the physical properties of the original bilayer will be better suited at maintaining the structural integrity of a membrane protein following its extraction from its natural lipidic environment. Indeed, specific lipids are sometimes required to maintain the structural stability of membrane proteins. Conversely, a "harsh" surfactant that cannot act as an effective membrane substitute is unlikely to maintain protein function.

As for the physical characteristics of surfactant, these compounds differ from biological lipids in that they self-assemble into micelles instead of the bilayer structures. Another property is the minimal concentration required for the formation of micelles, termed the critical micelle concentration (CMC). Because of their amphipathic properties, the surfactant monomers have limited solubility in water, and they partition into a micellar phase at concentrations above the CMC. Surfactants exist in water in a monomer-micelle equilibrium, and the amount of free monomer in solution remains essentially constant at surfactant concentrations greater than the CMC. In general, a surfactant must be used at concentrations above its CMC to act as an effective solubilizer. A general observation is that the head group has a strong influence on the interactions of surfactants with proteins, while the length of the alkyl chain affects the surfactant CMC and aggregation number. Furthermore, not all surfactants are useful over the same range of alkyl lengths, however, short chain surfactants may be hampered by very high CMC values, and long chain surfactants may be poorly soluble.

From these considerations, we may assume that in extraction and solubilization of proteins by surfactants a key concept is the critical solubilization concentration (CSC), which is the minimal surfactant concentration required to disrupt a membrane system into a predominantly micellar dispersion. The CSC is normally slightly higher than the CMC and very dependent on the exact starting conditions, especially the lipid concentration, and there may or may not be a sharp phase transition depending on the membrane system and the surfactant. The selective solubilization of proteins at surfactant concentrations below the CSC can be a very effective purification strategy. However, in many cases, is better to choose to completely solubilize the membrane preparation because this ensures that the target protein is fully extracted from the membrane.

In general terms, sugar-based surfactants have been the most successful in the crystallization of membrane proteins. The increasing interest in studying these amphiphiles is due to their preparation from renewable raw materials, ready biodegradability, mildness to the skin, and biocompatibility. An important structural feature of these surfactants is the sugar head-group, a voluminous and relatively rigid moiety that can be functionalized by a myriad of reagents and synthetic schemes [4–9]. Some of the more common surfactants used in functional and structural studies of membrane proteins are the alkyl glycosides, in particular the octyl glucoside (OG). These sugar surfactants have two interesting properties. On the one hand, are "mild" surfactants with respect to its denaturing effect on proteins [10–17] and on the other are not very efficient lipid solubilizing agents. Thus, the initial surfactant interaction destabilize the bilayer yielding mixed lipid-surfactant fragments and further surfactant addition leads to bilayer dissolution and protein solubilization [1,18]. This characteristic is essential in these studies given that extracted proteins must be purified in such a way that some native lipids remained bound to the protein. Under a practical view point, short chain alkyl maltosides and glucosides have been successful in the crystallization of membrane proteins [19–23], whereas longer-chain glycosides have been used to stabilize various oligomeric states of the G protein coupled receptor (GPCR), to study the unfolding of the 4-transmembrane helix protein DsbB from the inner membrane of *E. coli*, and to study the light-induced structural changes in mammalian rhodopsin by ^{19}F NMR [24–27].

The need to find effective and predictable means to solubilize and reconstitute these membranes, and to scale the reconstitution protocols for biological research or pharmacological applications, is one reason for interest in the nuances of membrane-surfactant interactions. A number of studies have been devoted to the understanding of the principles governing the interaction of surfactants with simplified membrane models such as phospholipid bilayers [28–30]. This interaction leads to the breakdown of lamellar structures and the formation of lipid-surfactant mixed micelles. A significant contribution in this area has been made by Lichtenberg [31], who postulated that the critical effective surfactant/lipid molar ratio (Re) producing saturation and solubilization of liposomes depends on the surfactant CMC and on the bilayer/aqueous medium distribution coefficients (K) rather than on the nature of the surfactants. Other fields in which surfactants have acquired great interest are those related with dermatology and with the cosmetic industry. The skin and specifically the stratum corneum (SC), the outermost layer of the epidermis, are the main target for a number of products containing surfactants. It is known that frequent exposure of SC to surfactants in household and personal care products can damage the skin in susceptible individuals provoking scaling, dryness, and erythema. Then research regarding the effect of surfactants on the skin is essential to guarantee preservation of this tissue and to retain the cohesion and the barrier function of the skin [32].

In this chapter, we shall describe three essential aspects related to the successful use of the sugar-base surfactants alkyl glucosides (alkyl chain lengths ranging from C_8 to C_{12}) in the solubilization and crystallization of membrane proteins. In the first section, we quantitatively characterize the sublytic and lytic alterations caused by a series of alkyl glucosides (alkyl chain lengths ranging from C_8 to C_{12}) in liposomes formed by PC and by lipids building the intercellular spaces of the SC. To this end, and according to Lichtenberg [31], we reported the Re and K parameters of these surfactants when interacted with PC and SC lipid liposomes both at sublytic and lytic levels. Knowledge of these parameters is useful in establishing a criterion for the evaluation of the surfactant activity and applicability in specific biological membranes. In the second section, it is described the transition states associated to the solubilization and reconstitution of phosphatidylcholine (PC) liposomes by OG. Thus, we describe the complex double step process involved in the formation of mixed micelles and the more simply one, the reconstitution of vesicles from micellar solutions. Finally, in the last section of this chapter it is described the use of this surfactant as a very selective tool to study the structure and function of the SC tissue. The use of this specific surfactant opens the possibility to investigate the physicochemical properties of the SC lipid assembly to better understanding its function as transcutaneous permeability barrier.

15.2   GENERAL METHODOLOGY

15.2.1   Preparation and Characterization of Liposomes

Liposomes with different lipid compositions and concentrations are usually prepared by reverse phase evaporation or by the method of the lipidic film [33,34]. Unilamellar vesicles are obtained by extrusion of these liposomes through 800–200 nm polycarbonate membranes to achieve a uniform size distribution [33]. The size distribution and the polydispersity index (PI) of liposomes and similar nanostructures, such as surfactant-PC aggregates resulting in the interaction liposome-surfactant, are frequently determined by dynamic light scattering techniques operating at different temperatures and reading angles [35]. Other liposome parameters such as the internal volume, (volume enclosed by a given amount of lipid, expressed in mL mmol^{-1}) and the encapsulation efficiency (fraction of the aqueous compartment sequestered by bilayers, expressed in percentage with respect to the original volume) are evaluated by measuring the concentration of a fluorescent probe encapsulated into liposomes after chromatographic separation of unencapsulated material (using resins) and subsequent destruction of liposomes by addition of Triton X-100 [36].

15.2.2 SOLUBILIZING PARAMETERS

In the analysis of the equilibrium partition model proposed by Schurtenberger [37] for bile salt/lecithin systems, Lichtenberg [31] and Almog et al. [38] have shown that for a mixing of lipids (at a concentration PL (mM)) and surfactant (at a concentration S_T (mM)), in dilute aqueous media, the distribution of surfactant between lipid bilayers and aqueous media obeys a partition coefficient K, given (in mM^{-1}) by:

$$K = \frac{S_B}{[(PL + S_B) \, S_W]} \qquad (15.1)$$

where S_B is the concentration of surfactant in the bilayers (mM) and S_W is the surfactant concentration in the aqueous medium (mM). For PL $\gg S_B$, the definition of K, as given by Schurtenberger, applies:

$$K = S_B/(PL \cdot S_W) = Re/S_W \qquad (15.2)$$

where Re is the effective molar ratio of surfactant to phospholipid in the bilayer: (Re $= S_B$/PL). Under any other conditions, Equation 15.1 has to be employed to define K; this yields:

$$K = Re/[S_W (1 + Re)] \qquad (15.3)$$

This approach is consistent with the experimental data offered by Lichtenberg [31] and Almog [38] for different surfactant phospholipid mixtures over wide ranges of Re values. Given that the range of lipid concentrations used in our investigation is similar to that used by Almog to test his equilibrium partition model, the K parameter has been determined using this equation. The determination of these parameters can be carried out on the basis of the linear dependence existing between the surfactant concentrations required to achieve these parameters and the lipid concentration in liposomes, which can be described by the equation:

$$S_T = S_W + Re \cdot PL \qquad (15.4)$$

where Re and the aqueous surfactant concentration (S_W) are in each curve respectively the slope and the ordinate at the origin (zero lipid concentration).

15.2.3 Permeability Alterations and Solubilization of Liposomes

To study bilayer permeability, vesicles containing a fluorescent probe are often used. A number of studies have been performed using 5(6)-carboxyfluorescein (CF) [33,39–42]. The permeability alterations caused by the surfactants in CF-containing liposomes are determined by monitoring the increase in the fluorescence intensity of the liposome suspensions due to the CF released from the interior of vesicles to the bulk aqueous phase [30,33]. The percentage of CF released is calculated by means of the equation:

$$\%CF \text{ release} = (I_T - I_0)/(I_\infty - I_0)100 \tag{15.5}$$

where I_0 is the initial fluorescence intensity of CF-loaded liposomes in the absence of surfactant, I_T is the fluorescence intensity measured 40 min after adding equal volumes of appropriate surfactant solutions to liposome suspensions. I_∞ corresponds to the fluorescence intensity resulting after the complete destruction of liposomes by the addition of Triton X-100 aqueous solution [30].

With regard to liposome solubilization, it has been previously demonstrated that static light-scattering (SLS) constituted a very convenient technique for the quantitative study of the bilayer solubilization by surfactants [30,41–43]. Accordingly, the solubilizing perturbation produced by surfactants in PC and SC lipid liposomes is monitored using this technique.

The overall solubilization can be mainly characterized by two parameters termed Re_{SAT} and Re_{SOL}, according to the nomenclature adopted by Lichtenberg [31] corresponding to the Re ratios at which SLS starts to decrease with respect to the initial value and shows no further decrease. These parameters correspond to the surfactant/lipid molar ratios at which the surfactant: (1) saturated liposomes and (2) led to a complete solubilization of these structures.

In order to characterize the aggregates formed during the solubilization, the technique of freeze fracture electron microscopy (FFEM) has resulted very useful [44–47]. Samples are cryofixed by dipping into nitrogen-cooled liquid propane and fractured. The replicas of the samples obtained by unidirectional shadowing with Pt/C and C are visualized by TEM.

15.2.4 Characterization of Stratum Corneum

15.2.4.1 Chemical Analysis

The separation of hairless pig epidermis and the isolation of SC sheets have been described in previous works [48–50]. When SC is treated with OG, part of the tissue components are disaggregated and part of the SC sheets remains visually almost unaltered [50]. The disaggregated material free of OG surfactant, concentrated to dryness and successively extracted with mixtures of chloroform–methanol can be analyzed by thin layer chromatography flame ionization detector (TLC/FID) [48,50]. This analysis gives information about the intercellular lipids of SC. To analyze lipids from the envelope of the SC corneocytes, additional mild alkaline hydrolysis are needed [50]. As for the protein analysis, the disaggregated material has to be hydrolyzed with HCl (6.0 M) at 110°C for 24 h. The amino acid analysis is carried out by HPLC.

15.2.4.2   TEM and Cryo-SEM for Stratum Corneum Visualization

In order to visualize SC by TEM, samples are fixed in glutaraldehyde and para-formaldehyde and postfixed in RuO_4 [50]. This method is based on a RuO_4 post-fixation protocol method described by Swatzendruber et al. [51]. The specimens, dehydrated in acetone are embedded in Spurr's resin and the ultrathin sections, are examined [50,51]. Similar procedure is used to observe corneocyte envelopes in the material disaggregated.

Freeze-substitution transmission electron microscopy (FSTEM) studies in combination with RuO_4 fixation, provides a different visualization of the SC [52]. The methodology used is based on that described by Van den Bergh et al. [53]. The SC is also fixed in glutaraldehyde and postfixed in RuO_4 with potassium fer-rocyanide. Afterwards, the tissue samples are cryofixed, by rapid freezing on a liquid-nitrogen cooled metal mirror. The freeze-substitution involves a gradual and low-temperature embedding procedure of the sample in the resin that reduces the formation of artefacts [52,53].

Additional visualization of SC can be carried out using high resolution low-temperature scanning electron microscopy (HRLTSEM) also called Cryo-SEM [54,55]. SC samples are fixed, cryoprotected with 30% (v:v) glycerol, frozen by plunging into liquid propane, and fractured. The fracture plane is coated with Pt/C and C, and resulting replicas are immediately cryotransferred on a Gatan cryoholder to be observed in a field emission scanning electron microscope (SEM). Images are usually obtained with the backscattered electron signal and recorded digitally.

15.3   RESULTS AND DISCUSSION

15.3.1   Sublytic and Lytic Effects Caused by the Alkyl Glucosides as a Function of Their Alkyl Chain Length in Liposomes Formed by PC or SC lipids

15.3.1.1   PC Liposomes

First of all, it is needed to determine the CMC of the surfactants used. These data are shown in Figure 15.1. It may be seen that the surface tension values decreased as the surfactant concentration rose, showing in each case a charac-teristic change in the slope, which corresponded to the surfactant CMC. From the CMC values obtained it is possible to assume that the increase in the alkyl chain length drastically reduced the surfactant CMCs and slightly reduced the surface tensions for these surfactants concentrations in all cases (from 31 mN m^{-1} for C_8-Glu to 28.5 mN m^{-1} for C_{12}-Glu). Given that the surface tension at the

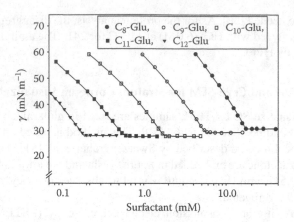

FIGURE 15.1   Variation of the surface tension versus total surfactant concentration for the alkyl glucosides tested. (●) C_8-Glu, (○) C_9-Glu, (□) C_{10}-Glu, (■) C_{11}-Glu, and (▼) C_{12}-Glu. (From De la Maza, A., Coderch, L., Gonzalez, P., and Parra, J.L., *J. Control. Release*, 52, 159, 1998. With permission.)

CMC (γ_{CMC}) is used as one of the criteria of surface activity of the system, (the lower the γ_{CMC}, the higher the surface activity [56]. It is possible to assume that the longer the surfactant hydrophobic tail, the higher its surface activity in the system.

The validity of the equilibrium partition model proposed by Lichtenberg and Almog et al. [31,38] based on the Equation 15.1 was investigated. According to these authors this equation may be expressed by: $L/S_B = (1/K)(1/S_W) - 1$. Hence, this validity requires a linear dependence between L/S_B and $1/S_W$; this line should have a slope of $1/K$, intersect with the L/S_B axis at -1 and intersect with the $1/S_W$ at K. These authors demonstrated the validity of this model for C_8-Glu in the range of PC and surfactant concentration used in other works [31,57]. To test the validity of the model for the other alkyl glucosides unilamellar PC liposomes were mixed with varying subsolubilizing concentrations of these surfactants (S_T). The resultant surfactant-containing vesicles were then spun at 140000 g at 20°C for 4 h to remove the vesicles. No PC was detected in the supernatants. The concentration of each surfactant in the supernatants (S_W) was determined by HPLC [58,59], and their concentration in the lipid bilayers was calculated ($S_B = S_T - S_W$). The results of the experiments in which S_B and S_W were measured (carried out for each surfactant in the same range of PC and surfactant concentrations used to determine K) were plotted in terms of the dependence of L/S_B on $1/S_W$. Straight lines were obtained for each surfactant tested ($r^2 = 0.990, 0.991, 0.990, 0.989$, and 0.992 for C_9-Glu, C_{10}-Glu, C_{11}-Glu, and C_{12}-Glu, respectively), which were dependent on L and intersected with the L/S_B axis always at -0.96 ± 0.10. Both the linearity of these dependences and the proximity of the intercept to -1 support the validity of this model to determine K for these surfactants at the two sublytic and lytic interaction levels investigated.

15.3.1.1.1 Sublytic Alterations

The sublytic alterations (permeability alterations) caused by the surfactants in CF-containing liposomes (lipid concentration ranging from 0.5 to 5.0 mM) were determined by monitoring the increase in the fluorescence intensity of the liposome suspensions due to the CF released from the interior of vesicles to the bulk aqueous phase.

It is known that, in surfactant/lipid systems, complete equilibrium may take several hours [43]. However, in subsolubilizing interactions a substantial part of the surfactant effect takes place within approximately 30 min after its addition to the liposomes [60]. To determine the time in which the leakage ceased, a kinetic study of the interaction of alkyl glucosides with these bilayer structures was carried out. Liposomes were treated with surfactants at subsolubilizing concentrations and subsequent changes in CF release were studied as a function of time. The results obtained for a constant C_8-Glu concentration (14.0 mM) the PC concentration ranging from 1.0 to 10.0 are indicated in Figure 15.2. It may be seen that the CF release showed always a transient state of enhanced permeability of the liposomal bilayers, in which about 40 min was needed to achieve CF release plateaux in all cases. This behavior was possibly due to the release of the fluorescent dye encapsulated into the vesicles through holes, or channels, created in the membrane and not to bilayer fusion. The incorporation of surfactant monomers into membranes may directly induce the formation of hydrophilic pores or merely

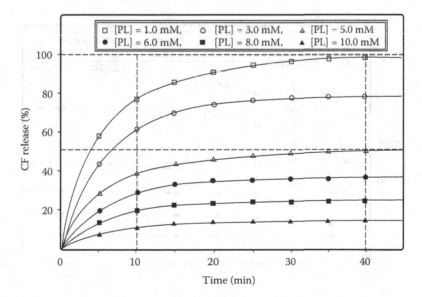

FIGURE 15.2 Time curves of the release of 5(6) carboxyfluorescein (CF) trapped into PC liposomes caused by the addition of C_8-Glu (14.0 mM). PC concentrations: (□) 1.0 mM, (○) 3.0 mM, (△) 5.0 mM, (•) 6.0 mM, (■) 8.0 mM, and (▲) 10.0 mM. (From De la Maza, A., Coderch, L., Gonzalez, P., and Parra, J.L., *J. Control. Release*, 52, 159, 1998. With permission.)

stabilize transient holes, in agreement with the concept of transient channels suggested by Schubert et al. [61]. The spontaneous release of the CF from the liposomes in the absence of surfactant and in this period was negligible. Hence, changes in CF release were studied 40 min after the addition of surfactants to the liposomes at 25°C.

To determine the Re and S_W parameters at two sublytic levels (50% CF release: $Re_{50\%CF}$, and 100% of CF release: $Re_{100\%CF}$), a systematic study of the sublytic liposome alteration caused by the addition of surfactants was carried out for various PC concentrations. The curves obtained for C_8-Glu (PC concentration ranging from 1.0 to 10.0 mM) are given in Figure 15.3. The surfactant concentrations resulting in 50% and 100% of CF release (for all the surfactants tested) were graphically obtained and plotted versus PC concentration. The r-squared statistic indicated that the straight lines obtained explained more than 98.9% of the surfactant concentration variability versus PC concentration. Therefore a very good linear fit was established in each case. These findings confirm that the straight lines for the Equations 15.6 and 15.7 were appropriate to determine Re and S_W for the surfactants tested.

$$S_{T,50\%CF} = S_{W,50\%CF} + Re_{50\%CF}. [L] \qquad (15.6)$$

$$S_{T,100\%CF} = S_{W,100\%CF} + Re_{100\%CF}. [L] \qquad (15.7)$$

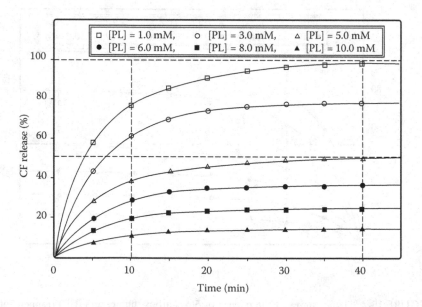

FIGURE 15.3 Percentage changes in CF release of PC liposomes, (lipid concentration ranging from 1.0 to 10.0 mM), induced by the presence of increasing concentrations of C_8-Glu. PC concentrations: (□) 1.0 mM, (○) 3.0 mM, (△) 5.0 mM, (●) 6.0 mM, (■) 8.0 mM, and (▲) 10.0 mM. (From De la Maza, A., Coderch, L., Gonzalez, P., and Parra, J.L., *J. Control. Release*, 52, 159, 1998. With permission.)

where $S_{T,50\%CF}$ and $S_{T,100\%CF}$ are the total surfactant concentrations. The surfactant to PC molar ratios $Re_{50\%CF}$ and $Re_{100\%CF}$ and the aqueous concentration of surfactant $S_{W,50\%CF}$ and $S_{W,100\%CF}$ are in each curve, respectively the slope and the ordinate at the origin (zero lipid concentration). The $K_{50\%CF}$ and $K_{100\%CF}$ parameters (bilayer/aqueous phase surfactant partition coefficient for 50% and 100% of CF release) were calculated using the Equation 15.3.

This method has also demonstrated to be valid for the study of the interactions of different surfactants with PC liposomes [41,42,62–65]. The Re, K, and S_W values obtained are given in Table 15.1 [41]. The CMCs for each surfactant together with the regression coefficients r^2 of the straight lines obtained are also included.

The S_W values increased as the amount of surfactant in the system rose regardless of the surfactant alkyl chain length, although showing always smaller values than those for the surfactant CMCs. Given that S_W is the surfactant concentration in water, these data indicate that the surfactant-liposome interaction studied was mainly ruled by the action of surfactant monomer. These findings are in accordance with the results reported for the overall interaction of C_8-Glu with PC liposomes [33,66]. As for the Re parameter, this value increased as the CF release percent rose (from $Re_{50\%CF}$ to $Re_{100\%CF}$) and as the surfactant alkyl chain length rose (from C_8 to C_{12}) showing at the interaction step for $Re_{50\%CF}$ a minimum in the range C_{10}–C_8. At the two interaction steps studied the variation in Re was more pronounced for the surfactants with longer alkyl chain lengths.

Given that the surfactant capacity to alter the release of the CF encapsulated into bilayers is inversely related to the Re value, the maximum activity corresponded to the C_9-Glu and to the C_8-Glu, respectively, whereas the minimum to the C_{12}-Glu. Thus, for 100% of CF release, the lower the surfactant alkyl

---

TABLE 15.1

Surfactant to Lipid Molar Ratios (Re), Partition Coefficients (K), Surfactant Concentrations in the Aqueous Medium (S_W) Resulting in the Subsolubilizing Interaction (50% and 100% of CF Release) of Alkyl Glucosides with PC Liposomes

	CMC (mM)	$S_{W,50\%CF}$ (mM)	$S_{W,100\%FC}$ (mM)	$Re_{50\%CF}$ (mole/ mole)	$Re_{100\%CF}$ (mole/ mole)	$K_{50\%CF}$	$K_{100\%CF}$	r^2 50%CF	r^2 100%CF
C_8-Glu	18.0	9.1	13.9	1.0	1.18	0.054	0.038	0.995	0.989
C_9-Glu	5.6	2.25	4.0	0.95	1.25	0.21	0.138	0.996	0.992
C_{10}-Glu	1.80	0.80	1.45	1.0	1.40	0.62	0.40	0.997	0.994
C_{11}-Glu	0.58	0.25	0.45	1.3	1.80	2.26	1.42	0.995	0.998
C_{12}-Glu	0.18	0.075	0.152	1.72	2.60	8.43	4.75	0.998	0.996

Source: de la Maza, A., Coderch, L., Gonzalez, P., and Parra, J.L., *J. Control. Release*, 52, 159, 1998. With permission.

Note: CMC of each surfactant tested is also included together with the regression coefficients of the straight lines obtained.

chain length (or the hydrophobic moiety) the higher its ability to alter the permeability of these bilayer structures. This finding is surprising taking into account that the C_{12}-Glu exhibits the lowest γ_{CMC} and, hence, the highest surface activity [64].

The structural changes in the molecular surfactant structure due to the increase in its alkyl chain length also resulted in an abrupt increase in the $K_{50\%CF}$ and $K_{100\%CF}$ values, which was also especially noticeable for the surfactants with longer alkyl chain lengths. Thus, the degree of partitioning of these surfactants into bilayers (or affinity with these structures) drastically increased as the surfactant alkyl chain length increased or its CMC decreased. The fact that these surfactants showed at 100% CF release lower K values than those for 50% may be attributed to the progressive saturation of the bilayers by the surfactants (the amounts of surfactants in the aqueous phase increased more than in the bilayers). These findings are in agreement with those reported by Paternostre et al. when studying the interaction of C_8-Glu with PC liposomes [66].

Hence, two opposite trends may be observed when comparing the variation of Re and K versus the surfactant alkyl chain length. The increase in the surfactant alkyl chain length resulted in a progressive decrease in the surfactant ability to alter the permeability of liposomes and inversely in an abrupt increase in its affinity with this bilayer structure. The overall balance of these two tendencies indicates that both the C_8-Glu and the C_9-Glu are the more appropriate to obtain an improved surfactant-induced release of PC liposomes.

15.3.1.1.2   Lytic Alterations

To determine the Re and S_W parameters at lytic level a systematic investigation of SLS variations in PC liposomes caused by the addition of surfactants have been carried out for various PC concentrations [67]. The curves obtained for C_8-Glu (PC concentration ranging from 1.0 to 10.0 mM) are given in Figure 15.4. The addition of surfactant led in all cases to an initial increase and a subsequent fall in the scattered intensity of the system until a low constant SLS value was achieved for complete bilayer solubilization via mixed micelle formation [31]. Surfactant concentrations producing 100% (S_{SAT}) and 0% (S_{SOL}) of SLS were obtained for each lipid concentration by graphical methods. The arrows A and B (curve for PC liposomes, concentration 10.0 mM in Figure 15.4) correspond to these two values.

When plotting the surfactant concentrations thus obtained versus PC concentration, curves were obtained in which an acceptable linear relationship was established in each case. The straight lines obtained corresponded to the Equations 15.8 and 15.9 from which the Re and S_W parameters were determined.

$$S_{SAT} = S_{W,SAT} + Re_{SAT} \cdot [L] \tag{15.8}$$

$$S_{SOL} = S_{W,SOL} + Re_{SOL} \cdot [L] \tag{15.9}$$

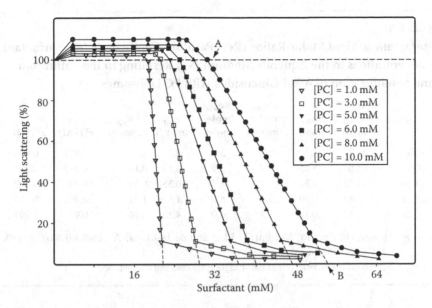

FIGURE 15.4 Percentage change in SLS of PC liposomes, (PC concentration ranging from 1.0 to 10.0 mM), induced by the presence of increasing amounts of C_8-Glu. Symbols: PC concentration ($\triangledown$) 1.0 mM, ($\square$) 3.0 mM, ($\blacktriangledown$) 5.0 mM, ($\blacksquare$) 6.0 mM, ($\blacktriangle$) 8.0 mM, and ($\bullet$) 10.0 mM. (From López, O., Cocera, M., Parra, J.L., and de la Maza, A., *Colloids Surf. A*, 193, 221, 2001. With permission.)

In these two equations S_{SAT} and S_{SOL} are the total surfactant concentrations for liposome saturation and completely solubilization. The surfactant to lipid molar ratios Re_{SAT} and Re_{SOL} and the aqueous surfactant concentrations $S_{W,SAT}$ and $S_{W,SOL}$ are in each curve, respectively the slope and the ordinate at the origin (zero lipid concentration). K_{SAT} and K_{SOL} (bilayer/aqueous phase surfactant partition coefficients for saturation and complete solubilization of liposomes) were determined from the Equation 15.3. The results obtained for each surfactant tested including the regression coefficients (r^2) of the straight lines are given in Table 15.2. [67].

Free surfactant concentrations ($S_{W,SAT}$, $S_{W,SOL}$) were always comparable to the surfactant CMCs, (see Table 15.1). This indicates that the liposome solubilization was mainly ruled by formation of mixed micelles. These results extend to the alkyl glucosides studied the generally admitted assumption that the free surfactant concentration (S_W) must reach the CMC before solubilization starts to occur [31,33].

The Re values clearly increased from liposome saturation (Re_{SAT}) to complete solubilization (Re_{SOL}), regardless of the surfactant alkyl chain length. The maximum activity at these two levels corresponded in all cases to the C_8-Glu (lowest

TABLE 15.2

Surfactant to Lipid Molar Ratios (Re), Partition Coefficients (K), Surfactant Concentrations in the Aqueous Medium (S_W) Resulting in the Saturation and Solubilization of Alkyl Glucosides with PC Liposomes

	$S_{W,SAT}$ (mM)	$S_{W,SOL}$ (mM)	Re_{SAT} (mole/mole)	Re_{SOL} (mole/mole)	K_{SAT} (mM^{-1})	K_{SOL} (mM^{-1})	r^2 (SAT)	r^2 (SOL)
C_8-Glu	17.8	18.3	1.3	3.6	0.03	0.04	0.992	0.990
C_9-Glu	5.6	5.65	1.4	3.9	0.10	0.14	0.997	0.998
C_{10}-Glu	1.75	1.81	1.6	4.5	0.35	0.45	0.996	0.997
C_{11}-Glu	0.57	0.59	2.0	5.0	1.17	1.41	0.999	0.993
C_{12}-Glu	0.17	0.18	2.8	6.0	4.33	4.76	0.992	0.994

Source: López, O., Cocera, M., Parra, J.L., and de la Maza, A., *Colloids Surf. A*, 193, 221, 2001. With permission.

Note: Regression coefficients (r^2) of the straight lines obtained are included.

Re values), whereas the minimum to the C_{12}-Glu (highest Re values). The surfactant partition coefficients for saturation (K_{SAT}) and complete liposome solubilization (K_{SOL}) show that the C_{12}-Glu molecules had the highest degree of partitioning into bilayers (maximum K values), whereas the C_8-Glu showed the lowest (minimum K values). These findings are correlated with the aforementioned higher surface activity of the alkyl glucosides with higher hydrophobic tails (lower γ_{CMC} values) [56]. The fact that the K values always increased from bilayer saturation to complete liposome solubilization means that the affinity of the surfactant molecules with the lipids building liposomes was greater in the complete bilayer solubilization (micellization process) than during the previous step of bilayer saturation (formation of mixed vesicles).

The relationship between the K_{SAT} and K_{SOL} for each surfactant tested may be correlated with the dynamic surfactant/PC equilibrium existing between the transition steps from mixed vesicles to mixed micelles. Comparison of these two parameters reveals that the higher the surfactant alkyl chain length the higher the quotient between both values (K_{SAT}/K_{SOL}). Hence, at the interaction level for bilayer saturation the degree of surfactant partitioning into liposomes relatively increased with respect to that for complete liposome solubilization as the alkyl chain length rose. Hence, the rise in the length of the hydrophobic tail in addition to improve the partitioning of surfactant molecules into bilayers (increasing K values) also resulted in a relative decrease in their ability to be associated with the molecules building liposomes to form mixed micelles. Possibly, the first order phase transition from mixed vesicles into mixed micelles appears to be relatively hampered by the increasing surfactant hydrophobic tail.

As occurred in the sublytic interaction, two opposite trends may be observed when comparing the variation of Re and K versus the surfactant alkyl chain length.

The rise in the alkyl chain length resulted (during the transition steps from mixed vesicles to mixed micelles) in a progressive decrease in the surfactant ability to saturate and solubilize liposomes and inversely in an abrupt increase in the surfactant partitioning between the bilayers and water. Thus, for C_{12}-Glu although a higher number of surfactant molecules were needed to saturate or solubilize liposomes, these molecules showed an increased affinity with these structures with respect to that exhibited by the more active C_8-Glu. As described in the sublytic interaction, the balance of these two opposite tendencies leads to the conclusion that the C_8-Glu was the more equilibrated surfactant in terms of hydrophilic–lipophilic balance, followed by the C_9-Glu, to obtain an improved interaction and, hence, a maximum solubilizing effect on PC liposomes.

15.3.1.2 Stratum Corneum Lipid Liposomes

In a parallel way to that used to describe the interaction of PC liposomes, in this section it is described the sublytic and lytic alterations caused by a series of alkyl glucosides (alkyl chain lengths ranging from C_8 to C_{12}) in liposomes formed by the intercellular lipids of the SC.

The SC of vertebrates is a major structural compartment that provides mechanical protection and prevents skin desiccation. This tissue is formed by corneocytes that are separated by an intercellular matrix mainly composed of lipids. The SC lipids, which has been described as approximately 40% ceramides (Cer), 25% cholesterol (Chol), 15% fatty acids, and small amounts of cholesterol sulfate (chol.sulf) and cholesterol esters, results in a very exceptional composition and organization not observed in biological membranes [68] These lipids are organized into bilayers that have been postulated both to account for the permeability properties of SC and to ensure the cohesiveness between corneocytes [69–72]. In all cellular and intercellular membranes, such bilayer-forming lipids consist predominantly of phospholipids. However, SC has been shown to be virtually devoid of phospholipids, as a result of which its ability to form bilayers has proved to be somewhat surprising. To find out whether SC lipids could form bilayers, Wertz et al. prepared liposomes from lipid mixtures approximating the composition of the SC lipids and characterized these bilayer structures [73].

We previously reported the formation of liposomes using a mixture of lipids modeling the composition of the SC (40% Cer, 25% Chol, 25% PA, and 10% Chol-sulf) [74], which was prepared following the method described by Wertz et al. [73]. After preparation, vesicles were annealed at 60°C for 30 min and incubated at 25°C under nitrogen atmosphere. The range of lipid concentrations in liposomes was 0.1–10.0 mM.

In order to find out whether all the components of the lipid mixture formed liposomes, vesicular dispersions were analyzed for these lipids [75]. The dispersions were then spun at 140,000 g at 25°C for 4 h to remove the vesicles. The supernatants were tested again for these components. No lipids were detected in any of the supernatants.

The phase transition temperature of the lipid mixture forming liposomes was determined by proton magnetic resonance (^{1}H NMR) showing a value of 55°C–56°C [76].

As occurred with PC liposomes the sublytic and lytic changes caused by the addition of alkyl glucosides on SC lipid liposomes were studied by measuring the corresponding CF release and static light-scattering changes during the process. In addition, and following the same scheme than for PC liposomes, the validity of the equilibrium partition model proposed by Lichtenberg and Almog et al. [31,38] has been studied for the system SC lipid liposomes/alkyl glucosides. This system demonstrated to be valid to determine K at the two sublytic and lytic interaction levels investigated.

15.3.1.2.1 Sublytic Alterations

It has been previously described that in sublytic interactions alkyl glucosides–PC liposomes a substantial part of the effect takes place within approximately 30 min after their addition to the liposomes [75]. Given, the specific biophysical characteristics of the SC lipids it was needed to determine previously the time in which the leakage ceased. To this end, a kinetic study of the interaction of alkyl glucosides with SC lipid liposomes was performed. Vesicles were treated with surfactants at sublytic concentrations and subsequent changes in CF release were studied as a function of time. The CF release showed always a transient state of enhanced permeability of bilayers, in which about 60 min was needed to achieve CF release plateaux in all cases. This biphasic behavior was similar to that exhibited by PC liposomes when interacted with the same surfactants However, the fact that almost a double period was needed to achieve the aforementioned CF release plateaux may be associated to the differences in the gel–liquid crystal phase transition temperature of the lipids building these two liposome models, which affects both the positional organization of lipids molecules and their polar heads as well as their mobility, also affecting the formation and stabilization of the aforementioned membrane holes. The spontaneous release of the CF from the liposomes in the absence of surfactants in this period was negligible. Hence, changes in CF release were studied 60 min after the addition of surfactants to the liposomes at 25°C.

A systematic study of the CF release changes caused by the addition of the surfactants to SC lipid liposomes was carried out for various lipid concentrations to determine the Re and S_W values. The surfactant concentrations for 50% and 100% CF release for each surfactant tested were graphically obtained and plotted versus lipid concentration. The r-squared statistic (regression coefficients r^2) indicated that the straight lines obtained explained more than 98.9% of the surfactant concentration variability versus lipid concentration. Therefore, a very good linear fit was established in all cases confirming than the straight lines for Equations 15.6 and 15.7 were suitable to determine Re and S_W for the surfactants and the SC liposomes tested. The Re, K, and S_W values obtained are given in Table 15.3 [42].

TABLE 15.3

Surfactant to Lipid Molar Ratios (Re), Partition Coefficients (K) and Surfactant Concentrations in the Aqueous Medium (S_W) Resulting in the Subsolubilizing Interaction (50% and 100% CF Release) of Alkyl Glucosides with SC Lipid Liposomes

	$S_{W,50\%CF}$ (mM)	$S_{W,100\%FC}$ (mM)	$Re_{50\%CF}$ (mole/mole)	$Re_{100\%CF}$ (mole/mole)	$K_{50\%CF}$ (mM^{-1})	$K_{100\%CF}$ (mM^{-1})	r^2 (50% CF)	r^2 (100% CF)
C_8-Glu	8.09	13.2	1.25	1.53	0.068	0.045	0.996	0.992
C_9-Glu	1.98	3.80	1.18	1.63	0.273	0.163	0.999	0.994
C_{10}-Glu	0.73	1.38	1.24	1.80	0.758	0.465	0.993	0.998
C_{11}-Glu	0.22	0.43	1.63	2.34	2.81	1.62	0.997	0.994
C_{12}-Glu	0.068	0.14	2.17	3.42	10.06	5.52	0.996	0.997

Source: Cócera, M., López, O., Coderch, L., Parra, J.L., and de la Maza, A., Colloids Surf. A, 176, 167, 2001. With permission.

Note: The regression coefficients of the straight lines obtained are included.

As occurred with PC liposomes, the S_W values always showed smaller values than those for the surfactant CMCs. This means that the sublytic interaction must be ruled mainly by the action of surfactant monomers, in line with the findings reported for the sublytic interaction of these surfactants with PC liposomes [41]. Furthermore, the C_9-Glu and the C_8-Glu showed the maximum sublytic activity against liposomes (lowest Re values), whereas the minimum corresponded to the C_{12}-Glu (highest Re values).

Comparison of the Re values for PC and SC lipid liposomes (Tables 15.1 and 15.3) shows that the surfactant ability to alter the permeability of SC liposomes was always less (higher Re values) than that exhibited with PC liposomes, although showing similar trends in the influence of the surfactant hydrophobic tail. Hence, the SC lipid vesicles exhibited more resistance to the surfactant perturbations than the PC vesicles. This different behavior may be explained bearing in mind the more hydrophilic nature of PC, which could facilitate the permeation of water and some other molecules (as surfactants) in PC vesicles either through the hydrophilic holes created on the PC polar heads or via formation of short-lived complexes surfactants-PC polar heads and subsequent transfer through the bilayers via flip-flop [74].

As it has been described for PC liposomes the K showed a maximum value for C_{12}-Glu, whereas the C_8-Glu exhibited the minimum. Thus, the C_{12}-Glu molecules had the highest degree of partitioning into bilayers. Comparison of the present K values of Tables 15.1 and 15.3 shows that the degree of surfactant partitioning into SC bilayers was always greater (higher K values) than that for PC ones.

However, the influence of the surfactant hydrophobic moiety in this affinity was also similar in both cases in spite of the different compositions and properties of these two bilayer structures. This increased degree of partitioning may be explained taking into account that the corresponding Re and S_W values were respectively higher and lower than those reported for PC ones in all cases (see Equation 15.3). The different chain length of lipid building these structures, degree of saturation and nature of polar heads appears to be responsible for these differences.

In general terms, it is possible to assume that although SC liposomes were more resistant to the action of surfactant monomers, the partitioning of these surfactants into SC structures was always greater than that for PC ones. Thus, although a greater number of surfactant molecules were needed to produce the same alterations in SC bilayers, these molecules showed increased affinity with these structures. However, a similar influence of the surfactant hydrophobic tail on the Re and K values is observed in both types of membrane models in spite of their different lipid compositions and physicochemical characteristics.

15.3.1.2.2 Lytic Alterations

As it has been described for PC liposomes, to determine the Re and S_W at lytic level, a systematic investigation of SLS variations due to the addition of the surfactants studied is performed. The curves obtained for C_8-Glu are given in Figure 15.5. Addition of surfactant to liposomes led to an initial increase and a subsequent fall

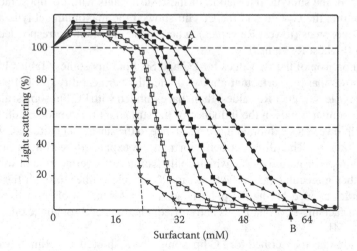

FIGURE 15.5 Percentage changes in SLS of SC lipid liposomes, (lipid concentration ranging from 1.0 to 10.0 mM), induced by the presence of increasing concentrations of C_8-Glu. Symbols for SC lipid concentrations: ($\triangledown$) 1.0 mM, ($\square$) 3.0 mM, ($\blacktriangledown$) 5.0 mM, ($\blacksquare$) 6.0 mM, ($\blacktriangle$) 8.0 mM, and ($\bullet$) 10.0 mM. (From López, O., Cocera, M., Parra, J.L., and de la Maza, A., *Colloid Polym. Sci.*, 279, 909, 2001. With permission.)

in the scattered intensity of the system until a low constant value was achieved, corresponding to the complete solubilization of liposomes. The curves obtained for the other alkyl glucosides investigated showed similar trends to those exhibited by C_8-Glu. This SLS behavior is similar to that reported for the interaction of C_8-Glu with PC liposomes although showing in all cases a more pronounced initial SLS increase. Surfactant concentrations producing 100% (S_{SAT}) and 0% (S_{SOL}) of SLS were obtained for each lipid concentration by graphical methods. The arrows A and B (curve for 10.0 mM lipid concentration, Figure 15.5) correspond to these two values.

When plotting surfactant concentrations thus obtained versus lipid concentration, curves were obtained in each case. The straight lines corresponded to Equations 15.8 and 15.9 from which the Re and S_W parameters were determined. The r-squared statistic (regression coefficients r^2 given in Table 15.4) indicated that the straight lines obtained explained more than 98.3% of the surfactant concentration variability versus lipid concentration. Therefore, a very good linear fit was established in each case. These findings confirm that these straight lines were appropriate to determine Re and S_W for the surfactants tested. The Re, K, and S_W values obtained and the surfactant CMCs are given in Table 15.4 [75].

The $S_{W,SAT}$ and $S_{W,SOL}$ values always showed similar values than those for the surfactant CMCs (Table 15.1). This finding extent to the SC lipid liposomes the observation made with PC liposomes, that the free surfactant concentration must reach the CMC for solubilization to commence and indicate that liposome solubilization was mainly ruled by the formation of mixed micelles.

As for the Re parameters, we may assume that the lower the surfactant alkyl chain length (or the hydrophobic moiety) the higher its ability to saturate

---

TABLE 15.4
Surfactant to Lipid Molar Ratios (Re), Partition Coefficients (K), and Surfactant Concentrations in the Aqueous Medium (S_W) Resulting in the Interaction of Alkyl Glucosides with SC Lipid Liposomes

	$S_{W,SAT}$ (mM)	$S_{W,SOL}$ (mM)	Re_{SAT} (mole/ mole)	Re_{SOL} (mole/ mole)	K_{SAT} (mM^{-1})	K_{SOL} (mM^{-1})	r^2 (SAT)	r^2 (SOL)
C_8-Glu	18.0	18.50	1.56	4.14	0.03	0.04	0.998	0.993
C_9-Glu	5.60	5.75	1.45	4.01	0.11	0.14	0.994	0.992
C_{10}-Glu	1.82	1.90	1.80	4.80	0.35	0.43	0.996	0.995
C_{11}-Glu	0.58	0.62	2.70	6.35	1.26	1.39	0.995	0.998
C_{12}-Glu	0.18	0.19	3.90	8.40	4.42	4.70	0.990	0.991

Source: López, O., Cocera, M., Parra, J.L., and de la Maza, A., *Colloid Polym. Sci.*, 279, 909, 2001. With permission.

Note: The regression coefficients of the straight lines obtained are also included.

and solubilize liposomes. Comparison of the Re values for PC and SC lipid liposomes (Tables 15.2 and 15.4) shows that the surfactant ability to saturate and solubilize SC liposomes was less (higher Re values) than that exhibited with PC liposomes. Hence, as occurred at sublytic level the SC lipid vesicles exhibited more resistance to the surfactant perturbations than the PC vesicles.

The K values increased from saturation to complete bilayer solubilization regardless of surfactant alkyl chain length and both K_{SAT} and K_{SOL} showed the highest values for C_{12}-Glu, and the lowest for C_8-Glu. This means that the affinity of surfactant molecules with the lipids building the liposome structure appears to be greater in the complete bilayer solubilization (micellization process) than during the previous step of bilayer saturation in line with the results reported for PC liposomes.

Comparison of the K values for C_8-Glu with those described for PC liposomes shows that the degree of partitioning of this surfactant into SC bilayers both at saturation and solubilization level was similar to that for PC vesicles in spite of the different degree of saturation and nature of polar heads of lipid building these two bilayer structures.

Two opposite trends may be also observed when comparing the variation of Re and K versus the surfactant alkyl chain length The structural changes corresponding to the increase in its CMC led to a rise in the surfactant ability to saturate or solubilize liposomes and inversely to an abrupt decrease in its affinity with these vesicles. From these findings, the molecular structure of the C_9-Glu was the more appropriate in terms of hydrophilic–lipophilic balance to obtain an improved activity against SC lipid vesicles.

From the overall information given in this section it is noteworthy that despite that both the C_8-Glu and the C_9-Glu have similar abilities to alter and solubilize PC and SC lipid liposomes, the use of C_9-Glu reduces approx three times the concentration of C_8-Glu needed for to produce the same effects in these bilayer structures. This effect is probably due to the fact that the CMC of the C_9-Glu is approximately three times lower than that of the C_8-Glu. This CMC value is very adequate to remove the surfactant by dialysis when reconstituting lipid-surfactant mixed micelles into liposomes. These findings open up new avenues in the application of C_9-Glu surfactant in the study of cell membranes, given its appropriate solubility in water, excellent hydrolysis stability, and may be considered as an alternative with respect to the use of the conventional OG.

15.3.2 OCTYL GLUCOSIDE-MEDIATED SOLUBILIZATION AND RECONSTITUTION OF LIPOSOMES

In this section, structural and kinetic aspects of the solubilization of PC liposomes induced by the action of OG and the reconstitution of PC vesicles by dilution of OG/PC mixed micelle systems are examined using dynamic light scattering and electron microscopy techniques.

15.3.2.1 Solubilization of Liposomes and Stability of the Systems

The size distribution curves for pure micelle OG solutions (OG concentration ranging from 20.0 to 40.0 mM) and for pure PC liposomes have been determined by DLS. The curves exhibit in both cases a monomodal distribution with a hydrodynamic diameter (HD) of 5 and 186 nm, respectively. This micellar size is in line with that reported by Lorber et al. using various techniques [76].

To study the solubilization of PC liposomes, increasing OG amounts are added to liposomes at different PC concentrations (from 0.5 to 5.0 mM) and the size of the resulting aggregates has been measured 10 h after mixing. At low OG concentrations only mixed vesicles were detected (monomodal distribution curves with a HD of about 186 nm). Increasing OG amounts led to the formation of mixed micelles, which were detected as small aggregates (bimodal distribution curves with a new peak at 11 nm). Higher OG concentrations resulted in the formation of systems, in which mixed vesicles and mixed micelles coexisted, with a progressive increase in the proportion of mixed micelles. Finally, only mixed micelles were detected (monomodal distribution curves with a peak of 11 nm), indicating the complete solubilization of liposomes. In addition, no peaks corresponding to intermediate complexes aggregates were found in equilibrated systems throughout the process [45].

When plotting the OG versus the PC concentrations for systems at which the formation of mixed micelles starts to occur (saturation of liposomes) and those with the only presence of mixed micelles (liposome solubilization) a linear dependence was obtained (Figure 15.6) with regression coefficients (r^2) 0.992 and 0.993, respectively. The slope of these two straight lines corresponded to the surfactant to PC molar ratio at which the surfactant saturated liposomes (Re_{SAT}) and led to

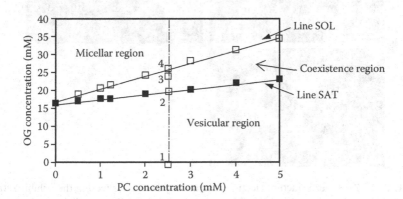

FIGURE 15.6 Phase diagram of the aggregates formed by mixtures of OG and PC in water, showing the phase boundary of the three domains; a "micellar" region formed by only mixed micelles, a "coexistence" region formed by mixed micelles and mixed vesicles, and a "vesicular" region formed by only mixed vesicles. Point 1 corresponds to pure liposomes (2.5 mM PC); points 2, 3 and 4 correspond to some representative systems of the "vesicular," "coexistence," and "micellar" regions, respectively. (From López, O., Cocera, M., Coderch, L., Parra, J.L., Barsukov, L., and de la Maza, A., *J. Phys. Chem. B*, 105, 9879, 2001. With permission.)

the complete liposome solubilization (Re_{SOL}), respectively [31,45]. These values were 1.4 and 3.2, respectively in agreement with previous solubilization studies [35,74,77]. Hence, these two straight lines delimited in the Figure 15.1 three regions; a "vesicular" region formed only by mixed vesicles, a "coexistence" region, in which mixed vesicles and mixed micelles coexisted, and a "micellar" region formed by only mixed micelles.

Under a structural viewpoint, some representative FFEM micrographs reflecting the solubilization of PC liposomes by rising OG amounts at the equilibrium (10 h after mixing) are shown in Figure 15.7. The four systems studied are indicated in Figure 15.6. Micrograph 1 for the system 1 (2.5 mM PC) shows some pure PC vesicles with a HD of about 186 nm. Micrograph 2 shows a system placed on the "vesicular phase boundary" (2.5 mM PC/20 mM OG, system 2) formed by mixed

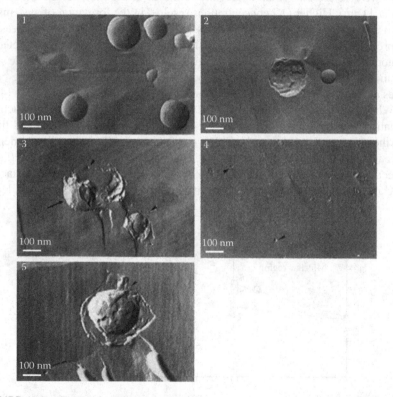

FIGURE 15.7 Freeze-fracture electron microscopy images reflecting the solubilization of 2.5 mM PC liposomes by rising OG amounts at the equilibrium (10 h after mixing). Micrograph 1 shows some pure PC vesicles (point 1, Figure 15.1). Micrograph 2 shows a system placed on the "vesicular phase boundary" (point 2, Figure 15.1). Micrograph 3 shows a system for the "coexistence" region (point 3, Figure 15.1). A system placed on the "micellar phase boundary" is shown in the micrograph 4 (point 4, Figure 15.1). Micrograph 5 shows the "coexistence" system of micrograph 3, but 24 h after. Structures are marked as follows: vesicles with arrows and mixed micelles with arrowheads. (From López, O., Cocera, M., Coderch, L., Parra, J.L., Barsukov, L., and de la Maza, A., *J. Phys. Chem. B*, 105, 9879, 2001. With permission.)

vesicles with a similar size to that found for pure PC vesicles (in accordance with the DLS data). However, some alterations in the topology of the vesicles were detected. Micrograph 3 shows a system for the "coexistence" region (2.5 mM PC/25 mM OG, system 3). Vesicles with clear morphological alterations together with small particles (arrows) corresponding to OG/PC mixed micelles were observed. It is noteworthy that no intermediate complex aggregates were observed, in accordance with the DLS data. A system placed on the "micellar phase boundary" (2.5 mM PC/26 mM OG, system 4) is shown in the micrograph 4, in which the only presence of mixed micelles was detected, in accordance with the DLS data. Micrograph 5 shows the system 3 (2.5 mM PC/25 mM OG) 24 h after of mixing. It can be noted that no changes in the appearance or size of the micelles (arrowheads) were detected; however an enlarging with changes in the topology of the mixed vesicles were observed reflecting the unstability of the system with time.

To study the solubilization kinetics three periods were considered: (1) the "micelle induction" as the time needed for micellization stars to occur (previously described by López et al. [78], 2) the "equilibration time" as the subsequent period needed for the formation of stable particles both in percentage and in size. This period will be associated to the end of the process for a given system, and 3) the "stability time" as the following period, in which the system remained without changes.

Figure 15.8 plots the particle size variations 30 s (A), 10 h (B), and 24 h (C) after mixing increasing OG concentrations (from 0 to 40 mM) with PC liposomes

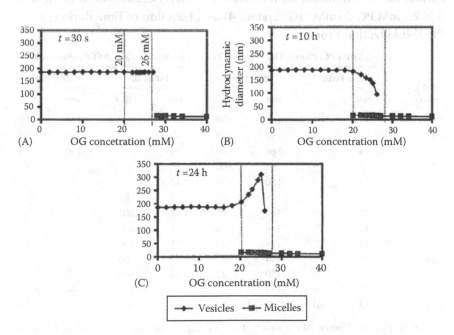

FIGURE 15.8 Particle size variation of the aggregates formed by 2.5 mM PC liposomes and increasing concentrations of OG, 30 s (A), 10 h (B), and 24 (C) h after mixing. (From López, O., Cocera, M., Coderch, L., Parra, J.L., Barsukov, L., and de la Maza, A., *J. Phys. Chem. B*, 105, 9879, 2001. With permission.)

(2.5 mM) in independent samples. After 30 s of mixing and up to 26 mM OG only one peak for mixed vesicles (HD of about 186 nm) was detected. Higher OG conc led to the only presence of mixed micelles (new peak for particles with a HD of 11 nm). After 10 and 24 h of mixing and up to 20 mM OG the size of the mixed vesicles remained almost constant. However, in a range of OG concentration from 20 to 26 mM mixed vesicles and mixed micelles coexisted with noticeable changes in the size of mixed vesicles. Higher OG concentrations showed the mixed micelles, which were already detected 30 s after mixing (stability higher than 24 h).

Table 15.5 shows the variation in size and in percentage of systems 3 (coexistence region) and 4 (micellar phase boundary) (Figure 15.6) with time [45]. Both systems 3 and 4 showed similar "micelle induction" and "equilibration" times, however, only in the system 4 was achieved the complete solubilization. Thus, the systems at the equilibrium (10 h after mixing) consisted in the first case (system 3) in a mixture of 27%–28% mixed vesicles and 72%–73% mixed micelles, and in the second (system 4) in 100% of mixed micelles. The "stability time" was in the first case of about 20 h (from this time the size of mixed vesicles rose approx two time and their proportion decreased), whereas in the second case the system was stable with time.

TABLE 15.5
Particle Size Distributions for the Systems 2.5 mM PC/25 mM OG (System 3) and 2.5 mM PC/26 mM OG (System 4) as a Function of Time during the Solubilization Process

| Time (min) | 2.5 mM PC/25 mM OG (System 3) | | | | 2.5 mM PC/26 mM OG (System 4) | | | |
| | 1st Peak | | 2nd Peak | | 1st Peak | | 2nd Peak | |
	nm	%	nm	%	nm	%	nm	%
0	—	—	186	100	—	—	190	100
0.5	—	—	187	100	—	—	189	100
1	—	—	186	100	—	—	190	100
1.5	—	—	186	100	—	—	188	100
3	—	—	150	100	12	60	190	40
3.5	11	57	145	43	12	85	190	15
4.5	11	73	144	27	11	100	—	—
10	11	70	144	30	11	100	—	—
60	12	72	145	28	11	100	—	—
300	12	73	144	27	12	100	—	—
800	11	73	150	27	11	100	—	—
1200	12	74	152	26	11	100	—	—
1440	12	80	295	20	11	100	—	—

Source: López, O., Cocera, M., Coderch, L., Para, J.L., Barsukov, L., and de la Maza, A., *J. Phys. Chem. B, 105*, 9879, 2001. With permission.

Note: The results are given as HD and the percentage corresponds to intensity values.

From all these data it can be noted that the solubilization of the system containing 26 mM OG was completed in 4.5 min (Table 15.5), whereas more concentrated systems achieved the total solubilization before 30 s (Figure 15.8). Thus, systems with higher OG concentration, in addition to led to stable micellar systems showed a faster solubilization kinetics.

15.3.2.2   Reconstitution of Vesicles by Dilution

From the diagram of Figure 15.6, it is expected that mixed micellar solutions resulted in the spontaneous formation of vesicles when the vesicular phase boundary was crossed by dilution. To confirm this finding various micellar systems, in which the molar ratio OG/PC corresponded to the Re_{SOL} (systems placed on the micellar phase boundary of Figure 15.6) were submitted to a fast one-step dilution. This dilution with buffer reduced the concentration of both components to the half being the phase boundary for the vesicular region crossed in all cases. Taking into account that no surfactant removal was performed during the dilution process it is assumed that the reconstituted vesicles included surfactant molecules and, hence, were mixed vesicles. The compositions of the three systems studied before dilution were 5.0 mM PC/35 mM OG, 2.5 mM PC/26 mM OG mM (system 4), and 1.25 mM PC/22 mM OG. The variation in the proportion of both components with respect to the Re_{SOL} (effective surfactant to PC molar ratio in bilayers for vesicle solubilization) is due to the fact that the total OG concentration (S_T) added is the sum of the surfactant in bilayers (S_B) and in the aqueous phase (S_W). The HD of the particles before dilution was 11 nm in all cases (mixed micelles). After dilution the size distribution curves for reconstituted vesicles showed also a monomodal distribution with HD values of 110, 60, and 45 nm, respectively, these sizes remaining stable with time. The size of the reconstituted vesicles is shown in Figure 15.9. A linear dependence was established between the size of the reconstituted vesicles and the OG and PC concentrations.

To study the kinetic of vesicle reconstitution by DLS, three periods of time were considered in a parallel way to the solubilization process: (1) "vesicle induction" as the time needed for vesicle reconstitution stars to occur, (2) "equilibration" time as the subsequent period needed for the complete formation of reconstituted vesicles (period associated with the end of the reconstitution process), and (3) "stability" time as the following period, in which the system remained without alterations [45].

Table 15.6 shows the kinetic of vesicle reconstitution for the systems 2.5 mM PC/26 mM OG (system 4) and 1.25 mM PC/22 mM OG. Although the "vesicle induction" time, was similar in both systems (about 30 s) the "equilibrium" time increased as the concentration of PC and OG decreased (2 and 2.5 min, respectively). The "stability time", however, was in both cases higher than 24 h indicating that in the period studied the stability of the vesicles formed was independent of the concentration of the components. In the system formed by 5.0 mM PC/35 mM OG the reconstitution process was too rapid to be followed by DLS.

To compare the kinetics of vesicle solubilization and reconstitution for the systems given in Table 15.6 the variation in the proportion of mixed micelles for

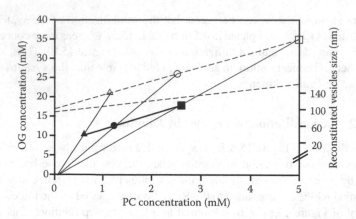

FIGURE 15.9 Phase diagram of Figure 15.6 plotting the systems formed by 5.0 mM PC/35 mM OG, 2.5 mM PC/26 mM OG mM (system 4), and 1.25 mM PC/22 mM OG before and after a fast one-step dilution and the size of the reconstituted vesicles. The molar ratio OG/PC corresponded to the Re_{SOL} (systems placed on the micellar phase boundary). (From López, O., Cocera, M., Coderch, L., Parra, J.L., Barsukov, L., and de la Maza, A., *J. Phys. Chem. B*, 105, 9879, 2001. With permission.)

TABLE 15.6

Particle Size Distributions for the Systems 2.5 mM PC/26 mM OG (System 4) and 1.25 mM PC/22 mM OG as a Function of Time during the Reconstitution Process by a Fast One-Step Dilution

	2.5 mM PC/26 mM OG (System 4)				1.25 mM PC/22 mM OG			
	1st Peak		2nd Peak		1st Peak		2nd Peak	
Time (min)	nm	%	nm	%	nm	%	nm	%
0	11	100	—	—	11	100	—	—
0.5	11	98	56	2	11	99	43	1
1	11	51	58	49	11	79	44	21
1.5	11	4	59	96	11	48	45	52
2.0	—	—	60	100	11	5	45	95
2.5	—	—	60	100	—	—	45	100
5.0	—	—	60	100	—	—	45	100
10	—	—	60	100	—	—	45	100
60	—	—	60	100	—	—	45	100
600	—	—	60	100	—	—	45	100
1440	—	—	60	100	—	—	45	100

Source: López, O., Cocera, M., Coderch, L., Parra, J.L., Barsukov, L., and de la Maza, A., *J. Phys. Chem. B*, 105, 9879, 2001. With permission.

Note: Results are given as HD and the percentage corresponds to intensity values.

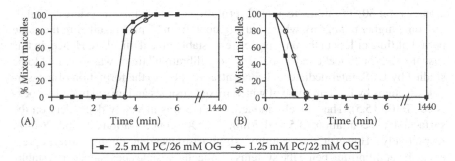

FIGURE 15.10   Variation in the percentage of mixed micelles as a function of time during the solubilization (A) and reconstitution by dilution (B) of PC vesicles for the systems formed by 2.5 mM PC/26 mM OG (■) (system 4) and by 1.25 mM PC/22 mM OG (○). (From López, O., Cocera, M., Coderch, L., Parra, J.L., Barsukov, L., and de la Maza, A., *J. Phys. Chem. B*, 105, 9879, 2001. With permission.)

these two processes are plotted versus time in Figure 15.10A and B, respectively. The second process was faster basically due to the short "induction time" associated to the formation of mixed micelles.

From the data obtained using DLS and FFEM techniques a simple mechanism of PC liposomes solubilization by OG in proposed. Although this mechanism is based on the three-stage model proposed by various authors [31,64,79,80] it differs in the following points: (1) up to the saturation of vesicles by the surfactant a direct formation of mixed micelles within the bilayer occurred "in situ micellization," (2) the progressive separation of the formed micelles from the liposome surface led to the complete solubilization of vesicles, and (3) this separation would take place without formation of stable complex intermediate aggregates with the only presence of mixed micelles and bilayer fragments. This mechanism is similar to that described for the nonionic surfactant Triton X-100 (T_{X-100}) and the anionic sodium dodecyl sulfate (SDS) [46,47]. However, some specific structural differences were observed depending on the physicochemical characteristics of each surfactant. Thus, using OG the size of saturated vesicles was unaffected, in the case of T_{X-100} this size increased, and using SDS the vesicle saturation occurred with a slight contraction. The preservation of the vesicular size after incorporation of the OG molecules seems to contradict many results previously published [81]. Moreover, it is seemingly in conflict with theoretical considerations, which indicate that the incorporation of OG molecules should result in a increase of vesicle diameter due to the increase in the total surface area of mixed vesicle. The only way to maintain the same vesicle size would be to change the bilayer topology. In this sense, micrograph 2 in Figure 15.7 presents evidence of such alteration, so it is clearly observed that OG-containing vesicles have a cramped wave-like appearance in contrast to the spherical smooth surface of pure PC vesicles.

DLS studies show that the level of liposome solubilization (PC conc 2.5 mM) and the kinetic of each process were dependent on the OG concentration (Figure 15.8). Thus, the use of a OG conc from 1 to 20 mM OG (OG CMC 18 mM) led already after 30 s of mixing to a rapid formation of stable mixed vesicles ("stability" time higher than 24 h), whereas using conc from 27 mM resulted in the same period of time or less in the only presence of stable mixed micelles. Hence, in this case the sum of "micelle induction" and "equilibration" times was too rapid to be studied by DLS. Intermediate OG concentrations, led to the formation of measurables systems by DLS, in some of which mixed vesicles and mixed micelles coexisted (Table 15.5). The "micelle induction" time was inversely dependent on the surfactant concentration (3.5 and 3 min for OG concentration 25 and 26 mM, respectively). However, the "stability" time was dependent on the particles present at the equilibrium, being the systems of micelle-vesicle coexistence less stable. The loss of stability of system 3 was reflected in the growth of mixed vesicles up to doubling the initial size and the reduction in its proportion (Table 15.5). These variations could be explained by fusion or aggregation of the mixed vesicles. Micrograph 3 and 6 of Figure 15.7 shows the system 3 (2.5 mM PC/25 mM OG) 10 and 24 h after of mixing. Comparison of these two micrographs in addition to confirm the growth of mixed vesicles with time (no signs of aggregation were detected) suggests a fusion mechanism, in which the resulting vesicles exhibited similar morphological alterations. It is a common observation that surfactants may induce vesicle fusion when interacted with liposomes at sublytic concentrations [66,80,82]. However, in our experiments [45] this phenomenon takes place in the systems of coexistence micellevesicle. This fusion may be explained because under these conditions the membranes are in a fluctuating dynamic state characterized by rapid exchange of phospholipid that favors fusion phenomena as reported by Kragh-Hansen et al. [83]. Another possible cause could be the formation of holes on the vesicle surface, which could act as critical points of fusion. This assumption is supported by the fact that the vesicle fusion occurred in the step of the separation of mixed micelles from the liposome surface.

Comparison of solubilization kinetics of PC liposomes by two nonionic surfactants, OG and T_{X-100}, shows that in systems in which vesicles and micelles coexisted the T_{X-100} had lower times of "micelle induction" and "equilibration" [84] indicating that the solubilization induced by T_{X-100} was faster than that induced by OG. Given the higher hydrophobicity of OG (hydrophilic–lipophilic balance of OG and T_{X-100} 12.6 and 13.6, respectively [85]) it would be expected that the OG exhibited a faster solubilization kinetics than the T_{X-100}, in agreement with Kragh-Hansen et al. [83] who associated the surfactant hydrophobicity with a higher rate of transbilayer movement (flip-flop) and, hence, with a faster solubilization kinetics. Taking into account the mechanism of micellization "in situ" proposed, the fact that the CMC of OG was approximately 120 times higher than that of T_{X-100} [88] indicates that a higher number of OG monomers should be incorporated in the bilayers for micellization starts to occur. This fact should increase the time of "micelle induction" in spite of the favorable hydrophobicity of the OG. Hence, the surfactant CMC should be considered as an important factor to explain the different solubilization kinetics of both surfactants. In this sense, Lasch et al. [65] compared the interaction of these two surfactants with phospholipid membranes demonstrating that T_{X-100} (lower CMC than the OG) induced an optimization of packing of the lipid molecules. This fact could favor the mechanism of "in situ" micellization proposed.

As for the reconstitution process, although the micellar size was initially the same (11 nm) a direct dependence was established between the size of the reconstituted vesicles and the concentrations of OG and PC. Furthermore, the lower the relative proportion of OG the higher the size of the reconstituted vesicles. These dependencies may be related to the reported formation of an equilibrium size of the vesicles associated with the curvature radius and the surfactant and lipid composition of the vesicles [66]. Under a practical viewpoint this finding opens the possibility to predict the size of reconstituted vesicles after a controlled one-step fast dilution of specific OG/PC micellar systems.

The fact that the kinetic of reconstitution was significantly shorter than that of solubilization ("vesicle induction" time 30 s and "micelle induction" time 3–3.5 min) contrasts with the results reported by various authors who claimed that the molecular rearrangements during the vesicle reconstitution follow the same pathway as the solubilization event [87,88]. However, the different "induction times" indicate different requirements for the formation of these two aggregates. Thus, the formation of mixed micelles requires first the initial release of the surfactant monomers from pure micelles governed by a micellar relaxation time τ_2 [89]. Secondly, the incorporation of these monomers into bilayers possibly through the hydrophilic holes created by these monomers on the PC polar heads or via formation short-lived complexes surfactants-PC polar heads [90]. Finally, the formation of mixed micelles would involve not only the adsorption of surfactant monomers into bilayers but also a desorption of mixed micelles from the bilayer surface [46]. However, the reconstitution of vesicles from micellar solutions only would involve the release of surfactants monomers from the mixed micelles and the reorganization of the remaining lipid and surfactant molecules in vesicles in a simpler way than that described until now [45].

15.3.3 USE OF THE OCTYL GLUCOSIDE IN STRATUM CORNEUM STUDIES

15.3.3.1 Octyl Glucoside as a Tool to Induce Modifications in Stratum Corneum

In this subsection it is described the use of the combined action of the OG surfactant and ultrasonic stimulation to induce structural modifications in the SC. The composition of the disaggregated material and the microstructure of the residual tissue after treatment are examined

15.3.3.1.1 Study of the Disaggregated Material
The treatments of SC with OG (10 and 20 mM) led to the disaggregation of 14% and 70% of the initial material, respectively [48,50]. The chemical analysis of this material (after the OG dialysis) shows the predominant presence of proteins (84%) and small amounts of lipids (9.1%). The proportions of these components were respectively higher and lower than those present in the native tissue (approximately

TABLE 15.7
Percentages of the Main Lipids and Amino Acids Disaggregated from the Stratum Corneum by OG at Two Concentrations, 10 mM and 20 mM (Expressed in Weight with Respect to the Initial Dry SC)

OG Concentration	10 mM		20 mM	
Percent of Disaggregation	14		70	
Main Amino Acid				
Glu	1.9		7.1	
Gly	1.2		5.3	
Ser	1.3		4.7	
Main Lipids				
	L. int.	**L. env.**	**L. int.**	**L. env.**
Ceramides	0.28	0.3	1.3	0.8
Cholesterol	0.24	—	0.7	—
FFA	0.3	—	0.6	—

Source: López, O., Cocera, M., Walther, P., de la Maza, A., Coderch, L., and Parra, J.L., *Colloids Surf. A*, 168, 115, 2000. With permission.

80% of proteins and 20% of lipids) [48,68]. The main amino acids detected were glutamic acid (Glu) (1.9% and 7.1%), glycine (Gly) (1.2% and 5.3%), and serine (Ser) (1.3% and 4.7%) (Table 15.7). Successive chloroform:methanol extractions were needed before TLC:FID lipid analysis to isolate and to characterize the inter-cellular lipids linked to the proteins by hydrophobic interactions indicating the existence of some kind of hydrophobic interactions between the removed compo-nents. In addition, a mild alkaline hydrolysis of the nonextracted residue (fol-lowed by chloroform: methanol extractions) was needed to detect the lipids linked to the amino acids belonging to the corneocyte envelopes by covalent bonds (envelope lipids), in agreement with Wertz et al. [69,73]. The percentages of the main lipids detected are given in Table 15.7. A dominant presence of ceramides (Cer) and cholesterol (Chol) was detected in the intercellular lipids (L. int.), whereas only Cer were found in the envelope lipids fraction (L. env.) (0.4% and 0.9%, respectively) [50]. TEM observations demonstrated the presence of corneocyte envelopes in the fraction disaggregated by OG 20mM (Figure 15.11). This figure shows a low-power electron micrograph of the disaggregated material in which some structures associated with corneocyte envelopes are visualized [50]. These pictures are similar to those of corneocyte envelopes reported by Swartzendruber et al. [91], despite the different surfactant and experimental conditions used. These results demonstrated the ability of the OG to disaggregate the SC tissue as

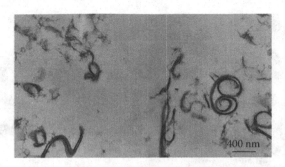

FIGURE 15.11 TEM micrograph of the disaggregated material showing corneocytes envelopes postfixed with RuO₄. The magnification is given in the micrograph. (From López, O., Cocera, M., Walther, P., de la Maza, A., Coderch, L., and Parra, J.L., *Colloids Surf.*, 168, 115, 2000. With permission.)

a function monomeric or micellar characteristics of the OG solutions. From an overall viewpoint the preferential removal of proteins (84%) together with the loss of cohesion (tissue disaggregation) indicates that the tissue cohesion against OG was mainly associated with the protein tissue components, i.e., the corneocyte and the corneocyte envelopes. These findings are in agreement with those reported by Wertz et al. and Chapman et al., who claimed that the corneocytes and their envelopes play a dominant role in this cohesion [69,92]. The fact that a proportion of the components in the disaggregated material (84% of proteins and 9.1% of lipids) were similar to that of the native tissue (80% of proteins and 20% of lipids) suggests a disaggregation process in layers. At a concentration 10 mM the OG was theoretically unable to solubilize "in the strict sense" lipid membranes given that the OG concentration must reach its CMC for solubilization to occur via mixed micelle formation [50]. However, some tissue disaggregation (14% of the initial tissue, see Table 15.7) occurred in these conditions possibly via formation of lipid:protein:surfactant complexes. This result is in line with those reported by various authors, who demonstrated the possibility of disaggregation of protein membranes by nonionic surfactants already at concentrations below their CMCs [50]. The use of an OG concentration slightly higher than its CMC (20 mM) increased the percentage of tissue disaggregation (70%, Table 15.7). The presence in the disaggregated material of Gly (dominant amino acid present in the corneocytes [68]) indicates that the treatment partially removed these structures with the subsequent SC disruption into individual corneocytes. Furthermore, the presence of high percentages of Glu (which has been described as the most abundant component of the corneocyte envelope [50,68]) indicates that this disaggregation process affects selectively these envelopes, in line with the action of this surfactant in other biological membranes [50,93].

15.3.3.1.2 Study of the SC Microstructure after OG Treatment
Macroscopic inspection of the SC pieces after OG treatment shows a loss in their cohesion, which was directly dependent on the surfactant concentration used.

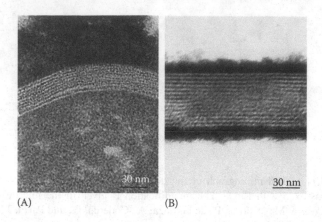

(A)   (B)

FIGURE 15.12 TEM micrographs corresponding to SC native (A) and treated with OG (B). The magnifications are given in the micrographs. (From López, O., Cocera, M., Walther, P., de la Maza, A., Coderch, L., and Parra, J.L., *Colloids Surf.*, 168, 115, 2000. With permission.)

Different microscopy techniques, (TEM, Cryo-SEM, and FSTEM) have been used to examine the microstructure of the residual tissue. The TEM micrograph of native SC (Figure 15.12A) showed the characteristic lipid lamellae assembly of the intercellular lipids [50]. A first electron-dense lamellae (outer lamellae for the lipids covalently bound to the corneocyte envelope) and a series of electron lucent-lamellae (polar portion of lamellae) alternating with paired electron dense-lamellae (intercellular lipid bilayers) were observed. Figure 15.12B shows a representative TEM image of treated SC sheets (20 mM OG), in which is observed that the lipid lamellae assembly remained almost unaffected. Comparison of both images indicates that the internal lamellar lipid structure was preserved even after a drastic removal of the proteinaceous material forming the external zone of corneocytes [50]. The Fourier transform of digitized SC images showed a lamellar spacing of 36 and 43 Å, for native and treated SC samples, respectively.

A Cryo-SEM micrograph obtained for native SC is shown in Figure 15.13A [50,54]. The corneocytes (letter C) are characterized by the particular pattern of keratin filaments entirely filling their interior and by the absence of cell organelles. Mostly, the fracture plane in the corneocyte lies vertically to the skin surface. In the zone corresponding to the intercellular spaces, however, the fracture plane goes along the lipid lamellae resulting in a clean view of the very smooth and relatively flat surface of the SC lipids (letter L). Occasionally, the bilayers were fractured straight across, resulting in sharp edges (open arrows). This observation indicates that the lipid matrix was constituted by multiple layers. These findings are in agreement with those reported by Dreher et al. [94], who found similar Cryo-SEM images for native human SC. The Figure 15.13B depicts a Cryo-SEM micrograph corresponding to the SC tissue after treatment with 20 mM

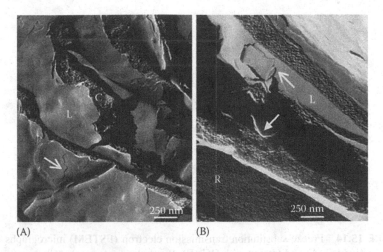

(A)   (B)

FIGURE 15.13   Cryo-SEM images of the SC native (A) and treated with the OG surfactant (B). The different ultrastructural characteristics are indicated as follows: corneocytes as letter C, fracture along lipid as letter L, fracture across lipid lamellae as open arrows, corneocyte envelope as arrows and rough structure as letter R. Magnifications are given in each micrograph. (From López, O., Cocera, M., Walther, P., Wehrli, E., Coderch, L., Parra, J.L., and de la Maza, A., *Micron*, 32, 201, 2001. With permission.)

OG [50,54]. In the intercellular lipid regions observed in this figure, the presence of sharps steps (open arrows) indicates that the lamellar structure remained partially unaltered after surfactant treatment. It is interesting to note the presence, in these lipid domains, of new rough structures (Figure 15.13B, letter R) which were not detected in the native tissue [50,94]. These structures are very similar to those reported by Hofland et al., and Van' Hal et al. [95,96] who associated their formation to a disorganization of the lipid lamellar structure. The formation of this "rough structure" could be attributed to the interaction of the protein material liberated from the corneocytes (when the corneocyte envelope was damaged) and the lipids of the intercellular spaces, with the subsequent disorder in their lamellar structure [50,54]. This alteration in the lipid domains could be associated with the disorder in these regions detected by x-ray diffraction in our previous work [97] and reported by various authors using microscopy techniques [94,96].

FSTEM was also able to visualize the SC tissue. Figure 15.14A clearly shows skin corneocytes (**C**), lipid bilayers (**L**) and corneosomes (**D**) (SC desmosomes representing contact areas between adjacent corneocytes). The technique of FSTEM showed that the treatment with the OG led to the general disaggregation of the SC structure and, consequently, to the disruption of part of the lipid and protein material. Figure 15.14B shows the loss of protein material from the corneocytes (**C**) and a certain disorganization in the lipid bilayers (**L**) that could be related to the rough structures visualized using Cryo-SEM in Figure 15.13B.

Both chemical characterization and microscopy techniques claim that although OG was always able to disaggregate part of the corneocyte envelopes,

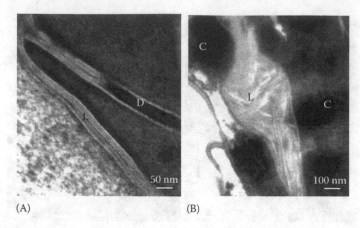

(A)   (B)

FIGURE 15.14 Freeze-substitution transmission electron (FSTEM) micrographs illustrating native SC (A) and treated with OG (B). Symbols: corneocytes (C), corneosomes (D) and intercellular lipid lamellae (L). Magnifications are given in each micrograph. (From López, O., Cocera, M., López-Iglesias, C., Walter, P., Coderch, L., Parra, J.L., and de la Maza, A., *Langmuir*, 18, 7002, 2002. With permission.)

the lamellar lipid structure in treated SC remained almost unaltered even when using the a OG concentration (20 mM) higher than its CMC. This preservation may be explained by the following arguments: the selectivity of OG to lipid membranes containing proteins such as the corneocyte envelopes and the complexity of the SC tissue, which may exert a self-protection of the underlying lipid lamellae structure. Besides, the fact that the highest OG concentration used was similar to that of its CMC would explain that only a small part of the SC lipid assembly was solubilized. From these findings a mechanism of tissue disaggregation in layers is proposed. This mechanism involves the disruption of SC into corneocytes, the disaggregation of the corneocyte envelopes, and the preservation of the underlying intercellular lamellae structure by means of a self-protection process. This mechanism is associated with the loss of SC cohesion and possibly to the desquamation process. From a practical viewpoint the ability of OG to induce modifications in the SC cohesion without noticeable changes in the lipid structure qualify this surfactant as a promising tool for selectively isolating the internal lipid lamellar domain. This possibility may be useful to investigate the physicochemical properties of this lipid assembly in order to better understand its function as transcutaneous permeability barrier.

15.3.3.2 Reconstitution of Vesicles in the SC by Use of OG

In this subsection, the interaction of PC liposomes with SC tissues is examined. To this end, SC tissue samples are treated with PC liposomes and subsequently with OG. A mechanism of vesicle penetration into the SC tissue is discussed.

15.3.3.2.1   Interaction between PC Liposomes and the SC

To study the effect of PC liposomes on the SC, that is, to investigate if PC liposomes were able to penetrate as intact vesicles into the SC, the tissue incubated with liposomes has been visualized using Cryo-SEM and FSTEM techniques [52]. No alterations or disorganization were detected in the intercellular lipids or in the corneocytes of the SC samples after treatment with liposomes. However, it is interesting to note that the technique of FSTEM, which allows visualization of the SC from the surface to the deeper layers, showed individual vesicles on the surface of the skin resembling lipid droplets (V) (Figure 15.15) [52]. These lipid structures are similar to those observed by Van den Bergh et al. in previous works, in which the SC was treated with elastic liquid-state vesicles [53]. These observations indicate that although the PC vesicles were not able to penetrate in the SC layers, these structures showed a certain affinity for the SC surface.

15.3.3.2.2   SC Samples Treated with PC Liposomes and Subsequently with OG

Figure 15.16 depicts a image using Cryo-SEM of SC previously incubated with PC liposomes (PC concentration, 15 mM) and immediately treated with OG. The lipid matrix exhibited the structure of multiple layers indicating the preservation of these structures as occurred in the SC treated with OG without previous incubation. However, the rough structures observed in the SC treated with OG were not observed in any micrograph of this sample [52]. Hence, the incubation with PC liposomes protected the SC against OG, this protection being associated with the incorporation of about 10% of PC into the SC [94]. The most striking result was the observation in the intercellular lipid spaces of vesicular structures of about 190 nm (Figure 15.16). Given that these structures were not observed either in the native SC (Figure. 15.13A) or in the tissue treated with OG without liposome incubation (Figure 15.13B) and taking into account their similarity in size and topology with PC vesicles, it is possible to associate these structures with PC

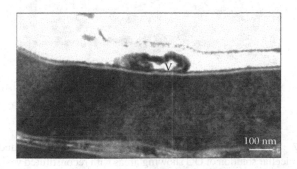

FIGURE 15.15   FSTEM micrograph of SC treated with PC liposomes. Individual vesicles (V) are visualized on the SC surface. Magnifications are given in each micrograph. (From López, O., Cocera, M., López-Iglesias, C., Walter, P., Coderch, L., Parra, J.L., and de la Maza, A., *Langmuir*, 18, 7002, 2002. With permission.)

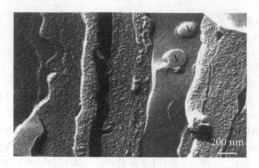

FIGURE 15.16   Cryo-SEM picture of SC incubated with PC liposomes and immediately treated with a micellar solution of OG showing detailed areas with reconstituted vesicles (V). Magnifications are given in each micrograph. (From López, O., Cocera, M., López-Iglesias, C., Walter, P., Coderch, L., Para, J.L., and de la Maza, A., *Langmuir*, 18, 7002, 2002. With permission.)

vesicles. The FSTEM micrographs from these SC samples treated with PC liposomes and OG also showed vesicles in the intercellular spaces (Figure 15.17) [52]. This fact confirms the results obtained by Cryo-SEM and demonstrates that the PC liposomes were able to penetrate in the SC by means of some mechanism induced by the OG, which will be explained in the next section. In any case, considering the size of these vesicles (190 nm) and the interlamellar distance of the ordered lipids (about 5–6 nm), the direct penetration of these vesicles into the SC appears to be difficult.

FIGURE 15.17   FSTEM picture of SC incubated with PC liposomes and immediately treated with a micellar solution of OG showing areas with reconstituted vesicles (V). Magnifications are given in each micrograph. (From López, O., Cocera, M., López-Iglesias, C., Walter, P., Coderch, L., Parra, J.L., and de la Maza, A., *Langmuir*, 18, 7002, 2002. With permission.)

15.3.3.2.3 Mechanism of Vesicle Penetration

To shed light on the mechanism of incorporation of PC vesicles inside the intercellular region of SC after the liposome incubation and subsequent OG treatment, the PC and OG amounts incorporated into the SC were determined [52]. The total weight of PC and OG incorporated in 15 mg of tissue was 2.04 and 6.84 µmol, respectively, the corresponding OG/PC molar ratio (Re) being 3.35. This Re corresponded to a system formed by 90% of OG-PC mixed micelles and 10% of OG-PC mixed vesicles, in agreement with our previous work [98]. In previous works, the formation of liposomes by continuous dilution of OG-PC micellar solutions has been reported [45]. Thus, we hypothesize that the presence of PC liposomes inside the SC is due to the reconstitution of these vesicles in the intercellular lipid spaces by dilution of OG-PC micellar aggregates incorporated into the SC [52]. To support this hypothesis, a series of DLS experiments in an aqueous medium were performed to know the variation of the mean HD of the OG/PC micellar solution (Re = 3.35) during its continuous dilution with buffer. The HD of particles increased with dilution up to a maximum (from 9 to 190 nm) indicating the reconstitution of the vesicles by dilution of the mixed micelles [52]. During the dilution process, the concentration of free OG in the buffer decreases, and for this reason the OG molecules forming mixed micelles tend to release to the buffer. Although in the experiments in the aqueous medium the total amount of OG molecules is always the same (partitioning into micelles and the medium), in the experiments using SC it is very probable that during the washing process some OG molecules were removed. However, this fact is independent of the formation of vesicles, which would occur by release of the OG molecules from mixed micelles regardless of the fate of the OG molecules.

From these findings, we assume that the presence of vesicles in the intercellular lipid region of the SC observed by the two microscopy techniques (Figures 15.16 and 15.17) is the result of the reconstitution of these structures inside the SC by means of a three-step process: first, formation in the SC surface of an OG-PC micellar system between the PC adsorbed during the incubation with liposomes and the OG added in the subsequent treatment; second, incorporation of this micellar system into the intercellular lipid region during the surfactant treatment; third, reconstitution of PC vesicles inside the lipid spaces by continuous dilution of this micellar system during the final washing of the SC [52]. This process is illustrated on Figure 15.18. In this sense, it is possible to assume that the washing resulted in a continuous release of buffer inside the SC. Furthermore, the driving force could be also generated by the large hydration gradient across the skin, varying from 15% to 29% in the SC to 70% in the stratum granulosum [68]. After reconstitution, the big size of the vesicles formed hinders their migration outside the tissue through the intercellular lipid spaces (distance of the ordered lipids of about 5–6 nm) by simple washing. This new strategy that favors the incorporation of PC vesicles inside the SC gives light on the controversy existing on the ability of PC liposomes to penetrate into this tissue and opens up new avenues in the therapeutic applications of liposomes in skin.

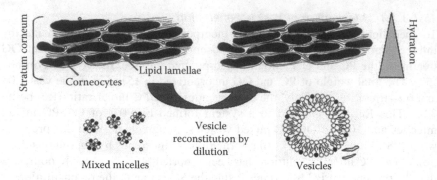

FIGURE 15.18 Cartoon showing the penetration in the SC of OG-PC mixed micelles and the reconstitution of vesicles by dilution of these micelles.

In general terms the application of alkyl glucosides in the solubilization and reconstitution of liposomes formed by different type of lipids and in the study of cell membranes has been successful in the last decade. In the future, the sugar-based surfactants and in particular the alkyl glucosides given their double characteristics of "mild" surfactants with respect to its denaturing effect on proteins and their low capacity to solubilize specific lipids must be considered as useful and precise tools to study biological protein structures.

ABBREVIATIONS

PC	phosphatidylcholine
SC	stratum corneum
Cer	ceramides type III
Chol	cholesterol
PA	palmitic acid
Chol-sulf	cholesteryl sulfate
PIPES	piperazine-1,4 bis(2-ethanesulfonic acid)
Re	effective surfactant/lipid molar ratio
$Re_{50\%CF}$	effective surfactant/lipid molar ratio for 50% of CF release
$Re_{100\%CF}$	effective surfactant/lipid molar ratio for 100% of CF release
Re_{SAT}	effective surfactant/lipid molar ratio for liposome saturation
Re_{SOL}	effective surfactant/lipid molar ratio for liposome solubilization
K	bilayer/aqueous phase surfactant partition coefficient
$K_{50\%CF}$	bilayer/aqueous phase surfactant partition coefficient for 50% of CF release
$K_{100\%CF}$	bilayer/aqueous phase surfactant partition coefficient for 100% of CF release
K_{SAT}	bilayer/aqueous phase surfactant partition coefficient for liposome saturation

K_{SOL}	bilayer/aqueous phase surfactant partition coefficient for liposome solubilization
S_W	surfactant concentration in the aqueous medium
$S_{W,50\%CF}$	surfactant concentration in the aqueous medium for 50% of CF release
$S_{W,100\%CF}$	surfactant concentration in the aqueous medium for 100% of CF release
S_{SAT}	surfactant concentration in the aqueous medium for liposome saturation
S_{SOL}	surfactant concentration in the aqueous medium for liposome solubilization
S_B	surfactant concentration in the bilayers
Cer	ceramides type III
Chol	cholesterol
PA	palmitic acid
Chol-sulf	cholesteryl sulfate
C_8-Glu	octyl glucoside (n-octyl β-D-glucopyranoside)
C_9-Glu	nonyl glucoside (n-nonyl β-D-glucopyranoside)
C_{10}-Glu	decyl glucoside (n-decyl β-D-glucopyranoside)
C_{11}-Glu	undecyl glucoside (n-undecyl β-D-glucopyranoside)
C_{12}-Glu	dodecyl glucoside (n-dodecyl β-D-glucopyranoside)
CF	5(6)-carboxyfluorescein
SLS	static light scattering
TLC-FID	thin-layer chromatography/automated flame ionization detection system
PI	polydispersity index
CMC	critical micellar concentration
r^2	regression coefficient

REFERENCES

1. Schuck, S., Honsho, M., Ekroos, K., Shevchenko, A., and Simons, K., *Proc. Natl. Acad. Sci. U S A*, 2003, *100*(10), 5795.
2. Jacobson, K., Mouritsen, O.G., and Anderson, R.G., *Nat Cell Biol*, 2007, *9*(1), 7.
3. Munro, S., *Cell*, 2003, *115*(4), 377.
4. Carnero Ruiz, C., Hierrezuelo, J.M., Aguiar, J., and Peula-Garcia, J.M., *Biomacromolecules*, 2007, *8*, 2497.
5. Fisicaro, E., Biemmi, M., Compari, C., Duce, E., Peroni, M., Viscardi, G., and Quagliotto, P., *Colloids Surf A*, 2007, *301*, 129.
6. Neimert-Andersson, K., Sauer, S., Panknin, O., Borg, T., Söderlind, E., and Somfai, P., *J. Org. Chem.*, 2006, *71*, 3623.
7. Yoshimura, T., Ishihara, K., and Esumi, K., *Langmuir*, 2005, *21*, 10409.
8. Kumpulainen, A.J., Persson, C.M., and Eriksson, J.C., *Langmuir*, 2004, *20*, 10935.
9. Johnsson, M., Wagenaar, A., Stuart, M.C.A., and Engberts, J.B.F.N., *Langmuir*, 2003, *19*, 4609.
10. Prive, G.G., *Methods*, 2007, *41*, 388.

11. Lu, S. and Somasundaran, P., *Langmuir*, 2007, *23*, 9188.
12. Raman, P., Cherezov, V., and Caffrey, M., *Cell Mol. Life Sci.*, 2006, *63*, 36.
13. Thiesen, P.H., Rosenfeld, H., Konidala, P., Garamus, V.M., He, L., Prange, A., and Niemeyer, B., *J. Biotech.*, 2006, *124*, 284.
14. Söderlind, E. and Karlsson, L., *Eur. J. Pharm. Biopharm.*, 2006, *62*, 254.
15. Iida, T., Kamo, M., Uozumi, N., Inui, T., and Imai, K., *J. Chrom. B*, 2005, *823*, 209.
16. Wiener, M.C., *Methods*, 2004, *34*(3), 364.
17. Ratnala, V.R.P., Swarts, H.G.P., VanOostrum, J., Leurs, R., DeGroot, H.J.M., Bakker, R.A., and DeGrip, W.J., *Eur. J. Biochem.*, 2004, *271*, 2636.
18. Ravaud, S., Do Cao, M.-A., Jidenko, M., Ebel, C., Le Maire, M., Jault, J.-M., Di Pietro, A., Haser, R., and Aghajari, N., *Biochem. J.*, 2006, *395*(2), 345.
19. Söderlind, E., Wollbratt, M., and von Corswant, M., *Int. J. Pharm.*, 2003, *252*, 61.
20. Rosenow, M.A., Brune, D., and Allen, J.P., *Acta Crystallogr D Biol Crystallogr.*, 2003, *59*, 1422.
21. Istvan, E.S., Hasemann, C.A., Kurumbail, R.G., Uyeda, K., and Deisenhofer, J., *Protein Sci.*, 1995, *4*(11), 2439.
22. Hirsch, A., Wacker, T., Weckesser, J., Diederichs, K., and Welte, W., *Proteins*, 1995, *23*(2), 282.
23. Scheffel, F., Fleischer, R., and Schneider, E., *Biochim. Biophys. Acta*, 2004, *1656*, 57.
24. Jastrzebska, B., Fotiadis, D., Jang, G.F., Stenkamp, R.E., Engel, A., and Palczewski, K., *J. Biol. Chem.*, 2006, *281*(17), 11917.
25. Qin, L., Hiser, C., Mulichak, A., Garavito, R.M., and Ferguson-Miller, S., *Proc. Natl. Acad. Sci. U S A*, 2006, *103*(44), 16117.
26. Sehgal, P. and Otzen, D.E., *Protein Sci.*, 2006, *15*(4), 890.
27. Klein-Seetharaman, J., Getmanova, E.V., Loewen, M.C., Reeves, P.J., and Khorana, H.G., *Proc. Natl. Acad. Sci. U S A*, 1999, *96*(24), 13744.
28. Morandat, S. and El Kirat, K., *Colloids Surf. B*, 2007, *55*, 179.
29. Cócera, M., López, O., Coderch, L., Parra, J.L., and de la Maza, A., *Langmuir*, 2002, *18*, 297.
30. Partearroyo, M.A., Alonso, A., Goñi, F.M., Tribout, M., and Paredes, S., *J. Colloid Interface Sci.*, 1996, *178*, 156.
31. Lichtenberg, D., *Biochim. Biophys. Acta*, 1985, *821*, 470.
32. Bouwstra, J., *Colloids Surf. A.*, 1997, *123–124*, 403.
33. De la Maza, A. and Parra, J.L., *Eur J. Biochem.*, 1994, *226*, 1029.
34. Wertz, P.W., In *Liposome Dermatics* (Griesbach Conference), Braun-Falco, O., Korting, H.C., and Maibach, H., (Eds.), Springer-Verlag, Berlin, Hedelberg, 1992, 38–43.
35. De la Maza, A. and Parra, J.L., *Langmuir*, 1996, *12*, 3393.
36. Deamer, D.W. and Uster, P.S., In *Liposomes*, Ostro, M.J. (Ed.), Marcel Dekker, Inc. New York, 1983, pp. 27–52.
37. Schurtenberger, P., Mazer, N., and Känzig, W., *J. Phys. Chem.*, 1985, *89*, 1042.
38. Almog, S., Litman, B.J., Wimley, W., Cohen, J., Wachtel, E.J., Barenholz, Y., Benshaul, A., and Lichtenberg, D., *Biochemistry*, 1990, *29*, 4582.
39. Weinstein, J.N., Ralston, E., Leserman, L.D., Klausner, R.D., Dragsten, P., and Blumenthal, R., In *Liposome technology*, Gregoriadis, G. (Ed.), CRC Press, Boca Raton, FL, 1986, Vol III, 183–204.
40. Singleton, W.S., Gray, M.S., Brown, M.L., and White, J.L., *J. Am. Oil Chem. Soc.*, 1965, *42*, 53.
41. De la Maza, A., Coderch, L., Gonzalez, P., and Parra, J.L., *J. Control. Release*, 1998, *52*, 159.
42. Cocera, M., López, O., Coderch, L., Parra, J.L., and de la Maza, A., *Colloids Surf A*, 2001, *176*, 167.

43. Ruiz, M.B., Prado, A., Goñi, F.M., and Alonso, A., *Biochim. Biophys. Acta*, 1994, *1193*, 301.
44. Egelhaaf, S.U., Wehrli, E., Muller, M., Adrian, M., and Schurtenberger, P., *J. Microsc.*, 1996, *184*, 214.
45. López, O., Cocera, M., Coderch, L., Parra, J.L., Barsukov, L., and de la Maza, A., *J. Phys Chem. B*, 2001, *105*, 9879.
46. López, O., de la Maza, A., Coderch, L., López-Iglesias, C., Wehrli, E., and Parra, J.L., *FEBS lett.*, 1998, *426*, 314.
47. López, O., Cocera, M., Wehrli, E., Parra, J.L., and de la Maza, A., *Arch. Biochem. Biophys.*, 1999, *367*, 153.
48. López, O., de la Maza, A., Coderch, L., and Parra, J.L., *Colloids Surf. A*, 1997, 123–124, 415.
49. Wertz, P.W. and Downing, D.T., *J. Lipid Res.*, 1983, *24*, 759.
50. López, O., Cocera, M., Walther, P., de la Maza, A., Coderch, L., and Parra, J.L., *Colloids Surf. A*, 2000, *168*, 115.
51. Swartzendruber, D.C., Burnett, I.H., Wertz, P.W., Madison, K.C., and Squier, C.A., *J. Invest. Dermatol.*, 1995, *104*, 417.
52. López, O., Cocera, M., López-Iglesias, C., Walter, P., Coderch, L., Parra, J.L., and de la Maza, A., *Langmuir*, 2002, *18*, 7002.
53. Van den Bergh, B.A.I., Vroom, J., Gerritsen, H., Junginger, H.E., and Bouwstra, J.A., *Biochim Biophys Acta,* 1999, *1461*, 155.
54. López, O., Cocera, M., Walther, P., Wehrli, E., Coderch, L., Parra, J.L., and de la Maza, A., *Micron*, 2001, *32*, 201.
55. Walther, P., Wehrli, E., Hermann, R., and Muller, M., *J. Microsc.*, 1995, *179*, 229.
56. Rosen, M.J., *J. Am Oil Chem. Soc.*, 1974, *51*, 461.
57. Vinson, P.K., Talmon, Y., and Walter, A., *Biophys. J.*, 1989, *56*, 669.
58. Seino, H., Uchibori, T., Nishitani, T., and Inamasu, S., *J. Am. Oil Chem. Soc.*, 1984, *61*, 1761.
59. García, M.T., Ribosa, I., Campos, E., and Sánchez Leal, J., *Chemosphere*, 1997, *35*, 545.
60. Ruiz, J., Goñi, F.M., and Alonso, A., *Biochim. Biophys. Acta,* 1988, *937*, 127.
61. Schubert, R., Beyer, K., Wolburg, H., and Schmidt, K.H., *Biochemistry*, 1986, *25*, 5263.
62. Inoue, T., Yamahata, T., and Shimozawa, R., *J. Colloid Interface Sci.,* 1992, *149*, 345.
63. De la Maza, A. and Parra, J.L., *J. Control. Release*, 1995, *37*, 33.
64. Jackson, M.L., Schmidt, C.F., Lichtenberg, D., Litman, B.J., and Albert, A.D., *Biochemistry*, 1982, *21*, 4576.
65. Lasch, J., *Biochim. Biophys. Acta*, 1990, *1022*, 171.
66. Paternostre, M., Meyer, O., Grabielle-Madelmont, C., Lesieur, S., Ghanam, M., and Ollivon, M., *Biophys. J.*, 1995, *69*, 2476.
67. López, O., Cocera, M., Parra, J.L., and de la Maza, A., *Colloid Surfaces A.*, 2001, *193*, 221.
68. Schäfer-Korting, M., In: *Liposome Dermatics*, Braun-Falco, O., Karting, H.C., and Maibach, H., (Eds.), (Griesbach Conference), Springer, Berlin, 1992, 299–307.
69. Wertz, P.W. and Downing, D.T., *Transdermal Drug Delivery. Developemental Issues and Research Iniciatives*, Marcel Dekker, New York, 1989.
70. Friberg, S.E., Goldsmith, L.B., Kayali, I., and Suhaimi, H., In: *Interfacial Phenomena in Biological Systems*, Bender, M. (Ed.), Surfactant Science Series, Marcel Dekker, New York, 1991, Vol 39, pp. 3–32.
71. Imokawa, G., Abe, A., Jin, K., Higaki, Y., Kawashima, M., and Hidano, A., *J. Invest. Dermatol.*, 1991, *96*, 523.

72. Bouwstra, J.A., Gooris, G.S., Bras, W., and Downing, D.T., *J. Lipid Res.*, 1995, *36*, 685.

73. Wertz, P.W., Abraham, W., Landmann, L., and Downing, D.T., *J. Invest. Dermatol.*, 1986, *87*, 582.

74. De la Maza, A., Manich, A.M., Coderch, L., Bosch, P., and Parra, J.L., *Colloids Surf. A*, 1995, *101*, 9.

75. López, O., Cocera, M., Parra, J.L., and de la Maza, A., *Colloid Polym. Sci.*, 2001, *279*, 909.

76. Lorber, B., Bishop, J., and DeLucas, L., *Biochim. Biophys. Acta*, 1990, *1023*, 254.

77. Keller, M., Kerth, A., and Blume, A., *Biochim. Biophys. Acta,* 1997, *1326*(2), 178.

78. López, O., Cocera, M., Pons, R., Azemar, N., and de la Maza, A., *Langmuir*, 1998, *14*, 4671.

79. Goñi, F.M., Urbaneja, M.A., Arrondo, J.L.R., Alonso, A., Durrani, A.A., and Chapman, D., *Eur. J. Biochem.*, 1986, *160*, 659.

80. Edwards, K. and Almgren, M., *Langmuir*, 1992, *8*, 824.

81. Edwards, K. and Almgren, M., *Progr. Colloid. Polym. Sci.*, 1990, *82*, 190.

82. Silvander, M., Karlsson, G., and Edwards, K., *J. Colloid Interface Sci.*, 1996, *179*, 104.

83. Kragh-Hansen, U., Le Marie, M., and Moller, J.V., *Biophys. J.*, 1998, *75*, 2932.

84. López, O., Cocera, M., Pons, R., Azemar, N., López-Iglesias, C., Wehrli, E., Parra, J.L., and de la Maza, A., *Langmuir*, 1999, *15*, 4678.

85. Ede, J., Nigmeijer, J.R.J., and Welling-Wester, S., *J. Chromatogr.*, 1989, *476*, 319.

86. De la Maza, A. and Parra, J.L., *Biochem. J.*, 1994, *303*, 907.

87. Levy, D., Gulik, A., Seigneuret, M., and Rigaud, J.L., *Biochemistry*, 1990, *29*, 9480.

88. Knol, J., Sjollena, K., and Poolman, B., *Biochemistry*, 1998, *37*(46), 16410.

89. Patist, A., Chhabra, V., Pagidipati, R., Shah, R., and Shah, D.O., *Langmuir*, 1997, *13*, 432.

90. Lasic, D.D., *Liposomes: From Physics to Applications,* Chapter 2, Elsevier Science Publishers B.V. Amsterdam, 1993.

91. Swartzendruber, D.C., Kitko, D.J., Wertz, P.W., Madison, K.C., and Downing, D.T., *Arch. Dermatol. Res.*, 1988, *280*, 424.

92. Chapman, S.J., Walsh, A., Jackson, S.M., and Friedmann, P.S., *Arch. Dermatol. Res.*, 1991, *283*, 167.

93. Stubbs, G.W. and Litman, B.J., *Biochemistry,* 1978, *17*, 215.

94. Dreher, F., Walde, P., Walther, P., and Wehrli, E., *J. Control. Release*, 1997, *47*, 131.

95. Hofland, H.E.J., Bouwstra, J.A., Spies, F., Boddé, H.E., Nagelkerke, J.F., Cullander, C., and Junginger, H.E., *J. Liposome Res.*, 1995, *5*, 241.

96. Van Hal, D.A., Jeremiasse, E., Junginger, H.E., Spies, F., and Bouwstra, J.A., *J. Invest. Dermatol.*, 1996, *106*, 89.

97. López, O., Cocera, M., Parra, J.L., and de la Maza, A., *Colloids Surf. A*, 2000, *162*, 123.

98. López, O., Cocera, M., Walther, P., Wehrli, E., Coderch, L., Parra, J.L., and de la Maza, A., *Colloids Surf. A*, 2001, *182*, 35.

Index

A

Adiabatic compressibility, 44–45
Adsorption
  hydrophilic solids
    critical micelle concentration, 208
    hydrocarbon chain length and glucose
      units, 210
    hydrogen bonding, 215, 218
    hydrophobicity, 210, 212
    pH effect, 213, 215
    sodium sulfate (Na_2SO_4) salt effect,
      212
    surface ion concentrations, 215–217
    three-stage adsorption process, 208
    urea and dimethyl sulfoxide (DMSO),
      218
    wettability, 209
    zeta-potential, 213–214
  hydrophobic solids
    hydrophobicity, 221
    stability, 220–221
    wettability, 220
  mixed n-dodecyl-β-D-maltoside(DM)/
    sodium dodecyl sulfonate (SDS_1)
    isoelectric point, 232
    molar ratios, 231–232
    synergism and antagonism, 232
  n-dodecyl-β-D-maltoside/dodecyltrimethyl
    ammonium bromide mixture
    anchor species, 235
    hydrophobic chain–chain interactions,
      234, 237
  n-dodecyl-β-D-maltoside/nonyl phenol
    ethoxylated decyl ether (NP-10)
    mixture
    antagonistic interaction, 238
    coadsorption, 240
    hydrophobic chain–chain interactions,
      237–238
    mixture behavior, 237
  n-dodecyl-β-D-maltoside/sodium
    dodecylsulfate mixture
    electrostatic interaction, 227
    hydrophobic chain–chain interactions,
      228
  orientation scheme, 222
  synergistic/antagonistic interactions, 230

Aggregation, bovine serum albumin
  8-anilinonaphthalene-1-sulfonate (ANS)
    definition, 553
    fluorescence intensity, 553–554
  number determination
    polymer–surfactant systems, 567
    protein–surfactant systems, 565–566
    pyrene fluorescence, 566
  pyrene 1:3 ratio method
    critical aggregation concentration,
      552–553
    micellar hydrophobic environment, 552
    vs. total surfactant concentration,
      550–551
  surface tension measurements, 554–555
Alkylammonium uronates, 338
Alkyl chain length
  glucose-based surfactants, 70
  nonionic surfactants, 71
Alkyl glucamides
  hydrophobic chain length effect
    CHAPS/n alkyl-N-methylglucamides,
      452–453
    lithium perfluorooctanesulfonate
      (LiFOS)/$C_{10}Glu_2$ system, 451–452
    MEGA-10/n-alkyltrimethylammonium
      bromides, 453–454
    micellar aggregation numbers,
      454, 456
  interaction parameters, 446
  ionic surfactants
    DBNMG/SDS system, 447–448
    MEGA-10/(α-SMy·Me), 448–449
    MEGA-10/CTPPB system, 449
    SDS and tetradecylmalono-bis-N-
      methylglucamide (TBNMG), 448
    SDS/MEGA-10 system, 449
  nonionic surfactants
    MEGA-10/$C_{16}E_{10}$ and $C_{12}E_{10}$ mixed
      system, 450–451
    MEGA-8/MEGA-9 mixture, 450
    polar head group effect, 451
Alkyl glycosides (AGs)
  classification
    galactoside-based surfactants, 25
    stereochemical configuration, 24–26
  foam films stability, 139–141
  hydrophilic–lipophilic balance (HLB), 24